Salinity Responses and Tolerance in Plants: Transport and Signaling Mechanisms

Salinity Responses and Tolerance in Plants: Transport and Signaling Mechanisms

Editor: Simon Vader

STATES
ACADEMIC PRESS
www.statesacademicpress.com

States Academic Press,
109 South 5th Street,
Brooklyn, NY 11249, USA

Visit us on the World Wide Web at:
www.statesacademicpress.com

ISBN: 978-1-63989-476-5 (Hardback)

Cataloging-in-Publication Data

Salinity responses and tolerance in plants : transport and signaling mechanisms / edited by Simon Vader.
p. cm.
Includes bibliographical references and index.
ISBN 978-1-63989-476-5
1. Plants--Effect of salts on. 2. Salt-tolerant crops--Physiology. 3. Plant physiology. 4. Plants--Hardiness. I. Vader, Simon.
QK753.S3 S25 2022
581.133 54--dc23

Table of Contents

Preface

Every book is initially just a concept; it takes months of research and hard work to give it the final shape in which the readers receive it. In its early stages, this book also went through rigorous reviewing. The notable contributions made by experts from across the globe were first molded into patterned chapters and then arranged in a sensibly sequential manner to bring out the best results.

Plants need to adapt to a variety of environmental conditions such as salinity, drought, light and temperature which affect their growth, flowering and fruiting. Plant cells communicate with each other in order to respond to changing external environment. Soil salinity is one of the major abiotic stresses which affect the plant growth. It induces cytotoxicity due to excessive uptake of sodium and chloride ions. Salinity is also accompanied by oxidative stress. It can be induced due to poor quality of water for irrigation and soil salinization. Some of the tolerance mechanisms adopted by plants are ion homeostasis, hormone modulation, synthesis of polyamines, compartmentalization, and activation of antioxidant enzymes. The book aims to shed light on some of the unexplored aspects of salinity response and tolerance in plants and the recent researches in this field. From theories to research to practical applications, case studies related to all contemporary topics of relevance to this field have been included herein. This book is a vital tool for all researching or studying salinity response and tolerance in plants as it gives incredible insights into emerging trends and concepts.

It has been my immense pleasure to be a part of this project and to contribute my years of learning in such a meaningful form. I would like to take this opportunity to thank all the people who have been associated with the completion of this book at any step.

Editor

1

Progress in Studying Salt Secretion from the Salt Glands in Recretohalophytes: How do Plants Secrete Salt?

Fang Yuan, Bingying Leng and Baoshan Wang*

Key Lab of Plant Stress Research, College of Life Science, Shandong Normal University, Ji'nan, China

Edited by:
Vadim Volkov,
London Metropolitan University, UK
Reviewed by:
Rosario Vera-Estrella, Universidad Nacional Autónoma de México, Mexico
Sergey Shabala, University of Tasmania, Australia
Wee Kee Tan,
National University of Singapore, Singapore
***Correspondence:**
Baoshan Wang
bswang@sdnu.edu.cn

To survive in a saline environment, halophytes have evolved many strategies to resist salt stress. The salt glands of recretohalophytes are exceptional features for directly secreting salt out of a plant. Knowledge of the pathway(s) of salt secretion in relation to the function of salt glands may help us to change the salt-tolerance of crops and to cultivate the extensive saline lands that are available. Recently, ultrastructural studies of salt glands and the mechanism of salt secretion, particularly the candidate genes involved in salt secretion, have been illustrated in detail. In this review, we summarize current researches on salt gland structure, salt secretion mechanism and candidate genes involved, and provide an overview of the salt secretion pathway and the asymmetric ion transport of the salt gland. A new model recretohalophyte is also proposed.

Keywords: asymmetric ion and water transport, recretohalophyte, salt gland, salt secretion mechanism, salt stress

INTRODUCTION

Soil salinization has long been known as an environmental problem (Jacobsen and Adams, 1958), and approximately 6% of the planet's total land, or more than 800 million ha, is affected (Shabala, 2013). Forty-five million ha (20%) of presently irrigated lands are also saline lands (Munns and Tester, 2008; FAO, 2015; Shelden et al., 2016). In China alone, more than one million acres of agricultural land are salt affected due to irrigation water containing high soluble salts[1]. Moreover, secondary salinization caused by inappropriate irrigation is increasing in many countries, and it is difficult to reclaim land once degraded in this way despite the availability of substantial funding for land recovery. Few crops can survive in salt-affected regions, leading to substantially reduced production and often further degradation and desertification (Flowers and Colmer, 2008). Halophytes are considered promising species for the use and improvement of saline land (Song and Wang, 2014).

Halophytes, which constitute 0.4% of the total plants in the world, are plants that can survive and complete their life cycle in media containing more than 200 mM NaCl (Flowers and Colmer, 2008; Santos et al., 2015). Amongst these halophytes there is a small number that are able to secrete salt from their leaves, the so-called recretohalophytes (Flowers et al., 2015). There are approximately 370 recretohalophyte species all over the world according to statistics from Breckle (1995), Zhou et al. (2001), Flowers and Colmer (2008), and Flowers et al. (2010). Recretohalophytes are distributed widely around the globe, inhabiting seawater and inland saline lands (eHALOPH[2]). Salt-secreting structures, namely salt bladders (Figure 1A) and salt glands (Figure 1B), are the unique structures that directly secrete these ions out of the plant, and they are also notable for their presence in recretohalophytes and absence from other halophytes and all non-halophytes (Shabala et al., 2014; Yuan et al., 2015).

Salt bladders and salt glands differ in their structure. Salt bladders are composed of one bladder cell, without or with one or more stalk cells while salt glands consist of either two-or multicellular structures (the details are discussed in the section, The Reported Recretohalophytes and the Structural Characteristics of Salt Glands). Single epidermal cells can function as a salt bladder, as seen in *Mesembryanthemum crystallinum* and bladders are often modified trichomes. Salt bladders once differentiated, expand rapidly and after exposure of the plant to salt may break up releasing ions to the environment. Salt glands form stable structures that directly secrete salt out of the plant to the external environment.

The earliest studies on salt secretion were performed on the salt bladders of *Hormosira banksii* (Bergquist, 1959) and the salt glands of *Spartina townsendii* (Skelding and Winterbotham, 1939). Since the latter half of the 20th century, more investigations on the ultrastructure and salt secretion of recretohalophytes have been performed. In recent decades, remarkable progress has been made in explaining salt exclusion and secretion mechanisms and the development of salt bladders and salt glands, with most studies concentrating on two plants, *Chenopodium quinoa* and *Limonium bicolor*.

C. quinoa is a typical recretohalophyte that possesses salt bladders, and its salt secretion mechanism and salt transport pathway were illustrated in detail in a recent review (Shabala et al., 2014). Comparison of metabolic changes in salt-treated relative to control samples without NaCl treatment showed that 352 different metabolites were identified in bladder cells of *M. crystallinum* under salt treatment (Barkla and Vera-Estrella, 2015). Recent studies of Oh et al. (2015) presented a transcriptomic analysis of bladder cells of *M. crystallinum* demonstrating cell-type-specific responses during adaptation to salt. The latest study of *Atriplex canescens* showed that the increasing of Na^+ accumulation in salt bladders can enhance the salt tolerance (Pan et al., 2016). *L. bicolor* has multicellular salt glands and the mechanisms of development and salt secretion, in particular the candidate genes, have been studied (Feng et al., 2014; Yuan et al., 2015): more detail is provided below.

The topics of salt glands and salt secretion have been previously reviewed and details of publications prior to 2010s can be found

[1] *http://www.fao.org/home/en/*

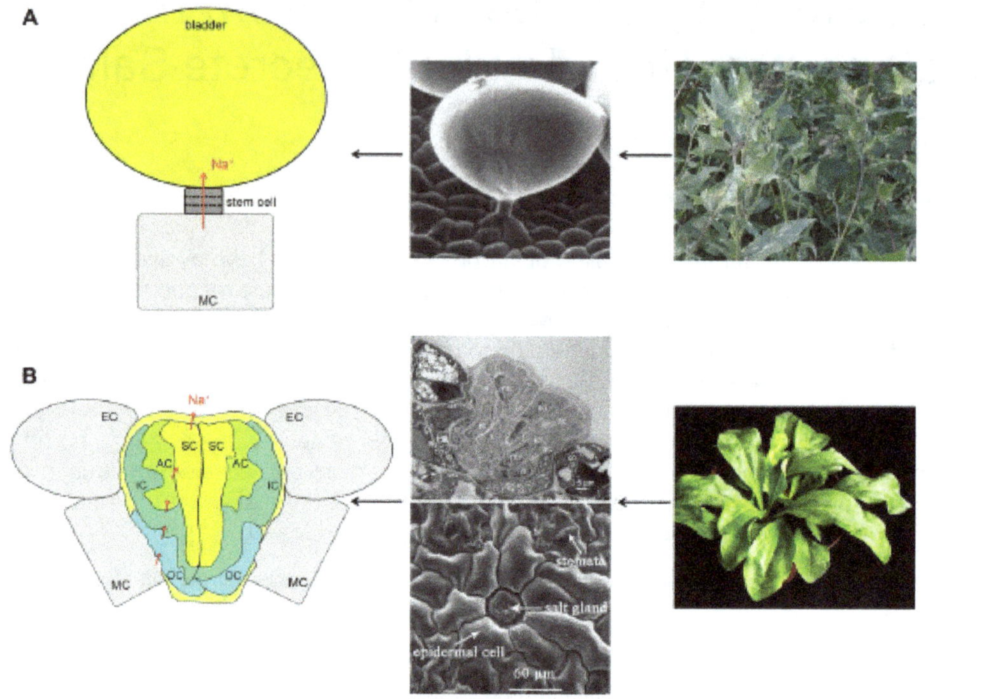

FIGURE 1 | The structure and Na^+ secretion pathway of a salt bladder (A) and a salt gland (B). (A) The large balloon represents the typical structure of the salt bladder. Na^+ can be transported into the balloon and released after bladder rupture. The representative plant is *Atriplex centralasiatica*. **(B)** The typical multi-cellular salt gland and the Na^+ pathway. The representative plant is *Limonium bicolor*. The photographs in **(B)** are reproduced from Feng et al. (2014) and Yuan et al. (2015) with some modifications. SC, secretory cell; AC, accessory cell; IC, inner cup cell; OC, outer cup cell; MC, mesophyll cell; EC, epidermal cell.

in Flowers and Colmer (2008), Ding et al. (2010b), and Flowers et al. (2010, 2015). In the current review, in addition to the basics before 2010s, we mainly focus on salt secretion mechanisms in salt gland in recretohalophyte that were published in the last 5 years.

THE REPORTED RECRETOHALOPHYTES ANDTHE STRUCTURAL CHARACTERISTICS OF SALT GLANDS

To date, the following 11 families (65 species) have been discove-red to have salt gland structures (**Figure 2**): Scrophulariaceae (one species; and the following number in parenthesis after each family represents the number of recretohalophyte species reported in that family), Frankeniaceae (1), Primulaceae (1), Myrsinaceae (2), Acanthaceae (2), Sonneratiaceae (3), Verbenaceae (5), Conovolvulaceae (8), Plumbagenaceae (12), Tamaricaceae (15), and Poaceae (15) according to the statistics of Zhou et al. (2001), Zhao et al. (2002), and Flowers et al. (2010). Most species were reported to have strong salt-secreting abilities as shown in **Figure 2**, e.g., *Frankenia grandifolia* of Frankeniaceae (Balsamo and Thomson, 1993), *Glaux maritima* of Primulaceae (Rozema and Riphagen, 1977), *Aegiceras corniculatum* of Myrsinaceae (Ball, 1988; Parida et al., 2004), *Acanthus ilicifolius* of Acanthaceae (Ye et al., 2005), *Sonneratia caseolaris* of Sonneratiaceae (Shan et al., 2008), *Avicennia marina* of Verbenaceae (Ball, 1988; Chen et al., 2010), *L. bicolor* of Plumbagenaceae (Ding et al., 2010a), and *Reaumuria trigyna* of Tamaricaceae (Dang et al., 2013). In the Poaceae, most genera showed low salt secretion ability except *Aeluropus* (Pollak and Waisel, 1970; Barhoumi et al., 2007), *Sporobolus* (Ramadan, 2001), and *Spartina* (Bradley and Morris, 1991). Attempts have been made to link the structure of salt gland (Zhao et al., 2002) to the salt secreting ability in two other families (Scrophulariaceae and Conovolvulaceae), and more findings involved in secretion ability will likely be discovered in both families in the near future.

The salt glands in different species possess various structural characteristics. The number of component cells has been used to separate multi-cellular salt gland and bi-cellular salt gland (**Figure 1**). In general, the salt glands in dicotyledonous recretohalophytes are multi-cellular and sunken into the epidermis. For instance, eight cells were identified in *Tamarix aphylla* with six secreting cells and two collecting cells in a symmetrical structure (Thomson and Platt-Aloia, 1985). Similarly, in *L. bicolor*, the salt glands consist of 16 cells, with four groups each of outer cup cells, inner cup cells, accessory cells and secretory cells (**Figure 1B**; Ding et al., 2010a; Feng et al., 2015; Yuan et al., 2015). Species of mangroves (of the Verbenaceae) grow in intertidal zone and possess salt glands with different numbers of secretory cells, e.g., 6–8 secretory cells in *A. officinalis* (Tan et al., 2010), 8–12 reported in *A. marina* (Shimony et al., 1973; Drennan et al., 1987), and eight found in *Avicennia germinans* (Balsamo and Thomson, 1993). In contrast to the multi-cellular glands, bi-cellular salt glands are found in the monocotyledonous recretohalophytes of the Poaceae, in species of *Aeluropus*, *Sporobolus*, *Spartina*, and *Zoysia* (see Ramadan and Flowers, 2004; Chen et al., 2009; Semenova et al., 2010; Céccoli et al., 2015). In all of the above examples, the innermost cells of the salt glands were positioned adjacent to the mesophyll cells, e.g., the collection cells in *Tamarix* and the outer cup cells in *Limonium*.

An interesting feature of salt glands is their autofluorescence

[2] *http://www.sussex.ac.uk/affiliates/halophytes/*

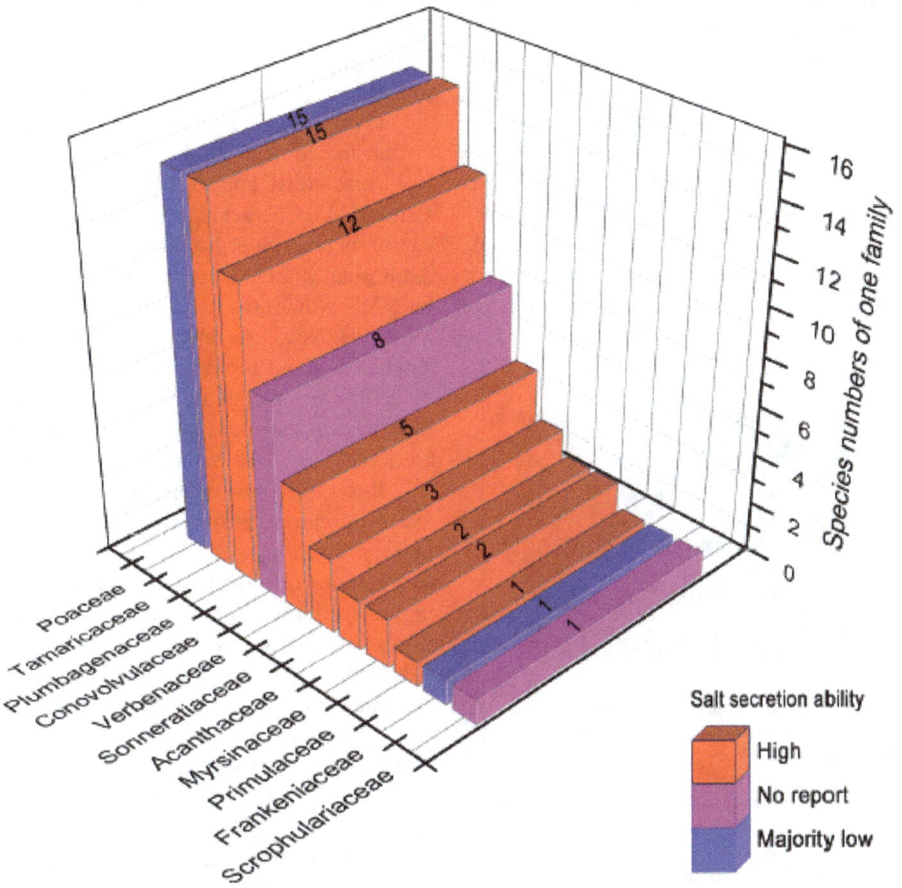

FIGURE 2 | The reported recretohalophytes possessing salt glands with different salt secretion ability. The numbers on the bars presented the species numbers of one family. Red, the species of these families showed strong salt secretion. Purple, there has been no report about the salt secretion in these families until now. Blue, majority of this family showed week salt secretion except *Aeluropus*, *Sporobolus*, and *Spartina*. The figures were drawn with reference to Zhou et al. (2001) and Zhao et al. (2002).

under UV excitation (e.g., of *L. bicolor;* Yuan et al., 2013). By successfully isolating the salt gland complex *in vitro*, Tan et al. (2010) showed that in *A. officinalis* this phenomenon was produced by the cuticles around the salt gland and that these complexes are acidic in nature. Recently, Deng et al. (2015) used Sudan IV staining to show that the autofluorescent substance was localized in the cuticle of the salt glands, and simultaneously suggested that the ferulic acid in the cuticle was directly involved in the salt secretion of the *L. bicolor* salt gland. The cuticle was considered an essential structure for preventing leakage of ions and for protecting the mesophyll from salt damage. The detailed roles of autofluorescencing substances and the cuticle in salt secretion are the subjects of ongoing study.

COMPARISON BETWEEN DIFFERENT METHODS FOR MEASURING SALT SECRETION

The salt secretion activity of a salt gland can be observed with the naked eye (e.g., **Figure 3**) or measured using leaf disks (Campbell et al., 1974; Thomson, 1975), a methodology that was recently improved by the use of oil (Yuan et al., 2013). In *L. bicolor*, in total 5 mg Na^+ secretion on single mature leaf in 24 h, treated with 92 mg NaCl (200 mM) each day. By brushing the salt bladders from the surface of leaves in *A. canescens*, the Na^+ concentration in bladders significantly raised with the increasing of external NaCl (Pan et al., 2016). However, these methods do not provide direct evidence of salt secretion by a single salt gland, so credible techniques are required in order to determine the salt secretion of a single salt gland. Such methods have been developed over the last 20 years (**Table 1**). X-ray microanalysis was first applied to *Porteresia coarctata* (Flowers et al., 1990) and *T. aphylla* (Storey and Thomson, 1994), and this method showed that the salt gland secreted Ca, Mg, and S as well as Na and Cl. Later, in the salt-secreting mangrove *A. marina*, X-ray fluorescence was used to determine the elemental composition with respect to Na, Cl, K, S, Ca, Br, and Zn (Sobrado and Greaves, 2000). SNP (sodium nitroprusside, a NO donor) significantly increased, particularly the Na^+-to-K^+ ratio in the salt glands of *A. marina* as measured by X-ray (Chen et al., 2010). Until recently, X-ray microanalysis was an accurate tool for measuring salt gland secretion and the position of the elements in the salt gland.

Subsequently, methods combining scanning electron micro-scopy (SEM) with X-ray were rapidly developed to measure salt secretion by single salt gland. *Chloris gayana* was found to secrete salt through its bicellular salt glands without rupturing the cuticle on the cap cell (Oi et al., 2013). Besides, salt glands can be identified in SEM images of intact leaves of *Cynodon dactylon* without any need for sectioning (Parthasarathy et al., 2015), and the salt glands observed by SEM in *L. bicolor* were apparently covered by exudates (Feng et al., 2014), which ensured the accuracy of observation *in situ*. Subcellular structures of the salt gland were also available by SEM and X-ray: in *Limoniastrum guyonianum*, where accumulating cells contained numerous, large and unshaped vacuoles (Zouhaier et al., 2015). SEM can also show the real-time secretion status of a salt gland and provide an in-depth understanding of structure–function relationships in multicellular salt glands (Zouhaier et al., 2015). Over a period of time, SEM has been recognized as the most precise and accurate means to observe salt secretion. However, the use of SEM and X-ray microanalysis in the estimation of salt secretion is beset by complicated sample preparation procedures. Consequently, researchers required more accurate and efficient methods in order to measure salt secretion rapidly and precisely.

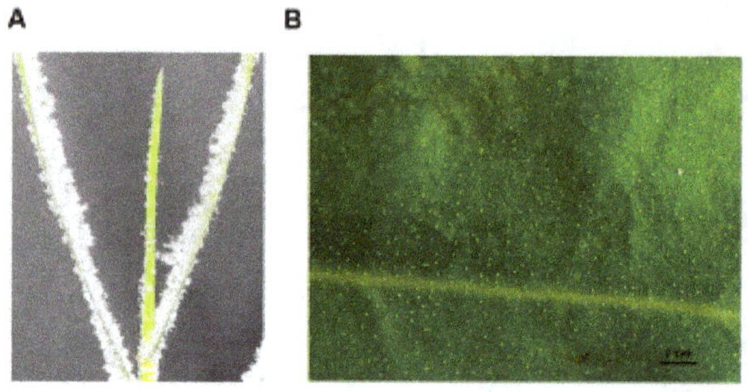

FIGURE 3 | The salt secretion of the salt gland with a distinct salt crystals on the leaves of *Distichlis spicata* (A) and *Limonium bicolor* (B). The photograph in **(A)** was reproduced from Semenova et al. (2010). The leaf blades of *D. spicata* exhibited salt crystals on their leaf surfaces after the plants were incubated in 0.55 mM NaCl solution for 20–24 h under conditions that prevented the slightest air flow (Semenova et al., 2010). The photograph in **(B)** was reproduced from Feng et al. (2014). Leaves of *L. bicolor* that were treated with 200 mM NaCl displayed distinct salt crystals on the surface of leaves, and the salt secretion rate per single salt gland was 8 ng h^{-1}, which is equivalent to a 5 mg Na^{+} secretion from a single mature leaf within 24 h after it was treated with 92 mg NaCl (200 mM) per day.

Over the past 5 years, a new method called Non-invasive Microsensing System (NMS)-BIO-001A (Younger USA Sci.&Tech) has been widely applied to measure the secretion of salt glands. Using NMS, Chen et al. (2010) found that SNP treatments significantly increased net Na^{+} efflux from the salt glands of *A. marina*. Additionally, the Na^{+} secretion rate from *L. bicolor* was obtained by NMS, which was greatly enhanced by NaCl treatment (Feng et al., 2014). NMS can realize the aim of measuring salt secretion of single salt gland *in vivo* and continuous measurement. NMS can detect the net flux of an ion of a salt gland. However, researchers have to peel the epidermis of leaves in order to detect the salt secretion of salt gland. Salt gland may be destroyed during peeling.

Recently, a new technology was introduced to determine the ion position in recretohalophytes, and it is known as nanoscale secondary ion mass spectrometry (SIMS) combined with high-pressure freezing (HPF) and freeze substitution (FS), which can achieve higher spatial resolution than NMT (Smart et al., 2010). SIMS was first performed in the recretohalophyte *L. bicolor*, and Feng et al. (2015) showed that K^{+}, which accumulated in both the cytoplasm and nucleus of salt gland cells under salinity, may play an important role in salt secretion, although the exact mechanism is unknown. The using of SIMS resolved the problems of ion position, distribution and content *in situ*, but complicated and strict operating procedures limited its application.

In conclusion, methods are now available for the elemental analysis of secretions from salt glands. Now, the question arises as to how the ions are transported into the salt gland and the role played by the unique structure of salt glands in the secretory process.

TABLE 1 | Comparative summary for different methods to measure salt secretion in recretohalophyte.

Method	Principle of the method	Advantages	Disadvantages	Secreted ions and concentration	Possible driving forces	Applications in recretohalophytes
Leaf disks secretion covered with oil	The method was first developed by Faraday et al. (1986) and Dschida et al. (1992), and recently improved by the use of oil (Yuan et al., 2013). The secreted ion solution by salt glands on single leaf can be isolated in oil droplets. The accurate ion concentration can be obtained in the further steps.	Simple method without a need of special equipment, and usually costs 12–24 h. Resolution is at the level of whole plant or the single leaf.	Cannot provide direct evidence of salt secretion by a single salt gland.	Na^{+} 20–200 mM, K^{+} 5–20 mM, Mg^{2+} 4–11 mM, Ca^{2+} 0–10 mM, Cl^{-} 20–200 mM, et al.	HKT1, CNGC, NSCC, H^{+}/Cl^{-} symporter	*Limonium bicolor* (Faraday et al., 1986; Yuan et al., 2013); *Avicennia germinans* (Dschida et al., 1992)
X-ray microanalysis	In X-ray analysis, the ratio of intensity of a target element of unknown concentration to that of an internal standard of known concentration is related to the concentration of the target element. The relative elemental sensitivity of the spectrometer was determined by the analysis of a multi-elemental aqueous standard solution containing V, Co, Zn, Se, and Sr (Sobrado and Greaves, 2000).	The method allows determinations in both materials from plants and animal species and is a useful tool because it provides multi-elemental analysis simultaneously.	Requires special equipment and strict procedure.	Na^{+} 20–200 mM, K^{+} 5–10 mM, Ca^{2+} 1–3 mM, Mg^{2+} 2–6 mM, $SO_4{}^{2-}$ 2–6 mM, et al.	HKT1, CNGC, NSCC	*Porteresia coarctata* (Flowers et al., 1990); *Tamarix aphylla* (Storey and Thomson, 1994); *Avicennia marina* (Sobrado and Greaves, 2000; Chen et al., 2010)

Method combined scanning electron microscopy (SEM) with X-ray	The method is based on SEM and X-ray analysis. SEM can give the clear component cells of salt glands and the secreted salts on the salt gland. In addition, X-ray can measure the ions by salt secretion in the specified locations. The structure–function relationships can be explored in-deep.	The method can give the intuitive secreted salt drops without any need for sectioning. Besides, this combined means may give the real-time secretion status of a salt gland. This is the most precise and accurate means to observe salt secretion and measure secreted ions.	Due to the complicated sample preparation procedures, this method required more accurate and efficient operations in order to measure salt secretion rapidly and precisely.	Na^+ 20–200 mM, K^+ 5–20 mM, et al.	HKT1, CNGC	*Chloris gayana* (Oi et al., 2013); *Cynodon dactylon* (Parthasarathy et al., 2015); *Limoniastrum guyonianum* (Zouhaier et al., 2015)
Non-invasive Microsensing System (NMS)-BIO-001A (Younger USA Sci.&Tech)	The ions can be measured by moving an electrode repeatedly between two positions in a predefined excursion (5–30 μm) at a programmable frequency in the range 0.01–10.00 Hz with a range of 0.3–0.5 Hz being typical for many types of electrodes.	The measurement of salt secretion of single salt gland is realized by this method. The salt secretion can be detected real-time. Most of inorganic ions can be measured, usually 5–20 min to measure one ion with simple operation.	NMS detects the net flux of an ion of a salt gland rather than ion efflux. Most of the organic acid anions cannot measure. Salt gland may be destroyed during peeling the epidermis.	Na^+ 20–200 mM, K^+ 5–20 mM, Cl^- 20–200 mM, et al.	HKT1, H^+/Cl^- symporter	*Avicennia marina* (Chen et al., 2010); *Limonium bicolor* (Feng et al., 2014)
Nanoscale secondary ion mass spectrometry (SIMS)	This method is operated combined with high-pressure freezing (HPF) and freeze substitution (FS). The bombardment of ions results in the ejection of charged atomic and molecular species from the surface layers of the sample. These secondary ions are then separated on the basis of their mass-to-charge ratio using a high-performance mass spectrometer, and are correlated with their spatial origin to form a chemical image (Smart et al., 2010).	The ions can be measured with spatial resolutions of better than 100 nm. The accurate ion distribution and concentration are showed on a chemical image based on TEM analysis. This method resolved the problems of ion position, distribution and content *in situ*.	More applications need to be attempted in salt secretion of salt glands. The complicated and strict operating procedures may limit the wide applications.	Na^+ 20–200 mM, K^+ 10–300 mM in nucleus of salt gland.	HKT1, CNGC	*Limonium bicolor* (Feng et al., 2015)

The distinction between the methods is partially arbitrary. The contents are partially referred to Volkov (2015). HKT1, high-affinity K^+ transporter 1; CNGC, cyclic nucleotide-gated cation channel; NSCC, non-selective cationic channel.

HOW IS SALTTRANSPORTED INTOTHE SALT GLAND?

Ion transport is the essential factor determining salinity tolerance in plants (Volkov, 2015). Salt gland can secrete various ions, and the secreted elements mainly depend on the environment (Kobayashi et al., 2007). Amounts of inorganic elements were reported to be excluded by salt glands in the late 5 years including various kinds cationic elements (Na, K, Ca, N, Mg, Fe, Mn, Si, and Zn) and anionic elements (Cl, O, Br, S, P, and C; Sobrado and Greaves, 2000; Semenova et al., 2010; Oi et al., 2013; Feng et al., 2014, 2015; Céccoli et al., 2015; Parthasarathy et al., 2015; Zouhaier et al., 2015), but secretion of Na^+ and Cl^- is higher than that of other ions (Ding et al., 2010b; Ma et al., 2011).

No direct evidence has been obtained to confirm the pathway of salt transport into the salt gland, but a number of possible paths have been suggested. The innermost cells of the salt glands were positioned adjacent to the mesophyll cells, the unique structures of salt glands determined the salt transported into the salt gland. A question then arises as to what caused salt to transfer from mesophyll cells into the salt gland directionally and without influence from the mesophyll cells. Which structure isolates the salt gland from the mesophyll? The cuticles surrounding the salt gland are likely to play this role.

Previous studies on the salt gland ultrastructure in *Spartina foliosa* (Levering and Thomson, 1971) and *T. aphylla* (Thomson et al., 1969) demonstrated that cuticles were present around the salt glands, and they formed a thick barrier from the mesophyll and the external environment. New findings of *Distichlis spicata* showed that these ions were transported into the salt gland through the bottom penetration area that was not covered by the cuticles of the salt gland, and the cuticles can prevent the ions from backflowing into the mesophyll (Semenova et al., 2010). Ions accumulated in the salt gland via the bottom penetration area and plasmodesmata generated fluid pressure due to the presence of the cuticle, and then secreted through salt gland pores. As stated in the former section, ferulic acid in the cuticle was proved to be directly involved in the salt secretion of salt glands (Deng et al., 2015). Therefore, cuticles play a significant role in salt secretion.

The plasmodesmata were considered as another typical feature in the ultrastructure of salt gland (Thomson and Platt-Aloia, 1985). Apparently, the plasmodesmata were present as a membrane-lined intercellular bridge between the mesophyll cells and the salt gland, which allowed these plant cells to communicate with virtually all the adjoining cells (Sager and Lee, 2014). The plasmodesmata were assumed to play an essential role in ion transport into the salt gland (Faraday et al., 1986). The callose and lipid are the most important composition of plasmodesmata, which are well-illustrated to be necessary for ion transport (Grison et al., 2015). Plasmodesmata showed dynamic structures that can actively and selectively transport very large molecules between cells (Overall and Blackman, 1996). However, the detailed role and mechanism of the plasmodesmata in salt gland ion transport required further verifications.

Of equal importance, large central vacuoles were not present in the salt gland cells; instead, many vesicles were detected (Thomson and Platt-Aloia, 1985; Tan et al., 2010, 2013). In particular, the neighboring mesophyll cells had much larger vacuoles than normal, and the secretory cells of the salt glands adjacent to the mesophyll cells possessed many small vesicles, such as *Limonium* (Ziegler and Lüttge, 1967; Yuan et al., 2015; Zouhaier et al., 2015). These small vesicles may play an important role in transporting salt into the salt gland.

In addition to the unique structural characteristics, the membrane-bound translocating proteins are likely to be involved in salt secretion. Another significant pathway passed through the ion carriers or channels inside the plasma membrane of the salt glands, which were similar to those in normal cells with possible exception that these transporters mainly distributed asymmetrically in the other side of plasma membranes of salt gland cells adjacent to the collecting chamber (**Figure 4**), e.g., the potassium transporter (HKT1 and KUP), the inward rectifier potassium channel (KEA), with the CNGC and NSCCs eventually increasing the Na^+ accumulation in the cytoplasm of the cells (Flowers et al., 1977; Flowers and Colmer, 2008). A low-affinity K^+ transporter called AlHKT2;1 from the recretohalophyte *Aeluropus lagopoides* played an important role in K^+ uptake during salt stress and in maintaining a high K^+/Na^+ ratio in the cytosol (Sanadhya et al., 2015b).

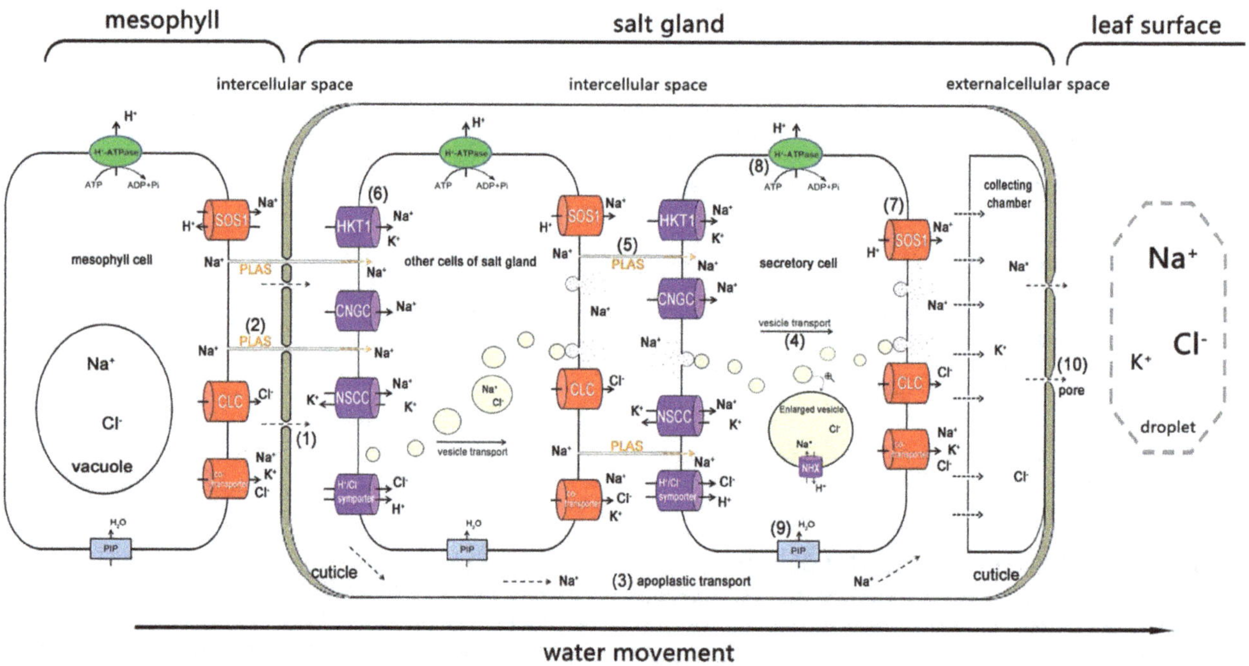

FIGURE 4 | The possible pathway of Na^+ transport from the mesophyll to the salt gland, and the way in which Na^+ is secreted out. Na^+ is transported into the salt gland through the bottom penetration area, which is not covered with cuticles (1), and the plasmodesmata (2). In the salt gland, the ions can be directly transported into the intercellular space (3).The ions arewrapped invesicles for transport from the cytosol to the plasma membrane and secreted out of the salt gland cells (4). Plasmodesmata play an important role in ion delivery among the salt gland component cells (5).Various ion transporters arewidely involved in moving salt in (6) and out (7). H^+-ATPase participated in salt secretion by establishing electric potential difference and proton motive force across plasma membrane (8). Water channels (e.g., PIP) also take part in all the processes as a medium for ion transport (9).The ions are eventually exuded from the secreting pores at the top of the salt gland due to high hydrostatic pressure (10). Ion transporters responsible for influx (dark blue) and efflux (red) are asymmetrically distributed in the plasma membrane of salt gland cells. PLAS, plasmodesmata; HKT1, high-affinity K^+ transporter 1; CNGC, cyclic nucleotide-gated cation channel; NSCC, non-selective cationic channel; PIP, plasma membrane intrinsic protein; NHX, Na^+/H^+ antiporter; SOS1, Na^+/H^+ antiporter; CLC, chloride channel.

Above all, salt was mainly transported into the salt gland through the plasmodesmata, vesicle transport and ion transpor-ters, which combined with both apoplastic and symplastic transport. Even so, there is still a question as to what mechanism drives the salt out of the salt gland.

A POSSIBLE PATHWAY FOR SECRETING SALT FROMTHE SALT GLAND

As stated above, salt secretion was believed to be an active physiological process of the plant. Various physiological data indicated that salt secretion activity requires large amounts of energy. Salt secretion was typically higher under light treatment than it was in the dark (Dschida et al., 1992). Under salinity, the expression of H^+-ATPase in particular increased with the same trend as that of the salt secretion (Chen et al., 2010). However, as reported in *T. aphylla* and *L. bicolor*, no chloroplast was present in the salt gland (Thomson and Liu, 1967; Shimony and Fahn, 1968; Yuan et al., 2015). Therefore, the plasmodesmata between the mesophyll cells and the salt glands in *L. bicolor* may provide the salt gland with photoassimilates and NADPH from the mesophyll cells, which were the original sources of energy for salt excretion. Moreover, during the differentiation of salt glands, mitochondria are the first subcellular organelle differentiated to provide the energy for differentiation of salt gland with small folding mitochondrial crista. The mitochondria in salt glands showed larger scale than that in mesophyll cells (Yuan et al., 2015).

Apparently, all ions are transported out of the salt gland by water; therefore, water efflux is indispensable for salt exclusion. Two aquaporin genes (PIP and TIP) were preferentially expressed in the salt gland cells of *A. officinalis,* and aquaporins were thought to contribute to the re-absorption of water during salt removal (Tan et al., 2013).

Three hypotheses have been proposed to explain the salt secretion of recretohalophytes on the basis of the ultrastructure, and Ding et al. (2010b) have discussed this progress in some detail which rely on (1) the role of the osmotic potential in salt secretion in *Limonium latifolium* (first suggested by Arisz et al., 1955), (2) a transfer system in *S. foliosa* that is similar to liquid flow in animals (raised by Levering and Thomson, 1971) and (3) salt solution excretion by vesicles in the plasma membrane (exocytosis; promulgated by Ziegler and Lüttge, 1967; Shimony and Fahn, 1968).

In the current paper, we concentrate on the newest findings on salt secretion that have been uncovered over the last 5 years; interested readers can obtain details of previous results in Ding et al. (2010b). The recent findings can be separated into three areas consistent with the previous studies. (1) As far as the first suggestion is concerned, recent finding from a study of *D. spicata* showed that the parietal layer of the cytoplasm is invaginated into the extracellular space (apoplast), which is separated only by a thin single membrane. A series of vacuolar-apoplastic continua were identified, and they look like a valve (Semenova et al., 2010). Ions are separated in the salt gland by cuticles and then secreted through secreting pores by fluid pressure. (2) Ouabain, a specific Na^+-ATPase inhibitor, can inhibit Na^+ efflux and enhance K^+ influx by combination with the outside K^+ binding sites (Kim et al., 2013; Mette et al., 2015). Comparisons with animal transport systems have shown that in *C. gayana* (Kobayashi et al., 2007) and the *Tamarix* species (Ma et al., 2011), salt secretion from their salt glands was significantly decreased in the presence of ouabain, which acts as a salt secretion inhibitor and indicates the existence of liquid flow in salt gland (possible Na^+-ATPase). (3) In terms of exocytosis, electrondense substances primarily accumulated in the vesicles of the salt gland in *L. bicolor,* which showed that under salt treatment, numerous vesicles fused with the plasma membrane (Feng et al., 2014; Yuan et al., 2015). While all three hypotheses have support from studies of different recretohalophytes, in the last few years, more and more evidence has supported the role of exocytosis (Echeverría, 2000). TEM images showed an identifiable vesicle-transporting pathway in the salt gland of an *L. bicolor* leaf that was prepared by HPF, followed by FS and staining (Feng et al., 2015).

Additionally, ion transporters were proposed as another important pathway for salt glands to secrete Na^+. H^+-ATPase may participate in salt secretion through establishing electrochemical potential difference and proton gradient across plasma membrane of salt gland. Leaf H^+-ATPase activity and the photosynthetic capacity of *Cakile maritima* can enhance salt resistance under increasing salinity (Debez et al., 2006). H^+-ATPase and the Na^+/H^+ antiporter are anticipated to have roles in the salt secretion and Na^+ sequestration of *A. marina* (Chen et al., 2010; Tan et al., 2010). Currently, high-throughput sequencing technology has broad and increasing applications for many recretohalophytes, and it has validated the view that various types of transporters may participate in salt secretion. Transcript profiling in *A. lagopoides* revealed that HAK, SOS1, and V-ATPase genes play a key role in regulating ion homeostasis (Sanadhya et al., 2015a). NHX and other members of the family play important roles in K^+ homeostasis, vesicle trafficking and cell expansion (Rodríguez-Rosales et al., 2009). The expression of SOS1 and NHX1 was proportional to the salt secretion in *A. marina* (Chen et al., 2010).

Recently, the application of transriptomics in recretohalo-phytes enhances the discovery of candidate genes involved in salt secretion. The transcription profiles of the recretohalophyte *R. trigyna* indicated that the genes related to ion transport were relevant to the salt secretion function of this species (Dang et al., 2013, 2014). Similarly, the candidate genes encoding ion transporters were also suggested in *Sporobolus virginicus* by RNA-seq (Yamamoto et al., 2015). An $NaHCO_3$-treated *L. bicolor* cDNA library was established (Wang et al., 2008), and a recent RNA-Seq library containing the different developmental stages of *L. bicolor* provided an accurate resource for identifying the genes encoding ion carriers and transporters (Yuan et al., 2015).

Nowadays, proteomics was gradually applied to investigate salinity responsive proteins in three recretohalophytes. A cell-type-specific proteomics approach was taken in individual epidermal bladder cells of *M. crystallinum* by shotgun peptide sequencing, with a high representation of proteins involved in H^+-transport, carbohydrate metabolism, and photosynthesis

(Barkla et al., 2012). Besides, Tan et al. (2015a) developed a reliable procedure for obtaining proteins from salt gland-rich tissues of *A. officinalis* and profiled the proteome through 1D-LC-MS/MS. Meanwhile, Tan et al. (2015b) used shotgun proteomics to identify proteins of *A. officinalis* that are present in the salt gland-enriched tissue by comparing the protein profiles of salt gland-enriched (isolated epidermal peels) and salt gland-deprived (mesophyll) tissues, which elucidated the molecular mechanism underlying secretion in plant salt glands. Using the suspension cell cultures of *Halogeton glomeratus*, iTRAQ-based proteomic approach was conducted to reveal several proteins involved in energy, carbohydrate metabolism, stress defense, protein metabolism, signal transduction, cell growth, and cytoskeleton metabolism (Wang et al., 2016). Transcriptomics combined with proteomics will play an essen-tial role in explaining salt secretion of salt gland of recretohalophytes.

However, all the known transcriptomes and proteomes described above are constructed on the basis of extractions from plant leaves, dominated by mesophyll cells so that genes differentially expressed on salt treatment were very likely derived from mesophyll cells; thus, we cannot determine that these genes are specific to the salt gland unless salt glands are first isolated for sequencing. As mentioned previously, isolating a single salt gland is feasible by enzymolysis in *A. officinalis* (Tan et al., 2010). Additionally, laser capture microdissection (LCM) technology has been used for separating target cells or tissues in many species (Li et al., 2010). Both methods can be introduced to isolate sufficient amounts of salt glands of sufficient purity for sequencing in the future.

Another approach to understanding salt secretion by recretohalophytes would be to analyze mutants whose salt secretion is in some way modified. Though many mutants in salt secretion have been obtained using gamma ray irradiation (Yuan et al., 2013), no recretohalophyte where specific genes have been silenced has been developed. Until now, the experiments suggesting candidate genes have only been reported in heterologous expression, such as *Lbchi32* of *L. bicolor* verified in pathogenic fungi (Liu et al., 2010). So the mutants in specific and candidate genes should be investigated in the future in order to accurately study molecular mechanism of salt secretion.

CONCLUSION AND PERSPECTIVES

It is widely accepted that salt secretion is closely aligned with the structure of salt glands (Ding et al., 2010b). Although salt glands in different species possess varying characteristics, their common characteristics can be summarized as follows: a thickened cuticle surrounding the salt gland, frequent plasmodesmata, a large number of developed mitochondria, no chloroplasts, and amounts of small vesicles in the cytoplasm (Thomson and Platt-Aloia, 1985; Ding et al., 2010b; Tan et al., 2013; Yuan et al., 2015). This special structure determines the functionality of bladders and glands; an ion can be rapidly transported from a mesophyll cells into a salt gland and secreted out of the salt gland with a force that is generated by mitochondrial activity and is then transported in vesicles, eventually being excluded through pores (Feng et al., 2014).

Using the muti-cellular salt gland such as *L. bicolor* as the model, **Figure 4** shows the possible pathway for an ion (e.g., Na^+) transport from the mesophyll cell into the salt gland and the pathway of ion secretion. Salts from roots are transported upward in the transpiration stream. The salt transport process hat moves ions in and out of the salt gland is thought to consist of a combination of both apoplastic and symplastic transport (Flowers and Yeo, 1986; Yeo and Flowers, 1986). Most of the Na^+ is transported into the salt gland through the basal penetration area that is not covered by cuticles (1), and prevented by the pull of transpiration stream and a valve structure only open inwardly (Semenova et al., 2010), so Na^+ cannot reflux through the basal penetration area. In other regions of the salt gland, the cuticles can also prevent ions from penetration into the mesophyll. Another possible pathway of the Na^+ flowing into the salt gland is the plasmodesmata (2). Once in the salt gland, there are two possible pathways for the movement of Na^+. The ions can be directly transported into the intercellular space that is separated from the mesophyll cells by the cuticles (3), where they can be temporarily collected in the collecting chamber. The ions are wrapped in vesicles for transport from the cytosol to the plasma membrane of salt gland cells and for secretion out of the salt gland; the membrane of vesicles can be recycled into plasma membrane. NHX may play an important role in transporting Na^+ into the vesicles (4). The plasmodesmata provide an important pathway for ion movement between the salt gland component cells (5). Various ion transporters and channels are widely involved in transporting Na^+ and Cl^- in (6) (e.g., HKT1, CNGC, NSCC and H+/Cl– symporter) and out (7) (e.g., SOS1, CLC and various types of co-transporters) of cells. Ions that are transported by symplastic transport are also temporarily stored in the collecting chamber. During the transport of ions in both import and export pathways, the chemical compounds such as photoassimilate is transported by the plasmodesmata, and ATP is directly generated from highly developed mitochondria (8). Water channels also participate as a medium for ion transport (9). The ions are eventually excluded from the secreting pores that are located in the cuticles in the top of the salt gland due to high hydrostatic pressure (10). The transport of ions from mesophyll cells into and out of salt glands is an asymmetric process, which is a common phenomenon in many biological systems, such as IAA polar transport (Estelle, 1998; Reinhardt et al., 2003), gravity-induced polar transport of calcium across root tips of maize (Lee et al., 1983), iron uptake of the root (Dubeaux et al., 2015), and etc. The asymmetric distribution of salt secretion components primarily involves the asymmetric ion uptake and efflux; asymmetry in the vesicle transport direction; and asymmetry in ion transporter distribution. Most ions are transported in the direction of the vesicles, and the ion transporters that are responsible for secreting ions out of the salt gland are primarily distributed along the outside of secretory cells.

The study of salt secretion in recretohalophytes has developed considerably in recent years. More evidence has been proposed to explain the salt secretion mechanism; however, uncertainties still exist. On one hand, the salt secretion activity is mediated by multi-cells and multi-genes, and thus, the underlying control of the salt gland development remains unclear. Although, Yuan et al. (2015) have illustrated five differentiation stages in the *L. bicolor* leaf and proposed that a series of candidate genes can participate in salt gland development and salt secretion (Yuan et al., 2016), the direct validations of the key genes are needed on the basis of these data. On the other hand, RNA-seq has been conducted in five recretohalophytes, but to date, no genome sequencing has been reported in any recretohalophyte. Unclear genetic background made molecular studies difficult, and a new proposed model recretohalophyte is urgently needed. Fortunately, as a typical recretohalophyte possessing salt glands, *L. bicolor* is a potential model plant for investigating the salt secretion mechanism by salt glands in the future. Partly deficient salt-gland mutants have been obtained by physical and chemical mutagenesis in *L. bicolor* (Yuan et al., 2013). Moreover, the mutant traits can be kept for long-term investigation because the seeds from these plants can be collected through self-pollination (Yuan et al., 2014b). Based on the established transformation system and leaf disk secretion model (Yuan et al., 2014a), the key genes involved in salt secretion can be transformed for validation. Efficient genome editing in plants using a CRISPR/Cas9 system in non-modal plants will also be used to study molecular mechanism of salt secretion of plant salt gland (Feng et al., 2013). For a decade, various ion transporters have been shown to participate in salt resistance as verified in *Arabidopsis* and *Eutrema/Thellungiella* spp. (Lee et al., 2016). However, the in-depth mechanism of halophyte salt tolerance is largely unknown. More and more scientists have begun to use halophytes as materials to study the salt tolerance mechanism. We believe that in the near future, the salt tolerance mechanism will be completely uncovered, and the dream of generating salt-tolerant crops will finally come true.

AUTHOR CONTRIBUTIONS

FY and BL wrote the manuscript. BW modified the article. All authors read and approved the final manuscript.

FUNDING

This work has been supported by the NSFC (National Natural Science Research Foundation of China, project No. 31570251), Programs Foundation of Ministry of Education of China (20123704130001), Natural Science Research Foundation of Shandong Province (ZR2014CZ002), Young Scientist Research Award Fund of Shandong Province (BS2015SW027), and Technology Plan of Universities of Shandong Province (J15LE08).

ACKNOWLEDGMENTS

We are very thankful to Professor Timothy J. Flowers (Depart-ment of Evolution, behaviour and environment, School of Life Sciences, University of Sussex) for his constructive suggestions and critical revision of the manuscript.

REFERENCES

Arisz, W., Camphuis, I., Heikens, H., and Tooren, A. V. (1955). The secretion of the salt glands of *Limonium latifolium* Ktze. *Acta Bot. Neerl.* 4, 322–338. doi: 10.1111/j.1438-8677.1955.tb00334.x

Ball, M. (1988). Salinity tolerance in the mangroves *Aegiceras corniculatum* and *Avicennia marina*. I. Water use in relation to growth, carbon partitioning, and salt balance. *Funct. Plant Biol.* 15, 447–464. doi: 10.1016/j.scitotenv.2013.11.106

Balsamo, R. A., and Thomson, W. W. (1993). Ultrastructural features associated with secretion in the salt glands of *Frankenia grandifolia* (Frankeniaceae) and *Avicennia germinans* (Avicenniaceae). *Am. J. Bot.* 80, 1276–1283. doi: 10.2307/2445711

Barhoumi, Z., Djebali, W., Smaoui, A., Chaïbi, W., and Abdelly, C. (2007). Contribution of NaCl excretion to salt resistance of *Aeluropus littoralis* (Willd) Parl. *J. Plant Physiol.* 164, 842–850. doi: 10.1016/j.jplph.2006.05.008

Barkla, B. J., and Vera-Estrella, R. (2015). Single cell-type comparative metabolomics of epidermal bladder cells from the halophyte *Mesem-bryanthemum crystallinum*. *Front. Plant Sci.* 6:435. doi: 10.3389/fpls. 2015.00435

Barkla, B. J., Vera-Estrella, R., and Pantoja, O. (2012). Protein profiling of epidermal bladder cells from the halophyte *Mesembryanthemum crystallinum*. *Proteomics* 12, 2862–2865. doi: 10.1002/pmic.201200152

Bergquist, P. L. (1959). The effect of cations and anions on the respiration rate of the Brown Alga, *Hormosira banksii*. *Physiol. Plant.* 12, 30–36. doi: 10.1111/j.1399-3054.1959.tb07882.x

Bradley, P. M., and Morris, J. T. (1991). Relative importance of ion exclusion, secretion and accumulation in *Spartina alterniflora* Loisel. *J. Exp. Bot.* 42, 1525–1532. doi: 10.1093/jxb/42.12.1525

Breckle, S. W. (1995). "How do halophytes overcome salinity?" in *Biology of Salt Tolerant Plants*, eds M. A. Khan and I. A. Ungar (Chelsea: Book Graffers), 199–213.

Campbell, N., Thomson, W. W., and Platt, K. (1974). The apoplastic pathway of transport to salt glands. *J. Exp. Bot.* 25, 61–69. doi: 10.1007/BF00385168

Céccoli, G., Ramos, J., Pilatti, V., Dellaferrera, I., Tivano, J. C., Taleisnik, E., et al. (2015). Salt glands in the Poaceae family and their relationship to salinity tolerance. *Bot. Rev.* 81, 162–178. doi: 10.1007/s12229-015-9153-7

Chen, J., Xiao, Q., Wu, F., Dong, X., He, J., Pei, Z., et al. (2010). Nitric oxide enhances salt secretion and Na+ sequestration in a mangrove plant, *Avicennia marina*, through increasing the expression of H^+-ATPase and Na+/H+ antiporter under high salinity. *Tree Physiol.* 30, 1570–1585. doi: 10.1093/treephys/tpq086

Chen, J., Yan, J., Qian, Y., Jiang, Y., Zhang, T., Guo, H., et al. (2009). Growth responses and ion regulation of four warm season turfgrasses to long-term salinity stress. *Sci. Hortic. (Amsterdam)* 122, 620–625. doi: 10.1016/j.scienta.2009.07.004

Dang, Z. H., Qi, Q., Zhang, H. R., Yu, L. H., Wu, S. B., and Wang, Y. C. (2014). Identification of salt-stress-induced genes from the RNA-Seq data of *Reaumuria trigyna* using differential-display reverse transcription PCR. *Int. J. Genomics* 2014:381501. doi: 10.1155/2014/381501

Dang, Z. H., Zheng, L. L., Wang, J., Gao, Z., Wu, S. B., Qi, Z., et al. (2013). Transcriptomic profiling of the salt stress response in the wild recretohalophyte *Reaumuria trigyna*. *BMC Genomics* 14:29. doi: 10.1186/1471-2164-14-29

Debez, A., Saadaoui, D., Ramani, B., Ouerghi, Z., Koyro, H. W., Huchzermeyer, B., et al. (2006). Leaf H+-ATPase activity and photosynthetic capacity of *Cakile maritima* under increasing salinity. *Environ. Exp. Bot.* 57, 285–295. doi: 10.1016/j.envexpbot.2005.06.009

Deng, Y., Feng, Z., Yuan, F., Guo, J., Suo, S., and Wang, B. (2015). Identification and functional analysis of the autofluorescent substance in *Limonium bicolor* salt glands. *Plant Physiol. Biochem.* 97, 20–27. doi: 10.1016/j.plaphy.2015.09.007

Ding, F., Chen, M., Sui, N., and Wang, B. S. (2010a). Ca2+ significantly enhanced development and salt-secretion rate of salt glands of *Limonium bicolor* under NaCl treatment. *S. Afr. J. Bot.* 76, 95–101. doi: 10.1016/j.sajb.2009.09.001

Ding, F., Yang, J. C., Yuan, F., and Wang, B. S. (2010b). Progress in mechanism of salt excretion in recretohalopytes. *Front. Biol.* 5:164–170. doi: 10.1007/s11515-010-0032-7

Drennan, P. M., Berjak, P., Lawton, J. R., and Pammenter, N. (1987). Ultrastructure of the salt glands of the mangrove, *Avicennia marina* (Forssk.) Vierh., as indicated by the use of selective membrane staining. *Planta* 172, 176–183. doi: 10.1007/BF00394585

Dschida, W., Platt-Aloia, K., and Thomson, W. (1992). Epidermal peels of *Avicennia germinans* (L.) Stearn: a useful system to study the function of salt glands. *Ann. Bot.* 70, 501–509.

Dubeaux, G., Zelazny, E., and Vert, G. (2015). Getting to the root of plant iron uptake and cell-cell transport: polarity matters! *Commun. Integr. Biol.* 8:e1038441. doi: 10.1080/19420889.2015.1038441

Echeverría, E. (2000). Vesicle-mediated solute transport between the vacuole and the plasma membrane. *Plant Physiol.* 123, 1217–1226. doi: 10.1104/pp.123.4.1217

Estelle, M. (1998). Polar auxin transport: new support for an old model. *Plant Cell* 10, 1775–1778. doi: 10.2307/3870902

FAO (2015). *FAO Land and Plant Nutrition Management Service*. Rome: Food and Agriculture Organization of the United Nations.

Faraday, C. D., Quinton, P. M., and Thomson, W. W. (1986). Ion fluxes across the transfusion zone of secreting *Limonium* salt glands determined from secretion rates, transfusion zone areas and plasmodesmatal frequencies. *J. Exp. Bot.* 37, 482–494. doi: 10.1093/jxb/37.4.482

Feng, Z., Deng, Y., Zhang, S., Liang, X., Yuan, F., Hao, J., et al. (2015). K^+ accumulation in the cytoplasm and nucleus of the salt gland cells of *Limonium bicolor* accompanies increased rates of salt secretion under NaCl treatment using NanoSIMS. *Plant Sci.* 238, 286–296. doi: 10.1016/j.plantsci.2015.06.021

Feng, Z., Sun, Q., Deng, Y., Sun, S., Zhang, J., and Wang, B. (2014). Study on pathway and characteristics of ion secretion of salt glands of *Limonium bicolor*. *Acta Physiol. Plant.* 36, 2729–2741. doi: 10.1007/s11738-014-1644-3

Feng, Z., Zhang, B., Ding, W., Liu, X., Yang, D., Wei, P., et al. (2013). Efficient genome editing in plants using a CRISPR/Cas system. *Cell Res.* 23, 1229–1232. doi: 10.1038/cr.2013.114

Flowers, T. J., and Colmer, T. D. (2008). Salinity tolerance in halophytes. *New Phytol.* 179, 945–963. doi: 10.1111/j.1469-8137.2008.02531.x

Flowers, T. J., Flowers, S. A., Hajibagheri, M. A., and Yeo, A. R. (1990). Salt tolerance in the halophytic wild rice, *Porteresia coarctata* Tateoka. *New Phytol.* 114, 675–684. doi: 10.1111/j.1469-8137.1990.tb00439.x

Flowers, T. J., Galal, H. K., and Bromham, L. (2010). Evolution of halophytes: multiple origins of salt tolerance in land plants. *Funct. Plant Biol.* 37, 604–612. doi: 10.1007/s10142-011-0218-3

Flowers, T. J., Munns, R., and Colmer, T. D. (2015). Sodium chloride toxicity and the cellular basis of salt tolerance in halophytes. *Ann. Bot.* 115, 419–431. doi: 10.1093/aob/mcu217

Flowers, T. J., Troke, P., and Yeo, A. (1977). The mechanism of salt tolerance in halophytes. *Annu. Rev. Plant Physiol.* 28, 89–121. doi: 10.1146/annurev.pp.28.060177.000513

Flowers, T. J., and Yeo, A. R. (1986). Ion relations of plants under drought and salinity. *Funct. Plant Biol.* 13, 75–91.

Grison, M. S., Brocard, L., Fouillen, L., Nicolas, W., Wewer, V., Dörmann, P., et al. (2015). Specific membrane lipid composition is important for plasmodesmata function in *Arabidopsis*. *Plant Cell* 27, 1228–1250. doi: 10.1105/tpc.114. 135731

Jacobsen, T., and Adams, R. M. (1958). Salt and silt in ancient mesopotamian agriculture. *Science* 128, 1251–1258. doi: 10.1126/science.128.3334.1251

Kim, S. H., Yu, H. S., Hong, G. P., Ha, K., Yong, S. K., Shin, S. Y., et al. (2013). Intracerebroventricular administration of ouabain, a Na/K-ATPase inhibitor, activates mTOR signal pathways and protein translation in the rat frontal cortex. *Prog. Neuropsychopharmacol. Biol. Psychiatry* 45, 73–82. doi: 10.1016/j.pnpbp.2013.04.018

Kobayashi, H., Masaoka, Y., Takahashi, Y., Ide, Y., and Sato, S. (2007). Ability of salt glands in Rhodes grass (*Chloris gayana* Kunth) to secrete Na^+ and K^+. *Soil Sci. Plant Nutr.* 53, 764–771. doi: 10.1111/j.1747-0765.2007.00192.x

Lee, J. S., Mulkey, T. J., and Evans, M. L. (1983). Gravity-induced polar transport of calcium across root tips of maize. *Plant Physiol.* 73, 874–876. doi: 10.1104/pp.73.4.874

Lee, Y. P., Funk, C., Erban, A., Kopka, J., Köhl, K. I., Zuther, E., et al. (2016). Salt stress responses in a geographically diverse collection of *Eutrema/Thellungiella* spp. accessions. *Funct. Plant Biol.* 43, 590–606. doi: 10.1071/FP15285

Levering, C. A., and Thomson, W. W. (1971). The ultrastructure of the salt gland of *Spartina foliosa*. *Planta* 97, 183–196. doi: 10.1007/BF00389200

Li, P., Ponnala, L., Gandotra, N., Wang, L., Si, Y., Tausta, S. L., et al. (2010). The developmental dynamics of the maize leaf transcriptome. *Nat. Genet.* 42, 1060–1067. doi: 10.1038/ng.703

Liu, Z. H., Yang, C. P., Xiao, T. Q., Li, L. X., and Yu, C. W. (2010). Cloning, heterologous expression, and functional characterization of a chitinase gene, lbchi32, from *Limonium bicolor*. *Biochem. Genet.* 48, 669–679. doi: 10.1007/s10528-010-9348-x

Ma, H., Tian, C., Feng, G., and Yuan, J. (2011). Ability of multicellular salt glands in *Tamarix* species to secrete Na^+ and K^+ selectively. *Sci. China Life Sci.* 54, 282–289. doi: 10.1007/s11427-011-4145-2

Mette, L., Jonas Lindholt, G., Laure, Y., Poul, N., and Fedosova, N. U. (2015). Structures and characterization of digoxin-and bufalin-bound Na^+,K^+-ATPase compared with the ouabain-bound complex. *Proc. Natl. Acad. Sci. U.S.A.* 112, 1755–1760. doi: 10.1073/pnas.1422997112

Munns, R., and Tester, M. (2008). Mechanisms of salinity tolerance. *Annu. Rev. Plant Biol.* 59, 651–681. doi: 10.1146/annurev.arplant.59.032607.092911

Oh, D., Barkla, B., Vera-Estrella, R., Pantoja, O., Lee, S., Bohnert, H., et al. (2015). Cell type-specific responses to salinity-the epidermal bladder cell transcriptome of *Mesembryanthemum crystallinum*. *New Phytol.* 207, 627–644. doi: 10.1111/nph.13414

Oi, T., Hirunagi, K., Taniguchi, M., and Miyake, H. (2013). Salt excretion from the salt glands in Rhodes grass (*Chloris gayana* Kunth) as evidenced by low-vacuum scanning electron microscopy. *Flora* 208, 52–57. doi: 10.1016/j.flora.2012.12.006

Overall, R. L., and Blackman, L. M. (1996). A model of the macromolecular structure of plasmodesmata. *Trends Plant Sci.* 1, 307–311. doi: 10.1016/S1360-1385(96)88177-0

Pan, Y., Guo, H., Wang, S., Zhao, B., Zhang, J., Ma, Q., et al. (2016). The photosynthesis, Na+/K+ homeostasis and osmotic adjustment of *Atriplex canescens* in response to salinity. *Front. Plant Sci.* 7:848. doi: 10.3389/fpls.2016.00848

Parida, A. K., Das, A. B., Sanada, Y., and Mohanty, P. (2004). Effects of salinity on biochemical components of the mangrove, *Aegiceras corniculatum*. *Aquat. Bot.* 80, 77–87. doi: 10.1016/j.aquabot.2004.07.005

Parthasarathy, M., Pemaiah, B., Natesan, R., Padmavathy, S. R., and Pachiappan, J. (2015). Real-time mapping of salt glands on the leaf surface of *Cynodon dactylon* L. using scanning electrochemical microscopy. *Bioelectrochemistry* 101, 159–164. doi: 10.1016/j.bioelechem.2014.10.004

Pollak, G., and Waisel, Y. (1970). Salt Secretion in *Aeluropus* litoralis (Willd.) Parl. *Ann. Bot.* 34, 879–888.

Ramadan, T. (2001). Dynamics of salt secretion by *Sporobolus spicatus* (Vahl) Kunth from sites of differing salinity. *Ann. Bot.* 87, 259–266. doi: 10.1006/anbo.2001.1326

Ramadan, T., and Flowers, T. J. (2004). Effects of salinity and benzyl adenine on development and function of microhairs of *Zea mays* L. *Planta* 219, 639–648. doi: 10.1007/s00425-004-1269-7

Reinhardt, D., Pesce, E. R., Stieger, P., Mandel, T., Baltensperger, K., Bennett, M., et al. (2003). Regulation of phyllotaxis by polar auxin transport. *Nature* 426, 255–260. doi: 10.1038/nature02081

Rodríguez-Rosales, M. P., Gálvez, F. J., Huertas, R., Aranda, M. N., Baghour, M., Cagnac, O., et al. (2009). Plant NHX cation/proton antiporters. *Plant Signal. Behav.* 4, 265–276. doi: 10.4161/psb.4.4.7919

Rozema, J., and Riphagen, I. (1977). Physiology and ecologic relevance of salt secretion by the salt gland of *Glaux maritima* L. *Oecologia* 29, 349–357. doi: 10.1007/BF00345808

Sager, R., and Lee, J. Y. (2014). Plasmodesmata in integrated cell signalling: insights from development and environmental signals and stresses. *J. Exp. Bot.* 65, 6337–6358. doi: 10.1093/jxb/eru365

Sanadhya, P., Agarwal, P., and Agarwal, P. K. (2015a). Ion homeostasis in a salt-secreting halophytic grass. *AoB Plants* 7:lv055. doi: 10.1093/aobpla/ plv055

Sanadhya, P., Agarwal, P., Khedia, J., and Agarwal, P. K. (2015b). A low-affinity K^+ transporter AlHKT2; 1 from recretohalophyte *Aeluropus lagopoides* confers salt tolerance in Yeast. *Mol. Biotechnol.* 57, 489–498. doi: 10.1007/s12033-015-9842-9

Santos, J., Al-Azzawi, M., Aronson, J., and Flowers, T. J. (2015). eHALOPH a database of salt-tolerant plants: helping put halophytes to work. *Plant Cell Physiol.* 57:e10. doi: 10.1093/pcp/pcv155

Semenova, G. A., Fomina, I. R., and Biel, K. Y. (2010). Structural features of the salt glands of the leaf of *Distichlis spicata* 'Yensen 4a' (Poaceae). *Protoplasma* 240, 75–82. doi: 10.1007/s00709-009-0092-1

Shabala, S. (2013). Learning from halophytes: physiological basis and strategies to improve abiotic stress tolerance in crops. *Ann. Bot.* 112, 1209–1221. doi: 10.1093/aob/mct205

Shabala, S., Bose, J., and Hedrich, R. (2014). Salt bladders: do they matter? *Trends Plant Sci.* 19, 687–691. doi: 10.1016/j.tplants.2014.09.001

Shan, L., Zhou, R. C., Dong, S. S., and Shi, S. H. (2008). Adaptation to salinity in mangroves: implication on the evolution of salt-tolerance. *Chin. Sci. Bull.* 11, 1708–1715.

Shelden, M. C., Dias, D. A., Jayasinghe, N. S., Bacic, A., and Roessner, U. (2016). Root spatial metabolite profiling of two genotypes of barley (*Hordeum vulgare* L.) reveals differences in response to short-term salt stress. *J. Exp. Bot.* 67, 3731–3745. doi: 10.1093/jxb/erw059

Shimony, C., and Fahn, A. (1968). Light and electron microscopical studies on the structure of salt glands of *Tamarix aphylla* L. *J. Linn. Soc. Lond. Bot.* 60, 283–288. doi: 10.1111/j.1095-8339.1968.tb00090.x

Shimony, C., Fahn, A., and Reinhold, L. (1973). Ultrastructure and ion gradients in the salt glands of *Avicennia marina* (Forssk.) Vierh. *New Phytol.* 72, 27–36. doi: 10.1111/j.1469-8137.1973.tb02006.x

Skelding, A., and Winterbotham, J. (1939). The structure and development of the hydathodes of *Spartina townsendii* Groves. *New Phytol.* 38, 69–79. doi: 10.1111/j.1469-8137.1939.tb07085.x

Smart, K. E., Smith, J. A. C., Kilburn, M. R., Martin, B. G., Hawes, C., and Grovenor, C. R. (2010). High-resolution elemental localization in vacuolate plant cells by nanoscale secondary ion mass spectrometry. *Plant J.* 63, 870–879. doi: 10.1111/j.1365-313X.2010.04279.x

Sobrado, M. A., and Greaves, E. D. (2000). Leaf secretion composition of the mangrove species *Avicennia germinans* (L.) in relation to salinity: a case study by using total-reflection X-ray fluorescence analysis. *Plant Sci.* 159, 1–5. doi: 10.1016/S0168-9452(00)00292-2

Song, J., and Wang, B. (2014). Using euhalophytes to understand salt tolerance and to develop saline agriculture: *Suaeda* salsa as a promising model. *Ann. Bot.* 115, 541–553. doi: 10.1093/aob/mcu194

Storey, R., and Thomson, W. W. (1994). An X-ray microanalysis study of the salt glands and intracellular calcium crystals of *Tamarix*. *Ann. Bot.* 73, 307–313. doi: 10.1006/anbo.1994.1036

Tan, W. K., Ang, Y., Lim, T. K., Lim, T. M., Kumar, P., Loh, C. S., et al. (2015a). Proteome profile of salt gland-rich epidermis extracted from a salt-tolerant tree species. *Electrophoresis* 36, 2473–2481. doi: 10.1002/elps.201500023

Tan, W. K., Lim, T. M., and Loh, C. S. (2010). A simple, rapid method to isolate salt glands for three-dimensional visualization, fluorescence imaging and cytological studies. *Plant Methods* 6:24. doi: 10.1186/1746-4811-6-24

Tan, W. K., Lim, T. K., Loh, C. S., Kumar, P., and Lin, Q. (2015b). Proteomic characterisation of the salt gland-enriched tissues of the mangrove tree species *Avicennia officinalis*. *PLoS ONE* 10:e0133386. doi: 10.1371/journal.pone.0133386

Tan, W. K., Lin, Q., Lim, T. M., Kumar, P., and Loh, C. S. (2013). Dynamic secretion changes in the salt glands of the mangrove tree species *Avicennia officinalis* in response to a changing saline environment. *Plant Cell Environ.* 36, 1410–1422. doi: 10.1111/pce.12068

Thomson, W., Berry, W., and Liu, L. (1969). Localization and secretion of salt by the salt glands of *Tamarix aphylla*. *Proc. Natl. Acad. Sci. U.S.A.* 63, 310–317. doi: 10.1073/pnas.63.2.310

Thomson, W., and Platt-Aloia, K. (1985). The ultrastructure of the plasmodesmata of the salt glands of *Tamarix* as revealed by transmission and freeze-fracture electron microscopy. *Protoplasma* 125, 13–23. doi: 10.1007/BF01297346

Thomson, W. W. (1975). "The structure and function of salt glands," in *Plants in Saline Environments*, Vol. 15, eds A. Poljakoff-Mayber and J. Gale (Heidelberg: Springer), 118–146.

Thomson, W. W., and Liu, L. L. (1967). Ultrastructural features of the salt gland of *Tamarix aphylla* L. *Planta* 73, 201–220. doi: 10.1007/BF00387033

Volkov, V. (2015). Salinity tolerance in plants. Quantitative approach to ion transport starting from halophytes and stepping to genetic and protein engineering for manipulating ion fluxes. *Front. Plant Sci.* 6:873. doi: 10.3389/fpls.2015.00873

Wang, J., Yao, L., Li, B., Meng, Y., Ma, X., Lai, Y., et al. (2016). Comparative proteomic analysis of cultured suspension cells of the halophyte *Halogeton glomeratus* by iTRAQ provides insights into response mechanisms to salt stress. *Front. Plant Sci.* 7:110. doi: 10.3389/fpls.2016.00110

Wang, Y., Ma, H., Liu, G., Xu, C., Zhang, D., and Ban, Q. (2008). Analysis of gene expression profile of *Limonium bicolor* under NaHCO3 stress using cDNA microarray. *Plant Mol. Biol. Rep.* 26, 241–254. doi: 10.1007/s11105-008-0037-4

Yamamoto, N., Takano, T., Tanaka, K., Ishige, T., Terashima, S., Endo, C., et al. (2015). Comprehensive analysis of transcriptome response to salinity stress in the halophytic turf grass *Sporobolus virginicus*. *Front. Plant Sci.* 6:241. doi: 10.3389/fpls.2015.00241

Ye, Y., Tam, N. F.-Y., Lu, C. Y., and Wong, Y. S. (2005). Effects of salinity on germination, seedling growth and physiology of three salt-secreting mangrove species. *Aquat. Bot.* 83, 193–205. doi: 10.1016/j.aquabot.2005.06.006

Yeo, A. R., and Flowers, T. J. (1986). Ion transport in *Suaeda maritima*: its relation to growth and implications for the pathway of radial transport of ions across the root. *J. Exp. Bot.* 37, 143–159. doi: 10.1093/jxb/37.2.143

Yuan, F., Chen, M., Leng, B. Y., and Wang, B. (2013). An efficient autofluorescence method for screening *Limonium bicolor* mutants for abnormal salt gland density and salt secretion. *S. Afr. J. Bot.* 88, 110–117. doi: 10.1016/j.sajb.2013.06.007

Yuan, F., Chen, M., Yang, J. C., Leng, B. Y., and Wang, B. S. (2014a). A system for the transformation and regeneration of the recretohalophyte *Limonium bicolor*. *In Vitro Cell. Dev. Biol. Plant* 50, 610–617. doi: 10.1007/s11627-014-9611-7

Yuan, F., Deng, Y. Q., and Wang, B. S. (2014b). The observation of pollen viability and pollen tube germination in *Limonium bicolor*. *WIT Trans. Biomed. Health* 18, 631–636. doi: 10.2495/HHME130841

Yuan, F., Lyu, M. J. A., Leng, B. Y., Zhu, X. G., and Wang, B. S. (2016). The transcriptome of NaCl-treated *Limonium bicolor* leaves reveals the genes controlling salt secretion of salt gland. *Plant Mol. Biol.* 91, 241–256. doi: 10.1007/s11103-016-0460-0

Yuan, F., Lyv, M. J., Leng, B. Y., Zheng, G. Y., Feng, Z. T., Li, P. H., et al. (2015). Comparative transcriptome analysis of developmental stages of the *Limonium bicolor* leaf generates insights into salt gland differentiation. *Plant Cell Environ.* 38, 1637–1657. doi: 10.1111/pce.12514

Zhao, K. F., Fan, H., and Ungar, I. (2002). Survey of halophyte species in China. *Plant Sci.* 163, 491–498. doi: 10.1186/1756-0500-7-927

Zhou, S., Han, J. L., and Zhao, K. F. (2001). Advance of study on recretohalophytes. *Chin. J. Appl. Environ. Biol.* 7, 496–501.

Ziegler, H., and Lüttge, U. (1967). Die Salzdrüsen von *Limonium vulgare*. *Planta* 74, 1–17. doi: 10.1007/BF00385168

Zouhaier, B., Abdallah, A., Najla, T., Wahbi, D., Wided, C., Aouatef, B. A., et al. (2015). Scanning and transmission electron microscopy and X-ray analysisof leaf salt glands of *Limoniastrum guyonianum* Boiss. under NaCl salinity. *Micron* 78, 1–9. doi: 10.1016/j.micron.2015.06.001

2

Regulation of Na^+ fluxes in plants

Frans J. M. Maathuis*, Izhar Ahmad and Juan Patishtan

Department of Biology, University of York, York, UK

Edited by:
Vadim Volkov, London Metropolitan University, UK

Reviewed by:
Lars Hendrik Wegner, Karlsruhe Institute of Technology, Germany
Vadim Volkov, London Metropolitan University, UK
Rana Munns, University of Western Australia, Australia

*Correspondence:
Frans J. M. Maathuis, Department of Biology, University of York, York YO10 5DD, UK
e-mail: frans.maathuis@york.ac.uk

When exposed to salt, every plant takes up Na^+ from the environment. Once in the symplast, Na^+ is distributed within cells and between different tissues and organs. There it can help to lower the cellular water potential but also exert potentially toxic effects. Control of Na^+ fluxes is therefore crucial and indeed, research shows that the divergence between salt tolerant and salt sensitive plants is not due to a variation in transporter types but rather originates in the control of uptake and internal Na^+ fluxes. A number of regulatory mechanisms has been identified based on signaling of Ca^{2+}, cyclic nucleotides, reactive oxygen species, hormones, or on transcriptional and post translational changes of gene and protein expression. This review will give an overview of intra- and intercellular movement of Na^+ in plants and will summarize our current ideas of how these fluxes are controlled and regulated in the early stages of salt stress.

Keywords: calcium, CBL, CIPK, cyclic nucleotides, flux, salinity, signaling, sodium

INTRODUCTION

Salinity, in the form of NaCl, is one of the main abiotic stresses that reduces plant growth and development. Saline soils are typically defined as soils with conductivity of 4 dS m^{-1} or more. Salinity has two major effects: a relative early osmotic stress and ionic stress, which is expressed after a longer period (Munns and Tester, 2008; Munns, 2010).

THE GLOBAL IMPACT OF SALT STRESS

Salt stress affects agriculture worldwide. More than 5% of arable soil is now salinized which is equivalent to 77 million hectares (Munns et al., 1999). Soil salinization can be the result of natural conditions and of human activities. Natural causes can include weathering of rocks or salt deposits through precipitation. To illustrate, rain containing 10 mg kg^{-1} of sodium chloride would deposit 10 kg ha^{-1} of salt for each 100 mm of rainfall per year (Munns and Tester, 2008). Man-made causes of salinization are usually related to irrigation, such as the use of water high in minerals, or disturbance of local hydrological configurations for example by removal of deep rooted vegetation causing saline ground waters to contaminate upper soil layers. To remedy saline arable soils, improved drainage and irrigation management can leach the excess of Na^+ away from the root:soil boundary.

THE ROLE OF Na^+ IN SALT STRESS

It has been argued many times that salt stress is largely due to Na^+ (rather than Cl^-) and that within cells, it is cytoplasmic Na^+ which is the main culprit. Indeed, generally low Na^+ levels in the cytoplasm and or high K^+/Na^+ ratios in the cytoplasm are considered to mitigate salt damage. The reason for Na^+ being more toxic than Cl^- stems from the notion that Na^+ inhibits enzyme activity, and this is particularly the case for the many enzymes that require K^+ for functioning (Maathuis, 2009). For example, the K^+ dependent pyruvate kinase has a Km (for K^+ binding) of around 5 mM (Sugiyama et al., 1968). Na^+ can also bind but has only 5–10% of the stimulating effect of K^+ and thus severely inhibits kinase action (Duggleby and Dennis, 1973).

The frequency and severity of such Na^+ toxicity effects depends on the cytoplasmic Na^+ concentration ($[Na^+]_{cyt}$). Unfortunately, accurate measurements of $[Na^+]_{cyt}$ are still relatively scarce and those that have been reported vary greatly. Carden et al. (2003), using microelectrodes, measured 10–30 mM steady state levels of Na^+ in the cytoplasm of barley cells. Kronzucker et al. (2006), using flux analysis, reported $[Na^+]_{cyt}$ values of over 300 mM while measurements with fluorescent dyes yielded estimates from 20 to 60 mM (e.g., Anil et al., 2007). It has to be pointed out that these values are likely to vary with experimental conditions and between species but in all, it is likely that $[Na^+]_{cyt}$ is in the tens of millimols and thus prone to negatively affect enzyme activity.

Na^+ UPTAKE AND DISTRIBUTION

The above considerations imply that $[Na^+]_{cyt}$ needs to be kept low to avoid toxicity which in turn requires the capacity to distribute Na^+ across tissues and to remove Na^+ from the cytosol (**Figure 1**). To achieve, this some control over Na^+ fluxes is likely to be essential. A flux is the movement of a substance across an area per unit of time, for example mol Na^+ sec^{-1} m^{-2} (e.g., Maathuis, 2006b). Na^+ flux is the movement of charge across the membrane which is equivalent to an electrical current. Because the cell has a negative interior of around −150 to 200 mV, Na^+ will be poised to enter the cell in almost all conditions. In contrast, Na^+ efflux (i.e., removal from the cell) is not spontaneous and will require the expenditure of energy.

The estimated values of $[Na^+]_{cyt}$ are far lower than the thermodynamic equilibrium concentration (e.g., with external Na^+ at 10 mM and a membrane potential of −120 mV $[Na^+]_{cyt}$ would be ~1 M at equilibrium) and implies that potent Na^+ extrusion mechanisms are present to keep $[Na^+]_{cyt}$ at permissible levels. To test this assertion, one can also compare the unidirectional and net (unidirectional minus efflux) Na^+ influx in intact tissues. A plant which contains ~200 mmol Na^+ per kg FW and has a RGR of 10%

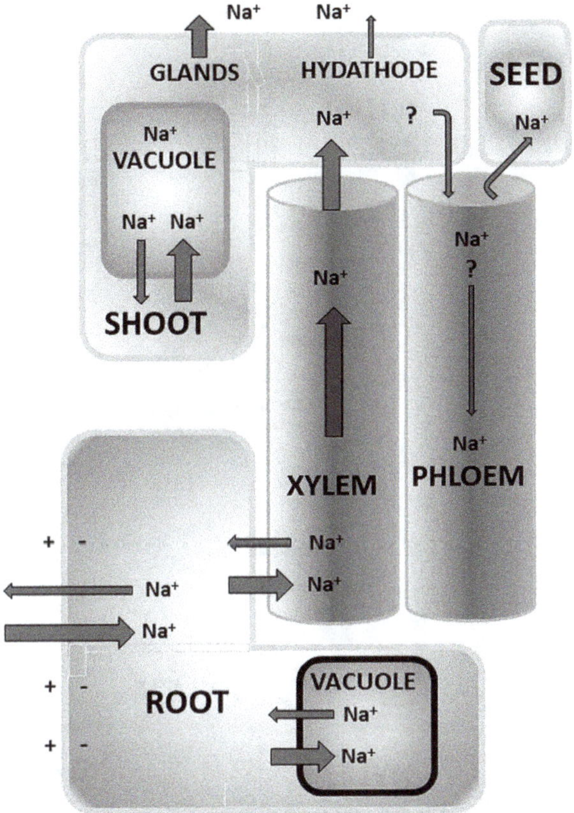

FIGURE 1 | Overview of the main Na^+ flux pathways that occur in terrestrial plants. Some mechanisms are still debated such as recycling of Na^+ from shoot to root via the phloem. Other mechanisms are anatomic adaptations found in a limited number of (halophytic) species only, such as extrusion via glands and hydathodes. The size of the arrow only provides a relative measure of various fluxes but is not meant to be quantitative.

day^{-1}, requires only a net Na^+ influx of around 800 nmol g^{-1} h^{-1} to stay at this level of cellular $[Na^+]$. However, experimentally obtained unidirectional Na^+ influx (e.g., measured in roots using radioactive Na^+) is typically 10 time higher than the above (e.g., Maathuis and Sanders, 2001). This implies that ~90% of Na^+ that initially entered the symplast is subsequently removed by Na^+ extrusion into the external medium. In the following section more detail is given about the mechanistic bases for these fluxes.

Na^+ UPTAKE

It is well established that Na^+ can enter the plant through ion channels and carrier type transporters. Na^+ permeable channels include glutamate like receptors (GLRs; Davenport, 2002) and cyclic nucleotide gated channels (CNGCs; Assmann, 1995; Bolwell, 1995; Trewavas, 1997; Newton and Smith, 2004) and possibly other, non-identified non-selective cation channels (NSCCs; Maathuis and Sanders, 1993; Tyerman et al., 1997; Demidchik and Tester, 2002; Essah et al., 2003). In addition, Na^+ uptake can be mediated by carrier type transporters on the plasma membrane, particularly those of the high affinity potassium transporters (HKT) family (Horie et al., 2001; Golldack et al., 2002; Mian et al., 2011).

Na^+ EFFLUX TO THE APOPLAST

Although some marine algae possess ATP driven Na^+ pumps (Wada et al., 1992) to extrude $[Na^+]_{cyt}$, higher plants rely on Na^+:H^+ antiporters (**Figure 2**). The latter are energized by the proton motive force (established by ATP and PPi driven H^+ pumps) which exists across plasma membrane and tonoplast. Biochemically, the presence of antiport systems was shown in the late 1980s using membrane vesicle and pH sensitive dyes such as acridine orange. However, in the late 1990s the first proteins and genes were identified using yeast complementation strategies (Apse et al., 1999) and forward mutant screens (Wu et al., 1996).

Removal of cytosolic Na^+ to the apoplast can be mediated by H^+ driven antiporters that are members of the NHA (Na^+:H^+ antiporter) family (**Figure 2**). Only one member of this family (SOS1) has been characterized in detail. SOS1 expression is prominent in root tip cells and also occurs in the xylem parenchyma (Wu et al., 1996). Root tip cells are predominantly evacuolate and hence incapable of vacuolar Na^+ compartmentation. Such tissues therefore must entirely rely on extrusion of cytoplasmic Na^+ into the apoplast, which is mediated by SOS1.

However, Na^+ extrusion into the apoplast is assumed to take place in most plant tissues, particularly at the root–soil boundary. Many of these tissues do not show SOS1 expression and it remains unclear which antiporters are involved. Other NHA isoforms may play a role but the NHA gene families in most plants only contain 1 or 2 isoforms. A further alternative is the CHX (cation H^+ exchange) family (**Figure 2**) which probably includes both K^+:H^+

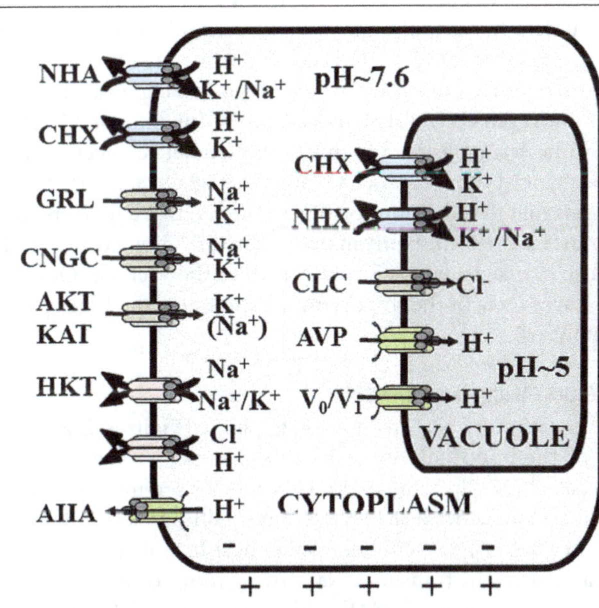

FIGURE 2 | Overview of the main membrane transporters that contribute to Na^+ and Cl^- uptake and distribution. Not necessarily all depicted proteins are expressed in one cell. AHA, H^+ pump; AKT/KAT, K^+ inward rectifying channel; AVP, vacuolar pyrophosphatase; CHX, cation-proton exchanger/antiport; CLC, chloride channel; CNGC, cyclic nucleotide gated channel; GRL, glutamate receptor like channel; HKT, K:Na and Na:Na, symporters; NHA, plasma membrane sodium-proton exchanger/antiport; NHX, tonoplast sodium-proton exchanger/antiport; Vo/V1, tonoplast H^+ ATPase.

and Na^+:H^+ antiporters and some CHXs have been implied in salinity tolerance (Evans et al., 2012).

VACUOLAR Na^+ SEQUESTRATION

Cytoplasmic Na^+ levels can also be kept low by exporting Na^+ into the vacuole, which typically makes up 80–90% of the cell volume. Although the vacuole may contain some Na^+-sensitive enzymes, it generally lacks any Na^+-sensitive biochemical machinery. The vacuolar [Na^+] can therefore rise much higher and often reaches several 100s of millimoles. Sequestration of Na^+ (and Cl^-) in the vacuole not only protects the cytoplasm, it also allows the plant to lower its cellular water potential and as such prevent water loss (in the cytoplasm so called compatible osmolytes fulfill a similar role but have to be synthesized at a considerable energetic cost).

One of the studies on plant antiporters discovered NHX1 (Na^+ H^+ exchanger 1) as a major player in the vacuolar cation movement. AtNHX1 overexpression significantly improved salinity tolerance in *Arabidopsis* (Apse et al., 1999) whereas its loss of function yielded the opposite effect. Manipulation of the expression of NHX1 orthologs in other species such as wheat (Xue et al., 2004), rice (Fukuda et al., 2004), and tomato (Zhang and Blumwald, 2001) showed the fundamental role this protein plays in salt tolerance and explains why it is a major focus for genetic engineering. However, more recent work has thrown some doubt on the molecular details of NHX activity and thus its physiological role. Most NHX isoforms that have been characterized can transport both K^+ and Na^+ and either have a similar Km for these substrates or even prefer K^+ (Jiang et al., 2010). This means that, unless the cytoplasmic Na^+ concentration is considerably higher than that for K^+, NHX exchangers mainly mediate K^+:H^+ exchange rather than Na^+:H^+ exchange (Zhang and Blumwald, 2001; Barragán et al., 2012). Indeed, loss of function of NHX1 and NHX2 in *Arabidopsis* led to impaired vacuolar K^+ accumulation but enhanced vacuolar Na^+ uptake (Barragán et al., 2012). Thus, it seems that the contribution of vacuolar NHX exchangers to salt tolerance is predominantly in maintaining K^+ homeostasis rather than in actual sequestration of Na^+ into the vacuole. Of course this leaves us with the important question how the latter process is catalyzed!

LONG DISTANCE Na^+ TRANSPORT

Translocation of Na^+ from root to shoot (**Figures 2** and **3**) is one of the important strategies in salt stress physiology (Flowers et al., 1977; Epstein, 1998). Glycophytes are mostly classified as salt excluders because they prevent significant accumulation of salts in photosynthetic tissues while most halophytes are includers and actively transport Na^+ from root to shoot (Flowers et al., 1977; Läuchli, 1984). This long distance transport has various points where the plant can exert control over salt distribution such as loading and translocation through the xylem and/or phloem mediated re-translocation from shoot to roots (**Figure 1**).

The Bypass flow solutes and water can reach the xylem via a symplastic or apoplastic route. The latter allows movement of solutes through the cell walls and intercellular spaces to the xylem without crossing plasma membranes and is sometimes called the

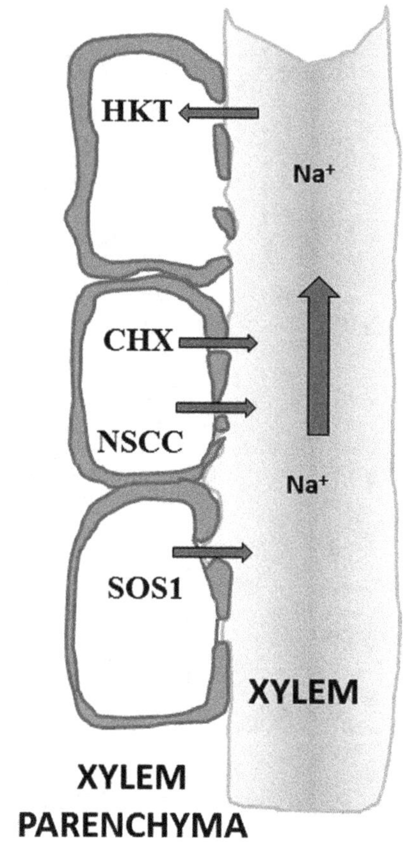

FIGURE 3 | Schematic of the xylem and xylem parenchyma anatomy. The xylem parenchyma consists of life cells that unload Na+ across their plasma membrane into the apoplast and xylem lumen (non-living tissue). Once in the xylem, bulk flow under positive and negative pressure ensures transport of minerals to shoot tissues.

"bypass flow" (Yeo et al., 1987; Anil et al., 2005; Krishnamurthy et al., 2009). Casparian strips and suberine layers in the root endodermal and exodermal layers provide barriers to apoplastic transport but in young roots and initiation sites of lateral roots these structures can be lacking or only partially effective. The efficacy of these anatomical features heavily depends on growth conditions such as the presence of silicon (Yeo et al., 1987) and Ca^{2+} (Anil et al., 2005). In many plants the apoplastic pathway is relatively limited but in other species such as rice the bypass flow can be substantial, especially at high levels of salinity (i.e., ~50% of total Na^+ uptake; Yeo et al., 1987) and therefore is responsible for significant amounts of Na^+ transport to the shoot.

The exact entry site for the Na^+ into the stele is not known. Yeo et al. (1987) proposed that the emerging sites of the lateral roots and cells walls near the root apices are the entry points. In rice, like other monocots, lateral roots arise from the pericycle through the endodermis breaking the casparian bands. Casparian bands are also often absent in the root tip regions. In contrast, Faiyue et al. (2010) showed that in rice the bypass flow significantly increases in the absence of lateral roots, by using mutant lines incapable of making lateral roots. These authors suggested that the higher

Na^+ content in the xylem sap and shoots of the mutant lines was caused by the different anatomical architecture and reduced suberine deposition on the exodermal and endodermal walls of the mutant. More recent data from Krishnamurthy et al. (2011) suggested the involvement of lateral root emergence in the leakage of tracer (trisodium-8-hydroxy-1,3,6-pyrenetrisulphonic acid) into the primary roots through the break created by the emergence of lateral roots.

Xylem loading Solutes delivered via the symplast have to cross the plasma membrane before they can be released into the xylem apoplast. This is another important step where plants can control solute translocation. Plasma membrane localized transporters are proposed to have a role in the xylem loading of Na^+ (Lacan and Durand, 1996) a process that involves the endodermis and xylem parenchyma cell layers (Epstein, 1998). The transport systems located at the xylem-parenchyma boundary may mediate both passive loading via Na^+ permeable channels and active loading through Na^+:H^+ exchangers (**Figure 3**). An example of the latter is SOS1, a plasma membrane antiporter that is expressed in root epidermis and root xylem parenchyma. The exact function of SOS1 is likely to depend on the severity of salinity stress and may include both xylem loading (low or moderate levels of salinity) and removal of Na^+ from the xylem (during high salinity). Members of CHX cation antiporter family are also implicated as playing a role in the loading of Na^+ into the xylem. For example, *Arabidopsis* CHX21 is mainly expressed in the root endodermis and loss of function in this protein reduced the level of Na^+ in the xylem sap (Hall et al., 2006).

The presence of non-selective ion channels (NSCs) in the plasma membrane of xylem parenchyma cells provides another pathway for Na^+ entry into the xylem. NSCs have been studied in xylem parenchyma cells of barley roots (Wegner and De Boer, 1997). The molecular identity of these NSCs is as yet unknown but could include members of the glutamate receptor like channels (GLRs) or CNGCs (Demidchik and Maathuis, 2007).

Na^+ retrieval Plants can reabsorb Na^+ from the xylem into the root cells as a mechanism to prevent large accumulation of Na^+ in the above ground tissues (Läuchli, 1984; Lacan and Durand, 1996). This retrieval mechanism was originally postulated in the 1970s (Lessani and Marschner, 1978) but now has a molecular basis (**Figure 3**). In *Arabidopsis*, disruption of HKT1 leads to hypersensitivity to salinity of the mutant lines with more Na^+ in the leaves (Mäser et al., 2002; Berthomieu et al., 2003; Davenport et al., 2007; Møller et al., 2009). The knockout lines showed higher Na^+ in the shoots but a lower level of K^+. These results favor the hypothesis that AtHKT1 is responsible for the retrieval of Na^+ from the xylem whilst directly stimulating K^+ loading. This is an ideal mechanism for plants to achieve a higher K^+/Na^+ ratio in shoots during salts stress (Horie et al., 2009). Similar reabsorption mechanisms were also found in rice and wheat. In rice, OsHKT1:5 is a plasma membrane Na^+ transporter expressed in xylem parenchyma cells that retrieves Na^+ from the xylem sap (Ren et al., 2005). The activity of OsHKT1;5 was significantly more robust in salt tolerant rice cultivars. In wheat, the HKTs NAX1 and NAX2 fulfill similar roles (Lindsay et al., 2004). Shi et al. (2002) suggested a similar role in xylem Na^+ reabsorption for the SOS1 transporter (depending on the level of salinity stress) but the evidence for this is less convincing.

Phloem recirculation of Na^+ Some Na^+ (and Cl^-) accumulation in shoot tissue is likely to be beneficial to plants because it provides "cheap" osmoticum to adjust the water potential. However, this is a risky strategy for most plants and indeed, in many glycophytes overall control of the Na^+ translocation breaks down, especially at higher salt levels (Maathuis, 2014). One mechanism to prevent shoot ion overaccumulation is an increased level of recirculation of Na^+ to the root via the phloem (**Figure 1**). At some stage, it was believed that HKT1 played an important role in this process (Berthomieu et al., 2003) but later work suggested this was not the case (Davenport et al., 2007). Indeed, the relevance of phloem Na^+ recirculation has yet to be firmly decided but one way of assessing this is to calculate the recirculation potential by comparing total xylem and phloem Na^+ flux. In saline conditions, xylem $[Na^+]$ can easily reach 100 mM (e.g., Mian et al., 2011). In contrast, phloem $[Na^+]$ does not appear to exceed ~20 mM, even on high salinity conditions (Munns and Fisher, 1986; Faiyue et al., 2010). This effective discrimination against Na^+ is one of the reasons why tissues that are phloem loaded, e.g., reproductive organs such as fruits and seeds or storage tissues such as tubers, typically show very low Na^+ contents even after exposure to saline conditions. The generally low phloem $[Na^+]$ coupled to the fact that phloem flow rates are typically threefold to fourfold smaller than xylem flow rates (e.g., Morandi et al., 2011), suggests that the overall potential to recirculate Na^+ would not exceed 5–7% of the total shoot Na^+ load and so is unlikely to play a major role in the reduction of shoot salt levels. However, accurate records of phloem flow rates would be very useful in this respect, especially during saline conditions.

Na^+ EXUDATION IN LEAVES

If the contribution of the phloem is small (see Long Distance Na^+ Transport) other mechanisms might help limit shoot ion levels (**Figure 1**). Guttation is common in many plants but more prominent in monocot plants. Guttation occurs through hydathodes, pore like structures that are often located along the margins and tips of leaves. Structurally, one can think of hydathodes as the end station of the xylem system where xylem sap exudes from the plant under influence of root pressure. Guttation fluid contains many inorganic ions but during salinity the Na^+ and Cl^- contents increase dramatically, a phenomenon that is often visible in the form of white salt crystals that precipitate on leaf and stem. Typical guttation rates are 5–10 ml h^{-1} m^{-2} (Shapira et al., 2013) but these will drop to 1–2 ml h^{-1} m^{-2} during osmotic stress or ~0.4–0.8 ml g^{-1} day^{-1}. If we assume a tissue $[Na^+]$ of 100 mM and a guttation solution with $[Na^+]$ of 5 mM (Chhabra et al., 1977), a total efflux of ~2–4 μmol g^{-1} day^{-1} would result, amounting to 2–4% of the total Na^+ content. Thus, guttation is not assumed to make any inroads in relieving the shoot of salt.

However, the principle of removing salt via exudation is taken further by some halophytes which have dedicated structures in the form of salt glands (**Figures 1** and **4**). Salt glands are usually modified hydathodes that contain specialized cells such as collecting cells that store salt and secretory cells that discharge the salt

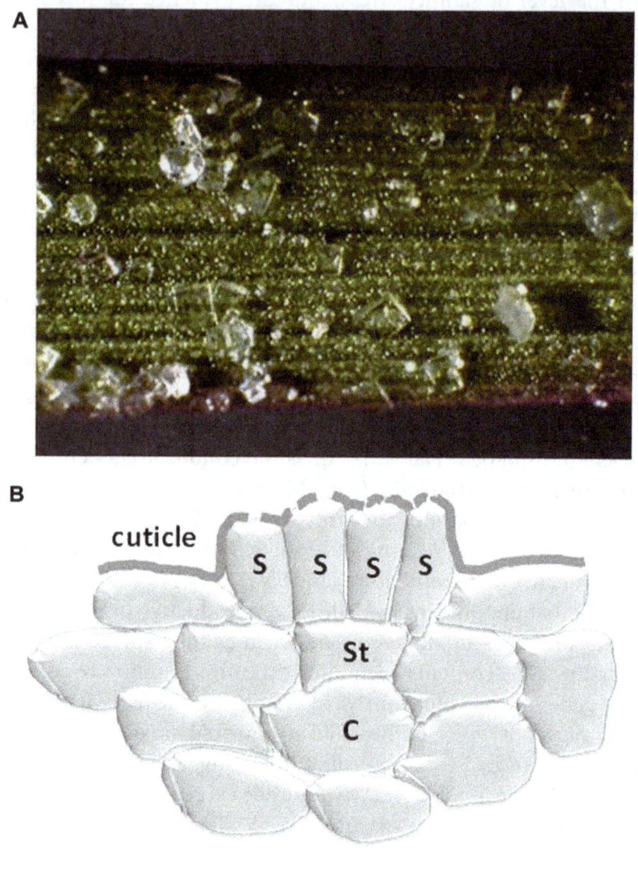

FIGURE 4 | (A) *Distichlis spicata* salt extrusion via multiple glands on its leaves. Note the formation of salt crystals (courtesy of Justin Klitzes, http://www.mbari.org/staff/conn/botany/seagrass/justin/salt_extrusion.htm). **(B)** Schematic diagram of a salt gland showing a collective cell (C), a stalk cell (St) and secretory cells (S). A cuticle is present with perforations to allow the excreted salt to exit the symplast.

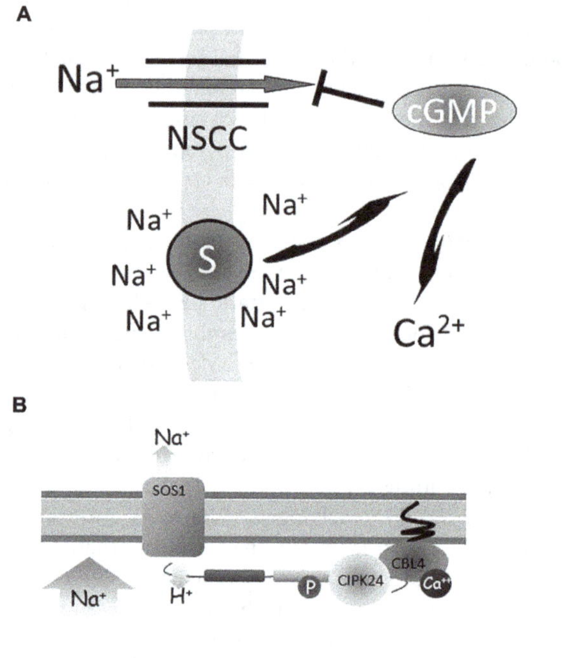

FIGURE 5 | Regulation of Na$^+$ flux. (A) A change in intra- or extracellular Na$^+$ may be reported by a (membrane–based) Na$^+$ sensor. This signal is translated into increased levels of cyclic GMP (cGMP) which in turn inhibits non-selective ion channel (NSCC) activity and thus reduces Na$^+$ influx. cGMP also affects transcription of many genes and also may be relayed downstream by calcium. **(B)** When salt is absent, the Na:H antiporter SOS1 is inactive. In the presence of salt a putative Ca^{2+} signal activates the calcineurin B like protein (CBL4) to interact with the CBL interacting protein kinase (CIPK24). The CBL-CIPK complex activates SOS1 via phosphorylation of its C-terminus leading to Na$^+$ extrusion.

through pores in the cuticle to the leaf surface. In contrast to guttation, salt excretion via glands can remove a large proportion of the shoot salt. For example, in mangroves 50–90% of salt that reaches the leaves can be lost again by this mechanism (Ball, 1988), showing excretion rates of 30 mmol/m^2 per day (Suarez and Medina, 2008). Excretion through salt glands has a large energetic cost and this may be the main reason why relatively few species show this adaptation.

SIGNALING PATHWAYS THAT REGULATE Na$^+$ FLUXES

When $[Na^+]_{cyt}$ exceeds a certain level, response mechanisms to reduce it are believed to initiate to prevent cellular damage. Response mechanisms could include vacuolar sequestration and/or extrusion of Na$^+$ to the apoplast, a reduction in Na$^+$ conductance of membranes at the root:soil boundary and for example increased Na$^+$ retrieval from the xylem. How such mechanisms are switched on and regulated is largely unknown especially where the early stages are concerned. Another, still largely unresolved question is whether plants do sense Na$^+$ at all or whether the response mechanisms we observe are mostly instigated by changes in osmotic potential.

REDUCING Na$^+$ CONDUCTANCE AT THE ROOT:SOIL BOUNDARY

Many pharmacological studies have shown that Na$^+$ influx is sensitive to various compounds such as ion channel blockers. Interestingly, the secondary messengers cyclic AMP and GMP also affect Na$^+$ influx (**Figure 5**). Studies on *Arabidopsis* seedlings (Maathuis and Sanders, 2001; Essah et al., 2003) and on mature pepper plants (Rubio et al., 2003) have shown that unidirectional Na$^+$ influx can be inhibited up to 35% by cGMP. Indeed, cyclic nucleotides appear to have a more general role in plant ion homeostasis, not only reducing Na$^+$ influx but also enhancing K$^+$ influx (Maathuis, 2006a) and as such may contribute to a high K$^+$:Na$^+$ ratio (Ahsan and Khalid, 1999; Maathuis and Amtmann, 1999). Further support for this mechanism was provided by later work that reported an increase in cellular cGMP within seconds after the onset of salt and osmotic stress (Donaldson et al., 2004). The latter study also implicated Ca^{2+} signaling as an intermediary in this process, downstream of the cGMP signal. cGMP also improves K$^+$ status during salt stress, possibly by improving K$^+$ uptake (Maathuis, 2006a) and by reducing K$^+$ efflux (Ordoñez et al., 2014). The latter work showed that reactive oxygen species (ROS) may be upstream of cyclic nucleotides.

Na$^+$ EXTRUSION TO THE APOPLAST

A rapid increase in cytosolic Ca^{2+} [Ca^{2+}]cyt concentration is a common response of plants exposed to stress conditions.

This increase of $[Ca^{2+}]$cyt can be sensed by several protein families such as calmodulins (CaMs; Assmann, 1995), calmodulin-binding protein kinases (CDPKs; Cheng et al., 2002) and calcineurin B-like (CBL) proteins (Batistič and Kudla, 2009). An increase in $[Ca^{2+}]$cyt can be detected by the CBL salt overly sensitive3 (SOS3/CBL4) protein (**Figure 5**). Activation of SOS3/CBL4 is followed by protein interaction with the serine/threonine protein kinase SOS2/CIPK24 (Halfter et al., 2000). The SOS3/CBL4-SOS2/CIPK24 complex migrates to the plasma membrane Na^+/H^+ antiporter SOS1 and increases its activity by phosphorylating the C-terminus of the SOS1 protein (Zhu, 2001).

REGULATION OF VACUOLAR Na^+ SEQUESTRATION

Some details are now known about how NHX1 activity may be upregulated (but see the above remarks about the uncertainty regarding the role of NHXs in Na^+ transport). Remarkably, one regulatory mechanism for NHX1 occurs in the vacuole: high vacuolar calcium may activate one of many Ca^{2+} binding proteins such as the calmodulin like protein CaM15. Binding of CaM15 to the (vacuolar) C-terminus of NHX1 is stimulated by acidic pH and inhibits Na^+:H^+ exchange by lowering the enzyme Vmax but, surprisingly, does not affect K^+:H^+ exchange (Yamaguchi et al., 2005). This scheme provides a regulatory framework to not only modulate vacuolar Na^+ deposition but also an elegant way to alter the cytoplasmic K^+:Na^+ ratio: a reduction in vacuolar pH, induced by increased proton exchange, would reduce CaM15 binding to NHX1 and therefore increase NHX1 Na^+:H^+ exchange.

Another potential Ca^{2+} dependent pathway to activate NHX activity is found in the cytoplasm. The calcium induced protein kinase CIPK24 is activated by an upstream protein called CBL10 (calcineurin-B-like). Interaction between CBL10 and CIPK24 is proposed to lead to phosphorylation of the NHX1 C-terminus and subsequent activation of the protein. Direct evidence for this has yet to be reported and it is difficult to envisage how this is consistent with the above model where the NHX1 C-terminus localizes to the vacuolar lumen (Reguera et al., 2014).

The above schemes evoke several intriguing questions. Firstly, if they are truly salt stress specific they are likely to be preceded by some type of sensor that reports on $[Na^+]$, either in the cytoplasm or other compartment. How plants register changes in ion levels is largely unknown except in the case of Ca^{2+}. In animals (where Na^+ plays an important physiological role) mechanisms for Na^+ sensing consist mostly of Na^+ selective ion channels. These can act as sensors (e.g., Watanabe et al., 2006) because the activity of these channels is directly correlated to the Na^+ concentration. Other mechanisms include proteins such as the protease thrombin which is modulated allosterically by Na^+ with a Kd of around 20 mM (Huntington, 2008). In thrombin, the Na^+ ion is coordinated by carbonyl oxygens from lysine and arginine residues and four water molecules and the Na^+ binding loop is directly connected to the active site (Huntington, 2008). Mammals also have Na^+ activated K^+ channels (expressed in neurons, kidney, heart, and skeletal muscle cells) that have Kd values for Na^+ binding between 50 and 70 mM. The Na^+ coordination domain in these channels has been located to the C-terminus (Zhang et al., 2010).

No Na^+ selective or Na^+ activated ion channels have been identified in plants, making it very unlikely that plant Na^+ sensing works in a similar fashion. However, like animals, plants may contain proteins that have regulatory Na^+ binding sites and on the basis of animal Na^+ binding sequences searches in plant genomes can be carried out to identify candidate proteins (Maathuis, 2014).

Another important question pertains to the selectivity of the Ca^{2+} signals. Ca^{2+} signals often occur in response to drought and salt stress with virtually identical signatures and convincing evidence for NaCl-specific rises in cytoplasmic Ca^{2+} are still lacking. Yeast studies showed that rapid Ca^{2+} transients (approximately 0–120 s) are entirely due to osmotic effects and did not vary for different salts or between ionic and non-ionic osmotica (Denis and Cyert, 2002; Matsumoto et al., 2002). Cell type specific Ca^{2+} responses to NaCl and equiosmolar sorbitol are also virtually identical (Kiegle et al., 2000) while a recent study by Moscatiello et al. (2013) showed no difference in Ca^{2+} signal between 600 sorb and 300 NaCl treated suspension cells. Thus, the observed transients most likely report changes in osmotic conditions. Similar doubts can be expressed regarding the specificity of cGMP signals that are recorded after both NaCl and osmotic stress (Donaldson et al., 2004). If signals report on changes in ionic status, then are they specific for Na^+ and/or Cl^- or do they report a generic response to variations in ionic strength such as changes in surface charge or membrane depolarization?

TRANSCRIPTIONAL REGULATION

Salinity can lead to rapid changes in transcript level. In most cases the involved genes are not salt specific and also respond to drought and osmotic stress. However, the transcription SERF (salt induced ethylene response factor) is upregulated after as little as 10 min of salt exposure (Schmidt et al., 2013). SERF is believed to be activated by ROS which have been shown to accumulate in salt exposed roots within minutes (Hong et al., 2009). The salt generated ROS may derive from membrane localized NADPH oxidases. Interestingly, the latter are themselves activated by Ca^{2+}. Downstream targets of SERF include MAP kinases which in turn modulate transcription of salt responsive genes.

CONCLUSION AND OUTLOOK

Salinity stress has, and continues, to put a major constraint on agriculture world-wide. Nevertheless, the existence of halophytic plants suggests that, at least in principle, crops can be adapted to grow in saline environments and a large effort has been spent over the past decades to pursue this strategy (Flowers et al., 1977). For example, many cereal crops have salt tolerant ancestors which can be exploited in breeding programs. Progress in this respect has been slow for many reasons but mainly because of the multigenic nature of salt tolerance. This is reflected in the wide dispersion of tolerance traits across the genome which in turn necessitates the introgression of many QTLs and genes to achieve a high yielding, tolerant variety.

The fact that salt tolerance relies on the activity of hundreds of genes makes a "gene specific" approach difficult as well (Munns, 2010). Manipulation of single genes, e.g., to increase

Na^+ sequestration or H^+ pumping capacity, yielded promising results in controlled conditions but applications in actual agricultural contexts have been limited so far. Engineering of tolerance is unlikely to be successful until robust methodologies are developed to alter expression of multiple genes, preferably in a tissue specific manner.

So where is progress going to come from? With more detailed insights into the genetic basis, signaling pathways and relevant proteins, both breeding and genetic engineering procedures should become more efficient. For example, molecular technology (Jia and Wang, 2014) increasingly facilitates the manipulating of multiple genes to either increase or suppress their expression. Altering the expression of key regulators that control suites of relevant genes may also be more promising in this respect. In the case of breeding, more targeted screens, e.g., at the organ or cellular level rather than based on broad whole plant phenotypes, should help to eliminate much noise and lack of reproducibility that previous analyses suffered from. In combination with molecular marker based programs the latter will accelerate the introgression of desired traits while minimizing "linkage drag."

REFERENCES

Ahsan, M., and Khalid, M. (1999). Effects of selecting for K^+/Na^+ and grain yield on salinity tolerance in spring wheat. *Pak. J. Biol. Sci.* 3, 679–681. doi: 10.3923/pjbs.1999.679.681

Anil, V. S., Krishnamurthy, P., Kuruvilla, S., Sucharitha, K., Thomas, G., and Mathew, M. (2005). Regulation of the uptake and distribution of Na+ in shoots of rice (*Oryza sativa*) variety Pokkali: role of Ca^{2+} in salt tolerance response. *Physiol. Plant* 124, 451–464. doi: 10.1111/j.1399-3054.2005.00529.x

Anil, V. S., Krishnamurthy, H., and Mathew, M. (2007). Limiting cytosolic Na^+ confers salt tolerance to rice cells in culture: a two-photon microscopy study of SBFI-loaded cells. *Physiol. Plant* 129, 607–621. doi: 10.1111/j.1399-3054.2006.00854.x

Apse, M. P., Aharon, G. S., Snedden, W. A., and Blumwald, E. (1999). Salt tolerance conferred by overexpression of a vacuolar Na^+/H^+ antiport in *Arabidopsis. Science* 285, 1256–1258. doi: 10.1126/science.285.5431.1256

Assmann, S. M. (1995). Cyclic AMP as a second messenger in higher plants (status and future prospects). *Plant Physiol.* 108, 885. doi: 10.1104/pp.108.3.885

Ball, M. (1988). Salinity tolerance in the mangroves *Aegiceras corniculatum* and *Avicennia marina*. I. Water use in relation to growth, carbon partitioning, and salt balance. *Funct. Plant Biol.* 15, 447–464. doi: 10.1071/PP9880447

Barragán, V., Leidi, E. O., Andrés, Z., Rubio, L., De Luca, A., Fernández, J. A., et al. (2012). Ion exchangers NHX1 and NHX2 mediate active potassium uptake into vacuoles to regulate cell turgor and stomatal function in *Arabidopsis. Plant Cell Online* 24, 1127–1142. doi: 10.1105/tpc.111.095273

Batistič, O., and Kudla, J. (2009). Plant calcineurin B-like proteins and their interacting protein kinases. *Biochim. Biophys. Acta* 1793, 985–992. doi: 10.1016/j.bbamcr.2008.10.006

Berthomieu, P., Conéjéro, G., Nublat, A., Brackenbury, W. J., Lambert, C., Savio, C., et al. (2003). Functional analysis of AtHKT1 in *Arabidopsis* shows that Na+ recirculation by the phloem is crucial for salt tolerance. *EMBO J.* 22, 2004–2014. doi: 10.1093/emboj/cdg207

Bolwell, G. P. (1995). Cyclic AMP, the reluctant messenger in plants. *Trends Biochem. Sci.* 20, 492–495. doi: 10.1016/S0968-0004(00)89114-8

Carden, D. E., Walker, D. J., Flowers, T. J., and Miller, A. J. (2003). Single-cell measurements of the contributions of cytosolic Na^+ and K^+ to salt tolerance. *Plant Physiol.* 131, 676–683. doi: 10.1104/pp.011445

Cheng, S.-H., Willmann, M. R., Chen, H.-C., and Sheen, J. (2002). Calcium signaling through protein kinases. The *Arabidopsis* calcium-dependent protein kinase gene family. *Plant Physiol.* 129, 469–485. doi: 10.1104/pp.005645

Chhabra, R., Ringoeta, A., Lambertsc, D., and Scheysc, I. (1977). Chloride losses from tomato plants (*Lycopersicon-esculentum*-mill). *Z. Pflanzenphysiol.* 81, 89–94. doi: 10.1016/S0044-328X(77)80042-1

Davenport, R. (2002). Glutamate receptors in plants. *Ann. Bot.* 90, 549–557. doi: 10.1093/aob/mcf228

Davenport, R. J., Muñoz-Mayor, A. A., Jha, D., Essah, P. A., Rus, A., and Tester, M. (2007). The Na+ transporter AtHKT1; 1 controls retrieval of Na+ from the xylem in *Arabidopsis. Plant Cell Environ.* 30, 497–507. doi: 10.1111/j.1365-3040.2007.01637.x

Demidchik, V., and Maathuis, F. J. (2007). Physiological roles of nonselective cation channels in plants: from salt stress to signalling and development. *New Phytol.* 175, 387–404. doi: 10.1111/j.1469-8137.2007.02128.x

Demidchik, V., and Tester, M. (2002). Sodium fluxes through nonselective cation channels in the plasma membrane of protoplasts from *Arabidopsis* roots. *Plant Physiol.* 128, 379–387. doi: 10.1104/pp.010524

Denis, V., and Cyert, M. S. (2002). Internal Ca^{2+} release in yeast is triggered by hypertonic shock and mediated by a TRP channel homologue. *J. Cell Biol.* 156, 29–34. doi: 10.1083/jcb.200111004

Donaldson, L., Ludidi, N., Knight, M. R., Gehring, C., and Denby, K. (2004). Salt and osmotic stress cause rapid increases in *Arabidopsis thaliana* cGMP levels. *FEBS Lett.* 569, 317–320. doi: 10.1016/j.febslet.2004.06.016

Duggleby, R. G., and Dennis, D. T. (1973). Pyruvate kinase, a possible regulatory enzyme in higher plants. *Plant Physiol.* 52, 312–317 doi: 10.1104/pp.52.4.312

Epstein, E. (1998). How calcium enhances plant salt tolerance. *Science* 280, 1906–1907. doi: 10.1126/science.280.5371.1906

Essah, P. A., Davenport, R., and Tester, M. (2003). Sodium influx and accumulation in *Arabidopsis. Plant Physiol.* 133, 307–318. doi: 10.1104/pp.103.022178

Evans, A., Hall, D., Pritchard, J., and Newbury, H. J. (2012). The roles of the cation transporters CHX21 and CHX23 in the development of *Arabidopsis thaliana. J. Exp. Bot.* 63, 59–67. doi: 10.1093/jxb/err271

Faiyue, B., Al-Azzawi, M. J., and Flowers, T. J. (2010). The role of lateral roots in bypass flow in rice (*Oryza sativa* L.). *Plant Cell Environ.* 33, 702–716. doi: 10.1111/j.1365-3040.2009.02078.x

Flowers, T., Troke, P., and Yeo, A. (1977). The mechanism of salt tolerance in halophytes. *Annu. Rev. Plant Physiol.* 28, 89–121. doi: 10.1146/annurev.pp.28.060177.000513

Fukuda, A., Nakamura, A., Tagiri, A., Tanaka, H., Miyao, A., Hirochika, H., et al. (2004). Function, intracellular localization and the importance in salt tolerance of a vacuolar Na^+/H^+ antiporter from rice. *Plant Cell Physiol.* 45, 146–159. doi: 10.1093/pcp/pch014

Golldack, D., Su, H., Quigley, F., Kamasani, U. R., Munoz-Garay, C., Balderas, E., et al. (2002). Characterization of a HKT-type transporter in rice as a general alkali cation transporter. *Plant J.* 31, 529–542. doi: 10.1046/j.1365-313X.2002.01374.x

Halfter, U., Ishitani, M., and Zhu, J.-K. (2000). The *Arabidopsis* SOS2 protein kinase physically interacts with and is activated by the calcium-binding protein SOS3. *Proc. Natl. Acad. Sci. U.S.A.* 97, 3735–3740. doi: 10.1073/pnas.97.7.3735

Hall, D., Evans, A., Newbury, H., and Pritchard, J. (2006). Functional analysis of CHX21: a putative sodium transporter in *Arabidopsis. J. Exp. Bot.* 57, 1201–1210. doi: 10.1093/jxb/erj092

Hong, C. Y., Chao, Y. Y., Yang, M. Y., Cheng, S. Y., Cho, S. C., and Kao, C. H. (2009). NaCl-induced expression of glutathione reductase in roots of rice (*Oryza sativa* L.) seedlings is mediated through hydrogen peroxide but not abscisic acid. *Plant Soil* 320, 103–115. doi: 10.1007/s11104-008-9874-z

Horie, T., Hauser, F., and Schroeder, J. I. (2009). HKT transporter-mediated salinity resistance mechanisms in *Arabidopsis* and monocot crop plants. *Trends Plant Sci.* 14, 660–668. doi: 10.1016/j.tplants.2009.08.009

Horie, T., Yoshida, K., Nakayama, H., Yamada, K., Oiki, S., and Shinmyo, A. (2001). Two types of HKT transporters with different properties of Na^+ and K^+ transport in *Oryza sativa. Plant J.* 27, 129–138. doi: 10.1046/j.1365-313x.2001.01077.x

Huntington, J. A. (2008). How Na^+ activates thrombin – a review of the functional and structural data. *Biol. Chem.* 389, 1025–1035. doi: 10.1515/BC.2008.113

Jia, H., and Wang, N. (2014). Targeted genome editing of sweet orange using Cas9/sgRNA. *PLoS ONE* 9:e93806. doi: 10.1371/journal.pone.0093806

Jiang, X., Leidi, E. O., and Pardo, J. M. (2010). How do vacuolar NHX exchangers function in plant salt tolerance? *Plant Signal. Behav.* 5, 792. doi: 10.4161/psb.5.7.11767

Kiegle, E., Moore, C. A., Haseloff, J., Tester, M. A., and Knight, M. R. (2000). Cell-type-specific calcium responses to drought, salt and cold in the *Arabidopsis* root. *Plant J.* 23, 267–278. doi: 10.1046/j.1365-313x.2000.00786.x

Krishnamurthy, P., Ranathunge, K., Franke, R., Prakash, H., Schreiber, L., and Mathew, M. (2009). The role of root apoplastic transport barriers in salt tolerance of rice (*Oryza sativa* L.). *Planta* 230, 119–134. doi: 10.1007/s00425-009-0930-6

Krishnamurthy, P., Ranathunge, K., Nayak, S., Schreiber, L., and Mathew, M. (2011). Root apoplastic barriers block Na^+ transport to shoots in rice (*Oryza sativa* L.). *J. Exp. Bot.* 62, 4215–4228. doi: 10.1093/jxb/err135

Kronzucker, H. J., Szczerba, M. W., Moazami-Goudarzi, M., and Britto, D. T. (2006). The cytosolic Na^+:K^+ ratio does not explain salinity-induced growth impairment in barley: a dual-tracer study using $42K^+$ and $24Na^+$. *Plant Cell Environ.* 29, 2228–2237. doi: 10.1111/j.1365-3040.2006.01597.x

Lacan, D., and Durand, M. (1996). Na^+-K^+ exchange at the xylem/symplast boundary (its significance in the salt sensitivity of soybean). *Plant Physiol.* 110, 705–711.

Läuchli, A. (1984). "Mechanisms of nutrient fluxes at membranes of the root surface and their regulation in the whole plant1," in *Roots, Nutrient and Water Influx, and Plant Growth*, eds S. A. Barber and D. R. Bouldin (Madison, WI: American Society of Agronomy, Crop Science Society of America, and Soil Science Society of America), 1–25.

Lessani, H., and Marschner, H. (1978). Relation between salt tolerance and long-distance transport of sodium and chloride in various crop species. *Funct. Plant Biol.* 5, 27–37. doi: 10.1071/PP9780027

Lindsay, M. P., Lagudah, E. S., Hare, R. A., and Munns, R. (2004). A locus for sodium exclusion (Nax1), a trait for salt tolerance, mapped in durum wheat. *Funct. Plant Biol.* 31, 1105–1114. doi: 10.1071/FP04111

Maathuis, F. J. M. (2006a). cGMP modulates gene transcription and cation transport in Arabidopsis roots. *Plant J.* 45, 700–711. doi: 10.1111/j.1365-313X.2005.02616.x

Maathuis, F. J. M. (2006b). "Transport across plant membranes," in *Plant Solute Transport*, eds T. J. Flowers and T. Yeo (Oxford: Blackwell), 75–98.

Maathuis, F. J. M. (2009). Physiological functions of mineral macronutrients. *Curr. Opin. Plant Biol.* 12, 250–258. doi: 10.1016/j.pbi.2009.04.003

Maathuis, F. J. M. (2014). Sodium in plants: perception, signalling, and regulation of sodium fluxes. *J. Exp. Bot.* 65, 849–858. doi: 10.1093/jxb/ert326

Maathuis, F. J. M., and Amtmann, A. (1999). K+ nutrition and Na+ toxicity: the basis of cellular K+/Na+ ratios. *Ann. Bot.* 84, 123–133. doi: 10.1006/anbo.1999.0912

Maathuis, F. J. M., and Sanders, D. (1993). Energization of potassium uptake in *Arabidopsis thaliana*. *Planta* 191, 302–307. doi: 10.1007/BF00195686

Maathuis, F. J. M., and Sanders, D. (2001). Sodium uptake in arabidopsis roots is regulated by cyclic nucleotides. *Plant Physiol.* 127, 1617–1625. doi: 10.1104/pp.010502

Mäser, P., Eckelman, B., Vaidyanathan, R., Horie, T., Fairbairn, D. J., Kubo, M., et al. (2002). Altered shoot/root Na+ distribution and bifurcating salt sensitivity in *Arabidopsis* by genetic disruption of the Na+ transporter AtHKT1. *FEBS Lett.* 531, 157–161. doi: 10.1016/S0014-5793(02)03488-9

Matsumoto, T. K., Ellsmore, A. J., Cessna, S. G., Low, P. S., Pardo, J. M., Bressan, R. A., et al. (2002). An osmotically induced cytosolic Ca^{2+} transient activates calcineurin signaling to mediate ion homeostasis and salt tolerance of *Saccharomyces cerevisiae*. *J. Biol. Chem.* 277, 33075–33080. doi: 10.1074/jbc.M205037200

Mian, A., Oomen, R. J., Isayenkov, S., Sentenac, H., Maathuis, F. J., and Véry, A. A. (2011). Over-expression of an Na+ and K+ permeable HKT transporter in barley improves salt tolerance. *Plant J.* 68, 468–479. doi: 10.1111/j.1365-313X.2011.04701.x

Møller, I. S., Gilliham, M., Jha, D., Mayo, G. M., Roy, S. J., Coates, J. C., et al. (2009). Shoot Na^+ exclusion and increased salinity tolerance engineered by cell type-specific alteration of Na^+ transport in *Arabidopsis*. *Plant Cell Online* 21, 2163–2178. doi: 10.1105/tpc.108.064568

Morandi, B., Zibordi, M., Losciale, P., Manfrini, L., Pierpaoli, E., and Grappadelli, L. C. (2011). Shading decreases the growth rate of young apple fruit by reducing their phloem import. *Sci. Horticult.* 127, 347–352. doi: 10.1016/j.scienta.2010.11.002

Moscatiello, R., Baldan, B., and Navazio, L. (2013). "Plant cell suspension cultures," in *Plant Mineral Nutrients*, Vol. 953, ed. F. J. M. Maathuis (Dordrecht: Humana Press), 77–93.

Munns, R. (2010). "Approaches to identifying genes for salinity tolerance and the importance of timescale," in *Plant Stress Tolerance* (Berlin: Springer), 25–38. doi: 10.1007/978-1-60761-702-0_2

Munns, R., Cramer, G. R., and Ball, M. C. (1999). Interactions between rising CO2, soil salinity and plant growth. Carbon dioxide and environmental stress. *Acad. Lond.* 139–167. doi: 10.1016/B978-012460370-7/50006-1

Munns, R., and Fisher, D. (1986). Na^+ and Cl^- transport in the phloem from leaves of NaCl-treated barley. *Funct. Plant Biol.* 13, 757–766. doi: 10.1071/PP9860757

Munns, R., and Tester, M. (2008). Mechanisms of salinity tolerance. *Annu. Rev. Plant Biol.* 59, 651–681. doi: 10.1146/annurev.arplant.59.032607.092911

Newton, R. P., and Smith, C. J. (2004). Cyclic nucleotides. *Phytochemistry* 65, 2423–2437. doi: 10.1016/j.phytochem.2004.07.026

Ordoñez, N. M., Marondedze, C., Thomas, L., Pasqualini, S., Shabala, L., Shabala, S., et al. (2014). Cyclic mononucleotides modulate potassium and calcium flux responses to H2O2 in Arabidopsis roots. *FEBS Lett.* 588, 1008–1015. doi: 10.1016/j.febslet.2014.01.062

Reguera, M., Bassil, E., and Blumwald, E. (2014). Intracellular NHX-type cation/H+ antiporters in plants. *Mol. Plant* 7, 261–226. doi: 10.1093/mp/sst091

Ren, Z.-H., Gao, J.-P., Li, L.-G., Cai, X.-L., Huang, W., Chao, D.-Y., et al. (2005). A rice quantitative trait locus for salt tolerance encodes a sodium transporter. *Nat. Genet.* 37, 1141–1146. doi: 10.1038/ng1643

Rubio, F., Flores, P., Navarro, J. M., and Martınez, V. (2003). Effects of Ca^{2+}, K^+ and cGMP on Na^+ uptake in pepper plants. *Plant Sci.* 165, 1043–1049. doi: 10.1016/S0168-9452(03)00297-8

Schmidt, R., Mieulet, D., Hubberten, H., Obata, T., Hoefgen, R., Fernie, A. R., et al. (2013). SALTRESPONSIVE ERF1 regulates reactive oxygen species–dependent signaling during the initial response to salt stress in rice. *Plant Cell* 25, 2115–2131. doi: 10.1105/tpc.113.113068

Shapira, O., Israeli, Y., Shani, U., and Schwartz, A. (2013). Salt stress aggravates boron toxicity symptoms in banana leaves by impairing guttation. *Plant Cell Environ.* 36, 275–287. doi: 10.1111/j.1365-3040.2012.02572.x

Shi, H., Quintero, F. J., Pardo, J. M., and Zhu, J.-K. (2002). The putative plasma membrane Na+/H+ antiporter SOS1 controls long-distance Na+ transport in plants. *Plant Cell Online* 14, 465–477. doi: 10.1104/pp.113.232629

Suarez, N., and Medina, E. (2008). Salinity effects on leaf ion composition and salt secretion rate in Avicennia germinans (L.). *Braz. J. Plant Physiol.* 20, 131–140. doi: 10.1590/S1677-04202008000200005

Sugiyama, T., Goto, Y., and Akazawa, T. (1968). Pyruvate kinase activity of wheat plants grown under potassium deficient conditions. *Plant Physiol.* 43, 730–734. doi: 10.1104/pp.43.5.730

Trewavas, A. J. (1997). Plant cyclic AMP comes in from the cold. *Nature* 390, 657–658. doi: 10.1038/37720

Tyerman, S. D., Skerrett, M., Garrill, A., Findlay, G. P., and Leigh, R. A. (1997). Pathways for the permeation of Na+ and Cl- into protoplasts derived from the cortex of wheat roots. *J. Exp. Bot.* 48, 459–480. doi: 10.1093/jxb/48.Special_Issue.459

Wada, M., Urayamab, O., Satoha, S., Harab, Y., Ikawab, Y., and Fujiia, T. (1992). A marine algal Na+-activated ATPase possesses an immunologically identical epitope to Na^+,K^+-ATPase. *FEBS Lett.* 309, 272–274. doi: 10.1016/0014-5793(92)80787-H

Watanabe, E., Hiyama, T. Y., Shimizu, H., Kodama, R., Hayashi, N., Miyata, S., et al. (2006). Sodium-level-sensitive sodium channel Nax is expressed in glial laminate processes in the sensory circumventricular organs. *Am. J. Physiol. Regul. Integr. Comp. Physiol.* 290, R568–R576. doi: 10.1152/ajpregu.00618.2005

Wegner, L., and De Boer, A. (1997). Properties of two outward-rectifying channels in root xylem parenchyma cells suggest a role in K^+ homeostasis and long-distance signaling. *Plant Physiol.* 115, 1707–1719.

Wu, S.-J., Ding, L., and Zhu, J.-K. (1996). SOS1, a genetic locus essential for salt tolerance and potassium acquisition. *Plant Cell Online* 8, 617–627. doi: 10.1105/tpc.8.4.617

Xue, Z.-Y., Zhi, D.-Y., Xue, G.-P., Zhang, H., Zhao, Y.-X., and Xia, G.-M. (2004). Enhanced salt tolerance of transgenic wheat (*Tritivum aestivum* L.) expressing a vacuolar Na^+/H^+ antiporter gene with improved grain yields in saline soils in the field and a reduced level of leaf Na^+. *Plant Sci.* 167, 849–859. doi: 10.1016/j.plantsci.2004.05.034

Yamaguchi, T., Aharon, G. S., Sottosanto, J. B., and Blumwald, E. (2005). Vacuolar Na+/H+ antiporter cation selectivity is regulated by calmodulin from within the vacuole in a Ca^{2+}- and pH-dependent manner. *Proc. Natl. Acad. Sci. U.S.A.* 102, 16107–16112. doi: 10.1073/pnas.0504437102

Yeo, A., Yeo, M., and Flowers, T. (1987). The contribution of an apoplastic pathway to sodium uptake by rice roots in saline conditions. *J. Exp. Bot.* 38, 1141–1153. doi: 10.1093/jxb/38.7.1141

Zhang, H.-X., and Blumwald, E. (2001). Transgenic salt-tolerant tomato plants accumulate salt in foliage but not in fruit. *Nat. Biotechnol.* 19, 765–768. doi: 10.1038/90824

Zhang, Z., Rosenhouse-Dantsker, A., Tang, Q. Y., Noskov, S., and Logothetis, D. E. (2010). The RCK2 domain uses a coordination site present in Kir channels to confer sodium sensitivity to Slo2.2 channels. *J. Neurosci.* 30, 7554–7562. doi: 10.1523/JNEUROSCI.0525-10.2010

Zhu, J.-K. (2001). Plant salt tolerance. *Trends Plant Sci.* 6, 66–71. doi: 10.1016/S1360-1385(00)01838-0

3

Quantitative description of ion transport via plasma membrane of yeast and small cells

Vadim Volkov*

Faculty of Life Sciences, School of Human Sciences, London Metropolitan University, London, UK

Edited by:
Hartmut Stützel, Leibniz Universität Hannover, Germany
Reviewed by:
Lars Hendrik Wegner, Karlsruhe Institute of Technology, Germany Dany Pascal Moualeu, Leibniz Universität Hannover, Germany
*Correspondence:
Vadim Volkov, Faculty of Life Sciences, School of Human Sciences, London Metropolitan University, 166-220 Holloway Road, London N7 8DB, UK
vadim.s.volkov@gmail.com

Modeling of ion transport via plasma membrane needs identification and quantitative understanding of the involved processes. Brief characterization of main ion transport systems of a yeast cell (Pma1, Ena1, TOK1, Nha1, Trk1, Trk2, non-selective cation conductance) and determining the exact number of molecules of each transporter per a typical cell allow us to predict the corresponding ion flows. In this review a comparison of ion transport in small yeast cell and several animal cell types is provided. The importance of cell volume to surface ratio is emphasized. The role of cell wall and lipid rafts is discussed in respect to required increase in spatial and temporary resolution of measurements. Conclusions are formulated to describe specific features of ion transport in a yeast cell. Potential directions of future research are outlined based on the assumptions.

Keywords: ion transport, yeast cell, erythrocyte, excitable membrane, lipid rafts, cell wall, salinity tolerance, systems biology

Introduction

The fungus *Saccharomyces cerevisiae* (Phylum Ascomycota) is a well-known baker's yeast. It is a small unicellular organism (**Figure 1**), which can grow in a wide range of pH, osmolality and various ion compositions of surrounding media. Yeast cells are among the best studied unicellular eukaryotic organisms with small sequenced genome, large available collections of mutants in specific genes, high growth rate in nutrient media. They are easy for genetic and molecular biological manipulation. Essential volume of accumulated knowledge about yeast facilitates further research in the area.

Yeast cells are widely used in the food industry, for baking and for brewing, for making wine and spirits. More recent and advanced applications include biotechnology, chemical industry and pharmacology where yeast cells are producing pharmaceutical and nutraceutical ingredients, commodity chemicals, biofuels and also heterologous proteins including different enzymes from other eukaryotic organisms. The commercial scale of production is achieved for the novel applications based on progress in synthetic biology and metabolic engineering (e.g., reviewed in Borodina and Nielsen, 2014). Yeast cells are invaluable for applications in biomedical research. Heterologous expression of mammalian proteins, especially membrane ones in yeast is an important means to understand their properties. Eukaryotic yeast cells with specific mutant phenotypes could be rescued after expressing homologous or complementing proteins from the other organisms, thus giving indications about the functions and interactions of the proteins. Amino acid mutations and substitutions within the proteins of interest allow detailed analysis of their structure and protein domains. Yeast two-hybrid screening is a technique in molecular biology to understand protein-protein interactions (Fields and Song, 1989; reviewed in Brückner et al., 2009); modifications of the method include split-ubiquitin system (Stagljar et al., 1998; reviewed in Thaminy et al., 2004) and several others for interacting membrane proteins *in vivo*. Yeast expression system helped to understand elements of calcium signaling in eukaryotic cells (reviewed in Ton and Rao, 2004). Yeast cells served for functional expression and characterization in more detail protein domains of an inward-rectifying mammalian K^+- channel mKir2.1 (Hasenbrink et al., 2005, 2007; Kolacna et al., 2005), rat neuronal potassium channel rEAG1 (Schwarzer et al., 2008), human estrogen alpha and beta receptors (Sievernich et al., 2004; Hasenbrink et al., 2006; Widschwendter et al., 2009), human receptors of the Hedgehog pathway (Joubert et al., 2010) and the other heterologous membrane proteins for functional and structural studies. The methods and applications are essentially based on the present and growing knowledge of yeast cells, their genetics, molecular biology and physiology.

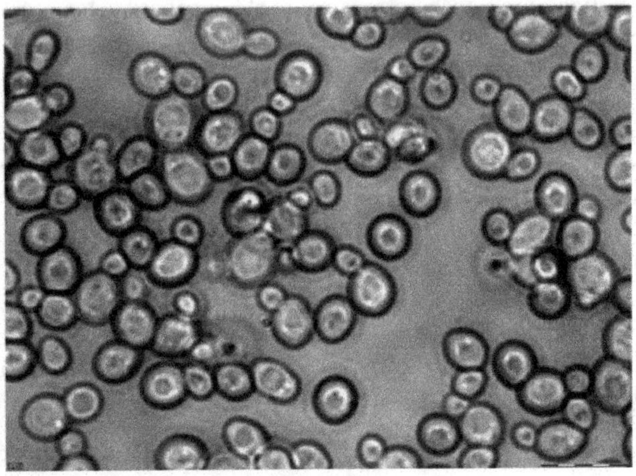

FIGURE 1 | Microscope image of mature yeast culture at high magnification. Yeast culture (BY4741) was grown for a day after reaching stationary phase. Scale bar is 12.5 μm.

Ion homeostasis is important for yeast growth (cell volume increases mainly by water uptake according to osmotic gradient against cell wall mechanical pressure) and also for the function of enzymes. Some proteins (e.g., yeast *HAL2* nucleotidase) may change conformation and lose activity under increased concentrations of Na^+ (Murguía et al., 1996; reviewed in Serrano, 1996; Albert et al., 2000). Understanding, describing and modeling ion transport is important for optimizing and improving growth conditions for yeast culture.

Initial assumptions for modeling seem oversimplified for a biologist; however, they are required for the basic biophysical description of the processes. The cell is considered to be a homogeneous spherical body consisting of viscous cytoplasm containing several ion species and surrounded by a lipid membrane. The lipid membrane contains a large number of incorporated proteins (ion pumps, channels, and transporters), which make pathways for selective and non-selective transport of ions. The cell is further surrounded by the cell wall.

Inner cell structures are present (e.g., nucleus, ATP producing mitochondria, clusters of so called lipid rafts within the plasma membrane, possible vacuolization and existing intracellular compartments etc.) and will be mentioned if necessary.

The numeric parameters of a yeast cell are—cell volume, membrane surface area, ion concentrations within and outside of the cell, yeast cell electric membrane potential, characteristics and number of ion transport systems of a yeast cell and also mechanical properties (elastic and plastic elasticity) of the cell wall. The presence of cell wall is a similarity between yeast, plant, algal and most of prokaryotic cells, while making them distinct from most of animal cells. Stretching cell wall balances hydrostatic turgor pressure, which is developed from the interior of the cell due to difference in osmotic pressures inside and outside of the cell. Positive turgor pressure is caused by water fluxes into the cell following higher concentration of osmotically active compounds inside. Ion gradients and partially the higher osmotic pressure are created by the concerted activity of ion pumps, channels and transporters, which also keep stable or ensure perturbed for signaling ion concentrations; ion transport systems are also responsible for negative membrane potential.

Exploring yeast with small size of their cells (several μm or around 10 wavelengths of red light) breaks trivial everyday experience about the world resembling to what is observed in microbiology (Beveridge, 1988) and cell biology (Albrecht- Buehler, 1990), hence requires special knowledge and equipment.

Quantitative Characteristics of Yeast Cells

Assuming an average diameter of yeast cell of about 6 μm and approximating the cell as a spherical body (**Figure 1**), we can calculate the **volume of the yeast cell** according to formula linking volume to diameter of sphere:

$$4/3^{*}\pi^{*}(6/2)^{3*}10^{-18}\text{m}^{3} = 4/3^{*}\pi^{*}27^{*}10^{-15}\text{L} \approx 100\,\text{fL} = 0.1\text{pL}. \quad (1)$$

For comparison, the volume of simpler prokaryotic microbe *Escherichia coli* is about 1,5-4,4 fL (e.g., Volkmer and Heinemann, 2011), the volume of a mammalian spermatozoon is about 20–30 fL (Curry et al., 1996). The volume of a human erythrocyte is also about 100 fL (Jay, 1975), but the volume of typical mammalian cardiomyocytes is 300 times larger (Satoh et al., 1996). The volume of giant squid axon is 108 times larger (5 cm long and 500 μm in diameter) (Lecar et al., 1967), the volume of barley leaf protoplasts is at least 100 times larger (Volkov et al., 2009).

It is important to mention that the basics of membrane ion transport were initially formulated in the outstanding works of Nobel Prize laureates Hodgkin and Huxley for large cells (squid axon) (Hodgkin and Huxley, 1952), so the principles for small cells with much higher surface/volume ratio may need to be scaled and amended. The surface/volume ratio is the surface per unit of volume; it consequently determines ion fluxes for the unit of volume and, *vice versa*, the potential metabolic activity per unit of surface (**Figure 2**). The surface/volume ratio is 125 times higher for a yeast cell than for a squid axon; this poses questions about potential principal differences in ion transport systems.

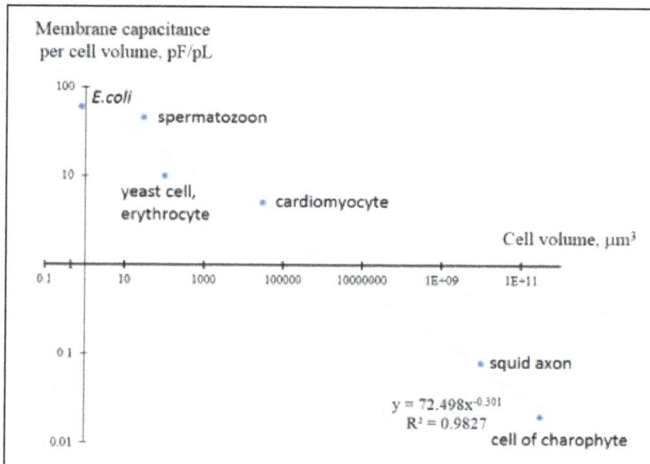

FIGURE 2 | Comparison of membrane electric capacitance per unit of volume in yeast and several small cells compared to larger giant squid axon and algal cells of charophytes. Membrane electric capacitance is linearly proportional to cell surface area with coefficient about $C_m = 1\mu F/cm^2$. The curve is approximated by inverse proportional function, both axes are in logarithmic scale. Higher values for spermatozoon are explained by the long tail of the cell; cardiomyocytes are flat cells with invaginations (T-tubules) to increase surface. More details are given in the text.

The cell surface area for a yeast cell is:

$$4^{*}\pi^{*}9\,\mu\text{m}^{2} \approx 100\,\mu\text{m}^{2}. \quad (2)$$

This value of surface area of biological membrane corresponds to electric capacitance 1 pF, since 100 μm^2 make up about 1 pF (the specific electric capacitance of biological membranes C_m is about C_m = 1μF/cm^2) (Hille, 2001). Similar results for capacitance are indeed found in patch clamp experiments (0.5–0.7 pF for yeast spheroplasts with diameters of 4–5 μm) (Roberts et al., 1999). This means that the electric charge q transfer of N = 600,000 monovalent cations (e.g., potassium, proton or sodium) out of cell will cause the change of voltage V from 0 to −100 mV according to the definition of electric capacitance C (q = C∗V) and taking into account the elementary charge $1.6*10^{-19}$ coulombs (equal to the electric charge of a monovalent cation; hence we have about $6.2^{*}10^{18}$ monovalent cations per coulomb):

$$N = (10^{-12}\text{F}^{*}0.1\,\text{V})/1.6^{*}10^{-19}\text{coulombs} \approx 6^{*}10^{5}. \quad (3)$$

This calculated number of cations is about 0.01% of K^+ ions in a yeast cell ($6*10^9$ ions per cell for 100 mM K^+, see below), however

100 times higher than the estimated number of free protons in the yeast cell (without a vacuole and without considering any pH buffering in the cell).

pH of the cytoplasm of a yeast cell is about 7.0, which means 10^{-7} M H^+/L, cell volume is 100 fL, therefore:

$$10^{-7}\ \mathrm{M\ H^+/L^*100^*10^{-15}\ L^*N_a} \approx 10^{-20}\ \mathrm{mole\ H^{+*}6.02^*10^{23}/mole} \approx 6,000\ \mathrm{protons/cell}, \quad (4)$$

where N_a = 6.02*10^{23}/mole is the Avogadro constant corresponding to the number of ions/molecules per mole of a chemical compound/substance.

Similar assumptions and calculations for the number of ions and voltage changes can be found also in Kahm (2011), for example, where several differential equations have been proposed and based on theoretical considerations to build a model of potassium homeostasis in *Saccharomyces cerevisiae*.

Ion concentrations in yeast cells were measured by several methods and gave readings about 100–150 mM for K^+ (Mulet and Serrano, 2002; Jennings and Cui, 2008; Zahrádka and Sychrova, 2012). However, concentrations of 50–300 mM of K^+ have been reported (depending on growth phase, external potassium etc., reviewed in Ariño et al., 2010; Kahm, 2011). Under low (0.1 mM) external K^+ at the background of 100– 300 mM NaCl internal K^+ concentration below 5 mM has been measured (Alemán et al., 2014). Na^+ depends more on the external concentration and may vary from a few mM to over 100 mM (García et al., 1997; Kolacna et al., 2005). For example, addition of external 1 M NaCl increased internal Na^+ concentration linearly from 0 to 150 mM after 100 min (García et al., 1997). Chloride concentration is relatively low and stable, about 0.1–1 mM, indicating that chloride is also a homeostatically controlled abundant ion in the cell (Jennings and Cui, 2008). Calcium concentration in yeast and eukaryotic cells is very low, 50–200 nM. This ion is essential for signaling, so specific pumps and transporters exist to ensure calcium homeostasis and signaling (e.g., Cui and Kaandorp, 2006).

These concentrations correspond to relatively small numbers of ions per cell. Even the number of K^+ ions is amenable for simple modeling using modern software and computers. Instead and in addition to using thermodynamic approach, location and properties of each ion could be potentially digitized and analyzed.

$$\begin{aligned} &100\ \mathrm{mM\ K^+\ in 100\ fL\ gives\ N_a^* 100\ mM^*100\ fL} \\ &\approx 6.02^*10^{23}/\mathrm{mole}^*100\ \mathrm{mmole/L}^*100\ 10^{-15}\ \mathrm{L} \\ &\approx 6^*10^9\ \mathrm{ions\ of\ K^+/cell}, \end{aligned} \quad (5)$$

where N_a = 6.02 * 10^{23}/mole is the Avogadro constant.

This number of ions is equivalent to an electrical current of 1.6*10^{-19} coulombs/ion *6*10^9 ions ≈ 10^{-9} A*seconds = 10 seconds * 100 pA (elementary charge of cation 1.6*10^{-19} coulombs multiplied by the number of ions and converted to Amperes; will be discussed later when considering ion currents via ion channels and transporters).

Direct measurements of **membrane potential** using glass microelectrodes is the only direct method available to measure membrane potential. However, for small cells of *Saccharomyces cerevisiae* the method may be inaccurate and can underestimate membrane potential considering that (1) membrane potential could be changed by the proton pump Pma1 within seconds (see discussion below) and (2) microelectrodes may have an effect on the measured values, for example, they cause KCl leak and mechanical stress (e.g., Blatt and Slayman, 1983). Therefore, indirect methods and comparative results for different yeast species and under several conditions are very important.

Microelectrode technique records membrane potentials of around −70 to −45 mV for S. *cerevisiae* (reviewed in Borst- Pauwels, 1981) and similar or lower values for other yeast species (summarized in **Table 1**).

TABLE 1 | Values of membrane potential of yeast and several fungal and bacterial cells recorded by microelectrodes.

Species	Membrane potential values, mv	References	Notes
Saccharomyces cerevisiae	Around −70 to −45	Reviewed in Borst-Pauwels, 1981	Values might be not realistic, fast decay in some experiments
Endomyces magnusii, (size of cells is about 15 × 30 μm)	(1) Around −40 (2) Around −150 to −120	(1) Vacata et al., 1981 (2) Bakker et al., 1986	(1) Measured in artificial pond water, pH is not indicated, over 1.5 mM K^+, 0.5 mM Na^+, 0.2 mM Ca^{2+} (2) −124 mV at pH 4.5 (0.1 mM KCl), preimpalement value estimated around −190 mV; −146 mV at pH 7.1 (0.1 mM KCl) and preimpalement value estimated around −275 mV
Pichia humboldtii (cell size is about 12–15 μm in diameter)	Around −90 to −50	Höfer and Novacky, 1986; Lichtenberg et al., 1988	100 mM KCl depolarized membrane of *Pichia humboldtii* from −48 to −21 mV, while 100 mM NaCl to −37 mV and 100 mM LiCl by 3 mV only; membrane potential depended on external pH increasing from −60 to −30 mV under external pH change from 6 to 8 or from 6 to 4 Höfer and Novacky, 1986
Neurospora. crassa (a) hyphae diameter over 10–15 μm or (b) spherical cells with diameter about 15–20 μm)	Below −300 to −120	(a) Slayman and Slayman, 1962; (b) Blatt and Slayman, 1983, 1987	Depended on concentration of external K^+ and internal pH
Giant *E. coli* cells (about 5 μm in diameter)	Below −140	Felle et al., 1980	−100 mV at pH 5.5 and −142 mV at pH 8.0

Indirect methods of membrane potential measurements are based on steady-state distribution of lipophilic cations and require further simple calculations; tetra[^{3}H]—phenylphosphonium (TPP^+) and tri[^{3}H]phenylmethylphosphonium ($TPMP^+$) are used to measure membrane potential in yeast cells. With this method values around −120 mV to −50 mV were estimated (Vacata et al., 1981). Measurements using 3,3′-dipropylthiacarbocyanine

gave estimates around −160 to −150 mV for *S. cerevisiae*; the membrane potential depolarized within 10 s by 25–30 mV in response to addition of 10 mM KCl (Peña et al., 2010). The effect of depolarization by KCl was observed in the presence of 20 mM glucose suggesting the implication of proton pump Pma1 in membrane energization (Peña et al., 2010). Recordings with 3,3′-dipropylthiacarbocyanine demonstrated depolarization for aerobic yeast *Rhodotorula glutinis*: the cells depolarized by 25 mV in 25 mM KCl, by 60 mV in 100 mM KCl and by 60 mV, when external pH changed from 6 to 3 (Plášek et al., 2012). In the presence of 100 mM glucose cells of *S. cerevisiae* depolarized by 30 mV in 100 mM KCl, by 20 mM in 100 mM NaCl and by 40 mV under pH decrease from 6 to 3 (Plášek et al., 2013). In the absence of glucose depolarization was larger for pH change and 100 mM KCl though smaller for 100 mM NaCl (Plášek et al., 2013).

This shows that indirectly measured membrane potential values are similar (Felle et al., 1980; Lichtenberg et al., 1988) or more negative (Vacata et al., 1981) than when measured directly by microelectrodes due to a variety of reasons, e.g., impalement problems and leak from microelectrodes, or non- specific absorption of cations.

Ion Transport Systems of Yeast Cells

Ion transport systems have been well studied for yeast cells within the last 30–40 years following the rise of molecular biology, electrophysiology and complete sequencing of the yeast genome. The new methods complement traditional experiments with yeast mutants and ion accumulation by yeast cells to elucidate the mechanisms of transport and role of specific proteins. The main ion transport systems include ion pumps, several transporters, one potassium channel and a non-selective cation current, which was recorded in electrophysiological experiments (**Figure 3**).

Ion Pumps

The most abundant protein in the yeast plasma membrane is the **proton pump Pma1** (P-type H^+-ATPase), which pumps protons out of the cytoplasm and shifts the membrane potential to more negative values. Pma1 transports one H^+ per ATP molecule (Ambesi et al., 2000) and makes up about 15–20% of membrane proteins (Eraso et al., 1987). Protein expression analysis (Ghaemmaghami et al., 2003) estimates the amount of Pma1 per cell to be 1,260,000 Pma1 molecules/cell. This is equivalent to 12,600 molecules/μm^2 and equates to one Pma1 molecule per 10∗10 nm^2 of membrane surface. The value is above the feasible limits, since the linear size of proton ATPases is large and, for example, the cross-section of similar H^+-ATPase from the plasma membrane of *Neurospora crassa* is nearly 10∗10 nm^2 (Auer et al., 1998). Assuming that only 10% of expressed Pma1 is present in the plasma membrane (many molecules could be in secretory vesicles in cytoplasm) we get 130,000 molecules per plasma membrane. This number coincides better with electron microphotographs of yeast plasma membranes labeled with antibodies for Pma1 (Permyakov et al., 2012).

One Pma1 molecule transports 20–100 H^+ per second (Serrano, 1988). The conclusion is confirmed by measurements of ATP hydrolysis activities (e.g., Perlin et al., 1989): assuming transport of one H^+ per one reaction of ATP hydrolysis and knowing ATPase activities for reconstituted mutant enzymes in μmol P_i mg^{-1}

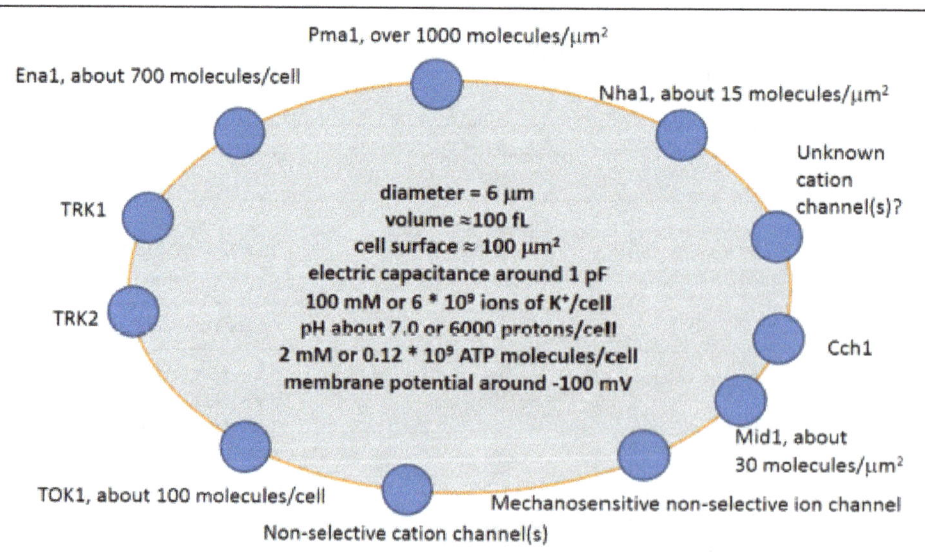

FIGURE 3 | Simplified scheme of cation transport systems in plasma membrane of a typical yeast cell. Pma1 is H^+-ATPase pumping protons out of the cell; Ena1 is Na^+-ATPase pumping Na^+ out of the cell; TRK1 and TRK2 are presumed potassium transporters; TOK1 is outward potassium-selective ion channel; Mid1 is a mechanosensitive cation-selective ion channel; Cch1 is calcium channel; Nha1 is presumed Na^+/H^+ antiporter. More explanation and controversy is provided in the text.

min-1 we can estimate the number of transported H^+/(Pma1 molecule∗second). The value is similar to animal Na^+/K^+-ATPase with a turnover of 160 (Skou, 1998). Taking the lower of these values it is simple to estimate the possible number of protons extruded by Pma1 within one second per yeast cell:

$$20H^+/(\text{second*pump molecule})*130,000\text{pump molecules/cell} = 2,600,000\ H^+/(\text{second*cell}). \quad (6)$$

This value exceeds by a factor of four the number of positive charges required to shift the membrane potential of the cell by −100 mV in one second (Equation 3). Obviously, the situation is more complex since one must also consider thermodynamics of transport and ATP hydrolysis. With a cytoplasmic pH similar to ambient medium Pma1 will not be able to pump protons against a membrane potential lower than -400 to -450 mV due to limitation by energy of ATP hydrolysis: −30 kJ/mole to be translated to the energy of transporting an elementary charge against the membrane voltage and would correspond to transport against −300 mV, slightly more negative hydrolysis energy could be achieved and result in more negative voltages (e.g., Bond and Russel, 1998). Indeed, the most negative membrane potentials recorded in fungal/autotrophic cells are around −300 to −350 mV for *Neurospora crassa* with equilibrium potential up to −450 mV (Blatt and Slayman, 1987).

These simple considerations show that the most important factors for membrane potential and driving force for translocation of protons (1μH⁺) in yeast cell are:

(1) ATP availability and cell metabolism;

(2) Pma1 regulation, not the potential maximal Pma1 activity;

(3) futile cycles of ion transport and inward electrogenic ion transport.

ATP concentration in yeast cells is about 2 mM in the presence of glucose (measured by aptamer-based ATP nanosensor) (Ozalp et al., 2010) or around: 250–350 attomole ATP per a cell (Ashe et al., 2000) estimated to 2.5–3.5 mM; 8.4 ± 0.4 μmole ATP/3∗10^7 cells (Ullah et al., 2013) recalculating to 2.8 mM. It is similar to the ATP concentration in bacteria, for example 3 mM ATP was measured in *Streptococcus bovis* (Bond and Russel, 1998). The total number of ATP molecules will be in that case:

$$2\ \text{mM ATP in } 100\ \text{fL gives } N_a^*2\ \text{mM}^*100\ \text{fL}$$
$$\approx 6.02^*10^{23}/\text{mole}^*2\ \text{mmole/L}^*100^*10^{-15}\text{L}$$
$$\approx 1.2^*10^8\ \text{ATP molecules/cell, where}$$
$$N_a = 6.02^*10^{23}/\text{mole is the Avogadro constant.}$$

This is at least 50 times higher than the potential minimal requirement of ATP consumed by Pma1 per a second (Equation 6); therefore ATP is likely not a limiting factor immediately linking metabolic activity of yeast cell with fast membrane transport processes (timescale of seconds). If the highest number of Pma1 in the membrane (1.3∗106 molecules) is assumed (not likely due to above mentioned steric calculations for the molecules) and the turnover of 100 H^+/(Pma1 molecule∗second) is taken, the amount of ATP is still sufficient for a second. There is a reasonable time lapse of 1–50 s (buffered by internal existing concentration of ATP) between (1) the proton transport and plasma membrane energization by Pma1 and (2) cell metabolism and ATP synthesis. It is worth to include for consideration ATP/ADP ratio, energy charge of yeast cells, concentrations of free phosphate and ADP. The parameters influence the change of the Gibbs free energy 1G for ATP hydrolysis (e.g., Bond and Russel, 1998) and also regulate many biochemical and physiological reactions in cells (e.g., Pradet and Raymond, 1983; Wilson et al., 1996; Gout et al., 2014) linking ion transport and cell metabolism. Total concentration of adenine nucleotides in cytoplasm of yeast cell was measured around 3–5 mM (Brindle et al., 1990; Nielsen et al., 2010), while ATP/ADP ratio varied from about 5 in high glucose to 0.3 after 15 min in low glucose, ATP/AMP ratio dropped from over 23 to 0.1 under the switch from high to low glucose (Wilson et al., 1996). Intracellular phosphate was similar to adenine nucleotides, 2.5–5.2 mM (Brindle et al., 1990). Complex and not always predictable changes of cytoplasmic nucleotide concentrations follow cell metabolism. For example, addition of 100 mM glucose at the background of 2% trehalose five-fold decreased ATP concentration less than in 30 s (Loret et al., 2007). Interestingly and unexpectedly, the reduction of ATP was due to increase of IMP and inosine without essential change of ADP and AMP; concentrations of adenine nucleotides recovered to initial levels in about 30 min (Loret et al., 2007). It is a complicated network of transcription factors, protein kinases and the other regulatory proteins and factors involved in nutritional, energy and metabolic control in yeast cells (e.g., Lee et al., 2008; Broach, 2012; Österlund et al., 2013). Pma1 is phosphorylated by several protein kinases and changes affinity to ATP in response to glucose metabolism; it is a way to modulate membrane potential (Goossens et al., 2000). Regulation of Pma1 involves the other numerous proteins and networks of events and is the subject of specific reviews (e.g., discussed in Ariño et al., 2010 and referenced therein; Babu et al., 2012), therefore it will not be discussed in more detail here.

The proton motive force of Pma1 is used for the transport of ions and small molecules by ion channels and transporters, dissipating the very negative membrane potential and proton gradient. Futile cycles of ion transport have been studied in bacteria (e.g., Bond and Russel, 1998); these show that over a third of the energy could be dissipated by cycling ions through the cell membrane. For yeast, futile cycles of protons and anions under adaptation to weak organic acids explained about 30–40% reduction in intracellular ATP compared to growth inhibitory conditions (e.g., Ullah et al., 2013), while the percentage of dissipated energy needs more investigation. Futile cycles of ion transport for Na^+, K^+ and the other major ions are also well known for plant root cells, the estimates predict over a third and more of energy consumed by the cycles (Britto and Kronzucker, 2006; Malagoli et al., 2008 and later publications from the laboratory). At a first glance, futile cycles seem to be waste of energy and pitfalls of tight regulation. On the opposite, from the point of systems biology futile cycles are an additional mechanism of control and signaling and also a prerequisite for non-linear complex behavior of a biological system (Newsholme and Crabtree, 1976; Samoilov et al., 2005; Qian and Beard, 2006; Tolla et al., 2015). In yeast it is assumed that 20–60% of cell ATP is used by Pma1 (reviewed in Kahm, 2011). There are over 250 predicted transporters in the yeast genome (Paulsen et al., 1998) and almost 100 plasma membrane transporters of known function (Van Belle and André, 2001). Many of them transport small organic molecules, not ions, and can use the proton gradients and membrane potential created by Pma1. The "leak" current via membrane (non-selective cation channels and several other potential routes) and activity of transporters are regulated by membrane potential, calcium, ion concentrations etc. (Bihler et al., 1998; Roberts et al., 1999; reviewed in Ariño et al., 2010) and so are linked to the activity of Pma1 generating a feedback loop. There are still many questions as to which parameter or parameters are controlled and which mechanisms are involved in the regulation of ion transport. Voltage sensors and ligand or ion-binding sites of ion channels are the simplest structures that directly regulate ion currents depending on membrane potential, ion and ligand concentrations (e.g., Hille, 2001), making up regulatory networks. The study of various mutants is helpful in elucidating the interactions. Mutants with about 80% reduced expression of plasma membrane H+-ATPase showed decreased rates of proton efflux and uptake of amino acids, while maintaining a stable membrane potential (measured by gradient of tetraphenylphosphonium) (Vallejo and Serrano, 1989). Mutants lacking outward potassium channel TOK1 had depolarized membrane potential (Maresova et al., 2006), while deletion of Trk1 and Trk2 potassium transporters hyperpolarized the membrane (Zahrádka and Sychrova, 2012).

Ena1 is another ion pump in the yeast plasma membrane recognized as a putative Na^+ pump (Haro et al., 1991) removing Na^+ out of cytoplasm. Around 688 molecules per cell were estimated by protein expression analysis (Ghaemmaghami et al., 2003), which means that the potential maximal efflux is extremely low. Assuming pumping activity of Ena1 similar to Pma1 with 20–100 H^+/second (Serrano, 1988) or animal Na^+/K^+-ATPase with a turnover of 160 (Skou, 1998), an estimate of 100 Na^+ pumped out/second seems reasonable. Then the expected drop in intracellular Na^+ due to activity of Ena1 would be:

$$\begin{aligned}&700\ \text{pumps/cell}^*100\ \text{Na}^+\text{ions/(second}^*\text{pump)}\\&= 70^*10^3\text{Na}^+\text{ions/(cell}^*\text{second) or about}\\&1\ \mu\text{M/(cell}^*\text{second)(converting to concentrations using}\\&\text{Avogadro constant and cell volume) or 3.6 mM/(cell}^*\text{h).}\end{aligned} \quad (7)$$

Thus, under control conditions Ena1 with a number of about 700 molecules/cell can hardly have an essential effect on sodium concentration. However, the transcription of Ena1 is activated by NaCl (Wieland et al., 1995) via multiple transduction pathways in dose-dependent manner (Márquez and Serrano, 1996). Microarray experiments revealed the complex pattern of activation (Posas et al., 2000 and http://transcriptome. ens.fr/ymgv/index.php for YDR040C), about 10-fold at most. Presumably the activation of transcription could be mirrored by increased number of protein molecules/cell. Interestingly, deletion of Ena1 in the yeast strain G19 resulted in dramatically retarded growth of yeast colonies on plates with added 0.5–1.3 M NaCl and even with added 1.8 M KCl especially at neutral pH 7.0 when compared with pH 3.6 and 5.5 (Bañuelos et al., 1998); therefore expression and interactions of Ena1 are expected to change more under long term salt stress.

Ion Channels and Transporters

Major channels and transporters of yeast plasma membrane include: (1) potassium-selective outward rectifying ion channel TOK1, (2) non-selective cation channels (registered in electrophysiological experiments, but gene sequence(s) and protein structure(s) responsible for the current had not been identified so far), (3) potassium transporters Trk1 and Trk2, (1) sodium/potassium-proton antiporter Nha1, (5) sodium- phosphate symporter Pho89 (reviewed in Ariño et al., 2010), (6) voltage-gated calcium channel Cch1 (Fischer et al., 1997; Peiter et al., 2005; Teng et al., 2008, 2013) probably interacting with mechanosensitive cation-selective channel Mid1 (Kanzaki et al., 1999; Peiter et al., 2005), and (7) mechanosensitive ion channel with similar selectivity for cations and anions (Gustin et al., 1988). A brief description concerning the number of molecules per cell and estimated ion currents carried by them may help to understand their role in membrane transport properties. As a first approximation: sodium/potassium-proton antiporter Nha1 was visualized in experiments on protein expression in yeast using both GFP and TAP tags which estimated 1480 molecules/cell (about 15 molecules/μm^2), Mid1 was detected at 3210 molecules/cell (about 30 molecules/μm^2), while Trk1, Trk2, TOK1, Pho89, and Cch1 were not visualized or below the detection limit of 50 molecules/cell (Ghaemmaghami et al., 2003). It is worth mentioning that the number of specific protein molecules per cell and the protein abundance variation among cells in large populations of *S. cerevisiae* are partially determined genetically (Albert et al., 2014).

Though **TOK1** was not visualized in the study on protein expression (Ghaemmaghami et al., 2003), earlier electrophysiological experiments predicted about 50–100 active ion channels per a cell (Bertl and Slayman, 1992; Bertl et al., 1998). Potassium-selective outward rectifying ion channel TOK1 is the best characterized ion channel in yeast cells. First single channels recordings with yeast protoplasts revealed dominant conductance with high selectivity for potassium over sodium.

The unitary conductance of about 13 pS (122 mM K^+ in pipetteand outside medium) was inhibited by 20 mM TEA^+ and 10 mM Mg^{2+} (Gustin et al., 1986). 17 pS potassium-selective and ATP-sensitive conductance was registered in H^+-ATPase mutants of *Saccharomyces cerevisiae* (150 KCl inside and outside configuration) (Ramirez et al., 1989). Several large (64 and 116 pS) potassium-selective conductances were revealed in yeast plasma membrane vesicles fused with planar bilayers. These conductances were inhibited by 10 mM TEA^+ and 10 mM BaCl2 and presumably correspond to TOK1 (Gómez-Lagunas et al., 1989). Single channel recordings have been studied in detail (Bertl and Slayman, 1992; Bertl et al., 1993). Flickering behavior was found for the channel. The ion channel demonstrated high selectivity for potassium over sodium and was inhibited by increasing cytoplasmic calcium and by cytoplasmic sodium. A kinetic model with several closed and open states was proposed for gating (transition between open and closed states) of the channel (Bertl and Slayman, 1992; Bertl et al., 1993).

Whole cell recordings with yeast protoplasts provided more information about TOK1. Outward current via TOK1 was about 500 pA at +100 mV (175 mM pipette = cytoplasmic KCl and 150 mM external KCl) and demonstrated ATP-dependence (the current exhibited strong rundown within 60 min without ATP in pipette medium). There was no effect of external protons within pH 5.0–8.0, but inhibition by cytosolic acidification; inhibition by TEA^+ and Ba^{2+} and gating by external potassium were also discovered (Bertl et al., 1998). Gating of the yeast potassium outward rectifier by external potassium was found for whole cell currents and confirmed for single channels. It was proposed (based on a previous model for the channel) that TOK1 has binding sites for potassium and plays a central role in osmoregulation and K^+-homeostasis in yeast (Bertl et al., 1998). Effects of TOK1 gating by external potassium and also amino acid residues responsible for the gating have been studied in detail using a heterologous expression system, oocytes of *Xenopus laevis* (Ketchum et al., 1995; Lesage et al., 1996; Loukin et al., 1997; Loukin and Saimi, 1999; Johansson and Blatt, 2006). It was an unexpected finding that TOK1 was slightly activated by volatile anaesthetic agents isoflurane, halothane and desflurane (Gray et al., 1998).

Simple estimates translate the current via the channel to ion concentrations: e.g., 10 pS correspond to 1.0 pA at 100 mV, the number of transported ions per second at 100 mV will be:

$$\approx 1*10^{-12}\text{coulombs/second}*1\text{second}*6.2*10^{18} \text{ monovalent ions/coulomb} \approx 6*10^{6} \text{ ions} \tag{8}$$

Assuming a change in K^+ concentration from 100 to 40 mM within 10 min (600 s), we estimate that only one TOK1 molecule (with extremely low conductivity, at the level of patch clamp resolution) is sufficient to ensure the potassium ion fluxes.

Questions about the physiological role of TOK1 arose together with electrophysiological characterization of the channel and were partially solved in experiments with yeast mutants. A yeast mutant with a deleted TOK1 gene had no potassium- selective outward current (Bertl et al., 2003). Deletion of TOK1 depolarized cell membrane, while overexpression hyperpolarized it. These results were obtained in an assay using 3,3′- dipropylthiacarbocyanine iodide fluorescence (Marcsova et al., 2006). Potential participation of TOK1 in potassium uptake by yeast cells was demonstrated in yeast mutants with overexpressed TOK1 and deleted Trk1 and Trk2 genes (Fairman et al., 1999). Overexpression of TOK1 restored growth on plates with low (1 mM) potassium concentration and more than doubled potassium contents in cells cultured on 5.0 mM K^+ (Fairman et al., 1999). The results seem unusual at first glance: overexpression of outward rectifying ion channel has a positive effect on ion accumulation. A possible explanation for this is due to small inward current below E reversal for K^+, which was recorded in TOK1-overexpressing cells (Fairman et al., 1999). However, *tok11* mutants also had slightly higher potassium contents and did not differ in growth (cited according to Kahm, 2011). It is therefore proposed that TOK1 can contribute to controlling membrane potential around reversal potential for potassium (Fairman et al., 1999): the number of ions transferred via TOK1 [(Equation 8) for 100 ion channels but multiplied by smaller voltage and open probability for channels] is comparable to the number of ions pumped by Pma1 per second (Equation 6); so both Pma1 and TOK1 may play a role in altering the membrane potential.

Non-selective cation current in yeast cells is well studied by electrophysiology. Surprisingly high inward currents often exceeding 1 nA at −200 mV have been registered in yeast protoplasts (from larger tetraploid cells) when external divalent ions were reduced below 10 μM (Bihler et al., 1998, 2002). The currents demonstrated inward rectification increasing at more negative voltages and included time-dependent and time-independent instantaneous components (Bihler et al., 1998, 2002). Slightly lower similar currents (about 300 pA at −140 mV) were recorded with 100 μM of external calcium and magnesium from spheroplasts of 4–5 μm in diameter (Roberts et al., 1999). The current was carried by monovalent cations and had similar selectivities for K^+ and Rb^+. Selectivity for Na^+ was about 50% less, while choline and TEA^+ were nearly impermeable cations (Roberts et al., 1999). Lowering intracellular potassium increased potassium was suggested (Roberts et al., 1999). While external pH within the range of 5.5–7.0 had no effect on the current (Roberts et al., 1999), pH 4.0 was found to be inhibitory (Bihler et al., 2002). Similar large inward currents have been recorded in another yeast, *Zygosaccharomyces bailii*; the conductance was not permeable to TEA^+ and was slightly inhibited by external 10 mM Ca^{2+} or pH 4.0 (Demidchik et al., 2005). Non-selective cation channels are better studied in plants, where they have an essential role in salt tolerance and development. Cyclic nucleotide-gated ion channels and amino acid-gated channels make up the total cation non-selective currents in plants (Demidchik and Maathuis, 2007), while no candidate genes or proteins responsible for the current have been identified so far in yeast.

Obviously, regulation of yeast non-selective cation currents and changes in membrane potential will essentially change the influx of ions by the pathway, since 1 nA current carries per second:

$$1.0\text{nA(measured at} - 200\text{ mV)corresponds to} \approx 1*10^{-9} \text{ coulombs/second}*1\text{second}*6.2*10^{18}\text{ ions/coulomb} = 6.2*10^{9}\text{ ions.} \tag{9}$$

That is about 100% of cell potassium concentration per second for 6 μm yeast cells or 25% for 10 μm tetraploid cells. Probably under realistic physiological conditions the fluxes of ions via ion channels are influenced by kinetic factors and diffusion of ions to ion channels.

Nha1 is an electroneutral or electrogenic cation/H^+ antiporter especially important for Na^+ efflux at low external pH values (Prior et al., 1996; Bañuelos et al., 1998; Kinclova- Zimmermannova et al., 2006). Assays with secretory vesicles evidence that during the transport cycle Nha1 transports with similar affinities one single ion of Na^+ or K^+ per more than one H^+ using energy of electrochemical gradient of H^+ (Ohgaki et al., 2005), but more experimental support is required. Though amino acids important for selectivity of Nha1 and protein domains essential for its regulation are known (Kinclová et al., 2001; Kinclova-Zimmermannova et al., 2005), mechanism and stoichiometry of ion transport by Nha1 still need to be elucidated. Potentially heterologous expression of Nha1 could be a step in the direction, however it was not successful so far (Volkov, personal communication).

Ion transporters have much lower transport rates compared to ion channels. Ion channels often transport over 106 ions/(second*molecule), while transporters about 100–10,000 ions per second with the estimated theoretical limit of 100,000, due to the speed of protein conformational changes (e.g., Levitan and Kaczmarek, 2001, p. 72; Chakrapani and Auerbach, 2005). Similar to the Nha1 transporter is NhaA found in *E. coli*, which has a turnover number of about 1500 ions per second at pH 8.5 (Taglicht et al., 1991), so an estimate of 1000 transported ions/(second∗transporter) in yeast seems reasonable to suggest.

Sodium transport out of cytoplasm by Nha1 would therefore be about:

$$\begin{aligned}&1500\text{Nha1 molecules/cell}*1000\text{ ions/(second*transporter)}\\&=1.5*10^6\text{ ions per second per a cell or (converting to}\\&\text{concentrations using Avogadro constant and cell volume)}\\&\text{about }25\,\mu\text{M/(second*cell)}=90\text{ mM/(hour*cell)}\end{aligned} \quad (10)$$

or from 9 to 900 mM/(hour∗cell), adding potential 10-fold variation in estimated unknown yet possible transport rates by Nha1. This value is still not very high compared to sodium uptake via non-selective cation currents (see above), Ena1 also has low sodium efflux capacity (Equation 7). Indeed, the rapid increase of intracellular sodium to 150 mM within 100 min was observed in yeast cells subjected to 1 M NaCl stress (García et al., 1997).

Several possible scenarios of cell behavior under high sodium stress may include in that case:

(1) a shift of membrane potential to more positive values and consequently essential drop of the non-selective cation conductance, which will reduce sodium accumulation in cells below the toxic levels. However, increasing the membrane potential to more positive values will have a negative effect on potassium uptake via non-selective cation conductance and consequently on growth;

(2) to induce expression of Ena1 and Nha1 increasing sodium efflux from yeast cell;

(3) to use other channels and transporters (which have yet been not discovered) to expel sodium from the cells while maintaining a high potassium uptake rate.

Experiments with yeast mutants with deleted Nha1 demonstrate, however, a more complex pattern. Deletion of Nha1 had no effect on membrane potential assayed by the fluorescent dye 3,3′-dipropylthiacarbocyanine iodide and no effect on growth on plates in the presence of 200 mM NaCl (Kinclova- Zimmermannova et al., 2006). Higher concentrations of sodium chloride (0.5 M) and in particular potassium chloride (1.8 M) inhibited growth of a yeast strain without the *nha1* gene on plates at pH 3.6 (Bañuelos et al., 1998). Overexpression of Nha1 increased NaCl tolerance and decreased intracellular sodium concentration by 150 mM within 20 min in yeast cells preloaded with sodium. However, the number of Nha1 transporters per cell was not known for the overexpressing cells (Bañuelos et al., 1998). Overexpression of Nha1 also hyperpolarized the plasma membrane (Kinclova-Zimmermannova et al., 2006), potentially involving proton pump Pma1 since Nha1 presumably transports inside more H^+ per Na^+ or K^+ out (Ohgaki et al., 2005).

Trk1 and Trk2 were isolated by a drop in potassium uptake by yeast cells from the external medium (Gaber et al., 1988; Ko and Gaber, 1991). Genes *trk1* and *trk2* are homologous to genes of HKT transporters, which are K^+ or Na^+ uniporters or K^+/Na^+ symporters (Mäser et al., 2002; Waters et al., 2013). However, some controversy appeared about their transport properties when both Trk1 and Trk2 potassium transporters were studied using *trk11 trk21* mutants in patch clamp experiments. Deletion of *trk1* and *trk2* nearly abolished inward current at −200 mV, the transporters are therefore responsible for about −40 pA in whole cell configuration with 150 mM KCl + 10 $CaCl_2$ outside/175 mM KCl inside. Deletion of trk1 alone reduced the inward current by 20 pA from −40 pA to −20 pA, while deletion of *trk2* abolished the current (Bertl et al., 2003). Surprisingly, in support of earlier electrophysiological experiments (Bihler et al., 1999) the current was initially attributed to H^+, not to K^+.

Simple estimates show that 40 pA per second will therefore be equivalent to:

$$\begin{aligned}&40\text{pA at}-200\text{ mV correspond to}\\&\approx 40*10^{-12}\text{ coulombs/second*1second*}6.2*10^{18}\\&\text{ions/coulomb}\approx 2.4*10^{8}\text{ protons}\end{aligned} \quad (11)$$

This calculation is nearly 105 times higher than the estimated number of 6000 protons per yeast cell at pH 7 (Equation 4) and will need robust and fast buffering system to maintain pH homeostasis. Complex dynamics for proton transport from the liquid layers adjacent to the membrane (including diffusion, kinetics of transport etc.) will also be required, since at pH 7.5 (Trk1/Trk2 currents measured in Bertl et al., 2003) the 1000 times larger volume (100 nL, 10 diameters of the yeast cell) around the cell would include the number of H^+ equal to:

$$\approx 6.02*10^{23}/\text{mole}*10^{-7}\text{L}*3*10^{-8}\text{ mole/L}\approx 18*10^{8}. \quad (12)$$

Only 7–8 times above the measured current per second. Fast transport of protons would need high-speed diffusion rates and a better understanding of the processes from the point of physical chemistry/biophysics of transport.

Further patch clamp studies with *S. cerevisiae* protoplasts revealed pH-dependent chloride conductance for both Trk1 and Trk2 rather than H^+ fluxes via the transporters (Kuroda et al., 2004; Rivetta et al., 2005, 2011). The inward ion current (Kuroda et al., 2004) did not depend on cation, but on chloride (or the other anion) concentrations in buffers for electrophysiological experiments. Small (below 5 pA) ion currents were also assumed to be associated with K^+-currents in protoplasts of *S. cerevisiae*. However, the current 5 pA at −200 mV would correspond per second to:

$$\begin{aligned}&\approx 5*10^{-12}\text{ coulombs/second*1second*}6.2*10^{18}\\&\text{ions/coulomb}\approx 3*10^{7}\text{ ions.}\end{aligned} \quad (13)$$

This would require (1) 10,000 Trk1 and Trk2 transporters (assuming turnover being 1,000 ions/second per transporter) per cell while they have not been found in protein expression essays so far (e.g., Ghaemmaghami et al., 2003); (2) more experimental investigation since small anion currents still could not be completely excluded when measuring small whole cell currents (within the range of few pA).

Chloride conductance by Trk1-type transporters was shown for another yeast-like species, human pathogenic fungus *Candida albicans*, where it was strongly inhibited by the chloride currents blocker DIDS (4,4′-diisothiocyanatostilbene-2,2′-disulfonic acid)

(Baev et al., 2004). Further experiments revealed that CaTrk1p transporter from C. *albicans* was homologous to Trk transporters from *S. cerevisiae* and CaTrk1p was responsible for low affinity K^+ uptake (indicated by Rb^+ uptake) in C. *albicans* (Baev et al., 2004; Miranda et al., 2009). However, the measured K^+ fluxes were estimated to be at the lower limit of resolution for patch clamp with protoplasts of *Candida albicans* (Miranda et al., 2009). Instead, large chloride conductance was measured in the protoplasts suggesting both chloride and potassium transport by CaTrk1p (Miranda et al., 2009).

The question remains as to whether Trk1 and Trk2 alone make up the proteins with a potassium-selective pore or if they are acting in cooperation with other proteins forming potassium uptake pathways (Ariño et al., 2010). Crystal structure of a similar type of transporter, TrkH from Vibrio parahaemolyticus, demonstrates a selectivity filter for K^+ and Rb^+ over Na^+ and Li^+ (Cao et al., 2011). Atomic scale and molecular dynamics modeling of Trk1 from *S. cerevisiae* confirmed homology with HKT transporters and revealed essential role of glycine residues within potassium selective filter region of the protein, sodium ions were inhibiting for K^+ transport (Zayats et al., 2015). However, still more experimental evidence is required and heterologous expression systems may also be useful for solving the puzzle. It is expected that the chloride conductance of Trk1 and Trk2 is masking the tiny potassium conductance which cannot therefore be easily measured electrophysiologically in yeast cells in whole cell configuration. Presumably chloride conductance is formed independently of K^+-conducting pathway and K^+ and Cl^- fluxes are not influencing each other, so Trk proteins rather have dual transport functions than symport/antiport K^+ and Cl^- ions.

Yeast mutants with deleted Trk1 and Trk2 transporters have hyperpolarized membrane potential assayed by fluorescent dyes 3,3′-dihexyloxacarbocyanin (Madrid et al., 1998) and 3,3′-dipropylthiacarbocyanide iodide, and also a more alkaline intracellular pH (by about 0.3 units) under high (50 mM) and low (15 μM) potassium in the liquid growth medium (Navarrete et al., 2010; Plášek et al., 2013). Proton efflux from the *trk11 trk21* mutants was higher under 50 mM potassium conditions (Navarrete et al., 2010). It is interesting to mention that the effect of hyperpolarization was associated with K^+ and Na^+, but not with Cl^- ions (according to results of their substitution by MES) (Madrid et al., 1998).

Basic Feedbacks for IonTransport Regulation in Yeast Cell

A simple sketch of participating elements, measured parameters and interactions between them for ion transport is given in **Figure 4**. A more detailed scheme will require complex models similar to metabolic models for yeast cell (e.g., Österlund et al., 2013 for metabolic control) with at least (1) determined and linking ion fluxes via distinct transport systems, (2) understanding and analysing non-linear interactions, and also (3) taking into account wide range of influencing physical factors (osmotic pressure, temperature, diffusion coefficients etc.) and chemical environment (pH, ion concentrations etc.) surrounding a yeast cell.

Effects of NaCl treatment under different external conditions with expected important pathways for ion transport are depicted at **Figure 5**. Among the present limitations are the knowledge gap between results obtained by different methods and extrapolation of experimental conclusions for higher ion concentrations. Electrophysiological patch clamp study provides evidence for fast kinetic changes; the data allow to reveal the role of individual types of ion channels or transporters, but yeast protoplasts are deprived of cell wall, do not have complete set of regulation feedbacks. Moreover, the patch clamp data are recorded under controlled conditions of experimental solutions, which traditionally do not use over 100–200 mM of Na^+, K^+ or other cations. Experiments with ion analysis of yeast cells or mutants in specific transport systems are with intact cells, the disadvantages are lower temporary resolution and net fluxes of definite ions; mutants may also have altered regulation of ion transport. It is reasonable to add that ion concentrations and pH in liquid layers adjacent to plasma membrane from the outside of cytoplasm are influenced by cell wall.

Cell Wall and Lipid Rafts Influence Ion Transport

The presence of a cell wall and the non-homogeneous structure of cell plasma membrane add more complexity for ion transport. Several as yet unexplored hypotheses should be considered based on the known physico-chemical properties of cell walls. They include (1) changes of cell membrane potential by fixed electric charges in cell walls; (2) effects of cell walls on ion buffering and ion concentrations; (3) expected ion-rectifying properties of cell walls; (4) assumed effect of shrinking and swelling of cell walls on mechanosensitive ion channels. Since yeast cell walls are similar to cell walls of plants though less studied, several facts and phenomena about plant cell walls will be also given.

The yeast cell wall consists mainly of polysaccharides (reviewed in Cabib et al., 2001), has a thickness of about 0.1 μm (90–250 nm according to different sources, 9–25-fold thickness of plasma membrane) (Bowen et al., 1992; Smith et al., 2000; Stenson, 2009; Dupres et al., 2010) and keeps cell volume stable under sudden fluctuations of surrounding medium. Under usual conditions of low osmotic pressure of medium and high osmotic pressure of cell solution the cell wall is stretched and balances the turgor pressure (positive hydrostatic pressure of several bars (1 bar = 0.1 MPa) inside the cell to equilibrate chemical potentials of water caused by the difference of osmotic pressures).

The yeast cell wall consists of glucans, mannose polysaccharides, a few proteins and several per cents of chitin (Lipke and Ovalle, 1998; Cabib et al., 2001). The surface electrical charge of the cell is determined by polymers in the cell wall (chitin has charged amino groups, proteins have several charged chemical groups), and changes depending on the pH and ion composition of medium. Titration curves of surface charges of the yeast cell wall indicated wave around pH 7.1, value of pKa2 for phosphate (Bowen et al., 1992). Much attention was then given to phosphomannans of the cell wall as they contain phosphate groups (Jayatissa and Rose, 1976).

Presumably the proton pump Pma1 and ion transporters have an effect on the charge distribution within the cell wall and *visa versa* ion transport is also influenced by the adjacent charged cell wall. The surface charge of a yeast cell can be measured by electrophoresis. For example, voltages 5–20 V of alternating current with frequencies in the range 0.03–5 Hz were applied to a suspension of yeast cells in distilled water. Cell of 5.7 μm in diameter was charged positively by about 10,000 elementary charges (personal communication of Ioan-Costin Stan in 2012: http://upcommons.upc.edu/pfc/bitstream/2099.1/ 10187/2/Master%20Thesis%20-%20Ioan.doc), which equates to about 100 positive elementary charges/μm2. Earlier results in buffered solutions, however, estimated a much higher surface charge (about 100 times larger in some cases) and demonstrated that electrophoretic mobility strongly depended on pH (negative charge at above pH 4), age of the culture and presence of phosphate in the growth medium (Eddy and Rudin, 1958). In these experiments isolated cell walls behaved similarly to whole yeast cells (Eddy and Rudin, 1958). In the other experiments electrophoretic mobility was also measured in buffered solutions; it dropped several times upon treatment with 60% hydrofluoric acid (HF) removing phosphate groups (Jayatissa and Rose, 1976).

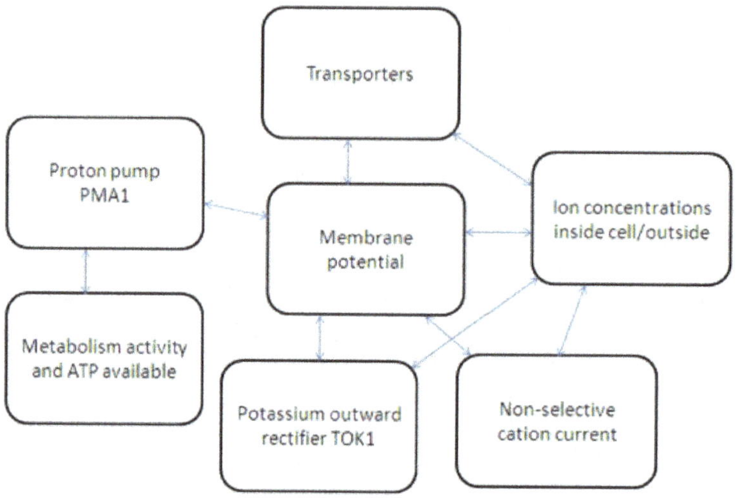

FIGURE 4 | Simplified scheme of several feedbacks for ion transport regulation in yeast cell. Metabolic activity and available ATP regulate proton pump Pma1, which influences membrane potential and also depends on membrane potential due to thermodynamic reasons. Non-selective cation currents and potassium current via potassium-selective outward rectifier TOK1 depend on cell membrane potential and ion concentration inside and outside yeast cell. Ion fluxes via transporters depend on ion concentrations; membrane potential is influencing them as well. The whole system of interacting factors and elements has non-linear feedbacks and links; therefore a system approach is required to describe the ion transport and its regulation.

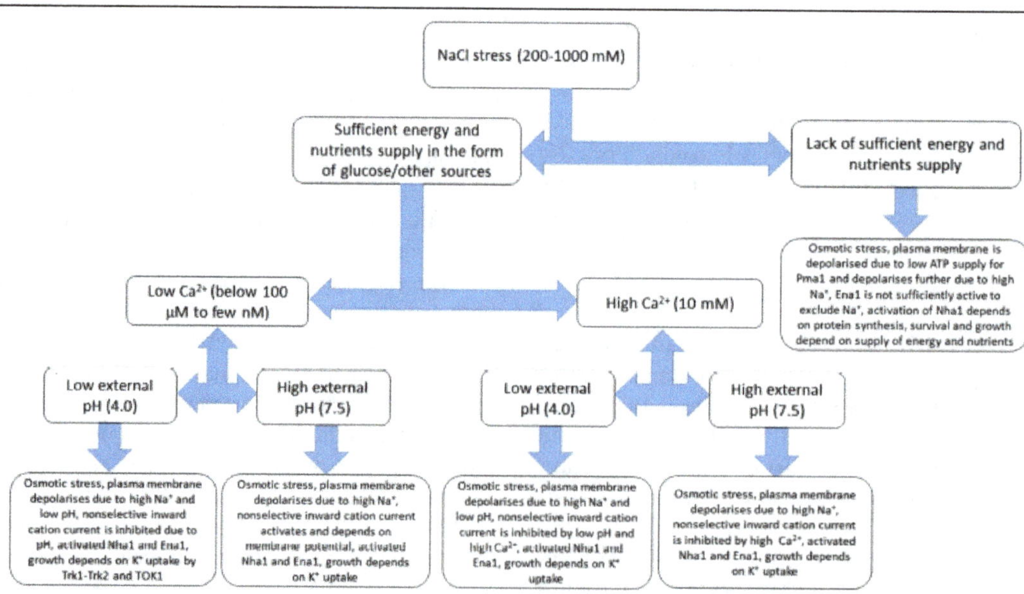

FIGURE 5 | Basic sketch of anticipated changes of ion transport in yeast cell under salinity treatment. Several additional factors including external calcium concentration, pH and sufficient energy supply are added. The scheme demonstrates complexity of responses and based on patch clamp and growth experiments, refers to membrane potential measurements and extrapolates some of the known results. Osmotic adaptation to compensate for rise of external osmotic pressure includes synthesis of compatible osmolytes and not shown at the Figure, while some responses could be more complex due to interactions between transport systems. More details are in the text, the given values are not exact and provided rather for indication. Continuous models with several parameters are outside the scope of the review being a special subject of study and research.

The surface charge is expressed as the electric potential between the cell surface and the surrounding liquid medium. Zeta potential had been measured for yeast cells, showing values varying between 0 mV and −40 mV, being more negative in distilled water and at higher pH values (pH range 3–8 was studied). Usually zeta potential was not below −10 mV for cells in culture medium (Thonart et al., 1982; Bowen et al., 1992; Tálos et al., 2012).

Apart from slightly changing the measured membrane potential, the cell walls also have ion exchange properties, which can influence/buffer ion concentrations in the vicinity of plasma membrane and create local ion gradients. More is known for algae and plants, where cells possess similar cell walls with slightly different chemical composition. For example, in plants four types of ionogenic groups were found in cell walls isolated from lupine roots. Maximal cation-exchange capacity exceeded one mmole per gram of dry weight of cell walls (at pH over 10.8), while anion-exchange capacity was about 20 times lower (at pH below 2.8) (Meychik and Yermakov, 2001c). Ion exchange properties of plant cell walls from roots of lupine and horse beans essentially depended on pH and ion concentrations (Sentenac and Grignon, 1981). Algae had been studied earlier and demonstrated calcium exchange capacity about 40–400 mmole/(L cell wall) for charophytes (Tyree, 1968; reviewed in Hope and Walker, 1975) and cation exchange capacity over 2,5 mmole/g^{-1} dry weight of cell walls for chlorophyte *Enteromorpha intestinalis* (Ritchie and Larkum, 1982). Cell walls of studied algae are thicker, averaging 0.3–1 μm (Wei and Lintilhac, 2007) and reported even up to 15 μm (Tyree, 1968).

Assuming similar range of cation-exchange capacity for yeast and then taking into account that cell wall makes up 15–30% of the dry weight of the cell and 25–50% of the volume based on calculations from electron micrographs (cited according to Lipke and Ovalle, 1998), it is simple to assess: about 1 mM per gram of dry weight is 1 M per kg of dry weight and about 200 mM per kg of fresh weight (1–5 is a reasonable fresh to dry ratio for yeast cell walls), similar to potassium concentration in yeast cell. Taking the volume of yeast cell wall to be about 20 fL (20% of 100 fL), the presumed maximal amount of cations bound by the cell wall would be:

$$20\ \text{fL} * 200\ \text{mM} = 4\ \text{fmole}. \tag{14}$$

The suggested amount may be released depending on the pH of medium within the cell wall (therefore the proton pump Pma1 activity may be involved) and buffer ion transport fluxes.

It is worth mentioning again that in plants (roots of lupine, wheat and pea) cell walls shrink at higher ion concentrations and at lower pH values (for example, dry cell walls of lupin roots bound up to 10 times more water at pH 9 compared to pH 4 at 0.3 mM ionic strength and the same amount at 1 M ionic strength) (Meychik and Yermakov, 2001a,b). Moreover, cation-exchange capacity of plant roots (spinach) increased by about 30% during growth in high salt conditions (250 mM NaCl) (Meychik et al., 2006). High cation-exchange capacity of cell walls of plant roots was visualized in experiments using small organic cation methylene blue, it resulted in 100–700-fold higher concentration of methylene blue in cell walls than in ambient solution (Meychik et al., 2007). Furthermore, diffusion coefficients of methylene blue into cell walls differed between species (10 times higher in mungbean compared to wheat) and positively correlated with the number of carboxyl groups in cell wall structure and swelling of cell walls (amount of water absorbed by dried cell walls) (Meychik et al., 2007). The comparisons might be useful for studying and predicting the behavior of yeast cell walls.

A deeper knowledge of yeast cell walls and the variation between yeast strains and under different growth conditions may lead to a better understanding of high affinity cation (especially potassium) transport in yeast cells. Cell-wall mutants of yeast (e.g., described in Soltanian et al., 2007) could be useful in the study of this. Yeast cell walls are heterogeneous within a cell (Rösch et al., 2005) and may potentially have electrically rectifying properties. Considering yeast cell walls as charged porous medium one may suggest the essential influence on ion fluxes via plasma membrane (Lemaire et al., 2010). The complexity of cell walls with potential analogs to micro- and nanofluidics in electrically charged medium under applied voltages needs to be studied in more detail. Several descriptions and models had already been developed for plant and algal cell walls. These include Donnan free space with micro- and nanochannels and voltages down to -100 mV within it and also water free space, both constituting cell wall (e.g., Dainty and Hope, 1961; Hope and Walker, 1975; Beilby and Casanova, 2014). The approach of considering cell wall and cell plasma membrane together as an interacting system may have implications on high affinity potassium transport in plants also. Inward rectifying potassium channels and transporters of HAK and HKT families are important for the pathway of transport in plants. For example, in *Arabidopsis thaliana* the potassium channel AKT1 is the high-affinity pathway for uptake of potassium from medium with concentrations as low as 10 μM while root cells had membrane potentials below −200 mV, sufficient for ensuring the electrochemical ion transport (Hirsch et al., 1998). Experimental evidence for effects of plant cell walls on calcium and proton fluxes was obtained for bean leaf mesophyll (Shabala and Newman, 2000). Isolated mesophyll protoplasts did not show NaCl-induced calcium efflux compared to mesophyll tissue; salt-induced H^+ efflux also differed between protoplasts compared to tissue. Presumably, all the Ca^{2+} efflux over an hour of measurement was from calcium ion exchange stores of cell walls (Shabala and Newman, 2000). Similarly, transient Ca^{2+} outward fluxes under certain pH changes were recorded for isolated cell walls of charophytes (Ryan et al., 1992). Further recent interesting observations have shown an essential step of iron binding to the cell wall in several algal species (though not in S. cerevisiae studied for comparison) before iron is taken up (Sutak et al., 2012).

Mechanical properties and turgor pressure of yeast cells have been measured by compression (Smith et al., 2000; Stenson, 2009; Stenson et al., 2011) and gives results around 100–180 MPa for mean Young elastic modulus. Turgor pressure in fission yeast was estimated at around 0.8 MPa (Minc et al., 2009). Elastic modulus of yeast cells looks higher at a first glance than usually measured for plant cells and their cell walls, the available studied similar system. For example, average volumetric elastic modulus of barley leaf cells was within 2–10 MPa and below 25 MPa, while turgor pressure was around 0.4–0.8 MPa (e.g., Volkov et al., 2007). However, methods of measurement are different for Young elastic modulus and volumetric elastic modulus even though both are expressed in units of pressure. Measurements of mechanical properties for oak leaf cells demonstrated a strong correlation between the two, with values of volumetric elastic modulus around 10 MPa corresponding to 100–200 MPa of Young elastic modulus (Saito et al., 2006). More detailed results concerning turgor pressure and elastic modulus of different plant cells (where water relations and mechanical properties of cell walls are studied better so far) can be found in numerous reviews and publications (e.g., Hüsken et al., 1978; Steudle, 1993; Fricke and Peters, 2002). Turgor pressure in magnetotactic bacteria of the gram-negative species *Magnetospirillum gryphiswaldense* was estimated in the range of 0.09–0.15 MPa using atomic force microscopy (Arnoldi et al., 2000); similar or slightly higher turgor pressure of 0.08–0.5 MPa was measured for several other gram-negative bacteria (Beveridge, 1988; Walsby et al., 1995). Gram-positive bacteria have 5–10 times higher turgor reaching 2–5 MPa (Beveridge, 1988; Doyle and Marquis, 1994), their cell walls are also much thicker being about 20–80 nm (e.g., Beveridge, 1988; Salton and Kim, 1996) compared to 5–10 nm in gram-negative bacteria (e.g., Salton and Kim, 1996).

Rapid changes in turgor pressure of yeast cells (influenced also by potential shrinking-swelling of cell wall) can activate mechanosensitive ion channels and induce ion flows via them, changing ion concentrations in different cell compartments. So far the well-studied yeast mechanosensitive ion channels include (1) mechanosensitive cation-selective channel Mid1 (Kanzaki et al., 1999; Peiter et al., 2005), (2) mechanosensitive ion channel with similar selectivity for cations and anions (Gustin et al., 1988), and also (3) the large (320 pS in 150 KCl in bath and 180 KCl in pipette solution) yeast vacuolar conductance resembling TRP-channels (Palmer et al., 2001; Zhou et al., 2003) with yet unknown functions (potentially the large trans-vacuolar currents under small changes in pressure could transport ions to cytoplasm for osmoregulation). Attempt to attribute activity of mechanosensitive ion channel with similar selectivity for cations and anions (Gustin et al., 1988) to Mid1 was not convincing (reviewed in Kung et al., 2010). The ion currents via mechanosensitive ion channels were registered in electrophysiological patch clamp experiments under slight positive (+) or negative (−) pressure applied to plasma or vacuolar membrane. The applied pressures were quite low compared to turgor pressure: −4 to −0.5 kPa for Mid1 (Kanzaki et al., 1999), +2.5 to +6.5 kPa and −8 to −0.5 kPa for mechanosensitive ion channel with similar selectivity for cations and anions (Gustin et al., 1988) and +0.6 to +6 kPa for mechanosensitive vacuolar Yvc1p channel (Zhou et al., 2003).

Presumably even small alterations below 1% of turgor pressure could be amplified by ion fluxes via mechanosensitive ion channels, include calcium signaling (Kanzaki et al., 1999; Palmer et al., 2001; Zhou et al., 2003; Peiter et al., 2005) and further osmotic stress signaling to restore turgor pressure (e.g., reviewed in Hohmann, 2002; modeled in Muzzey et al., 2009; Schaber et al., 2010). Rapid transients of ion fluxes with resolution of seconds under osmotic stress and variation in responses between different cells are still to be measured for yeast cells with intact cell walls. Recent experiments using technique of microelectrode ion flux estimation (MIFE) revealed fast activation of H^+ efflux by over 3–4 times, change from K^+ efflux to influx and slight increase in Mg^{2+} efflux under hyperosmotic stress (2 MPa by mannitol) in Gram-positive bacterium *Listeria monocytogenes* (Shabala et al., 2006), the MIFE technique is now being used for yeast cells (Ariño et al., 2014). The activity of mechanosensitive ion channels is also essentially determined by the rheology of membrane (Bavi et al., 2014) and so could be different in lipid raft compared to adjacent areas of plasma membrane.

Lipid rafts are found in yeast plasma membrane and the distribution of ion channels and transporters is not even within the membrane (e.g., reviewed in Malinsky et al., 2010). The Nha1 transporter was found to be associated with lipid rafts and requires sphingolipid for stable localization to the plasma membrane (Mitsui et al., 2009). The proton pump Pma1 was located to network structures, while proton transporters were in non-overlapping long-lived (over 10–30 min) 300 nm patches (Malínská et al., 2003; Grossmann et al., 2007), which changed upon membrane depolarization (Grossmann et al., 2007). More complex patchy pattern of plasma membrane had been discovered in yeast cells using total internal reflection fluorescence microscopy and following the location of 46 membrane proteins tagged with green fluorescent protein (GFP) (Spira et al., 2012). Another approach was to use atomic force microscopy (AFM) and the His-tagged transmembrane protein Wsc1. Scanning the cell surface with an AFM tip bearing Ni^{2+}- nitriloacetate revealed clusters with a diameter of around 200 nm (Heinisch et al., 2010). The membrane protein Wsc1 was located within these clusters and stress conditions (deionized water or heat shock) increased the number of clusters and their size (Heinisch et al., 2010).

Lipid rafts with patchy localization of transporters can interact with the membrane skeleton, cell wall, regulatory proteins and cell cytoskeleton and make up local temporary ion gradients at the scale of several nanometers. The local non-equilibrium ion states depend on activity of transport systems and diffusion coefficients, it finally creates dynamic and live structure with fast fluctuations of ion currents and their complex regulation (**Figure 6**).

Comparison ofYeast Cells and Large Excitable Cells

It is reasonable to ponder the difference between small yeast cells and excitable animal cells (**Figure 2**). The latter have a higher volume per unit of surface area (*vice versa* smaller surface per unit of volume), and use calcium and sodium ion channels with high conductance instead of immediate activity of electrogenic ion pumps for fast changes in membrane potential within hundreds of milliseconds and even much faster. Being more precise, the volume per unit of membrane electric capacitance has to be considered: the parameter is over 100 times higher for squid axon (Lecar et al., 1967) and about 5 times higher for cardiomyocytes (Satoh et al., 1996) compared to yeast cells. Moreover, the animal cells are functioning under controlled conditions of external medium (internal liquid of organisms) with buffered pH and relatively stable concentrations of K^+,

Na^+, Ca^{2+}, and Cl^-, but the limitations by cell geometry and by microscopic properties of ion pumps, transporters and ion channels are essential. Potentially to optimize ion fluxes the cells have choices of changing the number of membrane transport proteins and their regulation (e.g., yeast cells increase the density of plasma membrane proton pump Pma1 up to higher limits) or choosing specific transport proteins to fulfill the required physiological/biochemical demands. High density of sodium channels with high conductance is used in specialized segments of neurons (e.g., Kole et al., 2008). Imagining similar fluxes for a yeast cell with sodium channel density 2,500 pS μm^{-2}, it is simple to recalculate the theoretical sodium flux under the conditions of membrane potential equal −100 mV:

2500 pS μm^{-2} to 100 μm^2 of yeast cell is 250,000 pS/cell, under −100 mV it would correspond to −25 nA/cell. Taking into account that the number of monovalent cations per Ampere∗second = coulomb is equal to 6.2∗1018 ions/coulomb, the ion current per second is equivalent to ≈ 1.6∗10^{11} Na^+ ions. Note that it is Na+ current, but it is more than 25 times over the total number of, e.g., K+ ions in a yeast cell or 40 ms of the ion current are sufficient to carry the same number of cations which are present in the cell. Yeast cells do not have the sort of ion channels with high conductance and density, otherwise they will not be able to keep ion homeostasis under conditions of undetermined and changing ion composition of surrounding medium. Instead yeast cells use plasma membrane pump to keep required membrane potential and ion transporters with much lower transport rate to support smaller ion fluxes.

Another comparison is also interesting. Large cells of charophyte algae (**Figure 2**) have active plasma membrane H^+- ATPase and exhibit action potentials (reviewed in Beilby, 2007; Beilby and Casanova, 2014). The group of multicellular algae with cells up to 10 cm long and 2 mm in diameter (Johnson et al., 2002) demonstrates slow action potentials within tens of seconds based on changes in chloride and calcium electric conductances (Beilby, 2007). Similar to yeast cells the large charophytes use H^+-ATPase for shifting membrane potential to more negative values; charophytes have membrane potential down to -350 mV (comprehensively described in the book: Beilby and Casanova, 2014). Charophyte cells have several physiological "states" of plasma membrane depending on activity of ion transport systems. "Pump state" corresponds to active H^+-ATPase and membrane potential below -200 mV, "K^+ state" is dominated by K^+ conductances, while "H^+/OH^- state" is determined by passive H^+/OH^- conductances (Beilby and Casanova, 2014). It is questionable whether yeast cells may also have several physiological states of membrane. Slow compared to animal cells action potentials in charophytes are probably explained by their evolution and set of ion channels required for variable unstable ion environment despite the presence of shielding and ion buffering by thick cell walls; the situation needs further modeling and investigation.

The reasons for the difference in ion transport systems are of dual nature: (1) pure biophysical constrains determine the ion transport fluxes and also (2) the history of species and evolution. A few more comparisons with yeast cells are useful. Human erythrocytes have cell volume similar to yeast cells, possess an electrogenic Na^+/K^+-pump, have low membrane potential (about -10 mV when measured by glass electrode) (Jay and Burton, 1969) and several ion channels, but do not use the channels under usual conditions, instead relying on ion pump and transporters to keep ion homeostasis (reviewed in Thomas et al., 2011). Changing the membrane potential from -10 to -90 mV by sodium ionophore has no effect on swelling-activated potassium transport in erythrocytes (Kaji, 1993). Another example of small animals cells are spermatozoa. The cells have volume about 20–30 fL, their measured membrane potential is usually depolarized to around -40 mV and was reported from -80 to +13 mV depending on method of measurement and conditions (Rink, 1977; Guzmán-Grenfell et al., 2000; reviewed in Darszon et al., 2006; Navarro et al., 2007). The membrane potential of spermatozoa is easily shifted to more negative voltages by activated electrogenic Na^+/K^+- pump (Guzmán-Grenfell et al., 2000). Moreover, Na^+/K^+-pump is essential for lower membrane potential and also for fertility (Jimenez et al., 2011), though the cells are equipped with a range of ion channels and additionally depend on potassium channel to regulate the membrane potential (Darszon et al., 2006; Navarro et al., 2007).

Evolutionary aspects of differences in ion transport are quite complicated and could not be explained easily. Excitable cells usually have sodium channels, which have evolved from calcium channels and predated the origin of the nervous system in animals (Liebeskind et al., 2011). Interestingly, action potentials have been found in some marine algae, e.g., sodium/calcium- based action potential in *Odontella sinensis* (Taylor, 2009), which has a volume about 300 times larger than yeast. Chloride/calcium based action potentials in large charophytes are studied better (Beilby, 2007). Sodium channels and action potentials are not known in fungi so far, though sodium channels have surprisingly been found in the bacterium *Bacillus halodurans* (Ren et al., 2001). The bacterium, which is about 200 times smaller than the yeast cell by volume, can grow in a saline environment (2M NaCl) (Takami and Horikoshi, 1999). Opening of just one channel with 12 pS conductance (measured in 140 mM NaCl using expression in CHO-K1 cells) would carry large Na^+ currents and may essentially change sodium concentration in cytoplasm of the bacterium. The biological function of the channel is not known, but it may be related to driving the flagellar motor (Ren et al., 2001).

Biophysical limitations are easier for understanding, modeling and for an attempt of synthetic engineering (**Figure 2**), however, the more detailed description is outside the scope of the text.

Conclusions

The yeast plasma membrane is different from the plasma membrane of excitable animal cells: yeast cells have no abundant sodium and calcium channels with high conductance. The membrane potential of yeast cells can be easily controlled by copious plasma membrane ATPase Pma1, potentially it can change the membrane potential of the small cells by several hundred millivolts within a few seconds. Transporters are also involved in the regulation of membrane potential. Non-selective cation current and outward rectifying potassium selective channels may serve as safety valves against the sharp changes in membrane potential. The small volume of yeast cells, their relatively large surface/volume ratio and high transport capacity of Pma1 make the proton pump, not ion channels, the main element of ion transport. Therefore, the regulation of proton pump and ion transporters is vital for ion homeostasis in yeast cells. The basic principles of ion transport in yeast cells may be applicable to small cells and organelles. Rapid processes of ion transport, within scale of seconds, can essentially change ion concentrations in the cells taking into account their small volume. The long- term scale of minutes and hours needs other mechanisms and sensors for regulation and homeostasis. A high cation- exchange capacity of cells walls, location of ion channels and transporters within lipid rafts and changes in turgor pressure may essentially influence ion transport fluxes and make the situation more complex and yet unpredictable at smaller spatial scales.

Acknowledgments

The Author expresses sincere thanks to his colleagues Gareth Williams, Christopher Palmer, Leonid Krivitsky, Mary Beilby, and Louise Krska, Reviewers and the Editors for careful reading and comments on the manuscript and apologizes for not citing all the relevant literature sources due to limitations by the text frames of the Review.

References

Albert, A., Yenush, L., Gil-Mascarell, M. R., Rodriguez, P. L., Patel, S., Martínez-Ripoll, M., et al. (2000). X-ray structure of yeast Hal2p, a major target of lithium and sodium toxicity, and identification of framework interactions determining cation sensitivity. *J. Mol. Biol.* 295, 927–938. doi: 10.1006/jmbi. 1999.3408

Albert, F. W., Treusch, S., Shockley, A. H., Bloom, J. S., and Kruglyak, L. (2014). Genetics of single-cell protein abundance variation in large yeast populations. *Nature* 506, 494–497. doi: 10.1038/nature12904

Albrecht-Buehler, G. (1990). In defense of "nonmolecular" cell biology. *Int. Rev. Cytol.* 120, 191–241.

Alemán, F., Caballero, F., Ródenas, R., Rivero, R. M., Martínez, V., and Rubio, F. (2014). The F130S point mutation in the Arabidopsis high-affinity K^+ transporter AtHAK5 increases K^+ over Na^+ and Cs^+ selectivity and confers Na^+ and Cs^+ tolerance to yeast under heterologous expression. *Front. Plant Sci.* 5:430. doi: 10.3389/fpls.2014.00430

Ambesi, A., Miranda, M., Petrov, V. V., and Slayman, C. W. (2000). Biogenesis and function of the yeast plasma membrane H^+-ATPase. *J. Exp. Biol.* 203, 155–160. Available online at: http://jeb.biologists.org/content/203/1/155.long

Ariño, J., Aydar, E., Drulhe, S., Ganser, D., Jorrín, J., Kahm, M., et al. (2014). Systems biology of monovalent cation homeostasis in yeast: the Translucent contribution. *Adv. Microb. Physiol.* 64, 1–63. doi: 10.1016/B978-0-12-800143- 1.00001-4

Ariño, J., Ramos, J., and Sychrová, H. (2010). Alkali metal cation transport and homeostasis in yeasts. *Microbiol. Mol. Biol. Rev.* 74, 95–120. doi: 10.1128/MMBR.00042-09

Arnoldi, M., Fritz, M., Bäuerlein, E., Radmacher, M., Sackmann, E., and Boulbitch, A. (2000). Bacterial turgor pressure can be measured by atomic force microscopy. *Phys. Rev. E* 62, 1034–1044. doi: 10.1103/PhysRevE.62.1034

Ashe, M. P., De Long, S. K., and Sachs, A. B. (2000). Glucose depletion rapidly inhibits translation initiation in yeast. *Mol. Biol. Cell.* 11, 833–848. doi: 10.1091/mbc.11.3.833

Auer, M., Scarborough, G. A., and Kühlbrandt, W. (1998). Three-dimensional map of the plasma membrane H^+-ATPase in the open conformation. *Nature* 392, 840–843. doi: 10.1038/33967

Babu, M., Vlasblom, J., Pu, S., Guo, X., Graham, C., Bean, B. D., et al. (2012). Interaction landscape of membrane-protein complexes in *Saccharomyces cerevisiae*. *Nature* 489, 585–589. doi: 10.1038/nature11354

Baev, D., Rivetta, A., Vylkova, S., Sun, J. N., Zeng, G. F., Slayman, C. L., et al. (2004). The TRK1 potassium transporter is the critical effector for killing of *Candida albicans* by the cationic protein, Histatin 5. *J. Biol. Chem.* 279, 55060–55072. doi: 10.1074/jbc.M411031200

Bakker, R., Dobbelmann, J., and Borst-Pauwels, G. W. F. H. (1986). Membrane potential in the yeast *Endomyces magnusii* measured by microelectrodes and TPP^+ distribution. *Biochim. Biophys. Acta* 861, 205–209. doi: 10.1016/0005-2736(86)90421-9

Bañuelos, M. A., Sychrová, H., Bleykasten-Grosshans, C., Souciet, J. L., and Potier, S. (1998). The Nha1 antiporter of *Saccharomyces cerevisiae* mediates sodium and potassium efflux. *Microbiology* 144, 2749–2758.

Bavi, N., Nakayama, Y., Bavi, O., Cox, C. D., Qin, Q. H., and Martinac, B. (2014). Biophysical implications of lipid bilayer rheometry for mechanosensitive channels. *Proc. Natl. Acad. Sci. U.S.A.* 111, 13864–13869. doi: 10.1073/pnas.1409011111

Beilby, M. J. (2007). Action potential in charophytes. *Int. Rev. Cytol.* 257, 43–82. doi: 10.1016/s0074-7696(07)57002-6

Beilby, M. J., and Casanova, M. T. (2014). *The Physiology of Characean Cells.* Heidelberg; New York; Dordrecht; London: Springer, 205.

Bertl, A., Bihler, H., Reid, J. D., Kettner, C., and Slayman, C. L. (1998). Physiological characterization of the yeast plasma membrane outward rectifying K^+ channel, DUK1 (TOK1), *in situ*. *J. Membrane Biol.* 162, 67–80. doi: 10.1007/s002329900343

Bertl, A., Ramos, J., Ludwig, J., Lichtenberg-Fraté, H., Reid, J., Bihler, H., et al. (2003). Characterization of potassium transport in wild-type and isogenic yeast strains carrying all combinations of *trk1*, *trk2* and *tok1* null mutations. *Mol. Microbiol.* 47, 767–780. doi: 10.1046/j.1365-2958.2003.03335.x

Bertl, A., and Slayman, C. L. (1992). Complex modulation of cation channels in the tonoplast and plasma membrane of *Saccharomyces cerevisiae*: single-channel studies. *J. Exp. Biol.* 172, 271–287.

Bertl, A., Slayman, C. L., and Gradmann, D. (1993). Gating and conductance in an outward-rectifying K^+ channel from the plasma membrane of *Saccharomyces cerevisiae*. *J. Membr. Biol.* 132, 183–199. doi: 10.1007/BF00235737

Beveridge, T. J. (1988). The bacterial surface: general considerations towards design and function. *Can. J. Microbiol.* 34, 363–372. doi: 10.1139/m88-067

Bihler, H., Gaber, R. F., Slayman, C. L., and Bertl, A. (1999). The presumed potassium carrier Trk2p in *Saccharomyces cerevisiae* determines an H^+-dependent, K^+-independent current. *FEBS Lett.* 447, 115–120. doi: 10.1016/S0014-5793(99)00281-1

Bihler, H., Slayman, C. L., and Bertl, A. (1998). NSC1: a novel high-current inward rectifier for cations in the plasma membrane of *Saccharomyces cerevisiae*. *FEBS Lett.* 432, 59–64. doi: 10.1016/S0014-5793(98)00832-1

Bihler, H., Slayman, C. L., and Bertl, A. (2002). Low-affinity potassium uptake by *Saccharomyces cerevisiae* is mediated by NSC1, a calcium-blocked non-specific cation channel. *Biochim. Biophys. Acta* 1558, 109–118. doi: 10.1016/S0005-2736(01)00414-X

Blatt, M. R., and Slayman, C. L. (1983). KCl leakage from microelectrodes and its impact on the membrane parameters of a nonexcitable cell. *J. Membrane Biol.* 72, 223–234. doi: 10.1007/BF01870589

Blatt, M. R., and Slayman, C. L. (1987). Role of "active" potassium transport in the regulation of cytoplasmic pH by nonanimal cells. *Proc. Natl. Acad. Sci. U.S.A.* 84, 2737–2741. doi: 10.1073/pnas.84.9.2737

Bond, D. R., and Russel, J. B. (1998). Relationship between Intracellular Phosphate, Proton Motive Force, and Rate of Nongrowth Energy Dissipation (Energy Spilling) in *Streptococcus bovis* JB1. *Appl. Environ. Microbiol.* 64, 976–981.

Borodina, I., and Nielsen, J. (2014). Advances in metabolic engineering of yeast *Saccharomyces cerevisiae* for production of chemicals. *Biotechnol. J.* 9, 609–620. doi: 10.1002/biot.201300445

Borst-Pauwels, G. W. (1981). Ion transport in yeast. *Biochim. Biophys. Acta* 650, 88–127. doi: 10.1016/0304-4157(81)90002-2

Bowen, W. R., Sabuni, H. A., and Ventham, T. J. (1992). Studies of the cell-wall properties of *Saccharomyces cerevisiae* during fermentation. *Biotechnol. Bioeng.* 40, 1309–1318. doi: 10.1002/bit.260401104

Brindle, K., Braddock, P., and Fulton, S. (1990). 31 P NMR measurements of the ADP concentration in yeast cells genetically modified to express creatine kinase. *Biochemistry* 29, 3295–3302. doi: 10.1021/bi00465a021

Britto, D. T., and Kronzucker, H. J. (2006). Futile cycling at the plasma membrane: a hallmark of low-affinity nutrient transport. *Trends Plant Sci.* 11, 529–534. doi: 10.1016/j.tplants.2006.09.011

Broach, J. R. (2012). Nutritional control of growth and development in yeast. *Genetics* 192, 73–105. doi: 10.1534/genetics.111.135731

Brückner, A., Polge, C., Lentze, N., Auerbach, D., and Schlattner, U. (2009). Yeast two-hybrid, a powerful tool for systems biology. *Int. J. Mol. Sci.* 10, 2763–2788. doi: 10.3390/ijms10062763

Cabib, E., Roh, D. H., Schmidt, M., Crotti, L. B., and Varma, A. (2001). The yeast cell wall and septum as paradigms of cell growth and morphogenesis. *J. Biol. Chem.* 276, 19679–19682. doi: 10.1074/jbc.R000031200

Cao, Y., Jin, X., Huang, H., Derebe, M. G., Levin, E. J., Kabaleeswaran, V., et al. (2011). Crystal structure of a potassium ion transporter, TrkH. *Nature* 471, 336–340. doi: 10.1038/nature09731

Chakrapani, S., and Auerbach, A. (2005). A speed limit for conformational change of an allosteric membrane protein. *Proc. Natl. Acad. Sci. U.S.A.* 102, 87–92. doi: 10.1073/pnas.0406777102

Cui, J., and Kaandorp, J. A. (2006). Mathematical modeling of calcium homeostasis in yeast cells. *Cell Calcium.* 39, 337–348. doi: 10.1016/j.ceca.2005.12.001

Curry, M. R., Millar, J. D., Tamuli, S. M., and Watson, P. F. (1996). Surface area and volume measurements for ram and human spermatozoa. *Biol. Reprod.* 55, 1325–1332. doi: 10.1095/biolreprod55.6.1325

Dainty, J., and Hope, A. B. (1961). The electric double layer and the donnan equilibrium in relation to plant cell walls. *Aust. J. Biol. Sci.* 14, 541–551.

Darszon, A., Acevedo, J. J., Galindo, B. E., Hernández-González, E. O., Nishigaki, T., Treviño, C. L., et al. (2006). Sperm channel diversity and functional multiplicity. *Reproduction* 131, 977–988. doi: 10.1530/rep.1.00612

Demidchik, V., and Maathuis, F. J. (2007). Physiological roles of nonselective cation channels in plants: from salt stress to signalling and development. *New Phytol.* 175, 87–404. doi: 10.1111/j.1469-8137.2007.02128.x

Demidchik, V., Macpherson, N., and Davies, J. M. (2005). Potassium transport at the plasma membrane of the food spoilage yeast *Zygosaccharomyces bailii. Yeast* 22, 21–29. doi: 10.1002/yea.1194

Doyle, R. J., and Marquis, R. E. (1994). Elastic, flexible peptidoglycan and bacterial cell wall properties. *Trends Microbiol.* 2, 57–60. doi: 10.1016/0966-842X(94)90127-9

Dupres, V., Dufrêne, Y. F., and Heinisch, J. J. (2010). Measuring cell wall thickness in living yeast cells using single molecular rulers. *ACS Nano* 4, 5498–5504. doi: 10.1021/nn101598v

Eddy, A. A., and Rudin, A. D. (1958). The structure of the yeast cell wall. I. Identification of charged groups at the surface. *Proc. R. Soc. Lond. Ser. B Biol. Sci.* 148, 419–432. doi: 10.1098/rspb.1958.0035

Eraso, P., Cid, A., and Serrano, R. (1987). Tight control of the amount of yeast plasma membrane ATPase during changes in growth conditions and gene dosage. *FEBS Lett.* 224, 193–197. doi: 10.1016/0014-5793(87)80446-5

Fairman, C., Zhou, X., and Kung, C. (1999). Potassium uptake through the TOK1 K^+ channel in the budding yeast. *J. Membr. Biol.* 168, 149–157. doi: 10.1007/s002329900505

Felle, H., Porter, J. S., Slayman, C. L., and Kaback, H. R. (1980). Quantitative measurements of membrane potential in *Escherichia coli*. *Biochemistry* 19, 3585–3590. doi: 10.1021/bi00556a026

Fields, S., and Song, O. (1989). A novel genetic system to detect protein-protein interactions. *Nature* 340, 245–246. doi: 10.1038/340245a0

Fischer, M., Schnell, N., Chattaway, J., Davies, P., Dixon, G., and Sanders, D. (1997). The *Saccharomyces cerevisiae* CCH1 gene is involved in calcium influx and mating. *FEBS Lett.* 419, 259–262. doi: 10.1016/S0014-5793(97)01466-X

Fricke, W., and Peters, W. S. (2002). The biophysics of leaf growth in salt- stressed barley. A study at the cell level. *Plant Physiol.* 129, 374–388. doi: 10.1104/pp.001164

Gaber, R. F., Styles, C. A., and Fink, G. R. (1988). *TRK1* encodes a plasma membrane protein required for high-affinity potassium transport in *Saccharomyces cerevisiae. Mol. Cell. Biol.* 8, 2848–2859.

García, M. J., Ríos, G., Ali, R., Bellés, J. M., and Serrano, R. (1997). Comparative physiology of salt tolerance in *Candida tropicalis* and *Saccharomyces cerevisiae. Microbiology* 143, 1125–1131. doi: 10.1099/00221287-143-4-1125

Ghaemmaghami, S., Huh, W. K., Bower, K., Howson, R. W., Belle, A., Dephoure, N., et al. (2003). Global analysis of protein expression in yeast. *Nature* 425, 737–741. doi: 10.1038/nature02046

Gómez-Lagunas, F., Peña, A., Liévano, A., and Darszon, A. (1989). Incorporation of ionic channels from yeast plasma membranes into black lipid membranes. *Biophys. J.* 56, 115–119.

Goossens, A., de La Fuente, N., Forment, J., Serrano, R., and Portillo, F. (2000). Regulation of yeast H^+-ATPase by protein kinases belonging to a family dedicated to activation of plasma membrane transporters. *Mol. Cell. Biol.* 20, 7654–7661. doi: 10.1128/MCB.20.20.7654-7661.2000

Gout, E., Rébeillé, F., Douce, R., and Bligny, R. (2014). Interplay of $Mg2^+$, ADP, and ATP in the cytosol and mitochondria: unravelling the role of $Mg2^+$ in cell respiration. *Proc. Natl. Acad. Sci. U.S.A.* 111, E4560–E4567. doi: 10.1073/pnas.1406251111

Gray, A. T., Winegar, B. D., Leonoudakis, D. J., Forsayeth, J. R., and Yost, C. S. (1998). TOK1 is a volatile anesthetic stimulated K^+ channel. *Anesthesiology* 88, 1076–1084. doi: 10.1097/00000542-199804000-00029

Grossmann, G., Opekarova, M., Malinsky, J., Weig-Meckl, I., and Tanner, W. (2007). Membrane potential governs lateral segregation of plasma membrane proteins and lipids in yeast. *EMBO J.* 26, 1–8. doi: 10.1038/sj.emboj.7601466

Gustin, M. C., Martinac, B., Saimi, Y., Culbertson, M. R., and Kung, C. (1986). Ion channels in yeast. *Science* 233, 1195–1197. doi: 10.1126/science.2426783

Gustin, M. C., Zhou, X. L., Martinac, B., and Kung, C. (1988). A mechanosensitive ion channel in the yeast plasma membrane. *Science* 242, 762–765. doi: 10.1126/science.2460920

Guzmán-Grenfell, A. M., Bonilla-Hernández, M. A., and González-Martínez, M. T. (2000). Glucose induces a Na^+,K^+-ATPase-dependent transient hyperpolarization in human sperm. I. Induction of changes in plasma membrane potential by the proton ionophore CCCP. *Biochim. Biophys. Acta* 1464, 188–198. doi: 10.1016/S0005-2736(99)00247-3

Haro, R., Garciadeblas, B., and Rodríguez-Navarro, A. (1991). A novel P-type ATPase from yeast involved in sodium transport. *FEBS Lett.* 291, 189–191. doi: 10.1016/0014-5793(91)81280-L

Hasenbrink, G., Kolacna, L., Ludwig, J., Sychrova, H., Kschischo, M., and Lichtenberg-Fraté, H. (2007). Ring test assessment of the mKir2.1 growth based assay in *Saccharomyces cerevisiae* using parametric models and model- free fits. *Appl. Microbiol. Biotechnol.* 73, 1212–1221. doi: 10.1007/s00253-00 6-0589-x

Hasenbrink, G., Schwarzer, S., Kolacna, L., Ludwig, J., Sychrova, H., and Lichtenberg-Fraté, H. (2005). Analysis of the mKir2.1 channel activity in potassium influx defective *Saccharomyces cerevisiae* strains determined as changes in growth characteristics. *FEBS Lett.* 579, 1723–1731. doi: 10.1016/j.febslet.2005.02.025

Hasenbrink, G., Sievernich, A., Wildt, L., Ludwig, J., and Lichtenberg-Fraté, H. (2006). Estrogenic effects of natural and synthetic compounds including tibolone assessed in *Saccharomyces cerevisiae* expressing the human estrogen α and β receptors. *FASEB J.* 20, 1552–1554. doi: 10.1096/fj.05-5413fje

Heinisch, J. J., Dupres, V., Wilk, S., Jendretzki, A., and Dufrêne, Y. F. (2010). Single-molecule atomic force microscopy reveals clustering of the yeast plasma-membrane sensor Wscl. *PLoS ONE* 5:e11104. doi: 10.1371/journal.pone.0011104

Hille, B. (2001). *Ion Channels and Excitable Membranes, 3rd Edn.* Sunderland, MA: Sinauer Associates, 814.

Hirsch, R. E., Lewis, B. D., Spalding, E. P., and Sussman, M. R. (1998). A role for the AKT1 potassium channel in plant nutrition. *Science* 280, 918–921. doi: 10.1126/science.280.5365.918

Hodgkin, A. L., and Huxley, A. F. (1952). A quantitative description of membrane current and its application to conduction and excitation in nerve. *J. Physiol.* 117, 500–544. doi: 10.1113/jphysiol.1952.sp004764

Höfer, M., and Novacky, A. (1986). Measurement of plasma membrane potentials of yeast cells with glass microelectrodes. *Biochim. Biophys. Acta* 862, 372–378. doi: 10.1016/0005-2736(86)90240-3

Hohmann, S. (2002). Osmotic stress signaling and osmoadaptation in yeasts. *Microbiol. Mol. Biol. Rev.* 66, 300–372. doi: 10.1128/MMBR.66.2.300-372.2002

Hope, A. B., and Walker, N. A. (1975). *The Physiology of Giant Algal Cells.* Cambridge: Cambridge University Press, 226.

Hüsken, D., Steudle, E., and Zimmermann, U. (1978). Pressure probe technique for measuring water relations of cells in higher plants. *Plant Physiol.* 61, 158–163. doi: 10.1104/pp.61.2.158

Jay, A. W. (1975). Geometry of the human erythrocyte. I. Effect of albumin on cell geometry. *Biophys. J.* 15, 205–222. doi: 10.1016/S0006-3495(75)85812-7

Jay, A. W. L., and Burton, A. C. (1969). Direct measurement of potential difference across the human red blood cell membrane. *Biophys. J.* 9, 115–121. doi: 10.1016/S0006-3495(69)86372-1

Jayatissa, P. M., and Rose, A. H. (1976). Role of wall phosphomannan in flocculation of *Saccharomyces cerevisiae. J. Gen. Microbiol.* 96, 165–174. doi: 10.1099/00221287-96-1-165

Jennings, M. L., and Cui, J. (2008). Chloride homeostasis in *Saccharomyces cerevisiae*: high affinity influx, V-ATPase-dependent sequestration, and identification of a candidate Cl^- sensor. *J. Gen. Physiol.* 131, 379–391. doi: 10.1085/jgp.200709905

Jimenez, T., McDermott, J. P., Sánchez, G., and Blanco, G. (2011). Na,K-ATPase α4 isoform is essential for sperm fertility. *Proc. Natl. Acad. Sci. U.S.A.* 108, 644–649. doi: 10.1073/pnas.1016902108

Johansson, I., and Blatt, M. R. (2006). Interactive domains between pore loops of the yeast K+ channel TOK1 associate with extracellular K^+ sensitivity. *Biochem. J.* 393, 645–655. doi: 10.1042/BJ20051380

Johnson, B. R., Wyttenbach, R. A., Wayne, R., and Hoy, R. R. (2002). Action potentials in a giant algal cell: a comparative approach to mechanisms and evolution of excitability. *J. Undergrad. Neurosci. Educ.* 1, A23–A27.

Joubert, O., Nehmé, R., Bidet, M., and Mus-Veteau, I. (2010). Heterologous expression of human membrane receptors in the yeast *Saccharomyces cerevisiae. Methods Mol. Biol.* 601, 87–103. doi: 10.1007/978-1-60761-344-2_6

Kahm, M. (2011). *A Mathematical Model of the Potassium Homeostasis in Saccharomyces cerevisiae.* Dissertation zur Erlangung des Doktorgrades (Dr. rer. nat.) der Mathematisch-Naturwissenschaftlichen Fakultät der Rheinischen Friedrich-Wilhelms-Universität Bonn, 94. Available online at: http://hss.ulb.uni-bonn.de/2012/2903/2903.pdf

Kaji, D. M. (1993). Effect of membrane potential on K-Cl transport in human erythrocytes. *Am. J. Physiol.* 264, C376–C382.

Kanzaki, M., Nagasawa, M., Kojima, I., Sato, C., Naruse, K., Sokabe, M., et al. (1999). Molecular identification of a eukaryotic, stretch-activated nonselective cation channel. *Science* 285, 882–886. doi: 10.1126/science.285.5429.882

Ketchum, K. A., Joiner, W. J., Sellers, A. J., Kaczmarek, L. K., and Goldstein, S. A. (1995). A new family of outwardly rectifying potassium channel proteins with two pore domains in tandem. *Nature* 376, 690–695. doi: 10.1038/376690a0

Kinclová, O., Ramos, J., Potier, S., and Sychrová, H. (2001). Functional study of the *Saccharomyces cerevisiae* Nha1p C-terminus. *Mol. Microbiol.* 40, 656–668. doi: 10.1046/j.1365-2958.2001.02412.x

Kinclova-Zimmermannova, O., Gaskova, D., and Sychrova, H. (2006). The Na^+,K^+/H^+ -antiporter Nha1 influences the plasma membrane potential of *Saccharomyces cerevisiae. FEMS Yeast Res.* 6, 792–800. doi: 10.1111/j.1567- 1364.2006.00062.x

Kinclova-Zimmermannova, O., Zavrel, M., and Sychrova, H. (2005). Identification of conserved prolyl residue important for transport activity and the substrate specificity range of yeast plasma membrane Na^+/H^+ antiporters. *J. Biol. Chem.* 280, 30638–30647. doi: 10.1074/jbc.M506341200

Ko, C. H., and Gaber, R. F. (1991). *TRK1* and *TRK2* encode structurally related K^+ transporters in *Saccharomyces cerevisiae. Mol. Cell. Biol.* 11, 4266–4273.

Kolacna, L., Zimmermannova, O., Hasenbrink, G., Schwarzer, S., Ludwig, J., Lichtenberg-Fraté, H., et al. (2005). New phenotypes of functional expression of the mKir2.1 channel in potassium efflux-deficient *Saccharomyces cerevisiae* strains. *Yeast* 22, 1315–1323. doi: 10.1002/yea.1333

Kole, M. H., Ilschner, S. U., Kampa, B. M., Williams, S. R., Ruben, P. C., and Stuart, G. J. (2008). Action potential generation requires a high sodium channel density in the axon initial segment. *Nat. Neurosci.* 11, 178–186. doi: 10.1038/nn2040

Kung, C., Martinac, B., and Sukharev, S. (2010). Mechanosensitive channels in microbes. *Annu. Rev. Microbiol.* 64, 313–329. doi: 10.1146/annurev.micro.112408.134106

Kuroda, T., Bihler, H., Bashi, E., Slayman, C. L., and Rivetta, A. (2004). Chloride channel function in the yeast TRK-potassium transporters. *J. Membr. Biol.* 198, 177–192. doi: 10.1007/s00232-004-0671-1

Lecar, H., Ehrenstein, G., Binstock, L., and Taylor, R. E. (1967). Removal of potassium negative resistance in perfused squid giant axons. *J. Gen. Physiol.* 50, 1499–1515. doi: 10.1085/jgp.50.6.1499

Lee, S., Tong, L., and Denu, J. M. (2008). Quantification of endogenous sirtuin metabolite O-acetyl-ADP-ribose. *Anal. Biochem.* 383, 174–179. doi: 10.1016/j.ab.2008.08.033

Lemaire, T., Kaiser, J., Naili, S., and Sansalone, V. (2010). Modelling of the transport in electrically charged porous media including ionic exchanges. *Mech. Res. Commun.* 37, 495–499. doi: 10.1016/j.mechrescom.2010.05.009

Lesage, F., Guillemare, E., Fink, M., Duprat, F., Lazdunski, M., Romey, G., et al. (1996). A pH-sensitive yeast outward rectifier K^+ channel with two pore domains and novel gating properties. *J. Biol. Chem.* 271, 4183–4187. doi: 10.1074/jbc.271.8.4183

Levitan, I. B., and Kaczmarek, L. K. (2001). *The Neuron: Cell and Molecular Biology, 3rd Edn.* New York, NY; Oxford: Oxford University Press, 632.

Lichtenberg, H. C., Giebeler, H., and Höfer, M. (1988). Measurements of electrical potential differences across yeast plasma membranes with microelectrodes are consistent with values from steady-state distribution of tetraphenylphosphonium in *Pichia humboldtii. J. Membr. Biol.* 103, 255–261. doi: 10.1007/BF01993985

Liebeskind, B. J., Hillis, D. M., and Zakon, H. H. (2011). Evolution of sodium channels predates the origin of nervous systems in animals. *Proc. Natl. Acad. Sci. U.S.A.* 108, 9154–9159. doi: 10.1073/pnas.1106363108

Lipke, P. N., and Ovalle, R. (1998). Cell wall architecture in yeast: new structure and new challenges. *J. Bacteriol.* 180, 3735–3740.

Loret, M. O., Pedersen, L., and François, J. (2007). Revised procedures for yeast metabolites extraction: application to a glucose pulse to carbon-limited yeast cultures, which reveals a transient activation of the purine salvage pathway. *Yeast* 24, 47–60. doi: 10.1002/yea.1435

Loukin, S. H., and Saimi, Y. (1999). K^+-dependent composite gating of the yeast K^+ channel, Tok1. *Biophys. J.* 77, 3060–3070. doi: 10.1016/S0006- 3495(99)77137-7

Loukin, S. H., Vaillant, B., Zhou, X. L., Spalding, E. P., Kung, C., and Saimi, Y. (1997). Random mutagenesis reveals a region important for gating of the yeast K^+ channel Ykc1. *EMBO J.* 16, 4817–4825. doi: 10.1093/emboj/16.16.4817

Madrid, R., Gómez, M. J., Ramos, J., and Rodríguez-Navarro, A. (1998). Ectopic potassium uptake in *trk1 trk2* mutants of *Saccharomyces cerevisiae* correlates with a highly hyperpolarized membrane potential. *J. Biol. Chem.* 273, 14838–14844. doi: 10.1074/jbc.273.24.14838

Malagoli, P., Britto, D. T., Schulze, L. M., and Kronzucker, H. J. (2008). Futile Na^+ cycling at the root plasma membrane in rice (*Oryza sativa* L.): kinetics, energetics, and relationship to salinity tolerance. *J. Exp. Bot.* 59, 4109–4117. doi: 10.1093/jxb/ern249

Malínská, K., Malínskı, J., Opekarová, M., and Tanner, W. (2003). Visualization of protein compartmentation within the plasma membrane of living yeast cells. *Mol. Biol. Cell* 14, 4427–4436. doi: 10.1091/mbc.E03-04-0221

Malinsky, J., Opekarová, M., and Tanner, W. (2010). The lateral compartmentation of the yeast plasma membrane. *Yeast* 27, 473–478. doi: 10.1002/yea.1772

Maresova, L., Urbankova, E., Gaskova, D., and Sychrova, H. (2006). Measurements of plasma membrane potential changes in *Saccharomyces cerevisiae* cells reveal the importance of the Tok1 channel in membrane potential maintenance. *FEMS Yeast Res.* 6, 1039–1046. doi: 10.1111/j.1567-1364.2006.00140.x

Márquez, J. A., and Serrano, R. (1996). Multiple transduction pathways regulate the sodium-extrusion gene PMR2/ENA1 during salt stress in yeast. *FEBS Lett.* 382, 89–92. doi: 10.1016/0014-5793(96)00157-3

Mäser, P., Hosoo, Y., Goshima, S., Horie, T., Eckelman, B., Yamada, K., et al. (2002). Glycine residues in potassium channel-like selectivity filters determine potassium selectivity in four-loop-per-subunit HKT transporters from plants. *Proc. Natl. Acad. Sci. U.S.A.* 99, 6428–6433. doi: 10.1073/pnas.082123799

Meychik, N. R., Honarmand, S. J., and Yermakov, I. P. (2007). Organic cation diffusion in root cell walls of different plants. *Biologija* 53, 60–63.

Meychik, N. R., Nikolaeva, Y. I., and Yermakov, I. P. (2006). Ion-exchange properties of cell walls of *Spinacia oleracea* L. roots under different environmental salt conditions. *Biochem. Mosc.* 71, 781–789. doi: 10.1134/S000629790607011X

Meychik, N. R., and Yermakov, I. P. (2001a). Swelling of root cell walls as an indicator of their functional state. *Biochemistry (Mosc).* 66, 178–187. doi: 10.1023/A:1002843615188

Meychik, N. R., and Yermakov, I. P. (2001b). Ion exchange properties of plant root cell walls. *Plant Soil.* 234, 181–193. doi: 10.1023/A:1017936318435

Meychik, N. R., and Yermakov, I. P. (2001c). Ion-exchange properties of cell walls isolated from lupine roots. *Biochemistry (Mosc).* 66, 556–563. doi: 10.1023/A:1010219221077

Minc, N., Boudaoud, A., and Chang, F. (2009). Mechanical forces of fission yeast growth. *Curr. Biol.* 19, 1096–1101. doi: 10.1016/j.cub.2009.05.031

Miranda, M., Bashi, E., Vylkova, S., Edgerton, M., Slayman, C., and Rivetta, A. (2009). Conservation and dispersion of sequence and function in fungal TRK potassium transporters: focus on *Candida albicans. FEMS Yeast Res.* 9, 278–292. doi: 10.1111/j.1567-1364.2008.00471.x

Mitsui, K., Hatakeyama, K., Matsushita, M., and Kanazawa, H. (2009). *Saccharomyces cerevisiae* Na^+/H^+ antiporter Nha1p associates with lipid rafts and requires sphingolipid for stable localization to the plasma membrane. *J. Biochem.* 145, 709–720. doi: 10.1093/jb/mvp032

Mulet, J. M., and Serrano, R. (2002). Simultaneous determination of potassium and rubidium content in yeast. *Yeast* 19, 1295–1298. doi: 10.1002/yea.909

Murguía, J. R., Bellés, J. M., and Serrano, R. (1996). The yeast *HAL2* nucleotidase is an *in vivo* target of salt toxicity. *J. Biol. Chem.* 271, 29029–29033. doi: 10.1074/jbc.271.46.29029

Muzzey, D., Gómez-Uribe, C. A., Mettetal, J. T., and van Oudenaarden, A. (2009). A systems-level analysis of perfect adaptation in yeast osmoregulation. *Cell* 138, 160–171. doi: 10.1016/j.cell.2009.04.047

Navarrete, C., Petrezsélyová, S., Barreto, L., Martínez, J. L., Zahrádka, J., Ariño, J., et al. (2010). Lack of main K^+ uptake systems in *Saccharomyces cerevisiae* cells affects yeast cell physiological parameters both in potassium sufficient and limiting conditions. *FEMS Yeast Res.* 10, 508–517. doi: 10.1111/j.1567-1364.2010.00630.x

Navarro, B., Kirichok, Y., and Clapham, D. E. (2007). KSper, a pH-sensitive K^+ current that controls sperm membrane potential. *Proc. Natl. Acad. Sci. U.S.A.* 104, 7688–7692. doi: 10.1073/pnas.0702018104

Newsholme, E. A., and Crabtree, B. (1976). Substrate cycles in metabolic regulation and in heat generation. *Biochem. Soc. Symp.* 41, 61–109.

Nielsen, L. J., Olsen, L. F., and Ozalp, V. C. (2010). Aptamers embedded in polyacrylamide nanoparticles: a tool for *in Vivo* metabolite sensing. *ACS Nano* 4, 4361–4370. doi: 10.1021/nn100635j

Ohgaki, R., Nakamura, N., Mitsui, K., and Kanazawa, H. (2005). Characterization of the ion transport activity of the budding yeast Na^+/H^+ antiporter, Nha1p, using isolated secretory vesicles. *Biochim. Biophys. Acta* 1712, 185–196. doi: 10.1016/j.bbamem.2005.03.011

Österlund, T., Nookaew, I., Bordel, S., and Nielsen, J. (2013). Mapping condition-dependent regulation of metabolism in yeast through genome-scale modeling. *BMC Syst. Biol.* 7:36. doi: 10.1186/1752-0509-7-36

Ozalp, V. C., Pedersen, T. R., Nielsen, L. J., and Olsen, L. F. (2010). Time-resolved measurements of intracellular ATP in the yeast *Saccharomyces cerevisiae* using a new type of nanobiosensor. *J. Biol. Chem.* 285, 37579–37588. doi: 10.1074/jbc.M110.155119

Palmer, C., Zhou, X.-L., Lin, J., Loukin, S., Kung, C., and Saimi, Y. (2001). A TRP homolog in *Saccharomyces cerevisiae* forms an intracellular $Ca2^+$ permeable channel in the yeast vacuolar membrane. *Proc. Natl. Acad. Sci. U.S.A.* 98, 7801–7805. doi: 10.1073/pnas.141036198

Paulsen, I. T., Sliwinski, M. K., Nelissen, B., Goffeau, A., and Saier, M. Jr. (1998). Unified inventory of established and putative transporters encoded within the complete genome of *Saccharomyces cerevisiae. FEBS Lett.* 430, 116–125. doi: 10.1016/S0014-5793(98)00629-2

Peiter, E., Fischer, M., Sidaway, K., Roberts, S. K., and Sanders, D. (2005). The *Saccharomyces cerevisiae* $Ca2^+$ channel Cch1pMid1p is essential for tolerance to cold stress and iron toxicity. *FEBS Lett.* 579, 5697–5703. doi: 10.1016/j.febslet.2005.09.058

Peña, A., Sánchez, N. S., and Calahorra, M. (2010). Estimation of the electric plasma membrane potential difference in yeast with fluorescent dyes: comparative study of methods. *J. Bioenerg. Biomembr.* 42, 419–432. doi: 10.1007/s10863-010-9311-x

Perlin, D. S., Harris, S. L., Seto-Young, D., and Haber, J. E. (1989). Defective H^+-ATPase of hygromycin B-resistant *pma1* mutants from *Saccharomyces cerevisiae. J. Biol. Chem.* 264, 21857–21864.

Permyakov, S., Suzina, N., and Valiakhmetov, A. (2012). Activation of H^+- ATPase of the plasma membrane of *Saccharomyces cerevisiae* by glucose: the role of sphingolipid and lateral enzyme mobility. *PLoS ONE* 7:e30966. doi: 10.1371/journal.pone.0030966

Plášek, J., Gášková, D., Lichtenberg-Fraté, H., Ludwig, J., and Höfer, M. (2012). Monitoring of real changes of plasma membrane potential by diS-C3 (3). fluorescence in yeast cell suspensions. *J. Bioenerg. Biomembr.* 44, 559–569. doi: 10.1007/s10863-012-9458-8

Plášek, J., Gášková, D., Ludwig, J., and Höfer, M. (2013). Early changes in membrane potential of *Saccharomyces cerevisiae* induced by varying extracellular K^+, Na^+ or H^+ concentrations. *J. Bioenerg. Biomembr.* 45, 561–568. doi: 10.1007/s10863-013-9528-6

Posas, F., Chambers, J. R., Heyman, J. A., Hoeffler, J. P., de Nadal, E., and Ariño, J. (2000). The transcriptional response of yeast to saline stress. *J. Biol. Chem.* 275, 17249–17255. doi: 10.1074/jbc.M910016199

Pradet, A., and Raymond, P. (1983). Adenine nucleotide ratios and adenylate energy charge in energy metabolism. *Annu. Rev. Plant Physiol.* 34, 199–224. doi: 10.1146/annurev.pp.34.060183.001215

Prior, C., Potier, S., Souciet, J. L., and Sychrova, H. (1996). Characterization of the *NHA1* gene encoding Na^+/H^+-antiporter of the yeast *Saccharomyces cerevisiae. FEBS Lett.* 387, 89–93. doi: 10.1016/0014-5793(96) 00470-X

Qian, H., and Beard, D. A. (2006). Metabolic futile cycles and their functions: a systems analysis of energy and control. *IEE Proc. Syst. Biol.* 153, 192–200. doi: 10.1049/ip-syb:20050086

Ramirez, J. A., Vacata, V., McCusker, J. H., Haber, J. E., Mortimer, R. K., Owen, W. G., et al. (1989). ATP-sensitive K^+ channels in a plasma membrane H^+-ATPase mutant of the yeast *Saccharomyces cerevisiae. Proc. Natl. Acad. Sci. U.S.A.* 86, 7866–7870. doi: 10.1073/pnas.86.20.7866

Ren, D., Navarro, B., Xu, H., Yue, L., Shi, Q., and Clapham, D. E. (2001). A prokaryotic voltage-gated sodium channel. *Science* 294, 2372–2375. doi: 10.1126/science.1065635

Rink, T. J. (1977). Membrane potential of guinea-pig spermatozoa. *J. Reprod. Fertil.* 51, 155–157. doi: 10.1530/jrf.0.0510155

Ritchie, R. J., and Larkum, A. W. D.,(1982). Cation exchange properties of the cell walls of *Enteromorpha intestinalis* (L.) Link. *(Ulvales Chlorophyta). J. Exp. Bot.* 132, 125–139. doi: 10.1093/jxb/33.1.125

Rivetta, A., Kuroda, T., and Slayman, C. (2011). Anion currents in yeast K^+ transporters (TRK) characterize a structural homologue of ligand- gated ion channels. *Pflugers Arch.* 462, 315–330. doi: 10.1007/s00424-011- 0959-9

Rivetta, A., Slayman, C., and Kuroda, T. (2005). Quantitative modeling of chloride conductance in yeast TRK potassium transporters. *Biophys. J.* 89, 2412–2426. doi: 10.1529/biophysj.105.066712

Roberts, S. K., Fischer, M., Dixon, G. K., and Sanders, D. (1999). Divalent Cation Block of Inward Currents and Low-Affinity K^+ Uptake in *Saccharomyces cerevisiae. J. Bacteriol.* 181, 291–297.

Rösch, P., Harz, M., Schmitt, M., and Popp, J. (2005). Raman spectroscopic identification of single yeast cells. *J. Raman Spectrosc.* 36, 377–379. doi: 10.1002/jrs.1312

Ryan, P. R., Newman, I. A., and Arif, I. (1992). Rapid calcium exchange for protons and potassium in cell walls of *Chara. Plant Cell Environ.* 15, 675–683. doi: 10.1111/j.1365-3040.1992.tb01009.x

Saito, T., Soga, K., Hoson, T., and Terashima, I. (2006). The bulk elastic modulus and the reversible properties of cell walls in developing *Quercus* leaves. *Plant Cell Physiol.* 47, 715–725. doi: 10.1093/pcp/pcj042

Salton, M. R. J., and Kim, K.-S. (1996). "Structure," in *Medical Microbiology, 4th Edn.*, ed S. Baron (Galveston, TX: University of Texas Medical Branch at Galveston), Chapter 2.

Samoilov, M., Plyasunov, S., and Arkin, A. (2005). Stochastic amplification and signaling in enzymatic futile cycles through noise-induced bistability with oscillations. *Proc. Natl. Acad. Sci. U.S.A.* 102, 2310–2315. doi: 10.1073/pnas.0406841102

Satoh, H., Delbridge, L. M., Blatter, L. A., and Bers, D. M. (1996). Surface:volume relationship in cardiac myocytes studied with confocal microscopy and membrane capacitance measurements: species-dependence and developmental effects. *Biophys. J.* 70, 1494–1504. doi: 10.1016/S0006-3495(96)79711-4

Schaber, J., Adrover, M. A., Eriksson, E., Pelet, S., Petelenz-Kurdziel, E., Klein, D., et al. (2010). Biophysical properties of *Saccharomyces cerevisiae* and their relationship with HOG pathway activation. *Eur. Biophys. J.* 39, 1547–1556. doi: 10.1007/s00249-010-0612-0

Schwarzer, S., Kolacna, L., Lichtenberg-Fraté, H., Sychrova, H., and Ludwig, J. (2008). Functional expression of the voltage-gated neuronal mammalian potassium channel rat *ether à go-go1* in yeast. *FEMS Yeast Res.* 8, 405–413. doi: 10.1111/j.1567-1364.2007.00351.x

Sentenac, H., and Grignon, C. (1981). A model for predicting ionic equilibrium concentrations in cell walls. *Plant Physiol.* 68, 415–419. doi: 10.1104/pp.68.2.415

Serrano, R. (1988). Structure and function of proton translocating ATPase in plasma membranes of plants and fungi. *Biochim. Biophys. Acta* 947, 1–28. doi: 10.1016/0304-4157(88)90017-2

Serrano, R. (1996). Salt tolerance in plants and microorganisms: toxicity targets and defense mechanisms. *Int. Rev. Cytol.* 165, 1–52.

Shabala, L., Ross, T., McMeekin, T., and Shabala, S. (2006). Non-invasive microelectrode ion flux measurements to study adaptive responses of microorganisms to the environment. *FEMS Microbiol. Rev.* 30, 472–486. doi: 10.1111/j.1574-6976.2006.00019.x

Shabala, S., and Newman, I. (2000). Salinity effects on the activity of plasma membrane H^+ and $Ca2^+$ transporters in bean leaf mesophyll: masking role of the cell wall. *Ann. Bot.* 85, 681–686. doi: 10.1006/anbo.2000.1131

Sievernich, A., Wildt, L., and Lichtenberg-Fraté, H. (2004). *In vitro* bioactivity of 17α-estradiol. *J. Steroid Biochem. Mol. Biol.* 92, 455–463. doi: 10.1016/j.jsbmb.2004.09.004

Skou, J. C. (1998). Nobel lecture. The identification of the sodium pump. *Biosci. Rep.* 18, 155–169. doi: 10.1023/A:1020196612909

Slayman, C. L., and Slayman, C. W. (1962). Measurement of membrane potentials in Neurospora. *Science* 136, 876–877. doi: 10.1126/science.136.3519.876

Smith, A. E., Zhang, Z., Thomas, C. R., Kennith, E., Moxham, K. E., and Middelberg, A. P. J. (2000). The mechanical properties of *Saccharomyces cerevisiae. Proc. Natl. Acad. Sci. U.S.A.* 97, 9871–9874. doi: 10.1073/pnas.97.18.9871

Soltanian, S., Dhont, J., Sorgeloos, P., and Bossier, P. (2007). Influence of different yeast cell-wall mutants on performance and protection against pathogenic bacteria (*Vibrio campbellii*) in gnotobiotically-grown *Artemia. Fish Shellfish Immunol.* 23, 141–153. doi: 10.1016/j.fsi.2006.09.013

Spira, F., Mueller, N. S., Beck, G., von Olshausen, P., Beig, J., and Wedlich- Söldner, R. (2012). Patchwork organization of the yeast plasma membrane into numerous coexisting domains. *Nature Cell Biol.* 14, 640–648. doi: 10.1038/ncb2487

Stagljar, I., Korostensky, C., Johnsson, N., and te Heesen, S. (1998). A genetic system based on split-ubiquitin for the analysis of interactions between membrane proteins *in vivo. Proc. Natl. Acad. Sci. U.S.A.* 95, 5187–5192. doi: 10.1073/pnas.95.9.5187

Stenson, J. D. (2009). *Investigating the Mechanical Properties of Yeast Cells.* Ph.D. thesis, University of Birmingham, 253. Available online at: http://etheses.bham.ac.uk/304/1/Stenson09PhD.pdf

Stenson, J. D., Hartley, P., Wang, C., and Thomas, C. R. (2011). Determining the Mechanical Properties of Yeast Cell Walls. *Biotechnol. Prog.* 27, 505–512. doi: 10.1002/btpr.554

Steudle, E. (1993). "Pressure probe techniques: basic principles and application to studies of water and solute relations at the cell, tissue and organ level," in *Water Deficits: Plant Responses from Cell to Community*, eds J. A. C. Smith and H. Griffiths (Oxford: Bios Scientific Publishers), 5–36.

Sutak, R., Botebol, H., Blaiseau, P. L., Léger, T., Bouget, F. Y., Camadro, J. M., et al. (2012). A comparative study of iron uptake mechanisms in marine micro-algae: iron binding at the cell surface is a critical step. *Plant Physiol.* 160, 2271–2284. doi: 10.1104/pp.112.204156

Taglicht, D., Padan, E., and Schuldiner, S. (1991). Overproduction and purification of a functional Na^+/H^+ antiporter coded by nhaA (ant) from *Escherichia coli. J. Biol. Chem.* 266, 11289–11294.

Takami, H., and Horikoshi, K. (1999). Reidentification of facultatively alkaliphilic Bacillus sp. C-125 to Bacillus halodurans. *Biosci. Biotechnol. Biochem.* 63, 943–945. doi: 10.1271/bbb.63.943

Tálos, K., Pernyeszi, T., Majdik, C., Hegedûsova, A., and Páger, C. (2012). Cadmium biosorption by baker's yeast in aqueous suspension. *J. Serb. Chem. Soc.* 77, 549–561. doi: 10.2298/JSC110520181T

Taylor, A. R. (2009). A fast $Na^+/Ca2^+$-based action potential in a marine diatom. *PLoS ONE* 4:e4966. doi: 10.1371/journal.pone.0004966

Teng, J., Goto, R., Iida, K., Kojima, I., and Iida, H. (2008). Ion-channel blocker sensitivity of voltage-gated calcium-channel homologue Cch1 in *Saccharomyces cerevisiae. Microbiology* 154, 3775–3781. doi: 10.1099/mic.0.2008/021089-0

Teng, J., Iida, K., Imai, A., Nakano, M., Tada, T., and Iida, H. (2013). Hyperactive and hypoactive mutations in Cch1, a yeast homologue of the voltage- gated calcium-channel pore-forming subunit. *Microbiology* 159, 970–979. doi: 10.1099/mic.0.064030-0

Thaminy, S., Miller, J., and Stagljar, I. (2004). The split-ubiquitin membrane-based yeast two-hybrid system. *Methods Mol. Biol.* 261, 297–312. doi: 10.1385/1-59259-762-9:297

Thomas, S. L., Bouyer, G., Cueff, A., Egée, S., Glogowska, E., and Ollivaux, C. (2011).

Ion channels in human red blood cell membrane: actors or relics? *Blood Cells Mol. Dis.* 46, 261–265. doi: 10.1016/j.bcmd.2011.02.007

Thonart, P., Custinne, M., and Paquot, M. (1982). Zeta potential of yeast cells: application in cell immobilization. *Enzyme Microb. Technol.* 4, 191–194. doi: 10.1016/0141-0229(82)90116-8

Tolla, D. A., Kiley, P. J., Lomnitz, J. G., and Savageau, M. A. (2015). Design principles of a conditional futile cycle exploited for regulation. *Mol. Biosyst.* doi: 10.1039/c5mb00055f. [Epub ahead of print].

Ton, V. K., and Rao, R. (2004). Functional expression of heterologous proteins in yeast: insights into $Ca2^+$ signaling and $Ca2^+$-transporting ATPases. *Am. J. Physiol. Cell Physiol.* 287, C580–C589. doi: 10.1152/ajpcell.00135.2004

Tyree, M. T. (1968). Determination of transport constants of isolated *Nitella* cell walls. *Can. J. Bot.* 46, 317–327. doi: 10.1139/b68-054

Ullah, A., Chandrasekaran, G., Brul, S., and Smits, G. J. (2013). Yeast adaptation to weak acids prevents futile energy expenditure. *Front. Microbiol.* 4:142. doi: 10.3389/fmicb.2013.00142

Vacata, V., Kotyk, A., and Sigler, K. (1981). Membrane potentials in yeast cells measured by direct and indirect methods. *Biochim. Biophys. Acta* 643, 265–268. doi: 10.1016/0005-2736(81)90241-8

Vallejo, C. G., and Serrano, R. (1989). Physiology of mutants with reduced expression of plasma membrane H^+-ATPase. *Yeast* 5, 307–319. doi: 10.1002/yea.320050411

Van Belle, D., and André, B. (2001). A genomic view of yeast membrane transporters. *Curr. Opin. Cell Biol.* 13, 389–398. doi: 10.1016/S0955- 0674(00)00226-X

Volkmer, B., and Heinemann, M. (2011). Condition-dependent cell volume and concentration of Escherichia coli to facilitate data conversion for systems biology modeling. *PLoS ONE* 6:e23126. doi: 10.1371/journal.pone.0023126

Volkov, V., Boscari, A., Clement, M., Miller, A. J., Amtmann, A., and Fricke, W. (2009). Electrophysiological characterization of pathways for K^+ uptake into growing and non-growing leaf cells of barley. *Plant Cell Environ.* 32, 1778–1790. doi: 10.1111/j.1365-3040.2009.02034.x

Volkov, V., Hachez, C., Moshelion, M., Draye, X., Chaumont, F., and Fricke, W. (2007). Osmotic water permeability differs between growing and non-growing barley leaf tissues. *J. Exp. Bot.* 58, 377–390. doi: 10.1093/jxb/erl203

Walsby, A. E., Hayes, P. K., and Boje, R. (1995). The gas vesicles, buoyancy and vertical distribution of cyanobacteria in the Baltic Sea. *Eur. J. Phycol.* 30, 87–94, doi: 10.1080/09670269500650851

Waters, S., Gilliham, M., and Hrmova, M. (2013). Plant high-affinity potassium (HKT) transporters involved in salinity tolerance: structural insights to probe differences in ion selectivity. *Int. J. Mol. Sci.* 14, 7660–7680. doi: 10.3390/ijms14047660

Wei, C., and Lintilhac, P. M. (2007). Loss of stability: a new look at the physics of cell wall behavior during plant cell growth. *Plant Physiol.* 145, 763–772. doi: 10.1104/pp.107.101964

Widschwendter, M., Lichtenberg-Fraté, H., Hasenbrink, G., Schwarzer, S., Dawnay, A., Lam, A., et al. (2009). Serum oestrogen receptor a and b bioactivity are independently associated with breast cancer: a proof of principle study. *Br. J. Cancer.* 101, 160–165. doi: 10.1038/sj.bjc.6605106

Wieland, J., Nitsche, A. M., Strayle, J., Steiner, H., and Rudolph, H. K. (1995). The PMR2 gene cluster encodes functionally distinct isoforms of a putative Na^+ pump in the yeast plasma membrane. *EMBO J.* 14, 3870–3882.

Wilson, W. A., Hawley, S. A., and Hardie, D. G. (1996). Glucose repression/derepression in budding yeast: SNF1 protein kinase is activated by phosphorylation under derepressing conditions, and this correlates with a high AMP:ATP ratio. *Curr. Biol.* 6, 1426–1434. doi: 10.1016/S0960-9822(96) 00747-6

Zahrádka, J., and Sychrova, H. (2012). Plasma-membrane hyperpolarization diminishes the cation efflux via Nha1 antiporter and Ena ATPase under potassium-limiting conditions. *FEMS Yeast Res.* 12, 439–446. doi: 10.1111/j.1567-1364.2012.00793.x

Zayats, V., Stockner, T., Pandey, S. K., Wörz, K., Ettrich, R., and Ludwig, J. (2015). A refined atomic scale model of the *Saccharomyces cerevisiae* K^+-translocation protein Trk1p combined with experimental evidence confirms the role of selectivity filter glycines and other key residues. *Biochim. Biophys. Acta* 1848, 1183–1195. doi: 10.1016/j.bbamem.2015.02.007

Zhou, X. L., Batiza, A. F., Loukin, S. H., Palmer, C. P., Kung, C., and Saimi, Y. (2003). The transient receptor potential channel on the yeast vacuole is mechanosensitive. *Proc. Natl. Acad. Sci. U.S.A.* 100, 7105–7110. doi: 10.1073/pnas.1230540100

4

Biophysical and biochemical constraints imposed by salt stress: Learning from halophytes

***Bernardo Duarte*[1,2] *, *Noomene Sleimi*[3] *and Isabel Caçador*[1,2]**

[1] Centre of Oceanography, Faculty of Sciences, University of Lisbon, Lisbon, Portugal
[2] Marine and Environmental Sciences Centre, Faculty of Sciences, University of Lisbon, Lisbon, Portugal
[3] UR: MaNE, Faculté des Sciences de Bizerte, Université de Carthage, Bizerte, Tunisie

Edited by:
Vadim Volkov, London Metropolitan University, UK

Reviewed by:
Enrique Mateos-Naranjo, University of Seville, Spain
Muhammad Ajmal Khan, Qatar University, Qatar
Janine Adams, Nelson Mandela Metropolitan University, South Africa

****Correspondence:***
Bernardo Duarte, Centre of Oceanography, Faculty of Sciences, University of Lisbon, Lisbon, Portugal
e-mail: baduarte@fc.ul.pt

Soil salinization is one of the most important factors impacting plant productivity. About 3.6 billion of the world's 5.2 billion ha of agricultural dry land, have already suffered erosion, degradation, and salinization. Halophytes are typically considered as plants able to complete their life cycle in environments where the salt concentration is above 200 mM NaCl. Salinity adjustment is a complex phenomenon but essential mechanism to overcome salt stress, with both biophysical and biochemical implications. At this level, halophytes evolved in several directions, adopting different strategies. Otherwise, the lack of adaptation to a salt environment would negatively affect their electron transduction pathways and the entire energetic metabolism, the foundation of every plant photosynthesis and biomass production. The maintenance of ionic homeostasis is in the basis of all cellular counteractive measures, in particular in terms of redox potential and energy transduction. In the present work the biophysical mechanisms underlying energy capture and transduction in halophytes are discussed alongside with their relation with biochemical counteractive mechanisms, integrating data from photosynthetic light harvesting complexes, electron transport chains to the quinone pools, carbon fixation, and energy dissipation metabolism.

Keywords: photochemistry, halophytes, oxidative stress, osmoregulation, stress, physiological

INTRODUCTION

If we take a good look to our planet we will conclude that it is in fact a salt planet. About 70% of its surface is covered by salt water, with concentrations of Na^+ around 500 mM and contrasting low K^+ concentrations of 9 mM (Flowers, 2004). Alongside, the remaining 30% of the Earth's surface is severely affected by increased salinization, enhanced by improper agricultural soil use and irrigation practices (Zhang and Shi, 2014). We live in a time of changes, the ongoing climate-driven changes must also be considered as well as their consequences, such as increasing drought frequency and intensity, air temperature, and salt water intrusion in coastal soils (Duarte et al., 2013a). All these aspects impose severe constraints to the primary production of Earth, namely crop production. Salinity-induced constraints in plants are associated with reductions in leaf expansion, stomatal closure, reduced primary production, biomass losses, and nutritional deficiencies, like K^+ deficiency (Mahajan and Tuteja, 2005; Rahnama et al., 2010; James et al., 2011). Halophytes are an exception, being highly productive under saline conditions.

The typical definition of halophyte is a plant species that can survive and reproduce under growth conditions with more than 200 mM NaCl (Flowers and Colmer, 2008). Some of these species can be classified as 'obligatory halophytes' like *Suaeda maritima* and *Mesembryanthemum crystallinum* requiring saline environments for optimal growth, while other species like *Puccinellia maritima* and *Thellungiella halophila* are included in the group of the so-called "facultative halophytes" with optimal growth without salt in the substrate though tolerating high NaCl concentrations (Flowers, 1972; Gong et al., 2005; Gao et al., 2006; Agarie et al., 2007; Wang et al., 2007, 2009; Guo et al., 2012). The survival and productivity of these species outcomes from a complex network of mechanisms involving multiple biochemical and physiological traits of salt tolerance. Over the last decades, this issue attracted several research groups since the global soil salinization problem became more and more widespread, increasing the need to understand these mechanisms with the main objective of transposing this knowledge to economically relevant crops. Simultaneously, some halophytes were identified as potential nutritional sources with high nutritional value and possibilities to be cultivated in arid environments of the poorer regions of the planet. Several halophytes were already identified and used commercially as food sources like *Aster tripolium* (Ventura et al., 2013), *Chenopodium quinoa* (Eisa et al., 2012), and *Salicornia sp.* (Ventura and Sagi, 2013).

Several reviews have been published focusing at salt stress in halophytes and glycophytes from all over the world and in several different ecosystems. More than describing the anatomical and biochemical adaptations of different halophytes and halophytic strategies to salinity, the present work intends to connect these traits with the most recent biophysical approaches, relating adaptions, and stress signs with the cellular redox homeostasis and bioenergetics. The features are at the basis of the primary production, so knowledge of these bioenergetics traits can provide powerful insights for understanding salt stress in glycophytic crops as well as new opportunities for the improvement of their salinity tolerance.

ANATOMICAL MODIFICATIONS

Some of the evident adaptations to salt environments can be immediately detected just observing halophyte morphology. Typically, there are two mechanisms that halophytes use in order to overcome high salinity: secretion and exclusion. The secretion-based strategy implies the existence of specialized salt glands (**Figure 1**), located at the leaf surface. The main function of salt glands is the excretion of excessive Na^+ (Shabala et al., 2014) as a way to reduce its negative effects on cell metabolism. This is probably the most studied tolerance adaptation mechanism in halophytes (Rozema et al., 1981; Waisel et al., 1986; Shabala et al., 2014). The excreted salt crystals on the leaf surface are then washed out by rain or tidal waters, preventing its reabsorption to the leaf cells (Balsamo et al., 1995). On the contrary a typical halophyte excluder retains high amounts of K^+ and Ca^{2+} inside its cells to avoid Na^+ uptake, enabling survival in soils with very high salt concentrations (**Figure 2**). The increased Ca^{2+} concentrations allow the cell membrane to maintain the K^+/Na^+ selectivity and thus maintain the ionic balance of the cell (Cramer et al., 1987). Alongside with this shoot-exclusion, there is often an observable increase in root Ca^{2+} concentration accompanied by a decrease of the Na^+ root concentration. This exclusion strategy is well studied in *Sarcocornia fruticosa,* frequently followed by a dilution strategy, implying an increased cellular water uptake and thus decreasing the ionic concentration inside the cell (**Figure 3**). *T. halophila* also evidences a very similar strategy, retaining higher K^+ and lower Na^+ concentrations, while increasing its water uptake (Volkov and Amtmann, 2006). This differential ionic absorption is mediated by specific protein ionic channels, with a total of 32 salt induced differentially expressed proteins already identified in *T. halophila* (Pang et al., 2010). Under stress, K^+ transporter proteins are preferentially expressed alongside with changes in membrane potential and ion selectivity, counteracting the elevated extracellular Na^+ concentrations. Nevertheless, all these morphological adaptations have implications at both biophysical and biochemical levels.

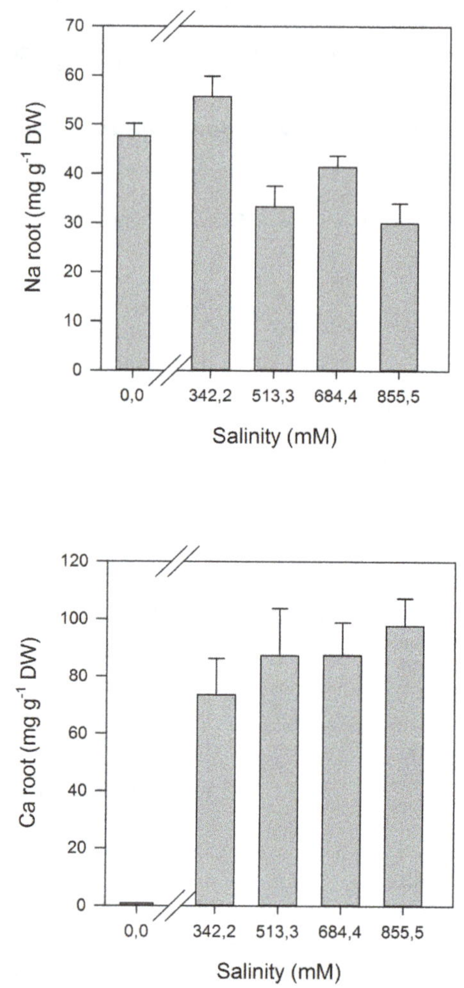

FIGURE 2 | Na^+ and Ca^{2+} ionome in the roots of *Sarcocornia fruticosa* exposed to increased salinity levels (average ± SE, *N* = 5).

BIOPHYSICAL FEEDBACK

As all other excessive ionic accumulation, excessive salinity has also redox implications at the cellular level, unbalancing the cellular electron fluxes. A decrease in the photosynthetic capacity is very common in salt stressed plants (Munns and Termaat, 1986; Munns, 1993; Qiu et al., 2003; Jaleel et al., 2007), mostly due to a low

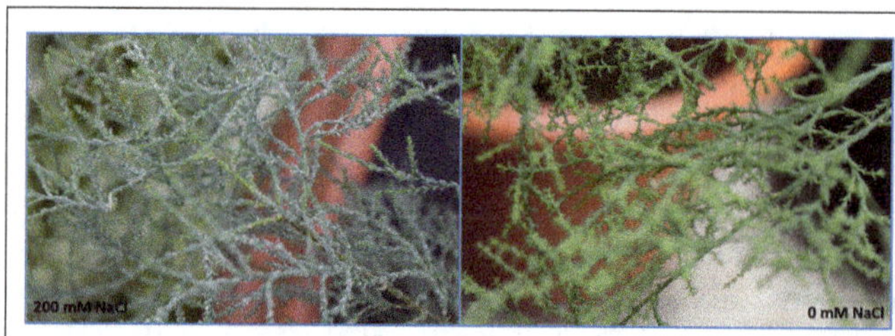

FIGURE 1 | *Tamarix gallica* leaves of individuals subjected to 200 and 0 mM NaCl Photo by B. Duarte (2012). Plants were originally collected in Tunisia and transplanted to the Centre of Oceanography greenhouse, where they were subjected to different salinity levels.

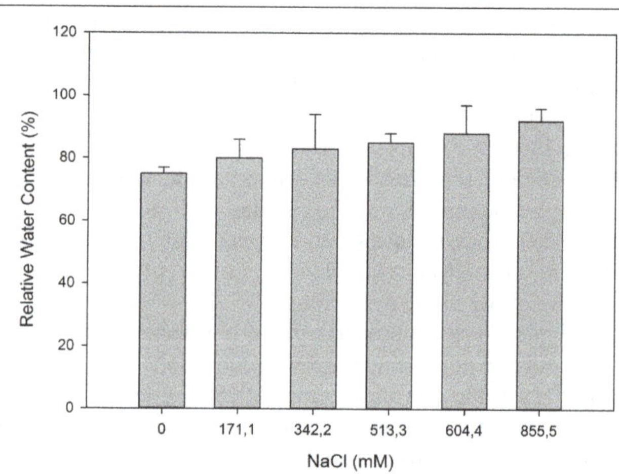

FIGURE 3 | Relative water content in photosynthetic stems of *S. fruticosa*, which was exposed to increased salinity levels (average ± SE, *N* = 5).

osmotic potential of the soil solution (osmotic stress), specific ion effects (salt stress), nutritional imbalances, or more usually, a combination of all these factors (Zhu, 2003). One of the consequences of salinity-induced photosynthetic impairment is the exposure of plants to excess of light energy and its inevitable consequences for the photosystem II (PSII). Plants under salt stress use less light energy for photosynthesis (Megdiche et al., 2008). Therefore the presence of efficient energy dissipation mechanisms is essential in order to prevent the accumulation of excessive energy within the cells in the form of excessive reducing potential (Demmig-Adams and Adams, 1992; Qiu et al., 2003). Salinity constraints for photosynthesis are not restricted to the light harvesting processes. Also the photosynthetic carbon fixation reactions are affected under salt stress, mostly due to disturbances of leaf osmotic potential, of the chloroplast membrane systems and of pigment composition (Munns, 2002; Zhao et al., 2007). To avoid damage in the PSII, plants have developed several strategies to dissipate excessive energy. Comparing the PSII activity of glycophytes (*Cyperus longus* for example) with halophytes (*Spartina versicolor* for example) in a salt medium, the differences are evident (**Figure 4**). In glycophyte species, both real (operational) and maximum PSII activities suffer drastic decreases under salt stress. On the other hand, halophytic species, well adapted to salt environments, show almost no differences along a salinity gradient even under oceanic salt concentrations. PSII quantum yield provides rapid and valuable insights on the overall PSII energetic processes. Nevertheless, in order to understand the causes behind these changes, as well as the mechanisms that allow halophytes to overcome salt stress, we need to take a closer look into the biophysics and energetics of the chloroplast. PSII efficiency relies essentially on two major processes: (1) photon harvesting, entrapment and energy transfer throughout the transport chain and (2) dissipation of excessive reducing power. The delicate balance between both these processes is important for all the electron transduction pathway and evidently for energy production. Overlooking the first one, and focusing especially in the electron transport processes, two strategies can be observed depending on the plant tolerance and mechanisms of the salt tolerance (**Figure 5**). Observing the rapid light curves obtained for *Halimione portulacoides* (excretion strategy) and *S. fruticosa* (exclusion strategy), the differences are evident. Although the exclusion strategy of *S. fruticosa* takes place in the roots, this will condition the Na^+ translocation for the aboveground organs. Nevertheless excessive Na^+ translocation can still happen and in this case the swelled photosynthetic steams will act as sinks, storing Na^+ in their vacuoles (Flowers and Colmer, 2008). In *S. fruticosa* the maximum electron transport rate (ETR_{max}), photosynthetic efficiency and the onset of light saturation are very similar between control and stressed individuals, with only small differences in the ETR at some light levels. On the other hand, *H. portulacoides* stressed and control individuals exhibited very distinct photosynthetic parameters. Not only the photosynthetic efficiency and the onset of light saturation were reduced to nearly zero, but also the ETR_{max} was severely decreased in stressed individuals. Observing *S. fruticosa* control and stressed individuals we found no major differences neither between the ETR nor in the onset of light saturation, indicating a normal functioning in the ETC. As for *H. portulacoides*, not only the ETR was rather decreased in stressed individuals, but these individuals also have a smaller onset for light saturation, indicating an incapacity to use the absorbed photons for primary photochemical purposes. This inevitably leads to an accumulation of large amounts of reducing power with a high potential for reactive oxygen species (ROS) generation that, as stated before, can destroy the D1 protein, impairing the photochemical apparatus (Rintamäki et al., 1995). Again, two tolerance mechanisms are evidenced between these two *Amaranthaceae* species. *S. fruticosa* presents a salinity tolerance mechanism that allows the PSs to absorb light even under high Na^+ concentrations. On the other hand, in *H. portulacoides* these mechanisms appear to be absent or inactivated, leading to lower light harvesting and carbon fixation efficiencies. In fact *S. fruticosa* exhibits a common feature among halophytes with an improvement of some energy conversion mechanisms

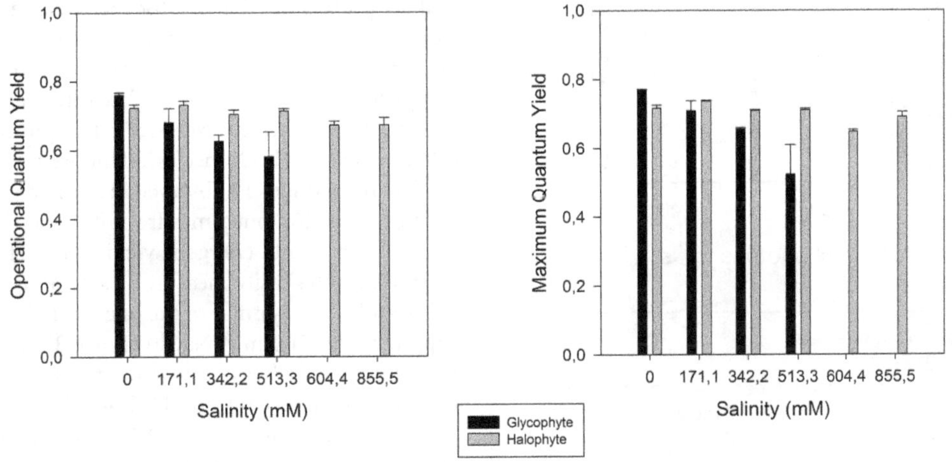

FIGURE 4 | Operational and maximum PSII efficiency in glycophyte *Cyperus longus* and in halophyte *Spartina patens* along a salinity gradient (average ± SE, *N* = 5).

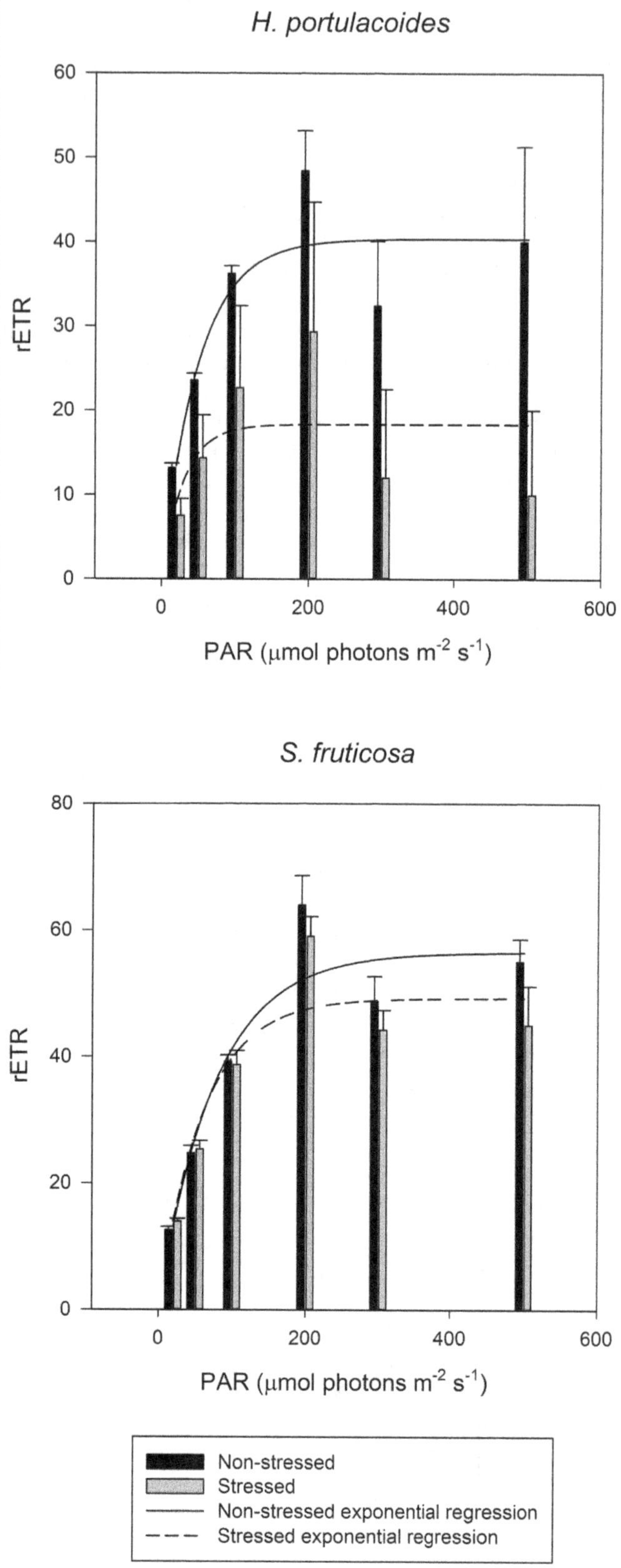

FIGURE 5 | Electron transport rate (rETR) at different light intensities in field stressed (gray) and non-stress individuals (black) of *Halimione portulacoides* and *S. fruticosa* (average ± SE, *N* = 5).

under elevated salt concentrations (Mateos-Naranjo et al., 2010; Rabhi et al., 2012). Diving even deeper in the electron transfer processes, it is possible to understand how the energy fluxes, which result in the total overall PSII activity, are affected by salt stress.

A closer investigation of the photochemical mechanisms (**Figure 6**) shows that in *S. fruticosa* the salinity adverse effects are mostly felt at the quinone level, affecting both the electron flow from reduced quinone to the electron transport chain (ETC) and also the quinone pool (Sm). Sm and the quinone reduction turnover rate (N) were severely reduced (**Figure 6**), leading to an excessive accumulation of reduced compounds and low redox potential (Kalaji et al., 2011). In *H. portulacoides*, the negative effects driven by salt stress result in lower light use efficiencies (LUE) due to high amounts of dissipated energy (Rintamäki et al., 1995). In these individuals, alongside with a lower probability that an incident photon can initiate an electron transfer via the ETC there is also a reduced efficiency for a trapped electron to move further than the oxidized quinone. This leads to an inevitable reduction in the maximum yield of primary photochemical processes (Kalaji et al., 2011).

A special group of fluorescence parameters derived from high-resolution measurements analysis of the chlorophyll *a* fluorescence kinetics, can offer detailed information on the structure and function of plant photosynthetic apparatus, mainly PSII. Analysis of O-J-I-P fluorescence transient by the JIP-test (Strasser et al., 1995) can be applied to derive a number of parameters quantifying the flow of energy through the PS II both at the reaction centre (RC) and at excited cross-section (CS) levels. This approach is far more sensitive that the traditional PSII quantum yields, being able to detect stress symptoms even before they are visible (Force et al., 2003; Christen et al., 2007). Strasser et al. (2000) also created a Performance Index (PI) to sum all the major processes within the JIP-test in order to express the plant vitality. This integrative parameters includes three independent variables: density of fully active RCs, efficiency of electron transfer generated by an exciton into the ETC and beyond the oxidized quinone pool (Q_A), and the probability that an absorbed quanta is trapped within the RCs. This way, PI reflects the functionality of both PSI and II and produces quantitative information of the plant performance, especially under stress conditions (Strasser et al., 2004). In the present case, although excessive salt produces negative effects at different levels in both species, all these effects can be well summarized in the reduced PI observed in stressed individuals (**Figure 7**). This PI reduction outcomes from its dependence on the primary photochemical and energetic yields. The behavior exhibited by *S. fruticosa* can be easily measured using a rapid induction Kautsky curve and is very similar to the one found in *Tamarix gallica* when supplied with 200 mM NaCl (**Figure 8**). This type of analysis is very quick and allows a rapid interpretation of the overall energetic fluxes underlying the PSII activity. In this assessment two phases can be distinguished: O-J step or photochemical phase and the J-I-P step or thermal phase. The first one is considered to be a good proxy of the photochemical energy production realized inside the chloroplasts, while the second one reflects the ability to dissipate excessive amounts of energy throughout thermal

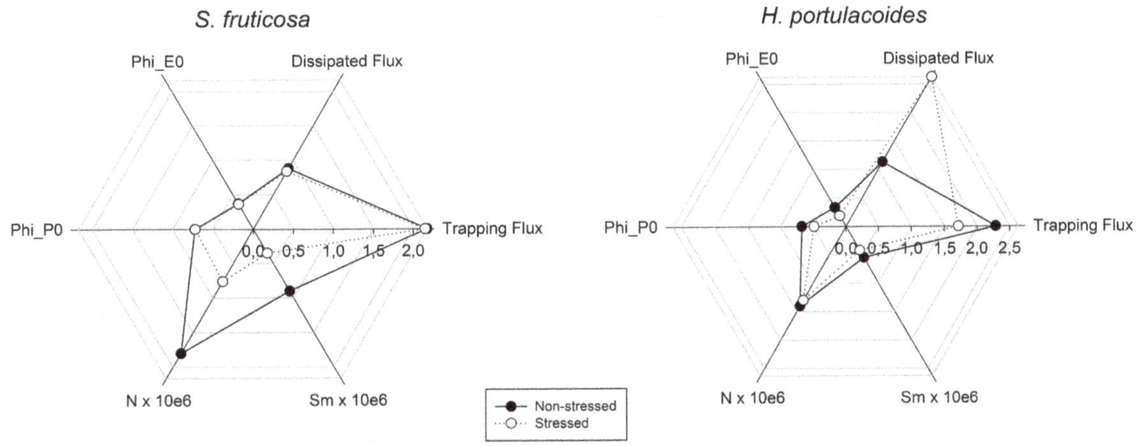

FIGURE 6 | Rapid transient OJIP curve calculated parameters (Phi P_0, maximum yield of primary photochemistry; Phi E_0, probability that an absorbed photon will move an electron into the ETC; N, the number of quinone turnovers until maximum fluorescence is attained; Sm, the number of electrons that flow from the quinones to the ETC) radar plot in field stressed (white) and non-stressed individuals (black) of *S. fruticosa* and *H. portulacoides* species (average values, *N* = 5).

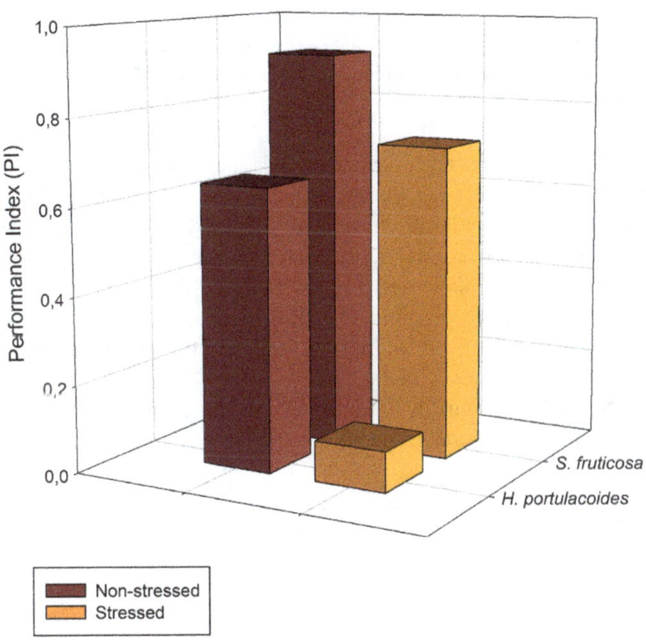

FIGURE 7 | Performance Index (PI) derived from the OJIP-test, in field stressed (orange) and non-stressed individuals (brown) of *S. fruticosa* and *H. portulacoides* species (average values, *N* = 5).

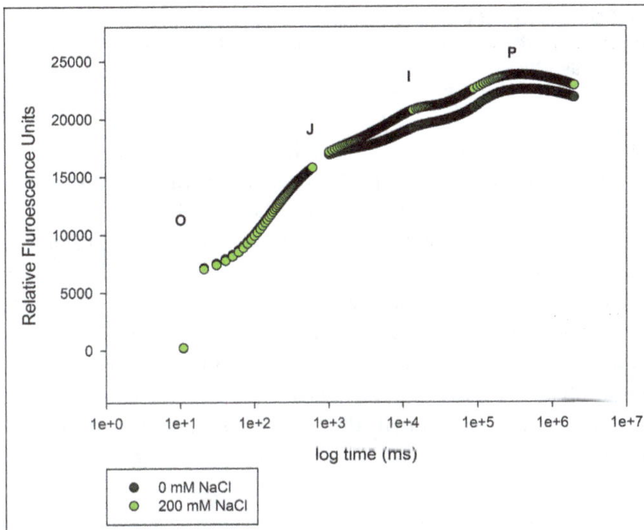

FIGURE 8 | Kautsky curve from *T. gallica* individuals exposed to 0 and 200 mM NaCl, during 20 days (average values, *N* = 5).

dissipation. It is possible to observe that *T. gallica* individuals have similar photochemical activity both with and without salt, but the individuals supplemented with 200 mM NaCl have a higher ability to dissipate excessive energy. This is one of the most common mechanisms by which halophytes overcome the accumulation of excessive reducing power, the primary source of ROS, avoiding this way the photo-destruction of the photosynthetic apparatus (Duarte et al., 2013b). Another interesting phenomenon observable while analyzing the Kautsky curves, is the appearance of a new phase, called K-step at 300 μs (**Figure 9**). The appearance of this K-step with salt stress is associated with damage in the PSII donor side mostly at the level of the oxygen-evolving complexes (Srivastava et al., 1997; Strasser and Stirbet, 2001; Strasser et al., 2004; Chen and Cheng, 2009). This is evident in *A. tripolium* exposed to different salt concentration and is normally indicative of a low stability of the oxygen evolving complexes (OECs) under excessive salt concentration, similarly to what was previously observed in plants subjected to thermal stresses (Wen et al., 2005).

BIOCHEMICAL RESPONSES

Beyond the biophysical processes, halophytes also have a battery of biochemical adjustments to counteract, at the molecular level, the cellular stress imposed by excessive ionic concentrations,

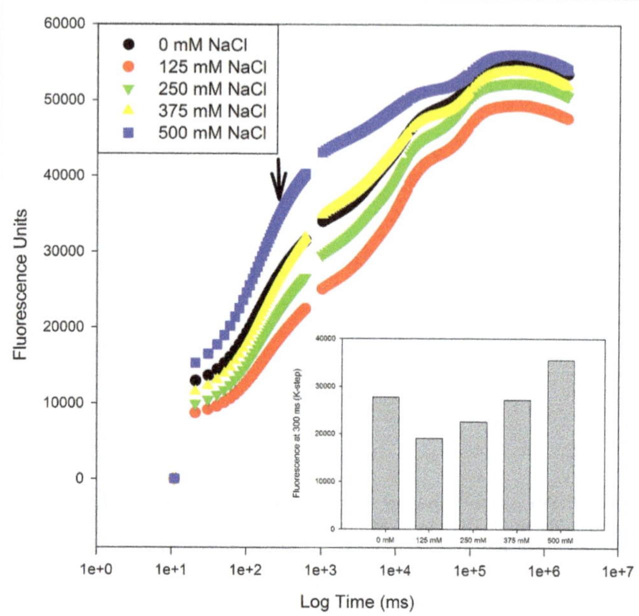

FIGURE 9 | Kautsky curve and K-step fluorescence (arrow) from *Aster tripolium* individuals exposed to a salinity gradient during 15 days (average values, *N* = 5).

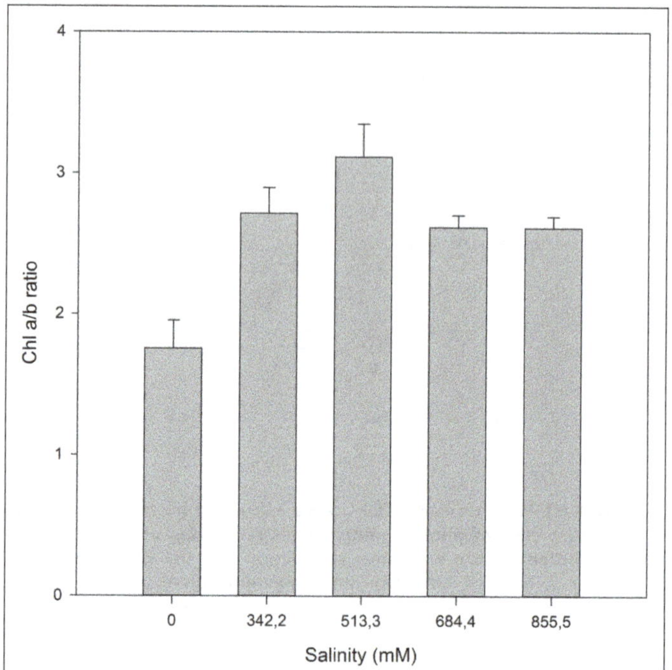

FIGURE 10 | Chlorophyll a/b ratio in *H. portulacoides* leaves from individuals exposed to a salinity gradient (average ± SE, *N* = 5).

namely Na^+. Still discussing the photosynthetic light harvesting mechanisms: the pigment profiles are frequently affected by elevated salt concentrations. On the other hand, under favorable conditions, the increased PS efficiency, driven by optimal salt concentrations is accompanied by a decrease of the PSII antenna size. Due to the lower requirements for light harvesting at optimum conditions, there is a reduction in the plant needs for larger light harvesting complexes (LHC) oppositely to the observed under stress conditions (Rabhi et al., 2012). This can be evaluated using the chlorophyll a/b ratio as proxy (**Figure 10**). An increase in the chlorophyll a/b ratio is directly related to higher number of active light harvesting RCs, being commonly used as indicator of an enhancement in the plant photochemical capacity. On the other hand, when the halophyte is out of its saline comfort concentrations, the excessive energy reaching the photosystems must be dissipated (Duarte et al., 2013b). *H. portulacoides* appears to have a physiological optimum at median NaCl concentrations (513.3 mM) similar to those observed in its natural habitat (estuarine salt marshes).

Nevertheless, this increase in LHC is sometimes not sufficient to sustain all the incoming solar radiation. At this moment, the

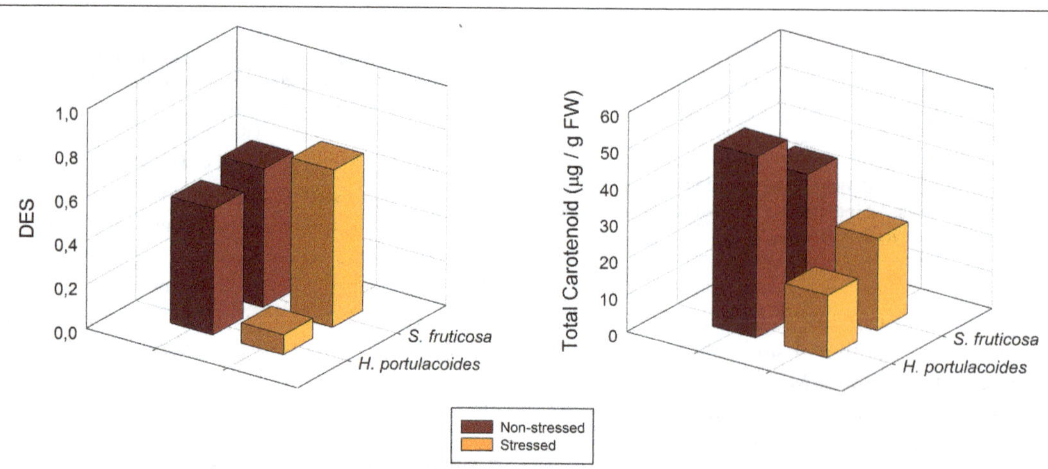

FIGURE 11 | De-epoxidation state (DES) and total carotenoids in *S. fruticosa* and *H. portulacoides* field stressed (orange) and non-stressed individuals (brown; average ± SE, *N* = 5).

FIGURE 12 | *Sarcocornia fruticosa* exhibiting a red coloration during the summer due to excessive salt concentrations in the sediments Photo by B. Duarte (2012).

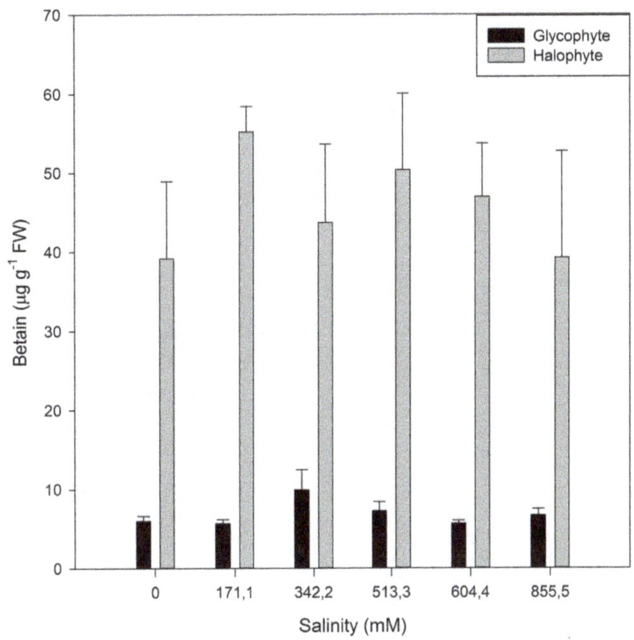

FIGURE 13 | Betain concentration in the leaves of a glycophyte (*Cyperus longus*) and of a halophyte (*Spartina patens*) exposed to a salinity gradient during 1 week (average ± SE, *N* = 5).

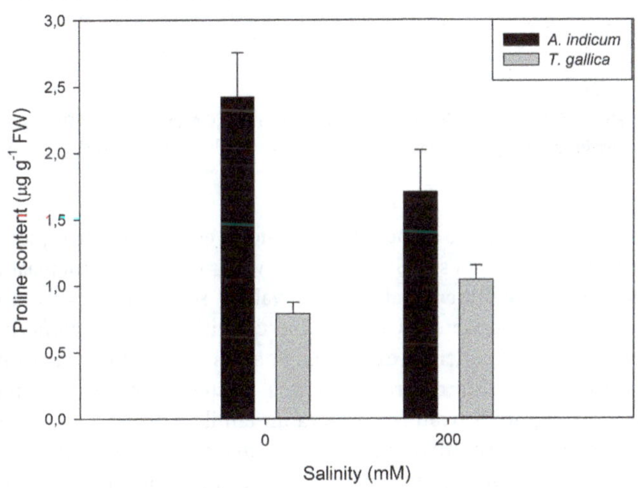

FIGURE 14 | Proline content in aboveground organs of *Arthrocnemum indicum* and *T. gallica* exposed to a salinity gradient during 20 days (average ± SE, *N* = 5).

plant needs to dissipate the energy in excess, either by fluorescence quenching or throughout a pigment metabolic pathway involving a class of carotenoids called xanthophylls (Demmig-Adams and Adams, 1992). As abovementioned, the salt stressed plants cannot withstand a usual dose of light as in a normal situation, and thus even at low solar radiances it undergoes photoinhibition increasing the energy dissipation needs. An evident signal of environmental stress is enhanced activation of the xanthophyll cycle, revealed by an increase in the De-Epoxidation State (DES) index (**Figure 11**). When the absorbed light exceeds the plant photochemical capacity (as revealed above by the decrease in the chl a/b ratio), this excessive energy may be transferred to the ever-present oxygen, generating ROS. These molecules affect many cellular functions by damaging nucleic acids, oxidizing proteins, and causing lipid peroxidation (Gill and Tuteja, 2010). Under steady state conditions, the ROS molecules are scavenged by various antioxidative enzymatic and non-enzymatic defense mechanisms (Foyer and Noctor, 2005). In this context, the conversion of violaxanthin to zeaxanthin throughout the xanthophyll cycle is considered to be one of the most effective energy dissipation mechanisms (Demmig-Adams and Adams, 1992). Zeaxanthin may be an important antioxidant in the thylakoid membrane bilayer itself, where it could scavenge ROS and/or terminate lipid peroxidation chain reactions (Muller et al., 2001).

Also the total chlorophyll to total carotenoids ratio, points out in the same direction. In stressed individuals it is common to observe an increase in this ratio, indicating lower chlorophyll concentration, enhancing photo-protection in detriment of light harvest (**Figure 11**).

Although this shift toward the carotenoid production is not evident by the naked eye, sometimes another phenomenon can

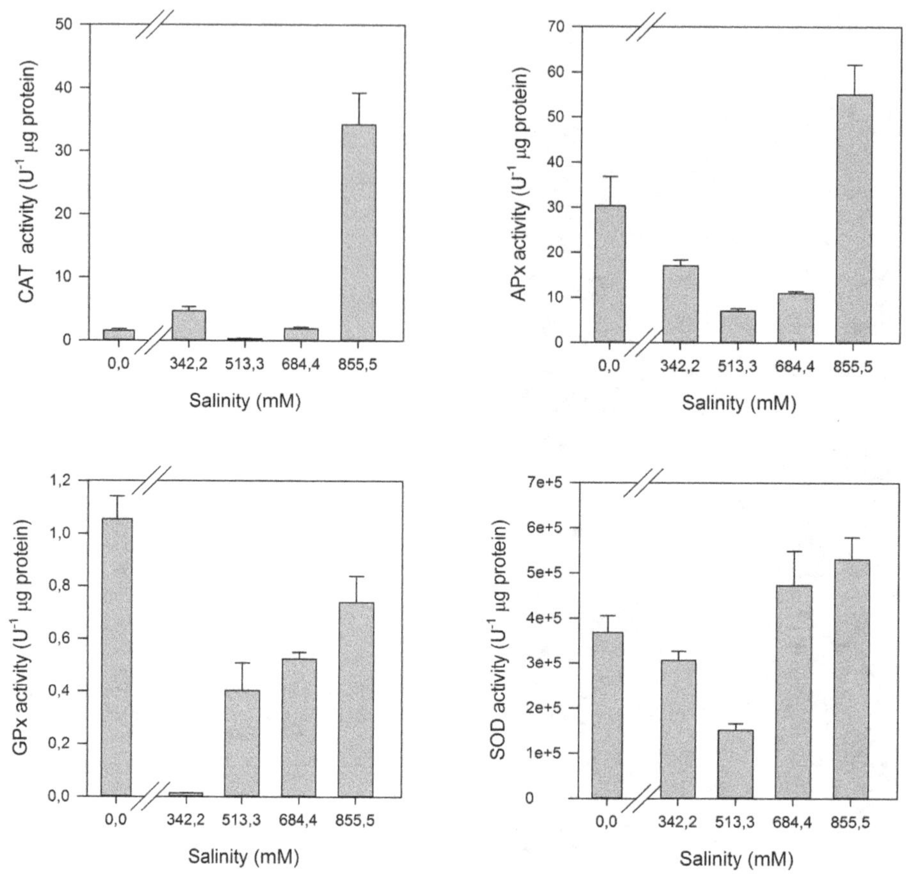

FIGURE 15 | Anti-oxidant enzymatic activities (CAT, Catalase; APx, Ascorbate Peroxidase; GPx, Guaiacol Peroxidase; SOD, Superoxide Dismutase) in the leaves of *H. portulacoides* exposed to a salinity gradient during 1 week (average ± SE, *N* = 5).

be observed in large halophytic extensions, especially during summer. During warm seasons, sediment water evaporates increasing greatly the sediment salinity, to values sometimes twice the observed in seawater. Under these conditions, *Amaranthaceae* salt marshes frequently exhibit large areas of red-colored plants (**Figure 12**). This coloration is due to the presence of water-soluble pigments from the betacyanin family, normally produced as response to salinity, anoxia, or thermal stresses (Chang-Quan et al., 2006). Betacyanins play an important role in scavenging ROS, generated under environmental stress conditions (Stintzing and Carle, 2004). Chang-Quan et al. (2006) found similar results for other *Amaranthaceae* species (*Suaeda salsa*), suggesting that this betacyanin production is part of a common defense mechanism against environmental stresses, namely salinity. Commonly, these pigments are also related to a high betain production, a quaternary ammonium compound, mainly accumulated in the chloroplast in order to counteract high Na^+ concentrations in this compartment (Rhodes and Hanson, 1993; McNeil et al., 1999). Again, comparing glycophytes (e.g., *Cyperus longus*) with halophytes (e.g., *Spartina patens*), the differences are evident (**Figure 13**). Halophytes are highly adapted to salinity, with an enormous production of betain in order to balance and regulate the osmotic potential inside its photosynthetic compartments. In glycophytes, these pathways are not well developed and thus the osmoregulation mechanisms are only adapted to small salinity fluctuations within an extremely low salinity range. Regarding the cytosol, the plant tends to accumulate proline, an amino acid with also a quaternary ammonium-based structure. In this cellular compartment, proline acts as an effective osmoregulator of the ionic pressure exerted by excessive salt concentrations. The use of this compatible solute can also reflect the salt tolerance strategy of a species. Comparing, e.g., an obligatory halophyte (*Arthrocnemum indicum*) with a salt-excreting facultative one (*T. gallica*) the differences are evident (**Figure 14**). While for *A. indicum* the absence of salt is an osmotic stress factor, in *T. gallica* the presence of salt, even at reduced concentration triggers the cytosolic accumulation of proline to counteract the osmotic imbalance. Allied with this compatible solute accumulation, *T. gallica* excretes the excessive salt from its leaves. In this case, the function of proline accumulation has a counteractive measure against the external medium osmotic pressure.

Halophytes are often classified as extremophile species, inhabiting extremely salinized and arid environments under extreme abiotic adverse conditions for life development. Another interesting adaptation developed by this group of plants was the acquisition and development of highly efficient battery of anti-oxidant enzymes. The interaction of high Na^+ concentrations, as well as any other excessive cation concentrations, with the cell organelles lead to generated ROS resulting to reactions with proteins and the cellular biological compounds in membranes (Duarte et al., 2013c). Halophytes developed a highly efficient enzymatic rapid response system toward salinity changes, quickly activated when the medium conditions shift aside from the saline comfort zone of a halophyte (**Figure 15**).

This battery has its higher expression at the first line of defense, superoxide dismutase (SOD). This enzyme catalyzes the conversion of the highly toxic superoxide anions to hydrogen peroxide. In the second line of defense, peroxidase-class enzymes, such as catalase (CAT), ascorbate peroxidase (APx), and guaiacol peroxidase (GPx) play key functions in the hydrogen peroxide detoxification, and thus in the reduction of ROS to non-damaging concentrations. While for glycophytes it would be expectable that these defense mechanisms are activated with the increasing salinity doses, in halophytes the lack of salt can also be a stress factor, especially if we are dealing with obligate halophytes. Some authors suggest that obligate halophytes not only exhibit optimum growing in salt mediums, but in fact they require salt as part of their nutrition in order to activate or de-activate several salt sensitive enzymes (Wang et al., 2011). These species frequently exhibit an activation of these enzymes at both very low Na^+ concentrations (below the physiological optimum) and at seawater Na^+ concentrations (considered excessive), pointing out to a physiological Na^+ dependence of certain halophytes, such as *H. portulacoides* (**Figure 13**).

LEARNING FROM HALOPHYTES: FINAL REMARKS

Halophytes are extremely plastic species with a high degree of adaptation to saline habitats, being therefore excellent models to study salt resistance and tolerance mechanisms. Alongside, some halophytes have recently been pointed out as potential alternative cash crops for replacing usual crops in soils with excessive salt concentrations. Their tolerance to salt goes from simple morphological adjustments, like increasing turgescence or specific salt glands, to efficient energy dissipation mechanisms based on electron fluxes adjustment inside the chloroplast or to the production of specific molecules with the main objective to counteract the osmotic unbalance driven by excessive salt. Nowadays, the metabolic biophysical and biochemical mechanisms underlying these processes are relatively well described for several halophytes. This opens a new door where physiology can be allied to biotechnology, identifying the key genes underlying these processes and introducing them into non-tolerant crops. This will allow glycophytic species to be cultured in arid and saline lands maintaining the food supply in some of the poorest regions of the planet.

ACKNOWLEDGMENTS

The authors would like to thank to the "Fundação para a Ciência e Tecnologia (FCT)" for funding the research in the Centre of Oceanography (CO) throughout the project PEst-OE/MAR/UI0199/2011 and this specific work throughout the ECOSAM project (PTDC/AAC-CLI/104085/2008). Bernardo Duarte investigation was supported by FCT throughout a PhD grant (SFRH/BD/75951/2011). The authors would also like to thank Dr. Vadim Volkov for the helpful comments and suggestions that greatly improved the present manuscript.

REFERENCES

Agarie, S., Shimoda, T., Shimizu, Y., Baumann, K., Sunagawa, H., Kondo, A., et al. (2007). Salt tolerance, salt accumulation, and ionic homeostasis in an epidermal bladder-cell-less mutant of the common ice plant *Mesembryanthemum crystallinum*. *J. Exp. Bot.* 58, 1957–1967. doi: 10.1093/jxb/erm057

Balsamo, R. A., Adams, M. E., and Thomson, W. W. (1995). Electrophysiology of the salt glands of *Avicennia germinans*. *Int. J. Plant Sci.* 156, 658–667. doi: 10.1086/297288

Chang-Quan, W., Zhao, J.-Q., Chen, M., and Wang, B.-S. (2006). Identification of betacyanin and effects of environmental factos on its accumulation in halophyte *Suaeda salsa*. *J. Plant Physiol. Mol. Biol.* 32, 195–201.

Chen, L. S., and Cheng, L. (2009). Photosystem 2 is more tolerant to high temperature in apple (*Malus domestica* Borkh.) leaves than in fruit peel. *Photosynthetica* 47, 112–120. doi: 10.1007/s11099-009-0017-4

Christen, D., Schonmann, S., Jermini, M., Strasser, R. J., and Dèfago, G. (2007). Characterization and early detection of grapevine (*Vitis vinifera*) stress responses to Esca disease by in situ chlorophyll florescence and comparision with drought stress. *Environ. Exp. Bot.* 60, 504–514. doi: 10.1016/j.envexpbot.2007.02.003

Cramer, G. R., Lynch, J., Lauchli, A., and Epstein, E. (1987). Effects of supplemental $Ca2^+$. *Plant Physiol.* 83, 510–516. doi: 10.1104/pp.83.3.510

Demmig-Adams, B., and Adams, W. W. II. (1992). Photoprotection and other responses of plants to light stress. *Annu. Rev. Plant Physiol. Plant Mol. Biol.* 43, 599–626. doi: 10.1146/annurev.pp.43.060192.003123

Duarte, B., Caçador, I., Marques, J. C., and Croudace, I. (2013a). Tagus Estuary salt marshes feedback to sea level rise over a 40-year period: insights from the application of geochemical indices. *Ecol. Indic.* 34, 268–276. doi: 10.1016/j.ecolind.2013.05.015

Duarte, B., Santos, D., Marques, J. C., and Caçador, I. (2013b). Ecophysiological adaptations of two halophytes to salt stress: photosynthesis, PS II photochemistry and anti-oxidant feedback – implications for resilience in climate change. *Plant Physiol. Biochem.* 67, 178–188. doi: 10.1016/j.plaphy.2013.03.004

Duarte, B., Santos, D., and Caçador, I. (2013c). Halophyte anti-oxidant feedback seasonality in two salt marshes with different degrees of metal contamination: search for an efficient biomarker. *Funct. Plant Biol.* 40, 922–930. doi: 10.1071/FP12315

Eisa, S., Hussein, S., Geissler, N., and Koyro, H.-W. (2012). Effect of NaCl salinity on water relations, photosynthesis and chemical composition of Quinoa (*Chenopodium quinoa* Willd.) as a potential cash crop halophyte. *Aust. J. Crop Sci.* 6, 357–368.

Flowers, T. J. (1972). Salt tolerance in *Suaeda maritima* (L.) Dum: the effect of sodium chloride on growth, respiration, and soluble enzymes in a comparative study with *Pisum sativum* L. *J. Exp. Bot.* 23, 310–321. doi: 10.1093/jxb/23.2.310

Flowers, T. J. (2004). Improving crop salt tolerance. *J. Exp. Bot.* 55, 307–319. doi: 10.1093/jxb/erh003

Flowers, T. J., and Colmer, T. D. (2008). Salinity tolerance in halophytes. *New Phytol.* 179, 945–963. doi: 10.1111/j.1469-8137.2008.02531.x

Force, L., Critchley, C., and Van Rensen, J. J. (2003). New fluorescence parameters for monitoring photosynthesis in plants. *Photosyn. Res.* 78, 17–33. doi: 10.1023/A:1026012116709

Foyer, C. H., and Noctor, G. (2005). Redox homeostis and antioxidant signaling: a metabolic interface between stress perception and physiological responses. *Plant Cell* 17, 1866–1875. doi: 10.1105/tpc.105.033589

Gao, F., Gao, Q., Duan, X., Yue, G., Yang, A., and Zhang, J. (2006). Cloning of an H^+-PPase gene from *Thellungiella halophila* and its heterologous expression to improve tobacco salt tolerance. *J. Exp. Bot.* 57, 3259–3270. doi: 10.1093/jxb/erl090

Gill, S. S., and Tuteja, N. (2010). Reactive oxygen species and antioxidant machinery in abiotic stress tolerance in crop plants. *Plant Physiol. Biochem.* 48, 909–930. doi: 10.1016/j.plaphy.2010.08.016

Gong, Q., Li, P., Ma, S., Rupassara, S. I., and Bohnert, H. J. (2005). Salinity stress adaptation competence in the extremophile *Thellungiella halophila* in comparison with its relative *Arabidopsis thaliana*. *Plant J.* 44, 826–839. doi: 10.1111/j.1365-313X.2005.02587.x

Guo, Q., Wang, P., Ma, Q., Zhang, J.-L., Bao, A.-K., and Wang, S.-M. (2012). Selective transport capacity for K^+ over Na^+ is linked to the expression levels of PtSOS1 in halophyte *Puccinellia tenuiflora*. *Funct. Plant Biol.* 39, 1047–1057. doi: 10.1071/FP12174

Jaleel, C. A., Gopi, R., Manivannan, P., and Panneerselvam, R. (2007). Antioxidative potentials as a protective mechanism in *Catharanthus roseus* (L.) G. Don. plants under salinity stress. *Turk. J. Bot.* 31, 245–251.

James, R. A., Blake, C., Byrt, C. S., and Munns, R. (2011). Major genes for Na^+ exclusion, Nax1 and Nax2 (wheat HKT1;4 and HKT1;5), decrease Na^+ accumulation in bread wheat leaves under saline and waterlogged conditions. *J. Exp. Bot.* 62, 2939–2947. doi: 10.1093/jxb/err003

Kalaji, H. M., Govindjee, Karolina, B., Janusz, K., and Zuk-Golaszewska, K. (2011). Effects of salt stress on photosystem II efficiency and CO2 assimilation of two Syrian barley landraces. *Environ. Exp. Bot.* 73, 64–72. doi: 10.1016/j.envexpbot.2010.10.009

Mahajan, S., and Tuteja, N. (2005). Cold, salinity and drought stresses: an overview. *Arch. Biochem. Biophys.* 444, 139–158. doi: 10.1016/j.abb.2005.10.018

Mateos-Naranjo, E., Gómez-Redondo, S., Cambrollé, J., and Figueroa, M. E. (2010). Growth and photosynthetic resposes of the cordgrass *Spartina maritima* to CO2 enrichment and salinity. *Chemosphere* 81, 725–731. doi: 10.1016/j.chemosphere.2010.07.047

McNeil, S. D., Nuccio, M. L., and Hanson, A. D. (1999). Betaines and related osmoprotectants. Targets for metabolic engineering of stress resistance. *Plant Physiol.* 120, 945–949. doi: 10.1104/pp.120.4.945

Megdiche, W., Hessini, K., Gharbi, F., Jaleel, C., Ksouri, R., and Abdelly, C. (2008). Photosynthesis and photosystem II efficiency of two salt-adapted halophytic seashore *Cakile maritima* ecotypes. *Photosynthetica* 46, 410–419. doi: 10.1007/s11099-008-0073-1

Muller, P., Li, X.-P., and Niyogi, K. K. (2001). Non-photochemical quenching. A response to excess light energy. *Plant Physiol.* 125, 1558–1566. doi: 10.1104/pp.125.4.1558

Munns, R. (1993). Physiological processes limiting plant growth in saline soil. Some dogmas and hypothesis. *Plant Cell Environ.* 16, 15–24. doi: 10.1111/j.1365-3040.1993.tb00840.x

Munns, R. (2002). Comparative physiology of salt and water stress. *Plant Cell Environ.* 25, 239–250. doi: 10.1046/j.0016-8025.2001.00808.x

Munns, R., and Termaat, A. (1986). Whole-plant responses to salinity. *Aust. J. Plant Physiol.* 13, 143–160. doi: 10.1071/PP9860143

Pang, Q., Chen, S., Dai, S., Chen, Y., Wang, Y., and Yan, X. (2010). Comparative proteomics of salt tolerance in *Arabidopsis thaliana* and *Thellungiella halophila. J. Proteome Res.* 9, 2584–2599. doi: 10.1021/pr100034f

Qiu, N., Lu, Q., and Lu, C. (2003). Photosynthesis, photosystem II efficiency and the xanthophyll cycle in the salt-adapted halophyte *Atriplex centralasiatica. New Phytol.* 159, 479–486. doi: 10.1046/j.1469-8137.2003.00825.x

Rabhi, M., Castagna, A., Remorini, D., Scattino, C., Smaoui, A., Ranieri, A., et al. (2012). Photosynthetic responses to salinity in two obligate halophytes: *Sesuvium portulacastrum* and *Tecticornia indica. S. Afr. J. Bot.* 79, 39–47. doi: 10.1016/j.sajb.2011.11.007

Rahnama, A., James, R. A., Poutini, K., and Munns, R. (2010). Stomatal conductance as a screen for osmotic stress tolerance in durum wheat growing in saline soil. *Funct. Plant Biol.* 37, 255–263. doi: 10.1071/FP09148

Rhodes, D., and Hanson, A. D. (1993). Quaternary ammonium and tertiary sulfonium compounds in higher plants. *Annu. Rev. Plant Physiol. Plant Mol. Biol.* 44, 357–384. doi: 10.1146/annurev.pp.44.060193.002041

Rintamäki, E., Salo, R., Lehtonen, E., and Aro, E.-M. (1995). Regulation of D1 protein degradation during photoinhibition of photosystem II in vivo: phosphorylation of the D1 protein in various plant groups. *Planta* 195, 379–386. doi: 10.1007/BF00202595

Rozema, J., Gude, H., and Pollack, G. (1981). An ecophysiological study of the salt secretion of four halophytes. *New Phytol.* 89, 207–217. doi: 10.1111/j.1469-8137.1981.tb07483.x

Shabala, S., Bose, J., and Hedrich, R., (2014). Salt bladders: do they matter? *Trends Plant Sci.* 19, 687–691. doi: 10.1016/j.tplants.2014.09.001

Srivastava, A., Guisse, B., Greppin, H., and Strasser, R. J. (1997). Regulation of antenna structure and electron transport in PS II of *Pisum sativum* under elevated temperature probed by the fast polyphasic chlorophyll a fluorescence transient: OKJIP. *Biochim. Biophys. Acta* 1320, 95–106. doi: 10.1016/S0005-2728(97)00017-0

Stintzing, F. C., and Carle, R. (2004). Functional properties of anthocyanins and betalains in plants, food, and in human nutrition. *Trends Food Sci. Technol.* 15, 19–38. doi: 10.1016/j.tifs.2003.07.004

Strasser, R. J., Srivastava, A., and Govindjee. (1995). Polyphasic chlorophyll a fluorescence transients in plants and cyanobacteria. *Photochem. Photobiol.* 61, 32–42. doi: 10.1111/j.1751-1097.1995.tb09240.x

Strasser, R. J., Srivastava, A., and Tsimilli-Michael, M. (2000). "The fluorescence transient as a tool to characterize and screen photosynthetic samples," in *Probing Photosynthesis: Mechanisms, Regulation and Adaptation,* eds M. Yunus, U. Pathre, and P. Mohanty (London: Taylor and Francis), 445–483.

Strasser, R. J., and Stirbet, A. D. (2001). Estimation of the energetic connectivity of PS II centres in plants using the fluorescence rise O–J–I–P. Fitting of experimental data to three different PS II models. *Math. Comput. Simul.* 56, 451–461. doi: 10.1016/S0378-4754(01)00314-7

Strasser, R. J., Tsimilli-Michael, M., and Srisvastava, A. (2004). "Analysis of the chlorophyll-a fluorescence transient," in *Advances in Photosynthesis and Respiration,* eds G. C. Papageorgiou and Govindjee (Berlin: Springer), 321–362.

Ventura, Y., Myrzabayeva, M., Alikulov, Z., Cohen, S., Shemer, Z., and Sagi, M. (2013). The importance of iron supply during repetitive harvesting of *Aster tripolium. Funct. Plant Biol.* 40, 968–976.

Ventura, Y., and Sagi, M. (2013). Halophyte crop cultivation: the case for *Salicornia* and *Sarcocornia. Environ. Exp. Bot.* 92, 144–153. doi: 10.1016/j.envexpbot.2012.07.010

Volkov, V., and Amtmann, A. (2006). *Thellungiella halophila,* a salt tolerant relative of *Arabidopsis thaliana,* has specific root ion-channel features supporting K^+/Na^+ homeostasis under salinity stress. *Plant J.* 48, 342–353. doi: 10.1111/j.1365-313X.2006.02876.x

Waisel, Y., Eshel, A., and Agami, M. (1986). Salt balance of leaves pf the mangrove *Avicennia marina. Physiol. Plant.* 67, 67–72. doi: 10.1111/j.1399-3054.1986.tb01264.x

Wang, C.-M., Zhang, J.-L., Liu, X.-S., Li, Z., Wu, G.-Q., Cai, J.-Y., et al. (2009). *Puccinellia tenuiflora* maintains a low Na^+ level under salinity by limiting unidirectional Na^+ influx resulting in a high selectivity for K^+ over Na^+. *Plant Cell Environ.* 32, 486–496. doi: 10.1111/j.1365-3040.2009.01942.x

Wang, S.-M., Zhang, J.-L., and Flowers, T. J. (2007). Low affinity Na^+ uptake in the halophyte *Suaeda maritima. Plant Physiol.* 145, 559–571. doi: 10.1104/pp.107.104315

Wang, W., Yan, Z., You, S., Zhang, Y., Chen, L., and Lin, G. (2011). Mangroves: obligate or facultative halophytes? A review. *Trees* 25, 953–963. doi: 10.1007/s00468-011-0570-x

Wen, X. G., Qiu, N. W., Lu, Q. T., and Lu, C. M. (2005). Enhanced thermotolerance of photosystem II in salt-adapted plants of the halophyte *Artemisia anethifolia. Planta* 220, 486–497. doi: 10.1007/s00425-004-1382-7

Zhang, J.-L., and Shi, H. (2014). Physiological and molecular mechanisms of plant salt tolerance. *Photosyn. Res.* 115, 1–22. doi: 10.1007/s11120-013-9813-6

Zhao, G. Q., Ma, B. L., and Ren, C. Z. (2007). Growth, gas exchange, chlorophyll fluorescence, and ion content of naked oat in response to salinity. *Crop Sci.* 47, 123–131. doi: 10.2135/cropsci2006.06.0371

Zhu, J.-K. (2003). Regulation of ion homeostasis under salt stress. *Curr. Opin. Plant Biol.* 6, 441–445. doi: 10.1016/S1369-5266(03)00085-2

5

Assessment of natural variation in the first pore domain of the tomato *HKT1;2* transporter and characterization of mutated versions of *SlHKT1;2* expressed in *Xenopus laevis* oocytes and via complementation of the salt sensitive *athkt1;1* mutant

Pedro M. F. Almeida[1]*, Gert-Jan de Boer[2] and Albertus H. de Boer[1]

[1] *Department of Structural Biology, Faculty Earth and Life Sciences, Vrije Universiteit Amsterdam, Amsterdam, Netherlands*
[2] *R&D Department, Enza Zaden, Enkhuizen, Netherlands*

Edited by:
Andreas P. M. Weber, Heinrich-Heine-Universität, Germany

Reviewed by:
Vadim Volkov, London Metropolitan University, UK
Alexandre Boscari, Institut National de la Recherche Agronomique, France

***Correspondence:**
Pedro M. F. Almeida, Department of Structural Biology, Faculty Earth and Life Sciences, Vrije Universiteit Amsterdam, De Boelelaan 1085, 1081 HV Amsterdam, Netherlands
e-mail: p.m.fidalgodealmeida@vu.nl

Single Nucleotide Polymorphisms (SNPs) within the coding sequence of HKT transporters are important for the functioning of these transporters in several plant species. To unravel the functioning of HKT transporters analysis of natural variation and multiple site-directed mutations studies are crucial. Also the *in vivo* functioning of HKT proteins, via complementation studies performed with *athkt1;1* plants, could provide essential information about these transporters. In this work, we analyzed the natural variation present in the first pore domain of the *HKT1;2* coding sequence of 93 different tomato accessions, which revealed that this region was conserved among all accessions analyzed. Analysis of mutations introduced in the first pore domain of the *SlHKT1;2* gene showed, when heterologous expressed in *Xenopus laevis* oocytes, that the replacement of S70 by a G allowed SlHKT2;1 to transport K^+, but also caused a large reduction in both Na^+ and K^+ mediated currents. The study of the transport characteristics of SlHKT1;2 revealed that Na^+-transport by the tomato SlHKT1;2 protein was inhibited by the presence of K^+ at the outside of the membrane. *GUS* expression under the *AtHKT1;1* promoter gave blue staining in the vascular system of transgenic *Arabidopsis*. *athkt1;1* mutant plants transformed with *AtHKT1;1, SlHKT1;2, AtHKT1;1S68G,* and *SlHKT1;2S70G* indicated that both AtHKT1;1 and SlHKT1;2 were able to restore the accumulation of K^+ in the shoot, although the low accumulation of Na^+ as shown by WT plants was only partially restored. The inhibition of Na^+ transport by K^+, shown by the SlHKT1;2 transporter in oocytes (and not by AtHKT1;1), was not reflected in Na^+ accumulation in the plants transformed with *SlHKT1;2*. Both AtHKT1;1-S68G and SlHKT1;2-S70G were not able to restore the phenotype of *athkt1;1* mutant plants.

Keywords: natural variation, tomato, *HKT1* genes, mutation, pore domain

INTRODUCTION

Salinity stress negatively affects crop yield (Munns and Tester, 2008). In order to sustain the growing human population it is necessary to increase the salt tolerance of crop plants, as the human population is growing faster than the area of agricultural land (FAO, 2009). To tolerate salinity plants rely on three different mechanisms: osmotic tolerance, ionic tolerance and Na^+ exclusion from the shoots (Munns and Tester, 2008). Na^+ exclusion from the shoots is the most studied and best understood mechanism, therefore, it is a promising candidate for an approach of genetic modification to enhance plant salt tolerance (Plett et al., 2010).

HKT transporters are often studied with regard to Na^+ exclusion from the shoots. HKT transporters belong to a superfamily of transporters including bacterial KtrBs transporters (Tholema et al., 1999) and yeast TRKs transporters (Rodriguez-Navarro, 2000). The *HKT* gene family is divided in two classes based on their gene structure and in the presence of either a glycine (G) or a serine (S) residue in the first pore domain of the transporter (Maser et al., 2002). Members of class 1 have an S at this position, whereas members of class 2, with the exception of OsHKT2;1, have a G at this position (Platten et al., 2006). HKT transporters are implicated in Na^+ transport in wheat (Davenport et al., 2005; James et al., 2006; Byrt et al., 2007; Munns et al., 2012), rice (Ren et al., 2005; Horie et al., 2007; Jabnoune et al., 2009) and *Arabidopsis* (Uozumi et al., 2000; Berthomieu et al., 2003; Rus et al., 2004; Sunarpi et al., 2005; Moller et al., 2009). Class I HKT transporters are low affinity transporters with specificity for Na^+ (Munns and Tester, 2008). Some of these members are located at the plasma membrane of root stele cells, in particular in the xylem parenchyma cells (XPC). They function in retrieving Na^+ from the xylem sap, and prevent Na^+ from reaching the shoots and damaging photosynthetic cells. The number of class I HKT members varies between mono- and dicotyledonous plants (Garciadeblas et al., 2003; Ren et al., 2005; Huang et al., 2006; Jabnoune et al., 2009). When first characterized, *athkt1;1* and wild type (WT) seedlings showed no difference in root and shoot growth after growing 6 days in a medium with (150 mM) or without NaCl (Rus et al., 2004). However, on the long term medium supplemented with 75 mM NaCl reduced the shoot growth and increased tip senescence of mature leaves of *athkt1;1* mutants (Rus et al., 2004). Due to the higher Na^+ accumulation in the shoots, *athkt1;1* mutant plants display Na^+ sensitivity, showing the role of HKT transporters in preventing Na^+ from reaching the shoots (Rus et al., 2001, 2004; Berthomieu et al., 2003; Sunarpi et al., 2005).

The discovery of genetic polymorphisms in *HKT* genes (Diatloff et al., 1998; Rubio et al., 1999; Ren et al., 2005; Cotsaftis et al., 2012) underlying the adaptation to salinity stress brought new insights in the understanding of the functions of genes involved in the adaptation mechanisms. Information on genetic polymorphisms in *HKT* genes will provide tools for the development of crops more tolerant to salinity stress. Within the same genus, the diversity of phenotypes across environmental gradients of stress can indicate the suitability for selection, and the study of the genotypes responsible for those phenotypes might lead to the discovery of genetic polymorphisms responsible for these adaptive responses (Baxter et al., 2010). Analysis of *OsHKT1;5* in two rice cultivars differing in their salinity tolerance showed that both the cellular location and expression patterns of *OsHKT1;5* were identical. Nevertheless, differences in the coding region producing four amino acid substitutions were linked to the functional variation of these two alleles (Ren et al., 2005). Recently, it was concluded that the V395L substitution present in Nona Bokra could directly affect the Na^+ transport rates and that it was responsible for the tolerant and sensitive behavior of Nona Bokra and Koshihikari, respectively (Cotsaftis et al., 2012).

Besides natural variation, studies of single (Diatloff et al., 1998; Maser et al., 2002) or multiple (Kato et al., 2007) site-directed mutations of HKT transporters (Diatloff et al., 1998; Maser et al., 2002; Kato et al., 2007) are also crucial to understand how these transporters function. The replacement of the S by a G in the first pore domain of AtHKT1;1 changed AtHKT1;1 from a Na^+ uniporter to a Na^+/K^+ symporter (Maser et al., 2002).

Data on the role of HKT transporters is often generated from heterologous expression of HKT transporters in, mainly, *Xenopus laevis* oocytes (Schachtman and Schroeder, 1994; Fairbairn et al., 2000; Uozumi et al., 2000; Horie et al., 2001; Liu et al., 2001; Golldack et al., 2002; Berthomieu et al., 2003; Su et al., 2003; Ren et al., 2005; Jabnoune et al., 2009; Lan et al., 2010; Yao et al., 2010; Horie et al., 2011) and, to a lesser extent, in *Saccharomyces cerevisiae* cells (Horie et al., 2001, 2011; Golldack et al., 2002; Garciadeblas et al., 2003; Su et al., 2003; Takahashi et al., 2007; Ali et al., 2012). However, in a report by Haro et al. (2005) a very important question was addressed: are results obtained for HKT transporters when expressed in heterologous systems of physiological importance *in planta*? Or, are heterologous expression systems too artificial to have physiological meaning *in planta*? We performed several experiments to investigate if there is natural variation present in the first pore domain of HKT1;2 in tomato and how the replacement of the S residue by a G residue in the first pore domain in HKT1;2 of tomato affects the function of the transporter when expressed in *Xenopus laevis* oocytes. Moreover, we investigated if the results obtained with *Xenopus oocytes* could be replicated *in planta*.

In this study we analyzed the presence of natural variation in the first pore domain of HKT1;2 of 93 tomato accessions. We also replaced the S residue of the first pore domain of SlHKT1;2 with a G residue, and we analyzed the effect of this mutation via expression of these mutated genes in *Xenopus laevis* oocytes. We generated stable lines of *Arabidopsis thaliana athkt1;1* plants expressing each of these constructs, and we characterized their biomass production, shoot water-content and ion content after 2 weeks of salt treatment.

MATERIALS AND METHODS

PLANT MATERIAL AND GROWTH CONDITIONS

Genomic DNA used in this experiment was extracted from 93 different tomato accessions (Supplementary File Table 1). Tomato seeds were surface sterilized by soaking in 1% (V/V) commercial sodium hypochlorite solution for 15 min and rinsed three times with sterile water. After sterilization, seeds were sown in rock wool plugs soaked with half-strength Hoagland solution (one seed per rock wool plug). Plugs were covered with dry vermiculite to avoid dehydration. A randomized design consisting of three biological replicates and 93 different accessions was used. Each biological replica consisted of a pool of seven to 10 plants. On alternate days plants were irrigated with half-strength Hoagland solution. Plants were kept in a climate chamber under a 14/10 h photoperiod and a 20/18°C day/night temperature. Plants were harvested after 5 weeks. Seeds of homozygous *athkt1;1* mutant (Columbia-0) were obtained from the NASC stock center (N6531) and were sown along with WT *Arabidopsis thaliana* Col-0. Plants were grown on a mix of sand and peat (1:1) at 24°C in a 16 h light/8 h dark cycle in a greenhouse. Plants were watered every 2 days. Selected transgenic lines (See the section "Cloning of *HKT* genes and generation of *Arabidopsis* transgenic lines" for a full description of how these lines were generated) were grown under the same conditions. Four-week old transgenic lines (T2 lines) were treated with 100 mM NaCl every 2 days during 2 weeks before harvesting of shoot material.

GENOMIC DNA EXTRACTION

For the extraction of genomic DNA approximately 25 mg of dried root material of each tomato accession was weighed and inserted in a 96 deep well plate. Samples were freeze dried for 1 week using a freeze dryer (Christ Alpha 1-4 LD plus, Germany). For the extraction of DNA a Nucleospin 96 Plant II kit (Macherey-Nagel, Germany) was used and the manufacturer's protocol was followed. The quality of the gDNA was checked on a 1.5% agarose gel. The concentration of the gDNA was calculated using the Quanti-iTTM PicoGREEN dsDNA assay kit (Invitrogen). gDNA was diluted in TE buffer pH 8.0 (10 mM Tris and 1 mM EDTA) and stored at 4°C.

PRIMER AND PROBE DESIGN AND ANALYSIS OF NATURAL VARIATION

Primers and probes (Supplementary File Table 2) were designed to have a Tm between 60 and 67°C with DNASIS MAX v3.0 software. Pairs of primers were designed to flank the SNP under study. The size of amplicons was designed to be smaller than 150 nucleotides. Unlabeled probes were blocked at the 3′ end to prevent extension in PCR reactions and designed to anneal over an SNP of interest. Reactions were performed in 384-well plates. 20 ng of gDNA per sample were used to study natural variation in tomato *HKT1;2* nucleotides of different cultivars. Per reaction, 3.21 μl of MilliQ, 0.05 μl of FW primer (10 μmol/μl) and 0.25 μl RV primer (10 μmol/μl), 0.04 μl of PAL (5 U/μl) (KAPA Biosystems, Boston, USA), 0.25 μl LC Green (Idaho Technology, Salt Lake City, USA), 0.20 μl dNTP (5 mM) and 1 μl PAL buffer (KAPA Biosystems, Boston, USA) supplemented with 12.5 mM $MgCl_2$, were added. The amplification reaction started with a denaturation step of 95°C for 10 min and continued with 14 cycles of 95°C for 15 sec and 60°C for 4 min. Samples were cooled to room temperature and a first melting curve analysis was performed to assess the quality of the amplification. Samples were again cooled down to room temperature and 2 μl (10 pmol/μl) of probe was added. Samples were heated up to 96°C for 3 min and cooled down to room temperature before analyzing the melting curve of the probe. The mix of PCR products and probes was heated up to 96°C for 3 min and then cooled down to room temperature to allow hetero-duplex formation. Thereafter, the mix of PCR products and probes was re-heated to 95°C at 0.3°C/s. Data were acquired between 50 and 95°C. Data acquisition was made with a Light Scanner HR384 (Idaho Technology Inc. Salt Lake City, USA). High Resolution Melting (HRM) curve analysis was performed using the "Unlabeled Probes" module in the "genotyping" mode of the software. This mode involves negative filter, normalization and grouping. *SlHKT1;2* coding sequence isolated from *S. lycopersicum* Arbasson F1 was used as a reference. Amplicons that showed a melting curve different from the reference melting-curve were selected, amplified and sequenced. Sequencing of the amplicons was performed at Macrogen Europe Laboratories, Amsterdam, The Netherlands.

PLASMID CONSTRUCTION

Site-directed mutagenesis of *SlHKT1;2* was conducted using overlap extension PCR. All primers used to make mutated *HKT1* genes are listed in (Supplementary File Table 3). *pGEM-HESphI+SlHKT1;2* and *pGEM-HESphI+AtHKT1;1* (see Almeida et al., 2014 for a full description of how these constructs were obtained) were used as template and the corresponding mutated gene cloned into the BamHI and XbaI restriction sites and SphI and BamHI restriction sites of an empty *pGEM-HESphI* vector for tomato and *Arabidopsis HKT1*, respectively. The constructs obtained were called *pGEM-HESphI+AtHKT1;1-S68G* and *pGEM-HESphI+SlHKT1;2-S70G*. All PCR-derived DNA-fragments were confirmed by sequencing.

HETEROLOGOUS EXPRESSION OF *HKT1* GENES IN *XENOPUS LAEVIS* OOCYTES

Preparation of template DNA, *in vitro* transcription and capping of mRNA and Two-electrode voltage clamping of *Xenopus* oocytes was performed according to Almeida et al. (2014). All measurements were performed at least 3 times on oocytes isolated from two different batches.

CLONING OF *HKT* GENES, GENERATION OF *ARABIDOPSIS* TRANSGENIC LINES AND GUS STAINING

The *pGEM-HE* vector used in this work was modified to contain the *SphI* restriction site. This restriction site was introduced between the *BamHI* and *XbaI* restriction sites present in the vector. This was made by designing two complementary primers containing the *SphI* restriction site flanked by both *BamHI* and *XbaI* restriction sites, and introducing this piece of DNA between the *BamHI* and *XbaI* restriction sites of *pGEM-HE* vector. The *pGEM-HE* vector containing this new restriction site was selected and used in this work. To be able to distinguish the original *pGEM-HE* and *pGEM-HE* containing the *SphI* restriction site, we called our vector *pGEM-HESphI*.

pGEM-HESphI+AtHKT1;1, *pGEM-HESphI+AtHKT1;1-S68G*, *pGEM-HESphI+SlHKT1;2*, and *pGEM-HESphI+SlHKT1;2-S70G* were used as template in the amplification of the *HKT* genes flanked by the *attB* Gateway (Invitrogen) recombination sites. The *AtHKT1;1*, *AtHKT1;1-S68G*, *SlHKT1;2*, and *SlHKT1;2-S70G* genes were cloned into the *pDONR221 P5-P2* vector (Invitrogen) and were named *p5-2AtHKT1;1*, *p5-2AtHKT1;1-S68G*, *p5-2SlHKT1;2*, and *p5-2SlHKT1;2-S70G*, respectively. A 5 kb DNA fragment upstream of the *ATG* start codon of the *AtHKT1;1* gene containing the promoter region, the tandem repeat and the small RNA target region (Baek et al., 2011) was cloned into *pDONR221 P1-P5* and *pDONR221 P1–P2* (Invitrogen) and the resultant constructs were named *p1-5AtHKT1;1prom*, *p1-2AtHKT1;1prom*, respectively. Cloning of DNA fragments into *pDONR221* (Invitrogen) vectors was performed by BP reactions (Invitrogen). Cloning of either *AtHKT* or *SlHKT* genes into *pHGW* (Karimi et al., 2002) under the *AtHKT1;1* promoter was performed by LR reactions (Invitrogen). In this way, *pHGW+AtHKT1;1prom+AtHKT1;1*, *pHGW+AtHKT1;1prom+AtHKT1;1-S68G*, *pHGW+AtHKT1;1prom+SlHKT1;2*, and *pHGW+AtHKT1;1prom+SlHKT1;2-S70G* constructs were created. *p1-2AtHKT1;1prom* was incubated with *pKGWFS7* (Karimi et al., 2002) in an LR reaction (Invitrogen) to create the *pKGWFS7+AtHKT1;1prom* construct. All constructs were sequenced prior to the transformation of *Arabidopsis* plants. All primers used are listed (Supplementary File Table 4). All constructs were introduced into *Agrobacterium tumefaciens* strain *GV3101pMP90*, including the *pHGW* empty vector, and transformed into 3-week old *athkt1-1* mutant plants, except *pKGWFS7+AtHKT1;1prom* and the *pKGWFS7* empty vector, which were transformed into 3-week old *Arabidopsis* WT plants. Plant transformation was performed by the flower dipping method (Clough and Bent, 1998). *athkt1;1* mutant plants (N6531) (Rus et al., 2001) were transformed with *pHGW+AtHKT1;1prom+AtHKT1;1* or *pHGW+AtHKT1;1prom+AtHKT1;1-S68G* or *pHGW+AtHKT1;1prom+SlHKT1;2* or *pHGW+AtHKT1;1prom+SlHKT1;2-S70G*. *Arabidopsis* WT plants were transformed with *pKGWFS7+AtHKT1;1prom* or *pKGWFS7* empty vector. Four weeks after transformation, seeds were harvested and surface sterilized. Surface sterilization was performed by washing seeds during 1 min with a 80% ethanol solution with 0.1% Tween-20, followed by a 20 min washing step with 1% commercial bleach, and three washing steps with sterile MilliQ. MilliQ was then replaced by warm half-strength Murashige and Skoog (MS) medium supplemented with 1% sucrose, 0.8% agar and 10 mg/L hygromycin or 50 mg/L kanamycin. Seeds were placed in round plates containing solid MS medium with the same composition. Seedlings showing hygromycin or kanamycin resistance were selected and transferred to pots containing a mix of soil and peat (1:1). They were grown for 4 weeks under the growing conditions described above, after which the seeds were harvested. T2 seeds were tested for antibiotic resistance and used to either investigate their growth response under different NaCl concentrations or for GUS staining. Histochemical assays of GUS activity (biological replicates $n = 3$) were conducted using samples that were stained according to Jefferson et al. (1987). Seven days old plantlets (plantlets $n = 7$) were incubated at 37°C for 6 h in the staining solution containing 0.5 mM X-Gluc (5-bromo-4-chloro-3-indolyl-b-D-glucoronide).

ANALYSIS OF TRANSCRIPTS LEVELS

Tissue specific expression of all *HKT* genes was tested by extracting total RNA from roots and leaves of three-week old WT and transformed *Arabidopsis* lines. RNA from transformed plants was extracted and immediately frozen in liquid nitrogen. RNA extraction was performed using the Innuprep Plant RNA kit (Westburg, the Netherlands). First strand cDNA was synthesized using one microgram of total RNA, random hexamers and SuperScript II Reverse Transcriptase (Invitrogen Life Technologies). First strand cDNA was used as a template for quantitative real-time PCR (qRT-PCR). Samples were measured from each transformed plant using a sequence detector system (7300 Real Time PCR System from Applied Biosystems). Primers used to test the expression of *AtHKT1;1* and *AtHKT1;1-S68G* according to (Jha et al., 2010). Primers used to test the expression of *SlHKT1;2* and *SlHKT1;2-S70G* according to Asins et al. (2012). *ß-Actin* transcript levels were used as an internal standard. The mean normalized expression was calculated according to Livak and Schmittgen (2001).

STATISTICAL ANALYSES OF DATA

Two-way ANOVA was used to assess the effect of salt treatment on *HKT1* gene expression, fresh weight, Na^+ and K^+ accumulation. Normality and homogeneity assumptions were checked. Gene expression data were rank-transformed if these assumptions were violated. If ANOVA was significant, a *post hoc* test (Tukey's test) was used to evaluate differences among treatment and plant lines. All tests were performed using SPSS 17.0.

SEQUENCE DATA

The *SlHKT1;2* sequence used in this study has the EMBL/GenBank accession number HG530660 (Almeida et al., 2014).

RESULTS

SLHKT1;2 S70 IS CONSERVED THROUGHOUT *SOLANUM* ACCESSIONS

From all 93 tomato accessions tested only three (PI 126435, LA 2931 and LA 1401) showed different melting curves (**Figure 1A**).

A

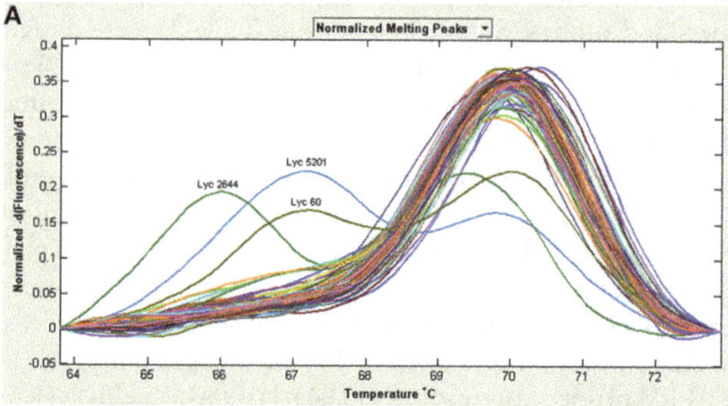

B SlHKT1;2 CCACAGTTTC TAGTATGTCC ACTATTGAAA TGG
PI 126435 CCACTGTTTC TAGTATGTCC ACTATTGAAA TGG
LA 2931 CCACAATTTC TAGTATGTCC ACTATTGAAA TGG
LA 1401 CCACTGTTTC TAGTATGTCC ACTATTGAAA TGG

FIGURE 1 | Tomato accessions show SNPs close to the S at the filter position of the first pore domain, but not in the nucleotides coding the S residue. (A) *Solanum HKT1;2* melting curve derivative plots from 93 tomato accessions showing the two different SNP's identified close to the 1st pore domain. **(B)** Nucleotide sequences of the *HKT1;2* gene of different *Solanum* accessions show that both SNPs (red letters) situate close to, but do not coincide with the S70 residue of the first pore domain (green letters). Black colored letters represent conserved nucleotides.

To identify whether the SNPs responsible for these different melting curves were located at our target region, we amplified and sequenced the region of the first pore domain of these three accessions (**Figure 1B**). Sequencing results revealed that each of these three accessions have a single SNP (PI 126435 and LA 1401 have the same SNP and LA 2931 has a different SNP), although none of them in the position of interest. The SNP in the accession LA 2931 resulted in an amino acid change from valine (V) to isoleucine (I); however, both amino acids are hydrophobic. In the case of accessions PI 126435 and LA 1401 the SNP did not result in any amino acid change, as both ACA and ACT code for a threonine (T) residue.

SITE-DIRECTED MUTAGENESIS OF S70 TO G OF SLHKT1;2

To test the hypothesis that S70 of the tomato SlHKT1;2 protein is crucial for the Na^+ selectivity we replaced S70 by a G (SlHKT1;2-S70G). cRNA of *SlHKT1;2-S70G* was injected in *Xenopus laevis* oocytes. After 2 days of incubation, currents produced in the presence of Na^+ and K^+ ions were recorded in oocytes expressing both WT and mutated HKT1 transporters from *Arabidopsis thaliana* and *Solanum lycopersicum* ($n = 3$) (**Figure 2**). *AtHKT1;1-S68G* expressing oocytes (Maser et al., 2002) were used as a positive control of SlHKT1;2-S70G. Currents produced by oocytes expressing either *AtHKT1;1* (**Figure 2A**) or *SlHKT1;2* (**Figure 2I**) increased when the oocytes were bathed in higher Na^+ concentrations (as seen by a more negative current). Increasing external K^+ concentration did not result in any change in the current levels produced by *AtHKT1;1* expressing oocytes (**Figure 2B**). In contrast, *SlHKT1;2* mediated inward and outward currents were sensitive to external K^+ concentration as both currents decreased with increasing bath K^+ concentration (**Figure 2J**). Increased concentrations of K^+ result in an inhibition on the transport of Na^+ by SlHKT1;2 but not by

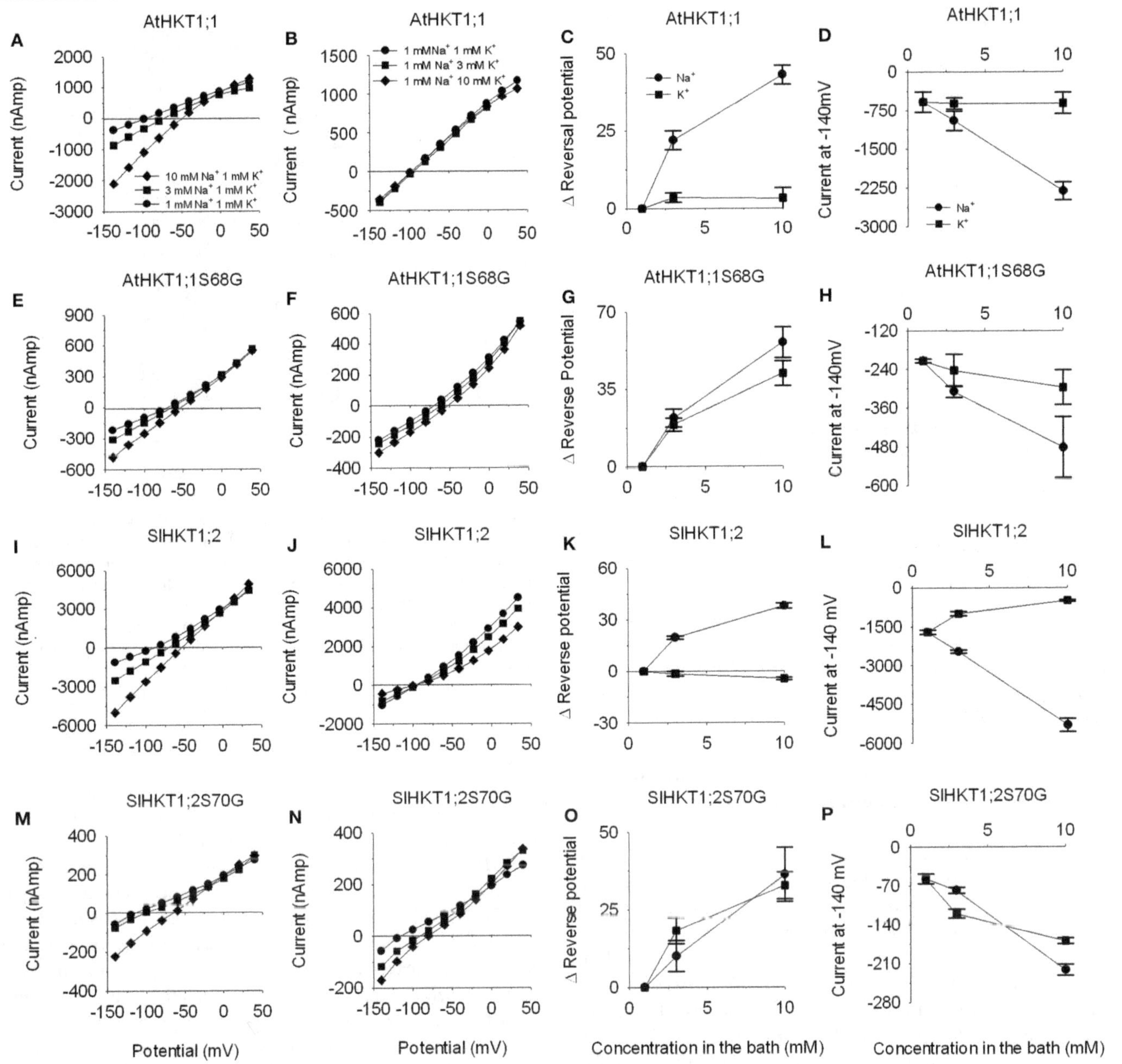

FIGURE 2 | Transport activity of *AtHKT1;1, AtHKT1;1-S68G, SlHKT1;2* and *SlHKT1;2-S70G* constructs expressed in *Xenopus laevis* oocytes. Oocytes were injected with *AtHKT1;1* and *SlHKT1;2* and the indicated mutated transporters. **(A,E,I,M)** Currents recorded at three Na^+ concentrations (1, 3, and 10 mM) with 1 mM K^+ as background; **(B,F,J,N)** Currents recorded at three K^+ concentrations (1, 3, and 10 mM) with 1 mM Na^+ as background; **(C,G,K,O)** Reversal potential shifts as a function of ion concentration. Only transporters where the S of the 1st pore domain was mutated to a G were permeable to K^+ as indicated by the large positive shifts in the reversal potential with increasing concentrations of K^+ in the bath; **(D,H,L,P)** Absolute currents as a function of ion concentration. Transporters where the S of the 1st pore domain was mutated to a G showed an increase in current with increasing K^+ concentration in the bath. Data are means ± *SE* ($n = 3$, experiments done with at least two different oocyte batches).

AtHKT1;1. When oocytes expressing either *AtHKT1;1-S68G* or *SlHKT1;2-S70G* were bathed with either increasing Na^+ concentration (**Figures 2E,M**) or K^+ concentration (**Figures 2F,N**) currents increased for both cations tested. For both AtHKT1;1- and SlHKT1;2-mediated currents a higher Na^+ concentration but not a higher K^+ concentration resulted in positive shifts in the reversal potential (**Figures 2C,K**), which is indicative of Na^+ permeation. Reversal potentials obtained with oocytes expressing either *AtHKT1;1-S68G* (**Figure 2G**) or *SlHKT1;2-S70G* (**Figure 2O**) showed positive shifts when both Na^+ concentration or K^+ concentration increased, indicating that the presence of a G residue at the filter position of the first pore domain allows the transport of both Na^+ and K^+ ions. **Figures 2D,H,L,P** show the currents recorded at −140 mV for AtHKT1;1, AtHKT1;1-S68G,

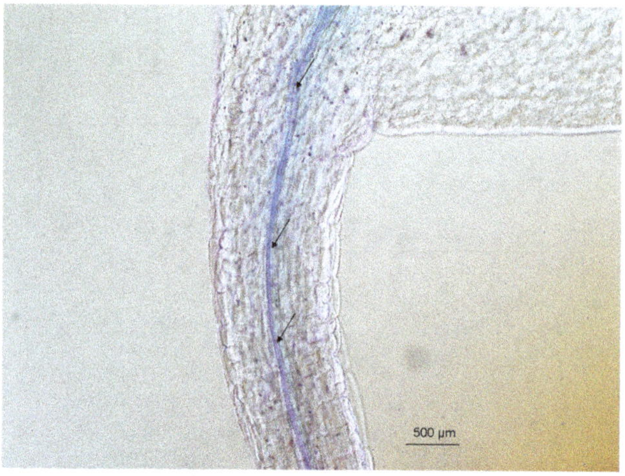

FIGURE 3 | Detection of GUS activity in the vascular system of transgenic *Arabidopsis thaliana* plants. *Arabidopsis* plants expressed the *GUS* gene under the control of the *AtHKT1;1* promoter were grown. Strong blue GUS staining was detected in the vicinity of the xylem and phloem in leaves. Arrows point to the xylem vessels. Biological replicates, $n = 3$; plantlets, $n = 7$.

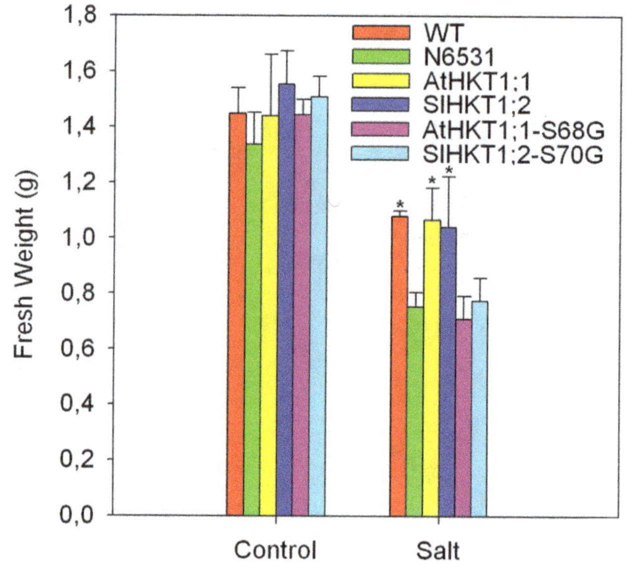

FIGURE 4 | Presence of 100 mM NaCl in the irrigation water during 2 weeks significantly reduced the fresh weight of transformed, WT and *athkt1;1* (N6531) *Arabidopsis* plants in comparison to control plants irrigated with water not supplemented with NaCl. Different inhibitions on the fresh weight production are observed amongst different transformed plant lines. *Arabidopsis* WT and *athkt1;1* plants were used as positive and negative controls, respectively. Treatment: $p < 0.05$; plant lines: $p < 0.05$; treatment*plant lines (n.s.); Two-Way ANOVA. Values indicate the means ± *SE* of three to seven biological replicates.

SlHKT1;2 and SlHKT1;2-S70G, respectively. These results show that the Na^+-mediated current of SlHKT1;2 is reduced by increased concentrations of K^+ in the bath. The presence of K^+ ions affects the transport of Na^+ by SlHKT1;2. This effect is not observed with AtHKT1;1.

SELECTION AND MOLECULAR ANALYSIS OF T2 PLANTS AND GUS EXPRESSION IN *A. THALIANA* UNDER THE *A. THALIANA* HKT1;1

HKT1 gene expression was analyzed in T2 *Arabidopsis* plantlets carrying the *SlHKT1;2* or *AtHKT1;1*, *SlHKT1;2-S70G*, or *AtHKT1;1-S68G* genes (data not shown). All genes were expressed under the *AtHKT1;1* endogenous promoter. All lines tested showed *HKT1* expression levels higher than the *athkt1;1* (N6531) mutant plants. Expression of *AtHKT1;1* was not detected in *athkt1;1* mutant plants. The 5 kb *AtHKT1;1* promoter fragment was able to drive the expression of *GUS* in the vascular tissues of transformed *Arabidopsis* WT plants (**Figure 3**). GUS activity driven by *AtHKT1;1* promoter was only observed in the vascular tissues of leaves of transgenic *Arabidopsis* plants (**Figure 3**).

ANALYSIS OF FW, Na^+, AND K^+ ACCUMULATION OF TRANSGENIC *ARABIDOPSIS* LINES

There was a marked difference in K^+-sensitivity of Na^+-transport mediated by the AtHKT1;1 and SlHKT1;2 proteins, as analyzed in *Xenopus laevis* oocytes. To see whether this difference was reflected in Na^+/K^+ homeostasis *in planta*, we expressed the *AtHKT1;1* and *SlHKT1;2* genes driven by a 5 kb long *AtHKT1;1* promoter in the *athkt1;1* mutant plants and studied how they complemented the mutant phenotype. Since the analysis of *SlHKT1;2-S70G* and *AtHKT1;1-S68G* in *Xenopus* oocytes showed interesting effects on transport activity and ion selectivity we transformed *athkt1;1* plants with *SlHKT1;2*-S70G and *AtHKT1-S68G*, and tested the functioning of these mutated genes *in planta* during salinity stress. **Figure 4** shows the effect of salinity treatment on the leaf fresh weight of WT plants, *athkt1;1* mutants and transformed plants. *athkt1;1*-mutant plants were more sensitive to salt than the WT plants. Both the *AtHKT1;1* and the *SlHKT1;2* gene were able to complement the *athkt1;1* mutant growth-phenotype on salt (**Figure 4**). Plants transformed with *SlHKT1;2*-S70G or *AtHKT1-S68G* were just as sensitive to salt as the *athkt1;1*-mutant plants. The analysis of the relative water content of the leaves did not show statistically significant differences between control and salt treated plants (data not shown) nor within transformed lines. The effect of the *athkt1;1* mutation on Na^+ and K^+ homeostasis was pronounced (**Figure 5**). The *athkt1;1* plants accumulated almost four-fold more Na^+ and two-fold less K^+, resulting in an eight-fold higher Na^+/K^+-ratio in the shoot of the mutant plants as compared to the WT plants. The transgenic lines that complemented the *athkt1;1* growth-phenotype (*AtHKT1;1* and *SlHKT1;2*) showed effects on ion accumulation: the *athkt1;1* K^+-phenotype during salt stress (i.e., strong reduction in K^+ accumulation) was completely restored, but Na^+ accumulation in the leaves of these transgenic lines was still significantly higher than that of the WT plants.

DISCUSSION

Several studies have shown that naturally occurring SNPs in genes involved in Na^+ and K^+ homeostasis can have a dramatic effect on the salinity tolerance of several plant species. This is the case for HKT transporters (Rubio et al., 1995; Ren et al., 2005; Baxter et al., 2010; Ali et al., 2012; Cotsaftis et al., 2012). Due to the

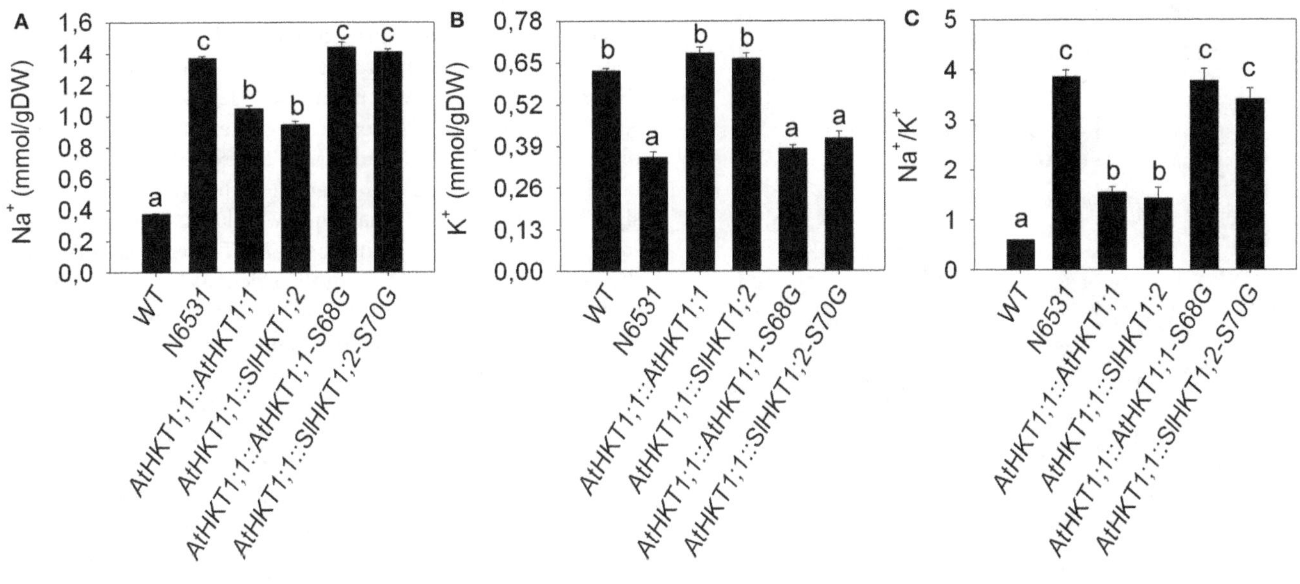

FIGURE 5 | Differences in ion accumulation between all plants analyzed. Na^+ **(A)** and K^+ **(B)** accumulation and Na^+/K^+ **(C)** ratio in the shoot of WT, *athkt1;1* mutant (N6531) and transgenic lines expressing different *HKT1* genes. Na^+/K^+ ratio was normalized for WT. The effect of the mutation in the *AtHKT1;1* gene is very clear, showing a strong increase in shoot Na^+ accompanied by a decrease in K^+, which resulted in a very strong increase in the Na^+/K^+ ratio. Both lines expressing *AtHKT1;1::AtHKT1;1* or *AtHKT1;1::SlHKT1;2* were able to reduce the accumulation of Na^+ and increase the accumulation of K^+ in comparison to *athkt1;1*. Lines expressing *AtHKT1;1::AtHKT1;1-S68G* or *AtHKT1;1::SlHKT1;2-S70G* were not able to ameliorate the phenotype of the *athkt1;1* plants. Different letters above bars indicate statistically significant differences. Values indicate the means ± *SE* of three to seven biological replicates.

effects of SNPs on the transport properties of HKT proteins, and consequently the effect of these SNPs on the salinity tolerance of the plants, we decided to study the presence/absence of SNPs in the sequence of the first pore domain of the *HKT1;2* gene of several tomato accessions (**Figure 1**). The first pore domain in all accessions studied showed an S, which is naturally occurring in AtHKT1;1, implying that if during evolution the amino acid residue that we studied would have changed, these changes were not in favor of salinity tolerance of the tomato plants and, therefore, did not persist (Diatloff et al., 1998).

The analysis of the properties of the heterologous expressed *SlHKT1;2* showed that the transport characteristics were in accordance with the presence of an S in the first pore domain of the transporter (**Figure 2**). This is reflected in the SlHKT1;2 transport characteristics as measured in heterologous expression, where it was shown that tomato HKT1;2 transports Na^+ but not K^+ (Almeida et al., 2014). However, a striking difference between the transport properties of AtHKT1;1 and SlHKT1;2 expressed in oocytes was observed when currents were measured at constant Na^+ in the bath (1 mM) and increasing K^+ (1, 3, and 10 mM). Whereas the AtHKT1;1 mediated current was virtually insensitive to higher K^+, the SlHKT1;2 mediated current decreased by 60% at 10 mM K^+ (**Figure 2**). A similar inhibitory action of K^+ on HKT-mediated currents was reported for OsHKT2;1 (Jabnoune et al., 2009) and for TmHKT1;5-A (Munns et al., 2012). It was proposed that this inhibition is caused by the association of K^+ to the Na^+ binding region within the pore region of HKT transporters (Rubio et al., 1995; Gassmann et al., 1996). This inhibition has not been observed with AtHKT1;1 (Uozumi et al., 2000) nor OsHKT1;5 (Ren et al., 2005) in *Xenopus* oocytes, which indicates that the K^+-sensitivity of these transporters is different from OsHKT2;1, TmHKT1;5-A and tomato SlHKT1;2. Physiologically, this K^+-induced reduction of Na^+-influx might mean that the tomato plants maintain a certain K^+/Na^+-homeostasis in the transpiration sap. High concentrations of K^+ in the xylem sap might imply a reduced Na^+-uptake into the XPC, as the Na^+/K^+ ratio is in favor of Na^+. On the other hand, low concentrations of K^+ and high concentrations of Na^+ in the xylem sap imply an induced Na^+-uptake into the XPC, as the Na^+/K^+ ratio is in favor of K^+ in the xylem sap. In this model, the Na^+/K^+ ratio is more important than the absolute concentration of Na^+ in the xylem sap in determining the Na^+ uptake into the XPC by HKT1 (**Figure 6**). Although the S → G mutation in the first pore domain of both the AtHKT1;1 and SlHKT1;2 protein had an effect on the ion selectivity of both transporters, as deduced from the shift in reversal potential at increasing external K^+ (**Figure 2**), a major difference observed was the reduction in total current transported by SlHKT1;2-S70G and AtHKT1-S68G, which was 95 and 78% respectively of that transported by the WT proteins at 10 mM Na^+ and 1 mM K^+ in the bath (**Figure 2**). The reason for this difference is unclear as the results we obtained do not match the results of the Na^+ currents produced by AtHKT1;1-S68G reported by Maser et al. (2002). Their results did not show this reduction in total currents. Interestingly, when Maser et al. (2002) mutated AtHKT1;1 in M69L, a reduction in outward currents with increasing K^+ concentrations in the bath was observed, whereas in TaHKT1;2 the reverse mutation, L92M, abolished this inhibitory effect of K^+ on outward currents (Maser et al., 2002).

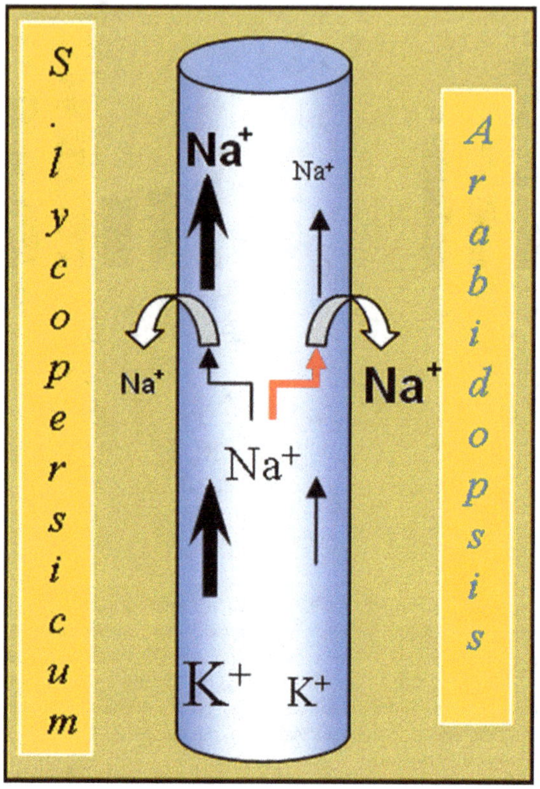

FIGURE 6 | Model depicting the difference in K$^+$-sensitivity of SlHKT1;2 from *S. esculentum* and AtHKT1;1 from *Arabidopsis*. When the K$^+$-concentration in the xylem sap is high, Na$^+$-uptake by the SlHKT1;2 transporter is reduced. As a result the amount of Na$^+$ in the xylem stream reaching the shoot of tomato is, at least partially, controlled by SlHKT1;2, which in turn depends on the concentration of K$^+$ present in the xylem sap. In *Arabidopsis*, Na$^+$-uptake in the XPCs by the AtHKT1;1 transporter is not affected by high K$^+$.

In their work, the leucine (L) adjacent to the G of the pore domain seems to confer K$^+$ sensitivity to outward currents in contrast with the methionine (M) at that same position, which abolished the effect of K$^+$ on Na$^+$ currents (Maser et al., 2002). Since in our study SlHKT1;2 has a M adjacent to the G of the pore domain, no sensitivity of outward currents was expected due to increasing K$^+$ concentrations in the bath. Nevertheless, in our study we observed outward currents, mediated by SlHKT1;2, sensitive to external K$^+$. In a future study it would be interesting to mutate SlHKT1;2-M71L, to check whether the presence of a L adjacent to the S of the first pore domain of SlHKT1;2 enhances or decreases this inhibitory effect of external K$^+$ on the outward Na$^+$-currents mediated by SlHKT1;2.

ATHKT1;1 AND *SLHKT1;2* EXPRESSING LINES

All plant lines studied showed a reduction in fresh weight caused by the salt treatment, and as expected the salt-induced growth reduction was stronger in the *athkt1;1*-mutant plants than in WT plants. With respect to growth on salt, *AtHKT1;1* and *SlHKT1;2* expression complemented the mutant, since the fresh weight of these lines was comparable to that of WT plants treated with salt and significantly higher than the growth of *athkt1;1* plants (**Figure 4**). Although expression of *AtHKT1;1* and *SlHKT1;2* completely restored the concentration of K$^+$ in the shoot to the same level of the WT plants, Na$^+$ levels were still higher in *AtHKT1;1* and *SlHKT1;2*-expressing lines in comparison to WT plants. Both *AtHKT1;1* and *SlHKT1;2* expressing lines accumulated significantly less Na$^+$ than the other transgenic lines, but significantly more Na$^+$ than the WT plants. This indicates that in these two transgenic lines both *HKT1* genes do not retrieve the same amount of Na$^+$ from the xylem as WT plants. This difference in Na$^+$ accumulation in lines expressing *AtHKT1;1* and *SlHKT1;2* in comparison to WT plants is surprising. In this study we used the native *AtHKT1;1* promoter (Baek et al., 2011) to drive the expression of all genes studied, avoiding non-native promoters as these are frequently referred to as the cause of unexpected results (Cominelli and Tonelli, 2010). The expression of *GUS* driven by the native *AtHKT1;1* promoter (**Figure 3**) showed that the expression patterns were similar to previous results (Sunarpi et al., 2005) and also the level of *HKT*-expression in the different transgenic lines was comparable to that in WT plants.

When expressed in *Xenopus laevis* oocytes *AtHKT1;1* and *SlHKT1;2* differed in two ways: first, SlHKT1;2 but not AtHKT1;1 showed an inhibition of Na$^+$ transport by K$^+$, and second, the total Na$^+$-mediated current (i.e., the turn-over) measured in *SlHKT1;2* expressing oocytes was considerably higher than the total Na$^+$-mediated current in *AtHKT1;1* expressing oocytes (**Figure 2**). The latter conclusion needs further testing, since it is based on the assumption that injection of the same amount of cRNA results in the same amount of protein expressed in the plasma membrane. However, it is clear that in these two transgenic lines both HKT1 transporters are involved in the retrieval of Na$^+$ ions from the xylem, as previously demonstrated for HKT1 transporters from several species (Liu et al., 2001; Laurie et al., 2002; Berthomieu et al., 2003; Garciadeblas et al., 2003; Ren et al., 2005; Horie et al., 2007; Jabnoune et al., 2009; Moller et al., 2009; Baxter et al., 2010; Plett et al., 2010; Baek et al., 2011; Mian et al., 2011; Ali et al., 2012; Munns et al., 2012).

As shown in **Figure 5**, mutating the *athkt1;1* gene not only results in an increase in shoot Na$^+$ concentration, but also in a strong reduction in shoot K$^+$ levels. This effect on K$^+$-accumulation has been reported before (Rus et al., 2004, 2005), but a good explanation for this effect is missing thus far. Moller et al. (2009) reported that the increase in K$^+$ concentration in the shoots from plants over-expressing *AtHKT1;1*, was a pleiotropic effect and a consequence of the reduced Na$^+$ shoot content. Another explanation is that the uptake of Na$^+$ from the xylem into the XPCs via HKT1 results in the depolarization of the membrane potential of XPCs and activation of the depolarization activated K$^+$-efflux channel SKOR, resulting in more K$^+$ release into the xylem (Sunarpi et al., 2005). A third explanation might be that AtHKT1;1 functionally interacts with a K$^+$-efflux transporter in the plasma membrane of XPCs. Support for this hypothesis is found in a recently published large-scale membrane interaction screen based on a yeast mating split-ubiquitin system (mbSUS) (Membrane-based Interactome Network Database, MIND: http://cas-biodb.cas.unt.edu/project/mind/index.php). In this screen, AtHKT1;1 was reported to

Network Database, MIND: http://cas-biodb.cas.unt.edu/project/mind/index.php). In this screen, AtHKT1;1 was reported to interact with KEA3, a putative K^+-efflux antiporter and member of the Proton Antiporter-2 (CPA2) family. However, this needs to be confirmed *in planta* in order to provide an explanation for the K^+-phenotype of the *athkt1;1* mutant. The absence of the AtHKT1;1 protein in the plasma membrane of XPCs of *athkt1;1* plants may have a negative effect on the KEA3 antiporter, resulting in reduced root to shoot K^+-transport. These hypotheses are certainly worth testing in view of the importance of Na^+/K^+-homeostasis during salinity stress.

In conclusion, the analysis of the presence of SNPs in the first pore domain of the *HKT1;2* gene in 93 tomato accessions showed that this amino acid is conserved in all these accessions. The replacement of S70 by a G in SlHKT1;2 proved to be sufficient to alter the transport specificity of this transporter, as analyzed by heterologous expression in *Xenopus laevis* oocytes. Moreover, when expressed under the *AtHKT1;1* native promoter, both *AtHKT1;1* and *SlHKT1;2* partially complemented the Na^+ accumulation phenotype and fully complemented the K^+ accumulation phenotype of *athkt1;1* mutant plants. *AtHKT1;1-S68G* and *SlHKT1;2-S70G* were unable to complement either the Na^+ or the K^+ phenotype of *athkt1;1* mutant plants. The transport activity of mutated HKT proteins, measured in *Xenopus* oocytes, was not reflected in altered Na^+/K^+-homeostasis of the *athkt1;1* mutants grown under soil conditions.

ACKNOWLEDGMENTS

The authors would like to thank Enza Zaden The Netherlands for the financial support. We thank the reviewers for their valuable comments, which helped to improve this manuscript.

REFERENCES

Ali, Z., Park, H. C., Ali, A., Oh, D. H., Aman, R., Kropornicka, A., et al. (2012). TsHKT1;2, a HKT1 homolog from the extremophile *Arabidopsis* relative *Thellungiella salsuginea*, shows K^+ specificity in the presence of NaCl. *Plant Physiol.* 158, 1463–1474. doi: 10.1104/pp.111.193110

Almeida, P., de Boer, G., and de Boer, A. H. (2014). Differences in shoot Na^+ accumulation between two tomato species are due to differences in ion affinity of HKT1;2. *J. Plant Physiol.* 171, 438–447. doi: 10.1016/j.jplph.2013.12.001

Asins, J. M., Villalta, I., Aly, M. M., Olias, R., Morales, P. A., Huertas, R., et al. (2012). Two closely linked tomato *HKT* coding genes are positional candidates for the major tomato QTL involved in Na^+/K^+ homeostasis. *Plant Cell Environ.* 36, 1171–1191. doi: 10.1111/pce.12051

Baek, D., Jiang, J. F., Chung, J. S., Wang, B. S., Chen, J. P., Xin, Z. G., et al. (2011). Regulated *AtHKT1;1* gene expression by a distal enhancer element and DNA methylation in the promoter plays an important role in salt tolerance. *Plant Cell Physiol.* 52, 149–161. doi: 10.1093/pcp/pcq182

Baxter, I., Brazelton, J. N., Yu, D. N., Huang, Y. S., Lahner, B., Yakubova, E., et al. (2010). A coastal cline in sodium accumulation in *Arabidopsis thaliana* is eriven by natural variation of the sodium transporter AtHKT1;1. *PLoS Genet.* 6:e1001193 doi: 10.1371/journal.pgen.1001193

Berthomieu, P., Conejero, G., Nublat, A., Brackenbury, W. J., Lambert, C., Savio, C., et al. (2003). Functional analysis of AtHKT1 in *Arabidopsis* shows that Na^+ recirculation by the phloem is crucial for salt tolerance. *EMBO J.* 22, 2004–2014. doi: 10.1093/emboj/cdg207

Byrt, C. S., Platten, J. D., Spielmeyer, W., James, R. A., Lagudah, E. S., Dennis, E. S., et al. (2007). HKT1;5-like cation transporters linked to Na^+ exclusion *loci* in wheat, *Nax2* and *Kna1*. *Plant Physiol.* 143, 1918–1928. doi: 10.1104/pp.106.093476

Clough, S. J., and Bent, A. F. (1998). Floral dip: a simplified method for *Agrobacterium*-mediated transformation of *Arabidopsis thaliana*. *Plant J.* 16, 735–743. doi: 10.1046/j.1365-313x.1998.00343.x

Cominelli, E., and Tonelli, C. (2010). Transgenic crops coping with water scarcity. *N. Biotechnol.* 27, 473–477. doi: 10.1016/j.nbt.2010.08.005

Cotsaftis, O., Plett, D., Shirley, N., Tester, M., and Hrmova, M. (2012). A two-staged model of Na^+ exclusion in rice explained by 3D modeling of HKT transporters and alternative splicing. *PLoS ONE* 7:e39865. doi: 10.1371/journal.pone.0039865

Davenport, R. J., James, R. A., Zakrisson-Plogander, A., Tester, M., and Munns, R. (2005). Control of sodium transport in durum wheat. *Plant Physiol.* 137, 807–818. doi: 10.1104/pp.104.057307

Diatloff, E., Kumar, R., and Schachtman, D. P. (1998). Site directed mutagenesis reduces the Na^+ affinity of HKT1, an Na^+ energized high affinity K^+ transporter. *FEBS Lett.* 432, 31–36. doi: 10.1016/S0014-5793(98)00833-3

Fairbairn, D. J., Liu, W. H., Schachtman, D. P., Gomez-Gallego, S., Day, S. R., and Teasdale, R. D. (2000). Characterisation of two distinct HKT1-like potassium transporters from *Eucalyptus camaldulensis*. *Plant Mol. Biol.* 43, 515–525. doi: 10.1023/A:1006496402463

FAO. (2009). *FAO Land and Plant Nutrition Management Service*. Available online at: http://www.fao.org/ag/agl/agll/spush/.

Garciadeblas, B., Senn, M. E., Banuelos, M. A., and Rodriguez-Navarro, A. (2003). Sodium transport and HKT transporters: the rice model. *Plant J.* 34, 788–801. doi: 10.1046/j.1365-313X.2003.01764.x

Gassmann, W., Rubio, F., and Schroeder, J. I. (1996). Alkali cation selectivity of the wheat root high-affinity potassium transporter HKT1. *Plant J.* 10, 869–882. doi: 10.1046/j.1365-313X.1996.10050869.x

Golldack, D., Su, H., Quigley, F., Kamasani, U. R., Munoz-Garay, C., Balderas, E., et al. (2002). Characterization of a HKT-type transporter in rice as a general alkali cation transporter. *Plant J.* 31, 529–542. doi: 10.1046/j.1365-313X.2002.01374.x

Haro, R., Banuelos, M. A., Senn, M. A. E., Barrero-Gil, J., and Rodriguez-Navarro, A. (2005). HKT1 mediates sodium uniport in roots. Pitfalls in the expression of *HKT1* in yeast. *Plant Physiol.* 139, 1495–1506. doi: 10.1104/pp.105.067553

Horie, T., Brodsky, D. E., Costa, A., Kaneko, T., Lo Schiavo, F., Katsuhara, M., et al. (2011). K^+ transport by the OsHKT2;4 transporter from rice with atypical Na^+ transport properties and competition in permeation of K^+ over Mg^{2+} and Ca^{2+} ions. *Plant Physiol.* 156, 1493–1507. doi: 10.1104/pp.110.168047

Horie, T., Costa, A., Kim, T. H., Han, M. J., Horie, R., Leung, H. Y., et al. (2007). Rice OsHKT2;1 transporter mediates large Na^+ influx components into K^+-starved roots for growth. *EMBO J.* 26, 3003–3014. doi: 10.1038/sj.emboj.7601732

Horie, T., Yoshida, K., Nakayama, H., Yamada, K., Oiki, S., and Shinmyo, A. (2001). Two types of HKT transporters with different properties of Na^+ and K^+ transport in *Oryza sativa*. *Plant J.* 27, 129–138. doi: 10.1046/j.1365-313x.2001.01077.x

Huang, S. B., Spielmeyer, W., Lagudah, E. S., James, R. A., Platten, J. D., Dennis, E. S., et al. (2006). A sodium transporter (HKT7) is a candidate for *Nax1*, a gene for salt tolerance in durum wheat. *Plant Physiol.* 142, 1718–1727. doi: 10.1104/pp.106.088864

Jabnoune, M., Espeout, S., Mieulet, D., Fizames, C., Verdeil, J. L., Conejero, G., et al. (2009). Diversity in expression patterns and functional properties in the rice HKT transporter family. *Plant Physiol.* 150, 1955–1971. doi: 10.1104/pp.109.138008

James, R. A., Davenport, R. J., and Munns, R. (2006). Physiological characterisation of two genes for Na^+ exclusion in durum wheat: *Nax1* and *Nax2*. *Plant Physiol.* 142, 1537–1547. doi: 10.1104/pp.106.086538

Jefferson, R. A., Kavanagh, T. A., and Bevan, M. W. (1987). GUS fusion: beta-glucuronidase as a sensitive and versatile gene fusion marker in higher plants. *EMBO J.* 20, 3901–3907.

Jha, D., Shirley, N., Tester, M., and Roy, S. J. (2010). Variation in salinity tolerance and shoot sodium accumulation in *Arabidopsis* ecotypes linked to differences in the natural expression levels of transporters involved in sodium transport. *Plant Cell Environ.* 33, 793–804. doi: 10.1111/j.1365-3040.2009.02105.x

Karimi, M., Inze, D., and Depicker, A. (2002). GATEWAY vectors for *Agrobacterium*-mediated plant transformation. *Trends Plant Sci.* 7, 193–195. doi: 10.1016/S1360-1385(02)02251-3

Kato, N., Akai, M., Zulkifli, L., Matsuda, N., Kato, Y., Goshima, S., et al. (2007). Role of positively charged amino acids in the M2(D) transmembrane helix of Ktr/Trk/HKT type cation transporters. *Channels* 1, 161–171. doi: 10.4161/chan.4374

Lan, W. Z., Wang, W., Wang, S. M., Li, L. G., Buchanan, B. B., Lin, H. X., et al. (2010). A rice high-affinity potassium transporter (HKT) conceals a calcium-permeable cation channel. *Proc. Natl. Acad. Sci. U.S.A.* 107, 7089–7094. doi: 10.1073/pnas.1000698107

Laurie, S., Feeney, K. A., Maathuis, F. J., Heard, P. J., Brown, S. J., and Leigh, R. A. (2002). A role for HKT1 in sodium uptake by wheat roots. *Plant J.* 32, 139–149. doi: 10.1046/j.1365-313X.2002.01410.x

Liu, W., Fairbairn, D. J., Reid, R. J., and Schachtman, D. P. (2001). Characterization of two HKT1 homologues from *Eucalyptus camaldulensis* that display intrinsic osmosensing capability. *Plant Physiol.* 127, 283–294. doi: 10.1104/pp.127.1.283

Livak, K. J., and Schmittgen, T. D. (2001). Analysis of relative gene expression data using real-time quantitative PCR and the 2-deltadeltaCT method. *Methods* 25, 402–408. doi: 10.1006/meth.2001.1262

Maser, P., Hosoo, Y., Goshima, S., Horie, T., Eckelman, B., Yamada, K., et al. (2002). Glycine residues in potassium channel-like selectivity filters determine potassium selectivity in four-loop-per-subunit HKT transporters from plants. *Proc. Natl. Acad. Sci. U.S.A.* 99, 6428–6433. doi: 10.1073/pnas.082123799

Mian, A., Oomen, R. J., Isayenkow, S., Sentenac, H., Maathuis, F. J., and Very, A. A. (2011). Overexpression of a Na^+ and K^+-permeable HKT transporter in barley improves salt tolerance. *Plant J.* 68, 468–479. doi: 10.1111/j.1365-313X.2011.04701.x

Moller, I. S., Gilliham, M., Jha, D., Mayo, G. M., Roy, S. J., Coates, J. C., et al. (2009). Shoot Na^+ exclusion and increased salinity tolerance engineered by cell type-specific alteration of Na^+ transport in *Arabidopsis*. *Plant Cell* 21, 2163–2178. doi: 10.1105/tpc.108.064568

Munns, R., James, R. A., Xu, B., Athman, A., Conn, J. S., Jordans, C., et al. (2012). Wheat grain yield on saline soils is improved by an ancestral Na^+ transporter gene. *Nat. Biotechnol.* 30, 360–364. doi: 10.1038/nbt.2120

Munns, R., and Tester, M. (2008). Mechanisms of salinity tolerance. *Ann. Rev. Plant Biol.* 59, 651–681. doi: 10.1146/annurev.arplant.59.032607.092911

Platten, J. D., Cotsaftis, O., Berthomieu, P., Bohnert, H., Davenport, R. J., Fairbairn, D. J., et al. (2006). Nomenclature for HKT transporters, key determinants of plant salinity tolerance. *Trends Plant Sci.* 11, 372–374. doi: 10.1016/j.tplants.2006.06.001

Plett, D., Safwat, G., Gilliham, M., Moller, I. S., Roy, S., Shirley, N., et al. (2010). Improved salinity tolerance of rice through cell type-specific expression of *AtHKT1;1*. *PLoS ONE* 5:e12571. doi: 10.1371/journal.pone.0012571

Ren, Z. H., Gao, J. P., Li, L. G., Cai, X. L., Huang, W., Chao, D. Y., et al. (2005). A rice quantitative trait locus for salt tolerance encodes a sodium transporter. *Nat. Genet.* 37, 1141–1146. doi: 10.1038/ng1643

Rodriguez-Navarro, A. (2000). Potassium transport in fungi and plants. *Biochim. Biophys. Acta* 1469, 1–30. doi: 10.1016/S0304-4157(99)00013-1

Rubio, F., Gassmann, W., and Schroeder, J. I. (1995). Sodium-driven potassium uptake by the plant potassium transporter HKT1 and mutations conferring salt tolerance. *Science* 270, 1660–1663. doi: 10.1126/science.270.5242.1660

Rubio, F., Schwarz, M., Gassmann, W., and Schroeder, J. I. (1999). Genetic selection of mutations in the high affinity K^+ transporter HKT1 that define functions of a loop site for reduced Na^+ permeability and increased Na^+ tolerance. *J. Biol. Chem.* 274, 6839–6847. doi: 10.1074/jbc.274.11.6839

Rus, A., Lee, B. H., Munoz-Mayor, A., Sharkhuu, A., Miura, K., Zhu, J. K., et al. (2004). AtHKT1 facilitates Na^+ homeostasis and K^+ nutrition in planta. *Plant Physiol.* 136, 2500–2511. doi: 10.1104/pp.104.042234

Rus, A., Yokoi, S., Sharkhuu, A., Reddy, M., Lee, B. H., Matsumoto, T. K., et al. (2001). AtHKT1 is a salt tolerance determinant that controls Na^+ entry into plant roots. *Proc. Natl. Acad. Sci. U.S.A.* 98, 14150–14155. doi: 10.1073/pnas.241501798

Rus, A. M., Bressan, R. A., and Hasegawa, P. M. (2005). Unraveling salt tolerance in crops. *Nat. Genet.* 37, 1029–1030. doi: 10.1038/ng1005-1029

Schachtman, D. P., and Schroeder, J. I. (1994). Structure and transport mechanism of a high-affinity potassium uptake transporter from higher-plants. *Nature* 370, 655–658. doi: 10.1038/370655a0

Su, H., Balderas, E., Vera-Estrella, R., Golldack, D., Quigley, F., Zhao, C. S., et al. (2003). Expression of the cation transporter McHKT1 in a halophyte. *Plant Mol. Biol.* 52, 967–980. doi: 10.1023/A:1025445612244

Sunarpi, Horie, T., Motoda, J., Kubo, M., Yang, H., Yoda, K., et al. (2005). Enhanced salt tolerance mediated by AtHKT1 transporter-induced Na^+ unloading from xylem vessels to xylem parenchyma cells. *Plant J.* 44, 928–938. doi: 10.1111/j.1365-313X.2005.02595.x

Takahashi, R., Liu, S., and Takano, T. (2007). Cloning and functional comparison of a high-affinity K^+ transporter gene *PhaHKT1* of salt-tolerant and salt-sensitive reed plants. *J. Exp. Bot.* 58, 4387–4395. doi: 10.1093/jxb/erm306

Tholema, N., Bakker, E. P., Bakker, E. P., Suzuki, A., and Nakamura, T. (1999). Change to alanine of one out of four selectivity filter glycines in KtrB causes a two orders of magnitude decrease in the affinities for both K^+ and Na^+ of the Na^+ dependent K^+ uptake system KtrAB from *Vibrio alginolyticus*. *FEBS Lett.* 450, 217–220. doi: 10.1016/S0014-5793(99)00504-9

Uozumi, N., Kim, E. J., Rubio, F., Yamaguchi, T., Muto, S., Tsuboi, A., et al. (2000). The *Arabidopsis HKT1* gene homolog mediates inward Na^+ currents in *Xenopus laevis* oocytes and Na^+ uptake in *Saccharomyces cerevisiae*. *Plant Physiol.* 122, 1249–1259. doi: 10.1104/pp.122.4.1249

Yao, X., Horie, T., Xue, A. W., Leung, H. Y., Katsuhara, M., Brodsky, D. E., et al. (2010). Differential sodium and potassium transport selectivities of the rice OsHKT2;1 and OsHKT2;2 transporters in plant cells. *Plant Physiol.* 152, 341–355. doi: 10.1104/pp.109.145722

6

The photosynthesis, Na^+/K^+ homeostasis and osmotic adjustment of atriplex canescens in response to salinity

Ya-Qing Pan[1†], Huan Guo[1†], Suo-Min Wang[1], Bingyu Zhao[2], Jin-Lin Zhang[1], Qing Ma[1], Hong-Ju Yin[1] and Ai-Ke Bao[1*]

1 State Key Laboratory of Grassland Agro-ecosystems, College of Pastoral Agriculture Science and Technology, Lanzhou University, Lanzhou, China,
2 Department of Horticulture, Virginia Polytechnic Institute and State University, Blacksburg, VA, USA

Edited by:
Vadim Volkov, London Metropolitan University, UK
Reviewed by:
Hong Zhang, Texas Tech University, USA
Marian Brestic, Slovak University of Agriculture, Slovakia
*Correspondence:
Ai-Ke Bao baoaik@lzu.edu.cn
† These authors have contributed equally to this work.

Atriplex canescens (fourwing saltbush) is a C4 perennial fodder shrub with excellent resistance to salinity. However, the mechanisms underlying the salt tolerance in A. canescens are poorly understood. In this study, 5-weeks-old A. canescens seedlings were treated with various concentrations of external NaCl (0–400 mM). The results showed that the growth of A. canescens seedlings was significantly stimulated by moderate salinity (100 mM NaCl) and unaffected by high salinity (200 or 400 mM NaCl). Furthermore, A. canescens seedlings showed higher photosynthetic capacity under NaCl treatments (except for 100 mM NaCl treatment) with significant increases in net photosynthetic rate and water use efficiency. Under saline conditions, the A. canescens seedlings accumulated more Na^+ in either plant tissues or salt bladders, and also retained relatively constant K^+ in leaf tissues and bladders by enhancing the selective transport capacity for K^+ over Na^+ (ST value) from stem to leaf and from leaf to bladder. External NaCl treatments on A. canescens seedlings had no adverse impact on leaf relative water content, and this resulted from lower leaf osmotic potential under the salinity conditions. The contribution of Na^+ to the leaf osmotic potential (Ψs) was sharply enhanced from 2% in control plants to 49% in plants subjected to 400 mM NaCl. However, the contribution of K^+ to Ψs showed a significant decrease from 34% (control) to 9% under 400 mM NaCl. Interestingly, concentrations of betaine and free proline showed significant increase in the leaves of A. canescens seedlings, these compatible solutes presented up to 12% of contribution to Ψs under high salinity. These findings suggest that, under saline environments, A. canescens is able to enhance photosynthetic capacity, increase Na^+ accumulation in tissues and salt bladders, maintain relative K^+ homeostasis in leaves, and use inorganic ions and compatible solutes for osmotic adjustment which may contribute to the improvement of water status in plant.

Keywords: Atriplex canescens, salt tolerance, photosynthesis, osmotic adjustment, salt bladder

INTRODUCTION

Salinity is one of the major environmental factors reducing the growth, development, and productivity of plants (Zhu, 2001; Flowers, 2004; Zhang et al., 2010; Shabala, 2013; Yan et al., 2013; Tang et al., 2015; Kalaji et al., 2016). It is estimated that about 10% of land area and half of irrigated land in the world are affected by salinity (Ruan et al., 2010; Shabala, 2013). Salt stress adversely reduces plant growth through ionic toxicity and osmotic stress (Hasegawa et al., 2000; Tester and Davenport, 2003; Flowers and Colmer, 2008), which influence a series of physiological processes, and finally suppress the photosynthesis (Zhu, 2001; Flowers and Colmer, 2008; Zhang and Shi, 2013; Flowers et al., 2015; Zhang et al., 2016). Some plants, such as halophytes, however, have evolved multiple adaptive strategies that ensure their survival and growth in a harsh environment (Flowers, 2004; Shabala, 2013). Therefore, to cope with the challenge of salinity for agriculture, there are increasing interests of studying the physiological responses underlying the salt resistance of halophytes, especially those species with high economic value and salt tolerance (Yamaguchi and Blumwald, 2005; Yan et al., 2013).

Salt resistance is a complex trait involving multiple mechanisms. One of the effective adaptations is reduction of Na^+ concentration in cytosol to alleviate Na^+ toxicity and maintain the intracellular ion homeostasis in a saline environment. Halophytes do achieve this goal by controlling net Na^+ uptake in the root, excreting Na^+ from the surface of stem or leaf, and sequestering Na^+ into the vacuole (Zhu, 2003; Flowers and Colmer, 2008; Hasegawa, 2013). Some species, such as *Thellungiella halophila* (Volkov and Amtmann, 2006; Wang et al., 2006) and *Puccinellia tenuiflora* (Wang et al., 2009, 2015; Guo et al., 2012; Niu et al., 2016) maintain a high K^+/Na^+ ratio in shoots by limiting net Na^+ influx into roots. Other species take up Na^+ from soil and then excrete large quantities of Na^+ via salt glands or bladders (Flowers and Colmer, 2008; Ding et al., 2010; Shabala, 2013; Shabala et al., 2014). Some succulent halophytes can sequester Na^+ into the vacuole to reduce the toxicity of excessive Na^+ in cytosol as well as regulate cellular osmotic potential by using Na^+ as an osmoregulation substance, and thus maintain cellular ion homeostasis and turgor (Zhu, 2001; Flowers, 2004; Wang et al., 2004, 2007; Zörb et al., 2005; Yamaguchi et al., 2013; Flowers et al., 2015). The osmotic adjustment

(OA) is another important physiological mechanism for plant adaptation to salinity, which involves the fall of osmotic potential (Ψs) in plant tissue resulting from the net accumulation of cellular solutes. It is essential for plants to maintain water uptake from a low water potential environment (Zhang et al., 1999; Munns and Tester, 2008; Ma et al., 2012). Inorganic ions, such as K^+ and vacuolar Na^+, can directly be engaged in decreasing the osmotic potential of cells (Flowers and Colmer, 2008). Moreover, higher plants also accumulate compatible solutes, such as betaine, free proline and soluble sugars, for OA under abiotic stress (Munns and Tester, 2008; Flowers et al., 2010).

Atriplex canescens (Pursh) Nutt. (fourwing saltbush), a C_4 perennial shrub native to saline and xeric deserts in North America, belongs to Chenopodiaceae with prominent resistance to salinity, drought, and cold (Glenn and Brown, 1998; Hao et al., 2013). This species is also an attractive fodder shrub for most livestock and large animals due to its high palatability as well as rich nutrition (Peterson et al., 1987; Kong, 2013). Moreover, it is especially useful for erosion control and reclamation of marginal lands due to its extensive root system and excellent adaptability (Benzarti et al., 2013; Kong, 2013). In 1989, *A. canescens* was introduced from USA to China and was widely used for soil and water conservation, sand-fixing and saline land restoration in north China (Kong, 2013). Previous studies showed that *A. canescens* accumulated more Na^+ under salinity conditions (Glenn et al., 1994, 1996; Glenn and Brown, 1998). On the other hand, recent studies suggested that the salt excretion via salt bladders (Ben Hassine et al., 2009; Belkheiri and Mulas, 2011; Shabala et al., 2014; Tsutsumi et al., 2015), the uptake and accumulation of K^+ (Bazihizina et al., 2009; Bose et al., 2015), photosynthetic responses (Redondo-Gómez et al., 2007; Bazihizina et al., 2009; Nemat-Alla et al., 2011), and OA by compatible solutes (Martínez et al., 2004, 2005; Ben Hassine et al., 2008; Bouchenak et al., 2012), may also contribute to the salt tolerance in some species of *Atriplex*. However, it is not clear if similar or different physiological mechanisms also contribute to the responses of *A. canescens* to saline environment.

Therefore, the aim of this study was to characterize the physiological responses of *A. canescens* to salinity by measuring various parameters related to photosynthesis, Na^+/K^+ homeostasis and OA under treatments with different concentrations of NaCl.

MATERIALS AND METHODS

Plant Growth Conditions and NaCl Treatments

Seeds of *A. canescens* were collected in Lingwu County (37.78° N, 106.25° E; elevation 1250 m) of Ningxia Autonomous Region, China. After corrosion of the hard coat with 75% H_2SO_4 (v/v) for 15 h, seeds were rinsed six times with distilled water and, germinated in vermiculite (moistened with distilled water) at 28°C in the dark for 6 days. Uniform seedlings were transplanted to plastic culture pots (5 cm × 5 cm × 5 cm; two plants/pot) containing vermiculite (with trace amounts of Na^+ and K^+, Ma et al., 2012) and watered with 1/2 strength Hoagland nutrient solution (Ma et al., 2012) at 2-days interval. The growth conditions in greenhouse were controlled to maintain a temperature of 28°C/25°C (day/night), a photoperiod of 16/8 $h^2 s^{-1}$) and (light/dark; the flux density was about 800 μmol m^- an approximate relative humidity of 65%.

After washing the leaves thoroughly with distilled water (to remove the salt from the surface of the leaves), 5-weeks-old seedlings were treated with 1/2 strength Hoagland nutrient solution containing additional 0, 100, 200, or 400 mM NaCl for 10 days, and the treatment solutions were renewed every 2 days to keep constant NaCl concentration. The treated and control plants were harvested for biomass measurement and physiological analysis.

Measurement of Parameters for Photosynthesis and Water Relations

Net photosynthesis rate (Pn), stomatal conductance (Gs), and transpiration rate (Tr) were measured by an automatic photosynthetic measuring apparatus (GFS-3000; Walz, Effeltrich, Germany) in the greenhouse under a light intensity of 900 μmol $m^{-2} s^{-1}$. The water use efficiency (WUE) was calculated by the following formula: WUE = Pn/Tr (Ma et al., 2012). Leaf areas were estimated using a photo scanner (Epson Perfection 4870; Epson America, Inc., Long Beach, CA, USA).

Measurement of Na^+ and K^+ Concentrations

Salt bladders were brushed from both adaxial and abaxial surfaces of leaves (approximately 1 g of fresh weight of leaves was used) into 20 mL deionized water using a hard nylon brush as described by Tsutsumi et al. (2015). Collected leaves were dried in an oven at 80°C for 72 h, and the dry weight (DW) was determined. Roots were rinsed with deionized water for 10 s, washed twice for 8 min in cold 20 mM $LiNO_3$ solution to exchange the cations in the apoplast. The DWs of stems and roots were determined after drying at 80°C for 72 h.

For determining the cation exclusion in bladders, the Na^+ and K^+ were extracted from brushed bladders under 90°C water bath for 1 h. The Na^+ and K^+ were determined using a flame photometer (Model 410 Flame; Sherwood Scientific, Ltd., Cambridge, UK), and cation concentration in salt bladders was calculated by the following formula: cation (Na^+ or K^+) concentration in salt bladders (mmol/g DW) = cation content in salt bladders (mmol)/ DW (g) of leaves. For measuring the cation accumulation in tissues, the Na^+ and K^+ were extracted from dried roots, stems and leaves with 100 mM acetic acid at 90°C for at least 2 h, the cation concentrations were then determined using flame spectrophotometer (Bao et al., 2016).

Selective transport (ST) capacity for K^+ over Na^+ between different parts (root, stem, leaf, and bladder) was calculated according to the following equation (Wang et al., 2002, 2009): $ST_{(A/B)}$ = (Na^+/K^+ in part A)/(Na^+/K^+ in part B). The higher ST value indicates the stronger net capacity of selection for transport of K^+ over Na^+ from part A to part B (Wang et al., 2002; Ma et al., 2014).

Scanning Electron Microscopic Observation of Leaf Surface

Segments of leaves before and after brushing the salt bladders were fixed on a stainless steel bracket and frozen with liquid nitrogen, then the samples were taken out the liquid nitrogen and the abaxial surfaces of leaves were observed quickly using scanning electron microscope (S-3400N; Hitachi, Tokyo, Japan). Meanwhile, the images were taken. The accelerating voltage was 15 kV.

Measurement of Betaine and Free Proline Concentration

For betaine determination, mature leaves from plants with different treatments were dried at 80°C for 1 day and ground to pass a 40-mesh sieve. The dried, finely ground sample (0.2 g) was shaken with 1 mL of 80% methanol (v/v) at 60°C for 30 min. The extracted solution was harvested after centrifugation at 11,000 × *g* under 25°C for 15 min. Then the betaine concentration was measured with a Reinecke salt Kit (Comin Biotechnology, Co. Ltd., Suzhou, China) following the manufacturer's instructions. Briefly, 0.25 mL of the extracted solution was mixed with 0.35 mL of Reinecke salt saturated solution (30 mg/L, pH = 1.0) and the reaction proceeded at 4°C for 2 h. The supernatant was discarded after centrifugation at 10,000 × *g* under 25°C for 15 min. The precipitate was washed with 0.3 mL of 99% ether (v/v) and then dissolved in 1 mL of 70% acetone (v/v). Finally, the absorbance was measured at 525 nm using a spectrophotometer (UV-6100PCS; Mapada Instruments, Co. Ltd., Shanghai, China). The betaine concentration was calculated in comparison with a standard sample in Kit.

For free proline measurement, 0.1 g of fresh leaf was homogenized with 1 mL of 5% salicylic acid (v/v) on the ice, then was extracted with shaking in boiling water for 10 min. The supernatant was collected after centrifugation at 10,000 × *g* under 25°C for 10 min. Finally, free proline concentration was determined according to the method described by Wang et al. (2004) using a spectrophotometer.

Measurement of Leaf Relative Water Content

The leaf relative water content (RWC) was calculated according to the following formula: RWC (%) = 100 × (FW – DW)/(TW — DW) (Bao et al., 2016). The leaves were excised from seedlings and the fresh weight (FW) was weighed immediately, then the turgid weight (TW) was measured after soaking the leaves in deionized water at 4°C overnight in the dark. Finally, leaves were dried in an oven at 80°C for 48 h and the DW were determined.

Measurement of Leaf Osmotic Potential (Ψs) and Evaluation of the Contributions of Solutes to Leaf Ψs

Leaves from each treatment were rinsed with deionized water and blotted on filter paper immediately, then were frozen in liquid nitrogen and thawed to extrude sap by a syringe, respectively. The resulting sap was used to determine the leaf osmotic potential (Ψs) according to the method described by Bao et al. (2014) using a cryoscopic osmometer (Osmomat-030, Gonotec GmbH, Berlin, Germany) at 25°C. To evaluate the contributions of solutes to leaf Ψs, the calculated osmotic potential (COP) values of Na^+, K^+, betaine, and free proline were calculated respectively by the Van't Hoff equation (Guerrier, 1996): COP = moles of solute × RK, where R = 0.008314 and K = 298°C. Then the contribution of each solute to leaf osmotic potential (C) was calculated by the following formula: C = COP/'-ls × 100% (Guerrier, 1996).

Statistical Analysis

Data were analyzed according to one-way analysis of variance (ANOVA) by SPSS statistical software (Ver. 19.0, SPSS, Inc., Chicago, IL, USA) and the significant differences among means were identified by Duncan's multiple range tests at a significance level of $P < 0.05$. All data were presented as mean ± SE ($n \geq 8$).

RESULTS

Atriplex canescens Seedlings Exhibited Strong Resistance to Salinity

After treatment with various concentrations of external NaCl for 10 days, all of the seedlings grew vigorously, especially, the seedlings under 100 mM NaCl exhibited larger and sturdier phenotypes than those under control (0 mM NaCl) and other NaCl treatments (**Figure 1a**), indicating that the addition of 100 mM NaCl might promote the growth of *A. canescens*. To further confirm above observations, plant height and biomass were measured. The data showed that the addition of 100 mM NaCl significantly increased plant height, FW and DW of *A. canescens* seedlings by 20, 13, and 15%, respectively, compared to control plants (**Figures 1b–d**). Furthermore, compared with the control, the addition of either 200 or 400 mM NaCl had no significant negative effects on plant height (**Figure 1b**) and DW (**Figure 1d**), although significantly reduced FW of plants (**Figure 1c**).

Effects of External NaCl on Photosynthesis of *A. canescens* Seedlings

To investigate the photosynthetic capacity of *A. canescens* seedlings under saline conditions, the net photosynthetic rate (Pn), stomatal conductance (Gs), transpiration rate (Tr), and WUE were measured. The results showed that the Pn of plants under NaCl treatments were significantly higher than that of control plants, and it actually increased within the measured range of NaCl concentrations. After 10 days of treatment, Pn of plants exposed to 100, 200, and 400 mM NaCl were 1.3, 1.5, and 2.3 fold higher than of control plants, respectively (**Figure 2A**). Compared to control, interestingly, Gs and Tr of *A. canescens* seedlings in the presence of additional 100 mM NaCl showed a sharp increase by 180 and 190%, respectively. However, both Gs and Tr were unaffected by 200 or 400 mM NaCl (**Figures 2B,C**). Correspondingly, the plant WUE was significantly reduced under 100 mM NaCl, but increased in the presence of 200 and 400 mM NaCl, which are 2.1 and 3.1 fold higher than for control plants, respectively (**Figure 2D**).

The Na^+/K^+ Homeostasis in *A. canescens* Seedlings Exposed to Salinity

To investigate the mechanism underlying salt resistance of *A. canescens* seedlings, we measured the amounts of Na^+ and K^+ accumulated in tissues and sequestered in salt bladders respectively, and also estimated the ST capacity for K^+ over Na^+ between different parts of *A. canescens* seedling.

With the increase of the external NaCl concentration, Na^+ accumulation exhibited a significant increase in different tissues of *A. canescens* seedlings. When treated with 400 mM NaCl for 10 days, the Na^+ concentrations in leaves, stems, and roots were 13.4, 17.2, and 3.4 fold higher than those in control plants, respectively (**Figures 3A–C**). Although, K^+ accumulations in all tissues of *A. canescens* seedlings were reduced by external NaCl (**Figures 3D–F**), the K^+ concentration in stems maintained a relative

FIGURE 1 | The growth status (a), plant height (b), fresh weight (c), and dry weight (d) of *Atriplex canescens* seedlings under different NaCl treatments for 10 days. Values in **(b–d)** are mean ± SE (n = 17–23). Columns with different letters indicate significant differences at P < 0.05 (Duncan test).

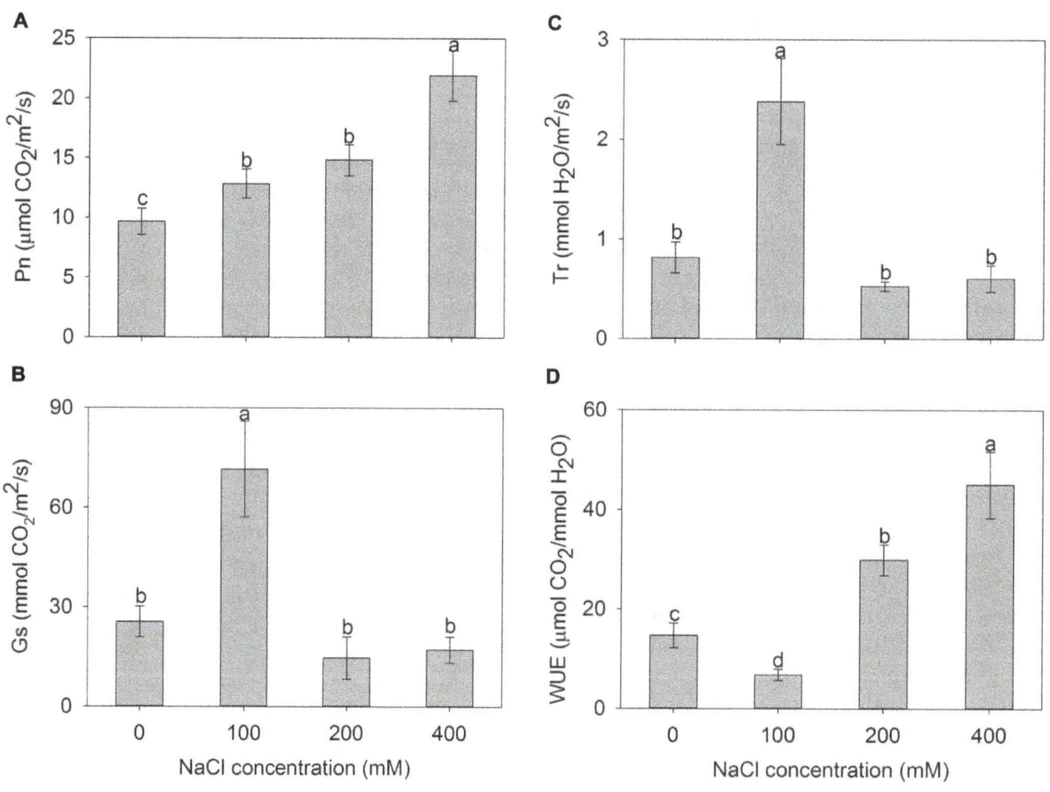

FIGURE 2 | Net photosynthesis rate (Pn) (A), stomatal conductance (Gs) (B), transpiration rate (Tr) (C), and water use efficiency (WUE) (D) of *A. canescens* seedlings under different NaCl treatments for 10 days. Values are mean ± SE (n = 10). Columns with different letters indicate significant differences at P < 0.05 (Duncan test).

stability among all external NaCl treatments (**Figure 3E**), and especially in leaves, it showed lesser decrease by 24 and 35% under 100 and 200 mM NaCl compared to control plants, respectively, even rebounded to control level under 400 mM NaCl (**Figure 3D**).

To investigate the Na^+ sequestration in salt bladders of *A. canescens* seedlings, we brushed the salt bladders from the surface of leaves (**Figure 4**). When *A. canescens* seedlings were grown in normal conditions (without NaCl supplement), only a small amount of Na^+ was measured in salt bladders. However, the bladder Na^+ concentration significantly raised with the increasing of external NaCl. Under 100, 200, and 400 mM NaCl for 10 days, the bladder Na^+ concentrations were 3.5, 3.6, and 5.9 fold higher than that of control, respectively (**Figure 5A**). On the other hand, compared to control, the bladder K^+ concentration was unaffected by 100 and 200 mM NaCl, and was significantly reduced by 33% under 400 mM NaCl (**Figure 5B**). Correspondingly, more Na^+ accumulation resulted in a significant increase of Na^+/K^+ ratio in salt bladders under various NaCl treatments. For example, the bladder Na^+/K^+ ratio of *A. canescens* seedlings under 400 mM NaCl (the value is 1.6) was 7.4 fold higher than that of the control plants (the value is 0.2; **Figure 5C**). These results indicate that sequestering more Na^+ into bladder may be one of important strategies for *A. canescens* seedlings to alleviate the toxicity of excessive Na^+ under saline conditions.

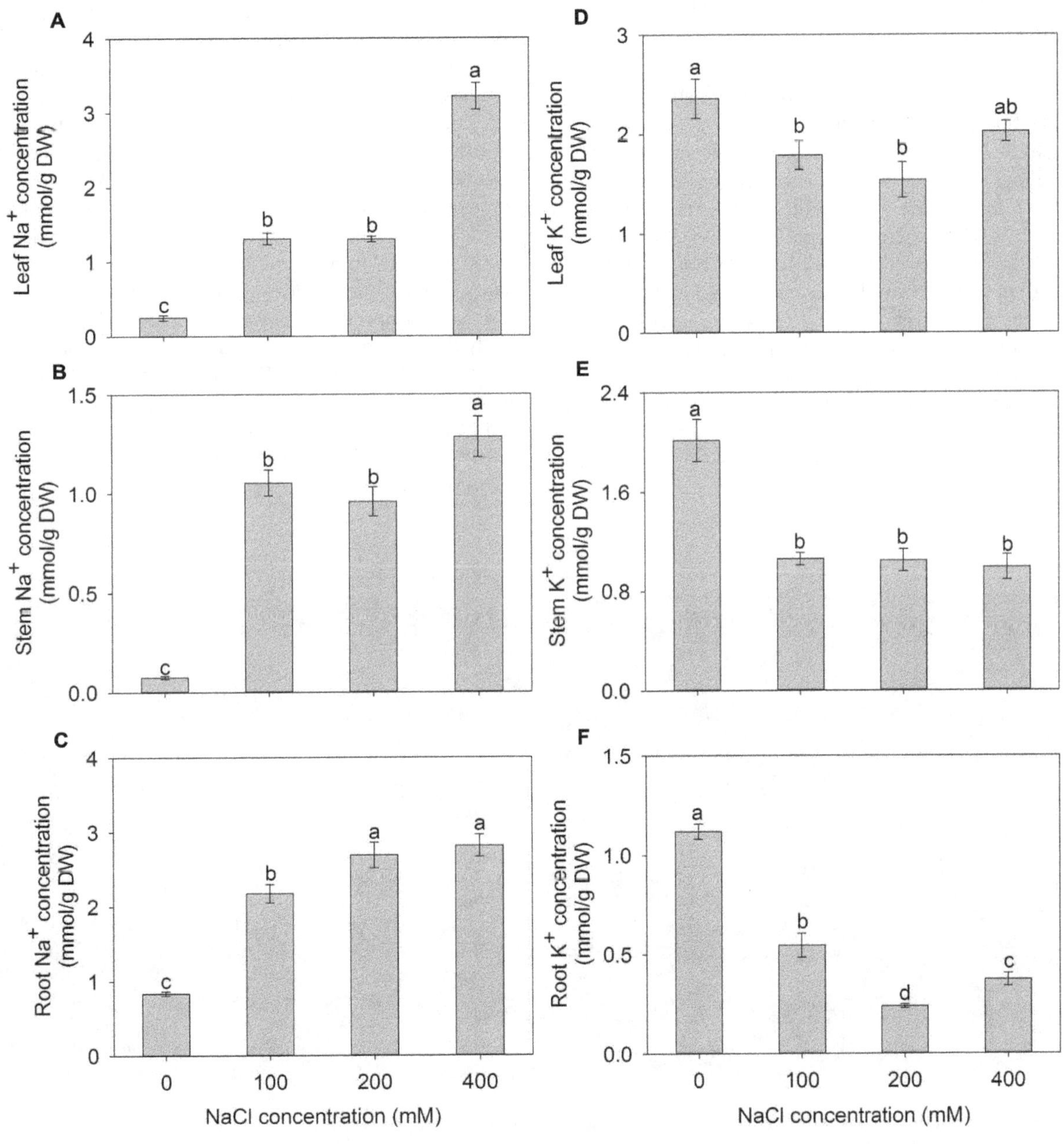

FIGURE 3 | The Na^+ (A–C) and K^+ (D–F) concentrations in the leaves, stems and roots of *A. canescens* seedlings under different NaCl treatments for 10 days. Values are mean ± SE (n = 10). Columns with different letters indicate significant differences at $P < 0.05$ (Duncan test).

The addition of external NaCl also influenced the ST capacity for K^+ over Na^+ (ST value) in *A. canescens* seedlings, but the change patterns of ST value varied among different parts (**Figure 6**). Compared to control, the ST values from root to stem were significantly decreased by 68, 45, and 65% under 100, 200, and 400 mM NaCl, respectively (**Figure 6A**). However, the ST values from stem to leaf of plants treated with 100–400 mM NaCl were significantly higher by 2.6, 1.8, and 1.3 fold than that in control plants, respectively (**Figure 6B**). Interestingly, the ST value from leaf to bladder showed significant increase by 1.2, 3.2, and 0.6 fold under 100, 200, 400 mM NaCl, respectively (**Figure 6C**). Most importantly, *A. canescens* seedlings showed highest ST value from root to stem under either control or saline conditions, suggesting that relatively more K^+ may be selectively loaded to xylem (**Figure 6**).

A. canescens Seedlings Accumulated More Betaine and Free Proline during NaCl Treatment

To investigate the effect of NaCl on compatible solute in *A. canescens* seedlings. We measure the concentrations of betaine and free proline in leaves. The leaf betaine concentration gradually increased with the increase of external NaCl concentration, and the highest value was detected under 400 mM NaCl, which was 66% higher than that of control plants (**Figure 7A**). Moreover, the leaf free proline concentrations of *A. canescens* seedlings were very low under either control or 100 mM NaCl conditions, but were sharply enhanced by 12 and 20 fold under 200 and 400 mM NaCl than in control, respectively (**Figure 7B**). These results indicate that salinity (especially at a high concentration) can induce more accumulation of compatible solutes in *A. canescens* seedlings.

A. canescens Seedlings Maintain Higher Leaf Relative Water Content by Effective Osmotic Adjustment Under Salinity Conditions

Maintaining water balance in plants is essential for their survival under saline conditions. Therefore, the leaf RWCs were determined for *A. canescens* seedlings after NaCl treatment. Compared with control, leaf RWC of *A. canescens* seedlings was not reduced by additional NaCl regardless of the concentrations, even showed a significant increase of 11% under 100 mM NaCl treatment (**Figure 8A**). These results are consistent with the growth data (**Figure 1**) and it implies that water status in plant may be one of key factors contributing to the survival and development of *A. canescens* seedlings under saline conditions.

To investigate the mechanism underlying high water retention capacity in *A. canescens* seedlings, the leaf osmotic potential (Ψs) was measured. As shown in **Figure 8B**, the addition of NaCl significantly decreased the leaf Ψs of *A. canescens* seedlings. The leaf Ψs continuously decreased in response to the increase of NaCl concentrations, which suggests that *A. canescens* seedlings could maintain a higher OA capacity in response to salinity. Finally, the contributions of different solutes in leaves to Ψs were further evaluated. With the increase of external NaCl concentrations, the contribution of Na^+ to Ψs significantly increased from 2% in control plants to 32, 35, and 49% in plants treated with 100, 200, and 400 mM NaCl, respectively. However, the contribution of K^+ significantly dropped from 34% in control plants to 9% in plants under 400 mM NaCl (**Table 1**). The contributions of both betaine and free proline showed significant increases under high concentration of NaCl treatments (200 and 400 mM) and accounted for 8 and 4% of contributions to Ψs under 400 mM NaCl, respectively (**Table 1**).

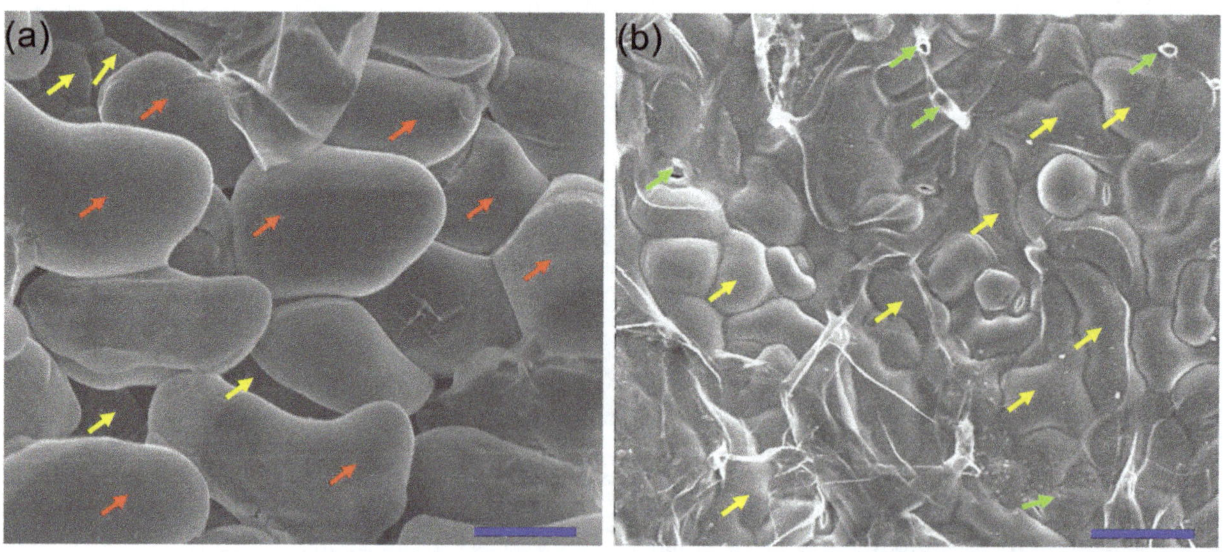

FIGURE 4 | Scanning electron microscopic observation of leaf surface. Before **(a)** and after **(b)** brushing the salt bladders, the abaxial surfaces of leaves were observed and photographed using scanning electron microscope and the images were taken subsequently. The accelerating voltage was 15 kV. Red arrows, salt bladders; yellow arrows, epidermal cells; green arrows, stalk cell. Bar = 0.1 mm.

DISCUSSION

It was proposed that halophytes such as *Suaeda* spp. grow better at moderate concentrations of NaCl, which is generally harmful to the growth of glycophyte species (Flowers, 2004; Shabala and Mackay, 2011). In the present work, the growth of *A. canescens* seedlings was stimulated by an external 100 mM NaCl. Plant height and biomass were significantly increased under 100 mM NaCl but were unaffected by external 200 or 400 mM NaCl treatments (**Figure 1**). Similar results were reported for other *Atriplex* species, such as *A. halimus* (Bajji et al., 1998; Martínez et al., 2004; Ben Hassine and Lutts, 2010; Nemat-Alla et al., 2011; Bouchenak et al., 2012), *A. gmelini* (Matoh et al., 1987; Tsutsumi et al., 2015), and *A. portulacoides* (Redondo-Gómez et al., 2007). Therefore, *A. canescens* is a typical halophytic species and highly tolerant to salinity.

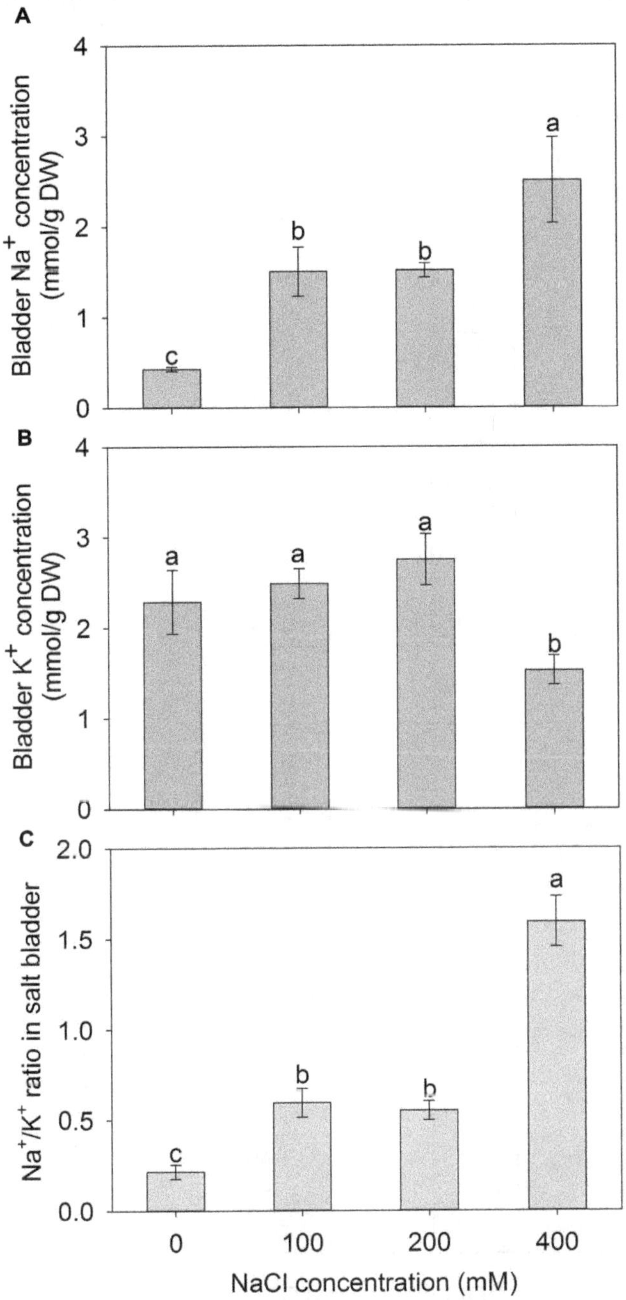

FIGURE 5 | The Na^+ (A) and K^+ (B) concentrations, and Na^+/K^+ ratio (C) in salt bladders of *A. canescens* seedlings under different NaCl treatments for 10 days. Values are mean ± SE ($n = 10$). Columns with different letters indicate significant differences at $P < 0.05$ (Duncan test).

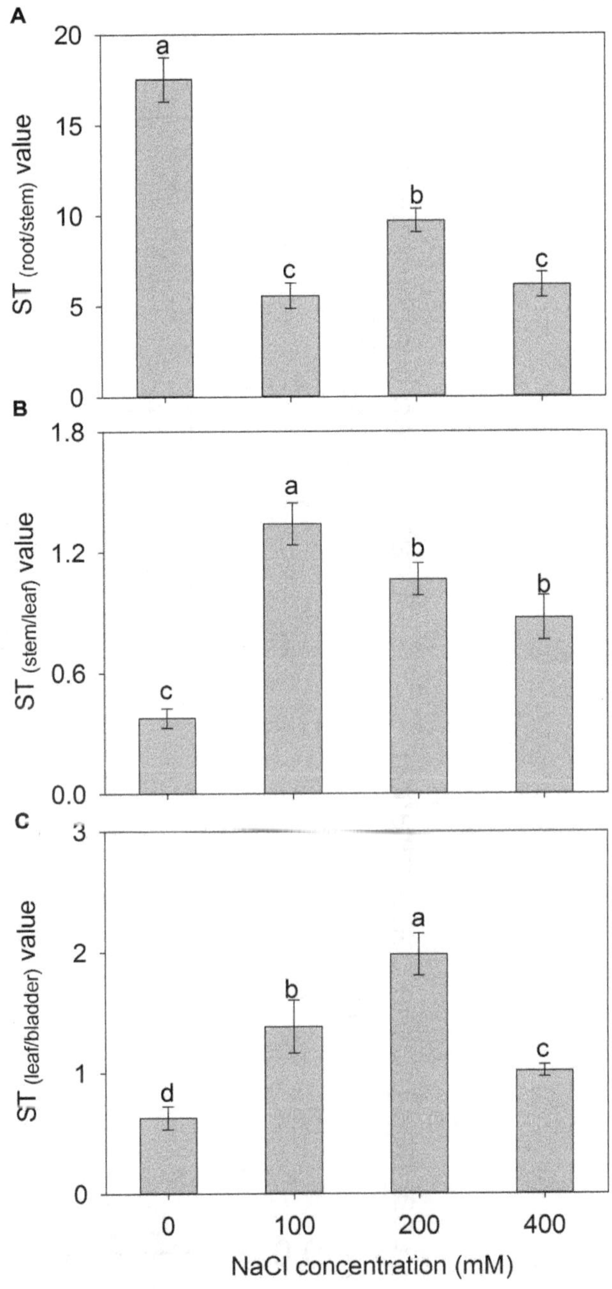

FIGURE 6 | The selective transport capacity for K^+ over Na^+ (ST value) from root to stem (A), from stem to leaf (B) and from leaf to salt bladder (C) in *A. canescens* seedlings under different NaCl treatments for 10 days. $ST_{(A/B)}$ value = (Na^+/K^+ in part A)/(Na^+/K^+ in part B). Values are mean ± SE ($n = 10$). Columns with different letters indicate significant differences at $P < 0.05$ (Duncan test).

The growth of higher plants depends directly on the photosynthetic capacity. In the present study, the net photosynthetic rate (Pn) of *A. canescens* was significantly increased by NaCl treatments (**Figure 2A**). However, stomatal conductance (Gs, **Figure 2B**) and transpiration rate (Tr, **Figure 2C**) were unaffected by 200 and 400 mM NaCl, which is different from the findings in a C_3 xero-halophyte *Zygophyllum xanthoxylum* that showed a positive correlations between Gs and Pn under salinity (Ma et al., 2012). This might be due to the C_4 properties of *A. canescens*. It was thought that Na^+ facilitates some biochemical processes in C_4 pathway photosynthesis such as the conversion of pyruvate into phosphoenolpyruvate (PEP) and the activity of photosystem II (PS II) in mesophyll chloroplasts (Chaves et al., 2011; Kronzucker et al., 2013). Therefore, the Na^+ might promote the C_4 photosynthetic process of *A. canescens* seedlings and thus improve the WUE under high salinity (**Figure 2D**).

Maintaining constant intracellular ion homeostasis, especially K^+ and Na^+ homeostasis, is essential for a series of physiological

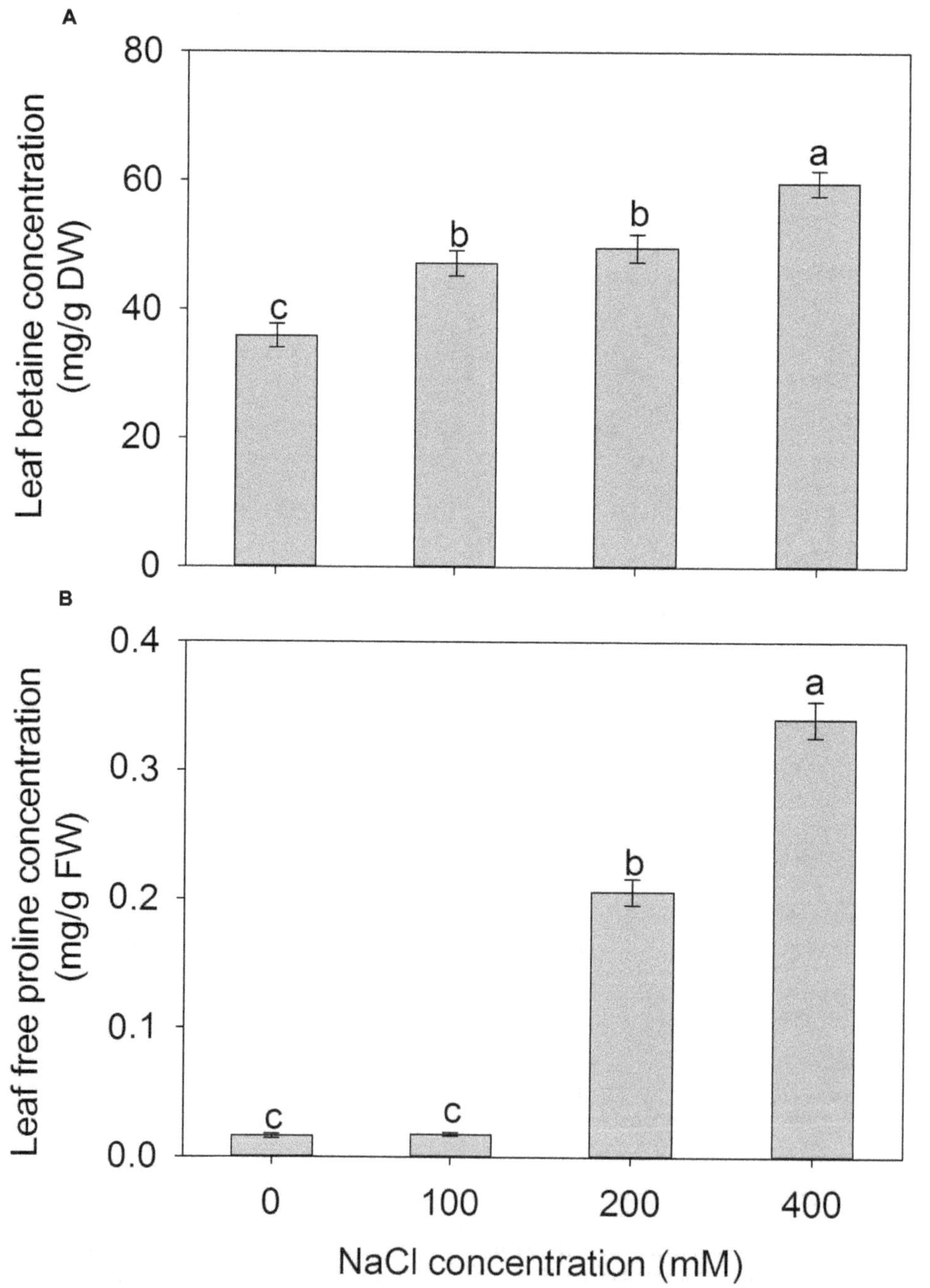

FIGURE 7 | The betaine (A) and free proline (B) concentrations in leaves of *A. canescens* seedlings under different NaCl treatments for 10 days. Values are mean ± SE ($n = 8$). Columns with different letters indicate significant differences at $P < 0.05$ (Duncan test).

processes in living cells, and is more crucial for plant adapting to saline environments (Zhu, 2003; Tang et al., 2015). Glenn et al. (1994, 1996) concluded that the tolerance of *A. canescens* to salinity was due to the accumulation of large amounts Na^+ in plants. This situation also was found in many species of genus *Atriplex* (Matoh et al., 1987; Bajji et al., 1998; Bose et al., 2015) and other succulent halophytes such as *Suaeda maritima* (Wang et al., 2007; Zhang et al., 2013) and *Z. xanthoxylum* (Wang et al., 2004; Ma et al., 2012, 2016; Yue et al., 2012). However, it has been demonstrated that excessive Na^+ in the cytosol is deleterious to cell through inhibiting functional enzymes, disrupting acquisition of K^+, inhibiting K^+-depending metabolic processes, and causing secondary stresses such as oxidative stress, regardless of species (Maathuis and Amtmann, 1999; Zhu, 2001; Flowers et al., 2015; Volkov,

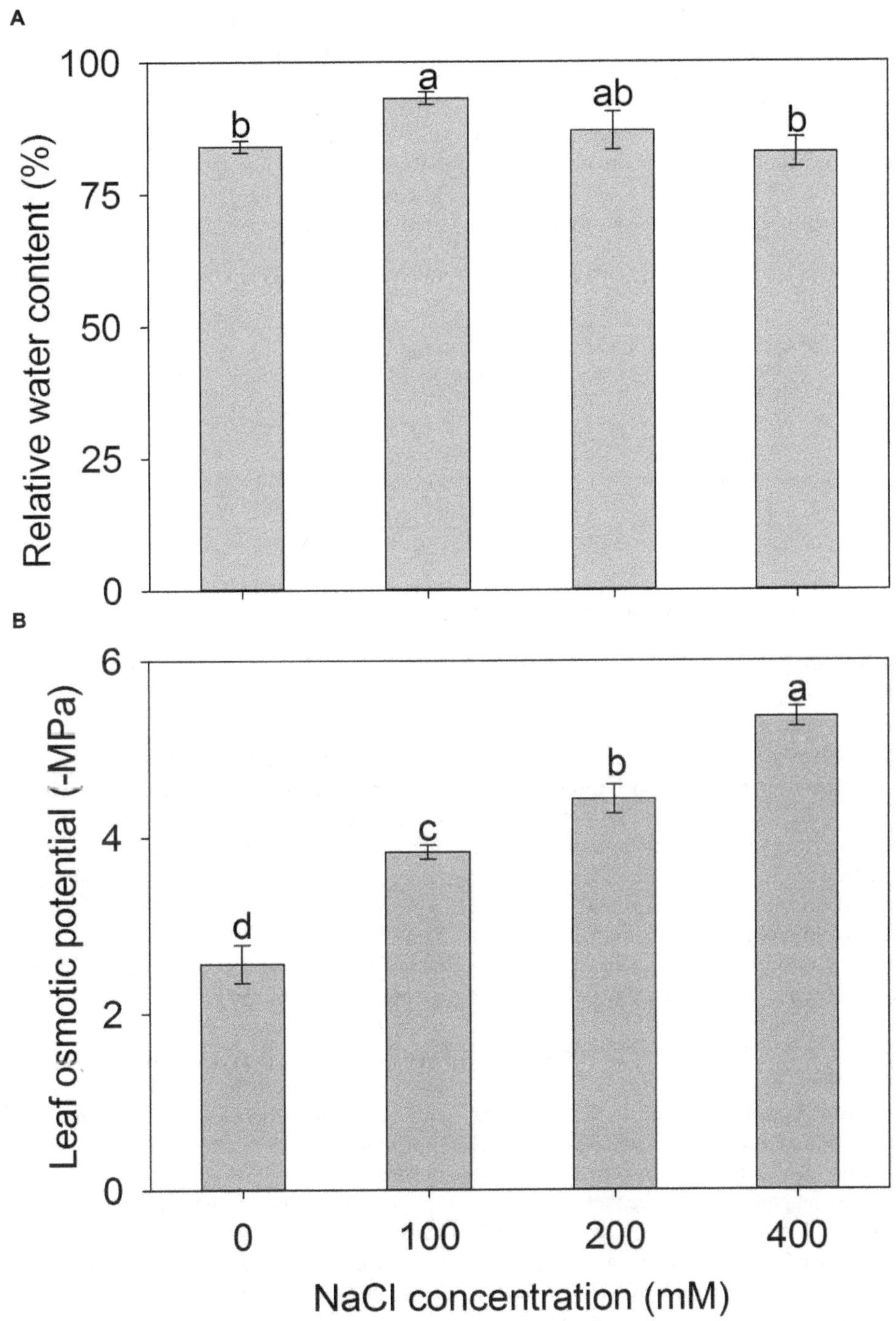

FIGURE 8 | Relative water content (A) and osmotic potential (B) in leaves of *A. canescens* seedlings under different NaCl treatments for 10 days. Values are mean ± SE ($n = 8$). Columns with different letters indicate significant differences at $P < 0.05$ (Duncan test).

2015). To reduce cytosolic Na^+ concentration, some halophytes developed a mechanism of ion compartmentation by sequestering excessive cytosolic Na^+ into the central vacuole, which alleviates the Na^+ toxicity, thus maintains ion homeostasis and OA of cell in saline conditions (Zhu, 2003; Yamaguchi et al., 2013; Flowers et al., 2015). In this study, the *A. canescens* seedlings showed less injury (**Figure 1**) although the accumulation of Na^+ in all tissues of *A. canescens* seedlings showed significant increase under external NaCl treatments (**Figures 3A–C**), suggesting that Na^+ might be sequestered into the vacuole by the strong capacity of ion compartmentation. This mechanism contributes to maintain a high cytosolic K^+/Na^+ ratio, which is one of the most important features that correlated with the salt tolerance of plants, since Na^+ shares similar physicochemical properties and competes with K^+ for the binding sites on enzymes in the cytoplasm and other key metabolic processes (Flowers and Colmer, 2008; Shabala and Cuin, 2008). On the other hand, the accumulation of K^+ was reduced in all tissues of *A. canescens* seedlings by external NaCl (**Figures 3D–F**). Similar results were also observed in many succulent halophytes (Ben Hassine et al., 2009; Nemat-Alla et al., 2011; Shabala and Mackay, 2011; Yue et al., 2012; Bose et al., 2015), and is due to the competition of Na^+ with K^+ for uptake into roots (Shabala and Mackay, 2011; Flowers et al., 2015). Interestingly, *A. canescens* seedlings maintained a relatively constant K^+ concentration in shoots, especially in leaves under saline conditions (**Figures 3D,E**), suggesting that *A. canescens* seedlings might struggle to retain more K^+ in shoot, especially in leaves, as a result, to maintain a relatively constant cytosolic K^+/Na^+ ratio in response to salinity. This conclusion was further supported by the fact that transport capacity for K^+ over Na^+ (ST value) from stem to leaf in *A. canescens* seedlings was significantly enhanced by external NaCl treatments (**Figure 5B**), since the higher $ST_{(A/B)}$ value implies the stronger capacity to selectively transport K^+ over Na^+ from tissue A to tissue B (Wang et al., 2002; Flowers and Colmer, 2008).

TABLE 1 | The contributions of Na^+, K^+, betaine, and free proline to leaf osmotic potential (\Ils) of Atriplex canescens seedlings under different NaCl treatments for 10 days.

NaCl treatments (mM)	Contribution of Na^+ to \Ils (%)	Contribution of K^+ to \Ils (%)	Contribution of betaine to \Ils (%)	Contribution of free proline to \Ils (%)
0	2.0 ± 0.3c	33.8 ± 5.2a	6.0 ± 0.3b	0.3 ± 0.1c
100	32.4 ± 3b	26.0 ± 3.2ab	6.4 ± 0.3b	0.3 ± 0.1c
200	34.7 ± 5.5b	19.9 ± 2.4b	7.3 ± 0.4a	2.8 ± 0.2b
400	48.8 ± 4.2a	9.2 ± 1.1c	8.1 ± 0.3a	3.8 ± 0.2a

Values are mean ± SE (n = 8). Columns with different letters indicate significant differences at P < 0.05 (Duncan test).

Almost all *Atriplex* species are regarded as salt-excreting plants since these species can sequester large quantities of absorbed Na^+ into epidermal bladder cells (EBCs) on their leaf surfaces and then release Na^+ from ruptured EBCs (Flowers and Colmer, 2008; Ding et al., 2010; Shabala, 2013; Shabala et al., 2014). In this study, we found that Na^+ sequestration in EBCs of *A. canescens* was significantly induced by external NaCl and showed a positive correlationship with the NaCl concentration (**Figures 4** and **5A**). This finding is consistent with the previous studies from other *Atriplex* spp. (Jeschke and Stelter, 1983; Ben Hassine et al., 2009; Tsutsumi et al., 2015), and it was proposed that each EBC could sequester about 1000 fold more Na^+ compared with leaf cell vacuoles because of its larger volume (Shabala et al., 2014). Similar with the process in 'traditional' mesophyll cells, indeed, large quantities of Na^+ in EBCs are transported into the huge central vacuoles, which will result in cytosolic K^+ and organic osmolytes accumulating for OA in EBCs (Shabala et al., 2014; Tsutsumi et al., 2015). This viewpoint was supported by our data. Under 100 or 200 mM external NaCl, the K^+ concentration in EBCs of *A. canescens* seedlings showed no change in comparison with control plants (**Figure 5B**). This may be partly due to the fact that *A. canescens* seedlings maintain high ST capacity for K^+ over Na^+ (ST value) from root to stem (**Figure 6A**), as well as increased ST value from leaf to salt bladder under salinity (**Figure 6C**). These results also suggest that there is less selectivity on Na^+ and K^+ in salt exclusion via EBC, which is different from the situation in most of other salt-excreting plants with multicellular salt gland such as *Limonium bicolor* (Ding et al., 2010; Feng et al., 2014), *Tamarix ramosissima* (Ma et al., 2011) and *Reaumuria soongarica* (Zhou et al., 2012), which have a high selectivity for the secretion of Na^+. This implies that there are the different salt-excreting mechanisms between salt bladder and salt gland.

Stable water status is essential for plants to survive from saline conditions. In this study, we found that the RWC in the leaf of *A. canescens* seedlings was increased by 100 mM NaCl, and was unaffected by 200 and 400 mM NaCl (**Figure 8A**), suggesting this species has a high water retention capacity that may result in better growth of plants under salinity. Similar phenotypes were also reported for other *Atriplex* species (Redondo-Gómez et al., 2007; Ben Hassine et al., 2009), and could be explained by decreased osmotic potential of cells (Flowers and Colmer, 2008; Munns and Tester, 2008; Yamaguchi et al., 2013). This viewpoint is supported by measurement of osmotic potential in leaves of *A. canescens* seedlings, which decreased significantly with the increase of external NaCl concentration (**Figure 8B**). Lower osmotic potential results in a higher OA capacity, which facilitates water uptake and thus may maintain the turgor in plants at low water potential conditions.

It is well-known that higher OA in plants subjected to salt stress mainly results from the accumulation of either inorganic ions or compatible solutes (or both; Flowers and Colmer, 2008; Kronzucker et al., 2013). In many cases, however, the importance of each solute to OA is controversial (Glenn et al., 1994, 1996; Bajji et al., 1998; Martínez et al., 2004, 2005; Ben Hassine et al., 2008). Therefore, the contributions of various solutes to OA were investigated under normal (no addition of external NaCl) and saline conditions (100–400 mM NaCl). In control plants, K^+ accounted for 34% of the leaf osmotic potential that was more than 16 fold higher than Na^+. The contribution of K^+ was significantly reduced by NaCl treatments while the contribution of Na^+ to leaf osmotic potential increased sharply to 49% under 400 mM NaCl (**Table 1**). These results suggest that Na^+ in mesophyll cell and EBCs of *A. canescens* seedlings can be used as an osmolyte contributing to OA in order to cope with osmotic stress under high salinity. In saline soil, *A. canescens* was able to absorb large quantities of Na^+ from soil and to accumulate in aboveground tissues (Glenn et al., 1994). Ma et al. (2012) found that *Z. xanthoxylum* can use Na^+ as an osmoregulatory substance by sequestering Na^+ in vacuoles of large cells mediated by the tonoplast Na^+/H^+ antiporter. In addition to inorganic ions, it was proposed that some compatible solutes, including betaine and free proline, may act as cytoplasmic osmoprotectant involved in OA and/or protection of cellular structures in plants under various abiotic stress conditions (Munns and Tester, 2008; Shabala and Mackay, 2011; Tsutsumi et al., 2015). In the present study, we found that the leaf betaine accumulation of *A. canescens* seedlings positively correlated with the concentration of external NaCl (**Figure 7A**) and its contribution to the leaf osmotic potential increased to 8% under 400 mM NaCl, which is close to the contribution of K^+ (**Table 1**). These results suggest that the betaine performed OA in *A. canescens* seedlings under higher salinity. On the other hand, it was reported that the betaine plays other roles in *Atriplex* genus species in response to salinity. For example, Tsutsumi et al. (2015) reported that high salinity induced the accumulation of betaine in the cytosol of the salt bladders of *A. gmelini*, which contributed to maintain membrane integrity and the enzyme activity and, as a result, ensured the bladder cells to load Na^+ into vacuole. In *A. halimus*, the accumulation of betaine helps to protect the photosynthetic machinery from salinity (Ben Hassine et al., 2008). Moreover, previous studies proposed that free proline is involved in the response to drought stress rather than to salinity in *A. halimus* (Ben Hassine et al., 2008, 2009). In *A. canescens*, however, the accumulation of leaf free proline was strongly induced by high salinity (200 and 400 mM NaCl) though it was unaffected by moderate concentrations (100 mM) of NaCl (**Figure 7B**), suggesting that the free proline may also be involved in physiological response of *A. canescens* to high salinity.

CONCLUSION

Our results demonstrate that the growth of *A. canescens* can be stimulated by moderate salinity (100 mM NaCl) and was not inhibited by higher salinity (200 and 400 mM NaCl). This adaptation is achieved through the following aspects: (i) to enhance the photosynthetic capacity by improving Pn and WUE. (ii) to increase Na^+ accumulation in tissues and salt bladders, as well as improve transport capacity for K^+ over Na^+ (ST value) from stem to leaf, which may maintain intracellular K^+ homeostasis. (iii) to maintain OA capacity and improve the water status in plant by accumulation of inorganic ions and compatible solutes.

AUTHOR CONTRIBUTIONS

Y-QP, S-MW, and A-KB conceived the study and designed the experiments; Y-QP and HG performed most of the work; BZ, J-LZ, H-JY, and QM provided technical assistance to experiments and data analysis, as well as, made revisions on the article. S-MW gave valuable suggestions on the article. Y-QP and A-KB wrote the article.

ACKNOWLEDGMENTS

We are very grateful to Professor Elizabeth A. Grabau from Virginia Tech, USA, for critically reviewing the manuscript and for valuable suggestions. This work was supported by the National Basic Research Program of China (2014CB138701), the National Natural Science Foundation of China (31372360, 31222053), the Key Technology R & D Program of Gansu Province (144FKCA058) and the Fundamental Research Funds for the Central Universities (lzujbky-2016-4, lzujbky-2015-250).

REFERENCES

Bajji, M., Kinet, J. M., and Lutts, S. (1998). Salt stress effects on roots and leaves of *Atriplex halimus* L. and their corresponding callus cultures. *Plant Sci.* 137, 131–142. doi: 10.1016/S0168-9452(98)00116-2

Bao, A. K., Du, B. Q., Touil, L., Kang, P., Wang, Q. L., and Wang, S. M. (2016). Co-expression of tonoplast Cation/H+ antiporter (NHX) and H+-pyrophosphatase (H+-PPase) from xerophyte *Zygophyllum xanthoxylum* improves alfalfa plant growth under salinity, drought, and field conditions. *Plant Biotechnol. J.* 14, 964–975. doi: 10.1111/pbi.12451

Bao, A. K., Wang, Y. W., Xi, J. J., Liu, C., Zhang, J. L., and Wang, S. M. (2014). Co-expression of xerophyte *Zygophyllum xanthoxylum* ZxNHX and ZxVP1-1 enhances salt and drought tolerance in transgenic *Lotus corniculatus* L. by increasing cations accumulation. *Funct. Plant Biol.* 41, 203–214. doi: 10.1071/FP13106

Bazihizina, N., Colmer, T. D., and Barrett-Lennard, E. G. (2009). Response to non- uniform salinity in the root zone of the halophyte *Atriplex nummularia*: growth, photosynthesis, water relations and tissue ion concentrations. *Ann. Bot.* 104, 737–745. doi: 10.1093/aob/mcp151

Belkheiri, O., and Mulas, M. (2011). The effects of salt stress on growth, water relations and ion accumulation in two halophyte *Atriplex* species. *Environ. Exp. Bot.* 86, 17–28. doi: 10.1016/j.envexpbot.2011.07.001

Ben Hassine, A., Ghanem, M. E., Bouzid, S., and Lutts, S. (2008). An inland and a coastal population of the Mediterranean xero-halophyte species *Atriplex halimus* L. differ in their ability to accumulate proline and glycinebetaine in response to salinity and water stress. *J. Exp. Bot.* 59, 1315–1326. doi: 10.1093/jxb/ern040

Ben Hassine, A., Ghanem, M. E., Bouzid, S., and Lutts, S. (2009). Abscisic acid has contrasting effects on salt excretion and polyamine concentrations of an inland and a coastal population of the Mediterranean xero-halophyte species *Atriplex halimus*. *Ann. Bot.* 104, 925–936. doi: 10.1093/aob/mcp174

Ben Hassine, A., and Lutts, S. (2010). Differential responses of saltbush *Atriplex halimus* L. exposed to salinity and water stress in relation to senescing hormones abscisic acid and ethylene. *J. Plant Physiol.* 167, 1448–1456. doi: 10.1016/j.jplph.2010.05.017

Benzarti, M., Ben Rejeb, K., Debez, A., and Abdelly, C. (2013). "Environmental and economical opportunities for the valorisation of the genus *Atriplex*: new insights," in *Crop Improvement*, eds K. R. Hakeem, P. Ahmad, and M. Ozturk (New York, NY: Springer), 441–457.

Bose, J., Rodrigo-Moreno, A., Lai, D., Xie, Y., Shen, W., and Shabala, S. (2015). Rapid regulation of the plasma membrane H+-ATPase activity is essential to salinity tolerance in two halophyte species, *Atriplex lentiformis* and *Chenopodium quinoa*. *Ann. Bot.* 115, 481–494. doi: 10.1093/aob/ mcu219

Bouchenak, F., Henri, P., Benrebiha, F. Z., and Rey, P. (2012). Differential responses to salinity of two *Atriplex halimus* populations in relation to organic solutes and antioxidant systems involving thiol reductases. *J. Plant Physiol.* 169, 1445–1453. doi: 10.1016/j.jplph.2012.06.009

Chaves, M. M., Costa, J. M., and Saibo, N. J. M. (2011). Recent advances in photosynthesis under drought and salinity. *Adv. Bot. Res.* 57, 49–104. doi: 10.1016/B978-0-12-387692-8.00003-5

Ding, F., Yang, J. C., Yuan, F., and Wang, B. S. (2010). Progress in mechanism of salt excretion in recretohalopytes. *Front. Biol.* 5:164–170. doi: 10.1007/s11515-010-0032-7

Feng, Z. T., Sun, Q., Deng, Y. Q., Sun, S. F., Zhang, J. G., and Wang, B. S. (2014). Study on pathway and characteristics of ion secretion of salt glands of *Limonium bicolor*. *Acta Physiol. Plant.* 36, 2729–2741. doi: 10.1007/s11738-014-1644-3

Flowers, T. J. (2004). Improving crop salt tolerance. *J. Exp. Bot.* 55, 307–319. doi: 10.1093/jxb/erh003

Flowers, T. J., and Colmer, T. D. (2008). Salinity tolerance in halophytes. *New Phytol.* 179, 945–963. doi: 10.1111/j.1469-8137.2008.02531.x

Flowers, T. J., Galal, H. K., and Bromham, L. (2010). Evolution of halophytes: multiple origins of salt tolerance in land plants. *Funct. Plant Biol.* 37, 604–612. doi: 10.1007/s10142-011-0218-3

Flowers, T. J., Munns, R., and Colmer, T. D. (2015). Sodium chloride toxicity and the cellular basis of salt tolerance in halophytes. *Ann. Bot.* 115, 419–431. doi: 10.1093/aob/mcu217

Glenn, E. P., and Brown, J. J. (1998). Effects of soil salt levels on the growth and water use efficiency of *Atriplex canescens* (Chenopodiaceae) varieties in drying soil. *Am. J. Bot.* 85, 10–16. doi: 10.2307/2446548

Glenn, E. P., Olsen, M., Frye, R., Moore, D., and Miyamoto, S. (1994). How much sodium accumulation is necessary for salt tolerance in subspecies of the halophyte *Atriplex canescens*? *Plant Cell Environ.* 17, 711–719. doi: 10.1111/j.1365-3040.1994.tb00163.x

Glenn, E. P., Pfister, R., Brown, J. J., Thompson, T. L., and O'leary, J. (1996). Na and K accumulation and salt tolerance of *Atriplex canescens* (Chenopodiaceae) genotypes. *Am. J. Bot.* 83, 997–1005. doi: 10.2307/2445988

Guerrier, G. (1996). Fluxes of Na^+, K^+ and Cl–, and osmotic adjustment in *Lycopersicon pimpinellifolium* and *L. esculentum* during short- and long- term exposures to NaCl. *Physiol. Plant.* 97, 583–591. doi: 10.1111/j.1399- 3054.1996.tb00519.x

Guo, Q., Wang, P., Ma, Q., Zhang, J. L., Bao, A. K., and Wang, S. M. (2012). Selective transport capacity for K^+ over Na^+ is linked to the expression levels of PtSOS1 in halophyte *Puccinellia tenuiflora*. *Funct. Plant Biol.* 39, 1047–1057. doi: 10.1071/FP12174

Hao, G. Y., Lucero, M. E., Sanderson, S. C., Zacharias, E. H., and Holbrook, N. M. (2013). Polyploidy enhances the occupation of heterogeneous environments through hydraulic related trade-offs in *Atriplex canescens* (Chenopodiaceae). *New Phytol.* 197, 970–978. doi: 10.1111/nph.12051

Hasegawa, P. M. (2013). Sodium (Na^+) homeostasis and salt tolerance of plants. *Environ. Exp. Bot.* 92, 19–31. doi: 10.1016/j.envexpbot.2013.03.001

Hasegawa, P. M., Bressan, R. A., Zhu, J. K., and Bohnert, H. J. (2000). Plant cellular and molecular responses to high salinity. *Annu. Rev. Plant Physiol. Plant Mol. Biol.* 51, 463–499. doi: 10.1146/annurev.arplant.51.1.463

Jeschke, W. D., and Stelter, W. (1983). Ion relations of garden orache, *Atriplex hortensis* L.: growthand ion distribution at moderate salinity and the function of bladder hairs. *J. Exp. Bot.* 34, 795–810. doi: 10.1093/jxb/ 34.7.795

Kalaji, H. M., Jajoo, A., Oukarroum, A., Brestic, M., Zivcak, M., Samborska, I. A., et al. (2016). Chlorophyll a fluorescence as a tool to monitor physiological status of plants under abiotic stress conditions. *Acta Physiol. Plant.* 38, 102. doi: 10.1007/s11738-016-2113-y

Kong, D. S. (2013). Morphological characteristics and eco-physiological adaptability of *Atriplex canescens*: a review. *Chin. J. Ecol.* 32, 210–216.

Kronzucker, H. J., Coskun, D., Schulze, L. M., Wong, J. R., and Britto, D. T. (2013). Sodium as nutrient and toxicant. *Plant Soil* 369, 1–23. doi: 10.1007/s11104-013- 1801-2

Ma, H., Tian, C., Feng, G., and Yuan, J. (2011). Ability of multicellular salt glands in *Tamarix* species to secrete Na^+ and K^+ selectively. *Sci. China Life Sci.* 54, 282–289. doi: 10.1007/s11427-011-4145-2

Ma, Q., Bao, A. K., Chai, W. W., Wang, W. Y., Zhang, J. L., Li, Y. X., et al. (2016). Transcriptomic analysis of the succulent xerophyte *Zygophyllum xanthoxylum* in response to salt treatment and osmotic stress. *Plant Soil* 402, 343–361. doi: 10.1007/s11104-016-2809-1

Ma, Q., Li, Y. X., Yuan, H. J., Hu, J., Wei, L., Bao, A. K., et al. (2014). ZxSOS1 is essential for long-distance transport and spatial distribution of Na^+ and K^+ in the xerophyte *Zygophyllum xanthoxylum*. *Plant Soil* 374, 661–676.

Ma, Q., Yue, L. J., Zhang, J. L., Wu, G. Q., Bao, A. K., and Wang, S. M. (2012). Sodium chloride improves photosynthesis and water status in the succulent xerophyte *Zygophyllum xanthoxylum*. *Tree Physiol.* 32, 4–13. doi: 10.1093/treephys/tpr098

Maathuis, F. J. M., and Amtmann, A. (1999). K^+ nutrition and Na^+ toxicity: the basis of cellular K^+/Na^+ ratios. *Ann. Bot.* 84, 123–133.

Martínez, J. P., Kinet, J. M., Bajji, M., and Lutts, S. (2005). NaCl alleviates polyethylene glycol-induced water stress in the halophyte species *Atriplex halimus* L. *J. Exp. Bot.* 56, 2421–2431. doi: 10.1093/jxb/eri235

Martínez, J. P., Lutts, S., Schanck, A., Bajji, M., and Kinet, J. M. (2004). Is osmotic adjustment required for water-stress resistance in the Mediterranean shrub *Atriplex halimus* L.? *J. Plant Physiol.* 161, 1041–1051. doi: 10.1016/j.jplph.2003.12.009

Matoh, T., Watanabe, J., and Takahashi, E. (1987). Sodium, potassium, chloride, and betaine concentrations in isolated vacuoles from salt-grown *Atriplex gmelini* leaves. *Plant Physiol.* 84, 173–177. doi: 10.1104/pp.84.1.173

Munns, R., and Tester, M. (2008). Mechanisms of salinity tolerance. *Annu. Rev. Plant Biol.* 59, 651–681. doi: 10.1146/annurev.arplant.59.032607.092911

Nemat-Alla, M. M., Khedr, A. A., Serag, M. M., Abu-Alnaga, A. Z., and Nada, R. M. (2011). Physiological aspects of tolerance in *Atriplex halimus* L. to NaCl and drought. *Acta Physiol. Plant.* 33, 547–557. doi: 10.1007/s11738-010- 0578-7

Niu, S. Q., Li, H. R., Guo, S. Q., Li, J., Guo, Q., Ma, Q., et al. (2016). Induced growth promotion and higher salt tolerance in the halophyte grass *Puccinellia tenuiflora* by beneficial rhizobacteria. *Plant Soil* doi: 10.1007/s11104-015-2767-z

Peterson, J. L., Ueckert, D. N., Potter, R. L., and Huston, J. E. (1987). Ecotypic variation in selected fourwing saltbush populations in western Texas. *J. Range Manage. Arch.* 40, 361–366. doi: 10.2307/3898738

Redondo-Gómez, S., Mateos-Naranjo, E., Davy, A. J., Fernández-Muñoz, F., Castellanos, E. M., Luque, T., et al. (2007). Growth and photosynthetic responses to salinity of the salt-marsh shrub *Atriplex portulacoides*. *Ann. Bot.* 100, 555–563. doi: 10.1093/aob/mcm119

Ruan, C. J., da Silva, J. A. T., Mopper, S., Qin, P., and Lutts, S. (2010). Halophyte improvement for a salinized world. *Crit. Rev. Plant Sci.* 29, 329–359. doi: 10.1080/07352689.2010.524517

Shabala, S. (2013). Learning from halophytes: physiological basis and strategies to improve abiotic stress tolerance in crops. *Ann. Bot.* 112, 1209–1221. doi: 10.1093/aob/mct205

Shabala, S., Bose, J., and Hedrich, R. (2014). Salt bladders: do they matter? *Trends Plant Sci.* 19, 687–691. doi: 10.1016/j.tplants.2014.09.001

Shabala, S., and Cuin, T. A. (2008). Potassium transport and plant salt tolerance. *Physiol. Plant.* 133, 651–669. doi: 10.1111/j.1399-3054.2007.01008.x

Shabala, S., and Mackay, A. (2011). Ion transport in halophytes. *Adv. Bot. Res.* 57, 151–199. doi: 10.1016/B978-0-12-387692-8.00005-9

Tang, X. L., Mu, X. M., Shao, H. B., Wang, H. Y., and Brestic, M. (2015). Global plant-responding mechanisms to salt stress: physiological and molecular levels and implications in biotechnology. *Crit. Rev. Biotechnol.* 35, 425–437. doi: 10.3109/07388551.2014.889080

Tester, M., and Davenport, R. (2003). Na^+ tolerance and Na^+ transport in higher plants. *Ann. Bot.* 91, 503–527. doi: 10.1093/aob/mcg058

Tsutsumi, K., Yamada, N., Cha-um, S., Tanaka, Y., and Takabe, T. (2015). Differential accumulation of glycinebetaine and choline monooxygenase in bladder hairs and lamina leaves of *Atriplex gmelini* under high salinity. *J. Plant Physiol.* 176, 101–107. doi: 10.1016/j.jplph.2014.12.009

Volkov, V. (2015). Salinity tolerance in plants. Quantitative approach to ion transport starting from halophytes and stepping to genetic and protein engineering for manipulating ion fluxes. *Front. Plant Sci.* 6:873. doi: 10.3389/fpls.2015.00873

Volkov, V., and Amtmann, A. (2006). *Thellungiella halophila*, a salt-tolerant relative of *Arabidopsis thaliana*, has specific root ion-channel features supporting K^+/Na^+ homeostasis under salinity stress. *Plant J.* 48, 342–353. doi: 10.1111/j.1365-313X.2006.02876.x

Wang, B., Davenport, R. J., Volkov, V., and Amtmann, A. (2006). Low unidirectional sodium influx into root cells restricts net sodium accumulation in *Thellungiella halophila*, a salt-tolerant relative of *Arabidopsis thaliana*. *J. Exp. Bot.* 57, 1161–1170. doi: 10.1093/jxb/erj116

Wang, C. M., Zhang, J. L., Liu, X. S., Li, Z., Wu, G. Q., Cai, J. Y., et al. (2009). *Puccinellia tenuiflora* maintains a low Na^+ level under salinity by limiting unidirectional Na^+ influx resulting in a high selectivity for K^+ over Na^+. *Plant Cell Environ.* 32, 486–496. doi: 10.1111/j.1365-3040.2009.01942.x

Wang, P., Guo, Q., Wang, Q., Zhou, X. R., and Wang, S. M. (2015). PtAKT1 maintains selective absorption capacity for K^+ over Na^+ in halophyte *Puccinellia tenuiflora* under salt stress. *Acta Physiol. Plant.* 37, 100. doi: 10.1007/s11738-015-1846-3

Wang, S. M., Wan, C. G., Wang, Y. R., Chen, H., Zhou, Z. Y., Fu, H., et al. (2004). The characteristics of Na^+, K^+ and free proline distribution in several drought-resistant plants of the Alxa Desert, China. *J. Arid Environ.* 56, 525–539. doi: 10.1016/S0140-1963(03)00063-6

Wang, S. M., Zhang, J. L., and Flowers, T. J. (2007). Low affinity Na^+ uptake in the halophyte *Suaeda maritima*. *Plant Physiol.* 145, 559–571. doi: 10.1104/pp.107.104315

Wang, S. M., Zheng, W. J., Ren, J. Z., and Zhang, C. L. (2002). Selectivity of various types of salt-resistant plants for K^+ over Na^+. *J. Arid Environ.* 52, 457–472. doi: 10.1006/jare.2002.1015

Yamaguchi, H., Hamamoto, S., and Uozumi, N. (2013). Sodium transport system in plant cells. *Front. Plant Sci.* 4:410. doi: 10.3389/fpls.2013.00410

Yamaguchi, T., and Blumwald, E. (2005). Developing salt-tolerant crop plants: challenges and opportunities. *Trends Plant Sci.* 10, 615–620. doi: 10.1016/j.tplants.2005.10.002

Yan, K., Shao, H., Shao, C., Chen, P., Zhao, S., Brestic, M., et al. (2013). Physiological adaptive mechanisms of plants grown in saline soiland implications for sustainable saline agriculture in coastal zone. *Acta Physiol. Plant.* 35, 2867–2878. doi: 10.1007/s11738-013-1325-7

Yue, L. J., Li, S. X., Ma, Q., Zhou, X. R., Wu, G. Q., Bao, A. K., et al. (2012). NaCl stimulates growth and alleviates water stress in the xerophyte *Zygophyllum xanthoxylum*. *J. Arid Environ.* 87, 153–160. doi: 10.1016/j.jaridenv.2012. 06.002

Zhang, D., Tong, J., He, X., Xu, Z., Xu, L., Wei, P., et al. (2016). A novel soybean intrinsic protein gene, GmTIP2;3, involved in responding to osmotic stress. *Front. Plant Sci.* 6:1237. doi: 10.3389/fpls.2015.01237

Zhang, J., Nguyen, H., and Blum, A. (1999). Genetic analysis of osmotic adjustment in crop plants. *J. Exp. Bot.* 50, 291–302. doi: 10.1093/jxb/50.332.291

Zhang, J. L., Flowers, T. J., and Wang, S. M. (2010). Mechanisms of sodium uptake by roots of higher plant. *Plant Soil* 326, 45–60. doi: 10.1007/s11104-009-0076-0

Zhang, J. L., and Shi, H. Z. (2013). Physiological and molecular mechanisms of plant salt tolerance. *Photosynth. Res.* 115, 1–22. doi: 10.1007/s11120-013-9813-6

Zhang, J. L., Wang, S. M., and Flowers, T. J. (2013). Differentiation of low-affinity Na^+ uptake pathways and kinetics of the effects of K^+ on Na^+ uptake in the halophyte *Suaeda maritima*. *Plant Soil* 368, 629–640. doi: 10.1007/s11104-012-1552-5

Zhou, H. Y., Bao, A. K., Du, B. Q., and Wang, S. M. (2012). The physiological mechanisms underlying how eremophyte *Reaumuria soongorica* responses to severe NaCl stress. *Pratacult. Sci.* 29, 71–75.

Zhu, J. K. (2001). Plant salt tolerance. *Trends Plant Sci.* 6, 66–71. doi: 10.1016/S1360-1385(00)01838-0

Zhu, J. K. (2003). Regulation of ion homeostasis under salt stress. *Curr. Opin. Plant Biol.* 6, 441–445. doi: 10.1016/S1369-5266(03)00085-2

Zörb, C., Noll, A., Karl, S., Leib, K., Yan, F., and Schubert, S. (2005). Molecular characterization of Na^+/H^+ antiporters (ZmNHX) of maize (*Zea mays* L.) and their expression under salt stress. *J. Plant Physiol.* 162, 55–66. doi: 10.1016/j.jplph.2004.03.010

7

The F130S point mutation in the Arabidopsis high-affinity K^+ transporter AtHAK5 increases K^+ over Na^+ and Cs^+ selectivity and confers Na^+ and Cs^+ tolerance to yeast under heterologous expression

Fernando Alemán †**, *Fernando Caballero*, *Reyes Ródenas*, *Rosa M. Rivero*, *Vicente Martínez* and *Francisco Rubio* ***

Centro de Edafología y Biología Aplicada del Segura-CSIC, Murcia, Spain

Edited by:
Vadim Volkov, London Metropolitan University, UK

Reviewed by:
Alonso Rodriguez-Navarro, Universidad Politecnica de Madrid, Spain
Vadim Volkov, London Metropolitan University, UK

****Correspondence:***
Francisco Rubio, Centro de Edafología y Biología Aplicada del Segura-CSIC, Campus de Espinardo, 30100 Murcia, Spain
e-mail: frubio@cebas.csic.es

† Present address:
Fernando Alemán, Division of Biological Sciences, University of California San Diego, La Jolla, USA

Potassium (K^+) is an essential macronutrient required for plant growth, development and high yield production of crops. Members of group I of the KT/HAK/KUP family of transporters, such as HAK5, are key components for K^+ acquisition by plant roots at low external K^+ concentrations. Certain abiotic stress conditions such as salinity or Cs^+-polluted soils may jeopardize plant K^+ nutrition because HAK5-mediated K^+ transport is inhibited by Na^+ and Cs^+. Here, by screening in yeast a randomly-mutated collection of AtHAK5 transporters, a new mutation in AtHAK5 sequence is identified that greatly increases Na^+ tolerance. The single point mutation F130S, affecting an amino acid residue conserved in HAK5 transporters from several species, confers high salt tolerance, as well as Cs^+ tolerance. This mutation increases more than 100-fold the affinity of AtHAK5 for K^+ and reduces the K_i values for Na^+ and Cs^+, suggesting that the F130 residue may contribute to the structure of the pore region involved in K^+ binding. In addition, this mutation increases the V_{max} for K^+. All this changes occur without increasing the amount of the AtHAK5 protein in yeast and support the idea that this residue is contributing to shape the selectivity filter of the AtHAK5 transporter.

Keywords: potassium, sodium, cesium, selectivity, HAK, Arabidopsis, random mutagenesis, point mutation

INTRODUCTION

Potassium (K^+) is an essential nutrient for plants, required for plant growth and development (Marschner, 2012). It fulfills important functions related to enzyme activation, protein synthesis, maintenance of cytoplasmic pH and transmembrane voltage gradients and neutralization of negative charges (Maathuis, 2009). It is the most abundant cationic component, accounting for more than a 10% of the plant dry weight (White and Karley, 2010). As it occurs with other nutrients, roots take up K^+ from the soil solution through specific transport systems at the plasma membrane of epidermal and cortical cells (Marschner, 2012). Physiological and molecular studies in the model plant *Arabidopsis thaliana* have described two systems, AtHAK5 and AKT1, a high-affinity K^+ transporter and an inward rectifying K^+ channel respectively, which are the major contributors to root K^+ uptake (Alemán et al., 2011; Nieves-Cordones et al., 2014). The relative contribution of these two systems to K^+ nutrition depends on the external K^+ concentration and several other factors such as other external ions or root cell membrane potential. At high K^+ concentrations (around 1 mM), within the low-affinity system described by Epstein (Epstein, 1966), AKT1 is the only system mediating K^+ uptake. Below 200 μM K^+, corresponding to the high-affinity range of concentrations, both transport systems can contribute. However, at very low K^+ concentrations (<10 μM) the only system involved in K^+ acquisition is AtHAK5 (Rubio et al., 2008, 2010; Pyo et al., 2010). This model may be extended to other plant species although the relative contribution of each of the two systems over the range of external K^+ concentration may vary among them. For example, in tomato or pepper plants, the AtHAK5 homologs LeHAK5 or CaHAK1 respectively, dominate K^+ uptake over the AKT1 homologs at low external concentrations (Martínez-Cordero et al., 2004, 2005; Nieves-Cordones et al., 2007).

Operation of the HAK5-type of transporters may be essential for K^+ nutrition in K^+ depleted soils. In addition, K^+ uptake under abiotic stress conditions such as salinity or Cs^+-polluted soils may also largely depend on HAK5-type transporters (Hampton et al., 2004; Qi et al., 2008; Alemán et al., 2009). Under salinity, uptake and accumulation of Na^+ in roots cells depolarize their plasma membrane (Volkov and Amtmann, 2006; Chen et al., 2007; Nieves-Cordones et al., 2008) and reduce the driving force for K^+ uptake. As a result, AKT1 function may be importantly impaired but, K^+ uptake through the K^+-H^+ symporter mechanism proposed for the HAK5-type transporters, may be still possible (Rodríguez-Navarro, 2000). Supporting this idea it has been reported that the Arabidopsis AtHAK5 is required for

K^+ uptake and plant growth at low K^+ in the presence of salinity, even though *AtHAK5* expression is reduced under high Na^+ concentrations (Nieves-Cordones et al., 2010). In Cs^+-polluted soils, K^+ uptake through AKT1 may be importantly inhibited because Cs^+ is a potent blocker of AKT1 channels (Bertl et al., 1997; White and Broadley, 2000).

Although HAK5-type transporters may be the major contributors to K^+ uptake under Na^+- or Cs^+-affected soils, they are inhibited by these cations. Both Na^+ and Cs^+ competitively inhibit K^+ uptake through HAK5-type transporters (Santa-María et al., 1997; Rubio et al., 2000). This leads to two important effects. On one hand, K^+ supply is reduced and in fact, it has been described that high external Na^+ or Cs^+ may lead to K^+ deficiency (Botella et al., 1997; Carden et al., 2003; Hampton et al., 2004; Shabala et al., 2006; Adams et al., 2013). On the other hand, Na^+ and Cs^+ may accumulate into the cell by entering through HAK5 transporters (Santa-María et al., 1997; Rubio et al., 2000; Qi et al., 2008). Overall, pivotal parameters related to Na^+ or Cs^+ tolerance such as the cytoplasmic K^+/Na^+ (Maathuis and Amtmann, 1999; Shabala and Cuin, 2008) or K^+/Cs^+ (Hampton et al., 2004; Adams et al., 2013) ratios are reduced and plant growth is inhibited.

These three stress conditions, K^+-depleted, salt-affected or Cs^+-polluted soils, that may make K^+ nutrition largely dependent on HAK5 operation, affect agriculture worldwide. Many agricultural lands are intrinsically deficient in K^+ (Rengel and Damon, 2008; Römheld and Kirkby, 2010) or become K^+ deficient because of an inadequate fertilization that prioritize N and P over K fertilization (Römheld and Kirkby, 2010). Salinity affects more than 20% of irrigated lands worldwide and, because most crop species are glycophytes, it reduces crop productivity (Munns and Tester, 2008). Finally, although the natural concentration of Cs^+ in soils is low, certain areas may contain Cs^+ concentrations that reduce plant growth because Cs^+ is toxic to plants. In addition, agricultural lands affected by discharges of nuclear power plants result in polluted soils with radiocesium, which readily enters the food chain (White and Broadley, 2000).

Obtaining plants with more selective HAK5 transporters may be a solution to these agricultural problems (Niu et al., 1995). This requires function-structure relationship studies, which are scarce in literature. Some studies reported the isolation of mutated plant transporters of the HAK5-type that when expressed in yeast showed altered kinetics of K^+ uptake (Rubio et al., 2000; Senn et al., 2001; Garciadeblas et al., 2007; Mangano et al., 2008; Nieves-Cordones et al., 2008). Here we have focused on the identification of amino acid residues that are important for K^+ over Na^+ or Cs^+ selectivity of the Arabidopsis AtHAK5 transporter. Mutagenic PCR followed by selection in yeast allowed us for the identification of a highly conserved amino acid residue in the family of HAK transporters, F130, that is probably involved in the selectivity of the transporter for K^+.

MATERIALS AND METHODS

YEAST AND BACTERIAL STRAINS AND GROWTH MEDIA

Saccharomyces cerevisiae yeast strain 9.3 (*MAT***a**, *ena1D:HIS3::ena4D, leu2, ura3–1, trp1–1, ade2–1, trk1D, trk2::pCK64*) (Bañuelos et al., 1995) with deleted TRK1 and TRK2 K^+ uptake systems and ENA1-ENA4 Na^+ extrusion Na^+-ATPases was used throughout the study. The *Escherichia coli* strain DH5α (Hanahan, 1985) was used for plasmid DNA amplification. For yeast growth YPD (1% yeast extract, 2% peptone, 2% glucose), SD (Sherman, 1991) and Arginine Phosphate medium (AP) (Rodríguez-Navarro and Ramos, 1984) supplemented with KCl, NaCl and requirements as indicated were used. For bacterial growth, LB media supplemented with ampicillin as required was used. Yeast transformation was carried out by the high-efficiency protocol of the lithium acetate/single-stranded carrier DNA/PEG method of transformation of *S. cerevisiae* (Gietz and Schiestl, 2007). Standard procedures were used for *E. coli* growth and transformation, and DNA manipulations (Sambrook and Russell, 2001).

For determining doubling times of yeast cells, liquid cultures were used. The optical density at 550 nm of the yeast cultures was determined in a spectrophotometer at different time points during 72 h. The optical density vs. time was plotted and the doubling time (the time required to duplicate the optical density) calculated during the exponential growth phase.

RANDOM AtHAK5 MUTAGENESIS, MUTANT SELECTION AND SITE-DIRECTED MUTAGENESIS

The AtHAK5 cDNA was mutagenized randomly by error prone PCR (epPCR) as described by Wong et al. (2007) with minor modifications. In short, two epPCRs biased toward transitions or transversions were carried out to amplify AtHAK5 cDNA to get a broader range of possible amino acid substitutions (Table S1). The primers used to amplify the AtHAK5 cDNA were Forward: 5′-AAAAAATATACCCCAGCCTCG-3′ and Reverse 5′-GGCGAAGAAGTCCAAAGCTGG-3, which can bind up- and downstream regions respectively of the NotI and BamHI pDR195 cloning sites (Rentsch et al., 1995), where the cDNA of AtHAK5 was inserted.

The amplified cDNA fragments and a NotI-BamHI digested pDR195 plasmid were cotransformed into yeast to allow homologous recombination (ratio 10:1 epPCR:vector) as previously described (Rubio et al., 1999). The amplified cDNA and the plasmid shared a region of more than 50 base pairs on each end to allow proper homologous recombination (Muhlrad et al., 1992). Transformants were first selected on SD medium lacking uracil and supplemented with 100 mM KCl and then replica-plated to AP medium supplemented with 0.1 mM KCl plus different concentrations of Na^+. The best growing colonies were selected, and their plasmids were isolated and reintroduced into yeast. Yeast retransformants that confirmed growth on high Na^+ were selected, their plasmids isolated and the AtHAK5 cDNA sequenced. To obtain single point mutations, PCR-based site directed mutagenesis was followed as described elsewhere (Cadwell and Joyce, 1992).

INTERNAL IONIC CONTENT DETERMINATION IN YEAST

Internal K^+, Na^+, and Cs^+ concentrations in yeast were determined after 24 h of growth in liquid AP media supplemented with K^+, Na^+, and Cs^+ as indicated. In short, samples of yeast cultures were filtered through a 0.8-μm-pore nitrocellulose membrane filter (Millipore, Bedford, MA) and washed with 20 mM $MgCl_2$.

Ionic contents were extracted after 24 h incubation in 0.1 M HCl. The K^+, Na^+, or Cs^+ concentrations were determined by atomic emission spectrophotometry of the acid extracts from the cells (Rodríguez-Navarro and Ramos, 1984).

Rb^+ UPTAKE EXPERIMENTS

For kinetic characterization of K^+ uptake, Rb^+ was used in uptake experiments (Rodríguez-Navarro, 2000). Yeast cells of the 9.3 strain transformed with plasmids harboring WT AtHAK5 and the different mutants were grown overnight in AP liquid media supplemented with 3 mM K^+ at 28°C with shaking. After overnight incubation, cell cultures reached an optical density (550) of 0.2–0.3. Then, cells were collected by centrifugation and suspended in fresh AP media without K^+, and incubated for 6 h. After this 6 h period of K^+ starvation cells were collected by centrifugation and suspended in uptake buffer that contained 10 mM MES, 0.1 mM $MgCl_2$ and 2% glucose brought to pH 6,0 with $Ca(OH)_2$. Cells were placed at 28°C in a shaker bath and at time zero Rb^+ was added. Samples were taken at different time points during 10 min and collected on nitrocellulose 0.8 μm pore filters that were incubated overnight in 5 mL of 0.1 M HCl to extract their internal Rb^+, which was determined by atomic emission spectrophotometry. The internal Rb^+ concentrations per unit of cell dry weight were plotted against time and the initial rates of Rb^+ uptake were calculated on a dry weight basis and per unit of time (Rodríguez-Navarro and Ramos, 1984). The rates of Rb^+ uptake were plotted at the different external concentrations of Rb^+ employed and fitted to Michaelis-Menten equations to calculate the kinetic parameters. These experiments were carried out with cultures from several independent transformants producing the same results.

K^+ AND Cs^+ DEPLETION EXPERIMENTS

For determining the V_{max} of K^+ and Cs^+ uptake, depletion experiments were performed. Yeast cells were grown and starved of K^+ as described for the Rb^+ uptake experiments and then transferred to AP media to reach an optical density of 2. At zero time, K^+ or Cs^+ was added to reach a 50 μM concentration. Samples of the cell suspensions were taken at different time points and centrifuged to spin down cells. K^+ or Cs^+ concentrations were determined in the supernatant by atomic emission spectrophotometry. The K^+ or Cs^+ uptake rates were calculated from the difference in the external cation concentration between two time points per unit of dry weight and unit of time.

PROTEIN EXTRACTION AND WESTERN BLOT

Yeast cells expressing the different AtHAK5 versions as well as the empty plasmid were grown overnight in YPD media supplemented with 100 mM K^+. Total membranes from yeast were isolated as described elsewhere (Serrano, 1988). 25 μg of proteins from total membranes of each yeast strain were subjected to SDS/PAGE and immunoblotting following standard procedures (Sambrook and Russell, 2001). AtHAK5 protein of a predicted mass of 87.85 kDa was immunodetected with a polyclonal antibody raised in rabbit against the YGYKEDIEEPDEFE peptide present in AtHAK5 amino acid sequence, synthesized by GenScript (Piscataway, NJ USA). After immunological detection, the PVDF membrane was washed and stained with comassie blue.

RESULTS

SELECTION OF AtHAK5 MUTANTS THAT INCREASE TOLERANCE TO NaCl

To isolate novel HAK5 mutants, the Na^+ sensitive yeast strain 9.3, lacking functional TRK1 and TKR2 K^+ uptake systems as well as ENA1-ENA4 Na^+ extruding ATPases, was used for heterologous expression (Bañuelos et al., 1995). Plasmids containing putative mutants of AtHAK5 were obtained by allowing homologous recombination in yeast of mutated cDNA and plasmid fragments that rescued the ura^- mutation of the 9.3 yeast strain, as described in Materials and Methods Section. Colonies growing in SD-URA selective media supplemented with 100 mM K^+ were replica plated to different selective minimal AP media. These selective media contained 0.1 mM K^+ with no added Na^+ or with 400 mM Na^+. Thirty five colonies that showed a better growth in the presence of Na^+ were selected for further analysis. To ensure that the best performance of the selected yeast colonies was promoted by the plasmid and not by a reversion of the yeast genetic background, the plasmids were recovered from yeast cells and retransformed into the 9.3 yeast strain. The capacity of the retransformants to grow in the presence of high Na^+ was checked. The 9.3 strain transformed with the WT AtHAK5 was used as a control. However, as previously shown (Rubio et al., 2000), the low V_{max} of K^+ uptake mediated by WT AtHAK5 in yeast prevented its functional complementation of *trk1, trk2* yeast mutants. Therefore, an additional control that consisted of yeast cells expressing the previously characterized AtHAK5mutant with the L776H mutation was used. This mutation increased the V_{max} of K^+ uptake and allowed growth of *trk1, trk2* yeasts at low K^+ (Rubio et al., 2000). None of the yeast cells harboring the putative AtHAK5 mutants grew better than those expressing the L776H mutant (not shown).

In order to obtain AtHAK5 mutations that enhanced Na^+ tolerance beyond that promoted by the L776H mutant, a similar approach was undertaken, but using the cDNA of the L776H mutant as a starting material. The selective pressure was also increased by using higher Na^+ concentrations in the selective media. In this case, 17 Na^+-tolerant colonies were selected on media containing 0.1 mM K^+ and 600 mM Na^+. After retransformation and selection for growth, one colony, that was named Mut11, was selected because it showed a better growth than the L776H mutant on plates with high Na^+. The plasmid containing the Mut11 mutant was isolated and the AtHAK5 cDNA sequenced. It was observed that in addition to the L776H mutation, the Mut11 mutant contained the F130S mutation. To study individual effects of each of these mutations, a cDNA that encoded the single mutant F130S was generated by site directed mutagenic PCR, cloned into the pDR195 plasmid and transformed into the 9.3 strain. Growth of yeast cells expressing double and single mutants as well as WT AtHAK5 and the empty plasmid was studied by drop tests on minimal media (**Figure 1**).

In the presence of 5 mM K^+, all yeast strains grew well. When the K^+ concentration was reduced to 0.1 mM K^+, yeast cells transformed with the empty plasmid (pDR195) or with WT AtHAK5 were unable to grow. Yeast cells expressing any of the mutated versions of AtHAK5 could grow. When 400 mM Na^+ was

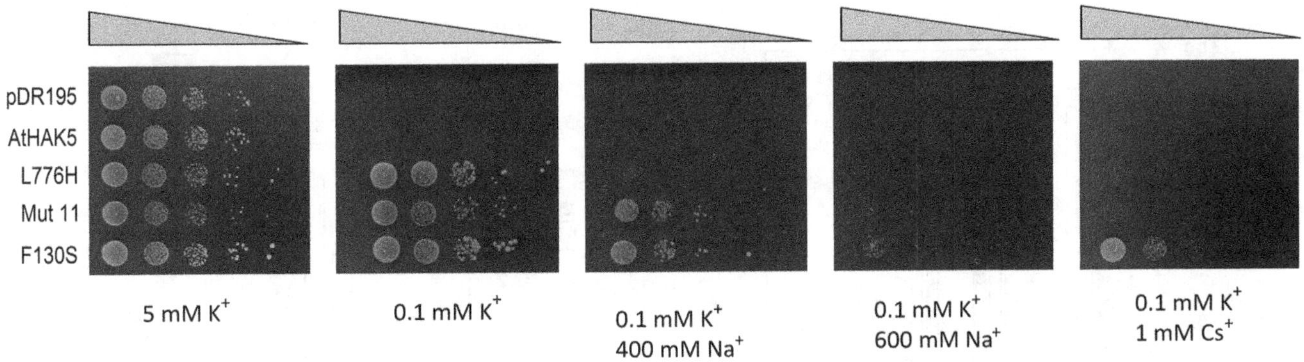

FIGURE 1 | Growth of 9.3 yeast cells expressing WT and mutated AtHAK5 K^+ transporters. Cells of the yeast strain 9.3 (*trk1, trk2, ena1-4*) expressing the empty plasmid pDR195, WT AtHAK5 and the mutants L776H, Mut11 (L776H, F130S) and F130S were used for growth assays on drop tests on solid media. Serial dilutions of cell suspensions were plated on minimal AP media supplemented with the indicated K^+, Na^+ and Cs^+ concentrations.

added to the 0.1 mM K^+ medium, only the double Mut11 mutant and the single F130S mutant were able to grow. Increasing Na^+ to 600 mM completely arrested growth of Mut11. However, at this high Na^+ concentration, the cells expressing the single mutant F130S were able to grow, although at low rates.

Tolerance to Cs^+ was also studied. The presence of 1 mM Cs^+ in the 0.1 mM K^+ medium completely inhibited growth of all yeast strains with the exception of those expressing the F130S mutation, that showed an outstanding growth capacity.

For a detailed study of growth rates and tolerance to Na^+ and Cs^+, liquid cultures were used. The doubling times of the different yeast strains growing in the presence of different levels of K^+, Na^+, and Cs^+ were determined (**Table 1**). At high K^+ (3 mM K^+) all strains grew although mutants did so faster than WT AtHAK5. At low K^+ (0.1 mM K^+) only the mutants grew with no significant differences among them. Including high Na^+ concentrations to the media gave rise to differences in growth rates among the different mutants. In the presence of 100 mM Na^+ new mutants, Mut11 and F130S, grew better than the L776H mutant. Further increase of Na^+ to 300 mM inhibited growth of L776H and reduced that of the new mutants. Importantly, the F130S showed a significant higher rate of growth than the double Mut11 mutant. When 1 mM Cs^+ was present in the growth media, a dramatic reduction of the growth rates was observed for all mutants, with the exception of the F130S that retained a higher rate of growth.

Table 1 | Doubling time (h) of 9.3 yeast strain expressing the indicated version of AtHAK5 under different external concentrations of K^+, Na^+, and Cs^+.

	Treatment				
Mutation	**3 mM K^+**	**0.1 mM K^+**	**0.1 mM K^+ 100 mM Na^+**	**0.1 mM K^+ 300 mM Na^+**	**0.1 mM K^+ 1 mM Cs^+**
WT	6.6 ± 0.2^b	–	–	–	–
L776H	3.6 ± 0.2^a	4.3 ± 0.2^a	9.5 ± 0.6^b	–	30.3 ± 0.9^b
Mut11	3.9 ± 0.3^a	3.7 ± 0.2^a	5.5 ± 0.2^a	12.4 ± 2^b	32.7 ± 1.5^b
F130S	3.4 ± 0.1^a	3.5 ± 0.1^a	5.9 ± 0.4^a	7.3 ± 1.8^a	6.8 ± 0.4^a

Shown are mean values of at least three replicates ± standard error. Different letters indicate significant differences within each column according to Tukey test ($p < 0.01$).

INTERNAL K^+ AND Na^+ CONCENTRATIONS OF YEAST EXPRESSING THE AtHAK5 MUTANTS UNDER SALT STRESS

Internal concentrations of K^+ and Na^+ after 24 h of yeast growth in media with different external concentrations of K^+ and Na^+ were determined (**Figure 2**). Under control conditions (3 mM K^+), all strains exhibited internal K^+ concentrations above 200 nmol mg^{-1} (**Figure 2A**). At low external K^+ (0.1 mM K^+), the K^+ concentration of yeast expressing WT AtHAK5 dropped to 52.9 ± 17.4 nmol mg^{-1}, whereas yeast expressing the mutants retained a high internal K^+ (292.4 ± 3.7, 304.8 ± 4.8, and 264.2 ± 3.4 nmol mg^{-1} for L776H, Mut11, and F130S respectively). The presence of 100 mM Na^+ reduced the internal K^+ concentrations for WT AtHAK5 to levels below 1 nmol mg^{-1}, and no differences among the different mutants were found in their K^+ concentrations (188.0.4 ± 3.3, 218.6 ± 1.5, and 146.9 ± 2.6 nmol mg^{-1} for L776H, Mut11, and F130S respectively). Further increase of Na^+ to 300 mM gave rise to differential results. Cells expressing the F130S mutation showed the highest K^+ concentration (85.6 ± 0.2 nmol mg^{-1}) followed by Mut11 (46.4 ± 1.8 nmol mg^{-1}) and L776H (38.1 ± 4.0 nmol mg^{-1}).

Internal Na^+ concentrations were also determined (**Figure 2B**). In the absence of added Na^+, internal Na^+ concentrations were very low. Addition of 100 mM Na^+ increased internal Na^+ in all cases, resulting the Na^+ concentrations of cells expressing WT AtHAK5 significantly higher than those expressing the F130S mutant. The presence of 300 mM Na^+ produced cells with similar high internal Na^+ concentrations in cells expressing WT AtHAK5, L776H, and Mut11. Importantly, cells expressing the F130S mutation showed a significant lower Na^+ concentration than the other strains.

The internal K^+/Na^+ ratios were calculated because this is one of the most important parameters for salt tolerance in yeast as well as in plants (Gaxiola et al., 1992; Maathuis and Amtmann, 1999; Gisbert et al., 2000). In the presence of 100 mM Na^+ (0.1 mM K^+), cells expressing the AtHAK5 mutants

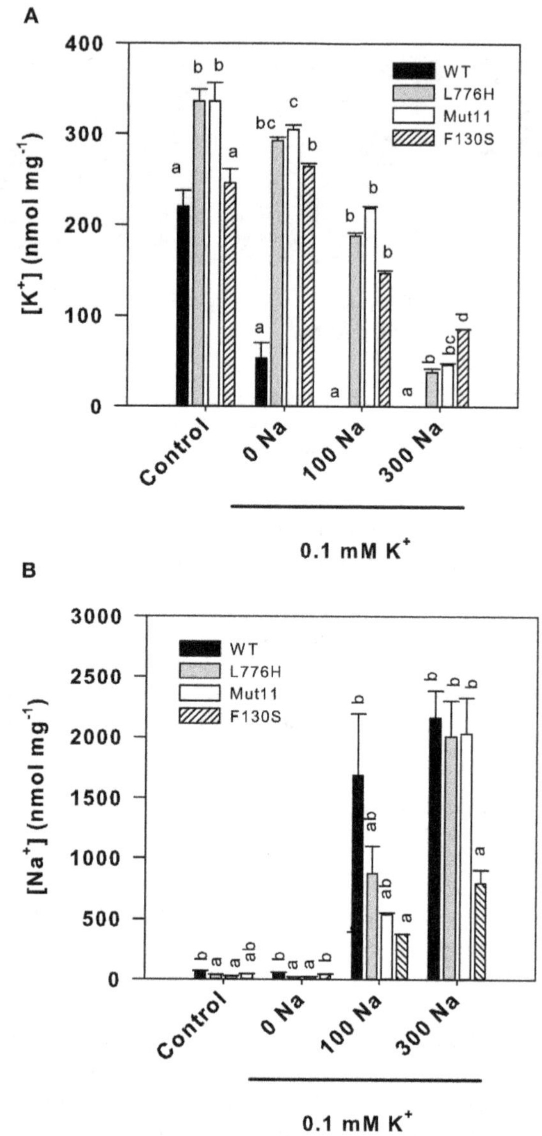

FIGURE 2 | Internal K^+ and Na^+ concentrations of 9.3 yeast cells expressing WT AtHAK5 and mutated AtHAK5 K^+ transporters. The same strains described in **Figure 1** were grown for 24 h in minimal AP liquid media supplemented with the indicated K^+ and Na^+ concentrations. The control medium contained 3 mM K^+. After 24 h cells were collected by filtration and their ionic content was acid extracted. The K^+ and Na^+ concentrations in the acid extracts were determined by atomic emission spectrometry and the internal K^+ and Na^+ concentrations calculated on a dry weight basis. Shown are averages of internal K^+ **(A)** and Na^+ **(B)** concentrations of at least 6 replicates and error bars denote standard error. Columns with different letters within each ionic treatment are significantly different according to Tukey test ($p < 0.05$).

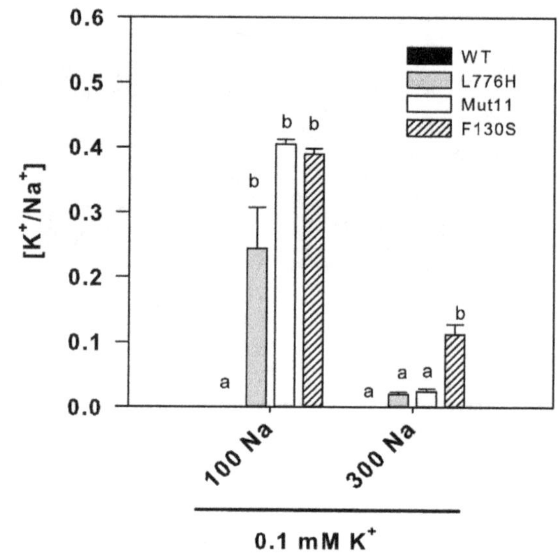

FIGURE 3 | Internal K^+/Na^+ ratios of 9.3 yeast cells expressing WT AtHAK5 and mutated AtHAK5 K^+ transporters. Cells were treated and analyzed as indicated in **Figure 2**. Shown are the ratios of internal K^+/Na^+ concentrations of the cells grown in the presence of 0.1 mM K^+ at 100 or 300 mM Na^+. Reported values are the average of at least 6 replicates and error bars denote standard error. Columns with different letters within each ionic treatment are significantly different according to Tukey test ($p < 0.05$).

showed similarly higher K^+/Na^+ ratios than those expressing WT AtHAK5 (**Figure 3**). The presence of 300 mM Na^+ importantly reduced the K^+/Na^+ ratio in all cases. WT AtHAK5 and the L776H and Mut11 mutants showed similar low K^+/Na^+ ratios. The F130S mutant showed a significantly higher K^+/Na^+ ratio than the rest of the yeast strains.

INTERNAL K^+ AND Cs^+ CONCENTRATIONS OF YEAST EXPRESSING THE MUTANTS UNDER Cs^+ STRESS

Since the F130S mutant showed a remarkable enhanced tolerance to Cs^+ (**Figure 1** and **Table 1**), we further examined the internal K^+ and Cs^+ concentrations in yeast grown for 24 h in liquid medium in the presence of at low K^+ (0.1 mM K^+) with 1 mM CsCl. The presence of 1 mM Cs^+ importantly reduced the internal K^+ concentrations to 36.6 ± 1.7, 26.7 ± 1.8, 38.4 ± 3.6, and 99.3 ± 9.8 nmol mg^{-1} for WT AtHAK5, L776H, Mut11, and F130S respectively (**Figure 4A**), far below the internal K^+ concentrations without Cs^+ (**Figure 2A**). However, cells expressing the F130S mutant retained a significantly higher K^+ concentration than the other yeast strains. When internal Cs^+ was determined, it was observed that cells expressing the F130S mutant accumulated a significantly lower Cs^+ concentration than the other strains (**Figure 4B**). As a result, the F130S mutation led to cells with significantly higher K^+/Cs^+ ratios (**Figure 5**), an important parameter for Cs^+ tolerance (Hampton et al., 2004).

KINETIC CHARACTERIZATION OF AtHAK5 MUTANTS

To gain further insights into the effects that the Mut11 and F130S mutations produced on AtHAK5 functionality, a kinetic characterization was performed. Yeast cells expressing the mutants were grown overnight on AP media supplemented with 3 mM K^+ and starved of K^+ for 6 h. Then, Rb^+ uptake experiments were performed as described in the Materials and Methods Section to calculate the initial rates of Rb^+ uptake at different external concentrations. The rates of Rb^+ uptake were plotted against the external Rb^+ concentration (**Figure 6**). Clear differences in the uptake kinetics could be observed among the mutants and WT

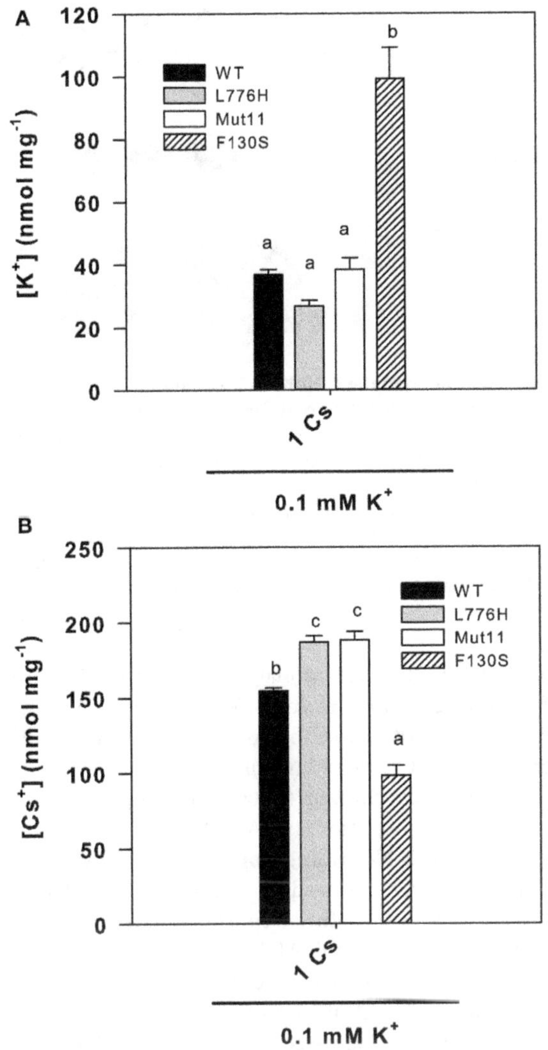

FIGURE 4 | Internal K^+ and Cs^+ concentrations of 9.3 yeast cells expressing WT AtHAK5 and mutated AtHAK5 K^+ transporters. The same strains described in **Figure 1** were grown for 24 h in minimal AP liquid media supplemented with 0.1 mM K^+ and 1 mM Cs^+. After 24 h cells were collected by filtration and their ionic content was acid extracted. The K^+ and Cs^+ concentrations in the acid extracts were determined by atomic emission spectrometry and the internal K^+ and Cs^+ concentrations calculated on a dry weight basis. Shown are averages of internal K^+ **(A)** and Cs^+ **(B)** concentrations of at least 6 replicates and error bars denote standard error. Columns with different letters within each ionic treatment are significant different according to Tukey test ($p < 0.01$).

AtHAK5. The data were fitted to Michaelis-Menten equations and V_{max} and K_m values determined (**Table 2**). All mutants increased the V_{max} for Rb^+ with respect to WT AtHAK5. However, whereas mutants containing the L776H mutation (L776H and Mut11) showed high V_{max} values around 10-fold the V_{max} of WT AtHAK5, the F130S only showed a 2-fold increase in this parameter. Mut11 showed a reduction of at least 3-fold in the K_m value with respect to WT AtHAK5. Importantly, the single F130S mutation led to a 100-fold increase in the affinity for Rb^+ in comparison to WT AtHAK5. The inhibition of Rb^+ uptake by

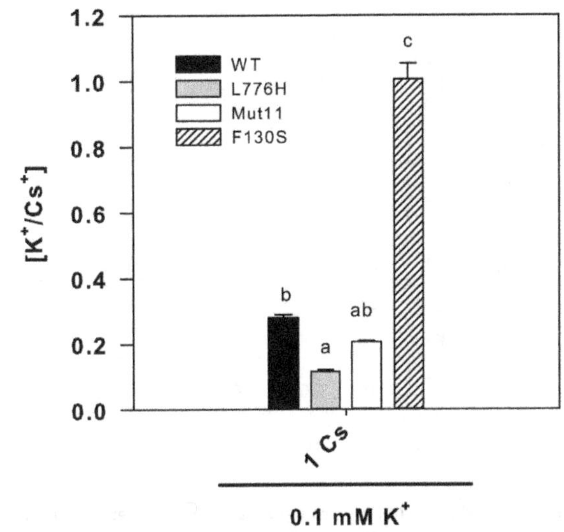

FIGURE 5 | Internal K^+/Cs^+ ratios of 9.3 yeast cells expressing WT AtHAK5 and mutated AtHAK5 K^+ transporters. Cells were treated and analyzed as indicated in **Figure 4**. Shown are the ratios of internal K^+/Cs^+ concentrations of the cells grown in the presence of 0.1 mM K^+ and 1 mM Cs^+. Reported values are the average of at least 6 replicates and error bars denote standard error. Columns with different letters within each ionic treatment are significantly different according to Tukey test ($p < 0.01$).

K^+, Na^+ and Cs^+ was also determined and the K_i values calculated (**Table 2**). As with the Rb^+ K_m values, the K_i constants were importantly reduced by the single F130S mutation. In addition, the rates of K^+ and Cs^+ depletion from a 50 μM solution were determined. It was observed that cells expressing the F130S depleted K^+ at higher rates than WT AtHAK5 cells (4.2 ± 0.5 vs 0.6 ± 0.3 nmol mg^{-1} min^{-1}) and they depleted Cs^+ at similar rates compared to WT AtHAK5 cells (2.6 ± 0.7 nmol mg^{-1} min^{-1}).

DETECTION OF AtHAK5 AND AtHAK5 MUTANT PROTEINS IN YEAST

The AtHAK5 protein of the different mutant versions as well as of WT AtHAK5 were immunologically detected. Proteins from total yeast membranes were extracted from liquid yeast cultures and resolved by SDS-PAGE. The AtHAK5 protein was detected by Western blot with a polyclonal antibody raised against AtHAK5. Yeast cells transformed with the empty plasmids did not produce any detectable signal (**Figure 7A**) whereas those expressing WT AtHAK5 and AtHAK5 mutants produced a signal band corresponding to the AtHAK5 size (predicted mass of 87.85 kDa). Differences in the amount of AtHAK5 with respect to WT AtHAK5 could be detected among the different yeast strains. L776H increased the amount of AtHAK5 protein, Mut11 showed a similar protein accumulation whereas the single F130S mutant showed a reduced amount of AtHAK5. Coomassie staining of the proteins transferred to the hybridized PVDF membrane showed that all lines contained the same amount of total protein (**Figure 7B**).

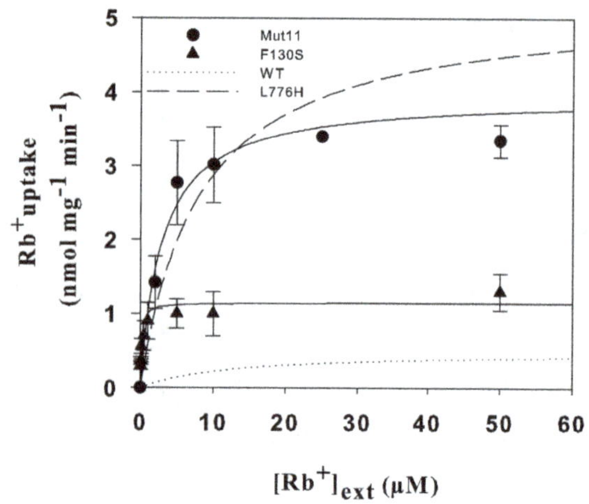

FIGURE 6 | Initial rates of Rb^+ uptake as a function of external Rb^+ in 9.3 cells expressing the AtHAK5 mutants. Cells of the 9.3 strain expressing the Mut11 and F130S mutants were grown overnight on minimal AP media supplemented with 30 mM K^+. After 6 h of incubation in K^+-free AP media cells were transferred to uptake buffer supplemented with different Rb^+ concentrations. The initial rates of Rb^+ uptake for each external Rb^+ concentration were determined and plotted for Mut11 (closed circles) and F130S (closed triangles). Values were fitted to Michaelis-Menten equations. The predicted curves for WT (dotted line) and the L776H mutation (dashed line) are also shown (Rubio et al., 2000).

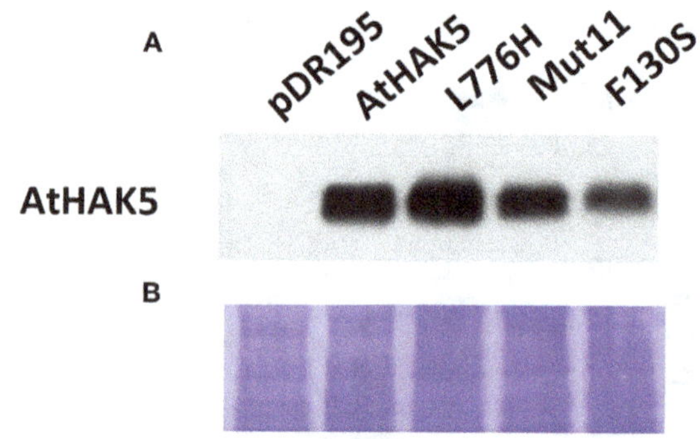

FIGURE 7 | Detection of the AtHAK5 WT and mutated proteins expressed in yeast by Western blot. The yeast strains described in **Figure 1** were grown overnight in YPD media supplemented with 100 mM K^+. Then, proteins of the total membrane fraction were isolated, resolved on polyacrylamide gel and transferred to a PVDF membrane for AtHAK5 detection by Western-bolt. A polyclonal antibody raised in rabbit against the YGYKEDIEEPDEFE peptide of AtHAK5 was used. **(A)** Shows the immunological detection of AtHAK5 in the different yeast strains. **(B)** Shows the coomasie blue staining of the hybridized PDVF membrane.

Table 2 | Kinetic parameters of Rb^+ uptake mediated by AtHAK5 and its mutant forms in yeast.

	V_{max} (nmol mg-1 min-1)	Rb^+ Km (µM)	K^+ Ki (µM)	Na^+ Ki(mM)	Cs^+ Ki (µM)
WT*	0.5 ± 0.3	12.6 ± 0.3	12	10	16
L776H*	5.2 ± 0.3	8.1 ± 1.0	12	10	16
Mut11	4.1 ± 0.2	2.5 ± 0.7	5.7 ± 0.1	12.3 ± 2.1	21.3 ± 0.8
F130S	1.0 ± 0.2	0.1 ± 0.1	0.3 ± 0.2	0.6 ± 0.3	0.3 ± 0.2

**From (Rubio et al., 2000).*

Values at the average of at least three replicates ± standard error. V_{max} (nmol mg^{-1} min^{-1}) and K_m (µM) values for Rb^+ uptake and K_i values for inhibition of Rb^+ uptake by K^+ (µM), Na^+ (mM), and and Cs^+ (µM).

DISCUSSION

In plants, four families of K^+ transport systems have been described: Shaker channels, KCO channels, HKT transporters and HAK transporters (Very and Sentenac, 2003). The members of the first three families bear putative pore forming regions derived from a K^+ channel ancestor subunit, region that is present in all K^+ channels among kingdoms (Mackinnon, 2003; Very and Sentenac, 2003; Corratgé-Faillie et al., 2010; Benito et al., 2014). This explains the high number of structure-function relationship studies on this type of transporters and the deep knowledge of their functional domains. However, this conserved pore forming region is absent in HAK transporters that, together with the lack of HAK representatives in the animal kingdom, place this family of transporters in a distinct group. This, and the scarce structure-function relationship studies on HAK transporters, make the identification of their functional domains an interesting and almost unexplored field of research. We have here undertaken the identification of mutants in the Arabidopsis AtHAK5 that help the identification of those unknown functional domains.

A previous study showed that introducing into the barley high-affinity K^+ transporter HvHAK1 (Santa-María et al., 1997) amino acids of the N-terminus of the barley low-affinity K^+ transporter HvHAK2, reduced the affinity of HvHAK1 for K^+ (Senn et al., 2001). Another report showed that the V336I and R591C mutants of HvHAK1 improved K^+ nutrition and increased Na^+ tolerance of yeast cells by increasing the V_{max} of K^+ uptake (Mangano et al., 2008). In Arabidopsis, the L776H mutation of AtHAK5 increased the V_{max} of Rb^+ uptake without affecting the K_m value, improving growth of yeast cells under low K^+ (Rubio et al., 2000). In *Physcomitrella patens*, the R443S, AQQP/VQP, L603H and G606E mutations in PpHAK1 decreased the K_m and, with the exception of R443S, increased the V_{max} for Rb^+ uptake in yeast. All these PpHAK1 mutants improved yeast growth at low K^+ (Garciadeblas et al., 2007). All these studies either used site-directed (Senn et al., 2001) or random mutagenic PCR (Mangano et al., 2008) or took advantage of the occurrence of spontaneous mutations in DNA (Rubio et al., 2000; Garciadeblas et al., 2007). Here, to increase the chances of isolating new mutations with higher transport selectivity, a modified mutagenic PCR approach was successfully used. According to it, different unbalanced mixtures of dNTPs were used in different reactions to favor the occurrence of transitions or transversions. To increase the selective pressure, mutants were selected on the presence of high Na^+ in the yeast strain 9.3, which is highly sensitive to Na^+ because of deletion of the endogenous K^+ uptake systems (TKR1, TRK2) as well the Na^+ extruding ATPases (ENA1-ENA4) (Bañuelos et al., 1995). In addition, the L776H mutant instead of WT AtHAK5 was used as a starting material, to generate new mutants that increase Na^+ tolerance beyond that of the L776H mutant.

Here we show a mutation, F130S, which importantly affects the kinetics of K^+ uptake of a HAK5 transporter, leading to improved growth at low K^+ as well as enhanced tolerance to Na^+ and Cs^+ (**Table 1** and **Figure 1**). The F130S mutation in the AtHAK5 transporter increases in yeast cells 100-fold the affinity for Rb^+ uptake, two-fold the V_{max} of Rb^+ uptake (**Table 2** and **Figure 6**) and 7-fold the maximal rate of K^+ depletion (4.2 ± 0.5 vs 0.6 ± 0.3 nmol mg^{-1} min^{-1} for F103S and WT respectively). In addition, important reductions in the K_i values for K^+, Na^+, and Cs^+ are observed (**Table 2**). The changes in the kinetic parameters promoted by the F130S mutation lead to yeast cells that were able to maintain higher concentrations of K^+ and lower concentrations of Na^+ at high external Na^+ concentrations (**Figure 2**), as well as higher concentrations of K^+ and lower of Cs^+ in the presence of Cs^+ (**Figure 4**). Overall, yeast cells expressing the mutant showed higher K^+/Na^+ (**Figure 3**) and K^+/Cs^+ (**Figure 5**) ratios, crucial parameters for salt and Cs^+ tolerances in yeast as well as in plants (Gaxiola et al., 1992; Maathuis and Amtmann, 1999; Gisbert et al., 2000; Hampton et al., 2004). Importantly, the changes mediated by the F130S amino acid substitution occurred without increasing the amount of transporter protein in yeast membranes (**Figure 7**).

The results summarized above suggest that the F130 residue of the AtHAK5 transporter may affect a domain that is involved in K^+ binding and/or transport from the external solution to the cytoplasm. K^+, Na^+, and Cs^+ competitively inhibit HAK5-mediated Rb^+ uptake and they are transported through HAK5 transporters (Santa-María et al., 1997; Rubio et al., 2000; Qi et al., 2008). Therefore, the domain involved in K^+ binding of this type of transporters is also probably involved in Rb^+, Na^+, and Cs^+ binding. Then, a mutation in the K^+ binding site that increases the affinity for K^+ is expected to produce a similar effect on the affinities for Rb^+, Na^+, and Cs^+, which is the effect that can be deduced from the data presented here (**Table 2**). The idea that the F130S mutation is affecting a domain involved in the binding of K^+, and therefore the binding of Na^+ and Cs^+, is further supported by the lower accumulation of Na^+ (**Figure 2B**) and Cs^+ (**Figure 4B**) observed in yeast expressing this mutant in comparison with those expressing WT AtHAK5.

Recently, the topology of a bacterial member of the HAK family of transporters, the *Escherichia coli* Kup system, has been determined (Sato et al., 2014). According to this study, this type of transporters is composed of 12 transmembrane domains with the N- and C-termini facing the cytoplasmic side. The reported structure confirms the initial model proposed for AtHAK5 (Rubio et al., 2000), which was based on algorithmic prediction of transmembrane domains. This structure may be very likely extended to all members of the HAK family (Rodríguez-Navarro, 2000). Although all HAK transporters may have a similar structure they do not show extensive conserved regions. However, a study that included plant, fungal and bacterial representatives found 13 conserved glycine residues (Rodríguez-Navarro, 2000), which have been found later in many other HAK transporters. Three out of these 13 glycine residues are located in the GEGGTFALY domain, highly conserved among HAK transporters and that contains the F130 residue proposed here as important for K^+ affinity (**Figure 8**). Importantly, the selectivity pore of K^+ channels contains the glycine-rich motif GYG (Heginbotham et al., 1992) and related motifs are found in the proposed pore regions of HKT-TRK transporters (Durell and Guy, 1999; Mäser et al., 2002).

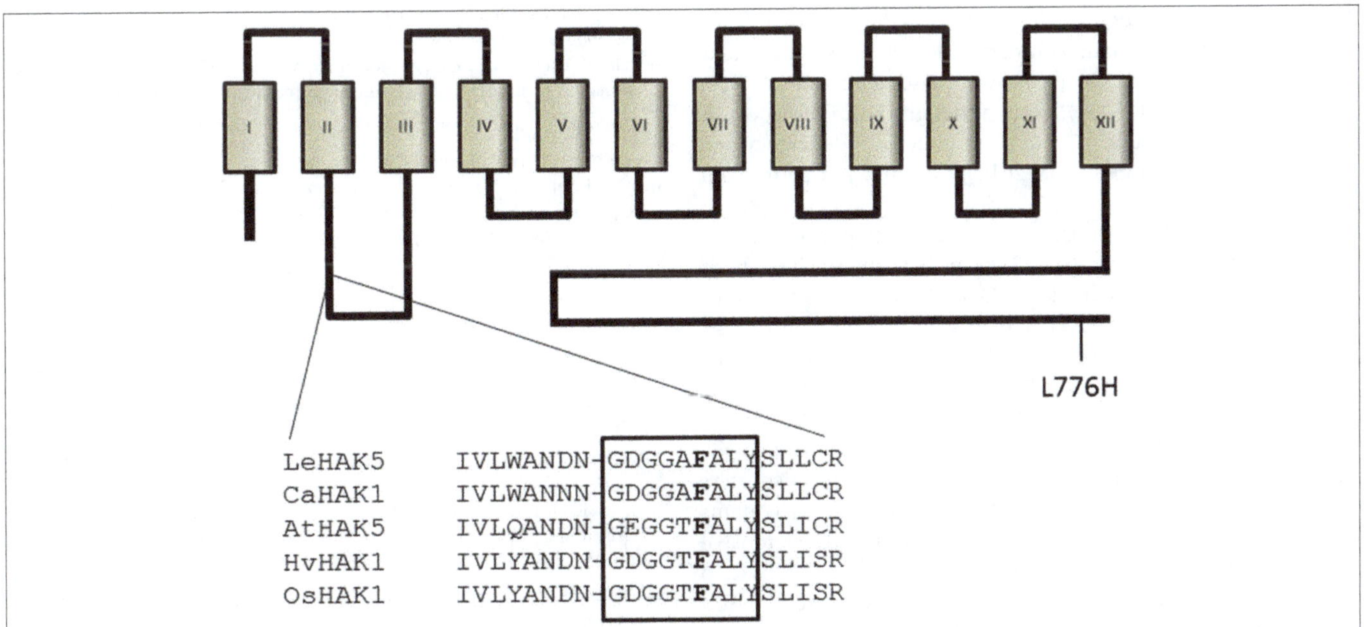

FIGURE 8 | Model of AtHAK5 topology and location of mutations. The proposed model for AtHAK5 protein is shown with 12 transmembrane domains and a long cytoplasmic C-terminus. The F130 residue is located between the second and the third transmembrane domains, in a highly glycine-rich conserved region among HAK transporters. Shown is the alignment of this conserved region in representative high-affinity HAK K^+ transporters of different species. The location of the L776H is also shown.

Interestingly enough, mutations that affect glycine residues in several K^+ transport systems that belong to different families, greatly affect their capacity to mediate K^+ transport. The paradigmatic example is the removal of the GYG motif of a K^+ channel that modifies its selectivity (Heginbotham et al., 1992). Other examples are the G91S mutation in TaHKT2.1 [former TaHKT1 (Platten et al., 2006)], that abrogates K^+ permeability (Mäser et al., 2002) or the G61S mutation in HvHAK1 that reduces the affinity and the V_{max} for K^+ uptake (Senn et al., 2001). In conclusion, it is tempting to speculate that the GEGGTFALY domain and other glycine-rich regions found along HAK proteins (Rodríguez-Navarro, 2000), are involved in shaping the pore region for K^+ binding, a very attractive hypothesis that needs further investigation. Modeling studies together with ion current measurements in heterologous systems should help to further characterize the structure-function relationships of this type of transporters. It is worth noting that the F130 residue studied here is conserved among almost all members of the HAK family identified so far in plants (Nieves-Cordones, personal communication). Importantly, this residue is present in all HAK transporters homologous to the high-affinity HAK5-type transporters identified in different plant species whose sequence has been deposited in the Genebank (Figure S1).

Interestingly, HAK5-type transporters can also mediate low-affinity Na^+ uptake (Santa-María et al., 1997) and high-affinity Cs^+ uptake (Rubio et al., 2000) and therefore, they may be involved in the accumulation of toxic Na^+ or Cs^+. The contribution of HAK5-type transporters to low-affinity Na^+ uptake and Na^+ accumulation may be low for two main reasons. Firstly, salinity reduces the number of HAK5 transporters in the plasma membrane because high external Na^+ reduces the expression of the encoding genes (Nieves-Cordones et al., 2008, 2010; Alemán et al., 2009). Secondly, although yet unidentified at the molecular level, it is proposed that the main pathways for low-affinity Na^+ uptake are non-selective cation channels (Amtmann and Sanders, 1999; Tester and Davenport, 2003). Thus, although HAK5 transporters may be crucial to sustain K^+ nutrition under K^+-deficiency and salinity (Nieves-Cordones et al., 2010), they may not be relevant in Na^+ accumulation.

However, HAK5 transporters may play an important role in the accumulation of toxic Cs^+. Two main pathways have been proposed for Cs^+ uptake in plants. One is a Ca^{2+}-sensitive pathway that operates in K^+-sufficient plants, and the other is insensitive to Ca^{2+} and it is only present in K^+-deprived plants (Heredia et al., 2002; Hampton et al., 2004; Caballero et al., 2012). This Ca^{2+}-insensitive pathway is very likely mediated by HAK5 transporters. Importantly, high Cs^+ concentrations may lead to K^+ starvation and induction of HAK5 genes, and HAK5 transporters may mediate Cs^+ uptake even in K^+-sufficient plants (Adams et al., 2013). Thus, furnishing plants with mutated versions of HAK5 that reduce the accumulation of Cs^+ could contribute to tolerance to this toxic cation. Since Cs^+ is a non-essential and rare cation in nature (White and Broadley, 2000), natural selection of HAK5 transporters with a high K^+/Cs^+ discrimination may have not been occurred. Thus, a biotechnological approach based on random mutagenesis and selection by function, should be a promising alternative.

The selection of HAK5 mutants in yeast, as shown here, results instrumental to identify functional domains of the transporter. However, a mutant may promote higher rates of K^+ uptake in yeast but not in plants. This can occur for example if the mutation produces a codon which is preferred in yeast and that leads to an increased amount of the protein in the fungus (Sharp et al., 1986; Campbell and Gowri, 1990). In this sense, mutants that show an increase in the affinity (**Table 2**) without importantly increasing the V_{max} nor the amount of protein in yeast (**Figure 7**) as the F130S, may provide more promising tools for biotechnological purposes.

Mutated versions of HAK5-type transporters, as the one described here, may provide biotechnological tools to improve K^+ nutrition of plants under K^+ deficiency or under certain abiotic stress conditions (Niu et al., 1995). Mutated HAK5 transporters can be expressed in plants to increase K^+ acquisition from low concentrations. In addition, the expression of these mutants can also provide a means of acquiring K^+ under salinity or Cs^+-polluted soils, conditions that are known to induced K^+ deficiency (Botella et al., 1997; Carden et al., 2003; Hampton et al., 2004; Shabala et al., 2006) and to reduce HAK-mediated accumulation of toxic Cs^+.

ACKNOWLEDGMENT

This work has been funded by grant AGL2012-33504 from Ministerio de Economía y Competitividad, Spain.

REFERENCES

Adams, E., Abdollahi, P., and Shin, R. (2013). Cesium inhibits plant growth through Jasmonate signaling in *Arabidopsis thaliana*. *Int. J. Mol. Sci.* 14, 4545–4559. doi: 10.3390/ijms14034545

Alemán, F., Nieves-Cordones, M., Martínez, V., and Rubio, F. (2009). Differential regulation of the HAK5 genes encoding the high-affinity K^+ transporters of *Thellungiella* halophila and *Arabidopsis thaliana*. *Environ. Exp. Bot.* 65, 263–269. doi: 10.1016/j.envexpbot.2008.09.011

Alemán, F., Nieves-Cordones, M., Martínez, V., and Rubio, F. (2011). Root K^+ acquisition in plants: the *Arabidopsis thaliana* model. *Plant Cell Physiol.* 52, 1603–1612. doi: 10.1093/pcp/pcr096

Amtmann, A., and Sanders, D. (1999). Mechanisms of Na^+ uptake by plant cells. *Adv. Bot. Res.* 29, 75–112. doi: 10.1016/S0065-2296(08)60310-9

Bañuelos, M. A., Klein, R. D., Alexander, S. J., and Rodriguez-Navarro, A. (1995). A potassium transporter of the yeast *Schwanniomyces* occidentalis homologous to the Kup system of *Escherichia* coli has a high concentrative capacity. *EMBO J.* 14, 3021–3027.

Benito, B., Haro, R., Amtmann, A., Cuin, T. A., and Dreyer, I. (2014). The twins K^+ and Na^+ in plants. *J. Plant Physiol.* 171, 723–731. doi: 10.1016/j.jplph.2013.10.014

Bertl, A., Reid, J. D., Sentenac, H., and Slayman, C. L. (1997). Functional comparison of plant inward-rectifier channels expressed in yeast. *J. Exp. Bot.* 48, 405–413. doi: 10.1093/jxb/48.Special_Issue.405

Botella, M. A., Martínez, V., Pardines, J., and Cerdá, A. (1997). Salinity induced potassium deficiency in maize plants. *J. Plant Physiol.* 150, 200–205. doi: 10.1016/S0176-1617(97)80203-9

Caballero, F., Botella, M. A., Rubio, L., Fernández, J. A., Martínez, V., and Rubio, F. (2012). A Ca^{2+}-sensitive system mediates low-affinity K^+ uptake in the absence of AKT1 in *Arabidopsis* plants. *Plant Cell Physiol.* 53, 2047–2059. doi: 10.1093/pcp/pcs140

Cadwell, R. C., and Joyce, G. F. (1992). Randomization of genes by PCR mutagenesis. *PCR Methods Appl.* 2, 28–33. doi: 10.1101/gr.2.1.28

Campbell, W. H., and Gowri, G. (1990). Codon usage in higher plants, green algae, and *Cyanobacteria*. *Plant Physiol.* 92, 1–11. doi: 10.1104/pp.92.1.1

Carden, D. E., Walker, D. J., Flowers, T. J., and Miller, A. J. (2003). Single-cell measurements of the contributions of cytosolic Na^+ and K^+ to salt tolerance. *Plant Physiol.* 131, 676–683. doi: 10.1104/pp.011445

Chen, Z., Pottosin, I. I., Cuin, T. A., Fuglsang, A. T., Tester, M., Jha, D., et al. (2007). Root plasma membrane transporters controlling K^+/Na^+ homeostasis in salt-stressed barley. *Plant Physiol.* 145, 1714–1725. doi: 10.1104/pp.107.110262

Corratgé-Faillie, C., Jabnoune, M., Zimmermann, S., Véry, A. A., Fizames, C., and Sentenac, H. (2010). Potassium and sodium transport in non-animal cells: the Trk/Ktr/HKT transporter family. *Cell. Mol. Life Sci.* 67, 2511–2532. doi: 10.1007/s00018-010-0317-7

Durell, S. R., and Guy, H. R. (1999). Structural models of the KtrB, TrkH, and Trk1,2 symporters based on the structure of the KcsA K^+ channel. *Biophys. J.* 77, 789–807. doi: 10.1016/S0006-3495(99)76932-8

Epstein, E. (1966). Dual pattern of ion absorption by plant cells and by plants. *Nature* 212, 1324–1327. doi: 10.1038/2121324a0

Garciadeblas, B., Barrero-Gil, J., Benito, B., and Rodriguez-Navarro, A. (2007). Potassium transport systems in the moss physcomitrella patens: pphak1 plants reveal the complexity of potassium uptake. *Plant J.* 52, 1080–1093. doi: 10.1111/j.1365-313X.2007.03297.x

Gaxiola, R., De Larrinoa, I. F., Villalba, J. M., and Serrano, R. (1992). A novel and conserved salt-induced protein is an important determinant of salt tolerance in yeast. *EMBO J.* 11, 3157–3164.

Gietz, R. D., and Schiestl, R. H. (2007). High-efficiency yeast transformation using the LiAc/SS carrier DNA/PEG method. *Nat. Protoc.* 2, 31–34. doi: 10.1038/nprot.2007.13

Gisbert, C., Rus, A. M., Bolarin, M. C., Lopez-Coronado, J. M., Arrillaga, I., Montesinos, C., et al. (2000). The yeast HAL1 gene improves salt tolerance of transgenic tomato. *Plant Physiol.* 123, 393–402. doi: 10.1104/pp.123.1.393

Hampton, C. R., Bowen, H. C., Broadley, M. R., Hammond, J. P., Mead, A., Payne, K. A., et al. (2004). Cesium toxicity in *Arabidopsis*. *Plant Physiol.* 136, 3824–3837. doi: 10.1104/pp.104.046672

Hanahan, D. (1985). "Techniques for transformation of *E. coli*," in *DNA Cloning: A Practical Approach*, Vol. 1, ed D. M. Glover (McLean, VA: IRL Press), 109.

Heginbotham, L., Abramson, T., and Mackinnon, R. (1992). A functional connection between the pores of distantly related ion channels as revealed by mutant K^+ channels. *Science* 258, 1152–1155. doi: 10.1126/science.1279807

Heredia, M. A., Zapico, R., García-Sánchez, M. A. J., and Fernández, J. A. (2002). Effect of calcium, sodium and pH on uptake and accumulation of radiocesium by riccia fluitans. *Aquatic Bot.* 74, 245–256. doi: 10.1016/S0304-3770(02)00107-9

Maathuis, F. J. M. (2009). Physiological functions of mineral macronutrients. *Curr. Opin. Plant Biol.* 12, 250–258. doi: 10.1016/j.pbi.2009.04.003

Maathuis, F. J. M., and Amtmann, A. (1999). K^+ Nutrition and Na^+ Toxicity: the basis of cellular K^+/Na^+ ratios. *Ann. Bot.* 84, 123–133. doi: 10.1006/anbo.1999.0912

Mackinnon, R. (2003). Potassium channels. *FEBS Lett.* 555, 62–65. doi: 10.1016/S0014-5793(03)01104-9

Mangano, S., Silberstein, S., and Santa-Maria, G. E. (2008). Point mutations in the barley HvHAK1 potassium transporter lead to improved K^+-nutrition and enhanced resistance to salt stress. *FEBS Lett.* 582, 3922–3928. doi: 10.1016/j.febslet.2008.10.036

Marschner, P. (2012). *Marschner's Mineral Nutrition of Higher Plants, 3rd Edn.* San Diego, CA: Academic Press.

Martínez-Cordero, M. A., Martinez, V., and Rubio, F. (2005). High-affinity K^+ uptake in pepper plants. *J. Exp. Bot.* 56, 1553–1562. doi: 10.1093/jxb/eri150

Martínez-Cordero, M. A., Martínez, V., and Rubio, F. (2004). Cloning and functional characterization of the high-affinity K^+ transporter HAK1 of pepper. *Plant Mol. Biol.* 56, 413–421. doi: 10.1007/s11103-004-3845-4

Mäser, P., Hosoo, Y., Goshima, S., Horie, T., Eckelman, B., Yamada, K., et al. (2002). Glycine residues in potassium channel-like selectivity filters determine potassium selectivity in four-loop-per-subunit HKT transporters from plants. *Proc. Natl. Acad. Sci. U.S.A.* 99, 6428–6433. doi: 10.1073/pnas.082123799

Muhlrad, D., Hunter, R., and Parker, R. (1992). A rapid method for localized mutagenesis of yeast. *Yeast* 8, 79–82. doi: 10.1002/yea.320080202

Munns, R., and Tester, M. (2008). Mechanisms of salinity tolerance. *Ann. Rev. Plant Biol.* 59, 651–681. doi: 10.1146/annurev.arplant.59.032607.092911

Nieves-Cordones, M., Aleman, F., Martinez, V., and Rubio, F. (2010). The *Arabidopsis thaliana* HAK5 K^+ transporter is required for plant growth and K+ acquisition from low K+ solutions under saline conditions. *Mol. Plant* 3, 326–333. doi: 10.1093/mp/ssp102

Nieves-Cordones, M., Alemán, F., Martínez, V., and Rubio, F. (2014). K^+ uptake in plant roots. the systems involved, their regulation and parallels in other organisms. *J. Plant Physiol.* 171, 688–695. doi: 10.1016/j.jplph.2013.09.021

Nieves-Cordones, M., Martinez-Cordero, M. A., Martinez, V., and Rubio, F. (2007). An NH_4^+-sensitive component dominates high-affinity K^+ uptake in tomato plants. *Plant Sci.* 172, 273–280. doi: 10.1016/j.plantsci.2006.09.003

Nieves-Cordones, M., Miller, A., Alemán, F., Martínez, V., and Rubio, F. (2008). A putative role for the plasma membrane potential in the control of the expression of the gene encoding the tomato high-affinity potassium transporter HAK5. *Plant Mol. Biol.* 68, 521–532. doi: 10.1007/s11103-008-9388-3

Niu, X., Bressan, R. A., Hasegawa, P. M., and Pardo, J. M. (1995). Ion homeostasis in NaCl stress environments. *Plant Physiol.* 109, 735–742.

Platten, J. D., Cotsaftis, O., Berthomieu, P., Bohnert, H., Davenport, R. J., Fairbairn, D. J., et al. (2006). Nomenclature for HKT transporters, key determinants of plant salinity tolerance. *Trends Plant Sci.* 11, 372–374. doi: 10.1016/j.tplants.2006.06.001

Pyo, Y. J., Gierth, M., Schroeder, J. I., and Cho, M. H. (2010). High-Affinity K^+ transport in *Arabidopsis*: AtHAK5 and AKT1 are vital for seedling establishment and postgermination growth under low-potassium conditions. *Plant Physiol.* 153, 863–875. doi: 10.1104/pp.110.154369

Qi, Z., Hampton, C. R., Shin, R., Barkla, B. J., White, P. J., and Schachtman, D. P. (2008). The high affinity K+ transporter AtHAK5 plays a physiological role in planta at very low K+ concentrations and provides a caesium uptake pathway in *Arabidopsis*. *J. Exp. Bot.* 59, 595–607. doi: 10.1093/jxb/erm330

Rengel, Z., and Damon, P. M. (2008). Crops and genotypes differ in efficiency of potassium uptake and use. *Physiol. Plant.* 133, 624–636. doi: 10.1111/j.1399-3054.2008.01079.x

Rentsch, D., Laloi, M., Rouhara, I., Schmelzer, E., Delrot, S., and Frommer, W. B. (1995). NTr1 encodes a high affinity oligopeptide transporter in *Arabidopsis*. *FEBS Lett.* 370, 264–268. doi: 10.1016/0014-5793(95)00853-2

Rodríguez-Navarro, A. (2000). Potassium transport in fungi and plants. *Biochim. Biophys. Acta* 1469, 1–30. doi: 10.1016/S0304-4157(99)00013-1

Rodríguez-Navarro, A., and Ramos, J. (1984). Dual system for potassium transport in *Saccharomyces cerevisiae*. *J. Bacteriol.* 159, 940–945.

Römheld, V., and Kirkby, E. (2010). Research on potassium in agriculture: needs and prospects. *Plant Soil* 335, 155–180. doi: 10.1007/s11104-010-0520-1

Rubio, F., Alemán, F., Nieves-Cordones, M., and Martínez, V. (2010). Studies on *Arabidopsis athak5, atakt1* double mutants disclose the range of concentrations at which AtHAK5, AtAKT1 and unknown systems mediate K^+ uptake. *Physiol. Plant.* 139, 220–228. doi: 10.1111/j.1399-3054.2010.01354.x

Rubio, F., Nieves-Cordones, M., Alemán, F., and Martinez, V. (2008). Relative contribution of AtHAK5 and AtAKT1 to K^+ uptake in the high-affinity range of concentrations. *Physiol. Plant.* 134, 598–608. doi: 10.1111/j.1399-3054.2008.01168.x

Rubio, F., Santa-Maria, G. E., and Rodríguez-Navarro, A. (2000). Cloning of *Arabidopsis* and barley cDNAs encoding HAK potassium transporters in root and shoot cells. *Physiol. Plant.* 109, 34–43. doi: 10.1034/j.1399-3054.2000.100106.x

Rubio, F., Schwarz, M., Gassmann, W., and Schroeder, J. I. (1999). Genetic selection of mutations in the high affinity K^+ transporter HKT1 that define functions of a loop site for reduced Na^+ permeability and increased Na^+ tolerance. *J. Biol. Chem.* 274, 6839–6847. doi: 10.1074/jbc.274.11.6839

Sambrook, J., and Russell, D. W. (2001). *Molecular Cloning: A Laboratory Manual.* New York, NY: Cold Spring Harbor Laboratory Press.

Santa-María, G. E., Rubio, F., Dubcovsky, J., and Rodriguez-Navarro, A. (1997). The HAK1 gene of barley is a member of a large gene family and encodes a high-affinity potassium transporter. *Plant Cell* 9, 2281–2289. doi: 10.1105/tpc.9.12.2281

Sato, Y., Nanatani, K., Hamamoto, S., Shimizu, M., Takahashi, M., Tabuchi-Kobayashi, M., et al. (2014). Defining membrane spanning domains and crucial membrane-localized acidic amino acid residues for K+ transport of a Kup/HAK/KT-type escherichia coli potassium transporter. *J. Biochem.* 155, 315–323. doi: 10.1093/jb/mvu007

Senn, M. E., Rubio, F., Bañuelos, M. A., and Rodriguez-Navarro, A. (2001). Comparative functional features of plant potassium HvHAK1 and HvHAK2 transporters. *J. Biol. Chem.* 276, 44563–44569. doi: 10.1074/jbc.M108129200

Serrano, R. (1988). H^+-ATPase from plasma membranes of *Saccharomyces cerevisiae* and avena sativa roots: purification and reconstitution. *Methods Enzymol.* 157, 533–544. doi: 10.1016/0076-6879(88)57102-1

Shabala, S., and Cuin, T. A. (2008). Potassium transport and plant salt tolerance. *Physiol. Plant.* 133, 651–669. doi: 10.1111/j.1399-3054.2007.01008.x

Shabala, S., Demidchik, V., Shabala, L., Cuin, T. A., Smith, S. J., Miller, A. J., et al. (2006). Extracellular Ca^{2+} ameliorates NaCl-induced K+ loss from *Arabidopsis* root and leaf cells by controlling plasma membrane K+-permeable channels. *Plant Physiol.* 141, 1653–1665. doi: 10.1104/pp.106.082388

Sharp, P. M., Tuohy, T. M. F., and Mosurski, K. R. (1986). Codon usage in yeast: cluster analysis clearly differentiates highly and lowly expressed genes. *Nucleic Acids Res.* 14, 5125–5143. doi: 10.1093/nar/14.13.5125

Sherman, F. (1991). Getting started with yeast. *Methods Enzymol.* 194, 3–21. doi: 10.1016/0076-6879(91)94004-V

Tester, M., and Davenport, R. J. (2003). Na^+ tolerance and Na^+ transport in higher plants. *Ann. Bot.* 91, 503–527. doi: 10.1093/aob/mcg058

Very, A. A., and Sentenac, H. (2003). Molecular mechanisms and regulation of K^+ transport in higher plants *Ann. Rev. Plant Biol.* 54, 575–603. doi: 10.1146/annurev.arplant.54.031902.134831

Volkov, V., and Amtmann, A. (2006). *Thellungiella halophila*, a salt-tolerant relative of *Arabidopsis thaliana*, has specific root ion-channel features supporting K+/Na+ homeostasis under salinity stress. *Plant J.* 48, 342–353. doi: 10.1111/j.1365-313X.2006.02876.x

White, P. J., and Broadley, M. R. (2000). Mechanisms of caesium uptake by plants. *New Phytol.* 147, 241–256. doi: 10.1046/j.1469-8137.2000.00704.x

White, P. J., and Karley, A. J. (2010). "Potassium," in *Cell Biology of Metals and Nutrients*, ed R. M. R. R. Hell (Heidelberg: Springer), 199–224.

Wong, T. S., Roccatano, D., and Schwaneberg, U. (2007). Are transversion mutations better? a mutagenesis assistant program analysis on P450 BM-3 heme domain. *Biotechnol. J.* 2, 133–142. doi: 10.1002/biot.200600201

8

Linking stomatal traits and expression of slow anion channel genes *HvSLAH1* and *HvSLAC1* with grain yield for increasing salinity tolerance in barley

Xiaohui Liu[1,2†], Michelle Mak[1†], Mohammad Babla[1], Feifei Wang[3], Guang Chen[4], Filip Veljanoski[1], Gang Wang[5], Sergey Shabala[3], Meixue Zhou[3] and Zhong-Hua Chen[1]*

[1] *School of Science and Health, University of Western Sydney, Penrith, NSW, Australia*
[2] *School of Chemical Engineering and Technology, Tianjin University, Tianjin, China*
[3] *School of Land and Food, University of Tasmania, Hobart, TAS, Australia*
[4] *College of Agriculture and Biotechnology, Zhejiang University, Hangzhou, China*
[5] *School of Environmental Science and Engineering, Tianjin University, Tianjin, China*

Edited by:
Vadim Volkov, London Metropolitan University, UK

Reviewed by:
Guoping Zhang, Zhejiang University, China
Yizhou Wang, University of Glasgow, UK

****Correspondence:***
Zhong-Hua Chen, School of Science and Health, University of Western Sydney, Bourke Street, Richmond, NSW 2753, Australia
e-mail: z.chen@uws.edu.au

† These authors have contributed equally to this work.

Soil salinity is an environmental and agricultural problem in many parts of the world. One of the keys to breeding barley for adaptation to salinity lies in a better understanding of the genetic control of stomatal regulation. We have employed a range of physiological (stomata assay, gas exchange, phylogenetic analysis, QTL analysis), and molecular techniques (RT-PCR and qPCR) to investigate stomatal behavior and genotypic variation in barley cultivars and a genetic population in four experimental trials. A set of relatively efficient and reliable methods were developed for the characterization of stomatal behavior of a large number of varieties and genetic lines. Furthermore, we found a large genetic variation of gas exchange and stomatal traits in barley in response to salinity stress. Salt-tolerant cultivar CM72 showed significantly larger stomatal aperture under 200 mM NaCl treatment than that of salt-sensitive cultivar Gairdner. Stomatal traits such as aperture width/length were found to significantly correlate with grain yield under salt treatment. Phenotypic characterization and QTL analysis of a segregating double haploid population of the CM72/Gairdner resulted in the identification of significant stomatal traits-related QTLs for salt tolerance. Moreover, expression analysis of the slow anion channel genes *HvSLAH1* and *HvSLAC1* demonstrated that their up-regulation is linked to higher barley grain yield in the field.

Keywords: soil salinity, stomata, gas exchange, quantitative trait loci, *HvSLAC1*, *HvSLAH1*, *Hordeum vulgare* L

INTRODUCTION

Soil salinity is one of the most difficult challenges facing global agriculture as it endeavors to increase productivity to meet world crop demands for human consumption and animal fodder. A third of the world's agricultural land will be significantly affected by salinity by 2050. Low precipitation, high evaporation, irrigation with saline water, and poor agricultural practice are among the major contributors to increased soil salinity (Pitman and Lähli, 2002; Zhu, 2002; Munns et al., 2006; Munns and Tester, 2008). The levels of salinity in some agricultural areas have exceeded the threshold of 50% yield reduction of many commercial crops even salt-tolerant barley. Therefore, breeding crops with higher salt-tolerance is a current and serious concern in agriculture (Mano and Takeda, 1997; Munns and Tester, 2008; Xue et al., 2009; Shabala and Mackay, 2011).

Understanding how plants respond to salinity stress has resulted in an improvement in crop yields and is seen as one of the key strategies to deliver continued crop improvements through genetic engineering (Munns et al., 2006; Ullrich, 2011; Schroeder et al., 2013). Non-halophytic crop species are usually unable to withstand saline soils. Firstly, the presence of salts cause water in soil to be more tightly bound, therefore reducing the water availability to plants thus consequently causing dehydration in plants. Secondly, most plants are not able to avoid salt absorption into their root and leaf tissues. This leads to high salt accumulation, disrupting normal physiological and biochemical functions in plants (Chen et al., 2005; Munns et al., 2006; Shabala and Mackay, 2011; Ullrich, 2011). Barley (*Hordeum vulgare L.*) is the fourth largest cereal crop grown worldwide and has one major advantage over wheat and many other crops: that is its ability to tolerate higher levels of soil salinity (Munns et al., 2006; Ullrich, 2011). Although barley is relatively tolerant to salt, there are large variations between genotypes in their salinity tolerance. This led to the utilization of genetic diversity to meet the increasing need to understand the physiological and genetic responses of barley to salt stress (Zhu, 2002; Chen et al., 2007a,b; Munns and Tester, 2008; Xue et al., 2009; Ullrich, 2011; Zhou et al., 2012).

As a powerful tool to link phenotypic traits and genotypic markers for salinity tolerance, Quantitative Trait Loci (QTLs) analysis is feasible to fine-map some genes involved in particular major quantitative traits (Kearsey, 1998; Qiu et al., 2011). In the last two decades, QTLs associated with saline tolerance have been mapped in barley at the germination, seedling, and late growth stages by using various genetic populations (Mano and Takeda, 1997; Xue et al., 2009; Siahsar and Narouei, 2010; Qiu et al., 2011; Zhou, 2011). Breeding barley for adaptation to saline soil lies in a better understanding of molecular mechanisms involved in QTLs of stomatal distribution and opening and genes encoding membrane transporters. Stomata consist of specialized guard cells (two guard cells and two subsidiary cells for monocots), which regulate photosynthetic CO_2 uptake and transpiration (Chen and Blatt, 2010; Kim et al., 2010; Chen et al., 2012; Hills et al., 2012). However, apart from correlating stomatal conductance to salinity stress, there is little systematic research for the identification of genetic control of stomatal behavior and its relationship to salinity tolerance in barley. Moreover, understanding the mechanisms of the complex network of regulatory genes of the control of stomata under salinity could be critical to reduce water loss and to maintain a high photosynthetic rate for better yield (Munns and Tester, 2008; Kim et al., 2010; Hedrich, 2012; Deinlein et al., 2014). Among the key regulatory genes, guard cell slow (or S-type) anion channels SLAC was found to be the "master switch" for stomatal closure (Vahisalu et al., 2008; Chen et al., 2010; Barbier-Brygoo et al., 2011; Geiger et al., 2011; Dreyer et al., 2012; Maierhofer et al., 2014; Zheng et al., 2014). The SLAC protein family comprised of SLAC1 and four SLAC1 homologs (SLAHs) in Arabidopsis (Negi et al., 2008; Vahisalu et al., 2008). The Arabidopsis genes *SLAH3* were found in mesophyll cells, but complementation of SLAC1 with SLAH3 recovered the anion channel function in guard cells. Therefore, it was proposed that SLAC1 and SLAH3 have an overlapping function (Negi et al., 2008; Geiger et al., 2011). Their difference is that SLAH3 predominately conducts nitrate but SLAC1 exhibited non-specific anion conductance. Given the key roles of SLAC and SLAH in stomatal closure, we propose that the *SLAC/SLAH* gene family might connect the stomatal response to salt stress with grain yield in barley.

The overarching hypothesis of this study was that physiological and molecular analysis of stomatal behavior contributes to the discovery of salt tolerance mechanisms in barley. The objectives of this study were to evaluate genetic variation of stomata behavior of barley under salinity stress and determine the links between physiological and molecular aspects of stomatal control and grain yield under saline condition in barley. It is likely that molecular markers and membrane transporter genes linked to stomatal traits could provide useful information for improving barley salinity tolerance in the future.

MATERIALS AND METHODS

PLANT MATERIALS

Barley varieties and a double haploid (DH) population (CM72/Gairdner) were used for the four experimental trials. The DH population of 108 lines, developed by another culture of the F1 hybrid between CM72 (California Mariout 72, six-rowed; salt-tolerant), and Gairdner (an Australian malt barley cultivar, two-rowed; salt-sensitive) was used in Glasshouse Trial 2.

EXPERIMENTAL TRIALS

Glasshouse Trial 1: Seeds of 10 parental barley cultivars were sown and seedlings were thinned to 5 plants per pot with 4 replicates for both control and NaCl treatment. *Glasshouse Trial 2:* Seeds of 108 DH lines, their two parental cultivars (CM72 and Gairdner) and two reference cultivars (Yerong and Franklin) were sown and seedlings were thinned to 4 plants per pot with 4 replicates for both control and NaCl treatment. *Field Trial*: Fifty-one barley varieties from around the world were selected to evaluate their salinity tolerance in the field in Tasmania, Australia. Each variety was sown in six 1.5 m × 0.2 m plots with half of them being treated with NaCl 1 week after germination. Salt treatment were achieved by gradually adding salt to the three of the treatment plots at a rate of 500 g NaCl m^{-2} over 3 consecutive days. The final electrical conductivity (EC) of the salt treated soils was ~10 dS m^{-1}. Normal pest and fertilizer application was employed. Leaf samples were collected for stomatal assay and gene expression analysis at Week 15. Plants were harvested for yield analysis at Week 20. *Glasshouse Trial 3:* The Field Trial was repeated in a glasshouse in New South Wales, Australia.

For the three Glasshouse Trials, plants were grown in two glasshouse rooms with identical conditions. All the plants were sown in 4-Litre pots containing potting mix and Osmocot® slow release fertilizer. Prior to salt treatment, all plants were watered and fed with full strength Hoagland's solution weekly. The glasshouse growth conditions were 24 ± 2°C and 60% relative humidity (RH) during the day, 22 ± 2°C and 70% RH at night. Broad spectrum growth lamps (600W HPS, GE Lighting, Smithfield, NSW, Australia) were used to provide a 12 h/12 h light/dark cycle. The average photosynthetically active radiation (PAR) received at the top of the plants was ~400 μmol m^{-2} s^{-1} over the duration of growth seasons. The plants were well watered to avoid drought stress during the experiments. Salt treatment was introduced 5 weeks after sowing at a concentration of 200 mM NaCl over 4 consecutive days at a rate of 50 mM (Munns et al., 2006; Chen et al., 2007c; Xue et al., 2009). All leached salt solutions were collected into the pot saucer and re-applied to ensure stability of NaCl concentrations across all treatment pots.

GAS EXCHANGE MEASUREMENT

Gas exchange was measurement as described in Chen et al. (2005); Mak et al. (2014); and O'carrigan et al. (2014). Net CO_2 assimilation (A), intercellular CO_2 concentration (Ci), stomata conductance (g_s), transpiration rate (T_r), leaf vapor pressure deficit (VPD), and leaf temperature (T_{leaf}) measurements were collected from third fully expanded leaves 4 week after salt treatment using a LI-6400XT infrared gas analyser (Li-Cor Inc., Lincoln, NE, USA). The conditions in the measuring chamber were controlled at a flow rate of 500 mol s^{-1}, a saturating PAR at 1500 μmol m^{-2} s^{-1}, a CO_2 level at 400 μmol mol^{-1} and a relative humidity of 65%. Gas exchange measurements were taken at the same time of day (approximately 10 a.m. to 4 p.m.) as for stomatal assay during full daylight for maximum photosynthesis. Each DH line was measured in its control and salt treatment pair to ensure comparative environmental and experimental conditions. Plants of each replication were randomly measured to minimize the effects of timing on gas exchange measurements.

STOMATAL ASSAY

Stomatal imaging was conducted as described in Mak et al. (2014) and O'carrigan et al. (2014) with some modification.

Third fully expanded leaves were collected from the glasshouse and transferred to the laboratory on tissue paper soaked in stomata stabilizing solution (50 mM KCl, 5 mM Na ± MES, pH 6.1) in petri dishes. The abaxial epidermal strips were then peeled and mounted on slides using a measuring solution (10 mM KCl, 5 mM Ca^2± MES, pH 6.1). Prompt peeling and mounting was used as an important quality control step to ensure aperture images are true representations of the stomata found naturally on the whole plant in the glasshouse. Images of the stomata were taken using a CCD camera (NIS-F1 Nikon, Tokyo, Japan) attached to a microscope (Leica Microsystems AG, Solms, Germany). All images were managed using a Nikon NIS Element imaging software (Nikon, Tokyo, Japan) and measured with Image J software (NIH, USA). The 12 stomatal parameters were aperture length and width, aperture width/length, stomatal pore area, guard cell length, width and volume, subsidiary cell length, width and volume, and stomatal density and index.

QTL ANALYSIS

The method for QTL analysis was essentially described in Xu et al. (2012). The DH population treated in 200 mM NaCl was used to identify QTLs controlling stomatal traits and gas exchange parameters using a map constructed with Diversity Array Technology (DArT) and Simple Sequence Repeat (SSR) markers. Using the software package MapQTL6.0 (Van Ooijen and Kyazma, 2009), QTLs were first analyzed by interval mapping (IM), followed by composite interval mapping (CIM). The closest marker at each putative QTL identified using interval mapping was selected as a cofactor and the selected markers were used as genetic background controls in the approximate multiple QTL model (MQM) of MapQTL6.0. Logarithm of the odds (LOD) threshold values for the presence of a QTL were estimated by performing the genome wide permutation tests implemented in MapQTL6.0 using at least 1000 permutations of the original data set for each trait, resulting in a 95% LOD threshold around 2.9. Two LOD support intervals around each QTL were established, by taking the two positions, left and right of the peak, that had LOD values of two less than the maximum (Van Ooijen and Kyazma, 2009), after performing restricted MQM mapping which does not use markers close to the QTL. The percentage of variance explained by each QTL (R^2) was obtained using restricted MQM mapping implemented with MapQTL5.0. Graphical representation of linkage groups and QTL was carried out using MapChart 2.2 (Voorrips, 2002).

PHYLOGENETIC AND ALIGNMENT ANALYSIS

The phylogenetic tree of the SLAC/SLAH family from Arabidopsis and selected cereal crops was generated using MEGA 6 program (Tamura et al., 2013). The Maximum Likelihood with WAG+Freqs (F) and Gamma distribution (G) model was employed. Names of organisms and accession numbers are from the National Centre for Biological Information (NCBI). Statistical values of phylogeny were estimated by the bootstrap method and were shown at the nodes. Alignment of the barley slow anion channel like 1 (HvSLAH1) and slow anion channel 1 (HvSLAC1) with homologous proteins in selected species was conducted using the CLASTALW tool in the BioEditor software (http://www.mbio.ncsu.edu/bioedit/bioedit.html) with 1000-bootstraps.

RT-PCR AND qPCR

Flag leaves of 16 barley varieties were collected from the Field Trial and immediately frozen in liquid nitrogen. Total RNA was extracted with TRIZOL® reagent (Life Technologies, Mulgrave, VIC, Australia) following the manufacturer's instructions. One microgram of total RNA sample was used to synthesize cDNA with the sensiFAST™ Kit (Bioline, Alexandria, NSW, Australia). The concentration of isolated total RNA and cDNA was determined using a NanoDrop ND-1000™ spectrophotometer (Thermo Scientific, Waltham, MA, USA).

PCR of *HvSLAH1* and *HvSLAC1* in cDNA was performed as described in the GoTaq® Flexi DNA Polymerase protocol (Promega, Alexandria, NSW, Australia). PCR thermal cycling conditions were 2 min at 95°C for initial denaturation, followed by 35 cycles of denaturation at 95°C for 30 s, annealing at 60°C for 30 s and an extension at 72°C for 11 s, and a final extension at 72°C for 5 min. For each experiment, 500 ng of cDNA was used for PCR amplification in a 20-μl reaction. PCR products were separated on a 0.9% (w/v) agarose gel and visualized with GelRed (Biotium, USA). Gel images were processed with Image J software (NIH, USA) to estimate the integrated fluorescence intensity of each band similar to cell imaging analysis in Bonales-Alatorre et al. (2013).

Transcript levels of *HvSLAH1* and *HvSLAC1* in leaves of 16 barley varieties were also determined by quantitative Real Time PCR (qPCR) using gene specific primers (Table S1). Purified cDNA samples were assayed using a Rotor-Gene® Q6000 (QIAGEN, Hilden, Germany) with the SensiFAST™ SYBR No-ROX Kit (Bioline, Alexandria NSW, Australia). qPCR conditions were consisted of 3-step cycling: 1.) polymerase activation at 95°C for 2 min; 2.) 40 cycles were set up for 5 s denaturation at 95°C, 10 s annealing at 63°C, 15 s extension at 72°C; 3.) SYBR green signal data were acquired at end. *HvACTIN* was used as the reference gene for normalization, which were selected from a number of candidates (Cao et al., 2014). Data are averages of three independent biological experiments, where each has two technical replicates.

STATISTICAL ANALYSIS

Statistical significance between control and the treatments was examined using the Student's *t*-test at $P < 0.05$ and $P < 0.01$. Pearson correlation analysis of all the parameters was conducted using SPSS 20 (IBM, New York, USA) and SigmaPlot 12 (Systat Software Inc., San Jose, CA, USA).

RESULTS

LARGE GENETIC DIVERSITY OF SALT TOLERANCE IN BARLEY

Barley genotypes showed contrasting response to salinity stress (**Figures 1A,B**). Salt-tolerant variety CM72 gained a significantly higher ($P < 0.05$) grain yield than salt-sensitive Franklin (**Figure 1B**). Some of the European elite varieties such as Bellini and Henley even performed better under salinity stress than CM72. In the Glasshouse Trial 1, similar trend was observed where CM72 and YYXT retained significantly more shoot dry

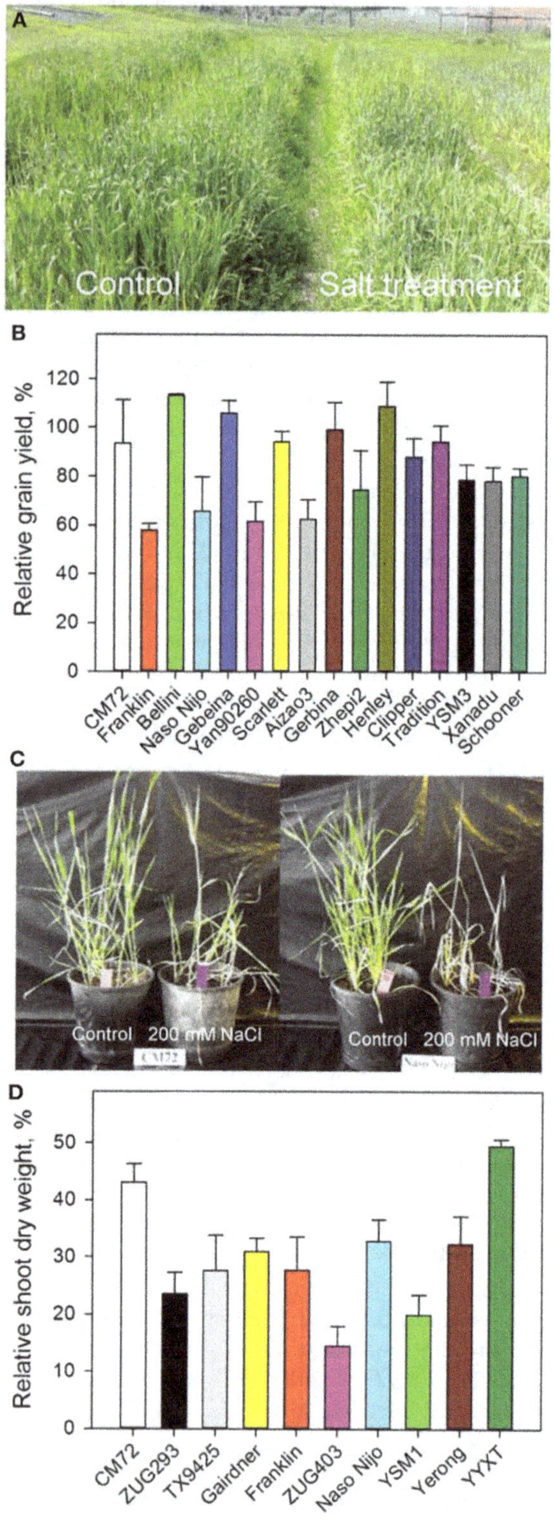

FIGURE 1 | Growth and yield of barley in the Field Trial and Glasshouse Trial 1. (A) Image shows 15-week old barley plants grown in control (left) and 10 dS m^{-1} NaCl (right) in the Field Trial in Launceston, Tasmania. **(B)** Relative grain yields of 16 randomly selected genotypes. Data are percentage of grain yields under the salt treatment compared to control. *(Continued)*

FIGURE 1 | Continued
Data are mean ± *SE* ($n = 3$). **(C)** Images of two representative varieties contrasting in their salinity tolerance. **(D)** Relative shoot dry weight of 10 varieties in Glasshouse Trial 1. Data shown are percentage of shoot dry weights under the salt treatment compared to control. Data are mean ± *SE* ($n = 4$).

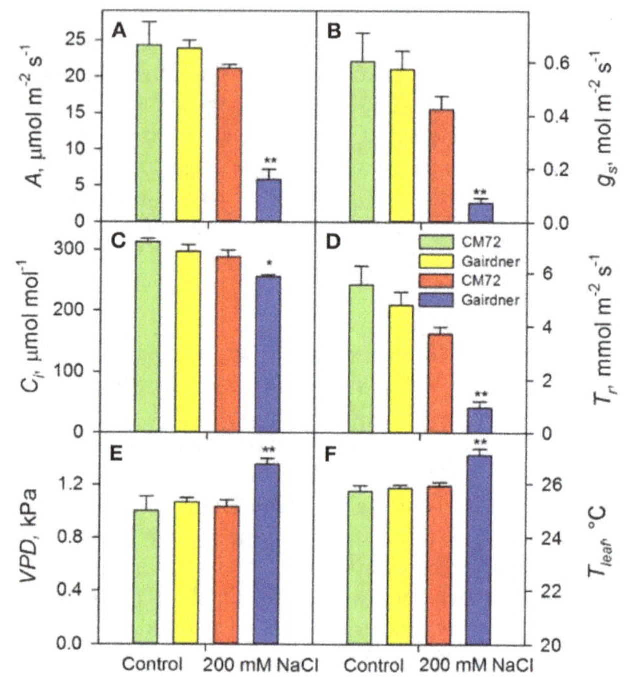

FIGURE 2 | Photosynthetic performance of two contrasting barley genotypes CM72 and Gairdner in the Glasshouse Trial 1. Data show net CO_2 assimilation **(A)**, stomatal conductance **(B)**, intracellular CO_2 concentration **(C)**, transpiration rate **(D)**, leaf vapor pressure deficit **(E)**, and leaf temperature **(F)** in the Control (green and yellow bars) and in 200 mM NaCl (red and blue bars) for 4 weeks. Data are mean ± *SE* ($n = 4$). * and ** indicate significant difference between CM72 and Gairdner under the salt treatment at $P < 0.05$ and $P < 0.01$, respectively.

weight than those salt-sensitive genotypes (**Figures 1C,D**). A large genetic diversity thus provided a foundation for further analysis of gas exchange, stomatal traits, gene expression and grain yield in contrasting cultivars and among the DH lines.

GAS EXCHANGE AND STOMATAL TRAITS DIFFER SIGNIFICANTLY BETWEEN SALT-TOLERANT AND SENSITIVE GENOTYPES

Distinct performance between salt-tolerant CM72 and salt-sensitive Gairdner was observed for gas exchange parameters in both Glasshouse Trial 1 (**Figure 2**) and Glasshouse Trial 2 (data not shown). Four weeks of salt treatment at 200 mM NaCl imposed no obvious change in these parameters in CM72, however, Gairdner exhibited huge reductions ($P < 0.01$) in A (72.5%), g_s (83.1%), and T_r (74.4%) as well as significant increases ($P < 0.01$) in *VPD* (30.9%) and T_{leaf} (4.4%) in comparison to these in CM72 (**Figures 2A–F**). We then further assessed the stomatal traits in these varieties in control (**Figures 3A,B**) and

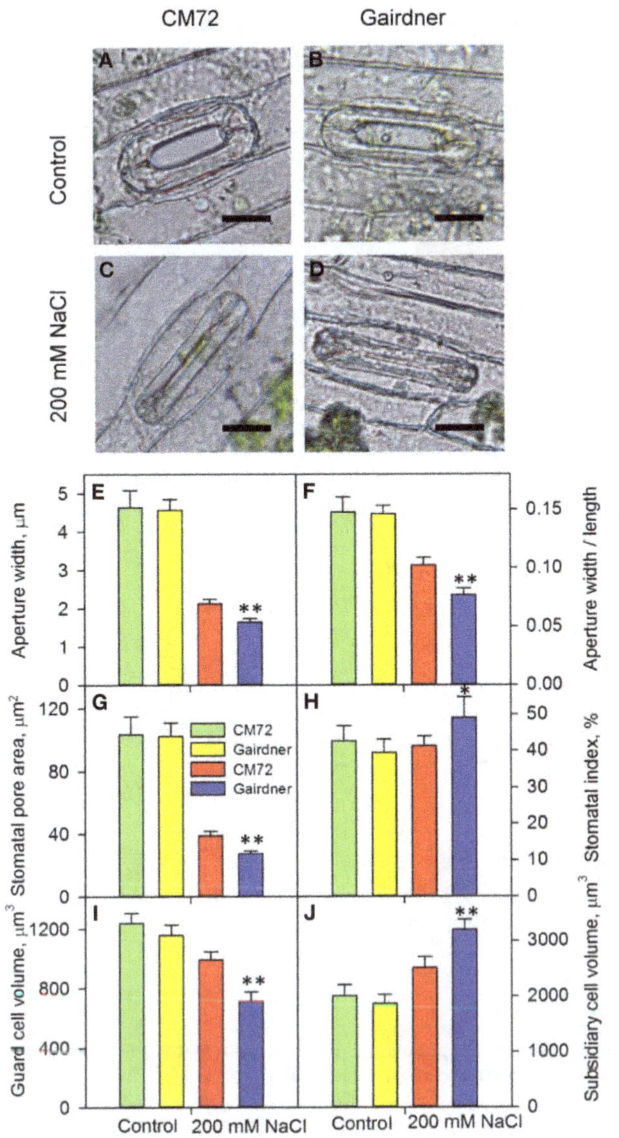

FIGURE 3 | Stomatal traits of two contrasting barley genotypes CM72 and Gairdner in the Glasshouse Trial 1. (A–D) Representative images of single stomatal complex for control and 200 mM NaCl treatment. Scale bars = 10 μm. Data show aperture width **(E)**, aperture width/length **(F)**, stomatal pore area **(G)**, stomatal index **(H)**, guard cell volume **(I)** and subsidiary cell volume **(J)** in control (green and yellow bars) and in 200 mM NaCl (red and blue bars) for 4 weeks. Data are mean + *SE* (n = 8–18 for stomatal index; n = 25–46 for rest of the stomatal traits). * and ** indicate significant difference between CM72 and Gairdner in the salt treatment at $P < 0.05$ and $P < 0.01$, respectively.

salinity treatment (**Figures 3C,D**) in Glasshouse Trial 1. There were negative impacts on the 11 stomatal traits 4 weeks after the imposition of salinity stress. Compared to Gairdner, CM72 gained 28.7, 32.9, 40.6, 39.6% larger aperture width, aperture width/length, stomatal pore area, guard cell volume and 19.1 and 27.4% lower stomatal index and subsidiary cell volume, respectively (**Figures 3E–J**). On the contrary, stomatal density was slightly increased ($P > 0.05$) by salinity treatment in both genotypes (**Figure S1**). Therefore, these stomatal traits showing large response to salt stress were employed for phenotypic characterization to identify potential salt tolerance-related QTLs in the CM72/Gairdner DH population.

STOMATAL TRAITS CONTRIBUTE SIGNIFICANTLY TO SALINITY TOLERANCE AND GRAIN YIELD IN BARLEY

The speed and accuracy of measuring stomatal traits is a major obstacle for its use in breeding program. In this study, we have refined this technique by standardizing the measurements for stomatal traits (**Figure 3**). Our frequency distribution results highlighted a distinct pattern of change in aperture width/length and grain yield at control (**Figures 4A,C**) and 200 mM NaCl (**Figures 4B,D**) of the DH lines. Salinity stress has led both traits to skewed distributions toward the lower values (**Figures 4B,D**). The potential contribution of 12 stomatal and 6 gas exchange traits to salinity tolerance in barley were demonstrated by significant correlations between these traits and biomass and grain yield in Glasshouse Trial 2 (data not shown). For instance, relative aperture length and relative aperture width/length showed significant correlation with relative grain yield (**Figures 5A,C**). Additionally, relative aperture width/length was correlated with relative biomass with statistical significance at $P < 0.01$ (**Figure 5B**). Most of the stomatal traits in the DH lines showed distinct response to the salt treatment. The five best performing DH lines were 2.3, 3.3, and 2.2-fold higher in aperture width/length, stomatal area, and guard cell volume in contrast to the five least performing ones (Table S3). Interestingly, one QTL was identified for this important stomatal trait—aperture width/length. This QTL explained 11.8% of the phenotypic variation with an LOD value of 2.90 (**Figure 6**). The nearest marker for this QTL is bPb-4564. This QTL is located on a similar position to one of the QTLs for salinity tolerance based on leaf injury in 240 mM NaCl treatment (Fan et al. unpublished data).

SLOW ANION CHANNEL GENES POSITIVELY REGULATE SALT TOLERANCE IN BARLEY

Slow anion channels SLAC and SLAH are the key regulators of stomatal closure in plants (Vahisalu et al., 2008; Geiger et al., 2011; Dreyer et al., 2012; Hedrich, 2012). Our comparative genomic study indicated that SLACs and SLAHs exist in a large range of plant species (**Figure 7**). There are 4, 14, 10, 5, 3, and 3 SLAH-like genes in Arabidopsis, rice, maize, sorghum, wheat and barley, respectively. The numbers of corresponding SLAC-like genes are much lower (1, 3, 2, 1, 0, and 1) in these species (**Figure 7**, **Figure S2**). Further analysis showed that coding DNA sequence of *HvSLAC1* and *HvSLAH1* identified in this study have 55.0 and 54.8% homology to the well-characterized Arabidopsis *SLAC1* and *SLAH3*, respectively. Alignment analysis of amino acids revealed that HvSLAC1 and HvSLAH belong to SLAC1-group and SLAH1/4-group, respectively, as proposed in Dreyer et al. (2012). Their protein sequences showed a clear difference as compared to Arabidopsis and other cereal crops (**Figure S2**). These data provide a valuable source of information for future studies into the understanding of mechanisms of the different SLAC/SLAH contributions to barley salt tolerance.

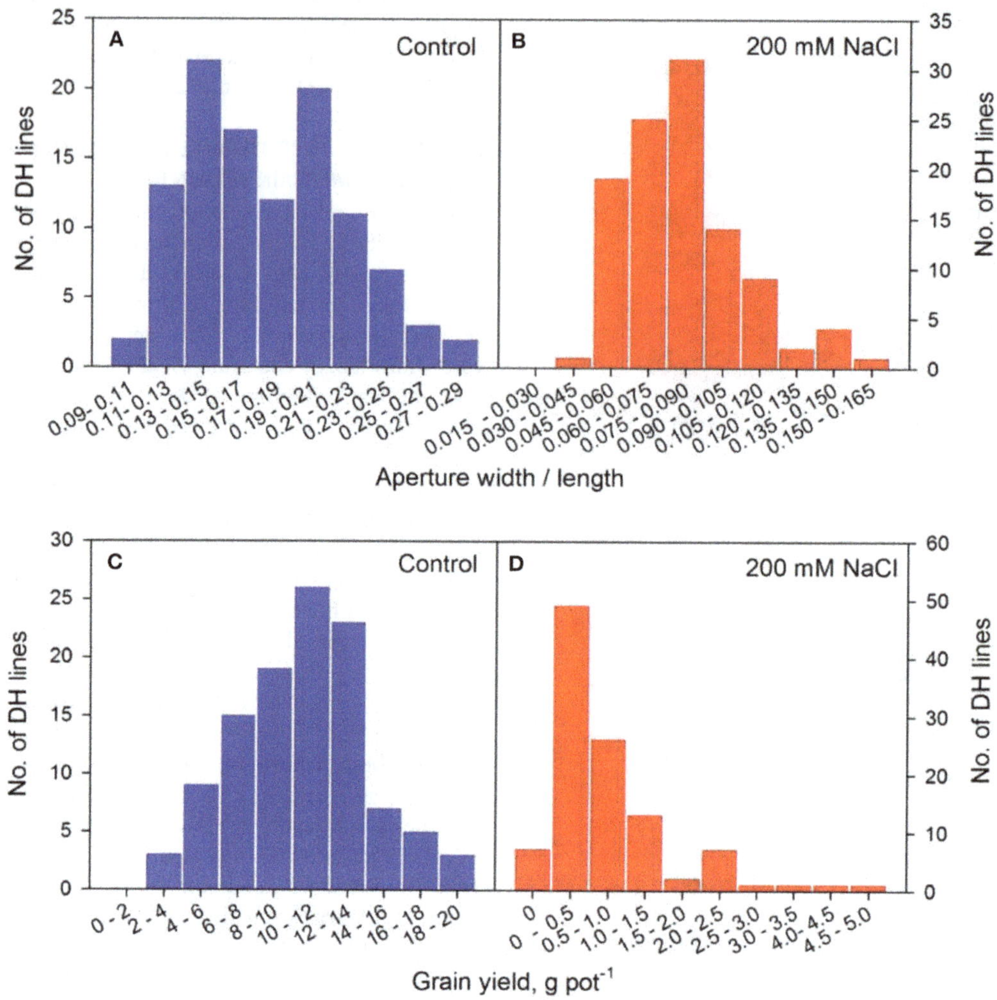

FIGURE 4 | Frequency distribution of representative phenotyping traits of the CM72/Gairdner DH population in Glasshouse Trial 2. Frequency distribution of aperture width/length at control **(A)** and 200 mM NaCl **(B)** and grain yield at Control **(C)** and 200 mM NaCl **(D)** of the DH lines.

We then conducted experiments to identify these barley *SLAC/SLAHs* in leaves from control and salt treatment in the field (**Figures 1A, 8**). As the first to report these genes in barley leaves, we named them as *HvSLAH1* and *HvSLAC1*. The expression of these two genes showed a large variation among the 16 genotypes in both control and salinity stress (**Figure 8**). Specifically, the salt-tolerant variety CM72 showed a NaCl-induced significant up-regulation of both *HvSLAH1* and *HvSLAC1*, which were significantly decreased or remained unchanged in salt-sensitive varieties Franklin and Naso Nijo (**Figures 8B,C**). Interestingly, the European varieties Bellini, Scarlett and Henley tended to have NaCl-induced up-regulation, but Asian varieties such as Naso Nijo, Yan90260, and Aizao3 showed down-regulation of *HvSLAH1* and *HvSLAC1* (**Figure 8**). Furthermore, highly significant ($P < 0.01$) positive correlations were found between the transcripts of two genes and salt tolerance of barley both in the field and the glasshouse (**Figure 9** and Table S2). These were consistent across the RT-PCR gel integrated intensity based results and qPCR analysis (**Figure 8**). Additionally, the transcripts of the two genes from the samples collected in the Field Trial were even significantly linked to the visual salt tolerance score and grain yield in the Glasshouse Trial 3 (Table S2).

DISCUSSION

EXPLORING THE GENETIC DIVERSITY OF BARLEY FOR SALT TOLERANCE

The Hordeum genus, to which modern cultivated barley belongs, contains species which have adapted to grow in a wide variety of environmental conditions from the sub-arctic Scandinavia to the subtropical North Africa. Species include annual and perennial varieties with a wide array of shapes and structures, phenology and reproductive variations, suggesting that the genus is highly diverse and adaptable (Ullrich, 2011; Dai et al., 2012, 2014). It would not come as a surprise that landrace species of barley are found in saline meadows and marshes along coastal regions, utilizing morphologies like deep root system to evolve into salt-tolerant varieties such as *Hordeum marinum* (Ullrich, 2011). Recently, Tibetan wild barley (*Hordeum vulgare spp. spontaneum*),

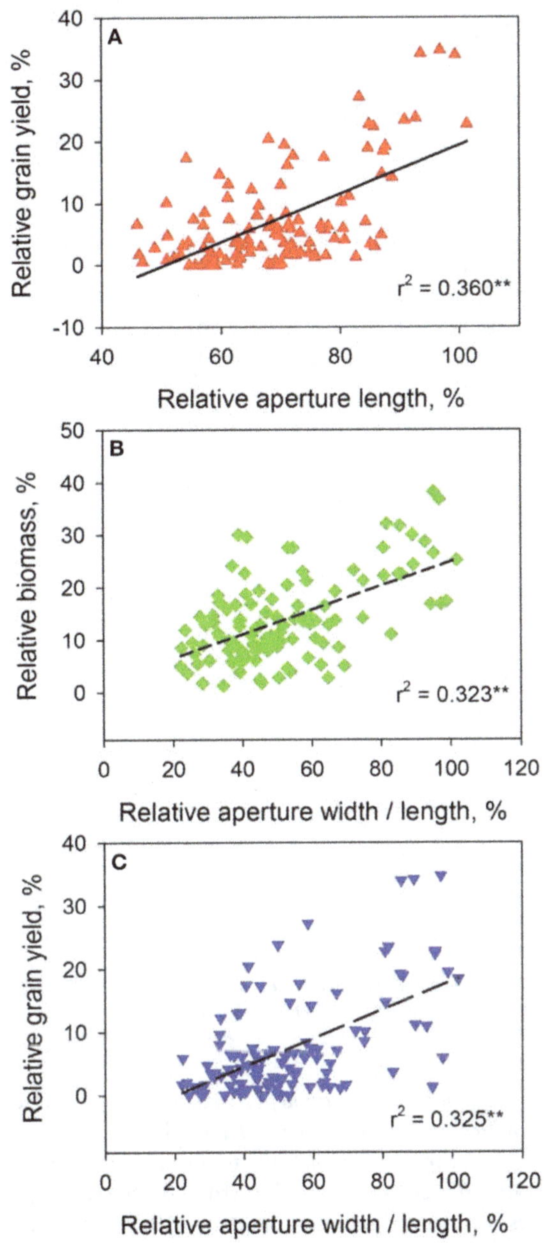

FIGURE 5 | Correlation analysis of representative stomatal and agronomical traits of the 108 DH lines. Data show correlations between relative grain yield and relative aperture length **(A)**, relative aperture width/length and relative biomass **(B)** and relative grain yield **(C)**. $^{**}P < 0.01$.

as one of the ancestors of cultivated barley (Dai et al., 2012, 2014), has been found to be even more tolerant to salinity (Qiu et al., 2011; Wu et al., 2011, 2013a,b). Furthermore, the level of salt tolerance in modern cultivated barley also differs significantly among genotypes. For example, seedlings of a barley cultivar California Mariout (CM) showed no growth retardation at 400 mM NaCl, while other barley genotypes were more affected (Epstein et al., 1980). Our results have further extended the large genetic variation to gas exchange (data not shown), stomatal

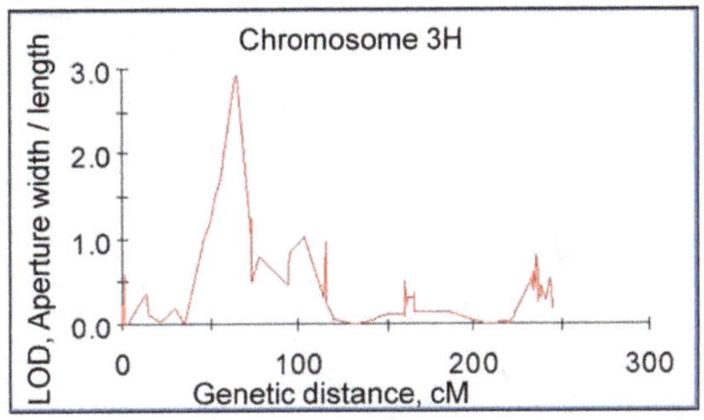

FIGURE 6 | A significant QTL for aperture width/length in salt tolerance. This QTL was mapped on barley chromosome 3H in the double haploid population from CM72/Gairdner using a map constructed with Diversity Array Technology (DArT) and Simple Sequence Repeat (SSR) markers.

traits (**Figures 3–6**) and the expression levels of *HvSLAH1* and *HvSLAC1* (**Figure 8**) in response to salinity stress in barley.

GAS EXCHANGE TRAITS MAY BE ACCURATE INDICATORS FOR SALINITY TOLERANCE BUT NOT FOR PHENOTYPING STUDY IN BARLEY

The effects of salinity stress on plants are largely due to osmotic stress and ion cytotoxicity (Zhu, 2002; Munns and Tester, 2008; Shabala and Mackay, 2011). Crop yield under salinity stress is a result of balancing the allocations of limited photosynthetic carbon gain toward both reproduction and growth. Photosynthesis is a multi-facetted process, which has dependencies on the availability of CO_2, water and light (Wong et al., 1979; Farquhar and Sharkey, 1982). Upon the NaCl-induced stomatal closure, the reduced CO_2 acquisition becomes the limiting factor in photosynthesis, resulting in leaf temperature increase, higher intracellular O_2:CO_2 ratio and oxidative stress (Farquhar and Sharkey, 1982; Chaves et al., 2009). For instance, salinity stress remarkably reduced A and g_s in sorghum (Yan et al., 2012), wheat (Zheng et al., 2009) and barley (Jiang et al., 2006), the extent being considerably larger in salt-sensitive genotypes than salt-tolerant ones. Hence, the salt-tolerant genotypes could harmonize the relationship between CO_2 assimilation (source) and the grain yield (sink) under the saline conditions (James et al., 2002). Measurement of gs provided the best information to assess genetic differences in barley for absolute performance when subjected to salinity stress (Jiang et al., 2006). In our experiments with gas exchange measurements (Glasshouse Trials 1, 2, and 3), some key traits (e.g., g_s, T_r, VPD) were found to be important indicators for salt tolerance after 4 weeks of 200 mM NaCl treatment. However, no significant QTLs were identified for the 6 gas exchange traits in the CM72/Gairdner DH population. We have reasoned that measuring these traits was highly variable over the course of the day and over a period up to 6 days. Also, many stomatal and non-stomatal limitations (James et al., 2002; Munns et al., 2006) are likely to make these measurements more difficult to control. In addition, there were limited replicates of gas exchange measurements as compared to up to 70 for the stomatal traits in some lines.

FIGURE 7 | Phylogenetic tree of SLAC and SLAH anion channel family in barley, Arabidopsis, and selected cereal crops. Statistical values of phylogeny estimated by bootstrap method are shown at the nodes. HvSLAC1 and HvSLAH1 are marked by a green square and a blue circle, respectively.

SYSTEMATIC INVESTIGATION OF STOMATAL TRAITS FOR BARLEY SALINITY TOLERANCE

Most of the previously studies have mainly evaluated a small numbers of genotypes or a small genetic population for their stomatal behavior in response to salinity stress. Among those, limited traits such as stomatal length and width, stomatal density and index measured from leaf imprint have been used (Pallaghy, 1971; Edwards and Meidner, 1979; Aryavand et al., 2003; Mumm et al., 2011). Here, we have conducted a comprehensive comparative study with large datasets on stomatal traits (e.g., 75,640 data points for 12 parameters in Glasshouse Trial 2 alone). In addition, we have developed an efficient technique to obtain viable single-layer epidermal peels for microscopic recording of high resolution images (**Figure S1**, **Figure 3**). All of these have allowed the assessment of a large number of genotypes and relatively big genetic populations in the glasshouse and even in the field.

Salt-tolerant plant species *Aeluropus lagopoides* had fewer and smaller stomata on both leaf surfaces and stomata of *Osyris compressa* were found only on the abaxial surface under salinity stress (Naz et al., 2010). Salt tolerance was related to lower stomatal density and decreased stomatal area in *Sporobolus ioclados* (Naz et al., 2010) and *Chenopodium quinoa* (Shabala et al., 2013). Here, the correlation between stomatal density and salt tolerance in the barley DH population is marginal due to a lack of difference in the parental lines (**Figure S1**). Another study on this issue using 51 genotypes is currently under investigation. However, the relationships between aperture length, aperture width/length (**Figure 5**), guard cell and subsidiary cell volumes (data not shown) and grain yield in response to salinity stress and significant stomatal trait-related QTLs (**Figure 6**) have provided some insights into the contribution of stomatal traits in salinity tolerance in barley. The salt-tolerant varieties and best performing DH lines showed significantly larger stomatal pore area and guard cell volume as compared to the salt-sensitive and worst performing DH lines (**Figures 3–6**). Therefore, the reduction in photosynthesis and grain yield could be identified by stomatal assay to differentiate the responses among genotypes and genetic populations.

FOCUS ON MEMBRANE TRANSPORTER REGULATION FOR FUTURE BREEDING FOR SALT-TOLERANT CROPS

Significant progress has been made on the molecular mechanisms of membrane transport for salt tolerance in plants. Combining

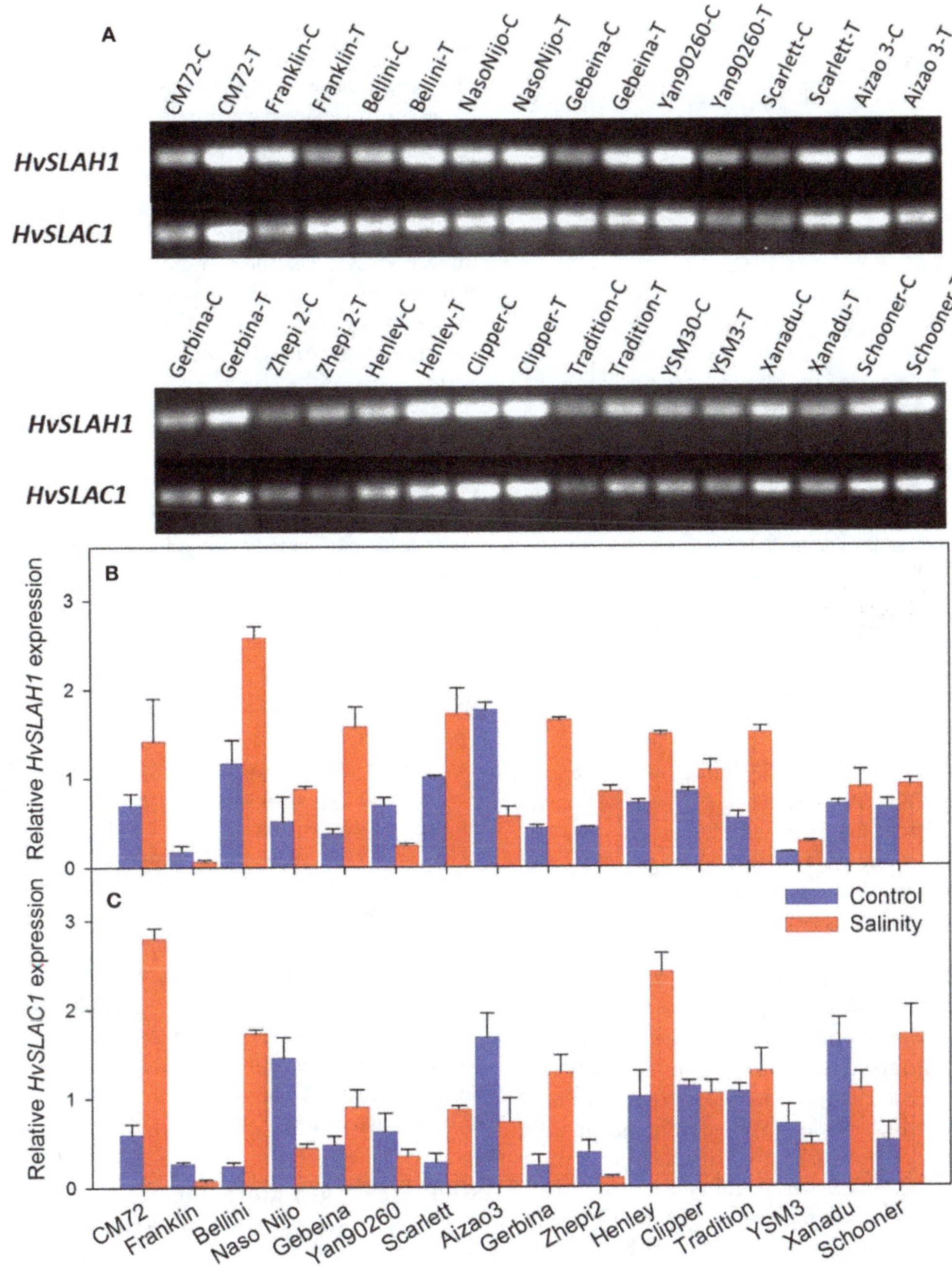

FIGURE 8 | Expression of *HvSLAH1* and *HvSLAC1* in leaves from the Field Trial. (A) Real time-PCR gel electrophoresis of the two slow anion channel genes *HvSLAH1* and *HvSLAC1* in control **(C)** and salt treatment (T). Quantitative RT-PCR of the two slow anion channel genes *HvSLAH1* **(B)** and *HvSLAC1* **(C)** in the Control (blue bars) and salt treatment (red bars). *HvACTIN* was used as a reference gene. Data are average ± SD with three biological replicates.

plant physiology, molecular breeding and genetic engineering have resulted in improvements in crop yields in saline soil (Munns and Tester, 2008; Tavakkoli et al., 2010; Schroeder et al., 2013; Deinlein et al., 2014). For instance, *TmHKT1;5-A* in the *Nax2* locus encodes a Na ± selective transporter located on the plasma membrane of root cells surrounding xylem vessels of bread wheat. It can remove Na^+ from the xylem and reduce transport of Na^+ to leaves. Field trials demonstrated that *TmHKT1;5-A* significantly reduces leaf Na^+ and increases durum wheat grain yield by 25% (Munns et al., 2012). More recently, the expression of *AVP1*, an Arabidopsis gene encoding a vacuolar proton pumping pyrophosphatase, has been shown to improve the salinity tolerance of transgenic plants in both greenhouse and in field trials (Schilling et al., 2014). It is also the case for our comparative

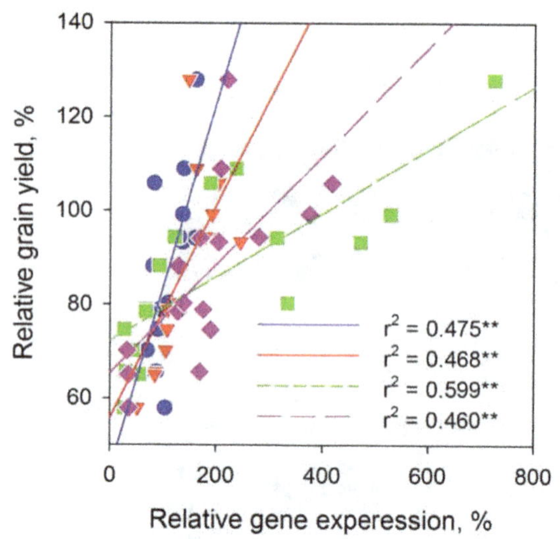

FIGURE 9 | Linking leaf *HvSLAH1* and *HvSLAC1* expression to barley grain yield in the Field Trial. Relative gene expression was expressed as the transcripts under the salt treatment over that in control. Data were collected from 16 barley cultivars and fitted separately for each gene in both RT-PCR (blue circles and red triangles) and q-PCR (green squares and pink diamonds) experiments. $^{**}P < 0.01$.

experimental trials in both field and glasshouse, where the grain yields and transcript of *HvSLAC1* and *HvSLAH1* are consistently and significantly correlated (**Figure 9**; Table S2).

Stomatal aperture in monocots is controlled by guard cells and their dynamic interactions and "shuttling" of ions and solutes with subsidiary cells. These processes are largely regulated by the concerted coordination of membrane transporters on both cell types (Raschke and Fellows, 1971; Edwards et al., 1976; Fairley-Grenot and Assmann, 1992; Philippar et al., 2003; Mumm et al., 2011). However, there are very few reports on guard cell transporters and salinity tolerance. Guard cell cation channels were found to be involved in Na ± induced stomatal closure in a halophyte *Aster tripolium*, which possesses no specific morphological adaptation to salinity. Short-term Na^+ treatment (~30 min) induced stomatal opening. The plasma membrane K^+ inward and outward rectifying channels were highly selective for K^+ against Na^+. Cytosolic Na^+ accumulation over long-term has led to a delayed and dramatic deactivation of the K^+ inward rectifying channels via an increase of cytosolic Ca^{2+} concentration (Very et al., 1998). In Arabidopsis, two SLAC/SLAH channels regulated anion efflux from the guard cells for stomatal closure and another two regulate anion/nitrate transport in roots (Vahisalu et al., 2008; Dreyer et al., 2012; Hedrich, 2012; Hills et al., 2012). Having higher transcripts of *HvSLAC1* and *HvSLAH1* or potentially higher number of these channels, salt-tolerant barley could have higher anion efflux for closure under salinity stress. However, the tolerant varieties showed larger stomatal aperture in salt treatment (**Figure 3**), which is seemingly contradicting to the up-regulation of *HvSLAC1* and *HvSLAH1*. Therefore, these genes could assist salt-tolerant varieties to have more rapid closure as an efficient tool to optimize water balance. On the other hand, one can argue that ion channels are closed most of the time and are not commonly considered as the limiting factor. Why do plants need to express more of *SLACs/SLAHs* if they can simply have available ones open for longer? Nevertheless, these controversies have provided new directions to investigate these genes in barley, Arabidopsis and other cereal crops using comparative genomic tool and the assembled barley genome (International Barley Genome Sequencing Consortium, 2012) in more depth in the future. It was also indicated that the other three HvSLAH channels might contribute to the regulation of stomata and nitrogen homeostasis. Further research is necessary to dissect the phylogenetic relationships and functional properties of the HvSLAHs and decipher their roles for salt tolerance in barley.

AUTHOR CONTRIBUTIONS

Zhong-Hua Chen, and Meixue Zhou designed research; Xiaohui Liu, Michelle Mak, Mohammad Babla, Feifei Wang, and Filip Veljanoski performed research; Xiaohui Liu, Michelle Mak, Mohammad Babla, Guang Chen, Zhong-Hua Chen. analyzed data; Zhong-Hua Chen, Meixue Zhou, Gang Wang, Sergey Shabala, Xiaohui Liu, and Michelle Mak wrote the paper.

ACKNOWLEDGMENTS

This project is supported by a Science and Innovation Award and a Minister's Award to Zhong-Hua Chen funded by Grains Research and Development Corporation (GRDC), Department of Agriculture, Australia. Zhong-Hua Chen is also supported by a University of Western Sydney Research Lectureship and an Australian Research Council (ARC) Discovery Early Career Research Award (DE140101143). Xiaohui Liu is the receipt of a China Scholarship Council award. Gang Wang is supported by a National Natural Science Foundation of China project (31271419). We thank Professor Chengdao Li at Department of Agriculture and Food Western Australia for providing the barley seeds. We thank Linda Westmoreland, Jennie Nelson, Renee Smith, Sumedha Dharmaratne, and Rosemary Freeman for their technical assistance.

SUPPLEMENTARY MATERIAL

The Supplementary Material for this article can be found online at: http://www.frontiersin.org/journal/10.3389/fpls.2014.00634/abstract

Figure S1 | Stomatal density of two contrasting barley genotypes CM72 and Gairdner in the Glasshouse Trial 1. (A-D) Representative stomatal images for control and 200 mM NaCl treatment. Scale bars = 40 μm. **(E)** Stomatal density in control (green and yellow bars) and in 200 mM NaCl (red and blue bars) for 4 weeks. Data are mean ± *SE* (n = 8–18).

Figure S2 | Alignment of HvSLAC1 and HvSLAH1 with homologous proteins in selected species. (A-C) Alignment of HvSLAC1 and some SLACs from monocot plant species and Arabidopsis. **(D-G)** Alignment of HvSLAH1 and some S-type anion channel SLAH from monocot plant species and Arabidopsis. Identical residues among the other proteins are indicated with asterisks. Colons indicate amino acids with high similarity among the proteins. Dashes indicate gaps.

REFERENCES

Aryavand, A., Ehdaie, B., Tran, B., and Waines, J. G. (2003). Stomatal frequency and size differentiate ploidy levels in *Aegilops neglecta*. *Genet. Resour. Crop Evol.* 50, 175–182. doi: 10.1023/A:1022941532372

Barbier-Brygoo, H., De Angeli, A., Filleur, S., Frachisse, J.-M., Gambale, F., Thomine, S., et al. (2011). Anion channels/transporters in plants: from molecular bases to regulatory networks. *Annu. Rev. Plant Biol.* 62, 25–51. doi: 10.1146/annurev-arplant-042110-103741

Bonales-Alatorre, E., Shabala, S., Chen, Z. H., and Pottosin, I. (2013). Reduced tonoplast fast-activating and slow-activating channel activity is essential for conferring salinity tolerance in a facultative halophyte, Quinoa. *Plant Physiol.* 162, 940–952. doi: 10.1104/pp.113.216572

Cao, F., Sun, H., Chen, F., Wang, F., Zhang, G., Chen, Z., et al. (2014). Genome-wide transcriptome analysis reveals cadmium-induced differential expression of key genes in Cd-tolerant and -sensitive barley genotypes. *BMC Genomics* 15:611. doi: 10.1186/1471-2164-15-611

Chaves, M., Flexas, J., and Pinheiro, C. (2009). Photosynthesis under drought and salt stress: regulation mechanisms from whole plant to cell. *Ann. Bot.* 103, 551–560. doi: 10.1093/aob/mcn125

Chen, Z., Cuin, T. A., Zhou, M., Twomey, A., Naidu, B. P., and Shabala, S. (2007a). Compatible solute accumulation and stress-mitigating effects in barley genotypes contrasting in their salt tolerance. *J. Exp. Bot.* 58, 4245–4255. doi: 10.1093/jxb/erm284

Chen, Z.-H., and Blatt, M. R. (2010). "Membrane transport in guard cells," in *Encyclopedia of Life Sciences* (Chichester: John Wiley & Sons, Ltd.), 1–13. doi: 10.1002/9780470015902.a0021630

Chen, Z. H., Hills, A., Batz, U., Amtmann, A., Lew, V. L., and Blatt, M. R. (2012). Systems dynamic modeling of the stomatal guard cell predicts emergent behaviors in transport, signaling, and volume control. *Plant Physiol.* 159, 1235–1251. doi: 10.1104/pp.112.197350

Chen, Z.-H., Hills, A., Lim, C. K., and Blatt, M. R. (2010). Dynamic regulation of guard cell anion channels by cytosolic free Ca^{2+}concentration and protein phosphorylation. *Plant J.* 61, 816–825. doi: 10.1111/j.1365-313X.2009.04108.x

Chen, Z., Newman, I., Zhou, M., Mendham, N., Zhang, G., and Shabala, S. (2005). Screening plants for salt tolerance by measuring K^+ flux: a case study for barley. *Plant Cell Environ.* 28, 1230–1246. doi: 10.1111/j.1365-3040.2005.01364.x

Chen, Z., Pottosin, I. I., Cuin, T. A., Fuglsang, A. T., Tester, M., Jha, D., et al. (2007b). Root plasma membrane transporters controlling K^+/Na^+ homeostasis in salt stressed barley. *Plant Physiol.* 145, 1714–1725. doi: 10.1104/pp.107.110262

Chen, Z., Zhou, M., Newman, I. A., Mendham, N. J., Zhang, G., and Shabala, S. (2007c). Potassium and sodium relations in salinised barley tissues as a basis of differential salt tolerance. *Funct. Plant Biol.* 34, 150–162. doi: 10.1071/FP06237

Dai, F., Chen, Z.-H., Li, Z., Wang, X., Cai, S., Wu, D., et al. (2014). Transcriptome profiling reveals mosaic genomic origins of modern cultivated barley. *Proc. Natl. Acad. Sci. U.S.A.* 111, 13403–13408. doi: 10.1073/pnas.1414335111

Dai, F., Nevo, E., Wu, D., Comadran, J., Zhou, M., Qiu, L., et al. (2012). Tibet is one of the centers of domestication of cultivated barley. *Proc. Natl. Acad. Sci. U.S.A.* 109, 16969–16973. doi: 10.1073/pnas.1215265109

Deinlein, U., Stephan, A. B., Horie, T., Luo, W., Xu, G., and Schroeder, J. I. (2014). Plant salt-tolerance mechanisms. *Trends Plant Sci.* 19, 371–379. doi: 10.1016/j.tplants.2014.02.001

Dreyer, I., Gomez-Porras, J. L., Riaño-Pachón, D. M., Hedrich, R., and Geiger, D. (2012). Molecular evolution of slow and quick anion channels (SLACs and QUACs/ALMTs). *Front. Plant Sci.* 3:263. doi: 10.3389/fpls.2012.00263

Edwards, M., and Meidner, H. (1979). Direct measurements of turgor pressure potentialsIV. Naturally occurring pressures in guard cells and their relation to solute and matric potentials in the epidermis. *J. Exp. Bot.* 30, 829–837. doi: 10.1093/jxb/30.4.829

Edwards, M., Meidner, H., and Sheriff, D. W. (1976). Direct measurements of turgor pressure potentials of guard cells II. The mechanical advantage of subsidiary cells, the spannunqsphase, and the optimum leaf water deficit. *J. Exp. Bot.* 27, 163–171. doi: 10.1093/jxb/27.1.163

Epstein, E., Norlyn, J. D., Rush, D. W., Kingsbur, Y. R. W., Kelley, D. B., Cunningham, G. A., et al. (1980). Saline culture of crops: a genetic approach. *Science* 210, 399–404. doi: 10.1126/science.210.4468.399

Fairley-Grenot, K. A., and Assmann, S. M. (1992). Whole-cell K^+ current across the plasma membrane of guard cells from a grass: *Zea mays*. *Planta* 186, 282–293. doi: 10.1007/BF00196258

Farquhar, G., and Sharkey, T. (1982). Stomatal conductance and photosynthesis. *Annu. Rev. Plant Physiol.* 33, 317–345. doi: 10.1146/annurev.pp.33.060182.001533

Geiger, D., Maierhofer, T., Al-Rasheid, K. A., Scherzer, S., Mumm, P., Liese, A., et al. (2011). Stomatal closure by fast abscisic acid signaling is mediated by the guard cell anion channel SLAH3 and the receptor RCAR1. *Sci. Signal.* 4, ra32. doi: 10.1126/scisignal.2001346

Hedrich, R. (2012). Ion channels in plants. *Physiol. Rev.* 92, 1777–1811. doi: 10.1152/physrev.00038.2011

Hills, A., Chen, Z. H., Amtmann, A., Blatt, M. R., and Lew, V. L. (2012). OnGuard, a computational platform for quantitative kinetic modeling of guard cell physiology. *Plant Physiol.* 159, 1026–1042. doi: 10.1104/pp.112.197244

International Barley Genome Sequencing Consortium. (2012). A physical, genetic and functional sequence assembly of the barley genome. *Nature* 491, 711–716. doi: 10.1038/nature11543

James, R. A., Rivelli, A. R., Munns, R., and Von Caemmerer, S. (2002). Factors affecting CO_2 assimilation, leaf injury and growth in salt-stressed durum wheat. *Funct. Plant Biol.* 29, 1393–1403. doi: 10.1071/FP02069

Jiang, Q., Roche, D., Monaco, T., and Durham, S. (2006). Gas exchange, chlorophyll fluorescence parameters and carbon isotope discrimination of 14 barley genetic lines in response to salinity. *Field Crops Res.* 96, 269–278. doi: 10.1016/j.fcr.2005.07.010

Kearsey, M. (1998). The principles of QTL analysis (a minimal mathematics approach). *J. Exp. Bot.* 49, 1619–1623. doi: 10.1093/jxb/49.327.1619

Kim, T. H., Böhmer, M., Hu, H., Nishimura, N., and Schroeder, J. I. (2010). Guard cell signal transduction network: advances in understanding abscisic acid, CO_2, and Ca^{2+} signaling *Annu. Rev. Plant Biol.* 61, 561–591. doi: 10.1146/annurev-arplant-042809-112226

Maierhofer, T., Lind, C., Hüttl, S., Scherzer, S., Papenfuß, M., Simon, J., et al. (2014). A single-pore residue renders the Arabidopsis root anion channel SLAH2 highly nitrate selective. *Plant Cell* 26, 2554–2567. doi: 10.1105/tpc.114.125849

Mak, M., Babla, M., Xu, S.-C., O'carrigan, A., Liu, X.-H., Gong, Y.-M., et al. (2014). Leaf mesophyll K^+, H^+ and Ca^{2+} fluxes are involved in drought-induced decrease in photosynthesis and stomatal closure in soybean. *Environ. Exp. Bot.* 98, 1–12. doi: 10.1016/j.envexpbot.2013.10.003

Mano, Y., and Takeda, K. (1997). Mapping quantitative trait loci for salt tolerance at germination and the seedling stage in barley (*Hordeum vulgare* L.). *Euphytica* 94, 263–272. doi: 10.1023/A:1002968207362

Mumm, P., Wolf, T., Fromm, J., Roelfsema, M. R. G., and Marten, I. (2011). Cell type-specific regulation of ion channels within the maize stomatal complex. *Plant Cell Physiol.* 52, 1365–1375. doi: 10.1093/pcp/pcr082

Munns, R., James, R. A., and Läuchli, A. (2006). Approaches to increasing the salt tolerance of wheat and other cereals. *J. Exp. Bot.* 57, 1025–1043. doi: 10.1093/jxb/erj100

Munns, R., James, R. A., Xu, B., Athman, A., Conn, S. J., Jordans, C., et al. (2012). Wheat grain yield on saline soils is improved by an ancestral Na^+ transporter gene. *Nat. Biotechnol.* 30, 360–364. doi: 10.1038/nbt.2120

Munns, R., and Tester, M. (2008). Mechanisms of salinity tolerance. *Annu. Rev. Plant Biol.* 59, 651–681. doi: 10.1146/annurev.arplant.59.032607.092911

Naz, N., Hameed, M., Ashraf, M., Al-Qurainy, F., and Arshad, M. (2010). Relationships between gas-exchange characteristics and stomatal structural modifications in some desert grasses under high salinity. *Photosynthetica* 48, 446–456. doi: 10.1007/s11099-010-0059-7

Negi, J., Matsuda, O., Nagasawa, T., Oba, Y., Takahashi, H., Kawai-Yamada, M., et al. (2008). CO_2 regulator SLAC1 and its homologues are essential for anion homeostasis in plant cells. *Nature* 452, 483–486. doi: 10.1038/nature06720

O'carrigan, A., Hinde, E., Lu, N., Xu, X.-Q., Duan, H., Huang, G., et al. (2014). Effects of light irradiance on stomatal regulation and growth of tomato. *Environ. Exp. Bot.* 98, 65–73. doi: 10.1016/j.envexpbot.2013.10.007

Pallaghy, C. K. (1971). Stomatal movement and potassium transport in epidermal strips of *Zea mays*. *Planta* 101, 287–295. doi: 10.1007/BF00398115

Philippar, K., Buchsenschutz, K., Abshagen, M., Fuchs, I., Geiger, D., Lacombe, B., et al. (2003). The K^+ channel KZM1 mediates potassium uptake into the phloem and guard cells of the C4 grass *Zea mays*. *J. Biol. Chem.* 278, 16973–16981. doi: 10.1074/jbc.M212720200

Pitman, M. G., and Lähli, A. (2002). "Global impact of salinity and agricultural ecosystems," in *Salinity: Environment-Plants-Molecules*, eds A. Läuchli and U. Lüttge (Dordrecht:Springer), 3–20.

Qiu, L., Wu, D., Ali, S., Cai, S., Dai, F., Jin, X., et al. (2011). Evaluation of salinity tolerance and analysis of allelic function of *HvHKT1* and *HvHKT2* in Tibetan wild barley. *Theor. Appl. Genet.* 122, 695–703. doi: 10.1007/s00122-010-1479-2

Raschke, K., and Fellows, M. P. (1971). Stomatal movement in *Zea mays* shuttle of potassium and chloride between guard cells and subsidiary cells. *Planta* 101, 296–316. doi: 10.1007/BF00398116

Schilling, R. K., Marschner, P., Shavrukov, Y., Berger, B., Tester, M., Roy, S. J., et al. (2014). Expression of the Arabidopsis vacuolar H^+-pyrophosphatase gene (AVP1) improves the shoot biomass of transgenic barley and increases grain yield in a saline field. *Plant Biotechnol. J.* 12, 378–386. doi: 10.1111/pbi.12145

Schroeder, J. I., Delhaize, E., Frommer, W. B., Guerinot, M. L., Harrison, M. J., Herrera-Estrella, L., et al. (2013). Using membrane transporters to improve crops for sustainable food production. *Nature* 497, 60–66. doi: 10.1038/nature11909

Shabala, S., Hariadi, Y., and Jacobsen, S.-E. (2013). Genotypic difference in salinity tolerance in quinoa is determined by differential control of xylem Na^+ loading and stomatal density. *J. Plant Physiol.* 170, 906–914. doi: 10.1016/j.jplph.2013.01.014

Shabala, S., and Mackay, A. (2011). Ion transport in halophytes. *Adv. Bot. Res.* 57, 151–199. doi: 10.1016/B978-0-12-387692-8.00005-9

Siahsar, B., and Narouei, M. (2010). Mapping QTLs of physiological traits associated with salt tolerance in 'Steptoe'×'Morex' doubled haploid lines of barley at seedling stage. *J. Food Agric. Environ.* 8, 751–759.

Tamura, K., Stecher, G., Peterson, D., Filipski, A., and Kumar, S. (2013). MEGA6: molecular evolutionary genetics analysis version 6.0. *Mol. Biol. Evol.* 30, 2725–2729. doi: 10.1093/molbev/mst197

Tavakkoli, E., Rengasamy, P., and McDonald, G. K. (2010). The response of barley to salinity stress differs between hydroponic and soil systems. *Funct. Plant Biol* 37, 621–633. doi: 10.1071/FP09202

Ullrich, S. E. (2011). *Barley: Production, Improvement and Uses.* West Sussex: Wiley Blackwell.

Vahisalu, T., Kollist, H., Wang, Y.-F., Nishimura, N., Chan, W.-Y., Valerio, G., et al. (2008). SLAC1 is required for plant guard cell S-type anion channel function in stomatal signalling. *Nature* 452, 487–491. doi: 10.1038/nature06608

Van Ooijen, J. W., and Kyazma, B. V. (2009). *"MapQTL 6" Software for the Mapping of Quantitative Trait Loci in Experimental Populations of Diploid Species.* Kyazma BV: Wageningen.

Very, A. A., Robinson, M. F., Mansfield, T. A., and Sanders, D. (1998). Guard cell cation channels are involved in Na^+-induced stomatal closure in a halophyte. *Plant J.* 14, 509–521. doi: 10.1046/j.1365-313X.1998.00147.x

Voorrips, R. E. (2002). MapChart: software for the graphical presentation of linkage maps and QTLs. *J. Hered.* 93, 77–78. doi: 10.1093/jhered/93.1.77

Wong, S. C., Cowan, I. R., and Farquhar, G. D. (1979). Stomatal conductance correlates with photosynthetic capacity. *Nature* 282, 424–426. doi: 10.1038/282424a0

Wu, D., Cai, S., Chen, M., Ye, L., Zhang, H., Dai, F., et al. (2013a). Tissue metabolic responses to salt stress in wild and cultivated barley. *PLoS ONE* 8:e55431. doi: 10.1371/journal.pone.0055431

Wu, D., Qiu, L., Xu, L., Ye, L., Chen, M., Sun, D., et al. (2011). Genetic variation of *HvCBF* genes and their association with salinity tolerance in Tibetan annual wild barley. *PLoS ONE* 6:e22938 doi: 10.1371/journal.pone.0022938

Wu, D., Shen, Q., Cai, S., Chen, Z.-H., Dai, F., and Zhang, G. (2013b). Ionomic responses and correlations between elements and metabolites under salt stress in wild and cultivated barley. *Plant Cell Physiol.* 54, 1976–1988. doi: 10.1093/pcp/pct134

Xu, R., Wang, J., Li, C., Johnson, P., Lu, C., and Zhou, M. (2012). A single locus is responsible for salinity tolerance in a Chinese landrace barley (*Hordeum vulgare* L.). *PLoS ONE* 7:e43079. doi: 10.1371/journal.pone.0043079

Xue, D., Huang, Y., Zhang, X., Wei, K., Westcott, S., Li, C., et al. (2009). Identification of QTLs associated with salinity tolerance at late growth stage in barley. *Euphytica* 169, 187–196. doi: 10.1007/s10681-009-9919-2

Yan, K., Chen, P., Shao, H., Zhao, S., Zhang, L., Zhang, L., et al. (2012). Responses of photosynthesis and photosystem II to higher temperature and salt stress in *Sorghum*. *J. Agron. Crop Sci.* 198, 218–226. doi: 10.1111/j.1439-037X.2011.00498.x

Zheng, X., He, K., Kleist, T., Chen, F., and Luan, S. (2014). Anion channel SLAH3 functions in nitrate–dependent alleviation of ammonium toxicity in Arabidopsis. *Plant Cell Environ.* doi: 10.1111/pce.12389

Zheng, Y. H., Xu, X. B., Wang, M. Y., Zheng, X. H., Li, Z. J., and Jiang, G. M. (2009). Responses of salt-tolerant and intolerant wheat genotypes to sodium chloride Photosynthesis, antioxidants activities, and yield. *Photosynthetica* 47, 87–94. doi: 10.1007/s11099-009-0014-7

Zhou, G., Johnson, P., Ryan, P. R., Delhaize, E., and Zhou, M. (2012). Quantitative trait loci for salinity tolerance in barley (*Hordeum vulgare* L.). *Mol. Breed.* 29, 427–436. doi: 10.1007/s11032-011-9559-9

Zhou, M. (2011). Accurate phenotyping reveals better QTL for waterlogging tolerance in barley. *Plant Breed.* 130, 203–208. doi: 10.1111/j.1439-0523.2010.01792.x

Zhu, J.-K. (2002). Salt and drought stress signal transduction in plants. *Annu. Rev. Plant Biol.* 53, 247–267. doi: 10.1146/annurev.arplant.53.091401.143329

9

Overexpression of copper/zinc superoxide dismutase from mangrove *Kandelia candel* in tobacco enhances salinity tolerance by the reduction of reactive oxygen species in chloroplast

Xiaoshu Jing[1], Peichen Hou[2], Yanjun Lu[1], Shurong Deng[1], Niya Li[3], Rui Zhao[1], Jian Sun[4], Yang Wang[1], Yansha Han[1], Tao Lang[1], Mingquan Ding[5], Xin Shen[1] and Shaoliang Chen[1]*

[1] Department of Plant Science, College of Biological Sciences and Technology, Beijing Forestry University, Beijing, China
[2] Department of Bio-Instruments, National Engineering Research Center for Information Technology in Agriculture, Beijing, China
[3] Department of Biology, College of Life Science, Hainan Normal University, Haikou, China
[4] Department of Plant Science, College of Life Science, Jiangsu Normal University, Xuzhou, China
[5] Department of Crop Science, College of Agricultural and Food Science, Zhejiang Agricultural and Forestry University, Hangzhou, China

Edited by:
Vadim Volkov, London Metropolitan University, UK

Reviewed by:
Dejuan Euring, Georg-August-Universität Göttingen, Germany
Kai Chen, Environment Canada, Canada
Isacco Beritognolo, CNR Istituto di Biologia Agroambientale e Forestale, Italy

***Correspondence:**
Shaoliang Chen, College of Biological Sciences and Technology, Beijing Forestry University, Qinghua-East Road 35, Haidian, Beijing 100083, China
e-mail: lschen@bjfu.edu.cn

Na^+ uptake and transport in *Kandelia candel* and antioxidative defense were investigated under rising NaCl stress from 100 to 300 mM. Salinized *K. candel* roots had a net Na^+ efflux with a declined flux rate during an extended NaCl exposure. Na^+ buildup in leaves enhanced H_2O_2 levels, superoxide dismutase (SOD) activity, and increased transcription of *CSD* gene encoding a Cu/Zn SOD. Sequence and subcellular localization analyses have revealed that KcCSD is a typical Cu/Zn SOD in chloroplast. The transgenic tobacco experimental system was used as a functional genetics model to test the effect of KcCSD on salinity tolerance. *KcCSD*-transgenic lines were more Na^+ tolerant than wild-type (WT) tobacco in terms of lipid peroxidation, root growth, and survival rate. In the latter, 100 mM NaCl led to a remarkable reduction in chlorophyll content and a/b ratio, decreased maximal chlorophyll *a* fluorescence, and photochemical efficiency of photosystem II. NaCl stress in WT resulted from H_2O_2 burst in chloroplast. Na^+ injury to chloroplast was less pronounced in *KcCSD*-transgenic plants due to upregulated antioxidant defense. *KcCSD*-transgenic tobacco enhanced SOD activity by an increment in SOD isoenzymes under 100 mM NaCl stress from 24 h to 7 day. Catalase activity rose in *KcCSD* overexpressing tobacco plants. *KcCSD*-transgenic plants better scavenged NaCl-elicited reactive oxygen species (ROS) compared to WT ones. In conclusion, *K. candel* effectively excluded Na^+ in roots during a short exposure; and increased *CSD* expression to reduce ROS in chloroplast in a long-term and high saline environment.

Keywords: *Kandelia candel*, Na^+ flux, superoxide anion, hydrogen peroxide, salt, catalase, superoxide dismutase

INTRODUCTION

NaCl-exposed plants accumulate a high level of Na^+ in roots and leaves regardless of Na^+-resistant or -sensitive species (Chen and Polle, 2010; Polle and Chen, 2014). Na^+ excess would lead to ionic imbalance, causing Na^+ injury (Volkov et al., 2004). To avoid excessive buildup of Na^+, non-secretor mangrove species (*Kandelia candel*) can maintain a high capacity to restrict Na^+ uptake and transport after NaCl exposure (Li et al., 2008). *K. candel* roots exhibited Na^+ efflux by increasing H^+ influx, indicating that Na^+ efflux resulted from active Na^+ exclusion across the plasma membrane (Lu et al., 2013; Lang et al., 2014). However, its roots and shoots could accumulate large amount of Na^+ under a long-term of increasing salinity (Li et al., 2008). This implies that the capacity for Na^+ exclusion decreased in salinized roots. However, this hypothesis needs further investigations.

In addition to ion-specific toxicity, Na^+ accumulation in leaves leads to oxidative stress by the production of reactive oxygen species (ROS) in trees (Wang et al., 2007, 2008). Superoxide anions (O_2^-) are generated as a byproduct of electron transport mainly in mitochondria or chloroplasts, which results in subsequent formation of hydrogen peroxide (H_2O_2) and hydroxyl radicals ($OH^{\cdot -}$) by a successive univalent reduction of oxygen (O_2) via chemical and enzymatic reactions (Asada, 1999; Apel and Hirt, 2004). Excessive ROS are potentially harmful to plant cells because of inactivating photosystem (PS) I and PS II (Jakob and Heber, 1996), and causing oxidative damage to proteins, lipids, and nucleic acids (Apel and Hirt, 2004). *K. candel* plants have an oxygen scavenging system against ROS under NaCl stress. Proteomic analysis of its leaves revealed that superoxide dismutase (SOD) abundance increased in response to high NaCl at 450–600 mM (Wang et al., 2014). SODs constitute the first line of cellular defense against ROS by rapidly converting O_2^- and water to H_2O_2 and O_2 (Bowler et al., 1992; Fridovich, 1995). Furthermore, SOD contributes to minimizing $OH^{\cdot -}$ formed by

Haber–Weiss or Fenton reactions (Bowler et al., 1992; Gutteridge and Halliwell, 2010). Wang et al. (2013) found that abiotic-stress proteins were up-regulated by NaCl in *K. candel* chloroplasts. However, the protection of chloroplast Cu/Zn SOD (CSD) to salt tolerance is still poorly understood.

The objectives of this study were to investigate Na^+ uptake and transport in *K. candel* plants and antioxidative defense under increasing NaCl salinity. Alterations of Na^+ flux were recorded in young roots using a non-invasive ion flux technique. Transcriptional response of *CSD* to salinity was examined in *K. candel* leaves. Microarray analysis has shown that NaCl stress increases *KcCSD* expression (Hou, 2010). To clarify the role of *KcCSD* in salinity tolerance, *KcCSD* gene of chloroplast from *K. candel* was cloned and transferred to *Nicotiana tabacum*, a model system for investigating novel genes in salt tolerance (Han et al., 2013; Shen et al., 2013). In wild-type (WT) tobacco and *KcCSD*-overexpressing lines, ROS accumulation in leaf cells and activities of SOD, catalase (CAT), and ascorbate peroxidase (APX) were examined under 100 mM NaCl stress. This study could provide scientific evidence of KcCSD protection for antioxidant defense in chloroplast.

MATERIALS AND METHODS

PLANT MATERIAL AND TREATMENTS

Uniform mature hypocotyls of *K. candel* were obtained from Dongzhai Harbor in Hainan Province of China (19°51′N, 110°24′E). Uniform hypocotyls from the same tree were planted in 5-L pots containing sand. Potted plants were placed in a greenhouse at Beijing Forestry University and fertilized with half-strength Hoagland's nutrient solution (Hoagland and Arnon, 1950) every 2 weeks at 20–25°C under a 12-h daily photoperiod with 200–300 $\mu mol\ m^{-2}s^{-1}$. Plants were 25–30 cm high and had four pair leaves after 3 months of culture (Supplementary Figure S1). These plants were raised in Hoagland's nutrient solution without the addition of NaCl. Mangrove plants were in good physiological state since shoots and roots retained abundant salts (Li et al., 2008; Lu et al., 2013). NaCl treatment was started from 100 mM and increased stepwise by 100 mM weekly, until reaching 300 mM, which could avoid osmotic shock effects of NaCl saline on the plants (Li et al., 2008). Na^+ flux in roots was recorded weekly and ion concentrations in roots and shoots were measured at low (100 mM) and high saline (300 mM) treatments, respectively. H_2O_2, *CSD* expression and SOD activity in leaves were measured after 8 h, 24 h, and 1, 2, and 3 weeks of NaCl treatment. The second pair leaves were harvested, quickly frozen in liquid nitrogen, and used for quantitative real-time PCR assays, SOD activity, and H_2O_2 measurements.

Na^+ FLUX RECORDING IN *K. CANDEL* ROOTS

Steady-state fluxes of Na^+ were measured using non-invasive micro-test technique (NMT-YG-100, Younger USA LLC, Amherst, MA, USA). Na^+ microelectrodes were prepared and calibrated as previously described (Lu et al., 2013; Lang et al., 2014). Length of primary roots of *K. candel* seedlings ranged from 1 to 5 cm (Supplementary Figure S1). Young roots with apices of 2–3 cm were excised from control and salinized plants for Na^+ flux determination. Steady flux profiles of Na^+ were measured along the root axis at the apical zones, where a vigorous flux was usually observed in woody plants (Sun et al., 2009a,b; Lu et al., 2013; Lang et al., 2014). Roots were rinsed with redistilled water and immediately incubated in measuring solution (with 0.1 mM Na^+) for 30 min equilibration. The basic Na^+ measuring solution (with low interfering ions of Ca^{2+} and K^+, Cuin et al., 2011) was 0.1 mM NaCl, $MgCl_2$, and $CaCl_2$, and 0.5 mM KCl at pH 6.0. Roots were immobilized on the bottom of the chamber. Then ion flux recordings started 200 μm from the apex and conducted along root axis until 2000 μm with an interval of 300 or 500 μm. Ionic flux rates were obtained using MageFlux developed by Yue Xu (1995) (http://www.youngerusa.com).

ION ANALYSIS

Roots, leaves, stems, and hypocotyls were harvested from control and NaCl-stressed *K. candel* plants, and oven-dried to constant weight at 65°C for 4 days, ground and passed through a 1.0-mm sieve. Samples were digested by H_2SO_4-H_2O_2, and Na^+ concentration was measured using an atomic absorption spectrophotometer (Perkin-Elmer 2280, PerkinElmer, Inc., Wellesley Hills, MA, USA) (Lu et al., 2013).

RNA EXTRACTION AND REAL-TIME PCR ANALYSIS

Total RNA was extracted from *K. candel* leaves (the second pair leaves) with a modified hot borate method (Wan and Wilkins, 1994). This protocol is commonly used for isolating RNA from plant tissues rich in polyphenols and polysaccharides. For tobacco, total RNA was extracted from the 3rd–4th mature leaves from the top using the Total RNA Extraction Kit (QBio Technologies Inc.). The integrity of total RNA was determined by running samples on 1.0% formaldehyde agarose gels stained with ethidium bromide. The quantity/yield of total RNA was estimated spectrophotometrically at 230, 260, 280 nm (NanoDrop 2000 spectrophotometer, Thermo Scientific, Wilmington, USA). The first strand was synthesized from 2 μg of total RNA using M-MLV reverse transcriptase (Promega, Madison, USA) and oligo (d_T) 12–15 primer at 42°C for 1 h. The real-time PCR conditions were 10 min at 95°C, 35 cycles of 95°C for 30 s, 55°C for 30 s, 72°C for 30 s, and 10 min at 72°C. The primers used for *KcCSD* were 5′-ATTAGCAACTATGTTTCCCA-3′ (forward), and 5′- CTACAACGGTGAATGTTC-3′ (reverse). The real-time PCR data in *K. candel* were normalized against *Tubulin: Tubulin* -F, 5′-TGCCCAAGGATGTGAACG-3′; and *Tubulin*-R, 5′-CCATACC CTCACCCACAT-3′. In tobacco, *EF1α* was used as the internal control (forward, 5′-GCTGTGAGGGACATGCGTCAAA-3′; and reverse, 5′-GTAGTAGATCGCGAGTACCACCA-3′).

The real-time PCR analysis was performed in a Real-time PCR System (MJ Opticon2 Bio-Rad). The relative level of expression was quantified using MJ Opticon Monitor software (Bio-Rad, Hercules, CA, USA). The expression of the target genes were normalized to the expression level of the reference gene (*Tubulin* in *K. candel* and *EF1α* in tobacco) using the $2^{-\Delta\Delta C_T}$ method (Livak and Schmittgen, 2001).

SEQUENCE ALIGNMENT AND PHYLOGENETIC TREE

Full-length amino acid sequences of Cu/Zn SOD were aligned using ClustalW2 online (http://www.ebi.ac.uk/Tools/msa/

clustalw2/). Amino acid sequences of chloroplast transit peptide were predicted by ChloroP (http://www.cbs.dtu.dk/services/TargetP/; http://www.cbs.dtu.dk/services/ChloroP/) (Emanuelsson et al., 1999, 2007). Phylogenetic tree was generated using the Neighbor–Joining (NJ) method in MEGA version 6.0 software (bootstrap analysis with 1000 replicates). The accession numbers of Cu/Zn SOD protein sequences used in multiple sequence alignment and phylogenetic analysis are provided in Supplementary Table S1.

PLASMID CONSTRUCTS

The open reading frame (ORF) of *KcCSD* (GenBank accession number KP143653) was amplified by PCR from *K. candel* cDNA using specific primers (forward, 5′-ATGCAAGCGGTAGTTGCG-3′; reverse, 5′-TAACACTGGTGTCAACCCAACAAC-3′). The PCR fragment was first cloned into the vector pMD18-T (Takara, Dalian, China) and verified by sequencing. The ORF of *KcCSD* was released by digestion with *Eco*RI and *Hin*dIII and introduced into the expression vector, pCAMBIA2300, under the control of the constitutive cauliflower mosaic virus 35S promoter.

OVEREXPRESSION OF *KcCSD* IN TRANSGENIC TOBACCO LINES

To further analyze the functions of KcCSD upon NaCl stress, *KcCSD* was transferred to tobacco plants via *Agrobacterium*-mediated gene transfer. The pCAMBIA2300-*KcCSD* construct was transferred to a strain of *Agrobacterium tumefaciens* (LBA4404) with a freeze–thaw method. Tobacco was infected by the *Agrobacterium tumefaciens* using the leaf disc method (Horsch et al., 1985). The infected leaves were placed on MS medium (Murashige and Skoog, 1962) with no antibiotic for 2–3 days and transferred to MS with 50 mg/L kanamycin. Individual kanamycin-resistant shoots were selected and shooted on MS medium with no growth regulators or antibiotic. More than 10 independently transformed plants were selected for *KcCSD* expression assay. Those overexpressing *KcCSD* were used for further study (L7 and L8; Supplementary Figure S2). WT and *KcCSD*-overexpressed plants were kept in the greenhouse to yield seeds. Using PCR, T2 generation of L7 and L8 was checked for the presence of the *KcCSD* gene.

T2 generation seeds of wild-type (WT) and transgenic lines (L7 and L8) were germinated on MS medium for 7 days, then subjected to 0 or 150 mM NaCl for 7 days. Root length, survival rates, and H_2O_2 level in chloroplast were measured in WT and transgenic plants. The capacity to control ROS was compared between WT and transgene tobacco under NaCl stress. Four-weeks old rooted plants of WT and transgenic lines were transferred to 1/4 Hoagland's nutrient solution for 2-weeks acclimation, then exposed to 0 or 100 mM NaCl for 7 days. O_2^- and H_2O_2 production, lipid peroxidation, and activity of antioxidant enzymes (SOD, CAT and APX), SOD isoenzymes were examined during NaCl treatment (100 mM, 7 day). Leaf photosynthesis, chlorophyll a fluorescence, chlorophyll content were measured in NaCl-stressed WT and transgenic plants.

SUBCELLULAR LOCALIZATION OF THE GFP FUSION PROTEINS

To generate a translational fusion of KcCSD-GFP, the *KcCSD* was obtained by PCR using specific primers (forward 5′-ATGCAAG CGGTAGTTGCG-3′; reverse 5′- CACTGGTGTCAACCCAACA AC -3′). PCR products were cloned into the pMD18-T vector and sequenced. The resulting construct was digested by *Eco*RI and *Pst*I and introduced into the constructed pGreen0029-GFP vector driven by the 35S promoter. Vector-carrying 35S-driven GFP was used as free GFP control.

KcCSD-GFP was transiently transformed to mesophyll protoplasts of *Arabidopsis thaliana* due to a high frequency of gene transformation, as compared to the tobacco. Isolation of *Arabidopsis* mesophyll protoplasts and polyethylene glycol–mediated transformation were performed according to Yoo et al. (2007). Confocal images were obtained after 16–20 h of incubation. Fluorescence was examined with a Leica inverted fluorescence microscope (Leica Microsystems GmbH) at 510–535 nm for GFP and at 650–750 nm for chlorophyll.

MALONDIALDEHYDE (MDA) CONTENT

Tobacco leaves (0.1 g, 3rd–5th mature leaves from the top) were homogenized in 1 ml of 0.1% trichloroacetic acid solution on ice. MDA content in WT and *KcCSD*-transgenic lines was determined by the thiobarbituric acid (TBA) reaction according to Heath and Packer (1968).

CHLOROPHYLL CONTENT, FLUORESCENCE PARAMETERS, AND NET PHOTOSYNTHETIC RATE

The 3rd–5th mature leaves from the top were used for chlorophyll contents, fluorescence, and photosynthesis measurements. Chlorophyll concentrations in WT and transgenic lines were measured according to Wellburn (1994) and Lichtenthaler (1987). Chlorophyll *a* fluorescence parameters were measured using a PAM fluorometer (Junior PAM, Walz, Germany). Plants were dark-adapted for 20 min to determine dark fluorescence yield (Fo), and then exposed to a single red pulse to determine maximal fluorescence yield (Fm) and Fv/Fm ratio using the formula (Fm-Fo)/Fm. ΦPSII, the PSII actual photochemical efficiency, was determined according to Wang et al. (2007). Net photosynthetic rate (Pn) was measured using a Li-6400 photosynthesis system (Li-Cor Inc., Lincoln, NE, USA) at 800 μmol photons m^{-2} s^{-1}. Chamber air temperature was maintained at 25°C and CO_2 concentration was 380 $\mu L\ L^{-1}$.

TOTAL PROTEIN EXTRACTION AND ENZYME ASSAYS

Antioxidant enzymes were extracted from *K. candel* (3rd–4th mature leaves from the top) and tobacco leaves (3rd–5th mature leaves from the top) and measured according to Wang et al. (2007, 2008). The total SOD activity was determined by monitoring super-radical–induced reduction of nitro blue tetrazolium (NBT) at 560 nm (Giannopolitis and Ries, 1977; Wang et al., 2008; Shen et al., 2013). One unit of SOD (relative unit) was defined as the amount of enzyme that causes 50% inhibition of the reaction compared with a blank sample (Giannopolitis and Ries, 1977). CAT activity was determined spectrophotometrically by monitoring the disappearance of H_2O_2 at 240 nm for 1 min (Aebi, 1984). APX activity was determined by monitoring the H_2O_2-dependent oxidation of ascorbate at 290 nm for 1.5 min (Nakano and Asada, 1981).

NATIVE PAGE OF SOD ISOENZYMES

Salt-elicited alterations in SOD isoenzymes were examined after 7 days of NaCl treatment (100 mM). SOD was extracted from transgenic and WT tobacco leaves (3rd–5th mature leaves from the top) at 4°C (Beauchamp and Fridovich, 1971; Wang et al., 2007, 2008). Native PAGE of SOD was performed on a 7.5% separating gel and 3.9% stacking gel at 4°C. After electrophoresis, the SOD isoenzymes were differentiated on a pre-equilibrating gel for 30 min in 50 mM potassium phosphate buffer and 1 mM EDTA at pH 7.8. The gel was incubated in dark for 30 min in fresh staining solution of 50 mM potassium phosphate buffer (pH 7.8), 0.24 mM NBT, 33.2 μM riboflavin, and 0.2% tetramethylethylenediamine. Then it was illuminated with 400 μmol m^{-2} s^{-1} fluorescence until uniformly blue except areas with SOD activity.

O_2^- AND H_2O_2 PRODUCTION IN LEAVES

In situ accumulations of O_2^- and H_2O_2 were examined with histochemical staining protocols. O_2^- was detected with NBT (Dutilleul et al., 2003) and H_2O_2 with 3-3′-diaminobenzidine (DAB; Thordal-Christensen et al., 1997), respectively. For *in situ* staining of O_2^-, leaf discs (2 cm in diameter) were sampled from the second fully developed leaf on the top, and immediately vacuum infiltrated in 0.5 mg/ml NBT and 10 mM potassium phosphate buffer at pH 7.8. For the negative control, the NBT solution was supplemented with 10 U ml^{-1} SOD and 10 mM $MnCl_2$ before the infiltration (Supplementary Figure S3). After being incubated in dark at room temperature for 1 h, samples were cleared in 90% ethanol at 70°C to remove chlorophyll. O_2^- was visualized as a blue color at the site of NBT precipitation.

Leaf discs (2 cm in diameter) were vacuum infiltrated in 1 mg/ml DAB at pH 3.8, incubated in dark at room temperature for 14 h and transferred to 90% ethanol at 70°C until complete removal of chlorophyll and visualization of H_2O_2 as brown color at the site of DAB polymerization. Samples were stored and examined in 70% glycerol. For the negative control, 10 mM ascorbic acid was added into DAB solution for infiltration. Total leaf H_2O_2 was determined according to Wang et al. (2008).

H_2O_2 DETECTION IN CHLOROPLAST

H_2O_2 in chloroplast was detected as described by Ramírez et al. (2013) with minor modifications. H_2O_2-specific fluorescent probe, H_2DCF-DA, was an indicator of H_2O_2 (Sun et al., 2010a,b, 2012). WT and *KcCSD*-transgenic tobacco seedlings were germinated on MS for 7 days, then transferred to MS supplemented with 0 or 150 mM NaCl. Seedlings were vacuum infiltrated for 10 min in 10 μM H_2DCF-DA (Sigma Aldrich) in 5 mM MES buffer at pH 5.7. After being incubated in dark for 20 min, leaf samples were washed with 5 mM MES for 30 min. Fluorescence was examined with a Leica inverted fluorescence microscope (Leica Microsystems GmbH) at 500–530 nm for H_2DCF-DA and 650–750 nm for chlorophyll, respectively.

STATISTICAL ANALYSIS

All experimental data were subjected to SPSS (SPSS Statistics 17.0, 2008) for statistical tests and analyses. When $P < 0.05$, differences between means were considered significant unless otherwise stated.

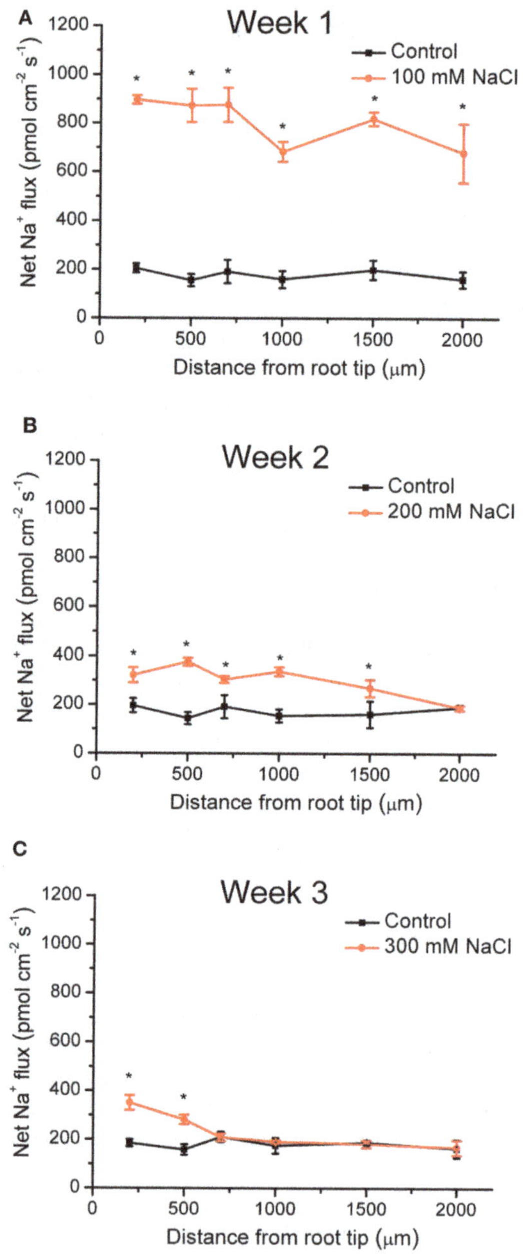

FIGURE 1 | Effects of NaCl on net Na^+ fluxes in young roots of *K. candel*. Plants were treated by a rising NaCl concentration from 100 to 300 mM weekly for 3 weeks. The control was not exposed to NaCl. Steady Na^+ flux profiles on week 1 **(A)**, week 2 **(B)**, and week 3 **(C)** were determined along root axis at apical zones, 200–2000 μm from the tip. Each point is the mean of 4–6 replicates. Bars represent standard error of the mean. *$P < 0.05$, control and NaCl treatment.

RESULTS

Na^+ FLUX AND Na^+ CONCENTRATIONS IN ROOTS AND SHOOTS

During NaCl treatment, Na^+ flux of *K. candel* was recorded along the root axes (200–2000 μm from the apex), in which a vigorous flux of Na^+ was usually observed. Under control conditions, *K. candel* roots exhibited a stable and constant Na^+ efflux with a mean flux of 176 pmol cm^{-2} s^{-1} (**Figure 1**). The Na^+ efflux in control plants was due to Na^+ previously accumulated by roots (**Figure 2**). However, Na^+ efflux along roots was significantly increased by 4.5-fold in the first week of NaCl exposure (**Figure 1**). Flux rate in NaCl-stressed roots remarkably decreased with the extension of the exposure (**Figure 1**). The mean Na^+ efflux in salinized roots was only 30% higher than the controls in the third week (**Figure 1**).

Na^+ content in roots, hypocotyls, stems, and leaves significantly rose after 3 weeks treatment. It was 54–400% higher than that in the control (**Figure 2**). The Na^+ accumulation was more pronounced in roots than in shoots (**Figure 2**).

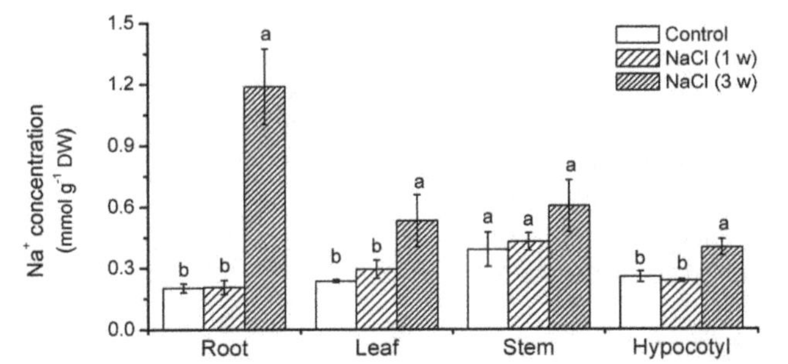

FIGURE 2 | Effects of NaCl on Na^+ concentrations in roots and shoots in *K. candel*. Plants were treated by a rising NaCl concentration from 100 to 300 mM weekly for 3 weeks. Each column is the mean of 4–6 replicates. Bars represent standard error of the mean. Different letters denote significant differences at $P < 0.05$ between treatments in each organ.

NaCl-ELICITED H_2O_2, SOD ACTIVITY AND EXPRESSION OF *KcCSD*

NaCl salinity increased H_2O_2 levels in *K. candel* leaves although H_2O_2 levels fluctuated during 3 weeks experiment (**Figure 3**). Salinized *K. candel* enhanced SOD activity in its leaves (**Figure 3**). Quantitative real-time PCR (qRT-PCR) showed *KcCSD* upregulation during NaCl treatment (**Figure 3**). These results indicate a molecular and biochemical change in expression of *KcCSD* in salinized *K. candel*.

KcCSD CLONING AND SEQUENCE ANALYSIS

The cDNA sequence of *KcCSD* contains 687 bp with a predicted open reading frame (ORF) of 228 amino acids (**Figure 4A**). KcCSD protein conserves Cu^{2+} or Zn^{2+} binding site and active site (amino acids from 84 to 219), which catalyzes the conversion of O_2^- to O_2 (**Figure 4A**). It also contains a chloroplast transit peptide with a potential cleavage site at amino acid position 73 (**Figure 4A**). Comparative phylogenetic analysis of KcCSD has revealed that KcCSD is homologous to *Arabidopsis* CSD2 and other chloroplast CSDs from different species (**Figure 4B**). Collectively, KcCSD can be classified as CSD2, a chloroplast-targeted protein.

SUBCELLULAR LOCATION OF KcCSD

A C-terminal translational construct was generated by the fusion of *KcCSD* to the green fluorescent protein (GFP) reporter gene. The construct was transiently expressed in *Arabidopsis* protoplasts (**Figure 5**). Fluorescence emitted by the GFP fusion of KcCSD overlapped chlorophyll autofluorescence, revealing that KcCSD was targeted to the chloroplast (**Figures 5D–F**). Fluorescence of the free GFP under the control of 35S promoter was distributed in cytoplasm of protoplasts, not merging with red autofluorescence from chloroplast (**Figures 5A–C**).

SALINITY TOLERANCE OF *KcCSD*-TRANSGENIC TOBACCO PLANTS

Analysis by qRT-PCR identified a strong overexpression of *KcCSD* in the transgenic lines L1, L7, L8, and L10, but *KcCSD* was not detectable in the WT plants (Supplementary Figure S2). L7 and L8 transgenic lines showed a remarkably higher transcript abundance, indicating that *KcCSD* driven by the 35S promoter

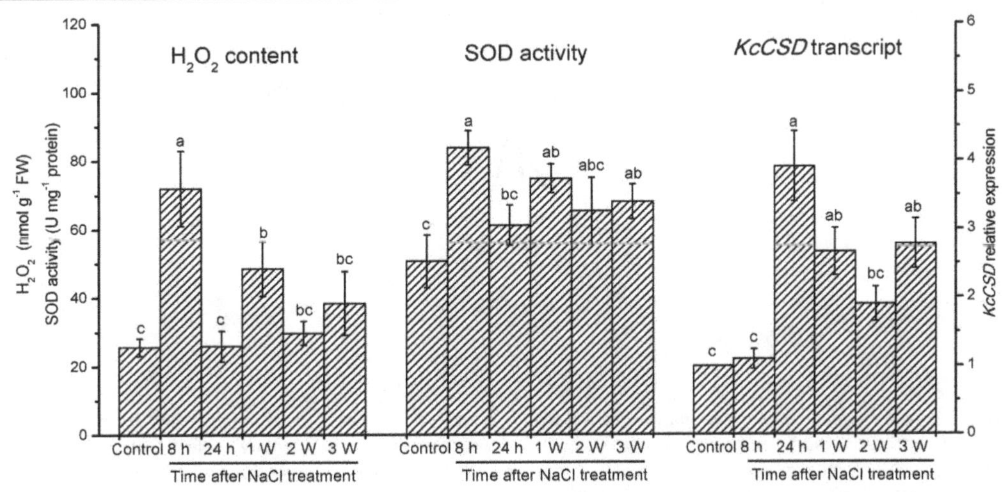

FIGURE 3 | Total concentration of H_2O_2, total SOD activity, and *KcCSD* expression in *K. candel* leaves under rising NaCl stress. Plants were treated by a rising NaCl concentration from 100 to 300 mM weekly for 3 weeks. Each column is the mean of 4–6 replicates. Bars represent standard error of the mean. Different letters above columns represent significant differences at $P < 0.05$ during treatment.

A

```
KcCSD      MQAVVAAAMAAHTILTASPLLSHYPRLSPPTAPIGHTSTLHSSFHGLSLK--FARQSLPL 58
BgchlCSD   MQAVVAAAMAAHTILTAPPLQSHCPFLSPTTPPFGHTPTLHSSFHGLSLK--LARHSLPL 58
DlchlCSD   MQAAIAA-MAAHTILTASPPFSHHPLLS--TFPSPNTLTRTSSFRGVSLN--PPQR---T 52
GhchlCSD   --------MAAPYFSRTTP--SHLALSFPSSTNPSNPPVLFSSFRGVSLK--LPRQS--L 46
AtCSD2     --------MAATNTILAFS--SPSRLLIP---PSSNPSTLRSSFRGVSLNNNNLHRLQSV 47
                   ***     : .  *            :. .  ***:*:**:    ::

KcCSD      SLAASAAPKKPLAVVAATRKGVAVLKGTSNVEGVVALTQEDEGPTTVNVRVTGLTPGPHG 118
BgchlCSD   SLAA-AAPKKPLAVVAATKKAVAVLKGTSDVEGVVTLTQEDEGPTTVNVHVSGLTPGPHG 117
DlchlCSD   TFTLTAVASKPFTVVAAVKKAVAVLKGNSNVEGVVSLTQENDGPTTVNVRITGLTPGPHG 112
GhchlCSD   SLAA-TIPKKAFSVFAVTKKAVAVLKGNSEVEGVVTLTQENDGPTTVNVRITGLTPGPHG 105
AtCSD2     SFAV-KAPSKALTVVSAAKKAVAVLKGTSDVEGVVTLTQDDSGPTTVNVRITGLTPGPHG 106
           :::   ..*.::*.:..:*.******.*:*****:***::.*******::********

KcCSD      FHLHEYGDTTNGCISTGAHFNPNKMTHGAPEDEIRHAGDLGNIVANADGVAEATIVDKQI 178
BgchlCSD   FHLHEYGDTTNGCISTGAHFNPNKMTHGAPEDETRHAGDLGNIVANADGVAEAKIVDKQI 177
DlchlCSD   FHLHEYGDTTNGCMSTGAHFNPNSMTHGAPEDEVRHAGDLGNVVANANGVAEATIVDNQI 172
GhchlCSD   FHLHEYGDTTNGCMSTGAHFNPNNMTHGAPEDEVRHAGDLGNIIANADGVAEATIVDNQI 165
AtCSD2     FHLHEFGDTTNGCISTGPHFNPNNMTHGAPEDECRHAGDLGNINANADGVAETTIVDNQI 166
           *****:*******:***.*****.********* ********: ***:****:.***:**

KcCSD      PLSGPNTVVGRAFVVHELEDDLGKGGHELSLTTGNAGGRLACGVVGLTPV           228
BgchlCSD   PLSGPNTVVGRAFVVHELEDDLGKGGHELSLTTGNAGGRLACGVVGFTPV           227
DlchlCSD   PLSGPNTVIGRALVVHELEDDLGKGGHELSLSTGNAGGRLACGVVGLTPV           222
GhchlCSD   PLSGPNAVVGRAFVVHELEDDLGKGGHELSLTTGNAGGRLACGVVGLTPV           215
AtCSD2     PLTGPNSVVGRAFVVHELKDDLGKGGHELSLTTGNAGGRLACGVIGLTPL           216
           **:***:*:***:*****:************:************:*:**:
```

B

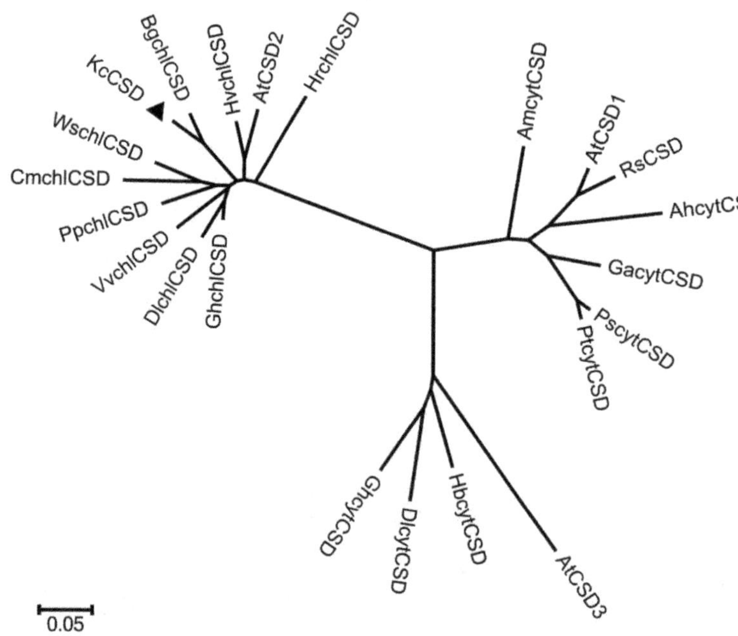

FIGURE 4 | Amino acid sequence and phylogenetic analysis of KcCSD. (A) Amino acid sequences of Cu/Zn SOD from *Kandelia candle* (KcCSD), *Bruguiera gymnorrhiza* (BgchlCSD), *Dimocarpus longan* (DlchlCSD), *Gossypium hirsutum* chloroplast (GhchlCSD), and *Arabidopsis thaliana* chloroplast (AtCSD2). Asterisks (*) and dots (·, :) indicate identical and conserved amino acid residues, respectively. Italics are chloroplast transit peptide. Bolds indicate the conserved Cu^{2+} or Zn^{2+} binding site. Activity sites are underlined. **(B)** Neighbor-joined phylogenetic tree for CSD protein sequence (chloroplast CSDs with no chloroplast transit peptides) in various species. The alignment used for this analysis is available from the database (Supplementary Table S1). Different species acronyms are: *Ah, Amaranthus hypochondriacus; Am, Avicennia marina; At, Arabidopsis thaliana; Bg, Bruguiera gymnorrhiza; Br, Brassica rapa subsp. Pekinensis; Cm, Chenopodiastrum murale; Dl, Dimocarpus longan; Ga, Gossypium arboreum; Gh, Gossypium hirsutum; Hb, Hevea brasiliensis; Hr, Haberlea rhodopenis; Hv, Hordeum vulgare; Kc, Kandelia candel; Pp, Prunus persica; Ps, Populus suaveolens; Pt, Populus tremuloides; Rs, Raphanus sativus; Vv, Vitis vinifera.*

was more efficiently expressed in these lines, as compared to other ones. To testify the importance of *KcCSD* in enhancing NaCl tolerance, L7 and L8 were used for further NaCl treatment studies.

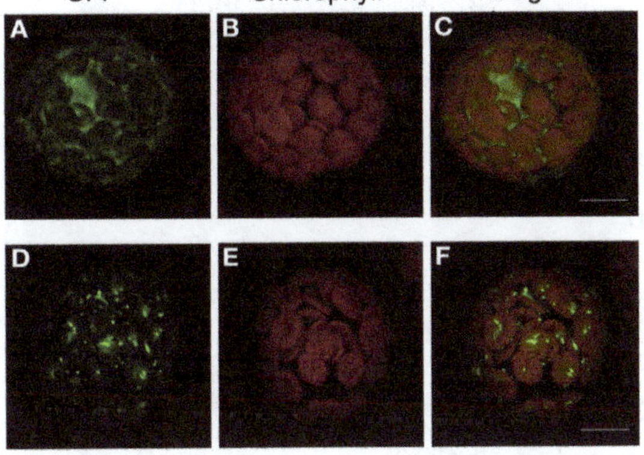

FIGURE 5 | Chloroplast subcellular localization of KcCSD. Determined by transient expression of GFP alone (vector control; **A–C**) and a fusion KcCSD-GFP protein (KcCSD-GFP; **D–F**) in *Arabidopsis* protoplasts. Green fluorescence of GFP **(A,D)** and red auto-fluorescence of chlorophyll **(B,E)** were monitored separately using a confocal laser scanning microscope, and the two color fluorescence images **(C,F)** were merged. Bars = 10 μm.

NaCl treatment (150 mM, 7 day) inhibited root growth and survival rate, but a more pronounced reduction occurred in WT (**Figures 6A,C,D**). Under hydroponic culture, WT plants showed a significant growth retardation compared to transgenic lines during a prolonged treatment (14 days; **Figure 6B**). Net photosynthetic rate (Pn) was decreased in WT and transgenic plants by salinity (**Figure 6E**). However, Pn was 74–93% higher in transgenic lines than in WT plants (**Figure 6E**). Under non-NaCl stress, both root and shoot growth of transgenic plants did not significantly differ from WT ones (**Figure 6**).

CHLOROPHYLL CONTENT AND FLUORESCENCE

NaCl treatment (100 mM) for 7 days resulted in a chlorophyll decline in tobacco plants (**Figure 7A**). However, it was more pronounced in WT than in transgenic lines L7 and L8 (**Figure 7A**). NaCl caused a decline in the chlorophyll a/b ratio by 14% in WT, which was greater than that in transgenic lines (**Figure 7B**).

After 100 mM NaCl treatment for a week, Fv/Fm was lowered in tobacco plants, but a more significant effect was observed in the WT ones (**Figure 7C**). ΦPSII, PSII actual photochemical efficiency (Maxwell and Johnson, 2000), in WT plants decreased remarkably by NaCl stress (**Figure 7D**). There was no similar change in *KcCSD*-transgenic lines (**Figure 7D**).

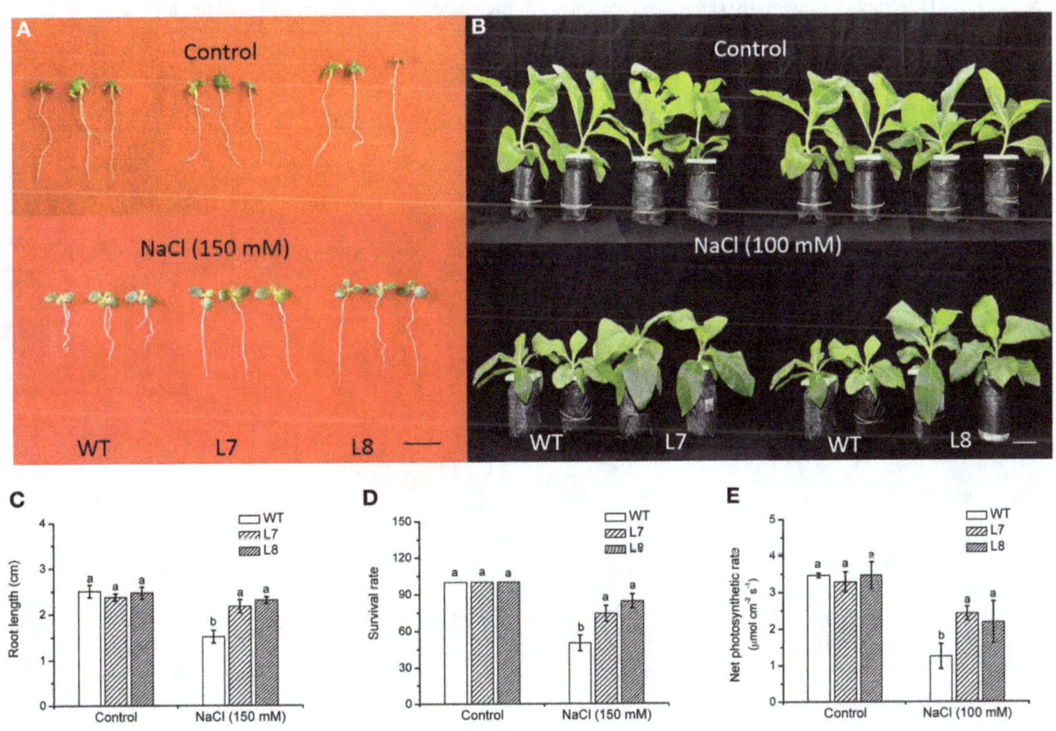

FIGURE 6 | NaCl tolerance of wild-type tobacco and *KcCSD*-transgenic lines. Seeds of wild-type (WT) and transgenic lines (L7 and L8) were germinated on MS medium for 7 days, then supplemented with 0 or 150 mM NaCl for another 7 days. **(A,C,D)** Show root growth and survival rate of tobacco seedlings. **(B,E)** Show plant performance and net photosynthetic rate of tobacco plants after exposure to 0 or 100 mM NaCl for 7 **(E)** or 14 **(B)** days. Prior to NaCl treatment, 4-weeks old rooted plants of WT and transgenic lines were transferred to 1/4 Hoagland's nutrient solution for 2-weeks acclimation. In **(C–E)**, each column is the mean of 4–6 replicates. Bars represent standard error of the mean. Different letters above columns represent significant differences at $P < 0.05$ between WT and transgenic lines in control and NaCl treatments. Scale bars: A: 0.5 cm, B: 5 cm.

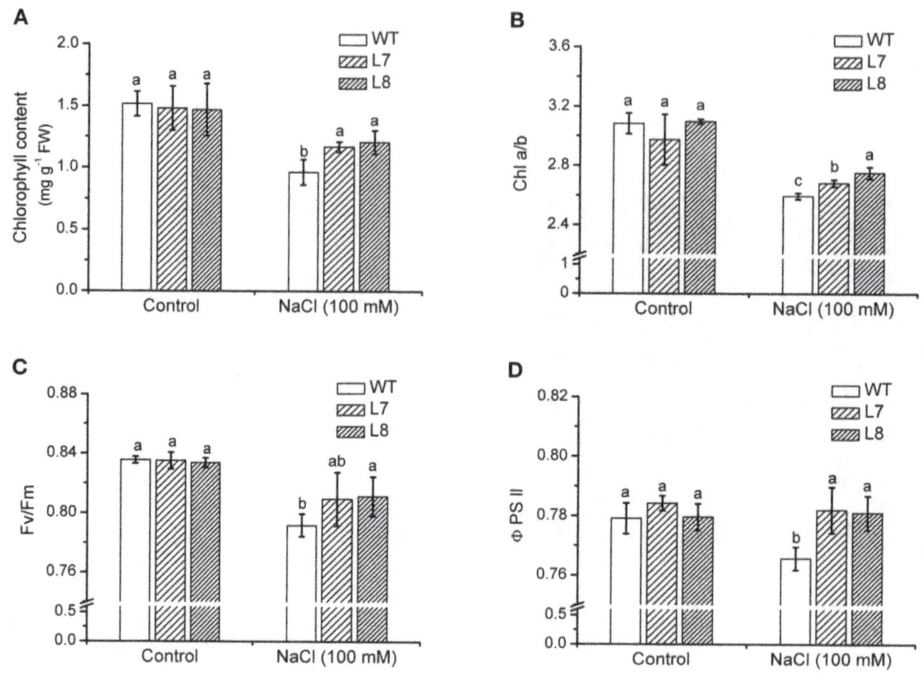

FIGURE 7 | Effects of NaCl on chlorophyll content and fluorescence in wild-type tobacco and *KcCSD*-transgenic lines. Four-weeks old rooted plants of wild-type (WT) and transgenic lines (L7 and L8) were transferred to 1/4 Hoagland's nutrient solution for 2-weeks acclimation, then exposed to 0 or 100 mM NaCl for 7 days. **(A)** Chlorophyll content, **(B)** Chlorophyll a/b ratio, **(C)** Ratio of variable to maximal chlorophyll fluorescence (Fv/Fm), **(D)** Actual photochemical efficiency of PSII (ΦPSII). Each column is the mean of 4–6 replicates. Bars represent standard error of the mean. Different letters above columns represent significant differences at $P < 0.05$ between WT and transgenic lines in control and NaCl treatments.

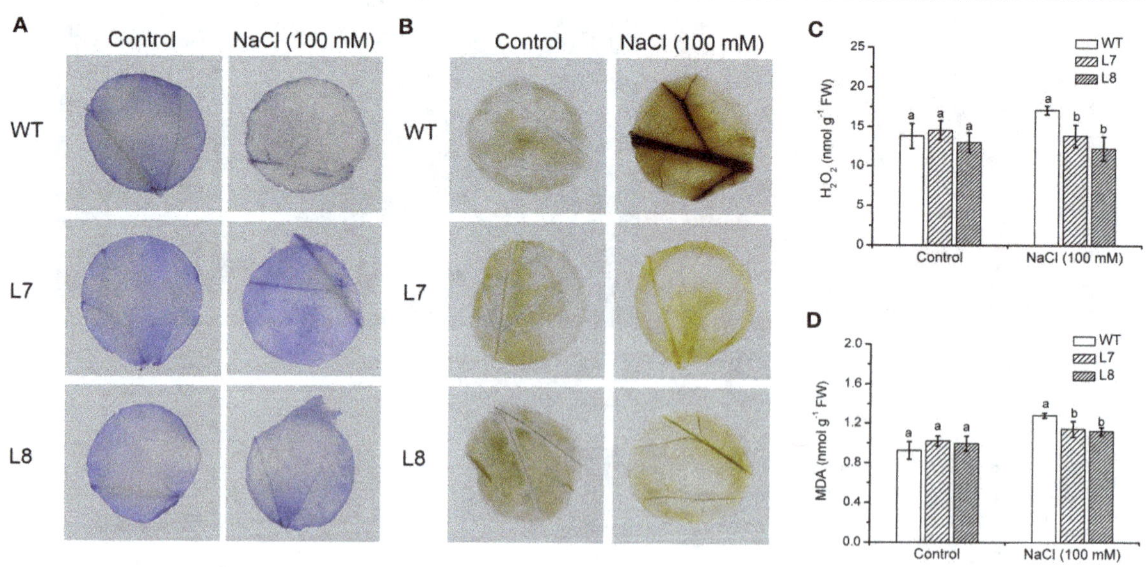

FIGURE 8 | Effects of NaCl on leaf ROS and malondialdehyde levels in wild-type tobacco and *KcCSD*-transgenic lines. Four-weeks old rooted plants of wild-type (WT) and transgenic lines (L7 and L8) were transferred to 1/4 Hoagland's nutrient solution for 2-weeks acclimation, then exposed to 0 or 100 mM NaCl for 7 days. Leaf discs (2 cm in diameter) were sampled. **(A)** *In situ* O_2^-, **(B)** *In situ* H_2O_2, **(C)** Total H_2O_2, and **(D)** Malondialdehyde (MDA) content. In **(C,D)**, each column is the mean of 4–6 replicates. Bars represent standard error of the mean. Different letters above columns represent significant differences at $P < 0.05$ between WT and transgenic lines in control and NaCl treatments.

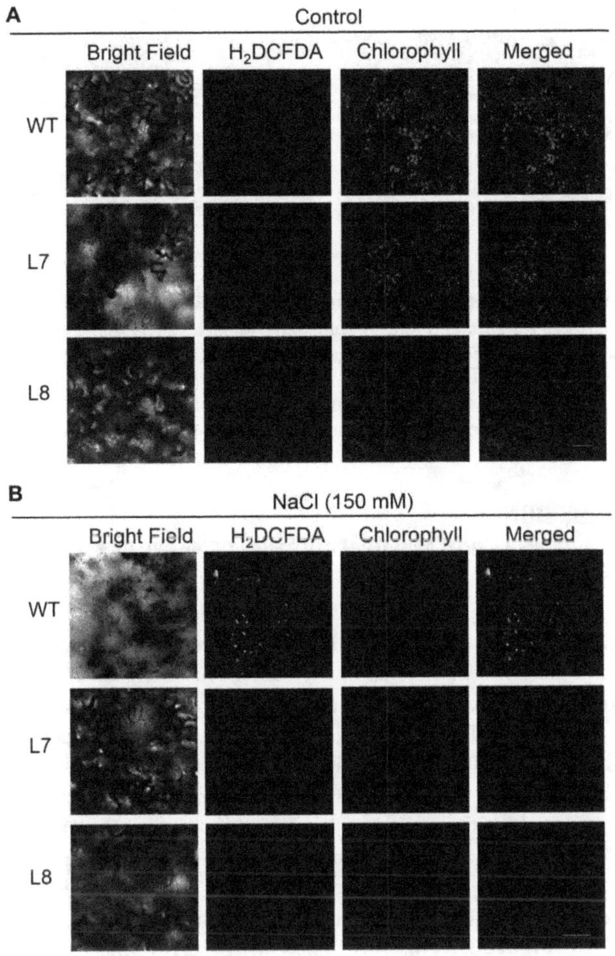

FIGURE 9 | Effect of NaCl on chloroplast H_2O_2 production in wild-type tobacco and *KcCSD*-transgenic lines. Seeds of wild-type (WT) and transgenic lines (L7 and L8) were germinated on MS medium for 7 days, then supplemented with 0 **(A)** or 150 mM NaCl **(B)** for 7 days. Seedlings treated with or without NaCl were stained with 10 μM H_2DCF-DA for H_2O_2 detection. Green fluorescence of H_2DCF-DA and red auto-fluorescence of chlorophyll were monitored separately using a confocal laser scanning microscope, and two color fluorescence images were merged. Bars = 20 μm.

O_2^- AND H_2O_2 PRODUCTION IN TOBACCO LEAVES

In situ O_2^- production in leaves was detected by the reduction of nitro blue tetrazolium (NBT). Formazan deposits were visualized in leaf discs of WT and transgenic lines under no-NaCl conditions (**Figure 8A**). After 1 week exposure to 100 mM NaCl, more formazan precipitates appeared in tobacco leaves, especially in transgenic plants (**Figure 8A**). However, formazan formation in WT and *KcCSD*-transgenic plants was suppressed by SOD, the scavenger of O_2^-, irrespective of NaCl and control treatments (Supplementary Figure S3). This indicates that NBT was reduced to formazan specifically by the superoxide anions in WT and *KcCSD*-transgenic plants.

By means of DAB staining, H_2O_2 levels were visible in WT and *KcCSD*-transgenic lines (**Figure 8B**). Compared to the control, the intensity of red–brown staining significantly increased in

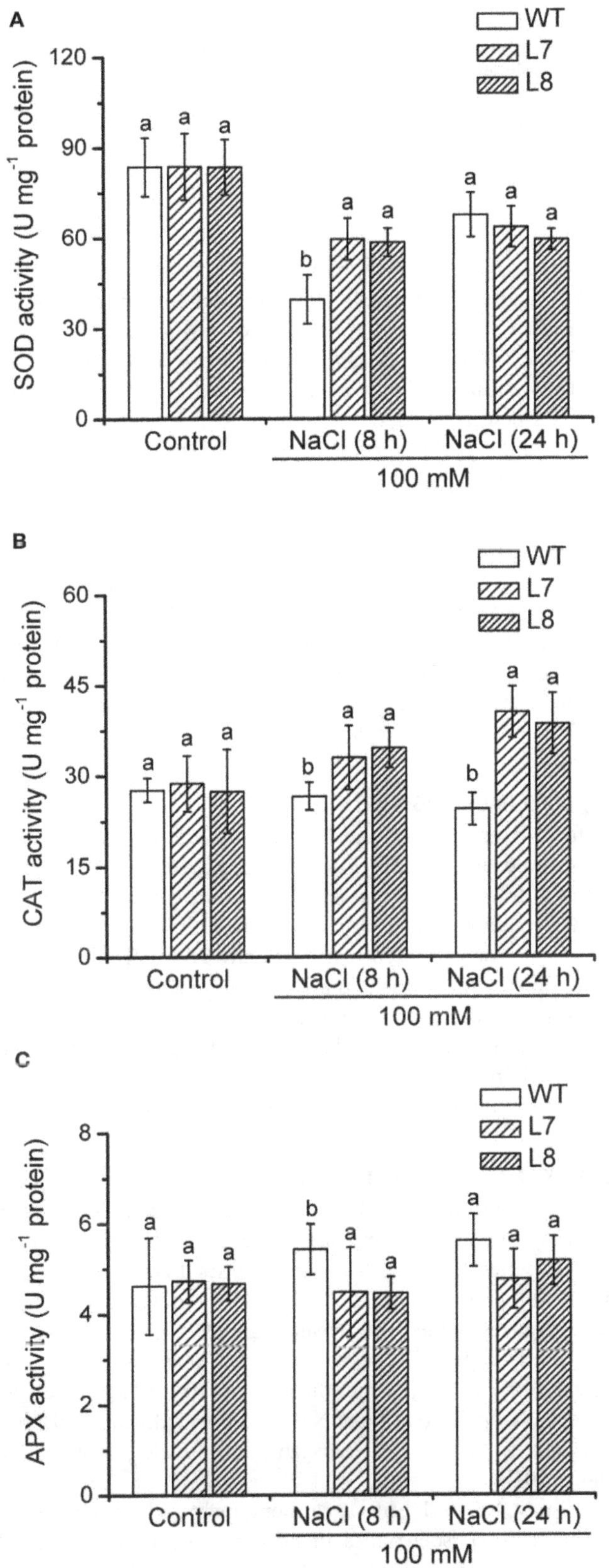

FIGURE 10 | Effect of NaCl on activities of antioxidant enzymes in wild-type tobacco and *KcCSD*-transgenic lines. Four-weeks old rooted plants of wild-type (WT) and transgenic lines (L7 and L8) were transferred

(Continued)

FIGURE 10 | Continued
to 1/4 Hoagland's nutrient solution for 2-weeks acclimation. Hydroponically acclimated plants were subjected to 0 or 100 mM NaCl for 24 h. **(A)** Total SOD activity, **(B)** CAT activity, and **(C)** APX activity. Each column is the mean of 4–6 replicates. Bars represent standard error of the mean. Different letters above columns represent significant differences at $P < 0.05$ between wild-type and transgenic lines in control and NaCl treatments.

NaCl-treated plants, and a more pronounced effect was observed in WT plants (**Figure 8B**). However, red–brown staining was absent in ascorbic acid-pretreated leaves (data not shown), indicating that the brownish staining was due to the reaction of DAB with H_2O_2. Total leaf H_2O_2 analysis showed a trend similar to that of *in situ* detection. H_2O_2 content in WT plants was increased by 16% under NaCl treatment, significantly higher than that in transgenic lines (**Figure 8C**).

MDA CONTENT IN LEAVES

NaCl-elicited increase in MDA, a marker of lipid peroxidation, was observed in both WT tobacco and *KcCSD*-transgenic lines (**Figure 8D**). However, WT showed 34% increase in MDA compared to transgenic lines L7 and L8 (10 and 17%; **Figure 8D**). This indicates that NaCl caused a more pronounced oxidative damage in WT than in transgene plants.

H_2O_2 LEVEL IN CHLOROPLASTS

Confocal laser scanning microscopy analysis of leaf epidermal cells showed the same level of chlorophyll red auto-fluorescence in the WT tobacco and transgenic lines of the control (**Figure 9A**). In NaCl treatment, DCF-dependent green fluorescence occurred in NaCl-stressed tobacco plants (**Figure 9B**). WT plants had a higher fluorescent intensity than the transgenic lines (**Figure 9B**). Furthermore, the green fluorescence overlapped the red auto-fluorescence (**Figure 9B**), indicating that the NaCl-elicited H_2O_2 mainly originated from chloroplast. Excessive H_2O_2 accumulation in chloroplast would cause oxidative damage to WT leaves.

ACTIVITY OF ANTIOXIDANT ENZYMES

In control conditions, SOD, CAT, and APX activities were similar in WT tobacco and transgenic plants (**Figure 10**). CAT and SOD activities of *KcCSD*-transgenic lines L7 and L8 were 20 and 50% higher than in WT after 8 h of NaCl stress, respectively (**Figures 10A,B**). Moreover, transgenic plants displayed a significantly higher CAT activity than WT plants after 24 h stress (**Figure 10B**), indicating that the capacity to scavenge H_2O_2 was enhanced by NaCl treatment. NaCl did not significantly decrease APX activity in WT and transgenic lines during short salinity (**Figure 10C**).

A 7-day 100 mM NaCl treatment increased activity of two dominant SOD isoenzymes in both transgenic lines and WT plants (**Figure 11**). Transgenic plants exhibited a higher increase in activity of one of four SOD isoenzymes than WT plants did under NaCl stress (arrowhead, **Figure 11**), indicating that overexpression of *KcCSD* in tobacco led to an increased antioxidant defense.

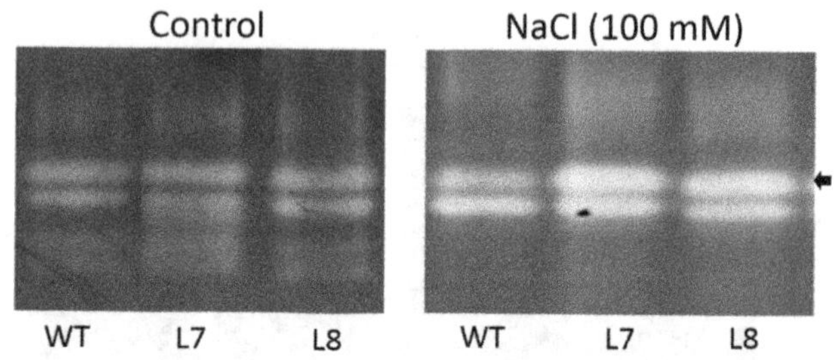

FIGURE 11 | Effect of NaCl on activities of SOD isoenzymes in leaves of wild-type tobacco and *KcCSD*-transgenic lines. Four-weeks old rooted plants of wild-type (WT) and transgenic lines (L7 and L8) were transferred to 1/4 Hoagland's nutrient solution for 2-weeks acclimation. Hydroponically acclimated plants were subjected to 0 or 100 mM NaCl for 7 days. The arrow indicates remarkably increased SOD isoform in transgenic lines.

DISCUSSION

Na^+ ACCUMULATION AND UPREGULATION OF *KcCSD* IN *K. CANDEL* LEAVES

Young roots of *K. candel* had a net Na^+ efflux under rising NaCl stress from 100 to 300 mM (**Figure 1**), which agreed with previous results in mangrove (Lu et al., 2013; Lang et al., 2014). Root Na^+ efflux resulted from active Na^+/H^+ antiport across the PM (Lu et al., 2013; Lang et al., 2014). Root Na^+ exclusion was more pronounced in a short treatment (1 week) than in a long-term stress (2–3 weeks) (**Figure 1**). This indicates that Na^+ extrusion capacity in *K. candel* roots declined with the exposure. As a result, large amount of Na^+ accumulated in roots was transported to other organs (hypocotyl, stem, and leaf) (**Figure 2**). Excessive Na^+ accumulation in *K. candel* leaves led to an increase of ROS (**Figure 3**). This was tallied with the findings in poplars (Wang et al., 2007, 2008). A limit in oxidative damage to photosynthetic apparatus and antioxidant enzyme activities could benefit for detoxifying Na^+-elicited ROS in *Populus euphratica* (Wang et al., 2007, 2008). In the present work, salinized *K. candel* promoted *KcCSD* expression and subsequently enhanced SOD activity in leaves (**Figure 3**). Protein abundance of SOD in *K. candel* leaves might increase under a high level of NaCl (Wang et al., 2014). Thus, it could be inferred that *K. candel* would upregulate the antioxidant enzymes to deal with a long-term saline stress. To investigate the role of KcCSD in salinity tolerance, *KcCSD* gene was transferred to the model species *Nicotiana tabacum*. The transgenic tobacco overexpressing *KcCSD* resulted in a greater root length and survival rate than WT plants under NaCl stress (**Figure 6**). This finding is consistent with other studies on transgenic Chinese cabbage plants (Tseng et al., 2007). Rice plants overexpressing a cytosolic Cu/Zn SOD gene (*Avicennia marina*) also conferred salinity tolerance in transgene plants (Prashanth et al., 2008). In this study, *KcCSD* overexpression in tobacco reduced ROS in chloroplasts under NaCl stress. The results confirmed the protection of chloroplast Cu/Zn SOD from NaCl stress.

KcCSD EXPRESSION AND ROS CONTROL IN CHLOROPLAST

Phylogenetic tree and sequence analyses have verified that KcCSD is a chloroplast CSD (**Figure 4**). The deduced protein sequence

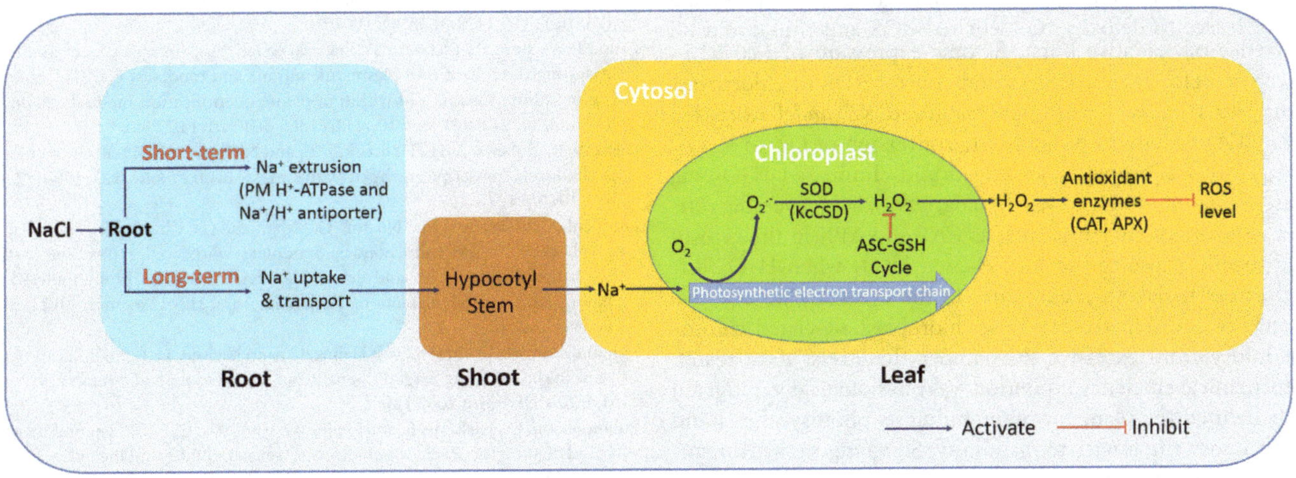

FIGURE 12 | Schematic model showing physiological mechanisms of *K. candel* adaptation to NaCl. Under a short term NaCl treatment, *K. candel* roots extrude Na^+ by the plasma membrane Na^+/H^+ antiport system driven by H^+-ATPase. During a prolonged stress, *K. candel* roots were unable to effectively limit Na^+ uptake and transport, leading to Na^+ accumulation in shoots. The buildup of Na^+ in leaves favors the formation of O_2^- via electron transport chain in chloroplasts. Salinized *K. candel* increases transcription of *CSD* gene encoding a copper/zinc superoxide dismutase (Cu/Zn SOD). An increased SOD activity in chloroplasts promotes the conversion of O_2^- to H_2O_2; subsequently, chloroplasts scavenge and eliminate H_2O_2 via the ascorbate-glutathione (ASC-GSH) cycle (Asada, 1999). Moreover, the activated antioxidant enzymes, such as catalase (CAT), and ascorbate peroxidase (APX), readily scavenge H_2O_2 when the molecule diffuses across thylakoid membranes to the cytosol. As a result, the efficient scavenging of Na^+-elicited ROS in chloroplasts avoids an occurrence of oxidative burst, and thus alleviates chloroplast injury from excessive ROS under the stress.

has confirmed a high similarity to chloroplast CSDs from other species, such as *Bruguiera gymnorrhiza*, *Dimocarpus longan*, *Gossypium hirsutum*, and *Arabidopsis thaliana* (**Figure 4**). In agreement with sequence analysis, a subcellular location assay revealed that KcCSD protein is targeted to chloroplast (**Figure 5**). Similarly, *Arabidopsis* CSD2 localized in chloroplast (Kliebenstein et al., 1998; Huang et al., 2012). ROS analysis in transgene tobacco plants indicated that KcCSD was involved in protecting chloroplasts from Na^+ damage. NaCl treatment caused a significant increase in leaf H_2O_2 (**Figure 8**) and WT tobacco chloroplast (**Figure 9**). The H_2O_2 burst in chloroplast resulted from the SOD-catalyzed conversion of O_2^-. It was formed predominantly at a high rate of electron transfer to O_2 (Asada, 1999; Apel and Hirt, 2004). The Na^+-induced oxidative damage occurred in WT tobacco leaves. It led to an increased MDA and declined Pn, chlorophyll content, chlorophyll a/b ratio, Fv/Fm, and ΦPSII (**Figures 6–8**). This was a result of high H_2O_2 in chloroplast (**Figure 9**; Stepien and Johnson, 2009). Excessive H_2O_2 has been shown to trigger membrane lipid peroxidation, and limit membrane lipid unsaturation and membrane protein polymerization (Bowler et al., 1992; Wang et al., 2007).

Compared to WT tobacco plants, *KcCSD*-transgenic plants accumulated less H_2O_2 in both leaves and chloroplast under NaCl stress (**Figures 8, 9**). Unexpectedly, NaCl-treated transgenic plants retained higher O_2^- production than WT (**Figure 8**). The high O_2^- in salinized transgenic plants was likely due to feedback activation of O_2^- production system. A high conversion of O_2^- to H_2O_2 in *KcCSD*-transgenic plants accelerated the electron transfer to O_2 via photosynthetic electron transport chain, thus activating positive feedback production of O_2^-. A lower H_2O_2 in *KcCSD*-transgenic plants mainly resulted from an increased activity of antioxidant enzymes (**Figures 10, 11**). Under either 24-h or 7-day NaCl stress, *KcCSD*-transgenic tobacco plants increased total activity of SOD due to an increment of SOD isoenzymes (**Figures 10, 11**). Moreover, in transgenic plants, NaCl treatment rapidly increased CAT after 8 h of salinity (**Figure 10**). Increased activities of CAT arose from an increased production of O_2^- and H_2O_2, as ROS are secondary messengers to induce antioxidant defenses (Desikan et al., 2001; Vranová et al., 2002). It is evidenced that the antioxidant enzymes SOD, CAT, and APX play a crucial role in maintaining O_2^- and H_2O_2 balance in the plants (Payton et al., 2001). Hence, increased activity of SOD in transgenic plants promoted the conversion of O_2^- to H_2O_2 (Bowler et al., 1992; Fridovich, 1995) and CAT activation. Consequently, increased CAT would assist transgenic plants in reducing NaCl-elicited H_2O_2 in leaf cells during an extended NaCl stress (**Figures 8, 9**). Similar findings were observed in *K. candel* under NaCl treatment. Li (2009) showed that *K. candel* increased activities of SOD, CAT, APX, and glutathione reductase (GR) in leaves to control ROS during NaCl stress.

CONCLUSIONS

This study has revealed that *K. candel* has different physiological mechanisms to adapt to NaCl stress (**Figure 12**). As shown in the schematic model, *K. candel* roots could maintain a high capacity to extrude Na^+ via a PM Na^+/H^+ antiport system driven by H^+-ATPase. Under a prolonged stress, *K. candel* could activate its antioxidant system when roots were unable to effectively limit Na^+ uptake and transport in the plant. The buildup of Na^+ in leaves would favor the formation of O_2^- via electron transport chain in chloroplast. *K. candel* could also upregulate *CSD* in leaves to detoxify Na^+-elicited ROS and thus avoid an occurrence of oxidative burst. Ectopic expression of *KcCSD* in tobacco revealed that KcCSD could control ROS in chloroplast during NaCl stress. Accordingly, salinized *K. candel* increased Cu/Zn SOD activity to promote the conversion of O_2^- to H_2O_2; subsequently, chloroplasts scavenged and eliminated H_2O_2 via the ascorbate-glutathione cycle (Asada, 1999). Moreover, the activated antioxidant enzymes, such as CAT and APX in the cytosol could readily scavenge the Na^+-elicited H_2O_2 when H_2O_2 diffused across thylakoid membranes to the cytosol. The effective scavenge of Na^+-elicited H_2O_2 in chloroplast alleviated chloroplast injury from excessive ROS under the stress. As a result, photochemical efficiency inhibition was physiologically mitigated and benefitted the plant for maintaining its photosynthesis and growth under the longer term salinity. Signaling network regulating *KcCSD* transcription under NaCl stress needs to be further investigated in the future.

AUTHOR CONTRIBUTIONS

Xiaoshu Jing designed and performed the experiments, analyzed the experimental data, and prepared the manuscript. Peichen Hou, Yanjun Lu, Shurong Deng, Niya Li, Yang Wang, Yansha Han, and Tao Lang partly participated in the experiments (Peichen Hou: gene cloning and expression analysis of *KcCSD* in *Kandelia candel*, Yanjun Lu: Na^+ fluxes recording in roots and Na^+ concentrations, Shurong Deng subcellular location of KcCSD). Rui Zhao designed KcCSD-GFP construct. Rui Zhao, Jian Sun, Mingquan Ding, and Xin Shen conceived research plan. Shaoliang Chen designed research work and revised the manuscript. All authors have read and approved the final version of this manuscript.

ACKNOWLEDGMENTS

This research was supported jointly by National Natural Science Foundation of China (Grant Nos. 31270654, 31160150), Research Project of Chinese Ministry of Education (Grant No. 113013A), key project for Overseas Scholars by Ministry of Human Resources and Social Security of P. R. China (Grant No. 2012001), Program of Introducing Talents of Discipline to Universities (Plan 111 Project, Grant No. B13007), and Program for Changjiang Scholars and Innovative Research Teams of the University (Grant No. IRT13047).

REFERENCES

Aebi, H. (1984). Catalase *in vitro*. *Meth. Enzymol.* 105, 121–126.

Apel, K., and Hirt, H. (2004). Reactive oxygen species: metabolism, oxidative stress, and signal transduction. *Annu. Rev. Plant Biol.* 55, 373–399. doi: 10.1146/annurev.arplant. 55.03-1903.141701

Asada, K. (1999). The water-water cycle in chloroplasts: scavenging of active oxygens and dissipation of excess photons. *Annu. Rev. Plant Biol.* 50, 601–639. doi: 10.1146/annurev. arplant.50.1.601

Beauchamp, C., and Fridovich, I. (1971). Superoxide dismutase: improved assays and an assay applicable to acrylamide gels. *Anal. Biochem.* 44, 276–287. doi: 10.1016/00032697-(71)90370-8

Bowler, C., Montagu, M. V., and Inze, D. (1992). Superoxide dismutase and stress tolerance. *Ann. Rev. Plant Biol.* 43, 83–116.

Chen, S., and Polle, A. (2010). Salinity tolerance of *Populus*. *Plant Biol.* 12, 317–333. doi: 10.1111/j.1438-8677.2009.00301.x

Cuin, T. A., Bose, J., Stefano, G., Jha, D., Tester, M., Mancuso, S., et al. (2011). Assessing the role of root plasma membrane and tonoplast Na^+/H^+ exchangers in salinity tolerance in wheat: in planta quantification methods. *Plant Cell Environ.* 34, 947–961. doi: 10.1111/j.1365-3040.2011.02296.x

Desikan, R., Soheila, A. H., Hancock, J. T., and Neill, S. J. (2001). Regulation of the Arabidopsis transcriptome by oxidative stress. *Plant Physiol.* 127, 159–172. doi: 10.1104/pp.127.1.159

Dutilleul, C., Garmier, M., Noctor, G., Mathieu, C., Chétrit, P., Foyer, C. H., et al. (2003). Leaf mitochondria modulate whole cell redox homeostasis, set antioxidant capacity, and determine stress resistance through altered signaling and diurnal regulation. *Plant Cell* 15, 1212–1226. doi: 10.1105/tpc. 009464

Emanuelsson, O., Brunak, S., von Heijne, G., and Nielsen, H. (2007). Locating proteins in the cell using TargetP, SignalP and related tools. *Nat. Protoc.* 2, 953–971. doi: 10.1038/nprot.2007.131

Emanuelsson, O., Nielsen, H., and Heijne, G. V. (1999). ChloroP, a neural network-based method for predicting chloroplast transit peptides and their cleavage sites. *Protein Sci.* 8, 978–984.

Fridovich, I. (1995). Superoxide radical and superoxide dismutases. *Annu. Rev. Plant Biochem.* 64, 97–112. doi: 10.1146/annurev.biochem.64.1.97

Giannopolitis, C. N., and Ries, S. K. (1977). Superoxide dismutases II. Purification and quantitative relationship with water-soluble protein in seedlings. *Plant Physiol.* 59, 315–318. doi: 10.1104/pp.59.2.315

Gutteridge, J., and Halliwell, B. (2010). Antioxidants: molecules, medicines, and myths. *Biochem. Biophy. Res. Commun.* 393, 561–564. doi: 10.1016/j.bbrc.2010.02.071

Han, Y., Wang, W., Sun, J., Ding, M., Zhao, R., Deng, S., et al. (2013). *Populus euphratica* XTH overexpression enhances salinity tolerance by the development of leaf succulence in transgenic tobacco plants. *J. Exp. Bot.* 64, 4225–4238. doi: 10.1093/jxb/ert229

Heath, R. L., and Packer, L. (1968). Photoperoxidation in isolated chloroplasts: I. Kinetics and stoichiometry of fatty acid peroxidation. *Arch. Biochem. Biophy.* 125, 189–198. doi: 10.1016/0003-9861(68)90654-1

Hoagland, D. R., and Arnon, D. I. (1950). *The Water-Culture Method for Growing Plants Without Soil.* Circular 347, 39. Berkeley, CA: California Agricultural Experiment Station.

Horsch, R. B., Fry, J. E., Hoffmann, N. L., Eichholtz, D., Rogers, S. A., and Fraley, R. T. (1985). A simple and general method for transferring genes into plants. *Science* 227, 1229–1231. doi: 10.1126/science.227.4691.1229

Hou, P. (2010). *Microarray Analysis of Gene Expression Patterns and Putative Functions of Differentially Expressed Genes in Mangrove Kandelia Candel Under High-Salinity Stress.* Ph.D. dissertation, 55–62, Beijing Forestry University, Beijing.

Huang, C. H., Kuo, W. Y., Weiss, C., and Jinn, T. L. (2012). Copper chaperone-dependent and-independent activation of three copper-zinc superoxide dismutase homologs localized in different cellular compartments in Arabidopsis. *Plant Physiol.* 158, 737–746. doi: 10.1104/pp.111

Jakob, B., and Heber, U. (1996). Photoproduction and detoxification of hydroxyl radicals in chloroplasts and leaves and relation to photoinactivation of photosystems I and II. *Plant Cell Physiol.* 37, 629–635.

Kliebenstein, D. J., Monde, R. A., and Last, R. L. (1998). Superoxide dismutase in Arabidopsis: an eclectic enzyme family with disparate regulation and protein localization. *Plant Physiol.* 118, 637–650. doi: 10.1104/pp.118.2.637

Lang, T., Sun, H., Li, N., Lu, Y., Shen, Z., Jing, X., et al. (2014). Multiple signaling networks of extracellular ATP, hydrogen peroxide, calcium, and nitric oxide in the mediation of root ion fluxes in secretor and non-secretor mangroves under salt stress. *Aquat. Bot.* 119, 33–43. doi: 10.1016/j.aquabot.2014. 06.009

Li, N., Chen, S., Zhou, X., Li, C., Shao, J., Wang, R., et al. (2008). Effect of NaCl on photosynthesis, salt accumulation and ion compartmentation in two mangrove species, *Kandelia candel* and *Bruguiera gymnorhiza*. *Aquat. Bot.* 88, 303–310. doi: 10.1016/j.aquabot.2007.12.003

Li, N. (2009). *Regulation of Ionic and Reactive Oxygen Species Homeostasis in Seedlings of Two Mangrove and the Relevance to Salinity.* Ph.D. dissertation, 69–76, Beijing Forestry University, Beijing.

Lichtenthaler, H. K. (1987). Chlorophylls and carotenoids: pigments of photosynthetic biomembranes. *Meth. Enzymol.* 148, 350–382.

Livak, K. J., and Schmittgen, T. D. (2001). Analysis of relative gene expression data using real-time quantitative PCR and the $2^{-\Delta\Delta C_T}$ method. *Methods* 25, 402–408. doi: 10.1006/meth.2001.1262

Lu, Y., Li, N., Sun, J., Hou, P., Jing, X., Zhu, H., et al. (2013). Exogenous hydrogen peroxide, nitric oxide and calcium mediate root ion fluxes in two non-secretor mangrove species subjected to NaCl stress. *Tree Physiol.* 33, 81–95. doi: 10.1093/treephys/tps119

Maxwell, K., and Johnson, G. N. (2000). Chlorophyll fluorescence-a practical guide. *J. Exp. Bot.* 51, 659–668. doi: 10.1093/jexbot/51.345.659

Murashige, T., and Skoog, F. (1962). A revised medium for rapid growth and bio assays with tobacco tissue cultures. *Physiol. Plant.* 15, 473–497. doi: 10.1111/j.1399-3054.1962.tb08052.x

Nakano, Y., and Asada, K. (1981). Hydrogen peroxide is scavenged by ascorbate-specific peroxidase in spinach chloroplasts. *Plant Cell Physiol.* 22, 867–880.

Payton, P., Webb, R., Kornyeyev, D., Allen, R., and Holaday, A. S. (2001). Protecting cotton photosynthesis during moderate chilling at high light intensity by increasing chloroplastic antioxidant enzyme activity. *J. Exp. Bot.* 52, 2345–2354. doi: 10.1093/jexbot/52.365.2345

Polle, A., and Chen, S. (2014). On the salty side of life: molecular, physiological and anatomical adaptation and acclimation of trees to extreme habitats. *Plant Cell Environ.* doi: 10.1111/pce.12440. [Epub ahead of print].

Prashanth, S. R., Sadhasivam, V., and Parida, A. (2008). Over expression of cytosolic copper/zinc superoxide dismutase from a mangrove plant *Avicennia marina* in *indica* Rice var Pusa Basmati-1 confers abiotic stress tolerance. *Transgenic Res.* 17, 281–291. doi: 10.1007/s11248-007-9099-6

Ramírez, L., Bartoli, C. G., and Lamattina, L. (2013). Glutathione and ascorbic acid protect *Arabidopsis* plants against detrimental effects of iron deficiency. *J. Exp. Bot.* 64, 3169–3178. doi: 10.1093/jxb/ert153

Shen, Z., Ding, M., Sun, J., Deng, S., Zhao, R., Wang, M., et al. (2013). Overexpression of *PeHSF* mediates leaf ROS homeostasis in transgenic tobacco lines grown under salt stress conditions. *Plant Cell Tiss. Org. Cult.* 115, 299–308. doi: 10.1007/s11240-013-0362-7

Stepien, P., and Johnson, G. N. (2009). Contrasting responses of photosynthesis to salt stress in the glycophyte *Arabidopsis* and the halophyte *Thellungiella*: role of the plastid terminal oxidase as an alternative electron sink. *Plant Physiol.* 149, 1154–1165. doi: 10.1104/pp.108

Sun, J., Chen, S., Dai, S., Wang, R., Li, N., Shen, X., et al. (2009a). NaCl-induced alternations of cellular and tissue ion fluxes in roots of salt-resistant and salt-sensitive poplar species. *Plant Physiol.* 149, 1141–1153. doi: 10.1104/pp.108

Sun, J., Dai, S., Wang, R., Chen, S., Li, N., Zhou, X., et al. (2009b). Calcium mediates root K^+/Na^+ homeostasis in poplar species differing in salt tolerance. *Tree Physiol.* 29, 1175–1186. doi: 10.1093/treephys/tpp04

Sun, J., Li, L., Liu, M., Wang, M., Ding, M., Deng, S., et al. (2010a). Hydrogen peroxide and nitric oxide mediate K^+/Na^+ homeostasis and antioxidant defense in NaCl-stressed callus cells of two contrasting poplars. *Plant Cell Tiss. Org. Cult.* 103, 205–215. doi: 10.1007/s11240-010-9768-7

Sun, J., Wang, M. J., Ding, M. Q., Deng, S. R., Liu, M. Q., Lu, C. F., et al. (2010b). H_2O_2 and cytosolic Ca^{2+} signals triggered by the PM H^+-coupled transport system mediate K^+/Na^+ homeostasis in NaCl-stressed *Populus euphratica* cells. *Plant Cell Environ.* 33, 943–958. doi: 10.1111/j.1365-3040.2010.02118.x

Sun, J., Zhang, X., Deng, S., Zhang, C., Wang, M., Ding, M., et al. (2012). Extracellular ATP signaling is mediated by H_2O_2 and cytosolic Ca^{2+} in the salt response of *Populus euphratica* cells. *PLoS ONE* 7:e53136. doi: 10.1371/journal.pone.0053136

Thordal-Christensen, H., Zhang, Z., Wei, Y., and Collinge, D. B. (1997). Subcellular localization of H_2O_2 in plants. H_2O_2 accumulation in papillae and hypersensitive response during the barley—powdery mildew interaction. *Plant J.* 11, 1187–1194. doi: 10.1046/j.1365-313X.1997.11061187.x

Tseng, M. J., Liu, C. W., and Yiu, J. C. (2007). Enhanced tolerance to sulfur dioxide and salt stress of transgenic Chinese cabbage plants expressing both superoxide dismutase and catalase in chloroplasts. *Plant Physiol. Biochem.* 45, 822–833. doi: 10.1016/j.plaphy.2007.07.011

Volkov, V., Wang, B., Dominy, P. J., Fricke, W., and Amtmann, A. (2004). *Thellungiella halophila*, a salt–tolerant relative of *Arabidopsis thaliana*, possesses effective mechanisms to discriminate between potassium and sodium. *Plant Cell Environ.* 27, 1–14. doi: 10.1046/j.0016-8025.2003.01116.x

Vranová, E., Inze, D., and Van Breusegem, F. (2002). Signal transduction during oxidative stress. *J. Exp. Bot.* 53, 1227–1236. doi: 10.1093/jexbot/53.372.1227

Wan, C. Y., and Wilkins, T. A. (1994). A modified hot borate method significantly enhances the yield of high-quality RNA from cotton (*Gossypium hirsutum L.*). *Anal. Biochem.* 223, 7–12. doi: 10.1006/abio.1994.1538

Wang, L., Liang, W., Xing, J., Tan, F., Chen, Y., Huang, L., et al. (2013). Dynamics of chloroplast proteome in salt-stressed mangrove *Kandelia candel* (L.) Druce. *J. Proteome Res.*12, 5124–5136. doi: 10.1021/pr4006469

Wang, L., Liu, X., Liang, M., Tan, F., Liang, W., Chen, Y., et al. (2014). Proteomic analysis of salt-responsive proteins in the leaves of mangrove *Kandelia candel* during short-term stress. *PLoS ONE* 9:e83141. doi: 10.1371/journal.pone.0083141

Wang, R., Chen, S., Deng, L., Fritz, E., Hüttermann, A., and Polle, A. (2007). Leaf photosynthesis, fluorescence response to salinity and the relevance to chloroplast salt compartmentation and anti-oxidative stress in two poplars. *Trees* 21, 581–591. doi: 10.1007/s00468-007-0154-y

Wang, R., Chen, S., Zhou, X., Shen, X., Deng, L., Zhu, H., et al. (2008). Ionic homeostasis and reactive oxygen species control in leaves and xylem sap of two poplars subjected to NaCl stress. *Tree Physiol.* 28, 947–957. doi: 10.1093/treephys/28.6.947

Wellburn, A. R. (1994). The spectral determination of chlorophylls a and b, as well as total carotenoids, using various solvents with spectrophotometers of different resolution. *J. Plant Physiol.* 144, 307–313. doi: 10.1016/S0176-1617(11)81192-2

Yoo, S. D., Cho, Y. H., and Sheen, J. (2007). *Arabidopsis* mesophyll protoplasts: a versatile cell system for transient gene expression analysis. *Nature Protoc.* 2, 1565–1572. doi: 10.1038/nprot.2007.199

Mechanisms of salt tolerance in habanero pepper plants (*Capsicum chinense* Jacq.): Proline accumulation, ions dynamics and sodium root-shoot partition and compartmentation

Emanuel Bojórquez-Quintal[1], Ana Velarde-Buendía[2], Ángela Ku-González[1], Mildred Carillo-Pech[1], Daniela Ortega-Camacho[3], Ileana Echevarría-Machado[1], Igor Pottosin[2] and Manuel Martínez-Estévez[1]*

[1] *Unidad de Bioquímica y Biología Molecular de Plantas, Centro de Investigación Científica de Yucatán, Yucatán, México*
[2] *Centro Universitario de Investigaciones Biomédicas, Universidad de Colima, Colima, México*
[3] *Unidad de Ciencias del Agua, Centro de Investigación Científica de Yucatán, Yucatán, México*

Edited by:
Vadim Volkov, London Metropolitan University, UK

Reviewed by:
Sergey Shabala, University of Tasmania, Australia
Vadim Volkov, London Metropolitan University, UK
Soledad Francisca Undurraga, Universidad Mayor, Chile

***Correspondence:**
Manuel Martínez-Estévez, Unidad de Bioquímica y Biología Molecular de Plantas, Centro de Investigación Científica de Yucatán (CICY), Calle 30 # 130, Col. Chuburná de Hidalgo, Mérida 97200, México
e-mail: luismanh@cicy.mx

Despite its economic relevance, little is known about salt tolerance mechanisms in pepper plants. To address this question, we compared differences in responses to NaCl in two *Capsicum chinense* varieties: Rex (tolerant) and Chichen-Itza (sensitive). Under salt stress (150 mM NaCl over 7 days) roots of Rex variety accumulated 50 times more compatible solutes such as proline compared to Chichen-Itza. Mineral analysis indicated that Na^+ is restricted to roots by preventing its transport to leaves. Fluorescence analysis suggested an efficient Na^+ compartmentalization in vacuole-like structures and in small intracellular compartments in roots of Rex variety. At the same time, Na^+ in Chichen-Itza plants was compartmentalized in the apoplast, suggesting substantial Na^+ extrusion. Rex variety was found to retain more K^+ in its roots under salt stress according to a mineral analysis and microelectrode ion flux estimation (MIFE). Vanadate-sensitive H^+ efflux was higher in Chichen-Itza variety plants, suggesting a higher activity of the plasma membrane H^+-ATPase, which fuels the extrusion of Na^+, and, possibly, also the re-uptake of K^+. Our results suggest a combination of stress tolerance mechanisms, in order to alleviate the salt-induced injury. Furthermore, Na^+ extrusion to apoplast does not appear to be an efficient strategy for salt tolerance in pepper plants.

Keywords: salt tolerance, pepper, roots, proline accumulation, sodium compartmentalization, potassium retention, ion fluxes, H^+-ATPase

INTRODUCTION

The excess of soluble salts in soil, particularly NaCl, causes three types of stresses in plants: osmotic, ionic, and oxidative. These stresses reduce absorption and induce a massive efflux of water and ions (K^+) in plant cells, resulting in water, and nutritional imbalances. The accumulation of Na^+ to toxic concentrations and the production of reactive oxygen species (ROS) reduce the growth, yield, and production of economically important crops, such as cereals and vegetables (Munns and Tester, 2008; Bojórquez-Quintal et al., 2012). Plants in relation to salt can be classified into two groups: halophytes (growth stimulated at moderate and tolerant to high salinity) and glycophytes, which display a suppressed growth in a saline environment (Flowers and Colmer, 2008; Ruan et al., 2010). In halophytes, various adaptive mechanisms to tolerate high levels of salt have been identified and intensively studied (Ruan et al., 2010; Adolf et al., 2013; Shabala, 2013). Unfortunately, some of these mechanisms may not be directly transferred to crop plants, which are mostly glycophytes (Zhang and Shi, 2013). Yet, several crops are relatively salt resistant, and there are also substantial differences in the salt tolerance between nearly isogenic varieties within the same plant species. Salt tolerance is a complex trait controlled by many genes and involves various biochemical and physiological mechanisms. The fine tuning of these mechanisms is necessary to achieve a significant increase in tolerance to salt (Zhang and Shi, 2013; Adem et al., 2014).

Proline is the most common compatible osmolyte in plants and has therefore been extensively studied. The accumulation of this amino acid is an important regulatory mechanism under osmotic stress (Huang et al., 2013). Proline is a multifunctional amino acid (Szabados and Savouré, 2009). In many plant species, the accumulation of proline has been associated with tolerance to salt stress and has even been used as a marker to select tolerant genotypes (Ashraf and Harris, 2004). However, a negative correlation between the accumulation of proline and salt tolerance has also been reported, indicating discrepancies in its function (Lutts et al., 1999; Chen et al., 2007a). Proline accumulation is made possible by the increase in the expression and activity of the synthesis enzymes (Δ-pyrroline-5-carboxylate synthetase, P5CS; Δ-pyrroline-5-carboxylate reductase, P5CR) or by the decrease in the degradation enzymes, proline dehydrogenase or proline oxidase (PDH or POX), and P5C dehydrogenase (P5CDH) (Huang et al., 2013). Under salt stress, the *P5CS1* and *PDH* genes are positively and negatively regulated, respectively (Kishor et al., 2005; Verslues and Sharma, 2010; Jaarsma et al., 2013). Similarly, the overexpression of the *P5CS* gene increases proline synthesis under salt stress and improves tolerance to salt (Kishore et al., 1995; Hmida-Sayari et al., 2005).

The roots are the first site of contact with high concentrations of Na^+ in the soil and therefore of the uptake or absorption of salt. Na^+ influx is mediated by non-selective cation channels (NSCC), high-affinity K^+ transporters (HKTs), and low-affinity cation transporters (LCT) in the root epidermal cells (Apse and Blumwald, 2007; Plett and Moller, 2010; Maathuis, 2014). Na^+ is then transported radially toward the root xylem via the apoplast and symplast. After being loaded into the xylem, Na^+ is finally transported to the shoots by xylem flow (Adams and Shin, 2014). In contrast to halophytes, Na^+ is not an essential element for most plants and becomes highly toxic at high concentrations, particularly in the aerial parts of the plant. Therefore, it is necessary to maintain efficient control of Na^+ content and intracellular compartmentalization in plant tissues. The high-affinity potassium transporters (HKTs), the Na^+/H^+ SOS1 (salt overly sensitive) antiporters on the plasma membrane and the intracellular NHX antiporters (Na^+/H^+) are transporters involved in the Na^+ homeostasis (Almeida et al., 2013; Adams and Shin, 2014).

The regulation of K^+ homeostasis is essential for plant adaptation to biotic and abiotic stresses. This adaptation is associated with the wide range of functions in which K^+ participates (Anschütz et al., 2014; Demidchik, 2014; Shabala and Pottosin, 2014). Recently, K^+ retention in the cells of roots and leaves has been identified as an important trait for salt tolerance. A strong negative correlation between the magnitude of salt-induced K^+ loss and salt tolerance, observed in various crop species, suggested K^+ retention as a selection criterion between salt tolerant and sensitive varieties (Chen et al., 2005, 2007b,c; Smethurst et al., 2008; Lu et al., 2013; Wu et al., 2013; Bonales-Alatorre et al., 2013a). Furthermore, it has been observed that the exogenous administration of organic compounds and divalent cations prevents K^+ efflux (Cuin and Shabala, 2005, 2007a,b; Shabala et al., 2006; Zhao et al., 2007; Chen et al., 2007a; Zepeda-Jazo et al., 2008). Efficient control of membrane potential due to the H^+-ATPase activity was shown to be important for the salt tolerance in several species (Chen et al., 2007b; Cuin et al., 2008; Hariadi et al., 2011; Bose et al., 2013, 2014). A more negative membrane potential during salt stress reduces the driving force for the K^+ loss and facilitates the K^+ absorption, thus allowing plants to retain K^+ in the cytosol (Chen et al., 2007b; Bose et al., 2013). Likewise, the H^+-ATPase activity is essential to fuel Na^+/H^+ exchangers in the plasma membrane (SOS1). At the same time, a higher activity of the H^+ pump consumes a large amount of ATP, hence has a higher energetic cost (Malagoli et al., 2008). Thus, keeping electrochemical gradients for physiologically important cations across the plasma membrane could present an energetic burden, so this tolerance mechanism cannot be considered permanent and may be used as a temporary solution at early times after the onset of the salt stress (Bose et al., 2013, 2014).

Peppers (*Capsicum* spp.) are an economically important genus of the *Solanaceae* family, which also includes tomatoes and potatoes. Among the 32 species native to America, *C. annuum L.*, *C. baccatum L.*, *C. frutescens L.*, *C. pubescens L.*, and *C. chinense* Jacq. are cultivated (Moscone et al., 2007; Perry et al., 2007). Overall, pepper plants are grown around the world because of their adaptation to different agro-climatic regions and their wide variety of shapes, sizes, colors, and pungencies of the fruit (Qin et al., 2014). However, these plants are sensitive to various biotic stresses, such as viruses and Oomycetes and abiotic factors such as drought and salinity. In fact, pepper plants are considered moderately sensitive, sensitive or highly susceptible to salt stress (Maas and Hoffman, 1977; Aktas et al., 2006). Nevertheless, despite their economic importance as a horticultural species, very little is known about the mechanisms of tolerance to high salt concentrations. To contribute to the understanding of salt stress in species of economic importance such as peppers, the difference in salt sensitivity of two varieties of the species *C. chinense* Jacq, commonly known as habanero pepper, was evaluated in this study. Furthermore, possible mechanisms of salt stress tolerance for the two varieties were addressed by electrophysiological studies using selective microelectrodes (MIFE) and by subcellular localization of Na^+ using fluorescent indicators.

MATERIALS AND METHODS

PLANT MATERIAL AND GROWTH CONDITIONS

Habanero pepper (*C. chinense* Jacq.) seeds of the Chichen-Itza (Seminis®) and Rex (Mayan Chan obtained in CICY) varieties were used in this study. To disinfect the seeds, they were rinsed in 80% ethanol (v/v) for 5 min and washed continuously with sterile water. Seeds were then incubated with 30% (v/v) sodium hypochlorite (Cloralex 5% NaOCl) and Tween (1 drop) for 15 min. Washes were continuous, and the seeds were kept in sterile water for 48 h at 4°C in the dark. After stratification, seeds of both varieties were incubated (in the dark) in Petri dishes with filter paper moistened with sterile water until the emergence of the radicle.

For the hydroponic experiments, germinated seeds were transferred to plastic containers with vermiculite moistened with a Hoagland solution to a fifth of its ionic strength (H1/5). Seeds were incubated under photoperiods of 16/8 h light/dark at 25°C. The light intensity was 123 μ mol m^{-2} s^{-1}. Seedlings were irrigated for 45 days with a sterile water solution and H1/5 with 7-day intervals. The modified Hoagland solution at one-fifth of its ionic strength (H1/5) contained the following: 1.2 mM KNO_3, 0.8 mM $Ca(NO_3)_2$, 0.2 mM KH_2PO_4, 0.2 mM $MgSO_4$, 50 μM $CaCl_2$, 12.5 μM H_3BO_3, 1 μM $MnSO_4$, 1 μM $ZnSO_4$, 0.5 μM $CuSO_4$, 0.1 μM $(NH_4)_6Mo_7O_{24}$, 0.1 μM NaCl, and 10 μM Fe-EDTA, pH 6.8.

For electrophysiological experiments and subcellular Na^+ localization, seeds with radicles were transferred to Petri dishes containing modified Gamborg-B5 growth medium (Sales B5, Sigma) at half ionic strength (B5/2). B5/2 medium was supplemented with 0.5% sucrose (w/v) and 1% agar (w/v). The pH was adjusted to 5.8. Seedlings 10 days of age with a primary root 8–10 cm in length were used for this experiment.

NaCl STRESS TREATMENT

Forty–five- to fifty-day-old seedlings (growing in vermiculite) of Rex and Chichen-Itza varieties were used. After pre-treatment with a solution of H1/5 for 7 days to avoid mechanical damage, seedlings homogeneous in size were selected. Three replicates of 10 seedlings of each variety were subjected to 7 days of treatment at concentrations of 0, 50, 100, and 150 mM NaCl in H1/5 solution. Treatments were performed in a culture room with photoperiods of 16/8 h light/dark at 25°C. At the end of the experiment, the seedlings were harvested and washed with sterile water to remove excess NaCl, and the fresh weight (FW), dry weight (DW), and the water content was determined by the formula (FW-DW)/FW. Each type of sample was dried in an electric oven at 60°C for 72 h. The leaves and roots were used for the determination of proline, Na^+, and K^+ content.

PROLINE CONTENT

A modification of the method described by Bates et al. (1973) was used to determine proline content. Briefly, dry leaf and root tissues were macerated and homogenized in 10 mL of boiling water. For each reaction, 2 mL of the supernatant was mixed with 2 mL of acetic acid and 2 mL of ninhydrin. The reaction mixture was heated in a water bath at 100°C for 60 min, and the reaction was stopped in an ice bath. For extraction, 4 mL of toluene was added, and samples were mixed vigorously for 15–20 s. Samples were then set aside to allow separation of the organic and aqueous phases. The organic phase containing the chromophore was collected in a new tube, and absorbance was read at 520 nm using toluene as a blank. Proline concentration was determined from a standard curve, and concentrations were calculated based on DW.

SODIUM AND POTASSIUM QUANTIFICATION

Samples of dried leaves and roots were weighed, and HNO_3:H_2O_2 was added at a 5:1 (v:v) ratio. Microwave digestion was performed at 1200 W using a ramp of 15 min to 200°C, 10 min to 200°C, and 5 min to 170°C. Subsequently, samples were adjusted to a volume of 25 mL with water (Milli Q), and Na^+ and K^+ content was quantified by inductively coupled plasma atomic emission spectroscopy (ICP-AES, Thermo IRIS Intrepid II XDL, New York, USA). Standard curves were used for each element.

MIFE TECHNIQUE

The net flux of K^+ and H^+ on the surface of the roots of the two varieties of *C. chinense* was measured non-invasively by the microelectrode ion flux estimation (MIFE) technique (Newman, 2001). For MIFE studies, seedlings grown *in vitro* with roots of 8–10 cm in length were transferred and fixed horizontally to a measuring chamber. Subsequently, 25 mL of measurement solutions were added for K^+ (0.5 mM KCl, 0.1 mM CaCl2, 5 mM MES, 2 mM Tris base, pH 6.0) and for H^+ (0.5 mM KCl and 0.1 mM $CaCl_2$, without pH-buffer) and incubated for 1 h to allow stabilization. Two electrodes selective for K^+ and H^+ were used in each experiment. For salt stress treatment, NaCl solution was added to the measuring chamber to a final concentration of 150 mM. Before the experiment, the microelectrodes were filled with 0.5 mM KCl for K^+ or 0.15 mM NaCl and 0.04 mM KH_2PO_4 for H^+, and the tip of each electrode was filled with the ion-selective resin (ion-liquid exchanger, LIX Fluka, Sigma-Aldrich) for the ion measured. Two electrodes were then mounted in a micromanipulator, and located perpendicular to the root axis 20–40 μm from the mature root zone, 1–2 cm from the root apex. Measurements were initiated by moving the electrodes 50 μm back and then forth and back in a cyclic manner every 8 s. The software CHART recorded potential differences between the two measurement positions and converted them into electrochemical potential differences using the Nernst slope. Net ion fluxes were calculated using the MIFEFLUX software for cylindrical diffusion geometry.

LOCALIZATION AND SUBCELLULAR ACCUMULATION OF SODIUM

Sodium Green™ indicator (S-6901, Molecular Probes, Life Technologies) was used to evaluate the subcellular localization and accumulation of sodium in the roots of *C. chinense.* Seedlings grown *in vitro* were incubated in a measurement solution (0.5 mM KCl, 0.1 mM $CaCl_2$, 5 mM MES, 2 mM Tris base, pH 6.0) supplemented or not with 150 mM NaCl. After 60 min of treatment, control, and NaCl treated seedlings were washed with distilled water and a solution of 0.5 mM $CaCl_2$. Root segments (1–1.5 cm long) were cut from the mature zone to 1–2 cm from the root apex and were incubated for 60 min in Eppendorf tubes of 500 μL (measurement solution) with 10 μM of Sodium Green™ (sodium staining) and 20 μM of FM®4-64 (membrane staining, T-13320, Molecular Probes, Life Technologies). Excess dye was removed, and the primary root segments were placed on a slide. Approximately 10 μL of Vectashield (H-1000, Vector Laboratories, Inc.) with DAPI (nuclei staining, D3571, Molecular Probes, Life Technologies) was added. The fluorescence was observed using the confocal microscope FluoView™ FV1000 (Olympus, Japan). A UPLFLN 40X0 (oil, NA: 1.3) lens was used with a scanning speed of 10 μs/pixel. DAPI, Sodium Green™, and FM4-64 exhibit excitation and emission wavelengths of 358–461 nm, 507–532 nm, and 515–670 nm, respectively. Z images had a resolution of 512 × 512 pixels and were projected as a single image. Fluorophores were merged using the software FV10-ASW 3.01b.

STATISTICAL ANALYSIS

Data were analyzed using a One-Way analysis of variance (ANOVA) (Sigma Stat Version 3.1). Treatment averages were compared using Tukey's range test.

RESULTS

VARIETIES OF *C. CHINENSE* DIFFER IN SENSITIVITY TO NaCl STRESS

Different varieties of *C. chinense* such as Rex and Chichen-Itza differ in their morphologic characteristics (fruit color, size, and root system architecture), as shown in **Figures 1A,C**. These two varieties exhibit differing sensitivities to salt stress, Rex being more tolerant than Chichen-Itza (**Figures 1B,C**). A concentration of 150 mM of NaCl over 7 days of culture in hydroponic conditions had a strong impact on the growth of the two varieties. Loss of turgor, leaf abscission, and darkening of the root system were observed, especially in the variety Chichen-Itza (**Figures 1B,C**).

A significant reduction of fresh and DWs was also induced by NaCl in both genotypes (**Figures 2A,B**). Under salt stress, the FW reduction was greater in Chichen-Itza, (75%) than in the Rex variety (50%) (**Figure 2A**). The water content in the Rex variety was identical to the water content in control seedlings with no salt treatments. However, the water content in the Chichen-Itza variety was significantly lower than in the untreated controls (**Figure 2C**). It is noteworthy that although symptoms of stress (wilting and senescence) were observed at concentrations below 150 mM NaCl, the effect on growth parameters was not significant between varieties by the end of the measurement period (data not shown). For this reason, a dose of 150 mM NaCl was selected for subsequent studies. This concentration has been used in various studies with glycophytes such as *A. thaliana*, *S. tuberosum*, and *S. lycopersicum* (Apse et al., 1999; Rodríguez-Rosales et al., 2008; Jaarsma et al., 2013).

FIGURE 1 | Morphologic and sensitivity differences to NaCl stress in two varieties of *C. chinense*. (A) Representative color of ripe habanero pepper fruits of the Rex (red) and Chichen-Itza (orange) varieties. **(B)** Forty-five-day-old seedlings of two varieties of *C. chinense* after 7 days of culture in H1/5 solution supplemented with 150 mM NaCl. **(C)** Appearance and sensitivity of Rex and Chichen-Itza seedlings after 7 days of treatment under control conditions (top) and 150 mM NaCl (bottom).

EFFECT OF NaCl ON PROLINE ACCUMULATION IN DIFFERENT VARIETIES OF *C. CHINENSE*

Many species of plants accumulate compatible solutes, such as proline, in response to abiotic stresses such as drought and salinity. In the varieties of *C. chinense* studied, proline accumulation was analyzed in the leaves and roots of seedlings grown in hydroponic cultures and subjected to 0 (control) or 150 mM NaCl for 7 days. Proline concentrations in the roots and leaves were similar in both varieties in the absence of salt. The proline concentration was higher in the leaves than in the roots (**Figure 3**). After 7 days of salt stress, the proline content in the Rex variety increased approximately 6.0-fold compared to control. However, in the Chichen-Itza variety, the values were similar to those in control seedlings growing without NaCl (**Figure 3A**). The same effect was observed in the roots as in the leaves. In the Rex variety, proline levels were 16-fold higher compared to control (**Figure 3B**). Surprisingly, the accumulation of proline in the roots was 50-fold higher in the Rex than in the Chichen-Itza variety after 7 days of treatment with NaCl (**Figure 3B**).

POTASSIUM RETENTION AND SODIUM ACCUMULATION IN ROOTS UNDER SALT STRESS

K^+ and Na^+ accumulation patterns in the two varieties of *C. chinense* are presented in **Figure 4**. In the absence of salt stress, K^+ content in the leaves and roots did not significantly differ between these varieties (**Figures 4A,D**). A higher K^+ concentration was observed in the leaves as compared to roots, due to the K^+ accumulation on the top of transpiration stream (Conn and Gilliham, 2010). NaCl stress did not modify K^+ content in the leaves in either variety of *C. chinense* (**Figure 4A**). However, a significant decrease in K^+ was observed in the roots of seedlings from the Chichen-Itza variety under salt stress. No significant reduction of root K^+ was observed in the Rex variety (**Figure 4D**). Furthermore, no differences in K^+ were observed for concentrations lower than 150 mM of NaCl in either variety (**Figure S1**).

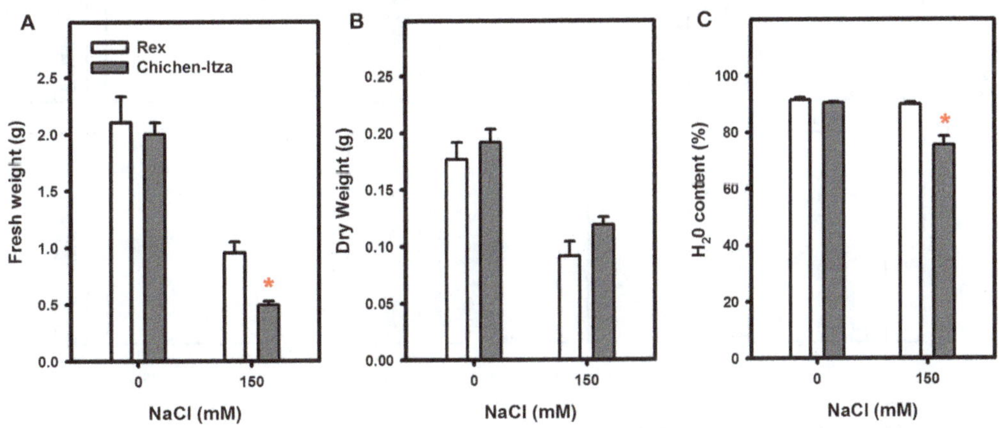

FIGURE 2 | Effect of saline stress (150 mM NaCl) on fresh weight (A), dry weight (B), and water content (C) of Chichen-Itza and Rex varieties seedlings after 7 days of culture with 150 mM NaCl. The white bars (Rex) and gray bars (Chichen-Itza) represent the mean FW, DW, and WC for the different treatments, ME ± SD ($n = 6$ seedlings). The asterisk indicates statistically significant differences between varieties for each treatment ($P < 0.050$, Tukey's test).

In absence of salt stress, Na^+ content in the leaves and roots of both varieties was minimal, with values of 0.02 mmol and 0.04 mmol g DW^{-1}, respectively. Treatment with 150 mM NaCl increased the Na^+ content in the leaves; however, no differences were observed between the two varieties (**Figure 4B**). Na^+ content increased in the roots treated with salt in both varieties and was surprisingly higher in Rex (twofold) compared to Chichen-Itza variety (**Figure 4E**). The Na^+/K^+ ratio in leaves and roots were very similar between pepper varieties when NaCl was not supplied (**Figures 4C,F**). As a consequence of increase in Na^+ and decreases in K^+ content by NaCl treatment, Chichen-Itza variety exhibited much higher Na^+/K^+ ratio in roots compared to the Rex variety (**Figure 4F**). In contrast, the Na^+/K^+ ratio in leaves at NaCl treatment did not differ between the genotypes (**Figure 4C**). Furthermore, at low and moderate concentrations of NaCl, the Rex variety exhibited higher Na^+ content in the roots and much lower Na^+ content in the leaves. The opposite effect was observed in the Chichen-Itza variety at 50 mM NaCl. Similar Na^+ content was observed between shoots and roots at a concentration of 100 mM (**Figure S2**).

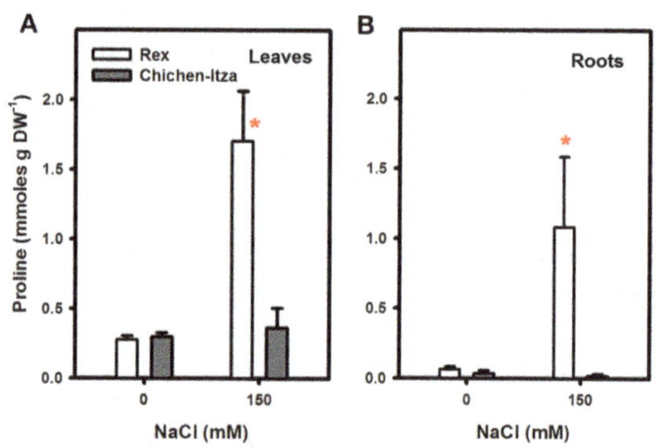

FIGURE 3 | Effect of NaCl treatment on proline accumulation in leaves (A) and roots (B) of two varieties of *C. chinense*. The bars (white: Rex, gray: Chichen-Itza) represent the means of treatments with or without NaCl (150 mM), ME ± SD ($n = 3$). The asterisk indicates statistically significant differences between varieties for each treatment ($P < 0.050$, Tukey's test).

NaCl INDUCES K^+ EFFLUX IN ROOTS OF *C. CHINENSE*

As described so far, Chichen-Itza and Rex varieties differ in their sensitivity to salt stress, the latter being less affected (**Figures 1, 2**). The Rex variety retains 55% of K^+ in salt-stress conditions, whereas Chichen-Itza looses about 90% of the root K^+ (**Figure 4**). To deepen the study of this response, the K^+ flux was measured using the non-invasive MIFE technique. The addition of 150 mM NaCl induced K^+ efflux from the epidermal cells in the mature root zone of the two varieties of *C. chinense* (**Figure 5**). This efflux started immediately following NaCl treatment. A higher K^+ efflux was observed in the roots of the Chichen-Itza variety compared to Rex variety (**Figure 5A**). The difference in K^+ efflux in the 1 min was double and the relative difference even increased with time (**Figure 5B**). In the Rex variety K^+ efflux was close zero after 35 min, whereas in Chichen-Itza a significant K^+ efflux of about 50 nmol m^{-2} s^{-1} was observed at late times.

EFFECT OF NaCl ON H^+ EFFLUX IN ROOTS OF *C. CHINENSE*

In the roots of habanero pepper, NaCl stress caused significant changes in the net flux of H^+ (**Figure 6**). Before starting the salt treatment (first 5 min), the net flux of H^+ was zero in both varieties. Application of 150 mM NaCl induced a substantial H^+ efflux (**Figure 6A**). In the roots of the Rex variety, NaCl induced H^+ efflux was much lower as compared the Chichen-Itza variety (**Figure 6B**). Furthermore, a pre-treatment

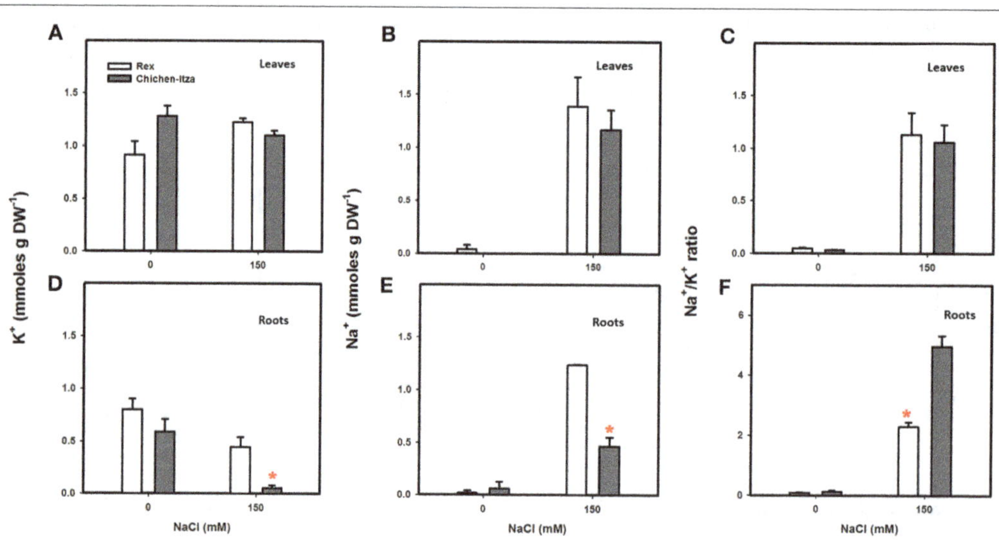

FIGURE 4 | Potassium and sodium content in two varieties of *C. chinense* under NaCl stress conditions. Forty-five-day-old seedlings cultivated in hydroponic cultures (H1/5) for 7 days with 0 and 150 mM of NaCl. K^+ content in the leaves **(A)** and roots **(D)** of Rex and Chichen-Itza varieties after treatment with salt. Na^+ content in the leaves **(B)** and roots **(E)** of the two *C. chinense* varieties after 7 days of treatment. The Na^+/K^+ ratios in leaves **(C)** and roots **(F)** in control and NaCl treatments. Bars represent the average of the treatments with or without NaCl (150 mM), ME ± SD ($n = 3$). The asterisk indicates statistically significant differences between varieties for each treatment ($P < 0.050$, Tukey's test).

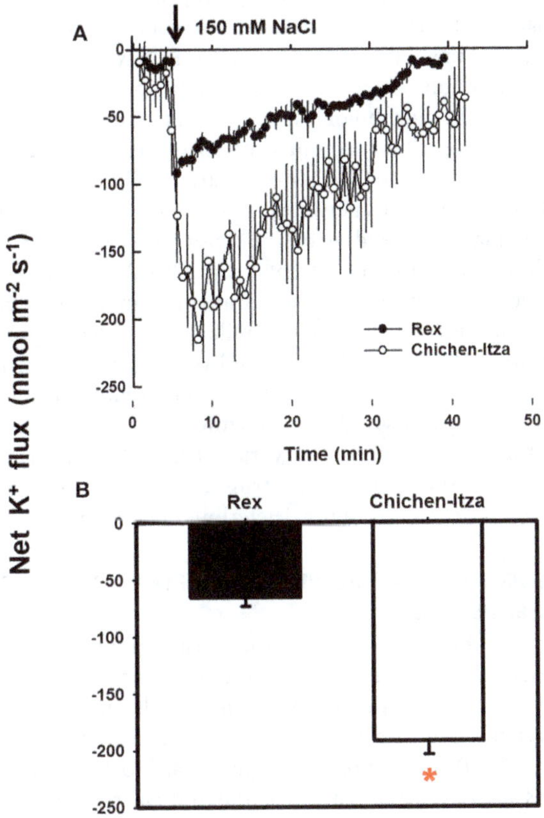

FIGURE 5 | Measurement of the net flow of K^+ in the roots of *C. chinense* in response to NaCl stress. (A) Kinetics of net K^+ flux measured in the mature root zone of seedlings (10 days old) of the Rex (closed circles) and Chichen-Itza (open circles) varieties in response to 150 mM NaCl (the arrow indicates the time of addition of the treatment). **(B)** Mean net flow of K^+ from the root of each variety in the first 10 min after application of 150 mM NaCl. ME ± SD (n = 4–5). The asterisk indicates statistically significant differences between varieties by treatment ($P < 0.001$, Tukey's test).

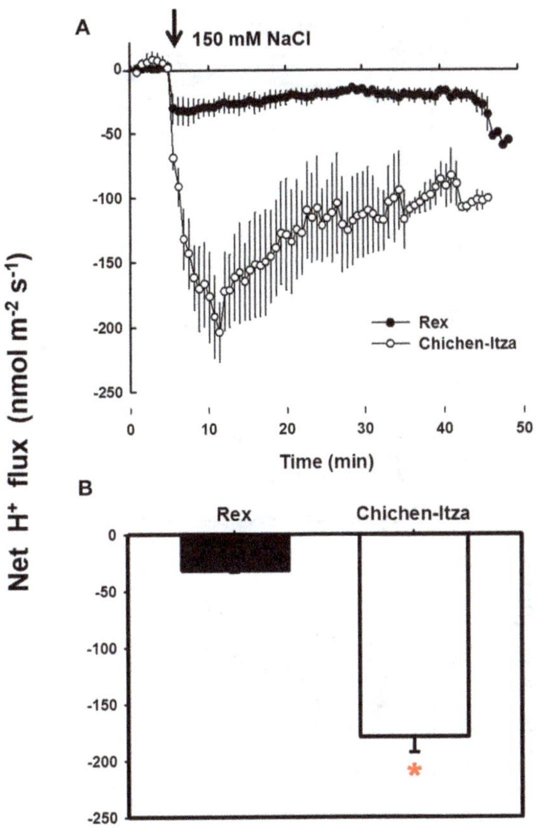

FIGURE 6 | Effect of NaCl on the net H^+ flux in the mature root area of *C. chinense*. (A) Kinetics of net flux of H^+ measured in the roots of 10-day-old seedlings of Rex (closed circles) and Chichen-Itza (open circles) varieties after adding 150 mM NaCl (arrow indicates the time addition of the treatment). **(B)** Mean net flow of H^+ from the root of each variety within the first 10 min of application of 150 mM NaCl. ME ± SD (n = 4–5). The asterisk indicates statistically significant differences between varieties by treatment ($P < 0.001$, Tukey's test).

of seedlings with 1 mM sodium orthovanadate (an inhibitor of P-type H^+-ATPases) strongly suppressed the H^+ efflux from the mature root zone of both genotypes (**Figure 7**). Thus, the NaCl-induced H^+ efflux was mediated by P-type H^+-ATPases.

Na^+ SUBCELLULAR LOCALIZATION IN ROOTS

After 60 min of treatment with 150 mM NaCl, marked differences were observed with respect to Na^+ localization in the mature root zone between two pepper genotypes (**Figure 8**). In the Rex variety, Sodium Green™ fluorescence was observed in vacuole-like structures of epidermal cells (red arrowheads, **Figure 8A**), suggesting an efficient mechanism for sodium compartmentalization in this variety. By contrast, in the Chichen-Itza variety, green fluorescence was observed around the epidermal cells in the mature root zone (white arrowheads, **Figure 8B**). Small endosomes stained with FM4-64 dye were observed in both pepper varieties under salt stress (yellow arrowheads, **Figure 8**). However, green fluorescence was not evident in these structures, indicating the absence of Na^+ accumulation (merge, **Figure 8**). In epidermal cells of the Rex variety, a conglomeration of small structures with green fluorescence was observed around a larger structure comparable to a vacuole (red arrowheads, **Figure 9A**); similar structures were previously observed in *Arabidopsis* (Hamaji et al., 2009). In contrast to the Rex variety, in the Chichen-Itza a less pronounced and diffuse pattern of small compartments stained with Sodium Green™ was observed (red arrowheads, **Figure 9A**). Overall, the highest levels of fluorescence were observed outside the cells in Chichen-Itza (white arrowheads, **Figure 9B**), demonstrating that Na^+ was mainly located in the apoplast. In control roots, Na^+ indicator did not report any change of fluorescence for either variety (**Figure S3**).

DISCUSSION

SENSITIVITY AND GROWTH IN *C. CHINENSE* SEEDLINGS UNDER NaCl STRESS

Most crop plants that provide food for the world population are glycophytes and are very sensitive to high concentrations of salts in the soil, mainly NaCl. Salt stress is the main abiotic factor that affects growth, yield, and quality. Peppers (*Capsicum* spp.)

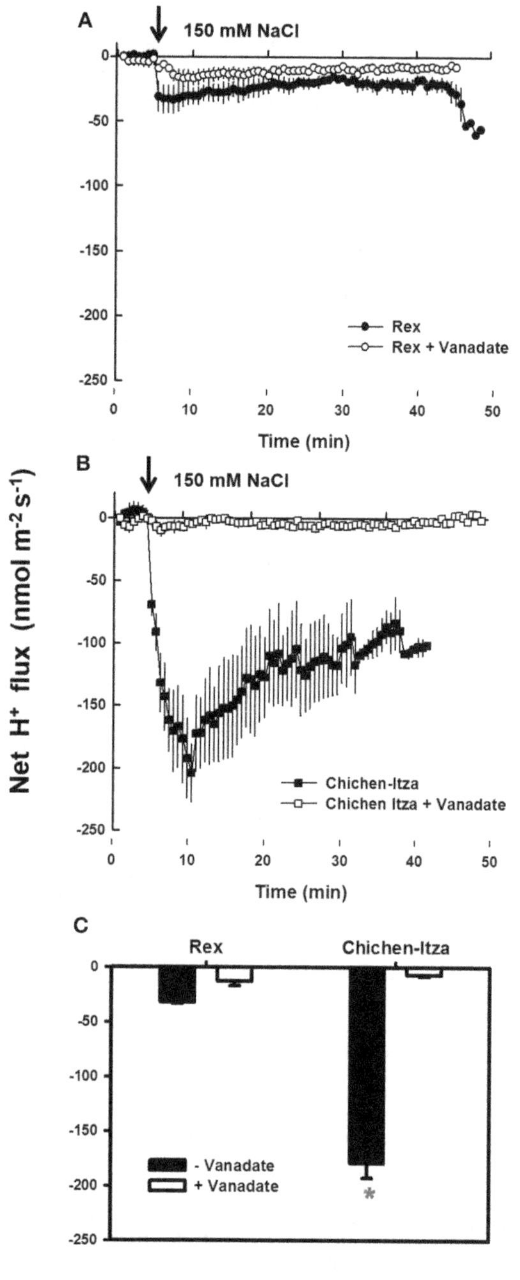

FIGURE 7 | Effect of vanadate pre-treatment on NaCl-induced H^+ efflux in roots of *C. chinense*. (A) Kinetics of net H^+ flux measured in the mature root zone of the Rex variety and **(B)** of the Chichen-Itza variety. 10-day-old roots of the two strains were pre-treated with 1 mM vanadate for 45–60 min before adding of 150 mM NaCl (marked by an arrow). Open circles and squares indicate pre-treatment with vanadate. Closed circles and squares indicate no pre-treatment. **(C)** Average effect of vanadate pre-treatment on the net flow of H^+ from the root of each variety in the first 10 min after the application of 150 mM NaCl, ME ± SD (n = 3–4). The asterisk indicates statistically significant differences between varieties by treatment ($P < 0.001$, Tukey's test).

are a major vegetable crop and are not exempt from the effect of salt throughout their ontogeny (Bojórquez-Quintal et al., 2012). Notably, pepper plants differ in their sensitivity to salt stress, including marked differences between varieties within the same species (Aktas et al., 2006). In this work, two varieties of *C. chinense* Jacq. were used as models. *C. chinense* Jacq. is a species in high demand in southeastern Mexico for its flavor and pungency and is commonly known as habanero pepper. These varieties exhibit different morphological characteristics and were shown to differ in their sensitivity to salt stress, with the Rex variety being more tolerant than the Chichen-Itza variety (**Figure 1**).

NaCl concentrations between 0 and 150 mM affect the growth of pepper plants, depending on the genotype, species, and condition of growth (Bojórquez-Quintal et al., 2012). In this study, the application of 150 mM NaCl had a dramatic impact on the growth of both varieties of *C. chinense*. Similar results were reported in *C. annuum* (Aktas et al., 2006). The FW, turgor, and water content of the Rex variety were less affected (**Figure 2**). In this variety, 70% of seedlings survived after treatment with NaCl, compared with 10% of the Chichen-Itza variety (data not shown). These results suggest the existence of intrinsic mechanisms of tolerance in the Rex variety to avoid the deleterious effect of NaCl.

DIFFERENCES IN PROLINE ACCUMULATION BETWEEN VARIETIES OF HABANERO PEPPERS

Proline accumulation is one of the most common and important responses of plants to adverse environmental factors such as drought and salt stress. Proline is a multifunctional amino acid participating in a wide range of functions (Szabados and Savouré, 2009) and represents a potential biochemical marker for the salt tolerance in plants (Ashraf and Harris, 2004). In our study, different proline content was observed in two pepper varieties, contrasting in their salt sensitivity (**Figure 3**). In leaves of the Rex variety, proline content increased six times in the presence of NaCl, as compared to a non-significant increase in Chichen-Itza leaves. The leaves are the major site of proline synthesis (source organ). It has been suggested that proline accumulation in this organ occurs to maintain chlorophyll content and turgor to protect the photosynthetic activity under salt stress conditions (Yildiztugay et al., 2011; Huang et al., 2013). Furthermore, treatment with NaCl induced a dramatic (16-fold) increase in proline content in the roots, but only of the tolerant variety Rex. At low water potential, proline is thought to be transported from the leaf (source) to the roots (sink) for growth processes or other functions, such as osmotic adjustment depending on proline content (Sharma et al., 2011). Exogenous administration of proline reduced NaCl- induced K^+ efflux in barley and Arabidopsis roots (Cuin and Shabala, 2005, 2007a,b). In *Solanaceae* such as the potato (*S. tuberosum*), increased proline content correlates with a higher expression of *P5CS* (synthesis gene) and decreased *PDH* gene expression (degradation) in tolerant but not in sensitive cultures (Jaarsma et al., 2013). Furthermore, overexpression of *P5CS* in *N. tabacum* and *S. tuberosum* stimulated proline accumulation under NaCl stress and improved the tolerance (Kishore et al., 1995; Hmida-Sayari et al., 2005). However, there is not always a good correlation between the accumulation of this osmolyte and tolerance to salt stress, and whether it is a symptom of damage or an indicator of tolerance is a matter of debates. For example, in rice (*O. sativa*), soybean (*G. max*), tomato (*S. lycopersicum*), and barley (*H. vulgare*), a negative correlation between proline accumulation and tolerance to stress has been reported. In these

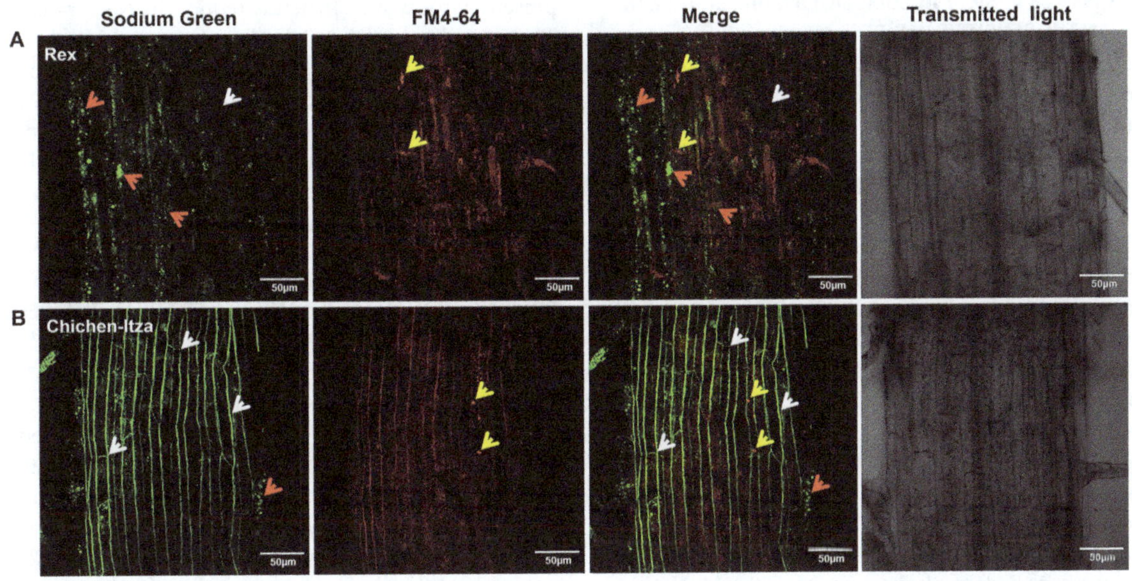

FIGURE 8 | Sodium localization in epidermal cells of the primary root of habanero pepper *C. chinense*). (A) Localization of Na^+ in the roots of the Rex and **(B)** Chichen-Itza varieties. 10-day-old habanero pepper roots were treated with 150 mM NaCl for 60 min and then stained with Sodium Green (Na^+ detection) and FM4-64 (membrane staining) before the confocal images were taken. White arrowheads indicate the location of the Na^+ around cells. Red arrowheads indicate the intracellular localization of Na^+, and yellow arrowheads indicate the presence of endosomes. Fluorescence analysis was determined in the mature root zone. Images are representative of the analysis of three roots per treatment and variety.

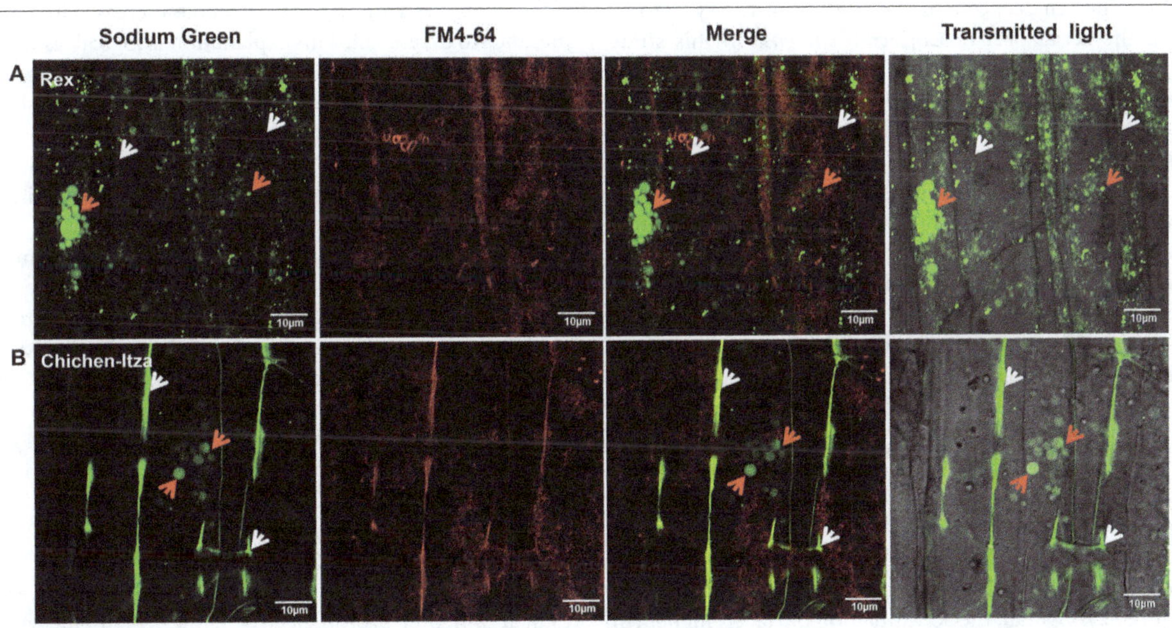

FIGURE 9 | Sodium subcellular localization in the apoplast and subcellular compartments. Magnified images of the mature root area shown in **Figure 8**. **(A)** Localization of Na^+ in the roots of the Rex and **(B)** Chichen-Itza varieties. The white arrowheads indicate the location of Na^+ in the apoplast, and red arrowheads indicate the location of Na^+ in subcellular structures. Images are representative of the analysis of three roots per treatment and variety.

studies, sensitive genotypes accumulated more proline (Moftah and Michel, 1987; Aziz et al., 1998; Lutts et al., 1999; Chen et al., 2007a). In addition, proline synthesis is a metabolically expensive strategy (Shabala and Cuin, 2007; Shabala and Shabala, 2011). Nevertheless, proline content has been shown to be higher in many plant varieties tolerant to salt relative to their susceptible counterparts (Ashraf and Harris, 2004), as also demonstrated by the results of this work (**Figure 3**).

THE TOLERANT PEPPER VARIETY ACCUMULATES MORE SODIUM IN THE ROOTS THAN THE SENSITIVE VARIETY

The hyperaccumulation of Na^+, particularly in the leaves, inhibits protein synthesis, enzymatic activity, and photosynthesis. Therefore, plants have the ability to control the transport and distribution of Na^+ to organs, tissues and cells where it causes less damage to protect against the accumulation of this cation. The most sensitive glycophytes (cereals or vegetables) are unable to control the transport of Na^+; therefore, large amounts of this ion are translocated to the shoot (Maathuis, 2014), inducing senescence, growth inhibition, and eventually death of the plant (Roy et al., 2014). In contrast, most halophyte and some glycophytes plants tend to accumulate large amounts of Na^+ in the leaves. It has been suggested that these plants use Na^+ in addition to K^+ to maintain turgor and growth (Hariadi et al., 2011; Adolf et al., 2013; Bonales-Alatorre et al., 2013a,b; Maathuis, 2014).

In our study (**Figure 4**), we observed that two varieties of habanero pepper (*C. chinense*) exhibit the same Na^+ content in the leaves but exhibit differences in Na^+ accumulation in the roots (after a minimum of 7 days of exposure to 150 mM NaCl). In plants of *C. annuum* treated with NaCl, a higher Na^+ content has been reported in the shoots of sensitive genotypes compared to tolerant genotypes. In fact, a positive correlation was observed between the severity of the symptoms in the leaves and Na^+ concentrations in the shoots (Aktas et al., 2006). This suggests an effect dependent on the accumulation of Na^+ as proposed Roy et al. (2014). The senescence and the severity of the symptoms observed in the aerial part of *Capsicum chinense* in this study (**Figure 1**) seem to be due to an osmotic effect rather than an ionic effect due to the presence of Na^+ in leaves (**Figure 4B**). In addition, this osmotic effect can be avoided by the accumulation of proline in the Rex variety (**Figure 3A**) as described in the previous section, as the effect of NaCl was less severe in that variety.

The root system is the first site of detection and the first line of defense against excess Na^+ in cells (Ji et al., 2013). NSCC are the principal source of Na^+ influx in the plant cell (Demidchik and Tester, 2002). As suggested for other species (Demidchik and Maathuis, 2007), NSCC can mediate Na^+ influx of pepper plants (Rubio et al., 2003). In our study, the Rex variety, less affected by salt, exhibited a two–fold higher Na^+ content in the root than in the Chichen-Itza variety (**Figure 4E**). Furthermore, the accumulation of Na^+ in the root system of the tolerant variety was higher at low NaCl concentrations (50 mM) where reached its peak and was maintained at higher NaCl concentrations (150 mM). The levels of Na^+ in the roots of the Chichen-Itza variety were lower and stable at all of the NaCl concentrations tested (**Figure S2**). Overall, the accumulation of Na^+ is higher (in both roots and leaves) in the tolerant variety than in the sensitive variety (**Figure S1** and **Figure 4**). At low and moderate salt concentration ranges, the Rex variety (tolerant) possesses more Na^+ in the roots and much less in the leaves. However, the opposite was observed in the Chichen-Itza variety (sensitive) at 50 mM NaCl. Furthermore, at 100 mM NaCl, the same Na^+ content was observed between the shoots and roots in the Chichen-Itza variety (**Figure S2**). The higher Na^+ content in roots than in leaves suggests that exclusion mechanisms (SOS1, antiporters, and HKT1 transporters) and compartmentalization (NHX, antiporters) are present in roots and efficiently reduce the Na^+ transport to leaves. These mechanisms have also been reported in other members of *Solanaceae*, such as tomatoes and potatoes (Queirós et al., 2009; Rodríguez-Rosales et al., 2009; Almeida et al., 2014). In particular, the HKTs (subfamily 1) exhibit an important role in the recovery of Na^+ from the xylem to prevent its transport to the aerial part, and recirculate Na^+ to the roots (Horie et al., 2009; Almeida et al., 2013; Adams and Shin, 2014).

Na^+ SUBCELULAR LOCALIZATION IN THE ROOTS: COMPARTMENTALIZATION AND Na^+ EFFLUX IN DIFFERENT VARIETIES OF *C. CHINENSE*

After 60 min of treatment with 150 mM NaCl, Na^+ was mainly located in vacuole-like structures in root epidermal cells in the Rex variety (**Figure 9**). This result suggests the existence of an efficient mechanism for Na^+ compartmentalization in this genotype. Similarly, it has been reported that Na^+ is confined in epidermal cells vacuoles and in the cortex in the roots of *Arabidopsis* and *Citrus* (Oh et al., 2009; Gonzalez et al., 2012). Furthermore, it is noteworthy that in the two varieties of habanero pepper, small compartments were stained with the fluorophore. These compartments were found in greater quantities in the roots of the Rex variety (**Figure 9**). Similar results were observed in the roots of *A. thaliana* under salt stress. Hamaji et al. (2009) reported that Na^+ accumulates in the vacuoles as well as in small vesicular compartments around vacuoles. These authors suggest that the fusion of these vesicles with the main vacuole increases its size and the tolerance to excess salt. This explanation is logical, as we observed a conglomeration of structures stained with the fluorophore in the roots of the tolerant variety after a short period of NaCl stress (**Figure 9**). In tolerant salt includes such as mangroves and barley, a rapid increase in vacuolar volume in response to salt stress was observed. This phenomenon was not reported in sensitive species such as peas and tomatoes (Mimura et al., 2003).

Recently, in tobacco salt-acclimated BY2 cells accumulation of Na^+ in vacuoles and small vesicles was reported. Interestingly, the putative VAMP711 (vesicle-associated membrane protein 711) and VPS46 (charged multivesicular body protein) proteins were highly induced in this BY2 cells, suggesting a Na^+ transport mechanism for vesicle trafficking (Garcia de la Garma et al., 2014). Furthermore, the increased Na^+ sequestration by vacuolar and small compartments in the Rex variety could be due to increased expression of NHX transporters as observed in tomato species with different sensitivity to salt (Galvez et al., 2012). Also, the up-regulation of V-ATPase (vacuolar-type H^+-ATPase) and H^+-PPase (vacuolar H^+-pyrophosphatase) in Rex variety may result in a higher proton electrochemical gradient, which facilitates enhanced sequestering of ions into the vacuole and endosomes, reducing water potential and resulting in increased salt tolerance (Gaxiola et al., 2001; Bassil et al., 2012; Pittman, 2012). In different species of plants, overexpression of AVP1 (vacuolar H^+-PPase) and co-overexpression with AtNHX1 enhances salt tolerance (Gaxiola et al., 2001; Pasapula et al., 2011; Shen et al., 2014). Finally, the reduction of Na^+ loss via non-selective vacuolar channels could assist efficient vacuolar Na^+ accumulation (Bonales-Alatorre et al., 2013b). In the Rex variety, all or some of these mechanisms may be involved in the Na^+ compartmentalization.

On the contrary, in the Chichen-Itza variety, Na^+ was mostly observed in the apoplastic area between cells of the root (**Figures 8, 9**). This result suggests an intensive Na^+ extrusion toward the outside of the cell, likely via the Na^+/H^+ antiporters (SOS1) of the plasma membrane (Shi et al., 2002). This is consistent with a low content of Na^+ (**Figure 4**) and with the large active H^+ efflux, mediated by vanadate-sensitive H^+-ATPase, in the roots of this variety. However, the high rate of Na^+ extrusion in sensitive cultivars could have a high energetic cost (Malagoli et al., 2008). This opts for the use of intracellular Na^+/H^+ (NHX-type) antiporters as compared to SOS1-type ones; the latter may be more useful in early responses to acute salt stress. Candidate genes for the intracellular (NHX) and plasma membrane (SOS1) exchangers were revealed in the genome of *Capsicum annuum* (Qin et al., 2014).

Furthermore, we may not exclude at this moment that accumulation of the indicator at cell walls, which possess high esterase activity, could contribute to the observed fluorescence signal and the difference between Rex and Chichen-Itza.

K^+ RETENTION IN THE ROOTS IS A TOLERANCE MECHANISM IN HABANERO PEPPER PLANTS

Potassium is an essential nutrient throughout the life cycle of plants, including the adaptation to hostile environments. The regulation of K^+ homeostasis plays a central role in tolerance to biotic and abiotic stresses in plants (Anschütz et al., 2014; Demidchik, 2014; Shabala and Pottosin, 2014). K^+ efflux from the root is a common physiological reaction that occurs under a wide range of stress conditions (Demidchik, 2014). K^+ retention in roots has been proved to confer salt tolerance in barley, wheat, lucerne, and poplar (Chen et al., 2007b,c; Cuin et al., 2008; Smethurst et al., 2008; Sun et al., 2009). In this study, the treatment with 150 mM NaCl significantly decreased the content of K^+ in the roots of the sensitive variety but not in the tolerant variety of habanero pepper (**Figure 4D**). In contrast, K^+ content in the leaves was not affected by treatment with NaCl in either genotype (**Figure 4A**). These data suggest that the ability to retain K^+ by roots is one of the salt tolerance mechanisms of *C. chinense*.

Initial NaCl induced K^+ efflux was higher in the Chichen-Itza as compared to the Rex variety and doubled after 10 min of exposure to NaCl. Similar differences in K^+ efflux have been reported in barley (*H. vulgare*), wheat (*T. aestivum*) and alfalfa (*M. sativa*). This difference has been used as a selection criterion to distinguish salt-tolerant from salt-sensitive genotypes (Chen et al., 2005, 2007b,c; Smethurst et al., 2008), yet some plants, like rice, did not show such a correlation (Coskun et al., 2013). In our results, content and K^+ efflux in roots (**Figures 4, 5**) were consistent with the observed differences in sensitivity between the varieties of habanero pepper (**Figures 1, 2**).

NaCl-induced K^+ efflux may be through outward-rectifying K^+ (KOR) channels activated by depolarization as demonstrated in barley (*H. vulgare*) and mangrove species (Chen et al., 2007b; Sun et al., 2009; Lu et al., 2013). The use of ion channels inhibitors in habanero pepper indicate that K^+ efflux from the roots are likely mediated by KOR channels rather than by NSCC channels (Bojorquez-Quintal et al., in review). Furthermore, ROS and K^+ deficiency have been associated with programmed cell-death (PCD) and necrosis (Anschütz et al., 2014; Demidchik, 2014; Shabala and Pottosin, 2014). Necrosis was observed in the roots of the Chichen-Itza variety (**Figure 1**), which have low ability to retain K^+ (**Figures 4, 5**).

DIFFERENCES IN H^+ EFFLUX IN THE ROOTS OF HABANERO PEPPER UNDER NaCl STRESS

H^+-ATPases generate an electrochemical gradient that maintains membrane potential and transports ions between the cytosol and the external medium. Under salt stress, NaCl universally induces H^+ efflux in the roots of cereals such as *H. vulgare* and *T. aestivum*, halophytes such as *C. quinoa* and even in the model plant *A. thaliana* (Chen et al., 2007b; Cuin et al., 2008; Hariadi et al., 2011; Bose et al., 2013, 2014). In *C. chinense*, NaCl also induced H^+ efflux in the mature root zone. H^+ flux differed significantly between the varieties (**Figure 6**). NaCl rapidly induced the H^+ efflux in both varieties of habanero peppers. It was suppressed by vanadate, the inhibitor of P-type H^+-ATPase (**Figure 7**). H^+ pumping activity of the plasma membrane H^+-ATPase is essential for salt tolerance (Palmgren and Nissen, 2010). It has been suggested that the ability to K^+ retain is related to the increase in H^+-ATPase activity, primarily through a membrane potential repolarization (Bose et al., 2013, 2014).

Maintaining a more negative membrane potential during salt stress prevents the loss of K^+ in the cytosol. In the salt-tolerant Arabidopsis relative species, *T. halophila*, a more negative membrane potential, correlated with a better K^+ retention, was observed during salt stress (Volkov and Amtmann, 2006). In transgenic *A. thaliana* (HO, heme oxygenase), K^+ retention was regulated by the increased H^+-ATPase activity (Bose et al., 2013). A higher H^+-ATPase activity maintained a more negative membrane potential and improved K^+ retention in tolerant genotypes of *H. vulgare* (Chen et al., 2007b). In contrast, higher H^+ transport activity was observed in the sensitive variety Chichen-Itza of *C. chinense* (**Figure 6**). Kinetics and magnitude of the H^+ efflux coincides with that of the K^+ efflux (**Figure 5**). Thus, it may be hypothesized that the descending phase of the H^+ efflux reflects the condition of membrane potential repolarization, so that the leakage of K^+ through KOR channels would be also reduced.

Furthermore, the plasma membrane H^+-ATPase activity provides the driving force the Na^+ extrusion via the Na^+/H^+ (SOS1) exchanger. In this work, we observed a massive accumulation of Na^+ in the apoplast of the Chichen-Itza variety roots (**Figure 9**). These data suggest that SOS1 antiporter may be very active in the Chichen-Itza variety. However, Na^+ efflux has a high energetic cost for the cell, especially keeping in mind a futile Na^+ cycling between cytosol and apoplast. It may recruit the ATP available for other metabolic processes, thus, being detrimental to growth and yield. For this reason, the activation of H^+-ATPase cannot be considered a permanent solution and might only be a temporal mechanism, as described by other studies (Ramani et al., 2006; Bose et al., 2013).

POSSIBLE TOLERANCE MECHANISM IN PEPPERS

Salt tolerance is a complex multigenic trait that involves many biochemical and physiological processes to achieve salt tolerance. In this paper, we demonstrate differences in salt sensitivity

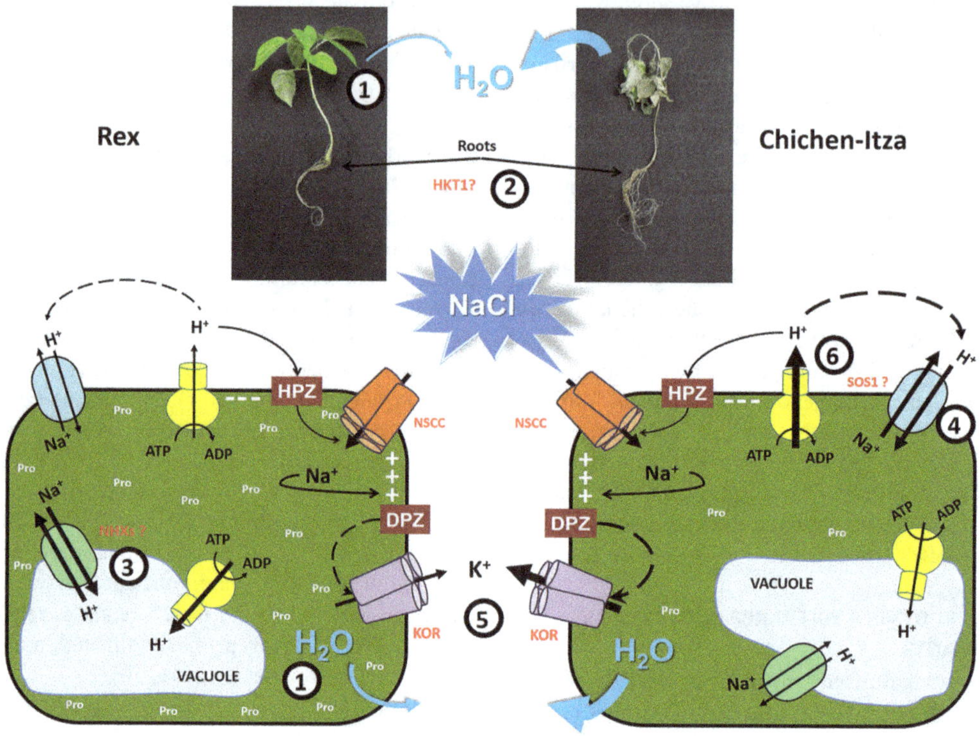

FIGURE 10 | Schematic model of habanero pepper plant tolerance mechanisms. (1) Osmotic adjustment through the accumulation of compatible solutes as proline in leaves and roots to maintain the absorption and prevent the loss of water. (2) Efficient control of Na^+ transport by confining this ion to roots, possibly through the recovery of Na^+ from the xylem by HKT1 transporters to avoid transport to photosynthetic tissues. (3) Na^+ compartmentalization in vacuole-like structures and small subcellular compartments, where it can act as an osmolyte, suggesting the involvement of vacuolar and endosomal NHX transporters. (4) Na^+ extrusion or efflux from the cytosol to the apoplast by SOS1 exchangers. (5) Regulation of K^+ homeostasis through K^+ retention in roots is crucial in habanero peppers. (6) Increase in the H^+-ATPase activity to generate the proton force for cell membrane repolarization or Na^+ extrusion. (KOR), outward-rectifying K^+ channels. HPZ is hyperpolarization and DPZ is depolarization.

between two varieties of *C. chinense* Jacq, one of the five domesticated pepper species. We have analyzed several parameters of salt stress responses in both genotypes and their differences that may underlie their differential salt tolerance (**Figure 10**). One of salt tolerance mechanisms is the osmotic adjustment through the accumulation of compatible solutes (in this case, proline) in roots and leaves to maintain the absorption and prevent the loss of water (1). A second tolerance mechanism is the efficient control of Na^+ transport by confining this ion to the roots, possibly through the recovery of Na^+ from the xylem by HKT1 transporters (at low, moderate and high concentration of NaCl) to avoid transport to photosynthetic tissues (2). Furthermore, if the Na^+ content in roots is high, this ion needs to be excluded from the cytosol to avoid toxicity. A third tolerance mechanism was observed in the tolerant variety (Rex), Na^+ was efficiently compartmentalized into vacuole-like structures and small compartments which can act as osmolytes. This mechanism is possibly mediated by vacuolar and endosomal NHX antiporters (3). Additional mechanism appears to be involved in salt-sensitive variety (Chichen-Itza), which extrudes large amounts of Na^+ into the apoplast (4). However, this mechanism appears to be less efficient due to its large energy cost. As in many other plant species, the regulation of K^+ homeostasis through its retention in roots is crucial in habanero peppers (5), demonstrating the universality of this mechanism to the salt stress tolerance of crops.

ACKNOWLEDGMENTS

For help with the confocal microscopy analysis we thank Luis Alberto Cruz Silva (Instituto de Ecología A.C.). We thank Lucila A. Sanchez Cach for their excellent technical assistance (Centro de Investigación Científica de Yucatán, A.C.). This work was supported by CONACYT project # 166621.

SUPPLEMENTARY MATERIAL

The Supplementary Material for this article can be found online at: http://www.frontiersin.org/journal/10.3389/fpls.2014.00605/abstract

Figure S1 | K^+ content of two varieties of *C. chinense* at different NaCl concentrations. K^+ content in the leaves **(A)** and roots **(B)** after treatment with salt. Forty-five-day-old seedlings cultivated in H1/5 for 7 days with 0, 50, 100 or 150 mM of NaCl. Bars represent averages for treatments with or without NaCl, ME $\pm$ SD ($n = 3$). The asterisk indicates statistically significant differences between varieties by treatment ($P < 0.050$, Tukey's test).

Figure S2 | Na^+ content in Rex and Chichen-Itza varieties at different NaCl concentrations. Na^+ content in the leaves **(A)** and roots **(B)** after 7 days under salt stress conditions. Forty-five-day-old seedlings were cultivated in H1/5 with 0, 50, 100, and 150 mM of NaCl. Bars represent the average effect of the treatments with or without NaCl, ME ± SD ($n = 3$). The asterisk indicates statistically significant differences between varieties by treatment ($P < 0.050$, Tukey's test).

Figure S3 | Roots of control seedlings exhibit an absence of Sodium Green™ fluorescence. Roots of untreated Rex **(A)** and Chichen-Itza **(B)** varieties stained with Sodium Green (SG), FM4-64, and DAPI. Images are representative of the analysis of three roots per treatment and variety.

REFERENCES

Adams, E., and Shin, S. (2014). Transport, signalling, and homeostasis of potassium and sodium in plants. *J. Integr. Plant Biol.* 56, 231–249. doi: 10.1111/jipb.12159

Adem, G. D., Roy, S. J., Zhou, M., Bowman, J., and Shabala, S. (2014). Evaluating contribution of ionic, osmotic and oxidative stress components towards salinity tolerance in barley. *BMC Plant Biol.* 14:113. doi: 10.1186/1471-2229-14-113

Adolf, V. I., Jacobsen, S. E., and Shabala, S. (2013). Salt tolerance mechanisms in quinoa (*Chenopodium quinoa* Willd.). *Environ. Exp. Bot.* 92, 43–54. doi: 10.1016/j.envexpbot.2012.07.004

Aktas, H., Abak, K., and Cakmak, I. (2006). Genotypic variation in the response of pepper to salinity. *Sci. Hortic.* 110, 260–266. doi: 10.1016/j.scienta.2006.07.017

Almeida, P., de Boer, G.-J., and de Boer, A. H. (2014). Differences in shoot Na^+ accumulation between two tomato species are due to differences in ion affinity of HKT1;2. *J. Plant Physiol.* 171, 438–447. doi: 10.1016/j.jplph.2013.12.001

Almeida, P., Katsching, D., and de Boer, A. H. (2013). HKT transporters-State of the art. *Int. J. Mol. Sci.* 14, 20359–20385. doi: 10.3390/ijms141020359

Anschütz, U., Becker, D., and Shabala, S. (2014). Going beyond nutrition: regulation of potassium homoeostasis as a common denominator of plant adaptive responses to environment. *J. Plant Physiol.* 171, 670–687. doi: 10.1016/j.jplph.2014.01.009

Apse, M. P., Aharon, G. S., Snedden, W. A., and Blumwald, E. (1999). Salt tolerance conferred by overexpression of a vacuolar Na^+/H^+ antiport in Arabidopsis. *Science* 285, 1256–1258. doi: 10.1126/science.285.5431.1256

Apse, M. P., and Blumwald, E. (2007). Na^+ transport in plants. *FEBS Lett.* 581, 2247–2254. doi: 10.1016/j.febslet.2007.04.014

Ashraf, M., and Harris, P. J. C. (2004). Potential biochemical indicators of salinity tolerance in plants. *Plant Sci.* 166, 3–16. doi: 10.1016/j.plantsci.2003.10.024

Aziz, A., Martin-Tanguy, J., and Larher, F. (1998). Stress-induced changes in polyamine and tyramine levels can regulate proline accumulation in tomato leaf discs treated with sodium chloride. *Physiol. Plant.* 104, 195–202. doi: 10.1034/j.1399-3054.1998.1040207.x

Bassil, E., Coku, A., and Blumwald, E. (2012). Cellular ion homeostasis: emerging roles of intracellular NHX Na^+/H^+ antiporters in plant growth and development. *J. Exp. Bot.* 63, 5727–5740. doi: 10.1093/jxb/ers250

Bates, L. S., Waldren, R. P., and Teare, D. (1973). Rapid determination of free proline for water-stress studies. *Plant Soil* 39, 2005–2007. doi: 10.1007/BF00018060

Bojórquez-Quintal, J. E., Echevarría-Machado, I., Medina-Lara, F., and Martínez-Estévez, M. (2012). Plants challenges in a salinized world: the case of Capsicum. *Afr. J. Biotechnol.* 11, 13614–13626. doi: 10.5897/AJB12.2145

Bonales-Alatorre, E., Pottosin, I., Shabala, L., Chen, Z.-H., Zeng, F., Jacobsen, S. E., et al. (2013a). Differential activity of plasma and vacuolar membrane transporters contributes to genotypic differences in salinity tolerance in a halophyte species, *Chenopodium quinoa*. *Int. J. Mol. Sci.* 14, 9267–9285. doi: 10.3390/ijms14059267

Bonales-Alatorre, E., Shabala, S., Chen, Z. H., and Pottosin, I. (2013b). Reduced tonoplast fast-activating and slow-activating channel activity is essential for conferring salinity tolerance in a facultative halophyte, quinoa. *Plant Physiol.* 162, 940–952. doi: 10.1104/pp.113.216572

Bose, J., Shabala, L., Pottosin, I., Zeng, F., Velarde-Buendía, A. M., Massart, A., et al. (2014). Kinetics of xylem loading, membrane potential maintenance, and sensitivity of K^+-permeable channels to reactive oxygen species: physiological traits that differentiate salinity tolerance between pea and barley. *Plant Cell Environ.* 37, 589–600. doi: 10.1111/pce.12180

Bose, J., Xie, Y., Shen, W., and Shabala, S. (2013). Haem oxygenase modifies salinity tolerance in Arabidopsis by controlling K^+ retention via regulation of the plasma membrane H^+-ATPase and by altering SOS1 transcript levels in roots. *J. Exp. Bot.* 64, 471–481. doi: 10.1093/jxb/ers343

Chen, Z., Cuin, T. A., Zhou, M., Twomey, A., Naidu, B. P., and Shabala, S. (2007a). Compatible solute accumulation and stress-mitigating effects in barley genotypes contrasting in their salt tolerance. *J. Exp. Bot.* 58, 4245–4255. doi: 10.1093/jxb/erm284

Chen, Z., Newman, I., Zhou, M., Mendham, N., Zhang, G., and Shabala, S. (2005). Screening plants for salt tolerance by measuring K+ flux: a case study for barley. *Plant Cell Environ.* 28, 1230–1246. doi: 10.1111/j.1365-3040.2005.01364.x

Chen, Z., Pottosin, I. I., Cuin, T. A., Fuglsang, A. T., Tester, M., Jha, D., et al. (2007b). Root plasma membrane transporters controlling K^+/Na^+ homeostasis in salt-stressed barley. *Plant Physiol.* 145, 1714–1725. doi: 10.1104/pp.107.110262

Chen, Z., Zhou, M., Newman, I. A., Mendham, N. J., Zhang, G., and Shabala, S. (2007c). Potassium and sodium relations in salinised barley tissues as a basis of differential salt tolerance. *Funct. Plant Biol.* 34, 150–162. doi: 10.1071/FP06237

Conn, S., and Gilliham, M. (2010). Comparative physiology of elemental distributions in plants. *Ann. Bot.* 105, 1081–1102. doi: 10.1093/aob/mcq027

Coskun, D., Britto, D. T., Jean, Y.-K., Kabir, I., Tolay, I., Torum, A. A., et al. (2013). K^+ efflux and retention in response to NaCl stress do not predict salt tolerance in contrasting genotypes of rice (*Oryza sativa* L.). *PLoS ONE* 8:e57767. doi: 10.1371/journal.pone.0057767

Cuin, T. A., Betts, S. A., Chalmandrier, R., and Shabala, S. (2008). A root's ability to retain K^+ correlates with salt tolerance in wheat. *J. Exp. Bot.* 59, 2697–2706. doi: 10.1093/jxb/ern128

Cuin, T. A., and Shabala, S. (2005). Exogenously supplied compatible solutes rapidly ameliorate NaCl-induced potassium efflux from barley roots. *Plant Cell Physiol.* 46, 1924–1933. doi: 10.1093/pcp/pci205

Cuin, T. A., and Shabala, S. (2007a). Amino acids regulate salinity-induced potassium efflux in barley root epidermis. *Planta* 225, 753–761. doi: 10.1007/s00425-006-0386-x

Cuin, T. A., and Shabala, S. (2007b). Compatible solutes reduce ROS-induced potassium efflux in Arabidopsis roots. *Plant Cell Environ.* 30, 875–885. doi: 10.1111/j.1365-3040.2007.01674.x

Demidchik, V. (2014). Mechanism and physiological roles of K^+ efflux from root cells. *J. Plant Physiol.* 171, 696–707. doi: 10.1016/j.jplph.2014.01.015

Demidchik, V., and Maathuis, F. J. M. (2007). Physiological roles of nonselective cation channels in plants: from salt stress to signaling and development. *New Phytol.* 175, 387–404. doi: 10.1111/j.1469-8137.2007.02128.x

Demidchik, V., and Tester, M. (2002). Sodium fluxes through nonselective cation channels in the plasma membrane of protoplasts from Arabidopsis roots. *Plant Physiol.* 128, 379–387. doi: 10.1104/pp.010524

Flowers, T. J., and Colmer, T. D. (2008). Salinity tolerance in halophytes. *New Phytol.* 179, 945–963. doi: 10.1111/j.1469-8137.2008.02531.x

Galvez, F. J., Baghour, M., Hao, G., Cagnac, O., Rodriguez-Rosales, M. P., and Venema, K. (2012). Expression of LeNHX isoforms in response to salt stress in salt sensitive and salt tolerant tomato species. *Plant Physiol. Biochem.* 51, 109–115. doi: 10.1016/j.plaphy.2011.10.012

Garcia de la Garma, J., Fernandez-Garcia, N., Bardisi, E., Pallol, B., Asencio-Rubio, J. S., Bru, R., et al. (2014). New insights into plant salt acclimation: the roles of vesicle trafficking and reactive oxygen species signalling in mitochondria and the endomembrane system. *New Phytol.* doi: 10.1111/nph.12997. [Epub ahead of print].

Gaxiola, R. A., Li, J., Undurraga, S., Dang, L. M., Allen, G. J., Alper, S. L., et al. (2001). Drought- and salt-tolerant plants result from overexpression of the AVP1 H+pump. *Proc. Natl. Acad. Sci. U.S.A.* 98, 11444 11449. doi: 10.1073/pnas.191389398

Gonzalez, P., Syvertsen, J. P., and Etxeberria, E. (2012). Sodium distribution in salt-stressed Citrus Rootstock seedlings. *HortScience* 47, 1504–1511.

Hamaji, K., Nagira, M., Yoshida, K., Ohnishi, M., Oda, Y., Uemura, T., et al. (2009). Dynamic aspects of ion accumulation by vesicle traffic under salt stress in Arabidopsis. *Plant Cell Physiol.* 50, 2023–2033. doi: 10.1093/pcp/pcp143

Hariadi, Y., Marandon, K., Tian, Y., Jacobsen, S. E., and Shabala, S. (2011). Ionic and osmotic relations in quinoa (*Chenopodium quinoa* Willd.) plants grown at various salinity levels. *J. Exp. Bot.* 62, 185–193. doi: 10.1093/jxb/erq257

Hmida-Sayari, A., Gargouri-Bouzid, R., Bidani, A., Jaoua, L., Savouré, A., and Jaoua, S. (2005). Overexpression of Δ1-pyrroline-5-carboxylate synthetase increases proline production and confers salt tolerance in transgenic potato plants. *Plant Sci.* 169, 746–752. doi: 10.1016/j.plantsci.2005.05.025

Horie, T., Hauser, F., and Schroeder, J. I. (2009). HKT transporter-mediated salinity resistance mechanisms in Arabidopsis and monocot crop plants. *Trends Plant Sci.* 14, 660–668. doi: 10.1016/j.tplants.2009.08.009

Huang, Z., Zhao, L., Chen, D., Liang, M., Liu, Z., Shao, H., et al. (2013). Salt stress encourages proline accumulation by regulating proline biosynthesis and degradation in Jerusalem Artichoke plantlets. *PLoS ONE* 8:e62085. doi: 10.1371/journal.pone.0062085

Jaarsma, R., de Vries, R. S. M., and de Boer, A. H. (2013). Effect of salt stress on growth, Na^+ accumulation and proline metabolism in potato (*Solanum tuberosum*) cultivars. *PLoS ONE* 8:e60183. doi: 10.1371/journal.pone.0060183

Ji, H., Pardo, J. M., Batelli, G., Van Oosten, M. J., Bressan, R. A., and Li, X. (2013). The salt overly sensitive (SOS) pathway: established and emerging roles. *Mol. Plant* 6, 275–286. doi: 10.1093/mp/sst017

Kishor, P. B. K., Sangam, S., Amrutha, R. N., Laxmi, P. S., Naidu, K. R., Rao, K. R. S. S., et al. (2005). Regulation of proline biosynthesis, degradation, uptake and transport in higher plants: its implications in plant growth and abiotic stress tolerance. *Curr. Sci.* 88, 424–438.

Kishore, B., Hong, Z., Miao, G., Hu, C., and Verma, D. (1995). Overexpression of delta-pyrroline-5-carboxylatesynthetase increase proline production and confers osmotolerance in transgenic plants. *Plant Physiol.* 108, 1387–1394.

Lu, Y., Li, N., Sun, J., Hou, P., Jing, X., Zhu, H., et al. (2013). Exogenous hydrogen peroxide, nitric oxide and calcium mediate root ion fluxes in two non-secretor mangrove species subjected to NaCl stress. *Tree Physiol.* 33, 81–95. doi: 10.1093/treephys/tps119

Lutts, S., Majerus, V., and Kinet, J. M. (1999). NaCl effects on proline metabolism in rice (*Oryza sativa* L.) seedlings. *Physiol. Plant.* 105, 450–458. doi: 10.1034/j.1399-3054.1999.105309.x

Maas, E. V., and Hoffman, G. J. (1977). Crop salt tolerance, current assessment. *ASCE J. Irrig. Drain. Div.* 103, 115–134.

Maathuis, F. J. M. (2014). Sodium in plants: perception, signalling, and regulation of sodium fluxes. *J. Exp. Bot.* 65, 849–858. doi: 10.1093/jxb/ert326

Malagoli, P., Britto, D. T., Schulze, L. M., and Kronzucker, H. J. (2008). Futile Na^+ cycling at the root plasma membrane in rice (*Oryza sativa* L.): kinetics, energetic, and relationship to salinity tolerance. *J. Exp. Bot.* 59, 4109–4117. doi: 10.1093/jxb/ern249

Mimura, T., Kura-Hotta, M., Tsujimura, T., Ohnishi, M., Miura, M., Okazaki, Y., et al. (2003). Rapid increase of vacuolar volumen in response to salt stress. *Planta* 216, 397–402. doi: 10.1007/s00425-002-0878-2

Moftah, A. E., and Michel, B. E. (1987). The effect of sodium chloride on solute potential and proline accumulation in soybean leaves. *Plant Physiol.* 83, 238–240. doi: 10.1104/pp.83.2.238

Moscone, E. A., Scaldaferro, M. A., Grabiele, M., Cecchini, N. M., Sánchez-García, Y., Jarret, R., et al. (2007). The evolution of chili peppers (*Capsicum-Solanaceae*): a cytogenetic perspective. *Acta Hortic.* 745, 137–170. Available online at: http://www.actahort.org/books/745/745_5.htm

Munns, R., and Tester, M. (2008). Mechanisms of salinity tolerance. *Annu. Rev. Plant Biol.* 59, 651–681. doi: 10.1146/annurev.arplant.59.032607.092911

Newman, I. A. (2001). Ion transport in roots: measurement of fluxes using ion-selective microelectrodes to characterize transporter function. *Plant Cell Environ.* 24, 1–14. doi: 10.1046/j.1365-3040.2001.00661.x

Oh, D.-H., Leidi, E., Zhang, Q., Hwang, S.-M., Li, Y., Quintero, F. J., et al. (2009). Loss of halophytism by interference with SOS1 expression. *Plant Physiol.* 151, 210–222. doi: 10.1104/pp.109.137802

Palmgren, M., and Nissen, P. (2010). P-type ATPases. *Annu. Rev. Biophys.* 40, 243–266. doi: 10.1146/annurev.biophys.093008.131331

Pasapula, V., Shen, G., Kuppu, S., Paez-Valencia, J., Mendoza, M., Hou, P., et al. (2011). Expression of an Arabidopsis vacuolar H+-pyrophosphatase gene (*AVP1*) in cotton improves drought and salt tolerance and increase fibre yield in the field conditions. *Plant Biotechnol. J.* 9, 88–99. doi: 10.1111/j.1467-7652.2010.00535.x

Perry, L., Dickau, R., Zarrillo, S., Holst, I., Pearsall, D. M., Piperno, D. R., et al. (2007). Starch fossils and the domestication and dispersal of chili peppers (*Capsicum* spp. *L.*) in the Americas. *Science* 325, 986–988. doi: 10.1126/science.1136914

Pittman, J. K. (2012). Multiple transport pathways for mediating intracellular pH homeostasis: the contribution of H+/ion exchangers. *Front. Plant Sci.* 3:11. doi: 10.3389/fpls.2012.00011

Plett, D. C., and Moller, I. S. (2010). Na^+ transport in glycophytic plants: what we know and would like to know. *Plant Cell Environ.* 33, 612–626. doi: 10.1111/j.1365-3040.2009.02086.x

Qin, C., Yu, C., Shen, Y., Fang, X., Chen, L., Min, J., et al. (2014). Whole-genome sequencing of cultivated and wild peppers provides insights into *Capsicum* domestication and specialization. *Proc. Natl. Acad. Sci. U.S.A.* 111, 5135–5140. doi: 10.1073/pnas.1400975111

Queirós, F., Fontes, N., Silva, P., Almeida, D., Maeshima, M., Gerós, H., et al. (2009). Activity of tonoplast proton pumps and Na^+/H^+ exchange in potato cell cultures is modulated by salt. *J. Exp. Bot.* 60, 1363–1374. doi: 10.1093/jxb/erp011

Ramani, B., Reeck, T., Debez, A., Stelzer, R., Huchzermeyer, B., Schmidt, A., et al. (2006). *Aster tripolium* L. and *Sesuvium portulacastrum* L.: two halophytes, two strategies to survive in saline habitats. *Plant Physiol. Biochem.* 44, 395–408. doi: 10.1016/j.plaphy.2006.06.007

Rodríguez-Rosales, M. P., Gálvez, F. J., Huertas, R., Aranda, M. N., Baghour, M., Cagnac, O., et al. (2009). Plant NHX catión/proton antiporters. *Plant Signal. Behav.* 4, 265–276. doi: 10.4161/psb.4.4.7919

Rodríguez-Rosales, M. P., Jiang, X., Gálvez, F. J., Aranda, M. N., Cubero, B., and Venema, K. (2008). Overexpression of the tomato K^+/H^+ antiporter LeNHX2 confers salt tolerance by improving potassium compartmentalization. *New Phytol.* 179, 366–377. doi: 10.1111/j.1469-8137.2008.02461.x

Roy, S. J., Negrão, S., and Tester, M. (2014). Salt resistant crop plants. *Curr. Opin. Biotechnol.* 26, 115–124. doi: 10.1016/j.copbio.2013.12.004

Ruan, C.-J., Teixeira da Silva, J. A., Mopper, S., Qin, P., and Lutts, S. (2010). Halophyte improvement for a salinized world. *CRC. Crit. Rev. Plant Sci.* 29, 329–359. doi: 10.1080/07352689.2010.524517

Rubio, F., Flores, P., Navarro, J. M., and Martínez, V. (2003). Effects of Ca^{2+}, K^+ and cGMP on Na^+ uptake in pepper plants. *Plant Sci.* 165, 1043–1049. doi: 10.1016/S0168-9452(03)00297-8

Shabala, S. (2013). Learning from halophytes: physiological basis and strategies to improve abiotic stress tolerance in crops. *Ann. Bot.* 112, 1209–1221. doi: 10.1093/aob/mct205

Shabala, S., and Cuin, T. A. (2007). Potassium transport and plant salt tolerance. *Physiol. Plant.* 133, 651–669. doi: 10.1111/j.1399-3054.2007.01008.x

Shabala, S., Demidchik, V., Shabala, L., Cuin, T. A., Smith, S. J., Miller, A. J., et al. (2006). Extracellular Ca^{2+} ameriorates NaCl-induced K^+ loss from Arabidopsis root and leaf cells by controlling plasma membrane K^+-permeable channels. *Plant Physiol.* 141, 1653–1665. doi: 10.1104/pp.106.082388

Shabala, S., and Pottosin, I. (2014). Regulation of potassium transport in plants under hostile conditions: implications for abiotic and biotic stress tolerance. *Physiol. Plant.* 151, 257–279. doi: 10.1111/ppl.12165

Shabala, S., and Shabala, L. (2011). Ion transport and osmotic adjustment in plants and bacteria. *Biomol. Concepts* 2, 407–419. doi: 10.1515/BMC.2011.032

Sharma, S., Villamor, J. G., and Verslues, P. E. (2011). Essential role of tissue-specific proline synthesis and catabolism in growth and redox balance at low water potential. *Plant Physiol.* 157, 292–304. doi: 10.1104/pp.111.183210

Shen, G., Wei, J., Qiu, X., Hu, R., Kuppu, S., Auld, D., et al. (2014). Co-overexpression of AVP1 and AtNHX1 in cotton further improves drought and salt tolerance in transgenic cotton plants. *Plant Mol. Biol. Rep.* doi: 10.1007/s11105-014-0739-8. [Epub ahead of print].

Shi, H., Quintero, F. J., Pardo, J. M., and Zhu, J.-K. (2002). The putative plasma membrane Na^+/H^+ antiporter SOS1 controls long-distance Na+ transport in plants. *Plant Cell* 14, 465–477. doi: 10.1105/tpc.010371

Smethurst, C. F., Rix, K., Garnett, T., Auricht, G., Bayart, A., Lane, P., et al. (2008). Multiple traits associated with salt tolerance in lucerne: revealing the underlying cellular mechanisms. *Funct. Plant Biol.* 35, 640–650. doi: 10.1071/FP08030

Sun, J., Dai, S., Wang, R., Chen, S., Li, N., Zhou, X., et al. (2009). Calcium mediates root K^+/Na^+ homeostasis in poplar species differing in salt tolerance. *Tree Physiol.* 29, 1175–1186. doi: 10.1093/treephys/tpp048

Szabados, L., and Savouré, A. (2009). Proline: a multifunctional amino acid. *Trends Plant Sci.* 15, 89–97. doi: 10.1016/j.tplants.2009.11.009

Verslues, P. E., and Sharma, S. (2010). Proline metabolism and its implications for plant-environmental interaction. *Arabidopsis Book* 8:e0140. doi: 10.1199/tab.0140

Volkov, V., and Amtmann, A. (2006). *Thellungiella halophila*, a salt-tolerant relative of *Arabidopsis thaliana*, has specific root ion-channel features supporting K^+/Na^+ homeostasis under salinity stress. *Plant J.* 48, 342–353. doi: 10.1111/j.1365-313X.2006.02876.x

Wu, H., Shabala, L., Barry, K., Zhou, M., and Shabala, S. (2013). Ability of leaf mesophyll to retain potassium correlates with salinity tolerance in wheat and barley. *Physiol. Plant.* 149, 515–527. doi: 10.1111/ppl.12056

Yildiztugay, E., Sekmen, A. H., Turkan, I., and Kucukoduk, M. (2011). Elucidation of physiological and biochemical mechanisms of an endemic halophyte *Centaurea tuzgoluensis* under salt stress. *Plant Physiol. Biochem.* 49, 816–824. doi: 10.1016/j.plaphy.2011.01.021

Zepeda-Jazo, I., Shabala, S., Chen, Z., and Pottosin, I. I. (2008). Na^+-K^+ transport in roots under salt stress. *Plant Signal. Behav.* 3, 401–403. doi: 10.4161/psb.3.6.5429

Zhang, J.-L., and Shi, H. (2013). Physiological and molecular mechanisms of plant salt tolerance. *Photosyn. Res.* 115, 1–22. doi: 10.1007/s11120-013-9813-6

Zhao, F., Song, C. P., He, J., and Zhu, H. (2007). Polyamines improve K^+/Na^+ homeostasis in barley seedlings by regulating root ion channel activities. *Plant Physiol.* 145, 1061–1072. doi: 10.1104/pp.107.105882

11

Salt tolerance at single cell level in giant-cell Characeae

Mary J. Beilby*

Plant Membrane Biophysics, Physics/Biophysics, School of Physics, University of New South Wales, Sydney, NSW, Australia

Edited by:
Richard Sayre, New Mexico Consortium at Los Alamos National Labs, USA
Reviewed by:
Suleyman I. Allakhverdiev, Russian Academy of Sciences, Russia
Martin Hagemann, University of Rostock, Germany
***Correspondence:**
Mary J. Beilby, Plant Membrane Biophysics, Physics/Biophysics (Visiting Fellow), School of Physics, University of New South Wales, Kensington 2052, Sydney, NSW, Australia m.j.beilby@unsw.edu.au

Characean plants provide an excellent experimental system for electrophysiology and physiology due to: (i) very large cell size, (ii) position on phylogenetic tree near the origin of land plants and (iii) continuous spectrum from very salt sensitive to very salt tolerant species. A range of experimental techniques is described, some unique to characean plants. Application of these methods provided electrical characteristics of membrane transporters, which dominate the membrane conductance under different outside conditions. With this considerable background knowledge the electrophysiology of salt sensitive and salt tolerant genera can be compared under salt and/or osmotic stress. Both salt tolerant and salt sensitive Characeae show a rise in membrane conductance and simultaneous increase in Na^+ influx upon exposure to saline medium. Salt tolerant Chara longifolia and Lamprothamnium sp. exhibit proton pump stimulation upon both turgor decrease and salinity increase, allowing the membrane PD to remain negative. The turgor is regulated through the inward K^+ rectifier and $2H^+/Cl$symporter. Lamprothamnium plants can survive in hypersaline media up to twice seawater strength and withstand large sudden changes in salinity. Salt sensitive C. australis succumbs to 50–100 mM NaCl in few days. Cells exhibit no pump stimulation upon turgor decrease and at best transient pump stimulation upon salinity increase. Turgor is not regulated. The membrane PD exhibits characteristic noise upon exposure to salinity. Depolarization of membrane PD to excitation threshold sets off trains of action potentials, leading to further loses of K^+ and Cl^-. In final stages of salt damage the H^+/OHchannels are thought to become the dominant transporter, dissipating the proton gradient and bringing the cell PD close to 0. The differences in transporter electrophysiology and their synergy under osmotic and/or saline stress in salt sensitive and salt tolerant characean cells are discussed in detail.

Keywords: Characeae, salt tolerance, electrophysiology, current-voltage characteristics, action potentials, proton pump, H^+/OH^- channels, non-selective cation channels

Introduction

Advantages of Characeae Experimental System for Salinity Studies

Large Cell Size and Simple Morphology

The thallus of characeaen plant consists of stems (axes), which are made of long multinucleate single cells interrupted by multicellular nodes. The nodes also give rise to branch-lets, which are similar to leaves of higher plants, but also consist of single cells (see **Figure 1**). The axial internode cell can be up to 1 mm in diameter and several cm long. The plants have colorless rhizoids instead of roots and these are also large cells joined end to end. The axial or leaf cells survive excision from the plant and can regenerate new plants from the adjacent nodal complexes. These excised cells can be used in prolonged experiments (up to 24 h). Pioneering electrical and transport measurements were performed on the characean plants (Walker, 1955; Hope and Walker, 1975; Beilby and Casanova, 2013).

Position on Phylogenetic Tree

Recent phylogenetic studies (Karol et al., 2001) have shown that charophytes (that contain the Characeae family) are the closest living relatives of the ancestors of all land plants. Land plants emerged onto land ~470 million years ago (Domozych et al., 2012), altering the atmosphere, reshaping the geology and enabling the evolution of terrestrial animals (Sorensen et al., 2010). While Characeae are now thought to be less closely related to land plants than another charophyte group Zygnematales (Wodniok et al., 2011; Timme et al., 2012), they are still positioned at the origin of land plants. Consequently, the large body of electrophysiological and physiological data provides valuable insights into many aspects of higher plants and into plant evolution (Beilby and Casanova, 2013). The question whether common ancestors of Characeae and land plants lived in freshwater or marine environments remains open (Graham and Gray, 2001; Kelman et al., 2004) as characean fossils were found in sediments from brackish and marine habitats (Martin et al., 2003). The transition of plants to land would have been less challenging from freshwater, as marine algae would have faced desiccation in air as well as hypersalinity in drying saline pools (Raven and Edwards, 2001). Further, fresh water plants would have already developed roots/rhizoids to acquire nutrients from the soil in the oligotrophic environment (Rodriguez-Navarro and Rubio, 2006).

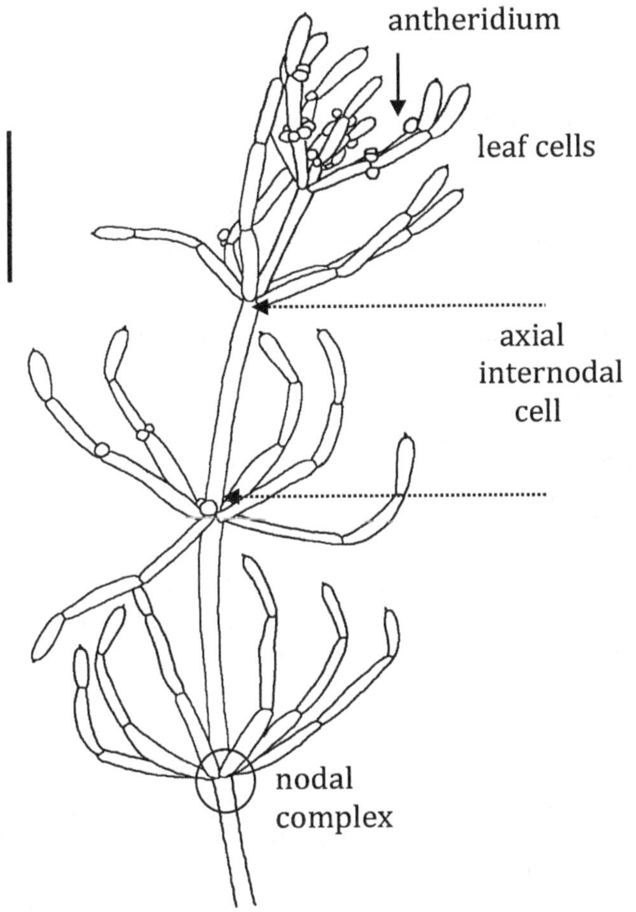

FIGURE 1 | *Chara australis* male plant: each segment is a single cell. Both leaf cells and axial internodes can be excised from the plant for experiments. The excised cells survive and new plants regenerate from small cells (not shown) in each nodal complex. Only the axial internodes near the top of the plant grow. The male reproductive structures, antheridia, can be found at the base of the whorl and branch-let nodes. The bar represents 10 mm (drawn by Michelle Casanova – adapted from Figure 1.6 of Beilby and Casanova, 2013).

SaltTolerant and Salt Sensitive Genera

The salt tolerance or sensitivity of the extant Characeae mirrors that of land plant glycophyte–halophyte distribution: majority live in fresh water and only few species are truly salt tolerant. The salt tolerant Characeae include some Tolypella, some Chara, and all Lamprothamnium species. The most salt tolerant species respond to salinity changes by complete turgor regulation through changing vacuolar concentrations of K^+, Cl^- and sometimes Na^+ or sucrose: Tolypella nidifica and glomerata (Winter et al., 1996), Chara longifolia (Hoffmann and Bisson, 1986), and all Lamprothamnium species (Bisson and Kirst, 1980a; Okazaki et al., 1984; Beilby et al., 1999; Casanova, 2013; Torn et al., 2014). The salt tolerance of Lamprothamnium is remarkable: plants with reproductive organs were found in Australian lakes at up to twice the salinity of seawater (Burne et al., 1980; Williams, 1998). C. australis or corallina and Nitella flexilis, on the other hand, are obligate freshwater species that regulate their internal osmotic pressure (Gutknecht et al., 1978; Sanders, 1981; Bisson and Bartholomew, 1984). C. australis plants exhibit 100% mortality after ~5 days in media containing 100 mM NaCl and 0.1 mM Ca^{2+} (Shepherd et al., 2008).

Components of saline stress

To resolve different components of salinity stress, cells can be exposed to a step up in osmolarity by employing sorbitol medium (for instance), followed by isotonic saline solution. Such experiments facilitate the measurement of short term defensive and stress responses to each component in dose dependent manner. The interpretation of results must allow for long term effects that might be due to slow acting mechanisms, such as compatible solute production and gene expression. For instance, Kanesaki et al. (2002) found that salt stress and hyperosmotic stress resulted in different gene expression in the ancient cyanobacteria Synechocystis.

The osmotic stress and Na^+ toxicity require different types of sensors and defensive mechanisms. The increase in osmolarity of the outside medium decreases the water potential and water flows out of the cell within seconds of exposure (Steudle and Zimmermann, 1974). The turgor of the cell drops, limiting growth, making cells prone to injury and affecting photosynthetic activity (Allakhverdiev et al., 2000; Allakhverdiev and Murata, 2008). To control turgor, cells have to be able to sense it. The turgor sensors are still being identified (Boudsocq and Lauriere, 2005).

The increase in NaCl concentration presents another problem. Characean cells are not very permeable to Cl^-, but Na^+ rapidly floods the cells through non-selective cation channels (NSCCs; Demidchik and Maathuis, 2007). Cell has to expend energy to move Na^+ from the cytoplasm, where it replaces K^+ and inhibits metabolic functions. In cyanobacteria Synechococcus increased Na^+ medium concentration caused slow and irreversible inactivation of the photosystems (Allakhverdiev et al., 2000).

The Characeae plants are totally submerged in the medium and cannot fight salinity by forming salt glands or exporting salt to sacrificial tissues, or blocking salt movement into the shoot: every single cell in the plant has to be salt tolerant. Thus, comparing the electrophysiology of salt tolerant and salt sensitive characean species is likely to identify a minimal ensemble of factors that bestow salt tolerance at cellular level. This review reports on the progress of such studies.

ExperimentalTechniques

Characeae experimental system is unique by providing the comparison between the intact cell and the preparations with escalating interventions. The methodology adapted to these large celled plants is briefly summarized below.

The cytoplasmic layer of a single cell is up to 10 μm thick and the vacuole occupies 95% of the cell volume (Raven, 1987; Beilby and Shepherd, 1989). Microelectrodes can be positioned in the cytoplasm as well as in the vacuole, and membrane potential difference (PD) is measured separately across plasma membrane and tonoplast (Findlay and Hope, 1964; Beilby, 1989). However, the cytoplasm often excludes microelectrodes (Walker, 1955) and the cells become more prone to damage upon electrode re-insertion. The access to the cytoplasmic compartment can be improved by gentle centrifugation that moves cytoplasm to one end of the internode (Hirono and Mitsui, 1981; Beilby and Shepherd, 1989). The electrode is impaled before the streaming cytoplasm redistributes along the cell, or the cell can be wilted in the air and the cytoplasmic plug tied off, preparing cytoplasm-enriched fragments (Hirono and Mitsui, 1981; Beilby and Shepherd, 1989). Alternatively, cell ends can be cut and perfusion medium replaces the vacuolar sap (Tazawa, 1964). Rapid perfusion rate or inclusion of EGTA in the perfusion medium disintegrates the tonoplast (Williamson, 1975; Tazawa et al., 1976). This preparation not only makes the plasma membrane accessible, but also surrounds it with media of known composition. To study the tonoplast the plasma membrane is permeabilized using EGTA (Shimmen and Tazawa, 1982; Tester et al., 1987). The tonoplast also surrounds cytoplasmic droplets formed by cutting an internodal cell and immersing the end in vacuolar sap like medium (Luhring, 1986). While some of the latter techniques allow experimenters greater control over the surroundings of each membrane, they also perturb the living cell and may introduce artifacts.

The size of characean cells facilitated early water permeability measurements (Wayne and Tazawa, 1990), using transcellular osmosis technique. The internodal cell was placed in two-compartment chamber with media of different osmolarity in each chamber. The reversible partial block by mercury derivatives suggested that some of the water moves across the membrane through water channels aquaporins. Pressure probe single cell measurements confirmed presence of aquaporins, although fraction of water and uncharged solutes permeating through them (as opposed to lipid bilayer) is still under consideration (Henzler and Steudle, 1995; Schutz and Tyerman, 1997). Ye et al. (2004, 2005) formulated tension/cohesion model for closure of water channels with increasing osmolarity. Henzler et al. (2004) found that water channels also close in response to oxidative stress.

The size of the cells can be a disadvantage when the membrane PD is controlled by voltage clamp. A longitudinal wire electrode or a small central compartment in a multi-compartment cell holder may be employed to space clamp the cell: to pass uniform currents and avoid conductance "hot spots" (Smith, 1984a; Beilby, 1990). To obtain current-voltage (I/V) characteristics of the plasma membrane (or both membranes in series), the membrane PD is clamped to bipolar staircase (Gradmann et al., 1978). The alternating short excursions to PD levels above and below the resting PD avoid prolonged large currents and transport number effects (accumulation of ions near the membrane outer surface – see Beilby, 1990). I/V profiles over wide PD windows (up to 500 mV) can be obtained in less than 10 s (Beilby, 1989). However, it is necessary to check the raw data to insure that the current has leveled at the end of each pulse and that it returned to near zero while the membrane PD was clamped at the resting level. Long time current dependencies need to be investigated separately by prolonged (seconds) voltage clamp to different PD levels.

Background of Characeae Electrophysiology

Thousands of I/V profiles have now been recorded on various Characeae species and under a range of conditions. The results are mostly consistent, suggesting that the plasma membrane can take on different states, depending on the outside conditions. This large body of data allows identification of ion transporter I/V profiles, their responses and synergies at the time of abiotic stress (such as salinity increase), as well as comparison to transporters of higher plants. The forthcoming sequencing of Characeae (Stefan Rensing, personal communication) will allow even more detailed comparisons on molecular level.

There are many ion transporters in both plasma membrane and the tonoplast and new ones are being discovered. However, only a small number of transporter types dominate the membrane conductance and the I/V profiles. The I/V characteristics can change substantially depending on the pH and K^+ concentration of the outside medium (see the central part of **Figure 2**).

Pump State

At neutral to slightly alkaline pHo with K^+ below 1 mM (a typical fresh water pond, where majority of characean species live), the plasma membrane resting PD is quite negative at −200 to −250 mV. The I/V characteristics exhibit a beautiful sigmoid shape generated by the energizing export of protons from the cytoplasm by the proton ATPase (see the bottom part of **Figure 2** and **Figure 3**). For history of the characean proton pump research and modeling with the cyclic enzymemediated HGSS model (Hansen et al., 1981) see chapter 2 of Beilby and Casanova (2013).

Background State

If the pump is turned off by some metabolic inhibitors or by circadian rhythms, the underlying null or background state is revealed: more depolarized resting PD near −100 mV and linear I/V profile in the PD window of ~ +50 to ~ −350 mV (see the left part of **Figure 2**). The background current was fitted by an empirical equation:

$$I_{background} = G_{background}(V - E_{background})$$

where $G_{background}$ is PD independent conductance and $E_{background}$ (E_L in **Figure 2**) is the reversal PD chosen as −100 (± 20) mV (−120 mV in **Figure 2**). This value was derived experimentally and its origin is still a puzzle. The background current is thought to flow through non-selective PD-independent cation channels (NSCC – Demidchik and Maathuis, 2007), which supply micronutrients to the plant and contribute to signaling (**Figure 3**).

At PD levels more negative than ~ −300 mV the background current is obscured by the inward rectifier current, K^+irc, while at PD levels more positive than ~ −50 mV the outward rectifier current, K^+orc, predominates. The inward and outward rectifier currents, mainly carried by K^+, were modeled by the Goldman-Hodgkin-Katz (GHK) equation, multiplied by the Boltzmann distribution of open probabilities to make the PDdependence stronger (Beilby and Walker, 1996; Amtmann and Sanders, 1999). The early activation of inward rectifier K^+irc can be observed near −400 mV on the pump state curves (bottom of **Figure 2**) and K^+ state curves (top of **Figure 2**). The early activation of outward rectifier, K^+orc, can be found near 0 PD in pump state curves and at +50 mV in the K^+ State curves.

The pump state and the underlaying background state are the native states for the salt sensitive Characeae in their low salt, slightly alkaline pond media.

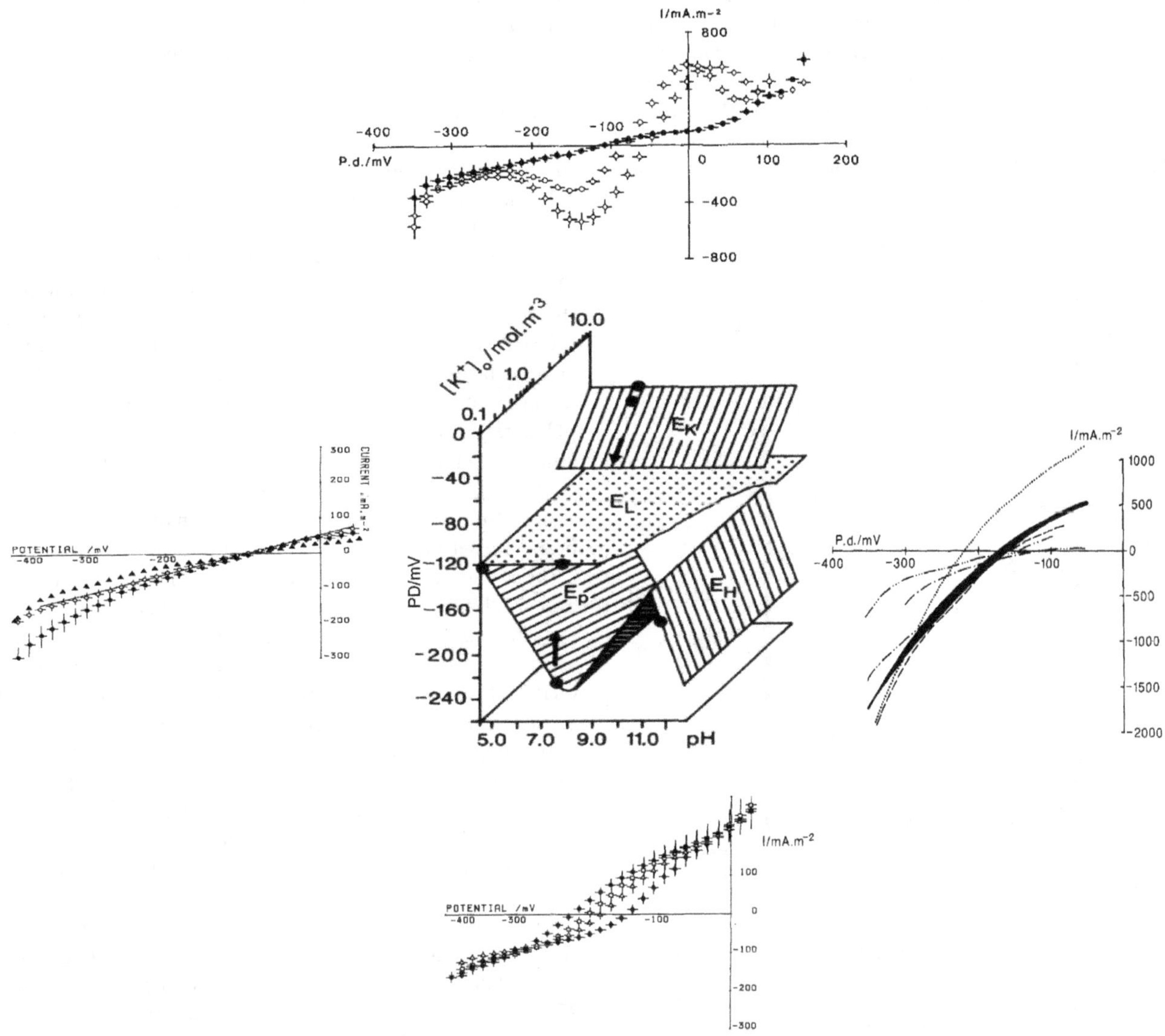

FIGURE 2 | Different states of plasma membrane: central panel shows the membrane PD in different states as function of pH$_o$ and [K$^+$]$_o$ (E$_p$ reversal PD of the proton pump, E$_L$ reversal PD for the background/leak current, E$_K$ Nernst PD for K$^+$, E$_H$ Nernst PD for H$^+$ or OH$^-$). The figure is based on **Figure 1** (Beilby, 1990) with the I/V characteristics updated from later publications. Bottom panel: proton pump-dominated state at pH$_o$ of 4.5 ♦, 5.5 **Δ**, 6.5 □, and 7.5 • (Beilby, 1984). E$_p$ is most negative at pH$_o$ 7–8 (shown as a dot on the "pump surface") and tends toward E$_L$ at pH 4.5 (a point where the pump and background surfaces meet). Left panel: background state with 10 I/V runs summarized from nine cells exposed to DES (diethyl stilbestrol). These cells stabilized after 30 min DES exposure (shown by a point on the background surface). Four La^{3+}-treated cells exposed to DES for 30 min, •, continued to change as shown by a single I/V run on La^{3+}-treated cell, with DES exposure of I hr 15 min (Beilby, 1984). Top panel (Beilby, 1986): I/V characteristics of cells in K$^+$ state, summary from seven cells in 5 mM K$^+$ APW ○, 10 mM K$^+$ APW ◇ (the two points on K$^+$ surface) and 0.1 mM K$^+$ APW • (here the K$^+$ channels closed revealing the background state, see the arrow in central part of figure). Right panel (Beilby and Bisson, 1992): high pH state: pH 11.5 (dotted line), pH 10.5 – two I/V runs in fast succession shown by black shading, pH 10.5 + 2.5 mM Na$_2$SO$_4$ (dashed line), back to pH 10.5 (dash, two dots, dash line), pH 10.5 + 10 mM Na$_2$SO$_4$ (dash dot dash line), and finally back to pH 10.5 (dash, three dots, dash line). The I/V characteristics in this state can be quite variable. As NaOH (5–30 mM) was used to bring the APW to high pH, the effect of Na$^+$ concentration increase was explored. Beilby and Bisson (1992) did not find a consistent effect, but high concentrations might affect the H$^+$/OH$^-$ channel activation via ROS response, see text.

K$^+$ State

As K$^+$ concentration in the medium rises above ~1 mM, large conductance K$^+$ (high G K$^+$) channels open, short-circuit the pump current and become the dominant membrane conductance (Oda, 1962; Smith and Walker, 1981; Sokolik and Yurin, 1981; Keifer and Lucas, 1982; Smith, 1984b; Beilby, 1985; Sokolik and Yurin, 1986; Tester, 1988a,b,c). The I/V characteristics of the K$^+$ state are very distinct: the two regions of negative conductance arise from the strong PD dependence of the channels (see the top part of **Figure 2**). The I/V profiles can also be modeled by GHK equation supplemented with Boltzmann distribution of open probabilities (Beilby and Shepherd, 2001b). The K$^+$ conductance increases with outside K$^+$ concentration until it becomes so large that the cells cannot be voltage-clamped. This property of K$^+$ state is exploited in K^+ anesthesia technique of measuring membrane PD without inserted electrodes (Hayama et al., 1979). K^+ channels are blocked totally and reversibly by tetraethylammonium (TEA) revealing the linear background state, which seems independent of K^+ concentration (see the top part of **Figure 2**). The activation of high G K^+ channels is observed in tissues, stomata and roots of land plants (Epstein, 1976; Cheeseman and Hanson, 1979; Blatt, 1988).

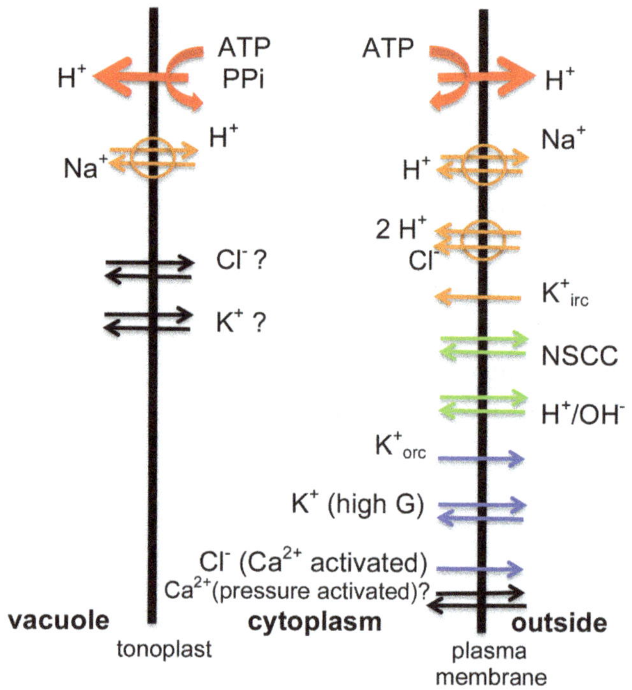

FIGURE 3 | The transporter scheme for the characean cell: same types of transporters are found in both salt tolerant and salt sensitive genera. The energizing ATPases and PPases (red arrows) drive H^+ out of the cytoplasm. The proton motive force is employed by Na^+/H^+ aniporters and $2H^+/Cl^-$ symporter, while negative PD opens K^+ inward rectifier (orange arrows). In salt sensitive Characeae the pumps are not stimulated by low turgor and fail in saline media. In normal pond water NSCC channels bring in nutrients and H^+/OH^- channels aid photosynthesis by exporting OH^- in alkaline bands (green arrows). In time of saline stress Na^+ enters through NSCC channels in all Characeae, but salt tolerant Characeae keep the Na^+/H^+ antiporters going and prevent global opening of H^+/OH^- channels. The outward rectifier, the high conductance K^+ channels or Ca^{2+} activated Cl^- channels are not active in steady state (blue arrows). The inflow of Ca^{2+} into the cytoplasm (black arrows) at the time of AP or hypoosmotic regulation is well documented, but the sources (outside, internal stores in the cytoplasm or the vacuole) are still disputed and beyond the scope of this article. Similarly, there are K^+ and Cl^- transporters on the tonoplast (black arrows), but discussion of these is beyond the scope of this article. The Ca^{2+}-activated Cl^- channels provide the depolarizing phase of the AP with the outward rectifier contributing to the recovery of resting PD. In salt tolerant Characeae the Ca^{2+}-activated Cl^- channels and high conductance K^+ channels mediate hypoosmotic regulation.

High pH State

If pH of the medium rises above 9.0, the resting PD starts to follow the equilibrium PD for H^+ or OH^- with resting PD more negative than −200 mV in some cells (Bisson and Walker, 1980). The membrane conductance can increase by up to 5 $S.m^{-2}$. This H^+/OH^- state is inhibited by darkness, photosynthesis inhibitor DCMU, various metabolic inhibitors (such as DES or DCCD) or lack of Ca^{2+} in the medium, which can be replaced by Mg^{2+} (Bisson, 1984). Beilby and Bisson (1992) found that the I/V characteristics exhibit slight downward curvature (see right part of **Figure 2**). However, the conductance and the reversal PD are quite variable. Al Khazaaly and Beilby (2012) modeled the I/V characteristics at high pH, using the GHK equation supplemented by the Boltzmann probability distribution. H^+ or OH^- was tested as the transported ion: OH^- seems more probable, as the channel number multiplied by the channel conductance remains relatively constant throughout the pH range, whereas for H^+ the same parameter had to be increased by several orders of magnitude to match the high conductance at very high pH (Beilby and Casanova, 2013). The high pH state is reversibly blocked by the zinc ion, potent inhibitor of animal H^+ channels (Al Khazaaly and Beilby, 2012). The H^+/OH^- channels participate in the pH banding pattern, where ring shaped zones of pump-dominated or H^+/OH^- channel dominated membrane are set up by the cells to acquire DIC (dissolved inorganic carbon) and make photosynthesis more efficient (Hope and Walker, 1975; Beilby and Bisson, 2012; Beilby and Casanova, 2013).

In their normal habitats, the whole Characeae cells and plants are not exposed to pH of 10 and higher. However, due to export of OH^- in the alkaline bands, the external pH rises to 10 and above. The exposure of cells to high pH highlighted mechanisms leading to banding formation (Beilby and Bisson, 2012) and facilitated detailed description and modeling of the I/V characteristics of the H^+/OH^- channels (Beilby and Bisson, 1992; Beilby and Al Khazaaly, 2009; Al Khazaaly and Beilby, 2012).

A detailed study of the I/V characteristics of Lamprothamnium cells acclimated to salinities ranging from 0.2 to full artificial sea water (ASW) showed that in the higher salinities (from ~1/3 ASW) the cells could be found in three different resting states: pump state, background state or K^+ state (Beilby and Shepherd, 2001a). Similarly to salt sensitive C. australis (left side of **Figure 2**), Lamprothamnium cells in the background state exhibited near linear I/V characteristics between −50 and −200 mV with reversal PD close to −100 mV.

Surviving in Saline Media: SaltTolerant Characeae

Hyperosmotic Adjustment Mechanisms

Upon increase of osmolarity in the external medium, the water potential drops, water flows out of the cell and the turgor pressure decreases. To restore turgor the osmolarity of the vacuole must be increased. This process requires energy. The energizing elements of the ion transport are proton ATPases and PPases on the plasma membrane and the tonoplast (**Figure 3**). The main osmotica, K^+ and Cl^- are imported through inward rectifier K^+irc and $2H^+/Cl^-$ symporter (**Figure 3**). Such turgor adjustment is well documented in higher plants (for instance in Arabidopsis roots – Shabala and Lew, 2002). The Na^+ inflow occurs through the NSCCs (Tester and Davenport, 2003; Demidchik and Maathuis, 2007). These channels can be partially blocked by high Ca^{2+} in the external medium (Tester and Davenport, 2003). The plant cells strive to keep low Na^+ in the cytoplasm employing Na^+/H^+ antiporters at both membranes (Tester and Davenport, 2003 and **Figure 3**). Several of these mechanisms were initially discovered in the Characeae experiments.

H^+ Pump Activation

Bisson and Kirst (1980a,b) and Okazaki et al. (1984) observed hyperpolarization of membrane PD in Lamprothamnium sp. upon increase in salinity while the vacuolar concentrations of K^+ and Cl^- increased to regulate turgor to ~300 mosmol/kg (Bisson and Kirst, 1980b). Reid et al. (1984) resolved the response further into a transient PD depolarization for ~10 min, followed by hyperpolarization. They also found a transient drop in ATP concentration paralleled by a rise in respiration. The cytoplasmic streaming speed diminished briefly and then increased, while the influxes of Na^+, K^+, and Cl^- increased over 800 min. Working on the only salt tolerant Chara species, C. longifolia, Yao et al. (1992) and Yao and Bisson (1993) demonstrated that the proton pumping also increased in more saline media and turgor was regulated: cell PD transiently hyperpolarized and conductance increased (95–144 h). Okazaki et al. (1984) challenged Lamprothamnium cells with a sorbitol hyperosmotic step and also observed membrane hyperpolarization. They suggested that it is the decrease in turgor, which initiates the turgor regulation observed by Bisson and Kirst (1980a). To confirm this hypothesis Al Khazaaly and Beilby (2007) compared the step from 1/6 ASW to 1/3 ASW in salinity and equivalent step in osmolarity using sorbitol (Sorbitol ASW). The I/V characteristics were measured for up to 7 h following either type of hyperosmotic step and the data were modeled to resolve the responses of various transporters. In both treatments the average cell PD hyperpolarized to similar level (~ −150 mV) through the activation of the proton pump and the inward rectifier channels were opened at more positive PDs. The authors also found that the cells in K^+ state were able to switch to pump state upon turgor decrease or salinity increase, and were able to regulate turgor.

Na^+ Transport

Kishimoto and Tazawa (1965), early pioneers of the I/V technique, measured increased Lamprothamnium cell membrane conductance with rising salinity of the medium. However, modeling of the currents through the different transporters was necessary to resolve the increase of pump conductance and the conductance due to Na^+ inflow. The Lamprothamnium cells in the background state showed clearly that the slope (conductance) of the Ibackground increased with medium salinity, while Ebackground remained close to −100 mV (Beilby and Shepherd, 2001a and **Figure 4C**). These cells confirmed the correct modeling of this underlying state for cells in pump state and K^+ state (Beilby and Shepherd, 2001a). Al Khazaaly and Beilby (2007) confirmed that after exposure to Sorbitol ASW the background conductance actually decreased slightly, while in 1/3 ASW the background conductance increased by about a third.

Working on C. longifolia the Bisson group found that the Na^+/Ca^{2+} ratio is important for turgor regulation (Hoffmann and Bisson, 1988; Hoffmann et al., 1989). The freshwater-incubated cells could only survive salinity increase and regulate turgor if the ratio was 10:1. Interestingly, the cells acclimated to their native medium (~100 mM Na^+ and 100 mM Mg^{2+} with Cl^- and SO_4^{2-} as main anions) did survive higher Na^+/Ca^{2+} ratios. Higher Ca^{2+} in the medium partially blocks the NSCC channels and diminishes the Na^+ inflow.

Tester and Davenport (2003) estimated that cytoplasm would equalize with the external medium of 50 mM NaCl in 3 min, if there was no Na^+ efflux. To frustrate the electrophysiologist Na^+/H^+ antiporter is electrically silent and consequently independent of membrane PD. Thermodynamically, the Na^+/H^+ anti-porter is an example of a Maxwell's demon. Protons are pumped out of the cytoplasm to create inward proton gradient and negative membrane PD, Na^+ flows in passively and is then removed by a "swap" for a proton. The primary energizing process is the proton pumping with ATP consumption as the cost to the cell (**Figure 3**).

Whittington and Bisson (1994) employed isotope $^{22}Na^+$ to measure Na^+ influx and efflux in salt sensitive C. corallina and fresh water-grown salt tolerant C. longifolia under mild salt stress of 20 mM NaCl. They found lower influx in salt tolerant C. longifolia. The addition of 1 mM Ca^{2+} to the medium greatly reduced influx into C. corallina. Na^+ efflux was greater in C. longifolia and increased in both Characeae as pHo changed from 7 to 5. In both species the efflux was unchanged by pH increase from 7 to 9. The efflux was not significantly inhibited by the Na^+/K^+ pump inhibitor amiloride, confirming the absence of such pump. C. longifolia grown in native saline medium (130 mM Na^+ and 110 mM MgSO4) exhibited greater Na^+ efflux, especially at pHo 5 (Kiegle and Bisson, 1996). At pH 9 the calculated L1μH was not sufficient to drive the efflux and the authors concluded that there might be other mechanisms for Na^+ efflux.

Davenport et al. (1996) found that despite surviving in salinities at seawater level and above, Lamprothamnium is also dependent on Ca^{2+} concentration of the medium. At low Ca^{2+} Na^+ influx increased and the cells died. For both Lamprothamnium and Chara australis influx of ~300 $nmolm^{-3}s^{-1}$ seems to be the limiting level for survival. The authors showed convincingly that low turgor promotes greater Na^+ influx. Reduction of turgor by addition of up to 100 mM mannitol doubled the influx even at low medium concentration of 3.5 mM NaCl. Conversely, increasing the turgor of Chara cells by soaking them in concentrated KCl decreased the Na^+ influx.

ModelingTransporter Response to Salinity

Resolving the different transporter populations by modeling the ITotal /V characteristics provides quantitative estimates of response to salinity/osmolarity increase. In Lamprothamnium plants acclimated to range of salinities the conductance of the background state increased with salinity from 0.5 $S.m^{-2}$ in 0.2 ASW to 22.0 $S.m^{-2}$ in full ASW (see **Figure 4C**). The cells in pump state increased pump currents in more saline media (see **Figure 4B**) with conductance maxima at 2 $S.m^{-2}$ in 0.2 ASW and 5 $S.m^{-2}$ in full ASW. But even with greater proton pumping the cell resting PDs became more depolarized with rising salinity: below −200 mV in 0.2 ASW to ~ −140 mV in full ASW (see **Figure 4A**). **Figures 4A–C** contain important message: Lamprothamnium cell can sense Na^+ concentration in the medium and adjust pump activity to counteract the increased background conductance. In higher salinity the cell has to expend more energy powering the proton pump. The cells in K^+ or background state could be "resting," saving energy temporarily, while the proton gradient runs down. Al Khazaaly and Beilby (2007) demonstrated that these cells are able to "switch" the pump back on. It might be interesting to find out if there is a periodic (perhaps circadian) pattern for the different states to dominate.

Beilby and Shepherd (2001a) also modeled transient changes of the Lamprothamnium proton pump after salinity step of 150 mOsmol.kg^{-1}, starting in dilute medium of 0.2 ASW (see **Figures 4D–F**). There was an initial decrease in the proton pump current and peak conductance (see **Figures 4D,E**, curve 1 magenta), coinciding with the low ATP concentration found by Reid et al. (1984). Then both the current and peak conductance increased, coming to a maximum after ~2 h of hyperosmotic challenge (see **Figures 4D,E**, curve 4, green). Inward rectifier cur-

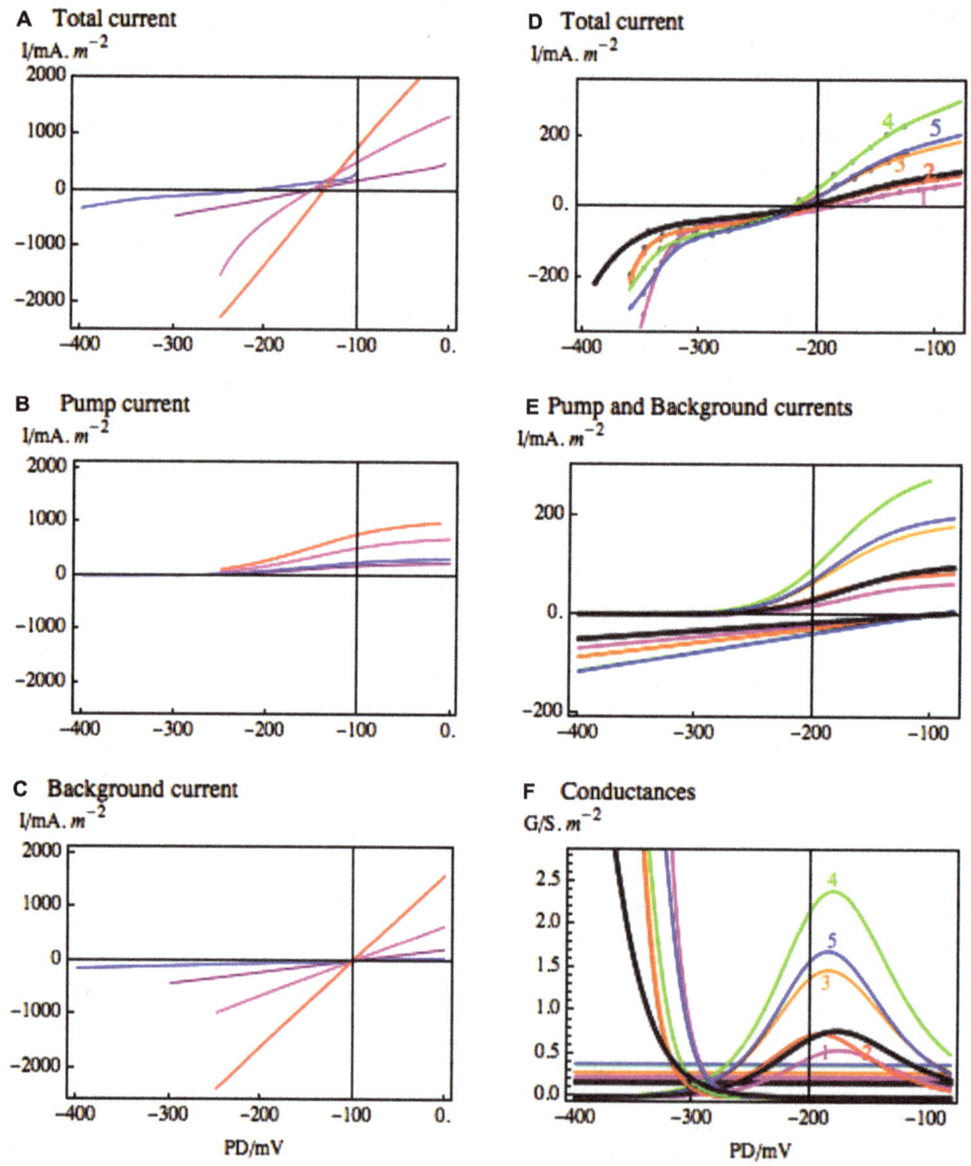

FIGURE 4 | The response of the most conductive transporters in *Lamprothamnium* to medium salinity. (A) I/V characteristics of *Lamprothamnium* sp. in steady state, grown in media of increasing salinity: 0.2 seawater (SW), blue; 0.4 SW, purple; 0.5 SW, magenta; full SW, red. The currents have been fitted to data from 6 to 8 cells from each medium (adapted from Figure 6 of Beilby and Shepherd, 2001a). The fitted pump **(B)** and the background **(C)** current components are shown in same colors (for the fit parameters see Table 2 of Beilby and Shepherd, 2001a). The fitting of background current in media 0.4 – full SW was supported by comparison with cells in background state in each medium. **(D)** Characteristics from one *Lamprothamnium* cell acclimated to 0.2 SW and challenged by doubling salinity to 0.4 SW. Steady state I/V in 0.2 SW: black thick line; curve 1, magenta, 5 min of 0.4 SW; curve 2, red, 21 min of 0.4 SW; curve 3, orange, 41 min of 0.4 SW; curve 4, green, 2 h 34 min of 0.4 SW; curve 5, blue, 3 h 30 min of 0.4 SW (adapted from Figure 7 of Beilby and Shepherd, 2001a). The fitted pump and background current components **(E)** are shown in same colors. The conductances of the fitted current components, including the inward rectifier, are displayed in **(F)** also in same colors (for fit parameters see Table 3 of Beilby and Shepherd, 2001a).

rent responded within minutes of salinity increase by activating at more positive PDs (**Figures 4D,F**).

Thus the more researched salt tolerant Characeae, Lamprothamnium sp. and Chara longifolia, both exhibit proton pump stimulation upon salinity increase to keep the membrane PD negative, despite partial short-circuit of greater background conductance of Na^+ inflow through NSCC channels. This response is crucial, as the proton pump provides the energy source for both up-regulation of turgor and prevention of toxic built up of Na^+ in the cytoplasm (**Figure 3**). The membrane PD more negative than E_K facilitates the import of K^+ through the inward rectifier channels to maintain K^+/Na^+ ratio supportive to normal enzyme function. K^+ is also transported into the vacuole for turgor regulation. The import of Cl^- into the vacuole upon salinity/osmolarity increase is mediated by the $2H^+/Cl^-$ symporter at plasma membrane, again powered by the proton electrochemical gradient (Sanders, 1980; Beilby and Walker, 1981). The Cl^- transporters at the tonoplast also have to work "uphill" to achieve the observed vacuolar concentrations (Teakle and Tyerman, 2010).

Hypoosmotic Adjustment Mechanisms

Plants living in saline media have to cope with salinity changes: osmolarity of a shallow pond can drop in minutes in a torrential downpour. Cells have been observed to explode if the turgor became too great. What are the mechanisms for such sudden downward turgor adjustment? Working with different starting media and different salinity/osmolarity decrease in Lamprothamnium or C. longifolia, several investigators found rapid membrane depolarization to ~ −70 mV for 30–60 min, accompanied by conductance rise up to an order of magnitude, with subsequent partial repolarization (Reid et al., 1984; Okazaki et al., 1984; Hoffmann and Bisson, 1990; Okazaki and Iwasaki, 1992; Beilby and Shepherd, 1996).

Okazaki and Tazawa (1986a,b) also observed streaming inhibition for up to 20 min upon hypoosmotic challenge. Low Ca^{2+} medium or presence of Ca^{2+} antagonist nifedipine abolished the depolarization, conductance increase and turgor regulation. Okazaki and Tazawa (1987) visualized the Ca^{2+} increase in the cytoplasm using fluorescence techniques: in ASW with normal high Ca^{2+} content fluorescence increased after about 1 min of hypoosmotic stress. If the cell was given a hypoosmotic shock in low Ca^{2+} ASW and Ca^{2+} increased later, the fluorescence rose immediately. The authors concluded that turgor mediated opening of Ca^{2+} channels has a small delay. Okazaki et al. (2002) made more detailed measurements of Ca^{2+} concentration in Lamprothamnium under hypoosmotic stress. After initial rapid increase from resting value of 100 nM to peak of 600 nM, the concentration dropped at 0.9 nM/s. They also estimated that maximum conductance was reached by 300 nM, but 400–600 nM was necessary for cytoplasmic streaming cessation.

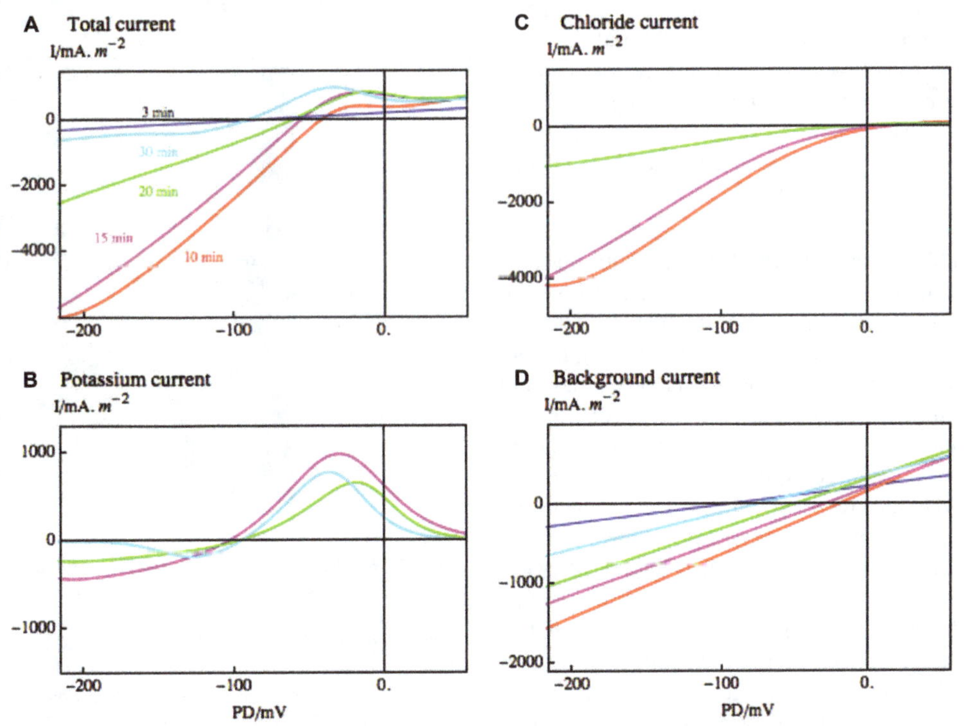

FIGURE 5 | Electrophysiology of the hypoosmotic regulation upon dilution step from 1/3 ASW to 1/6 ASW. (A) The I/V characteristics are compiled from currents fitted to cells with Cl^- current blocked by exposure to $LaCl_3$ or K^+ current blocked by TEA at times: 3 min (dark blue); 10 min (red); 15 min (magenta); 20 min (green); 30 min (blue; Beilby and Shepherd, 1996, 2001b). **(B)** The fitted Cl^- currents are shown with the same types of line as in **(A)**. These currents appear with a slight delay after hypoosmotic exposure at 10 and 15 min, and start to decline at 20 min. **(C)** The fitted K^+ currents are shown by the same types of line as in **(A)**. They appear with a greater delay at 15 min, 20 min, and 30 min. **(D)** The fitted background currents at same times as in **(A)**. Note the different scales in **(A)**, **(B)**, **(C)**, and **(D)**. While the K^+ currents are smaller than the Cl^- currents, they persist for a longer time, up to 60 min after hypoosmotic shock (the figure was adapted in color from Beilby and Casanova, 2013).

As turgor regulation in Lamprothamnium was mostly achieved by varying K^+ and Cl^- in the vacuole (Bisson and Kirst, 1980a), the increase of conductance must be due to the efflux of these ions. Okazaki and Iwasaki (1992) found a good correlation between Cl^- efflux and rise in conductance. Beilby and Shepherd (1996) employed the I/V technique and pharmacological dissection (TEA or LaCl3) to resolve the timing of the Cl^- and K^+ outflow (see **Figure 5**). The large G K^+ channels, identified by their typical I/V characteristics (compare top part of **Figures 2** and **5B**) and total block by TEA, mediate the K^+ outflow (Beilby and Shepherd, 2001b). The K^+ channel activation is clearly visible with the Cl^- currents blocked by La^{3+} (**Figure 5B**). The opening of K^+ channels is preceded by the Cl^- channel activation, but there is some overlap (Beilby and Shepherd, 1996). These results also suggest that K^+ channels do not require increased Ca^{2+} concentration in the cytoplasm. By Occam's razor argument the Cl^- channels are assumed to be the same Ca^{2+}-activated channels that participate in the AP (Beilby, 2007, see also **Figure 3**). With the Ca^{2+} concentration in the cytoplasm high for many minutes (Okazaki et al., 2002), the I/V characteristics of the Cl^- channels could be investigated (see **Figure 5C** and Beilby and Shepherd, 2006b). The channels are inwardly rectifying with maximum conductance near −100 mV and strong inactivation near 0 PD. Bisson et al. (1995) found that in C. longifolia K^+ channel activation preceded the Cl^- channel activation and that this second stage also required external Ca^{2+}.

In both Characeae, there was an initial depolarization upon the hypoosmotic step, which was independent of Ca^{2+} concentration in the medium, or presence of the blockers TEA and La^{3+} (Bisson et al., 1995; Beilby and Shepherd, 1996). This response could be modeled by changing the reversal PD for the background current (see **Figure 5D**). Thus at least some of this current might flow through stretch-activated (mechano-sensitive) channels (Beilby and Shepherd, 1996; Shepherd et al., 2002).

Hypoosmotic Effect Modulation by Cell Structure and Age

Bisson et al. (1995) found that small cells (less than 10 mm in length) of C. longifolia regulated turgor within 60 min, while longer (and mostly older) cells took up to 3 days for full regulation. Beilby et al. (1999) and Shepherd and Beilby (1999) discovered that older Lamprothamnium cells developed coating of sulphated polysaccharide mucilage, identified by staining with Toluidine Blue or Alcian Blue at pH 1.0 (see **Figure 6**). With thicker mucilage the cells exhibited graded response to hypoosmotic challenge. The inflow of Ca^{2+} and streaming stoppage was not observed, but the opening of K^+ channels was retained (such as **Figure 5B**). Very mucilaginous cells exhibited only a brief depolarization with linear I/V profiles, modeled as the background current with depolarizing reversal PD (**Figure 5D**; Shepherd et al., 1999). These cells still regulated turgor, but took 24 h or longer. Interestingly, Stento et al. (2000) were not able to find mucilage on cells of C. longifolia and it is possible that Chara genera do not produce it (Casanova, personal communication). On the other hand, Torn et al. (2014) collected six Lamprothamnium species from nine Australian locations and all species produced extracellular mucilage, with thickness and proportion of sulphated polysaccharides increasing with cell age, unrelated to the salinity of the environment.

After treatment of the mucilaginous cells with the heparinase enzyme, the cells responded to hypotonic shock with exaggerated depolarization, streaming stoppage and conductance increase (Shepherd and Beilby, 1999). Interestingly, the mucilage layer remained in place after heparinase treatment, but did not stain at low pH. Thus the effects of mucilage as an unstirred layer and a polyanionic layer could be separated. The removal of heparinase restored the muted response in the same cell. Young cells with no mucilage showed no change upon exposure to heparinase.

The young (fast regulating) and the old (mucilaginous and slow regulating) cells exhibited differences in sequestration of fluorochrome 6-carboxyfluorescein (6CF), which accumulated in the cytoplasm of the fast regulating cells and in the vacuole of the slow regulating cells. Beilby et al. (1999) speculated that the fast cells have more complex vacuole structure with canalicular elements, while slow cells had large central vacuoles. Patch clamp experiments performed on cytoplasmic droplets (thought to be bound by the tonoplast membrane) from slow and fast regulating cells exhibited K^+ channels and small conductance Cl^- channels. The Cl^- channels appeared more active in slow regulating cells.

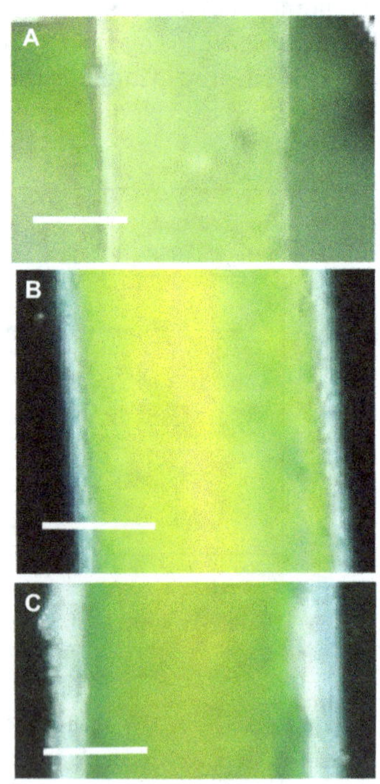

FIGURE 6 | The extracellular mucilage produced by *Lamprothamnium* plants. Cells were stained with Alcian Blue at pH 1 (Beilby et al., 1999). **(A)** Apical cell growing in ¡ ASW with mucilage ~7 μm thick. The staining is patchy, indicating that only fraction of the mucilage is sulphated. Bar = 50 μm. **(B)** Third internode from the apex from a plant growing in full ASW. Mucilage is ~28 μm. Bar = 100 μm. **(C)** Seventh internode of the same plant as in **(B)**, mucilage thickness is ~43 μm. Bar = 100 μm (from Beilby and Casanova, 2013).

The ability of Lamprothamnium sp. to make extracellular sulphated polysaccharide mucilage (similar to that found in many chlorophyte marine algae) highlights another important halophytic attribute. Aquino et al. (2011) found that glycophytes of agricultural importance (Zea mays L., Oryza sativa L., and Phaseolus vulgaris L.) were unable to synthesize sulphated mucilage in their roots and leaves, even when challenged by increased salinity. On the other hand, salt tolerant angiosperms, such as mangroves and sea grass, and salt tolerant fern Acrostichum aureum contained sulphated mucilage in roots and shoots. Further, the seagrass Ruppia maritima L. only produced sulphated mucilage in saline media, even if the plants were supplied with abundant sulfate in the freshwater growth medium.

Pathology of Salt Stress: Salt Sensitive Characeae

The salt sensitive C. australis was also exposed to the components of saline stress: sorbitol medium and saline medium of equivalent osmolarity. Shepherd et al. (2008) used 50–100 mM NaCl added to artificial pond water (APW) or 90–180 sorbitol APW. The proton pump in Chara cells does not respond to non-plasmolysing decrease in turgor, but can be transiently activated by an increase

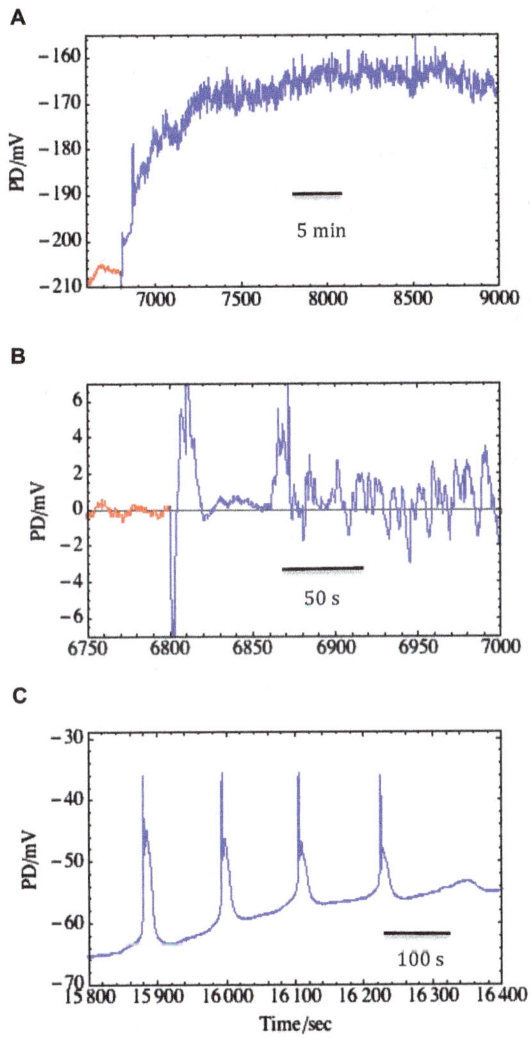

FIGURE 7 | Salinity-induced noise in membrane PD. (A) Transition from Sorbitol APW (red) to 50 mM NaCl saline APW (blue) resulted in depolarization and noisy membrane PD. There was no AP, but the resting PD measurement was disrupted by medium change and exhibited spikes between 6800 and 6900 s. **(B)** Using the same trace on expanded time scale, the higher frequency noise was isolated by subtracting the running average (fitted with $n = 100$): noise in sorbitol APW (red) and saline-induced noise (blue) that started promptly after ~50 s of the medium change (data from Beilby et al., 2014). **(C)** Spontaneous and repetitive APs after ~80 min of Saline APW (Shepherd et al., 2008).

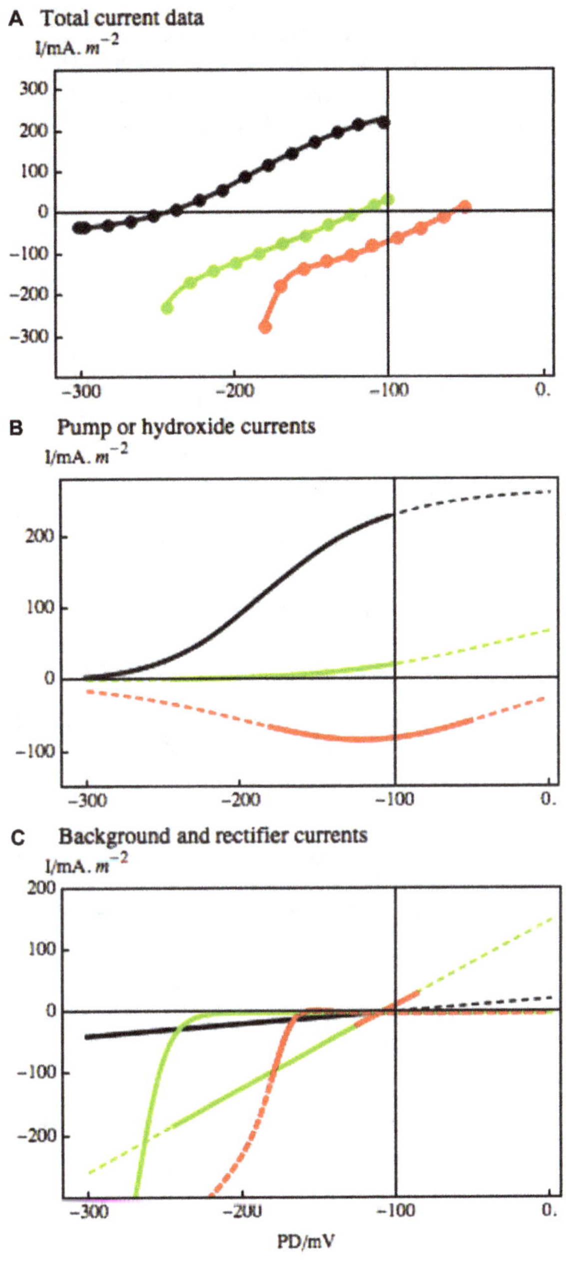

FIGURE 8 | The response of *Chara* I/V characteristics to salinity challenge. (A) Pump-dominated profile (black) in Sorbitol APW evolved to background-dominated profile after 67 min in 50 mM NaCl Saline artificial pond water (APW; green) and to upwardly concave profile (red) after 117 min of Saline APW. The experimental data, shown as points in **(A)**, were fitted by the pump or OH^- channel models **(B)** and background current and inward rectifier models **(C)**, using same line colors as in part **(A)**. The parameters are given in caption of Figure 1 of Beilby and Al Khazaaly (2009). The dashed lines in **(B)** and **(C)** extrapolate the models beyond the range of the data. The inward rectifier channels also activate at more depolarized PDs, but the membrane PDs are more depolarized for K^+ inflow to take place.

in Na^+ concentration if Na^+/Ca^{2+} ratio is not too high (Beilby and Shepherd, 2006a) and rapidly inactivated when Na^+/Ca^{2+} ratio increases (Shepherd et al., 2008). The increase in Ca^{2+} concentration in saline media exerts its protective influence by partially blocking NSCC channels and keeping the pump running (Bisson, 1984). The two calcium effects may be related through cytoplasmic Na^+ concentration, which may increase past some level critical to the pump. However, even at low external Na^+ (2–3 mM) the lack of Ca^{2+} caused decline of the pump activity within hours (Bisson, 1984). The inactivation of the pump brings the membrane potential to Ebackground, near −100 mV and at the excitation threshold. Spontaneous repetitive APs with long duration are often observed, further depleting the cell of K^+ and Cl^- (see **Figure 7C** and Shepherd et al., 2008).

The background conductance does not change upon nonplasmolysing turgor decrease (as in Lamprothamnium), but increases in a Ca^{2+} dependent manner in saline media (Shepherd et al., 2008). Given equivalent salt stress the background conductance is higher in C. australis than in Lamprothamnium (Beilby and Shepherd, 2006a), as low turgor was found to increase the Na^+ influx (see earlier section on Na^+ transport).

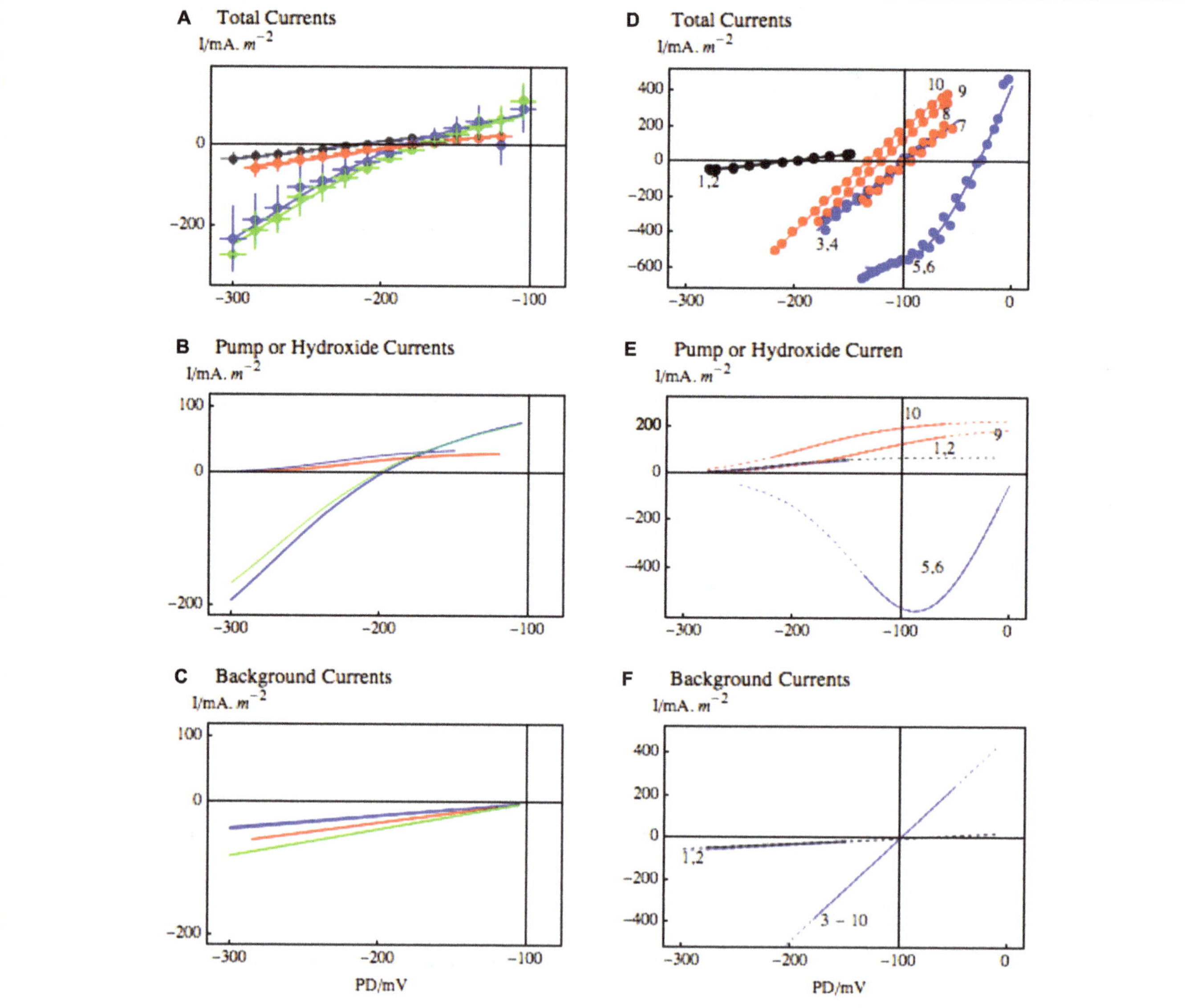

FIGURE 9 | The effect of zinc ion on the high pH state and putative H^+/OH^- channels at the time of salinity stress. (A) Statistics of 12 I/V profiles from 5 cells in APW (black), APW with pH increased to 11 (blue), 1.0 mM $ZnCl_2$ was then added to the high pH APW for average time of 36 min (red) and finally in three cells where 0.5 mM 2-mercaptoethanol (ME) replaced $ZnCl_2$ for average 35 min after the high pH state was inhibited (green). The data were fitted with the pump current or the OH^- current **(B)**. The background current increased slightly after application of $ZnCl_2$ **(C)**. The fit parameters are given in Al Khazaaly and Beilby (2012). **(D–F)** The effect of zinc ion on salinity induced I/V profiles. **(D)** The profiles 1 and 2 (black) were obtained in APW and APW with 90 mM sorbitol. After 15–30 min of APW + 50 mM NaCl, the I/V profiles 3,4 just showed the background current (blue). The resting PD then dropped further and the I/V profiles 5 and 6 were modeled with OH^- channels (blue). 1.0 mM $ZnCl_2$ was added to the saline APW and the PD repolarized to the background current (I/Vs 7 and 8, red) and later the pump was re-activated (I/Vs 9 and 10, red). For details see Figures 2 and 4 of Al Khazaaly and Beilby (2012).

Upon transfer from sorbitol to saline medium Chara exhibits salinity-induced noise in the membrane PD (see **Figures 7A,B** and Al Khazaaly et al., 2009). At frequencies between 1 and 500 mHz classical noise analysis revealed ($1/f^2$) rise of noise power as frequency falls, and a marked increase in noise power upon salinity challenge. Inspection of the time domain shows a continuous but random series of small rapid depolarizations followed by recovery (**Figure 7B**). As PD noise is unchanged if NaCl is exchanged for Na2SO4, we initially hypothesized that high Na^+ concentration activates H^+/OH^- channels. However, in recent experiments (Eremin et al., 2013) employed fluorescent probe dihydrodichlorofluorescein (DCHF) to trace reactive oxygen species (ROS) formation under strong spot illumination of Chara surface. The authors suggest that excess ROS formed in the chloroplasts was carried away by the cytoplasm, oxidizing either histidine or sulfhydryl (SH) groups on transport proteins, leading to opening the H^+/OH^- channels. In the animal kingdom, the voltage-gated H^+ channels in the brain microglia are activated by H2O2 (Wu, 2014). Beilby et al. (2014) were able to inhibit or postpone the saline-induced noise by presoaking the Chara cells in strong antioxidant melatonin prior to saline stress. Therefore, H^+/OH^- channels may be activated by an oxidative burst upon exposure to saline. Such a transient increase in ROS was observed by Li et al. (2007) upon exposing rice roots to salinity.

Once again the modeling of the I/V data allowed us to trace responses of ion transporter populations as function of exposure to saline. With longer exposure to high salinity, the membrane PD of Chara cells continues to depolarize toward zero, while the noise diminishes (suggesting that progressively larger numbers of H^+/OH^- channels were activated – Beilby et al., 2014). The shape of the I/V characteristics changes and could be simulated by H^+/OH^- channels and increased background current (see **Figure 8** and Beilby and Al Khazaaly, 2009). The global opening of these channels at the time of saline stress would be disastrous for the cells, as both the negative membrane potential and the pH gradients between the cytoplasm, vacuole and the medium are necessary for the cell survival (**Figure 3**). Further evidence for role of H^+/OH^- channels in the salt stress pathology is their blockage by zinc ion (Al Khazaaly and Beilby, 2012). Zinc ions are the most potent inhibitors of animal proton channels. While the permeating ion in Characeae is more likely OH^-, the channel proteins may still be closely related, as replacement of aspartate 112 by a neutral amino acid facilitated anion conduction in animal "proton" channels (DeCoursey, 2013). In Characeae the zinc ion reversibly inhibited the high pH state. The application of 0.5 mM 2-mercaptoethanol (ME) removed the zinc and restored the high pH state (**Figures 9A–C**). At the time of salt stress the depolarization to PD levels above −100 could be reversed by including 1.0 mM ZnCl2 in the saline APW. Even the function of the proton pump was temporarily restored (**Figures 9D–F**). The saline noise was also inhibited by zinc ion. However, as zinc has many functions in plant tissues, further proof of H^+/OH^- channel involvement in salt stress pathology is needed.

Interestingly, Kirst and Bisson (1982) found that salt tolerant Lamprothamnium exposed to pH above 9.5 suffered similar fate to salt sensitive C. australis: the turgor dropped, concentration of Na^+ in the cell compartments increased, while concentration of K^+ and Cl^- decreased leading to cell death. We assume that high pH opened the H^+/OH^- channels, placing the Lamprothamnium cell in the same electrophysiological state as Chara in the late stages of salt stress.

Conclusions, Relevance to Higher Plants and Future Research

The salt tolerant and salt sensitive Characeae contain the same types of ion transporters (**Figure 3**), but some of these transporters respond differently to osmotic and saline stress. The salt tolerant Lamprothamnium and C. longifolia cells sense both the decrease of turgor and the increase of Na^+, and respond by pumping protons faster to maintain a negative membrane potential while keeping H^+/OH^- channels closed in the acid bands. (The pH banding phenomenon as a function of salinity is under investigation at present. pH banding was observed in Lamprothamnium cells acclimated to fresh water – Beilby, unpublished). The cells regulate turgor by importing more K^+, Cl^- and Na^+. Salt sensitive C. australis does not respond to turgor decrease, does not regulate turgor, loses the energizing pump function and negative membrane potential and undergoes spontaneous repetitive APs. The global opening of H^+/OH^- channels speeds up the irreversible decline by further decreasing the membrane PD and promoting K^+ loss through outward rectifying channels. The proton gradient powering Na^+/H^+ antiporter and $2H^+/Cl^-$ symporter is dissipated.

Is the H^+/OH^- channels and the proton pump combination just peculiar to the ancient Characeae? This complex electrophysiological motif of pH banding can be observed in higher plants: aquatic angiosperms (Prins et al., 1980), pollen tubes (Feijo et al., 1999) and, most importantly, roots of land plants (Raven, 1991). In roots the source of the current is located in the acid root subapical zone with the sink at the alkaline tip (Raven, 1991). The author suggests that the acid and alkaline zones facilitate acquisition of molybdenum, phosphorus and iron as well as reduction of aluminum toxicity. Protoplasts from wheat roots were found to change from pump-dominated to H^+/OH^- channel dominated tate (Tyerman et al., 2001). Further, salinity induced noise was observed in wheat root protoplasts (Tyerman et al., 1997). Thus a future experiments should investigate the effect of salinity on the acid/alkaline zones of roots of both glycophytes and halophytes.

Being able to sense turgor is clearly important for salt tolerant cells. Staves et al. (1992), Shimmen (2008) suggest that it is the nodal complexes (see **Figure 1**) of the characean cells that sense difference in turgor. Future experiments are planned with nodeless constructs (Beilby and Shepherd, 1989, 1991) from Lamprothamnium cells to find if these can still sense turgor and regulate it.

How do the turgor sensors and Na^+ sensors communicate with the proton pump? Are the proton pumps of salt tolerant and salt sensitive Characeae different? Plant H^+ ATPase is encoded by a multi-gene family. In rice a new isoform of the proton pump genetic family was observed in response to salt stress (see a review by Janicka-Russak, 2011). This isoform was similar to that found in halophyte Suaeda maritima. Several post-translational modifications are also suggested, involving Cterminal domain and N-terminus of the protein. Phosphorylation is regulated by 14-3-3 proteins and pump molecules might form multimeric complexes. In electrophysiological experiments the pump current can be observed directly as function of time after osmotic or salinity increase. It might be possible to probe the post-translational modifications in response to salinity stress. The imminent sequencing of C. braunii genome (Stefan Rensing, personal communication) will provide molecular data of proton pump structure in salt tolerant and salt sensitive Characeae.

References

Al Khazaaly, S., and Beilby, M. J. (2007). Modeling ion transporters at the time of hypertonic regulation *Lamprothamnium succinctum* (Characeae, Charophyceae). *Charophytes* 1, 28–47.

Al Khazaaly, S., and Beilby, M. J. (2012). Zinc ion blocks H^+/OH− channels in *Chara australis*. *Plant Cell Environ.* 35, 1380–1392. doi: 10.1111/j.13653040.2012.02496.x

Al Khazaaly, S., Walker, N. A., Beilby, M. J., and Shepherd, V. A. (2009). Membrane potential fluctuations in *Chara australis*: a characteristic signature of high external sodium. *Eur. Biophys. J.* 39, 167–174. doi: 10.1007/s00249-009-0485-2

Allakhverdiev, S. I., and Murata, N. (2008). Salt stress inhibits photosystems II and I in cyanobacteria.*Photosynth. Res.* 98, 529–539. doi: 10.1007/s11120-0089334-x

Allakhverdiev, S. I., Sakamoto, A., Nishiyama, Y., Inaba, M., and Murata, N. (2000). Ionic and osmotic effects of NaCl-induced inactivation of photosystems I and II in *Synechococcus* sp. *Plant Physiol.* 123, 1047–1056. doi: 10.1104/pp.123.3.1047

Amtmann, A., and Sanders, D. (1999). Mechanism of Na^+ uptake by plant cells. *Adv. Bot. Res.* 29, 75–112. doi: 10.1016/S0065-2296(08)60310-9

Aquino, R. S., Grativol, C., and Mourão, P. A. S. (2011). Rising from the sea: correlations between sulfated polysaccharides and salinity in plants. *PLoS ONE* 6:e18862. doi: 10.1371/ journal.pone.0018862

Beilby, M. J. (1984). Current-voltage characteristics of the proton pump at *Chara* plasmalemma: I. pH dependence. *J. Membr. Biol.* 81, 113–125. doi: 10.1007/BF01868976

Beilby, M. J. (1985). Potassium channels at *Chara* plasmalemma. *J. Exp. Bot.* 36, 228–239. doi: 10.1093/jxb/36.2.228

Beilby, M. J. (1986). Factors controlling the K^+ conductance in *Chara*. *J. Membr. Biol.* 93, 187–193. doi: 10.1007/BF01870810

Beilby, M. J. (1989). "Electrophysiology of giant algal cells," in *Methods in Enzymology*, Vol. 174, eds S. Fleischer and B. Fleischer (San Diego, CA: Academic Press), 403–443.

Beilby, M. J. (1990). Current-voltage curves for plant membrane studies: a critical analysis of the method. *J. Exp. Bot.* 41, 165–182. doi: 10.1093/jxb/41.2.165

Beilby, M. J. (2007). "Action potential in charophytes," in *International Review of Cytology*, Vol. 257, ed. K. W. Jeon (San Diego, CA: Elsevier Inc.), 43–82.

Beilby, M. J., and Al Khazaaly, S. (2009). The role of H^+/OH− channels in salt stress response of *Chara australis*. *J. Membr. Biol.* 230, 21–34. doi: 10.1007/s00232009-9182-4

Beilby, M. J., Al Khazaaly, S., and Bisson, M. A. (2014). Salinity-induced noise in membrane potential of Characeae *Chara australis*: effect of exogenous Melatonin. *J. Membr. Biol.* 248, 93–102. doi: 10.1007/s00232-014-9746-9

Beilby, M. J., and Bisson, M. A. (1992). *Chara* plasmalemma at high pH: voltage dependence of the conductance at rest and during excitation. *J. Membr. Biol.* 125, 25–39. doi: 10.1007/BF00235795

Beilby, M. J., and Bisson, M. A. (2012). "pH banding in Charophyte algae," in *Plant Electrophysiology Methods and Cell Electrophysiology*, ed. A. G. Volkov (Berlin Heidelberg: Springer-Verlag), 247–271. doi: 10.1007/978-3-642-29119-7_11

Beilby, M. J., and Casanova, M. T. (2013). *The Physiology of Characean Cells*. Berlin: Springer.

Beilby, M. J., Cherry, C. A., and Shepherd, V. A. (1999). Dual regulation response to hypertonic stress in *Lamprothaminum papulosum*. *Plant Cell Environ.* 22, 347–359. doi: 10.1046/j.1365-3040.1999.00406.x

Beilby, M. J., and Shepherd, V. A. (1996). Turgor regulation in *Lamprothamnium papulosum*.1. I/V analysis and pharmacological dissection of the hypotonic effect. *Plant Cell Environ.* 19, 837–847. doi: 10.1111/j.1365-3040.1996.tb00420.x

Beilby, M. J., and Shepherd, V. A. (1989). Cytoplasm-enriched fragments of *Chara*: structure and electrophysiology. *Protoplasma* 148, 150–163. doi: 10.1007/BF02079334

Beilby, M. J., and Shepherd, V. A. (1991). Reassertion of morphology and physiology in cytoplasm enriched fragments of *Nitella*. *C. R. Acad. Sci.* 313, 265–271.

Beilby, M. J., and Shepherd, V. A. (2001a). Modeling the current-voltage characteristics of charophyte membranes: II. the effect of salinity on membranes of *Lamprothamnium papulosum*. *J. Membr. Biol.* 181, 77–89. doi: 10.1007/PL00020977

Beilby, M. J., and Shepherd, V. A. (2001b). Modeling the current–voltage characteristics of charophyte membranes III. K^+ state of *Lamprothamnium*. *Austr. J. Plant Physiol.* 28, 541–550. doi: 10.1071/PP01032

Beilby, M. J., and Shepherd, V. A. (2006a). The electrophysiology of salt tolerance in charophytes. *Cryptogam. Algol.* 27, 403–417.

Beilby, M. J., and Shepherd, V. A. (2006b). The characteristics of Ca^{++} -activated Cl− channels of the salt tolerant charophyte *Lamprothamnium*. *Plant Cell Environ.* 29, 764–777. doi: 10.1111/j.1365-3040.2005. 01437.x

Beilby, M. J., and Walker, N. A. (1981). Chloride transport in *Chara* I. Kinetics and current-voltage curves for a probable proton symport. *J. Exp. Bot.* 32, 43–54. doi: 10.1093/jxb/32.1.43

Beilby, M. J., and Walker, N. A. (1996). Modeling the current-voltage characteristics of *Chara* membranes: I. The effect of ATP and zero turgor. *J. Membr. Biol.* 149, 89–101. doi: 10.1007/s002329900010

Bisson, M. A. (1984). Calcium effects on electrogenic pump and passive permeability of the plasma membrane of *Chara corallina*. *J. Membr Biol.* 81, 59–67. doi: 10.1007/BF01868810

Bisson, M. A., and Bartholomew, D. (1984). Osmoregulation or turgor regulation in *Chara*? *Plant Physiol.* 74, 252–255. doi: 10.1104/pp.74.2.252

Bisson, M. A., Kiegle, E. A., Black, D., Kiyosawa, K., and Gerber, K. (1995). The role of calcium in turgor regulation in *Chara longifolia*. *Plant Cell Environ.* 18, 129–137. doi: 10.1111/j.1365-3040.1995.tb00346.x

Bisson, M. A., and Kirst, G. O. (1980a). *Lamprothamnium*, a euryhaline charophyte Osmotic relations and membrane potential at steady state. *J. Exp. Bot.* 31, 1223–1235. doi: 10.1093/jxb/31.5.1223

Bisson, M. A., and Kirst, G. O. (1980b). *Lamprothamnium*, a euryhaline charophyte II. time course of turgor regulation. *J. Exp. Bot.* 31, 1237–1244. doi: 10.1093/jxb/31.5.1237

Bisson, M. A., and Walker, N. A. (1980). The *Chara* plasmalemma at high pH. Electrical measurements show rapid specific passive uniport of H^+ or OH−. *J. Membr. Biol.* 56, 1–7. doi: 10.1007/BF01869346

Blatt, M. R. (1988). Potassium-dependent, bipolar gating of K^+ channels in guard cells. *J. Membr. Biol.* 102, 235–246. doi: 10.1007/BF01925717

Boudsocq, M., and Lauriere, C. (2005). Osmotic signaling in plants. Multiple pathways mediated by emerging kinase families. *Plant Physiol.* 138, 1185–1194. doi: 10.1104/pp.105.061275

Burne, R. V., Bauld, J., and de Dekker, P. (1980). Saline lake charophytes and their geological significance. *J. Sediment. Petrol.* 50, 281–294.

Casanova, M. T. (2013). *Lamprothamnium* in Australia (Characeae, Charophyceaea). *Austr. Syst. Bot.* 26, 268–290. doi: 10.1071/SB13026

Cheeseman, J. M., and Hanson, J. B. (1979). Energy-linked potassium influx as related to cell potential in corn roots. *Plant Physiol.* 64, 842–845. doi: 10.1104/pp.64.5.842

Davenport, R., Reid, R. J., and Smith, F. A. (1996). Control of sodium influx by calcium and turgor in two charophytes differing in salinity tolerance. *Plant Cell Environ.* 19, 721–728. doi: 10.1111/j.1365-3040.1996. tb00407.x

DeCoursey, T. E. (2013). Voltage-gated proton channels: molecular biology, physiology, and pathophysiology of the Hv family. *Physiol. Rev.* 93, 599–652. doi: 10.1152/physrev.00011.2012

Demidchik, V., and Maathuis, F. M. (2007). Physiological roles of non-selective cation channels in plants: from stress to signalling and development. *New Physiol.* 175, 387–404. doi: 10.1111/j.1469-8137.2007.02128.x

Domozych, D., Ciancia, M., Fangel, J. U., Mikkelsen, M. D., Ulvskov, P., and Willats, W. G. T. (2012). The plant walls of green algae: a journey through evolution and diversity. *Front. Plant Sci.* 3:82. doi: 10.3389/fpls.2012. 00082

Epstein, E. (1976). "Kinetics of ion transport and the carrier concept," in *Transport in Plants. IIB. Encyclopedia of Plant Physiology*, Vol. 2, eds U. Luttge and M. G. Pitman. (Berlin: Springer), 70–94.

Eremin, A., Bulychev, A., and Hauser, M. J. B. (2013). Cyclosis-mediated transfer of H2 O2 elicited by localized illumination of *Chara* cells and its relevance to the formation of pH bands. *Protoplasma* 250, 1339–1349. doi: 10.1007/s00709-0130517-8

Feijo, J., Sainhas, J., Hackett, G. R., Kunkel, J. G., and Hepler, P. K. (1999). Growing pollen tubes possess a constitutive alkaline band in the clear zone and a growth-dependent acidic tip. *J. Cell Biol.* 144, 483–496. doi: 10.1083/jcb.144.3.483

Findlay, G. P., and Hope, A. B. (1964). Ionic relations of cells of *Chara australis*: VII. The separate electrical characteristics of the plasmalemma and tonoplast. *Austr. J. Biol. Sci.* 17, 62–77.

Gradmann, D., Hansen, U.-P., Long, W. S., Slayman, C. L., and Warncke, J. (1978). Current-voltage relationship for the plasma membrane and its principal electrogenic pump in *Neurospora crassa*: I. Steady-state conditions. *J. Membr. Biol.* 39, 333–367. doi: 10.1007/BF01869898

Graham, L. E., and Gray, J. (2001). "The origin, morphology, and ecophysiology of early embryophytes: neontological and pleontological perspectives." in *Plants Invade the Land: Evolutionary and Environmental Perspectives*. eds P. G. Gensel and D. Edwards (New York, NY: Columbia University Press).

Gutknecht, J., Hastings, D. F., and Bisson, M. A. (1978). "Ion transport and turgor pressure regulation in giant algal cells," in *Membrane Transport in Biology*, Vol. III, *Transport Across Multimembrane Systems*, eds G. Giebisch, D. Tosteson, and G. Ussing (Berlin: Springer), 125–174.

Hansen, U. P., Gradmann, D., Sanders, D., and Slayman, C. L. (1981). "Interpretation of current-voltage relationships for "active" ion transport systems: I. Steady-state reaction-kinetic analysis of class-I mechanisms." *J. Membr. Biol.* 63, 165–190. doi: 10.1007/BF01870979

Hayama, T., Nakagava, S., and Tazawa, M. (1979). Membrane depolarization induced by transcellular osmosis in internodal cells of *Nitella flexilis*. *Protoplasma* 98, 73–90. doi: 10.1007/BF01676663

Henzler, T., and Steudle, E. (1995). Reversible closing of water channels in *Chara* internodes provides evidence for a composite transport model of the plasma membrane. *J. Exp. Bot.* 46, 199–209. doi: 10.1093/jxb/46.2.199

Henzler, T., Ye, Q., and Steudle, E. (2004). Oxidative gating of water channels (aquaporins) in *Chara* by hydroxyl radicals. *Plant Cell Environ.* 27, 1184–1195. doi: 10.1111/j.1365-3040.2004.01226.x

Hirono, C., and Mitsui, T. (1981). "The course of activation in plasmalemma of *Nitella axilliformis*," in *Nerve membrane*, eds A. G. Matsumoto and M. Kotani (Tokyo: University of Tokyo).

Hoffmann, R., and Bisson, M. A. (1986). *Chara buckelii*, a euryhaline charophyte from an unusual saline environment. I. Osmotic relations at steady state. *Can. J. Bot.* 64, 1599–1605. doi: 10.1139/b86-215

Hoffmann, R., and Bisson, M. A. (1988). The effect of divalent cations on Na^+ tolerance in Charophytes. I. *Chara buckelii*. *Plant Cell Environ.* 11, 461–472. doi: 10.1111/j.1365-3040.1988.tb01784.x

Hoffmann, R., and Bisson, M. A. (1990). *Chara buckelii*, a euryhaline charophyte from an unusual saline environment. III. Time course of turgor regulation. *Plant Physiol.* 93, 122–127. doi: 10.1104/pp.93.1.122

Hoffmann, R., Tufariello, J. M., and Bisson, M. A. (1989). Effect of divalent cations on Na^+ permeability of *Chara corallina* and fresh water grown *Chara buckelii*. *J. Exp. Bot.* 40, 875–881. doi: 10.1093/jxb/40.8.875

Hope, A. B., and Walker, N. A. (1975). *The Physiology of Giant Algal Cells*. London: Cambridge University Press.

Janicka-Russak, M. (2011). "Plant plasma memebrane H^+-ATPase in adaptation of plants to abiotic stresses," in *Abiotic Stress Response in Plants – Physiological, Biochemical and Genetic Perspectives*. ed. P. A. Shanker (Rijeka: INTECH), 197–218.

Kanesaki, Y., Suzuki, I., Allakhverdiev, S. I., Mikami, K., and Murata, N. (2002). Salt stress and hyperosmotic stress regulate the expression of different sets of genes in *Synechocystis* sp. *PCC* 6803. *Biochem. Biophys. Res. Commun.* 290, 339–348. doi: 10.1006/bbrc.2001.6201

Karol, K. G., McCourt, R. M., Cimino, M. T., and Delwiche, C. F. (2001). The closest living relatives of land plants. *Science* 294, 2351–2353. doi: 10.1126/science.1065156

Keifer, D. W., and Lucas, W. J. (1982). Potassium channels in *Chara corallina*. *Plant Physiol.* 69, 781–788. doi: 10.1104/pp.69.4.781

Kelman, R., Feist, M., Trewin, N. H., and Hass, H. (2004). Charophyte algae from the Rhynie chert. *Trans. R. Soc. Edinb. Earth Sci.* 94, 445–455.

Kiegle, E. A., and Bisson, M. A. (1996). Plasma membrane Na^+ transport in a salt-tolerant Charophyte. *Plant Physiol.* 111, 1191–1197. doi: 10.1104/pp.111.4.1191

Kirst, G. O., and Bisson, M. A. (1982). Vacuolar and cytoplasmic pH, ion composition and turgor pressure in *Lamprothamnium* as function of external pH. *Planta* 155, 287–295. doi: 10.1007/BF00429453

Kishimoto, U., and Tazawa, M. (1965). Ionic composition and electric response of *Lamprothamnium succinctum*. *Plant Cell Physiol.* 6, 529–536.

Li, J.-Y., Jiang, A.-L., and Zhang, W. (2007). Salt stress-induced programmed cll death in rice root tip cells. *J. Integr. Plant Biol.* 49, 481–486. doi: 10.1111/j.17447909.2007.00445.x

Luhring, H. (1986). Recording of single K^+ channels in the membrane of cyto-plasmic drop of *Chara australis*. *Protoplasma* 133, 19–27. doi: 10.1007/BF01 293183

Martin, G., Torn, K., Blindow, I., Schubert, H., Munsterhjelm, R., and Henricson, C. (2003). "Introduction to charophyte," in *Charophytes of the Baltic Sea*, Vol. 19, eds H. A. B. Schubert and I. Blindow (Königstein im Taunus: The Baltic Marine Biologists Publications), 3–14.

Oda, K. (1962). Polarized and depolarized states of the membrane in *Chara braunii*, with special reference to the transition between the two states." *Sci. Rep. Tohoku Univ. Fourth Ser. (Biology)* 28, 1–16.

Okazaki, Y., Ishigami, M., and Iwasaki, N. (2002). Temporal relationship between cytosolic free Ca2$^+$ and membrane potential during hypotonic turgor regulation in a brackish water charophyte *Lamprothamnium succinctum*. *Plant Cell Physiol.* 43, 1027–1035. doi: 10.1093/pcp/pcf127

Okazaki, Y., and Iwasaki, N. (1992). Net efflux of Cl– during hypotonic turgor regulation upon hypotonic treatment in internodal cells of *Lamprothamnium*. *Plant Cell Environ.* 15, 61–70. doi: 10.1111/j.1365-3040.1992. tb01458.x

Okazaki, Y., Shimment, T., and Tazawa, M. (1984). Turgor regulation in a brackish Charophyte, *Lamprothamnium succinctum* II. Changes in K^+, Na^+ and Cl–concentrations, membrane potential and membrane resistance during turgor regulation. *Plant Cell Physiol.* 25, 573–581.

Okazaki, Y., and Tazawa, M. (1986a). Ca2$^+$ antagonist nifedipine inhibits turgor regulation upon hypotonic treatment in internodal cells of *Lamprothamnium*. *Protoplasma* 135, 65–66. doi: 10.1007/BF01276378

Okazaki, Y., and Tazawa, M. (1986b). Involvement of calcium ion in turgor regulation upon hypotonic treatment in *Lamprothamnium succinctum*. *Plant Cell Environ.* 9, 185–190. doi: 10.1111/1365-3040.ep11611622

Okazaki, Y., and Tazawa, M. (1987). Increase in cytoplasmic calcium content in internodal cells of *Lamprothamnium* upon hypotonic treatment. *Plant Cell Environ.* 10, 619–621. doi: 10.1111/j.1365-3040.1987.tb01843.

Prins, H. B. A., Snel, J. F. H., Helder, R. J., and Zanstra, P. E. (1980). Photosynthetic HCO– utilization and OH– excretion in aquatic angiosperms. *Plant Physiol.* 66, 818–822. doi: 10.1104/pp.66.5.818

Raven, J. A. (1987). The role of vacuoles. *New Phytol.* 106, 357–422. doi: 10.1111/j.1469-8137.1987.tb00149.x

Raven, J. A. (1991). Terrestrial rhizophytes and and H^+ currents circulating over at least a millimeter: an obligate relationship? *New Phytol.* 117, 177–185. doi: 10.1111/j.1469-8137.1991.tb04899.x

Raven, J. A., and Edwards, D. (2001). Roots: evolutionary origins and biogeochemical significance. *J. Exp. Bot.* 52, 381–401. doi: 10.1093/jexbot/52.suppl_1.381

Reid, R. J., Jefferies, R., and Pitman, M. G. (1984). *Lamprothamnium*, a euryhaline charophyte. IV. Membrane potential, ionic fluxes and metabolic activity during turgor adjustment. *J. Exp. Bot.* 35, 925–937. doi: 10.1093/jxb/35. 6.925

Rodriguez-Navarro, A., and Rubio, F. (2006). High-affinity potassium and sodium transport systems in plants. *J. Exp. Bot.* 57, 1149–1160. doi: 10.1093/jxb/ erj068

Sanders, D. (1980). The mechanism of Cl– Transport at the plasma membrane of *Chara corallina* I. Cotransport with H^+. *J. Membr. Biol.* 53, 129–141. doi: 10.1007/BF01870581

Sanders, D. (1981). Physiological control of chloride transport in *Chara corallina*. I. Effects of low temperature, cell turgor pressure, and anions. *Plant Physiol.* 67, 1113–1118. doi: 10.1104/pp.67.6.1113

Schutz, K., and Tyerman, S. D. (1997). Water channels in *Chara corallina*. *J. Exp. Bot.* 48, 1511–1518. doi: 10.1093/jexbot/48.313.1511

Shabala, S., and Lew, R. (2002). Turgor regulation in osmotically stressed *Arabidopsis* epidermal root cells. Direct support for the role of inorganic ion uptake as revealed by concurrent flux and cell turgor measurements. *Plant Physiol.* 129, 290–299. doi: 10.1104/pp.020005

Shepherd, V. A., and Beilby, M. J. (1999). The Effect of an extracellular mucilage on the response to osmotic shock in the charophyte alga *Lamprothamnium papulosum*. *J. Membr. Biol.* 170, 229–242. doi: 10.1007/s002329900552

Shepherd, V. A., Beilby, M. J., and Heslop, D. J. (1999). Ecophysiology of the hypotonic response in the salt-tolerant charophyte alga *Lamprothamnium papulosum*. *Plant. Cell Environ.* 22, 333–346. doi: 10.1046/j.1365-3040.1999. 00414.x

Shepherd, V. A., Beilby, M. J., Al Khazaaly, S., and Shimmen, T. (2008). Mechanoperception in *Chara* cells: the influence of salinity and calcium on touch-activated receptor potentials, action potentials and ion transport. *Plant Cell Environ.* 31, 1575–1591. doi: 10.1111/j.1365-3040.2008.01866.x

Shepherd, V. A., Beilby, M. J., and Shimmen, T. (2002). Mechanosensory ion channels in charophyte cells: the response to touch and salinity stress. *Eur. Biophys. J.* 31, 341–355. doi: 10.1007/s00249-002-0222-6

Shimmen, T. (2008). Electrophysiological characterization of the node in *Chara corallina*: functional differentiation for wounding response. *Plant Cell Physiol.* 49, 264–272. doi: 10.1093/pcp/pcn002

Shimmen, T., and Tazawa, M. (1982). Reconstitution of cytoplasmic streaming in Characeae. *Protoplasma* 113, 127–131. doi: 10.1007/BF01282001

Smith, J. R. (1984a). The electrical properties of plant cell membranes. II. Distortion of non-linear current-voltage characteristics induced by the cable properties of *Chara*. *Austr. J. Plant Physiol.* 11, 211–224.

Smith, P. T. (1984b). Electrical evidence from perfused and intact cells for volt-agedependent K^+ channels in the plasmalemma of *Chara australis*. *Austr. J. Plant Physiol.* 11, 304–318.

Smith, P. T., and Walker, N. A. (1981). Studies on the perfused plasmalemma of *Chara corallina*: I. Current voltage curves: ATP and potassium dependence. *J. Membr. Biol.* 60, 223–236. doi: 10.1007/BF01992560

Sokolik, A. I., and Yurin, V. M. (1981). Transport properties of potassium channels of the plasmalemma in *Nitella* cells at rest. *Sov. Plant Physiol.* 28, 206–212.

Sokolik, A. I., and Yurin, V. M. (1986). Potassium channels in plasmalemma of *Nitella* cells at rest. *J. Membr. Biol.* 89, 9–22. doi: 10.1007/BF01870892

Sorensen, I., Domozych, D., and Willats, W. G. T. (2010). How have plant cell walls evolved? *Plant Physiol.* 153, 366–372. doi: 10.1104/pp.110.154427

Staves, M. P., Wayne, R., and Leopold, A. C. (1992). Hydrostatic pressure mimics gravitational pressure in characean cells. *Protoplasma* 168, 141–152. doi: 10.1007/BF01666260

Stento, N. A., Gerber, N. R., Kiegle, E. A., and Bisson, M. A. (2000). Turgor regulation in the salt-tolerant alga *Chara longifolia*. *Plant Cell Environ.* 23, 629–637. doi: 10.1046/j.1365-3040.2000.00571.x

Steudle, E., and Zimmermann, U. (1974). Determination of the hydraulic conductivity and of reflection coefficients in *Nitella flexilis* by means of direct cell-turgor pressure measurements. *Biochim. Biophys. Acta* 332, 399–412. doi: 10.1016/0005-2736(74)90362-9

Tazawa, M. (1964). Studies on *Nitella* having artificial cell sap. I Replacement of the cell sap with artificial solutions. *Plant Cell Physiol.* 5, 33–43.

Tazawa, M., Kikuyama, M., and Shimmen, T. (1976). Electric characteristics and cytoplasmic streaming of Characeae cells lacking tonoplast. *Cell Struct. Funct.* 1, 165–175. doi: 10.1247/csf.1.165

Teakle, N. L., and Tyerman, S. D. (2010). Mechanisms of Cl– transport contributing to salt tolerance. *Plant Cell Environ.* 33, 566–589. doi: 10.1111/j.13653040.2009.02060.x

Tester, M. (1988a). Pharmacology of K^+ channels in the plasmalemma of the green alga *Chara corallina*. *J. Membr. Biol.* 103, 159–169. doi: 10.1007/BF018 70946

Tester, M. (1988b). Blockade of potassium channels in the plasmalemma of *Chara corallina* by tetraethylammonium, $Ba2^+$, Na^+ and Cs^+. *J. Membr. Biol.* 105, 77–85. doi: 10.1007/BF01871108

Tester, M. (1988c). Potassium channels in the plasmalemma of *Chara corallina* are multi-ion pores: voltage -dependent blockade by Cs^+ and anomalous permeabilities. *J. Membr. Biol.* 105, 87–94. doi: 10.1007/BF01871109

Tester, M., and Davenport, R. (2003). Na^+ tolerance and Na^+ transport in higher plants. *Ann. Bot.* 91, 503–527. doi: 10.1093/aob/mcg058

Tester, M., Beilby, M. J., and Shimmen, T. (1987). Electrical characteristics of the tonoplast of *Chara corallina*: a study using permeabilised cells. *Plant Cell Physiol.* 28, 1555–1568.

Timme, R. E., Bachvaroff, T. R., and Delwiche, C. H. F. (2012). Broad phylogenomic sampling and the sister lineage of land plants. *PLoS ONE* 7:e29696. doi: 10.1371/journal.pone.0029696

Torn, K., Beilby, M. J., Casanova, M. T., and Al Khazaaly, S. (2014). Formation of extracellular sulphated polysaccharide mucilage on the salt tolerant Characeae. *Int. Rev. Hydrobiol.* 98, 1–9.

Tyerman, S., Beilby, M. J., Whittington, J., Juswono, U., Newman, I., and Shabala, S. (2001). Oscillations in proton transport revealed from simultaneous measurements of net current and net proton fluxes from isolated root protoplasts: MIFE meets patch-clamp. *Austr. J. Plant Physiol.* 28, 591–604.

Tyerman, S., Skerrett, M., Garrill, A., Findlay, G. P., and Leigh, R. A. (1997). Pathways for the permeation of Na^+, and Cl– into protoplasts derived from the cortex of wheat roots. *J. Exp. Bot.* 48, 459–480. doi: 10.1093/jxb/48.Special_Issue.459

Walker, N. A. (1955). Microelectrode experiments on *Nitella*. *Austr. J. Biol. Sci.* 8, 476–489. doi: 10.1071/BI9550476

Wayne, R., and Tazawa, M. (1990). Nature of the water channels in the internodal cells of *Nitellopsis*. *J. Membr. Biol.* 116, 31–39. doi: 10.1007/ BF01871669

Whittington, J., and Bisson, M. A. (1994). Na^+ fluxes in *Chara* under salt stress. *J. Exp. Bot.* 45, 657–665. doi: 10.1093/jxb/45.5.657

Williams, W. D. (1998). Salinity as a determinant of the structure of biological communities in salt lakes. *Hydrobiology* 381, 191–201. doi: 10.1023/A:1003287826503

Williamson, R. E. (1975). Cytoplasmic streaming in *Chara*: a cell model activated by ATP and inhibited by cytochalasin B. *J. Cell Sci.* 17, 655–668.

Winter, U., Soulie-Marsche, I., and Kirst, G. O. (1996). Effects of salinity on turgor pressure and fertility in *Tolypella* (Characeae). *Plant Cell Environ.* 19, 869–879. doi: 10.1111/j.1365-3040.1996.tb00423.x

Wodniok, S., Brinkmann, H., Glockner, G., Heidel, A. J., Philippe, H., Melkonian, M., et al. (2011). Origin of land plants: do conjugating green algae hold the key? *BMC Evol. Biol.* 11:104–114. doi: 10.1186/1471-2148-11-104

Wu, L.-J. (2014). Voltage-gated proton channel HV1 in microglia. *Neuroscientist* 20, 599–609. doi: 10.1177/1073858413519864

Yao, X., and Bisson, M. A. (1993). Passive proton conductance is the major reason for membrane depolarization and conductance increase in *Chara buckelii* in high-salt conditions. *Plant Physiol.* 103, 197–203.

Yao, X., Bisson, M. A., and Brzezicki, L. J. (1992). ATP-driven proton pumping in two species of *Chara* differing in their salt tolerance. *Plant Cell Environ.* 15, 199–210. doi: 10.1111/j.1365-3040.1992.tb01474.x

Ye, Q., Muhr, J., and Steudle, E. (2005). A cohesion/tension model for the gating of aquaporins allows estimation of water channel pore volumes in *Chara*. *Plant Cell Environ.* 28, 525–535. doi: 10.1111/j.1365-3040.2004. 01298.x

Ye, Q., Wiera, B., and Steudle, E. (2004). A cohesion/tension mechanism explains the gating of water channels (aquaporins) in *Chara* internodes by high concentration. *J. Exp. Bot.* 55, 449–461. doi: 10.1093/jxb/erh040

12

Linking salinity stress tolerance with tissue-specific Na^+ sequestration in wheat roots

Honghong Wu[1], Lana Shabala[1], Xiaohui Liu[2,3], Elisa Azzarello[4], Meixue Zhou[1], Camilla Pandolfi[4], Zhong-Hua Chen[2], Jayakumar Bose[1], Stefano Mancuso[4] and Sergey Shabala[1]*

[1] *Faculty of Science, Engineering and Technology, School of Land and Food, University of Tasmania, Hobart, TAS, Australia*
[2] *School of Science and Health, University of Western Sydney, Sydney, NSW, Australia*
[3] *School of Chemical Engineering and Technology, Tianjin University, Tianjin, China*
[4] *Department of Agrifood Production and Environmental Sciences, University of Florence, Florence, Italy*

Edited by:
Mary Jane Beilby, University of New South Wales, Australia

Reviewed by:
Sakiko Okumoto, Virginia Tech, USA
Suleyman I. Allakhverdiev, Russian Academy of Sciences, Russia

***Correspondence:**
Sergey Shabala, Faculty of Science, Engineering and Technology, School of Land and Food, University of Tasmania, Private Bag 54, Hobart, TAS 7001, Australia
e-mail: Sergey.Shabala@utas.edu.au

Salinity stress tolerance is a physiologically complex trait that is conferred by the large array of interacting mechanisms. Among these, vacuolar Na^+ sequestration has always been considered as one of the key components differentiating between sensitive and tolerant species and genotypes. However, vacuolar Na^+ sequestration has been rarely considered in the context of the tissue-specific expression and regulation of appropriate transporters contributing to Na^+ removal from the cytosol. In this work, six bread wheat varieties contrasting in their salinity tolerance (three tolerant and three sensitive) were used to understand the essentiality of vacuolar Na^+ sequestration between functionally different root tissues, and link it with the overall salinity stress tolerance in this species. Roots of 4-day old wheat seedlings were treated with 100 mM NaCl for 3 days, and then Na^+ distribution between cytosol and vacuole was quantified by CoroNa Green fluorescent dye imaging. Our major observations were as follows: (1) salinity stress tolerance correlated positively with vacuolar Na^+ sequestration ability in the mature root zone but not in the root apex; (2) contrary to expectations, cytosolic Na^+ levels in root meristem were significantly higher in salt tolerant than sensitive group, while vacuolar Na^+ levels showed an opposite trend. These results are interpreted as meristem cells playing a role of the "salt sensor;" (3) no significant difference in the vacuolar Na^+ sequestration ability was found between sensitive and tolerant groups in either transition or elongation zones; (4) the overall Na^+ accumulation was highest in the elongation zone, suggesting its role in osmotic adjustment and turgor maintenance required to drive root expansion growth. Overall, the reported results suggest high tissue-specificity of Na^+ uptake, signaling, and sequestration in wheat roots. The implications of these findings for plant breeding for salinity stress tolerance are discussed.

Keywords: bread wheat, cytosolic Na^+, Na^+ distribution, root zones, salinity stress tolerance, vacuolar Na^+ sequestration

INTRODUCTION

More than 800 million hectares (6%) of land are affected by salinity worldwide (Munns and Tester, 2008). As sodium is one of the most abundant metal elements, sodium salts dominate in many saline soils of the world (Rengasamy, 2006). High concentrations of salts in soils account for large decreases in the yields of a wide variety of crops all over the world (Tester and Davenport, 2003). In the light of predicted population growth to 9.3 billion by 2050 (Lee, 2011), global food requirements are expected to increase by 70–110% (Tilman et al., 2011). Wheat is one of the most important crops providing nearly 55% of the consumed carbohydrates world widely (Gupta et al., 1999) but is not highly salt tolerant, and its commercial production is substantially reduced as the soil salinity level rises to 100 mM NaCl and is not possible in soils containing more than 250 mM NaCl (Munns et al., 2006; Munns and Tester, 2008). Thus, improving salinity stress tolerance in wheat is an urgent task to cope with the possible shortage of food supply in the near future. This is especially true for the hexaploid bread wheat that composes about 95% of all wheat grown world wide (Shewry, 2009).

Sodium uptake and sequestration has always been in spotlight of researchers aimed at finding the traits or genes which can be selected to improving salinity tolerance in wheat. Early studies using $^{22}Na^+$ isotopes showed that salt tolerant wheat varieties have significantly lower Na^+ accumulation in the shoot (Davenport et al., 1997), suggesting an efficient Na^+ exclusion mechanism. The following studies by Munns and colleagues (Munns et al., 2002, 2003, 2006; Munns and James, 2003; Lindsay et al., 2004) suggested that targeting Na^+ exclusion from shoot was a promising way to improving salinity tolerance in this species. Indeed, under saline condition, flag leaf Na^+ was significantly reduced from 326 mM in commercial variety Tomaroi to

87 mM in transgenic Tamaroi plants that expressed *TmHKT1;5-A* gene enabling Na^+-retrieval from the xylem (Munns et al., 2012). This has resulted in about 20% increase in wheat yield under saline field conditions (from 1.30 to 1.61 tons per hectare). Using microelectrode ion flux measuring MIFE technique, Cuin et al. (2011) found that Kharchia 65 (accepted as a "standard" for salinity tolerance in wheat by most breeders) had the highest root Na^+ exclusion ability compared with other seven wheat varieties studied. Pharmacological experiments and experiments with transgenic Arabidopsis mutants have confirmed that this Na^+ efflux was mediated by the plasma membrane SOS1 Na^+/H^+ antiporter. Similar studies conducted on sorghum (Yang et al., 1990), maize (Fortmeier and Schubert, 1995), and tomato (Al-Karaki, 2000) have also suggested that plant's ability to exclude Na^+ from uptake in these species was positively correlated with the overall salinity tolerance. Even in lower plants the ability to avoid the accumulation of Na^+ in cytosol is critical for its salinity tolerance (e.g., cyanobacteria; Allakhverdiev et al., 2000; Allakhverdiev and Murata, 2008).

The above beneficial effects of Na^+ exclusion from uptake was always attributed to its toxic effect on cell metabolism (Maathuis and Amtmann, 1999; Munns and Tester, 2008) and essentiality to maintain low level of Na^+ in the cytosol. However, the same goal may be achieved by the efficient Na^+ sequestration in the vacuole. The latter trait is commonly employed by halophytes (naturally salt-loving plants; Flowers and Colmer, 2008; Shabala and Mackay, 2011; Shabala, 2013; Bonales-Alatorre et al., 2013a,b), and some evidences were presented that salt-tolerant wheat varieties may also possess better vacuolar Na^+ sequestration ability (e.g., Saqib et al., 2005). However, most of these studies were conducted on leaves, while the role of Na^+ sequestration in roots received less attention. In *Thellungiella salsuginea*, a halophytic relative of *Arabidopsis thaliana*, Oh et al. (2009) showed that vacuolar Na^+ fluorescence intensities in cortex cells of root tip region is higher in *thsos1-4* than in wild type. However, to the best of our knowledge, the issue of tissue-specificity of vacuolar Na^+ sequestration between different root zones, and its link with the overall salinity tolerance, has never received a proper attention, neither in wheat nor in any other crop species.

The root anatomy and functional structure can be generally divided into four different zones: (1) root meristem, (2) the distal elongation (or transition) zone, (3) elongation zone, and (4) mature zone (Verbelen et al., 2006; Baluška and Mancuso, 2013). So far, most studies of Na^+ distribution in plant roots under salt stress was conducted either at the level of whole root (e.g., Matsushita and Matoh, 1991; Flowers and Hajibagheri, 2001; Rus et al., 2001), or were focused on cell-type-specific Na^+ distribution in roots (Huang and van Steveninck, 1988; Storey et al., 2003; Oh et al., 2009, 2010). While these and some other (Cuin et al., 2011; Li et al., 2012) papers showed a heterogeneity of Na^+ distribution within the root, none of them discussed the difference in Na^+ patterning between intracellular compartments within functionally different root zones. In *T. salsuginea*, Na^+ accumulated inside the pericycle in *thsos1-4* mutant, while in the wild type it was confined in vacuoles of epidermal and cortical cells (Oh et al., 2009). Cell-type-specific Na^+ distribution patterns in hypodermis, cortex, endodermis, and pericycle were also studied in salinized grapevines using X-ray microscopy method (Storey et al., 2003). Here, vacuolar Na^+ was sequestered predominantly in endodermis and pericycle cells. However, to the best of our knowledge, no study has compared Na^+ distribution between cytosol and vacuole in functionally different root zones within the same tissue, at least in bread wheat.

In the present work, variability of vacuolar Na^+ sequestration in four different root zones under salt stress was studied using six bread wheat varieties contrasting in their salinity tolerance. Cytosolic and vacuolar Na^+ content in different root zones was quantified by CoroNa Green fluorescent dye imaging, and the link between tissue-specific vacuolar Na^+ sequestrations in specific root zones and the overall salinity stress tolerance was explored. We report that the overall salinity stress tolerance correlates positively with vacuolar Na^+ sequestration ability in the mature root zone but not in the root apex. At the same time, cytosolic Na^+ levels in root meristem were significantly higher in salt tolerant than sensitive group, suggesting that meristem cells may play a role of the "salt sensor." The overall Na^+ accumulation was highest in the elongation zone, suggesting its role in osmotic adjustment and turgor maintenance required to drive root expansion growth.

MATERIALS AND METHODS

PLANT MATERIALS AND GROWTH CONDITIONS

Six bread wheat (*Triticum aestivum*) varieties contrasting in their salinity tolerance (tolerant – Persia 118, Cranbrook, and Westonia; sensitive – Iran 118, Belgrade 3, and 340) were used in this study. All seeds were obtained from the Australian Winter Cereals Collection and multiplied in our laboratory. Plants were grown in February–March 2013 in the glasshouse facilities at the University of Tasmania essentially as described in Chen et al. (2007b). Twelve seeds for each variety were sown in 4.5 L PVC pots with the standard potting mix by triplicates. After emerging (roughly 6 days), salt treatment (300 mM NaCl) were applied for about 5 weeks. Plants were irrigated twice per day by an automatic watering system with dripper outlets, and were uniformly thinned to eight plants in each pot after roughly 10 days sowing. A saucer was placed under each pot. For confocal imaging experiments, seeds were sterilized with 5% commercial bleach for 15 min, and then washed thoroughly by the running tap water for half an hour. Seeds were then germinated in wet paper rolls in growth chambers at 23 ± 1°C (16 h light/8 h dark regime). Four days old wheat roots were treated with 100 mM NaCl for 72 h, and then stained with CoroNa Green dye for the LSCM (laser scanning confocal microscopy) measurements as described below.

WHOLE-PLANT PERFORMANCE ASSESSING

Before harvesting, plants "damage index" were scored on zero to 10 scale (0, no visual symptoms of stress; 10, dead plants; **Supplementary Figure S1**). The higher damage index score represents the lower salt tolerance. Then, the stem was cut 1 cm above the ground, and shoot fresh weight (FW) was measured.

CONFOCAL LASER SCANNING MICROSCOPY MEASUREMENTS

Measurements of cytosolic and vacuolar Na^+ content in wheat root cells using the green fluorescent Na^+ dye CoroNa Green acetoxymethyl ester were essentially as described in Bonales-Alatorre et al. (2013b). The dye has absorption and fluorescence emission maxima of approximately 492 and 516 nm, respectively. The dye was reconstituted as a stock with anhydrous dimethyl sulfoxide before use. The CoroNa Green indicator stock was added to 5 mL of measuring buffer (10 mM KCl, 5 mM Ca^{2+}-MES, pH 6.1) and diluted to a final concentration of 15 mM. Ten millimeters-long root segments were cut from the apical (the first 10 mm from the apex) and mature (30–40 mm from the apex) root zones and incubated for 2 h in the dark in a solution containing 20 μM CoroNa Green. After incubation, the samples were rinsed in a buffered MES solution and examined using confocal microscopy. Confocal imaging was performed using an upright Leica Laser Scanning Confocal Microscope SP5 (Leica Microsystems, Germany) equipped with a 40× oil immersion objective. To analyze sodium intracellular localisation CoronaNa Green AM (Molecular Probes, USA) was used. The excitation wavelength was set at 488 nm, and the emission was detected at 510–520 nm. Comparison of different levels of fluorescence between cells was carried out by visualizing cells with the identical imaging settings of the confocal microscope (i.e., exposure times, laser intensity, pinhole diameter and settings of the imaging detectors). Images were analyzed with LAS Lite software. Six to eight individual roots (each from different plants) were used; and at least two images were taken for each root zone. For analysis, several lines were drawn across the so-called "region of interest" (ROI; **Figure 1A**) in an appropriate root zone. Continuous fluorescence intensity distribution profiles (quantified in arbitrary units by LAS software) were then obtained and plotted in an Excel file (**Figure 1B**). The mean fluorescence intensity values for cytosolic and vacuolar compartments were then calculated for each cell by attributing signal profiles to root morphology (visualized by light microscopy images). The data was then averaged for all cells measured for the same treatment. The background signal was measured from the empty region and then subtracted from the readings, to obtain corrected fluorescence values. Depending on the root zone, readings from between 70 and 300 cells were averaged and reported for each genotype.

A further validation of the above protocol was conducted using roots co-stained with CoroNa Green-AM and FM4-64, a dye that stains both plasma and vacuolar membranes (Oh et al., 2010; Bassil et al., 2011) and allow a better resolution between intracellular compartments. After 1 h incubation with 20 μM CoroNa Green-AM (as described above), the same root samples were then incubated together with 20 μM FM4-64 for another 1 h to visualize tonoplast. Roots were then rinsed with buffer solution (10 mM KCl, 5 mM Ca^{2+}-MES, pH 6.1) for 3 min and analyzed using confocal imaging facilities. For FM4-64 fluorescence, the 488-nm excitation line was used and collected with a 615-nm long-pass filter. Results of this experiment are illustrated in **Figure 2** showing intracellular Na^+ distribution in the transition zone of cultivar Persia 118. **Figure 2A** shows distribution of CoroNa Green and FM4-64 in co-stained root samples, while **Figures 2B,C** show the same root stained with FM4-64 and CoroNa Green, respectively. Two cells having multiple vacuoles (circled) were then selected for analysis (depicted in **Figures 2D,E**). Two ROI lines were then drawn crossing two vacuoles in each of the cells. Intracellular Na^+ distribution was then quantified (**Figure 2F**). As one can see, two major peaks in each cell correspond to two vacuoles, while troughs report values for cytosolic Na^+.

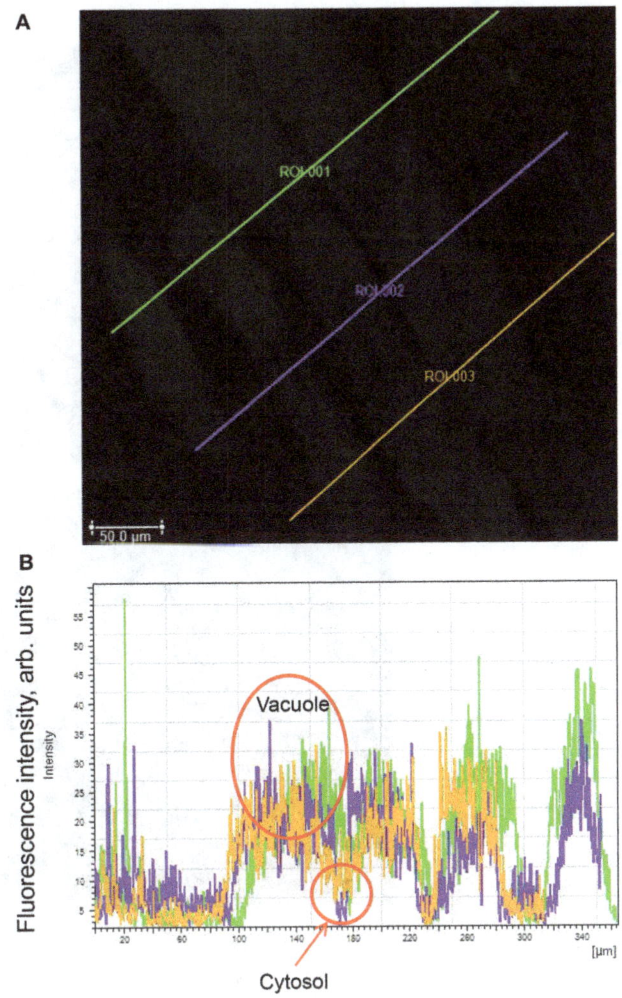

FIGURE 1 | Illustration of the quantification procedure for Na^+ distribution between the cytosol and the vacuole. (A) Several lines are drawn across the so-called "region of interest" (ROI) in an appropriate root zone. **(B)** Continuous fluorescence intensity distribution profiles obtained by LAS software. The mean fluorescence intensity values for cytosolic and vacuolar compartments are then calculated for each cell by attributing signal profiles to root morphology (visualized by light microscopy images; not shown in a figure). The shown image was obtained from mature region of Persia 118 cultivar.

STATISTICAL ANALYSIS

All data were analyzed by using SPSS 20.0 for windows (SPSS Inc., Chicago, IL, USA). Comparison of cytosolic or vacuolar Na^+ fluorescent intensity between different varieties in root zones was done by One-Way ANOVA based on Duncan's multiple range test. Different lowercase letters represent significant difference between varieties. Data with the same lowercase letters are not significantly different at $P < 0.05$. Comparison of cytosolic or vacuolar Na^+ fluorescent intensity between tolerant and sensitive groups in different root zones was done by independent samples

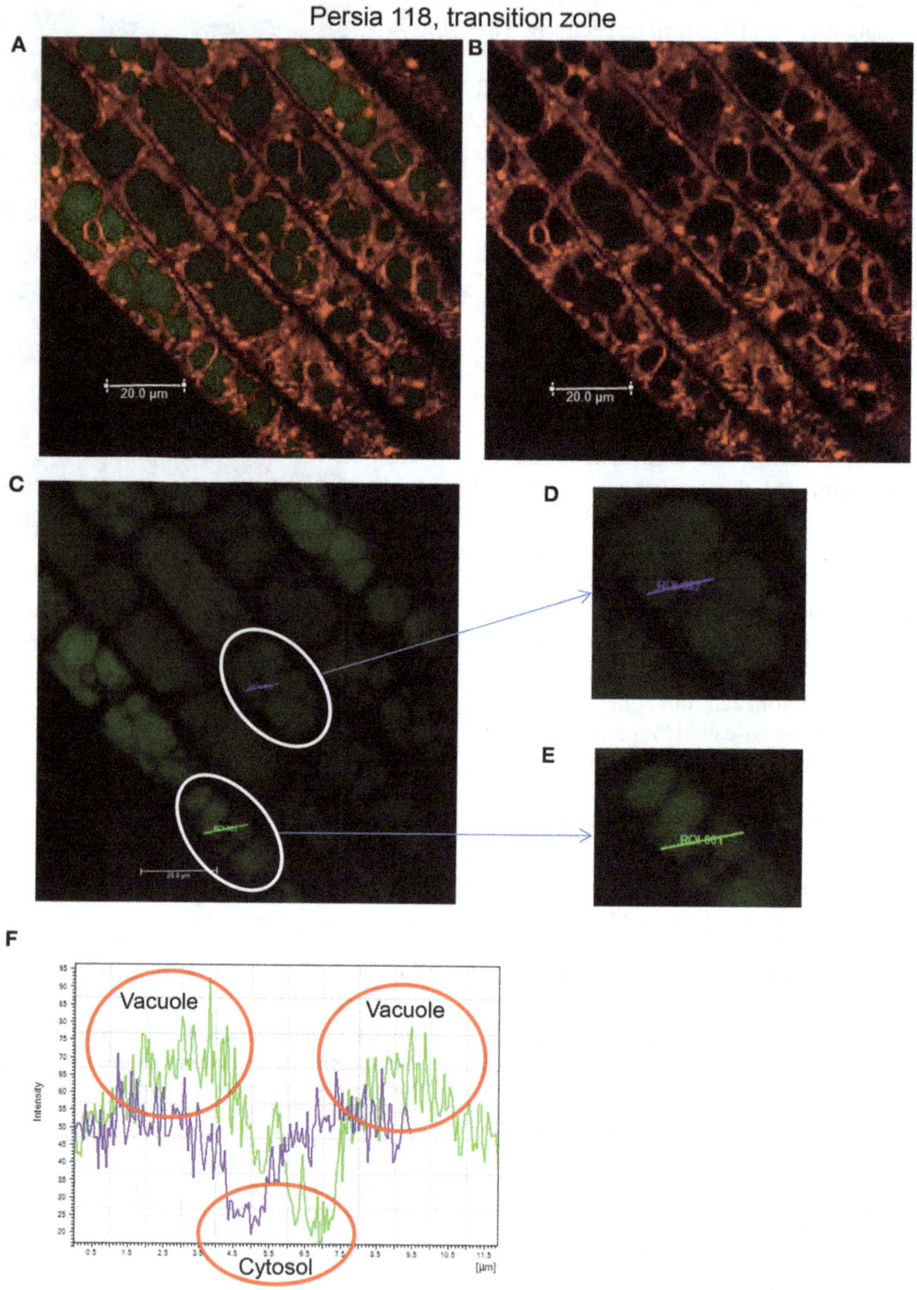

FIGURE 2 | Quantifying intracellular Na^+ distribution between the cytosol and the vacuole by double staining procedure. Intracellular Na^+ distribution is illustrated using the transition zone of cultivar Persia 118 as an example. **(A)** The root transition zone co-stained with Corona Green AM and FM4-64. The same root stained with FM4-64 dye **(B)** and CoroNa Green **(C)**. Two cells having multiple vacuoles (circled) were then selected for analysis (depicted in **D**,**E**). Two lines were then drawn across the "region of interest" (ROI) for each of these, crossing two vacuoles in each of cells. Intracellular Na^+ distribution was then quantified **(F)**.

t-test. The significance levels are $^{*}P < 0.05$, $^{**}P < 0.01$, and $^{***}P < 0.001$.

RESULTS

WHOLE-PLANT PERFORMANCE

Bread wheat varieties used in this study showed big variability in salinity stress tolerance. Salinity damage index (a measure of salt tolerance; see **Supplementary Figures S1A,B** for details) ranged from the highest (most sensitive) 5.5 ± 0.5 in variety 340 to the lowest (most tolerant) 2.3 ± 0.4 in variety Persia 118 (significant at $P < 0.01$; **Figure 3A**). Similarly, the highest shoot fresh weight (FW) was found in Persia 118 (1.5 ± 0.1 g/plant), while variety 340 showed the lowest shoot FW (0.6 ± 0.2 g/plant) (**Figure 3B**). Accordingly, all varieties were grouped into tolerant and sensitive clusters (**Figures 3C,D**). The tolerant cluster showed much less damage (about two-fold; $P < 0.05$, **Figure 3C**) and 70% higher

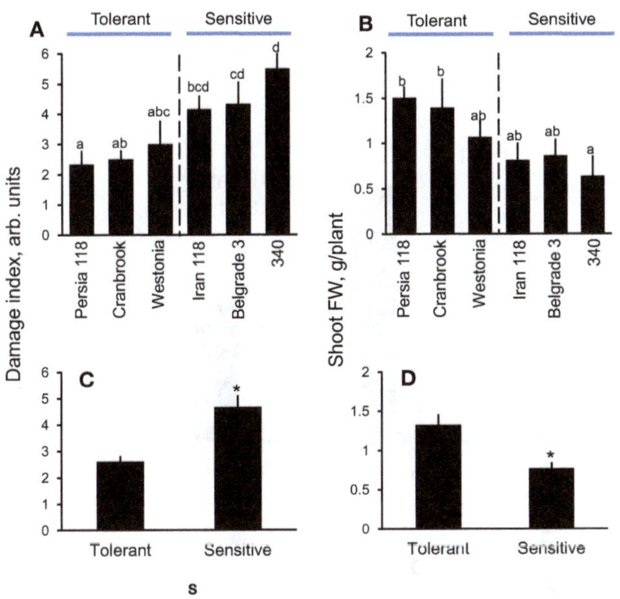

FIGURE 3 | Genetic variability of salinity stress tolerance in bread wheat. Damage index **(A)** and shoot fresh weight **(B)** of **six** bread wheat cultivars contrasting in their salinity tolerance after 38 days of 300 mM NaCl treatment in a glasshouse. Different lowercase letters represent significant differences between varieties at $P < 0.05$. **(C,D)** Average pooled data for tolerant and sensitive clusters shown in **A,B**. Mean ± SE ($n = 3$; 24 plants in total, eight plants in each pots). Asterisk indicates significant difference between clusters at $P < 0.05$.

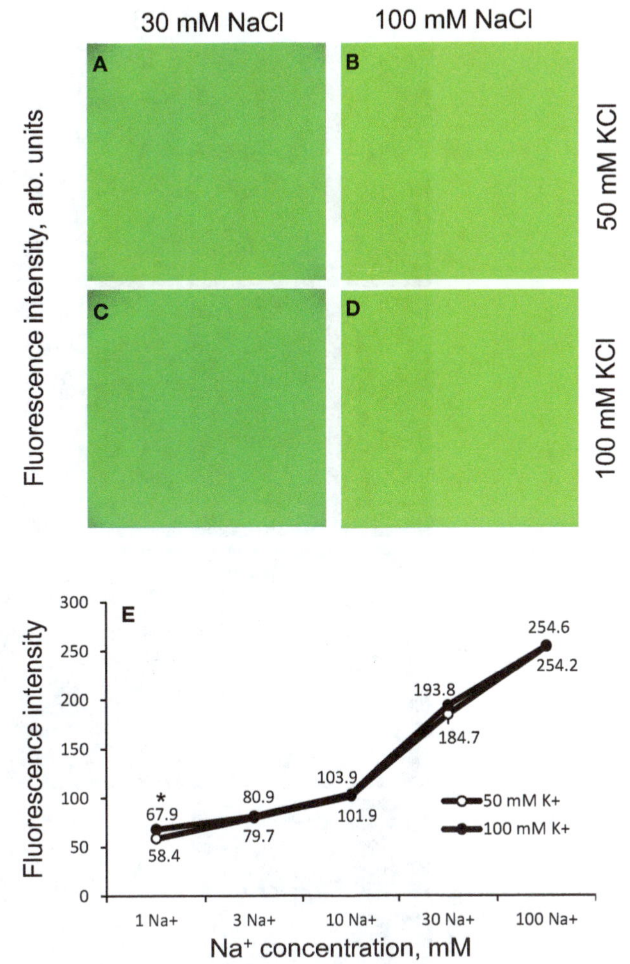

FIGURE 4 | Methodological aspects of CoroNa Green dye calibration. CoroNa Green dye was calibrated *in vitro* in a buffer solution containing different Na^+ and K^+ concentrations. **(A–D)** *In vitro* fluorescence images for two Na^+ and two K^+ concentrations used in experiment. One (of four) representative images is shown for each Na^+ and K^+ concentration; **(E)** dose-dependency of Na^+ fluorescence signals at 50 and 100 mM KCl background. Mean ± SE ($n = 60$; four images, 15 replicates recorded from each image). Asterisk indicates a significant (at $P < 0.05$) difference between K^+ treatments. Please note that in most cases the error bar is smaller than the symbol *per se*.

shoot biomass (significant at $P < 0.05$, **Figure 3D**) compared with the sensitive cluster. A significant negative correlation ($r^2 = 0.91$, $P < 0.01$) was found between shoot FW and damage index among all the varieties (**Supplementary Figure S1C**).

SODIUM ACCUMULATION PROFILES

Before cell- and genotype-specific Na^+ distribution was quantified, a series of methodological experiments was conducted to eliminate possible confounding effects of dye loading and stress-induced changes in intracellular ionic conditions on fluorescence measurements. First, CoroNa Green was calibrated in a cytosol-like solution (50–100 mM K^+; <1 μM Ca^{2+}; pH = 7.2) in a broad range of Na^+ concentrations (1, 3, 10, 30, and 100 mM) in *in vitro* experiments. As shown in **Figure 4**, a two-fold drop in background K^+ concentration from 100 to 50 mM (mimicking NaCl-induced reduction in cytosolic K^+ under stress conditions; Shabala et al., 2006) had no significant ($P < 0.05$) impact on fluorescence signal except the lowest (1 mM NaCl) concentration, which is well-below expected levels for cytosolic Na^+ (Munns and Tester, 2008). Calibration characteristics were also insensitive to pH in physiological (5–7.2) pH range (data not shown). Thus, the possible difference in K^+ retention ability or stress-induced changes in intracellular pH between genotypes had no confounding effects on CoroNa Green readings. Dye loading profiles were uniform between various Z-plains (**Supplementary Figure S2**) suggesting that 2 h of loading was sufficient to ensure its homogenous uptake by most cells.

Plants grown in Na-free solution (Milli-Q water) showed negligible small Na^+ fluorescence signals (**Figure 5**) while root exposure to 100 mM NaCl for 72 h resulted in massive accumulation of Na^+ in root tissues (**Figure 5**).

When the method was applied to all salt-grown plants, sodium distribution within the root showed a clear pronounced tissue- and genotype-specificity. The representative images for one tolerant (Persia 118) and one sensitive (Iran 118) varieties for each of four measured zones (meristem, transition, elongation, and mature zone) are shown in **Figure 6**, and the average amounts of Na^+ in cell vacuole and cytosol in each variety are quantified in **Figures 7–10**.

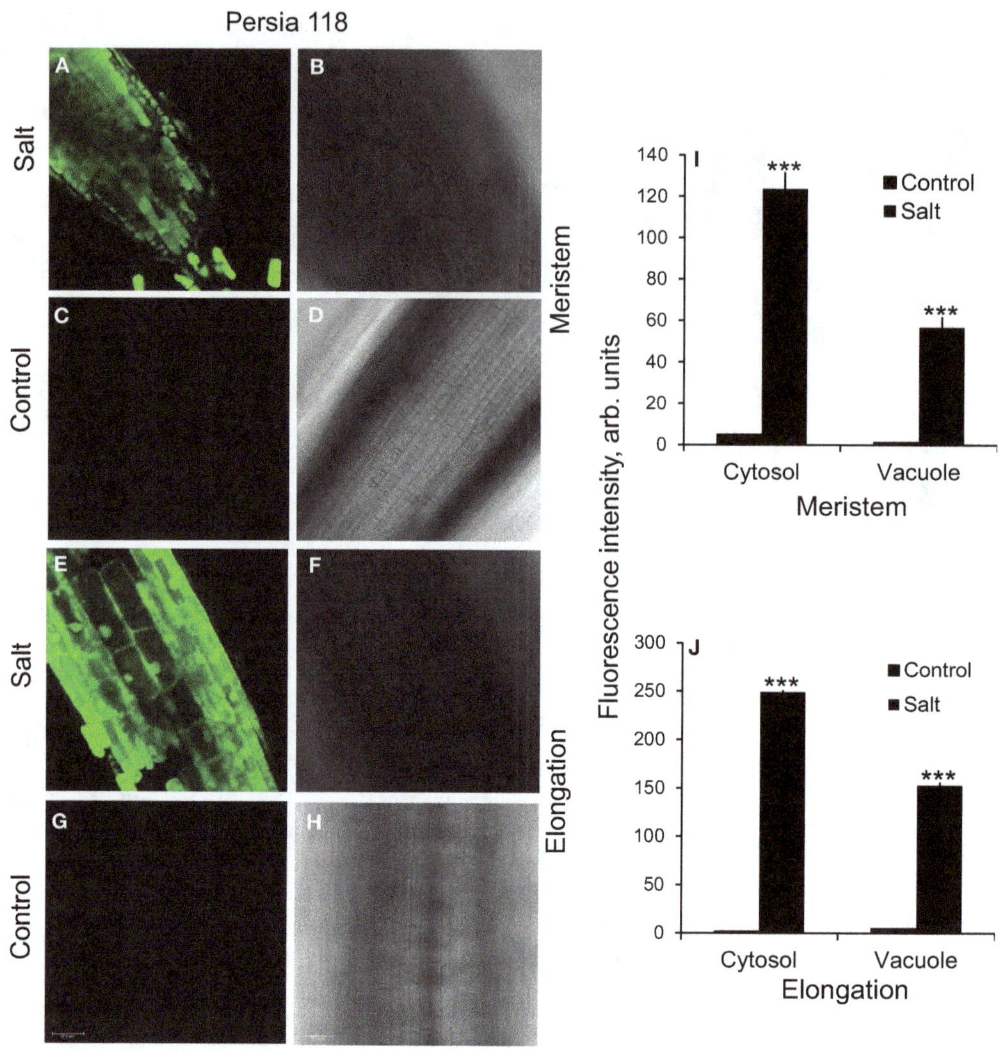

FIGURE 5 | The comparison of Na^+ fluorescence intensity in salt-grown and Na-free roots. (A,C,E,G) Na^+ fluorescence intensity was measured in root of Persia 118 cultivar grown in the presence (100 mM NaCl for 72 h) and absence (milliQ water) of Na^+ in the growth media. One (of 12) representative images is shown for each zone. **(B,D,F,H)** Light images on the appropriate slides shown on the left. **(I,J)** Quantification of cytosolic and vacuolar Na^+ in salt-grown and Na-free roots. Mean ± SE ($n = 42$–96 cells from six roots). ***indicates significant differences at $P < 0.001$.

ROOT MERISTEM

Significantly higher quantities of Na^+ were accumulated in the cytosol of meristematic cells in a tolerant compared with sensitive cluster (**Figure 7**). Here, cytosolic Na^+ intensity ranged from the highest 179.9 ± 7.7 (in tolerant Westonia) to the lowest 20.3 ± 1.2 (in sensitive Belgrade 3), declining in a sequence Westonia > Cranbrook > Persia 118 > Iran 118 > 340 > Belgrade 3 (**Figure 7A**). Overall, the amount of Na^+ stored in the cytosol of cells in the meristem region of the tolerant cluster was 4.3-fold higher compared with the sensitive cluster (159.2 ± 17.9 vs. 37.0 ± 8.4; $P < 0.01$; **Figure 7C**) and also significantly ($P < 0.05$) higher than vacuolar Na^+ content (**Figures 7A,B; Supplementary Figure S3**). At the same time, vacuolar Na^+ intensity in root meristem zone ranged from the highest 108.9 ± 4.4 (in sensitive Iran 118) to the lowest 56.7 ± 4.9 (in tolerant Persia 118), declining in a sequence Iran 118 > 340 > Cranbrook > Westonia > Belgrade 3 > Persia 118 (**Figure 7B; Supplementary Figure S3**). Overall, no significant (at $P < 0.05$ level) difference was found in vacuolar Na^+ in meristem cell vacuoles among two contrasting clusters (**Figure 7D**). Furthermore, a significant ($P = 0.05$) negative correlative relationship ($r^2 = 0.66$) was found between cytosolic Na^+ intensity in root meristem zone and salinity-induced damage index (**Supplementary Figure S4A**), while a weak positive correlation ($r^2 = 0.22$; $P > 0.05$) was found between the vacuolar Na^+ intensity in root meristem zone and the damage index (**Supplementary Figure S4E**).

TRANSITION ZONE

Vacuolar Na^+ intensity in the root transition zone was higher than cytosolic Na^+ intensity, in both tolerant and sensitive clusters (illustrated in **Figures 6E,G** and quantified in **Figure 8**).

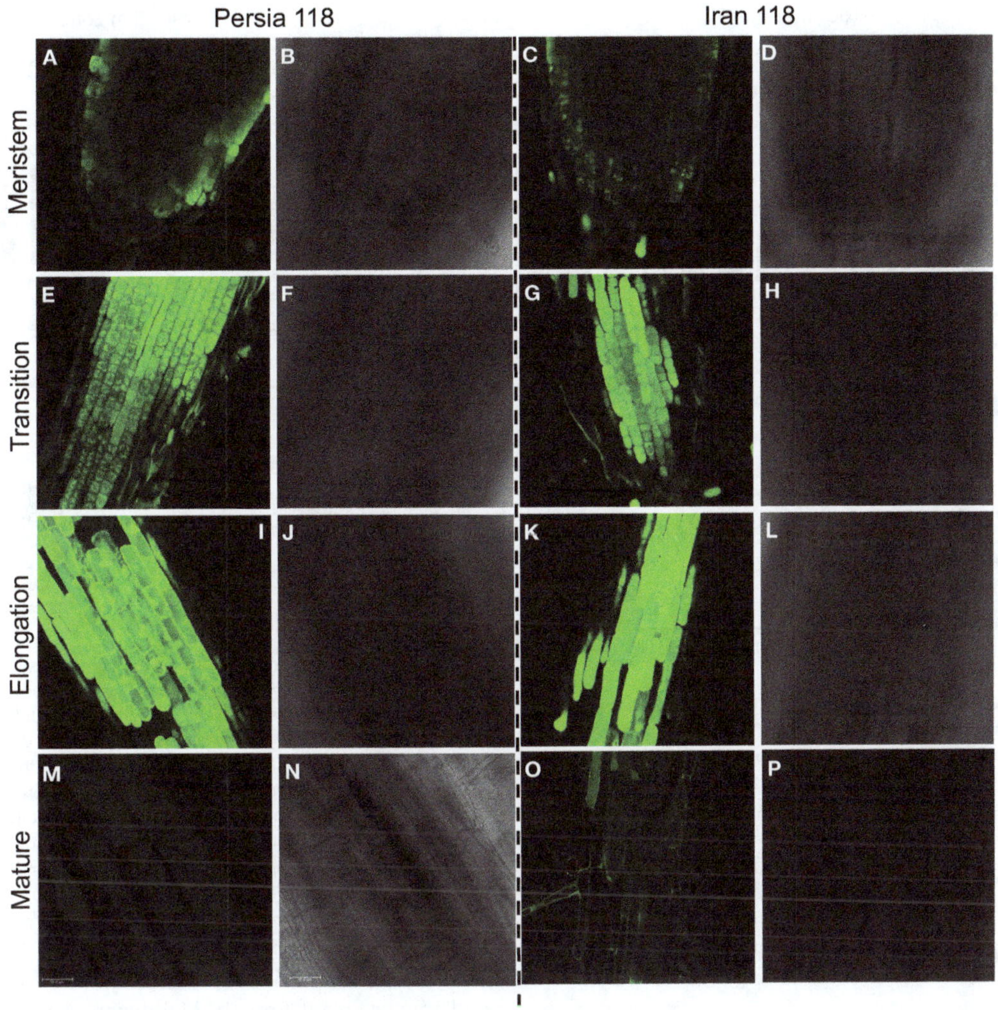

FIGURE 6 | Na^+ accumulation profiles in bread wheat visualized by CoroNa Green dye imaging. The representative images of Na^+ distribution within one tolerant (Persia 118) and one sensitive (Iran 118) varieties are shown for four different root zones: meristem **(A,C)**; transition zone **(E,G)**; elongation zone **(I,K)**, and mature zone **(M,O)**. One of 12 representative images is shown for each zone. Panels next to fluorescence images represent respective light images **(B,D,F,H,J,L,N,P)**.

Belgrade 3 showed the lowest (37.6 ± 2.3) cytosolic Na^+ intensity in root transition zones, while Cranbrook showed the highest (227.7 ± 5.0) (**Figure 8A**). Highest vacuolar Na^+ intensity was reported for Persia 118 (209.5 ± 6.2), and the lowest for Belgrade 3 (111.9 ± 5.8) (**Figure 8B**). Overall, no clear trend was observed in Na^+ distribution between cytosol and vacuole in the root transition zone. As a result, no significant difference in either cytosolic nor vacuolar Na^+ intensity was found between salt tolerant and sensitive clusters here (**Figures 8C,D**). Only very modest ($r^2 = 0.33$; $P = 0.23$) negative correlation was found between cytosolic Na^+ intensity and salt damage index (**Supplementary Figure S4B**), while no correlation was found between vacuolar Na^+ intensity in root transition zone and damage index ($r^2 = 0.06$, **Supplementary Figure S4F**).

ELONGATION ZONE

Cells in root elongation zone had slightly higher amounts of Na^+ in the cytosol compared with vacuoles (**Figures 6I,K**), both in tolerant and sensitive clusters (**Figure 9**). Belgrade 3 showed the lowest cytosolic Na^+ intensity (68.4 ± 4.7), while the highest values were reported for Persia 118 (248.9 ± 2.0) (**Figure 9A**). Vacuolar Na^+ intensity was the highest in a tolerant Westonia (250.4 ± 1.7) and the lowest in a sensitive variety 340 (145.4 ± 3.3) (**Figure 9B**). Overall, no clear trends were observed in Na^+ distribution between cytosol and vacuole in root elongation zone in six varieties (**Figure 9**), and no significant difference in either cytosolic or vacuolar Na^+ intensity was found between two contrasting clusters (**Figures 9C,D**). No significant (at $P <$ 0.05 level) correlation was found between the damage index and cytosolic Na^+ intensity in root elongation zone ($r^2 = 0.02$, **Supplementary Figure S4C**), as well as for vacuolar Na^+ intensity ($r^2 = 0.13$; $P = 0.48$) (**Supplementary Figure S4G**).

MATURE ZONE

Cytosolic Na^+ in root mature zone was significantly lower in salt-tolerant compared with salt-sensitive cluster while for vacuolar Na^+ this trend was inverse (**Figures 6M,O, 10**). Another important trend was that the overall amount of Na^+ accumulated

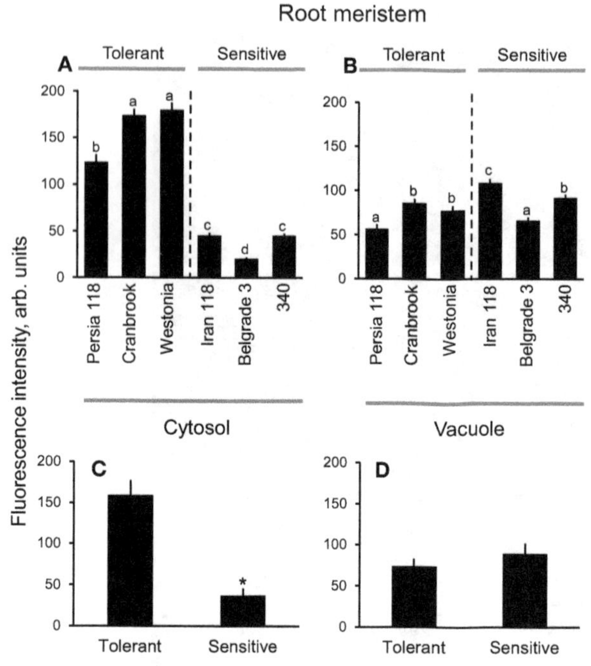

FIGURE 7 | Na^+ accumulation and patterning between cytosol and vacuole in meristematic root zone in bread wheat genotypes. Intensity of CoroNa Green fluorescence in cytosolic **(A)** and vacuolar **(B)** compartments (arb. units) in root meristem of six bread wheat varieties contrasting in their salinity tolerance. Mean ± SE ($n = 72$–96 cells from at least six individual plants). Different lowercase letters represents significant difference between varieties at $P < 0.05$. **(C,D)** Averaged pooled values for cytosolic **(C)** and vacuolar **(D)** Na^+ intensity for salt-tolerant and salt-sensitive clusters shown in **A**,**B**. Mean ± SE ($n = 215$ to 300; three varieties × 72–96 cells analyzed for each variety). Asterisk indicates significant difference between clusters at $P < 0.01$.

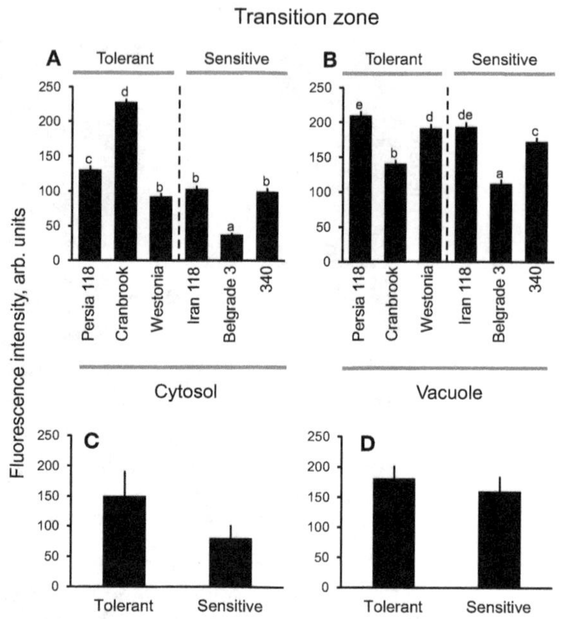

FIGURE 8 | Na^+ accumulation and patterning between cytosol and vacuole in transition root zone in bread wheat genotypes. Intensity of CoroNa Green fluorescence in cytosolic **(A)** and vacuolar **(B)** compartments (arb. units) in transition root zone of six bread wheat varieties contrasting in their salinity tolerance. Mean ± SE ($n = 72$–96 cells from at least six individual plants). Different lowercase letters represents significant difference between varieties at $P < 0.05$. **(C, D)** Averaged pooled values for cytosolic **(C)** and vacuolar **(D)** Na^+ intensity for salt-tolerant and salt-sensitive clusters shown in **A**,**B**. Mean ± SE ($n = 215$ to 300; three varieties × 72–96 cells analyzed for each variety).

in the cytosol was much less compared with any other zone (**Figure 6**). Cytosolic Na^+ intensity in root mature zone was highest in sensitive Iran 118 (99.6 ± 15.1) and lowest in tolerant Persia 118 (13.1 ± 3.5) (**Figure 10A**). Salt-tolerant cultivar Westonia had the highest vacuolar Na^+ intensity (149.0 ± 19.0), while salt-sensitive Belgrade 3 had the lowest (7.6 ± 1.3) (**Figure 10B**). On average, cytosolic Na^+ intensity in mature root zone was three-fold lower in tolerant (25.4 ± 7.7) than in sensitive (73.0 ± 14.6) variety (**Figure 10C**; significant at $P < 0.05$). At the same time, vacuolar Na^+ intensity in mature root zone in tolerant group was eight-fold higher compared with sensitive group (111.4 ± 22.2 vs. 13.9 ± 4.3; significant at $P < 0.01$; **Figure 10D**). Overall, a positive correlation ($r^2 = 0.57$; $P = 0.08$) was found between plant damage index and cytosolic Na^+ intensity in mature root zone (**Supplementary Figure S4D**), while for vacuolar Na^+ intensity this correlation was negative ($r^2 = 0.59$; $P = 0.08$; **Supplementary Figure S4H**).

DISCUSSION

VACUOLAR Na^+ SEQUESTRATION IN MATURE ROOT ZONE BUT NOT ROOT APEX CORRELATES WITH SALINITY TOLERANCE IN BREAD WHEAT

For glycophytes such as bread wheat, Na^+ is not considered to be an essential nutrient (Maathuis, 2014) and, when present in excessive quantities in soil, leads to cytosolic Na^+ toxicity, and impairs plant growth. Maintenance of the optimal cytosolic Na^+ level under saline conditions requires effective exclusion of Na^+ from the cytosol, either back to external media, or into vacuole (Maathuis and Amtmann, 1999; Blumwald, 2000).

In the present work, we have investigated Na^+ distribution between the cytosol and the vacuole in four different root zones in contrasting bread wheat varieties exposed to salinity stress. Surprisingly, vacuolar Na^+ sequestration was correlated positively with salinity tolerance *only* in mature root zone. In contrast, no significant difference in vacuolar Na^+ sequestration patterns was found between salt tolerant and sensitive clusters in either of other three zones: transition (**Figures 8B,D**), elongation (**Figures 9C,D**), or meristem (**Figures 7C,D**). At the same time, significantly lower cytosolic Na^+ intensity was found in salt tolerant compared with sensitive cluster in mature zone (**Figures 10A,B**). Taken together these results suggest that the ability of mature root cell vacuoles to sequester excessive Na^+ is one of the key determinants of salinity tolerance in bread wheat.

It should be noted that, given the non-linearity of the calibration curve (**Figure 4**), the fluorescent intensities measured might not linearly relate to the absolute sodium concentrations in plant tissues, so some caution is needed while interpreting the presented data in quantitative terms.

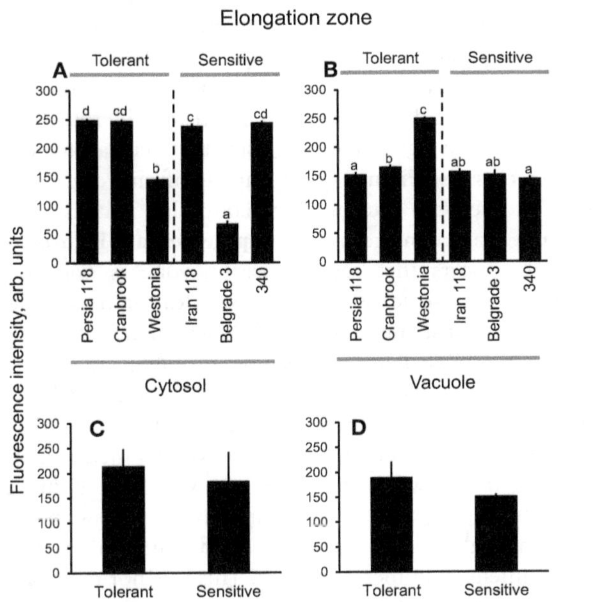

FIGURE 9 | Na^+ accumulation and patterning between cytosol and vacuole in elongation root zone in bread wheat genotypes. Intensity of CoroNa Green fluorescence in cytosolic **(A)** and vacuolar **(B)** compartments (arb. units) in root elongation zone of six bread wheat varieties contrasting in their salinity tolerance. Mean ± SE ($n = 72$–96 cells from at least six individual plants). Different lowercase letters represents significant difference between varieties at $P < 0.05$. **(C,D)** averaged pooled values for cytosolic **(C)** and vacuolar **(D)** Na^+ intensity for salt-tolerant and salt-sensitive clusters shown in **A,B**. Mean ± SE ($n = 215$–300; three varieties × 72–96 cells analyzed for each variety).

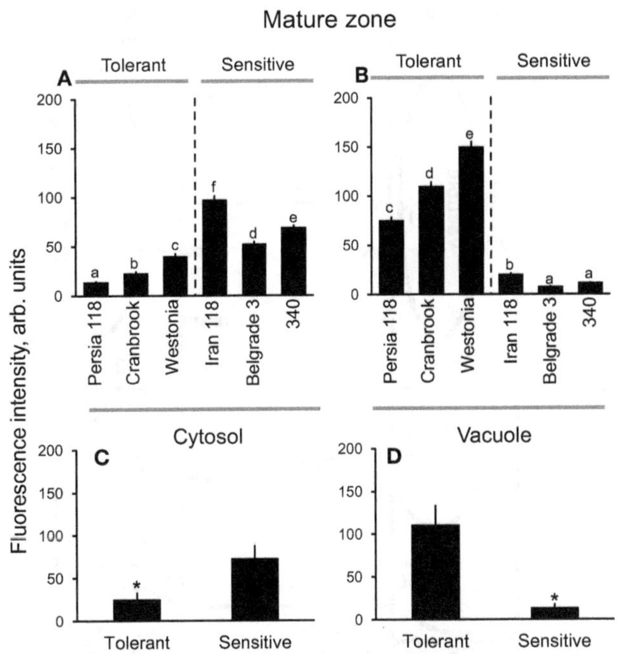

FIGURE 10 | Na^+ accumulation and patterning between cytosol and vacuole in mature root zone in bread wheat genotypes. Intensity of CoroNa Green fluorescence in cytosolic **(A)** and vacuolar **(B)** compartments (arb. units) in mature root zone of six bread wheat varieties contrasting in their salinity tolerance. Mean ± SE ($n = 72$–96 cells from at least six individual plants). Different lowercase letters represents significant difference between varieties at $P < 0.05$. **(C,D)** Averaged pooled values for cytosolic **(C)** and vacuolar **(D)** Na^+ intensity for salt-tolerant and salt-sensitive clusters shown in **A, B**. Mean ± SE ($n = 240$–320; three varieties × 72–96 cells analyzed for each variety). Asterisk indicates significant difference between clusters at $P < 0.05$.

We have used term "surprisingly" above because root apex is a house for most metabolically active cells and, as such, has often considered as a potential target for many abiotic stresses such as aluminium toxicity (Doncheva et al., 2005), oxidative stress (Demidchik et al., 2007), and heavy metal toxicity (Halušková et al., 2009). It was also unexpected as SOS1 Na^+/H^+ exchangers that remove Na^+ from uptake are believed to be expressed predominantly in the root apex (Shi et al., 2002). Given the fact that the functional expression of SOS1 exchangers was always considered as an important component of salinity tolerance trait (Zhu, 2003; Apse and Blumwald, 2007; Olías et al., 2009; Oh et al., 2009), the fact that cytosolic Na^+ levels in root meristem of tolerant group was six-fold higher compared with mature zone (159.2 ± 17.9 vs. 25.4 ± 7.7, respectively; **Figure 11**) was unexpected.

On the other hand, mature root zone represents a major bulk of the root and, thus, has to deal with the largest quantities of accumulated Na^+. This zone also has fully expanded cells, with large and well-formed vacuoles, while in meristematic or transition zones cells are much smaller and with small vacuoles (Verbelen et al., 2006). Thus, superior Na^+ sequestration ability in mature root zone make sense from both anatomical and physiological points of view. Another aspect to be considered is a need to maintain cell turgor pressure under hyperosmotic conditions caused by salinity. Indeed, cytosolic Na^+ intensity was *always* higher in elongation zone compared with all other zones in each variety studied; and so was the vacuolar Na^+ intensity (**Figure 11**). It may be suggested therefore that in this zone Na^+ might be utilized by roots as a cheap osmoticum to maintain turgor pressure and enable cell expansion. Consistent with this idea are findings that in maize root apical zone, the estimated turgor potential showed only a small decline although salt shock caused a rapid decrease in root water and solute potentials in the major bulk of the root (Rodríguez et al., 1997).

Na^+ sequestration into vacuole is believed to be mediated by the tonoplast NHX Na^+/H^+ antiporters (Apse et al., 1999). Indeed, salt tolerant bread wheat variety showed higher expression of *TaNHX* in roots (Saqib et al., 2005) and also higher vacuolar Na^+ sequestration in root mature zone compared with the sensitive one (Cuin et al., 2011). Similarly, relative expression of *ZmNHX* increased proportionally with increasing external NaCl concentration in maize in bred line roots (Zörb et al., 2005). Also, transgenic tobacco lines with *TNHXS1* (wheat Na^+/H^+ vacuolar antiporter gene) had greater Na^+/H^+ antiporter activity in root tonoplast vesicles and showed higher salt tolerance than the wild type (Gouiaa et al., 2012). Previous studies in our laboratory highlighted the importance of K^+ retention in mature root zone of various species (Chen et al., 2005, 2007a; Cuin et al., 2008; Smethurst et al., 2008). Here we provide the evidence that the vacuolar Na^+ sequestration in this zone correlates with

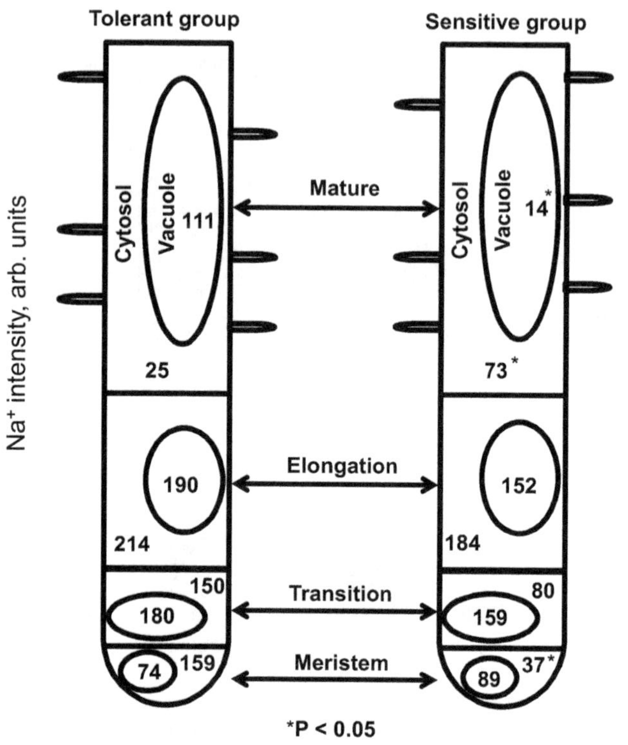

FIGURE 11 | A summary of Na^+ distribution between various intracellular compartments and functionally different root zones in bread wheat. Numbers represent mean pooled values for CoroNa Green fluorescence intensity (arbitrary units) for salt-sensitive and salt-tolerant clusters.

salinity tolerance. Taken together, cytosolic K^+ retention and Na^+ sequestration represent two major components of the tissue tolerance mechanism. Targeting these traits may be a promising way to improve salinity tolerance in bread wheat.

ROOT MERISTEM ZONE AS A SALT SENSOR?

How plants sense Na^+ to trigger the following signaling cascades to cope with the salt stress is a fundamental question which still need to be studied and clarified. In animals, Na^+ sensing mechanism appears to consist mostly of specific Na^+ selective ion channels and other Na^+ transporters; however, so far no Na^+ selective ion channels have been identified in plants (Maathuis, 2014). Displacement of Ca^{2+} by Na^+ from the plasmalemma of root cells was proposed as a primary response to salt stress (Cramer et al., 1985), but it has been deemed to be of a minor importance later (Kinraide, 1999). Plasma-membrane based SOS1 Na^+/H^+ antiporter was also suggested as a potential salt sensor (Zhu, 2003) but the explicit evidence is still lacking. Histidine kinases (such as Hik16/Hik41 in cyanobacterium, Marin et al., 2003; or AHK1/ATHK1 in Arabidopsis, Shinozaki and Yamaguchi-Shinozaki, 2000; Tran et al., 2007) were also named as potential NaCl- and/or osmo-sensors.

Generally, root is the first organ that perceives the salt stress signal. The transition zone in root apex has been suggested as a signaling-response nexus in the root (Baluška et al., 2010). This unique zone provides the root apices with an effective mechanism to reorient growth in response to many stimuli such as salinity, gravity, temperature, moisture, oxygen availability, electric fields, and heavy metals (Verbelen et al., 2006). In our investigation, both cytosolic and vacuolar Na^+ intensities in root transition zone showed no significant difference between salt tolerant and sensitive clusters (**Figures 8C,D**); no significant correlation was also found between cytosolic or vacuolar Na^+ intensities in this zone and plant damage index (**Supplementary Figures S4B,F**). Thus, it appears that root transition zone is *not* the main signaling-response nexus to salt stress, at least in bread wheat.

Salt stress causes nuclear and DNA degradation in root meristematic cells (Katsuhara and Kawasaki, 1996; Liu et al., 2000; Richardson et al., 2001), and NaCl-induced nuclear DNA fragmentation in the root meristem zone was higher in *sos1* mutant (lacking the ability to remove Na^+ to external media) compared with *Arabidopsis* wild type, when grown under saline conditions (Huh et al., 2002). From this point of view, one would expect that salt-tolerant varieties would maintain lower cytosolic Na^+ intensity in meristem cells. This was not the case in our study (**Figure 11**). On the contrary, salt tolerant group showed 4.3-fold higher cytosolic Na^+ intensity (159.2 ± 17.9 vs. 37.0 ± 8.4, respectively; **Figure 11**) than the sensitive cluster. At the same time, vacuolar Na^+ intensity was not significantly different between the groups (**Figure 7D**). It is tempting to suggest that higher cytosolic Na^+ intensity in salt tolerant cluster in the meristematic zone might be important to effectively convey or regulate signals during salt stress to other root zones even shoot after perceiving external salt stress. Hence, we suggest that, in addition to its role in cell division, root zone meristem also participates in, or executes, a role of the salt sensor. The specific details of this signaling mechanism should be revealed in further studies.

EVALUATING SALINITY TOLERANCE BY SCREENING VACUOLAR Na^+ SEQUESTRATION VIA LSCM TECHNIQUE

In addition to physiological and genetic complexity of the salinity tolerance trait, the progress in breeding was also hampered by the lack of convenient screening techniques. While agronomical (e.g., biomass/yield, plant survival, or leaf injury; Munns and James, 2003; Colmer et al., 2005) and biochemical (e.g., antioxidant activity or compatible solutes content; Ashraf and Harris, 2004) markers are convenient as rapid screening tools, they are not directly linked with major physiological mechanisms conferring salinity tolerance. Thus, the demand for a technique that is, on one hand, convenient enough to be used for the high throughput screening and, on another hand, was directly linked to a specific physiological trait in question, remains high. We believe that using CoroNa Green imaging may suit this purpose.

Na^+ measurement by CoroNa Green dye confocal imaging was firstly conducted in animals (Meier et al., 2006). It was then adopted by researchers to visualize Na^+ distribution in both root (Park et al., 2009; Oh et al., 2009, 2010; Li et al., 2012) and leaf (Bonales-Alatorre et al., 2013a,b) tissues under saline conditions. The present work narrows down salinity tolerance trait to vacuolar Na^+ sequestration in merely one specific root zone (**Figures 10, 11**) and shows a high ($r^2 = 0.59$, **Supplementary Figure S4H**) prognostic value as a screening tool.

In practical terms, imaging of one specific zone of a root takes only a few minutes, in addition to ~2 h required for staining. Thus, quantifying Na^+ fluorescent intensity in 100–120 roots per day by one operator is a realistic task. From our experience, 6–8 biological replicates are sufficient to get consistent results and eliminate out layers. Thus, even at the current stage, screening 15–20 genotypes per day may be feasible. The next practical step should be creating a DH population between Westonia and Belgrade 3 (varieties with highest and lowest vacuolar Na^+ sequestration ability), and then screening this DH population to determine QTL(s) associated with such sequestration ability. This work is next on agenda in our laboratories.

ACKNOWLEDGMENTS

This work was supported by the Grain Research and Development Corporation grant to SS and MZ, and by the Australian Research Council Discovery grant to SS. XL holds a Chinese Scholarship Council award.

SUPPLEMENTARY MATERIAL

The Supplementary Material for this article can be found online at: http://www.frontiersin.org/journal/10.3389/fpls.2015.00071/abstract

Supplementary Figure S1 | Quantifying salinity tolerance in bread wheat by the damage index. The extent of salinity damage to plants was quantified on 0 (no visual symptoms of stress) to 10 (all plants are dead) scoring scale. Two examples for tolerant variety Persia 118 **(A)** and sensitive variety 340 **(B)** are shown. **(C)** A correlation between damage index and shoot fresh weight in six bread wheat treated with 300 mM NaCl for about 5 weeks. Each point represents a variety.

Supplementary Figure S2 | Homogeneity of CoroNa Green fluorescence signal between various cell layers. (A,C) Representative (one of four) images of CoroNa Green fluorescence in the root apex of bread wheat cultivar Belgrade 3 taken at different focal depth (**A**, top cell layer; **C**, fifth cell layer; ~80 μm deeper inside the root). **(B,D)** Respective light images. **(E,F)** Mean fluorescence intensity values in cytosol **(E)** and vacuole **(F)** of cells in the first (shown in **A**) and fifth (shown in **C**) cell layers. Mean ± SE (n = 20–24). The difference between cell layers (different Z-plains) is not significant at $P < 0.05$, neither in vacuole nor in the cytosol.

Supplementary Figure S3 | Na^+ distribution between the cytosol and the vacuole in meristematic zone of wheat root. A representative images of the root meristem loaded with Corona Green AM is shown for salt-tolerant cultivar Persia 118 and salt-sensitive cultivar Iran 118. As one can see, in tolerant variety most of the Na^+ is located in the cytosol while vacuoles are dark and show not much fluorescent signal. The opposite is true for salt-sensitive genotype

Supplementary Figure S4 | Correlation between salinity stress tolerance and Na^+ distribution in cytosol and vacuole in different root zones. Correlation between salinity stress tolerance (quantified as a damage index) and cytosolic **(A–D)** and vacuolar **(E–H)** Na^+ intensities in four functional root zones. Each point represents a variety.

REFERENCES

Al-Karaki, G. N. (2000). Growth, water use efficiency, and sodium and potassium acquisition by tomato cultivars grown under salt stress. *J. Plant Nutr.* 23, 1-8. doi: 10.1080/01904160009381992

Allakhverdiev, A. I., and Murata, N. (2008). Salt stress inhibits photosystems II and I in cyanobacteria. *Photosyn. Res.* 98, 529-539. doi: 10.1007/s11120-008-9334-x

Allakhverdiev, A. I., Sakamoto, A., Nishiyama, Y., Inaba, M., and Murata, N. (2000). Ionic and osmotic effects of NaCl-induced inactivation of photosystem I and II in *Synechococcus* sp. *Plant Physiol.* 123, 1047-1056. doi: 10.1104/pp.123.3.1047

Apse, M. P., Aharon, G. S., Snedden, W. A., and Blumwald, E. (1999). Salt tolerance conferred by overexpression of a vacuolar Na^+/H^+ antiport in *Arabidopsis*. *Science* 258, 1256-1258. doi: 10.1126/science.285.5431.1256

Apse, M. P., and Blumwald, E. (2007). Na^+ transport in plants. *FEBS Lett.* 581, 2247-2254. doi: 10.1016/j.febslet.2007.04.014

Ashraf, M., and Harris, P. J. C. (2004). Potential biochemical indicators of salinity tolerance in plants. *Plant Sci.* 166, 3-16. doi: 10.1016/j.plantsci.2003.10.024

Baluška, F., and Mancuso, S. (2013). Root apex transition zone as oscillatory zone. *Front. Plant Sci.* 4:354. doi: 10.3389/fpls.2013.00354

Baluška, F., Mancuso, S., Volkmann, D., and Barlow, P. (2010). Root apex transition zone: a signalling-response nexus in the root. *Trends Plant Sci.* 15, 402–408. doi: 10.1016/j.tplants.2010.04.007

Bassil, E., Ohto, M., Esumi, T., Tajima, H., Zhu, Z., Cagnac, O., et al. (2011). The *Arabidopsis* intracellular Na^+/H^+ antiporters NHX5 and NHX6 are endosome associated and necessary for plant growth and development. *Plant Cell* 23, 224-239. doi: 10.1105/tpc.110.079426

Blumwald, E. (2000). Sodium transport and salt tolerance in plants. *Curr. Opin. Cell Biol.* 12, 431-434. doi: 10.1016/S0955-0674(00)00112-5

Bonales-Alatorre, E., Pottosin, I., Shabala, L., Chen, Z. H., Zeng, F., Jacobsen, S. E., et al. (2013a). Plasma and vacuolar membrane transporters conferring genotypic difference in salinity tolerance in a halophyte species, *Chenopodium quinoa*. *Int. J. Mol. Sci.* 14, 9267–9285 doi: 10.3390/ijms14059267

Bonales-Alatorre, E., Shabala, S., Chen, Z. H., and Pottosin, I. (2013b). Reduced tonoplast FV and SV channels activity is essential for conferring salinity tolerance in a facultative halophyte, *Chenopodium quinoa*. *Plant Physiol.* 162, 940–952 doi: 10.1104/pp.113.216572

Chen, Z., Newman, I., Zhou, M., Mendham, N., Zhang, G., and Shabala, S. (2005). Screening plants for salt tolerance by measuring K^+ flux: a case study for barley. *Plant Cell Environ.* 28, 1230-1246. doi: 10.1111/j.1365-3040.2005.01364.x

Chen, Z., Pottosin, I. I., Cuin, T. A., Fuglsang, A. T., Tester, M., Jha, D., et al. (2007a). Root plasma membrane transporters controlling K^+/Na^+ homeostasis in salt-stressed barley. *Plant Physiol.* 145, 1714-1725. doi: 10.1104/pp.107.110262

Chen, Z., Zhou, M., Newman, I. A., Mendham, N. J., Zhang, G. P., and Shabala, S. (2007b). Potassium and sodium relations in salinised barley tissues as a basis of differential salt tolerance. *Funct. Plant Biol.* 34, 150–162. doi: 10.1071/FP06237

Colmer, T. D., Munns, R., and Flowers, T. J. (2005). Improving salt tolerance of wheat and barley: future prospects. *Aust. J. Exp. Agric.* 45, 1425–1443. doi: 10.1071/EA04162

Cramer, G. R., Läuchli, A., and Polito, V. S. (1985). Displacement of Ca^{2+} by Na^+ from the plasmalemma of root cells. A primary response to salt stress? *Plant Physiol.* 79, 207–211. doi: 10.1104/pp.79.1.207

Cuin, T. A., Betts, S. A., Chalmandrier, R., and Shabala, S. (2008). A root's ability to retain K^+ correlates with salt tolerance in wheat. *J. Exp. Bot.* 59, 2697-2706. doi: 10.1093/jxb/ern128

Cuin, T. A., Bose, J., Stefano, G., Jha, D., Tester, M., Mancuso, S., et al. (2011). Assessing the role of root plasma membrane and tonoplast Na^+/H^+ exchangers in salinity tolerance in wheat: in *planta* quantification methods. *Plant Cell Environ.* 34, 947-961. doi: 10.1111/j.1365-3040.2011.02296.x

Davenport, R. J., Reid, R. J., and Smith, F. A. (1997). Sodium-calcium interactions in two wheat species differing in salinity tolerance. *Physiol. Plant.* 99, 323-327. doi: 10.1111/j.1399-3054.1997.tb05419.x

Demidchik, V., Shabala, S. N., and Davies, J. M. (2007). Spatial variation in H_2O_2 response of *Arabidopsis thaliana* root epidermal Ca^{2+} flux and plasma membrane Ca^{2+} channels. *Plant J.* 49, 377-386. doi: 10.1111/j.1365-313X.2006.02971.x

Doncheva, S., Amenos, M., Poschenrieder, C., and Barcelo, J. (2005). Root cell patterning: a primary target for aluminium toxicity in mazie. *J. Exp. Bot.* 56, 1213-1220. doi: 10.1093/jxb/eri115

Flowers, T. J., and Colmer, T. D. (2008). Salinity tolerance in halophytes. *New Phytol.* 179, 945-963. doi: 10.1111/j.1469-8137.2008.02531.x

Flowers, T. J., and Hajibagheri, M. A. (2001). Salinity tolerance in *Hordeum vulgare*: ion concentrations in roots cells of cultivars differing in salt tolerance. *Plant Soil* 231, 1–9. doi: 10.1023/A:1010372213938

Fortmeier, R., and Schubert, S. (1995). Salt tolerance of maize (*Zea mays* L.): the role of sodium exclusion. *Plant Cell Environ.* 18, 1041-1047. doi: 10.1111/j.1365-3040.1995.tb00615.x

Gouiaa, S., Khoudi, H., Leidi, E. O., Pardo, J. M., and Masmoudi, K. (2012). Expression of wheat Na^+/H^+ antiporter *TNHXS1* and H^+ – pyrophosphatase *TVP1* genes in tobacco from a bicistronic transcriptional unit improves salt tolerance. *Plant Mol. Biol.* 79, 137-155. doi: 10.1007/s11103-012-9901-6

Gupta, P. K., Varshney, R. K., Sharma, P. C., and Ramesh, B. (1999). Molecular markers and their applications in wheat breeding. *Plant Breed.* 118, 369–390. doi: 10.1046/j.1439-0523.1999.00401.x

Halušková, L., Valentovièová, K., Huttová, J., Mistrík, I., and Tamás, L. (2009). Effect of abiotic stresses on glutathione peroxidase and glutathione S-transferase activity in barley root tips. *Plant Physiol. Biochem.* 47, 1069–1074. doi: 10.1016/j.plaphy.2009.08.003

Huang, C. X., and van Steveninck, R. F. M. (1988). Effect of moderate salinity on patterns of potassium, sodium and chloride accumulation in cells near the root tip of barley: role of differentiating metaxylem vessels. *Physiol. Plant* 73, 525-533. doi: 10.1111/j.1399-3054.1988.tb05436.x

Huh, G. H., Damsz, B., Matsumoto, T. K., Reddy, M. P., Rus, A. M., Ibeas, J. I., et al. (2002). Salt causes ion disequilibrium-induced programmed cell death in yeast and plants. *Plant J.* 29, 649-659. doi: 10.1046/j.0960-7412.2001.01247.x

Katsuhara, M., and Kawasaki, T. (1996). Salt stress induced nuclear and DNA degradation in meristematic cells of barley roots. *Plant Cell Physiol.* 37, 169–173. doi: 10.1093/oxfordjournals.pcp.a028928

Kinraide, T. B. (1999). Interactions among Ca^{2+}, Na^+ and K^+ in salinity toxicity: quantitative resolution of multiple toxic and ameliorative effects. *J. Exp. Bot.* 50, 1495–1505. doi: 10.1093/jxb/50.338.1495

Lee, R. (2011). The outlook for population growth. *Science* 333, 569–573. doi: 10.1126/science.1208859

Li, R., Zhang, J., Wu, G., Wang, H., Chen, Y., and Wei, J. (2012). HbCIPK2, a novel CBL-interacting protein kinase from halophyte *Hordeum brevisubulatum*, confers salt and osmotic stress tolerance. *Plant Cell Environ.* 35, 1582–1600. doi: 10.1111/j.1365-3040.2012.02511.x

Lindsay, M. P., Lagudah, E. S., Hare, R. A., and Munns, R. (2004). A locus for sodium exclusion (*Nax1*), a trait for salt tolerance, mapped in durum wheat. *Funct. Plant Biol.* 31, 1105–1114. doi: 10.1071/FP04111

Liu, T., van Staden, J., and Cress, W. A. (2000). salinity induced nuclear and DNA degradation in meristematic cells of soybean (*Glycine max* (L.)) roots. *Plant Growth Regul.* 30, 49-54. doi: 10.1023/A:1006311619937

Maathuis, F. J. M. (2014). Sodium in plants: perception, signalling, and regulation of sodium fluxes. *J. Exp. Bot.* 65, 849-858. doi: 10.1093/jxb/ert326

Maathuis, F. J. M., and Amtmann, A. (1999). K^+ nutrition and Na^+ toxicity: the basis of cellular K^+/Na^+ ratios. *Ann. Bot.* 84, 123–133. doi: 10.1006/anbo.1999.0912

Marin, K., Suzuki, I., Yamaguchi, K., Ribbeck, K., Yamamoto, H., Kanesaki, Y., et al. (2003). Identification of histidine kinases that act as sensors in the perception of salt stress in *Synechocystis* sp. PCC 6803. *Proc. Natl. Acad. Sci. U.S.A.* 100, 9061–9066. doi: 10.1073/pnas.1532302100

Matsushita, N., and Matoh, T. (1991). Characterization of Na^+ exclusion mechanisms of salt-tolerant reed plants in comparison with salt-sensitive rice plants. *Physiol. Plant.* 83, 170–176. doi: 10.1111/j.1399-3054.1991.tb01298.x

Meier, S. D., Kovalchuk, Y., and Rose, C. R. (2006). Properties of the new fluorescent Na^+ indicator CoroNa Green: comparison with SBFI and confocal Na^+ imaging. *J. Neurosci. Meth*ods 155, 251–259. doi: 10.1016/j.jneumeth.2006.01.009

Munns, R., Husain, S., Rivelli, A. R., James, R. A., Condon, A. G. T., Lindsay, M. P., et al. (2002). Avenues for increasing salt tolerance of crops, and the role of physiologically based selection traits. *Plant Soil* 247, 93–105. doi: 10.1023/A:1021119414799

Munns, R., and James, R. A. (2003). Screening methods for salinity tolerance: a case study with tetraploid wheat. *Plant Soil* 253, 201–218. doi: 10.1023/A:1024553303144

Munns, R., James, R. A., and Läuchli, A. (2006). Approaches to increasing the salt tolerance of wheat and other cereals. *J. Exp. Bot.* 57, 1025–1043. doi: 10.1093/jxb/erj100

Munns, R., James, R. A., Xu, B., Athman, A., Conn, S. J., Jordans, C., et al. (2012). Wheat grain yield on saline soils is improved by an ancestral Na^+ transporter gene. *Nat. Biotechnol.* 30, 360–366. doi: 10.1038/nbt.2120

Munns, R., Rebetzke, G. J., Husain, S., James, R. A., and Hare, R. A. (2003). Genetic control of sodium exclusion in durum wheat. *Aust. J. Agric. Res.* 54, 627–635. doi: 10.1071/AR03027

Munns, R., and Tester, M. (2008). Mechanism of salinity tolerance. *Annu. Rev. Plant Biol.* 59, 651–681. doi: 10.1146/annurev.arplant.59.032607.092911

Oh, D. H., Lee, S. Y., Bressan, R. A., Yun, D. J., and Bohnert, H. J. (2010). Intracellular consequences of SOS1 deficiency during salt stress. *J. Exp. Bot.* 61, 1205–1213. doi: 10.1093/jxb/erp391

Oh, D. H., Leidi, E., Zhang, Q., Hwang, S. M., Li, Y., Quintero, F. J., et al. (2009). Loss of halophytism by interference with SOS1 expression. *Plant Physiol.* 151, 210–222. doi: 10.1104/pp.109.137802

Olías, R., Eljakaoui, Z., Li, J., de Morales, P. A., Marín-Manzano, M. C., Pardo, J. M., et al. (2009). The plasma membrane Na^+/H^+ antiporter SOS1 is essential for salt olerance in tomato and affects the patitioning of Na^+ between plant organs. *Plant Cell Environ.* 32, 904–916. doi: 10.1111/j.1365-3040.2009.01971.x

Park, M., Lee, H., Lee, J. S., Byun, M. O., and Kim, B. G. (2009). *In planta* measurements of Na^+ using fluorescent dye CoroNa Green. *J. Plant Biol.* 52, 298–302. doi: 10.1007/s12374-009-9036-8

Rengasamy, P. (2006). World salinization with emphasis in Australia. *J. Exp. Bot.* 57, 1017–1023. doi: 10.1093/jxb/erj108

Richardson, K. V. A., Wetten, A. C., and Caligari, P. D. S. (2001). Cell and nuclear degradation in root meristems following exposure of potatoes (*Solanum tuberosum* L.) to salinity. *Potato Res.* 44, 389–399. doi: 10.1007/BF02358598

Rodríguez, H. G., Roberts, J. K. M., Jordan, W. R., and Drew, M. C. (1997). Growth, water relations, and accumulation of organic and inorganic solutes in roots of maize seedlings during salt stress. *Plant Physiol.* 113, 881–893.

Rus, A., Yokoi, S., Sharkhuu, A., Reddy, M., Lee, B. H., Matsumoto, T. K., et al. (2001). AtHKT1 is a salt tolerance determinant that controls Na^+ entry into plant roots. *Proc. Natl. Acad. Sci. U.S.A.* 98, 14150-14155. doi: 10.1073/pnas.241501798

Saqib, M., Zörb, C., Rengel, Z., and Schubert, S. (2005). The expression of the endogenous vacuolar Na^+/H^+ antiporters in roots and shoots correlates positively with salt resistance of wheat (*Triticum aestivum* L.) *Plant Sci.* 169, 959-965. doi: 10.1016/j.plantsci.2005.07.001

Shabala, S. (2013). Learning from halophytes: physiological basis and strategies to improve abiotic stress tolerance in crops. *Ann. Bot.* 112, 1209-1221. doi: 10.1093/aob/mct205

Shabala, S., Demidchik, V., Shabala, L., Cuin, T. A., Smith, S. J., Miller, A. J., et al. (2006). Extracellular Ca^{2+} ameliorates NaCl-induced K^+ loss from Arabidopsis root and leaf cells by controlling plasma membrane K^+-permeable channels. *Plant Physiol.* 141, 1653–1665. doi: 10.1104/pp.106.082388

Shabala, S., and Mackay, A. (2011). Ion transport in halophytes. *Adv. Bot. Res.* 57, 151-199. doi: 10.1016/B978-0-12-387692-8.00005-9

Shewry, P. R. (2009). Wheat. *J. Exp. Bot.* 60, 1537–1553. doi: 10.1093/jxb/erp058

Shi, H., Quintero, F. J., Pardo, J. M., and Zhu, J. K. (2002). The putative plasma membrane Na^+/H^+ antiporter SOS1 controls long-distance Na^+ transport in plants. *Plant Cell* 14, 465-477. doi: 10.1105/tpc.010371

Shinozaki, K., and Yamaguchi-Shinozaki, K. (2000). Molecular response to dehydration and low temperature: differences and cross-talk between two stress signalling pathways. *Curr. Opin. Plant Biol.* 3, 217–223. doi: 10.1016/S1369-5266(00)80068-0

Smethurst, C. F., Rix, K., Garnett, T., Aurich, G., Bayart, A., Lane, P., et al. (2008). Multiple traits associated with salt tolerance in lucerne: revealing the underlying cellular mechanisms. *Funct. Plant. Biol.* 35, 640–650. doi: 10.1071/FP08030

Storey, R., Schachtman, D. P., and Thomas, M. R. (2003). Root structure and cellular chloride, sodium and potassium distribution in salinized grapevines. *Plant Cell Environ.* 26, 789–800. doi: 10.1046/j.1365-3040.2003.01005.x

Tester, M., and Davenport, R. (2003). Na^+ tolerance and Na^+ transport in higher plants. *Ann. Bot.* 91, 503–507. doi: 10.1093/aob/mcg058

Tilman, D., Balzer, C., Hill, J., and Befort, B. L. (2011). Global food demand and the sustainable intensification of agriculture. *Proc. Natl. Acad. Sci. U.S.A.* 108, 20260–20264. doi: 10.1073/pnas.1116437108

Tran, L. S. P., Urao, T., Qin, F., Maruyama, K., Kakimoto, T., Shinozaki, K., et al. (2007). Functional analysis of AHK1/ATHK1 and cytokinin receptor histidine kinases in response to abscisic acid, drought, and salt stress in *Arabidopsis*. *Proc. Natl. Acad. Sci. U.S.A.* 104, 20623–20628. doi: 10.1073/pnas.0706547105

Verbelen, J. P., De Cnodder, T., Le, J., Vissenberg, K., and Baluška, F. (2006). The root apex of *Arabidopsis thaliana* consists of four distinct zones of growth activities. Meristematic zone, transition zone, fast elongation zone and growth terminating zone. *Plant Signal. Behav.* 1, 296–304. doi: 10.4161/psb.1.6.3511

Yang, Y. W., Newton, R. J., and Miller, F. R. (1990). Salinity tolerance in sorghum. *I.* Whole plant response to sodium chloride in *S. bicolor* and *S. halepense*. *Crop Sci.* 30, 775-781. doi: 10.2135/cropsci1990.0011183X003000040003x

Zhu, J. K. (2003). Regulation of ion homeostasis under salt stress. *Curr. Opin. Plant Biol.* 6, 441–445. doi: 10.1016/S1369-5266(03)00085-2

Zörb, C., Noll, A., Karl, S., Leib, K., Yan, F., and Schubert, S. (2005). Molecular characterization of Na^+/H^+ antiporters (*ZmNHX*) of maize (*Zea mays* L.) and their expression under salt stress. *J. Plant Physiol.* 162, 55–66. doi: 10.1016/j.jplph.2004.03.010

Ultrastructural and physiological responses of potato (*Solanum tuberosum* L.) plantlets to gradient saline stress

Hui-Juan Gao[1†], Hong-Yu Yang[1†], Jiang-Ping Bai[1], Xin-Yue Liang[2], Yan Lou[1], Jun-Lian Zhang[1], Di Wang[1*], Jin-Lin Zhang[3*], Shu-Qi Niu[3] and Ying-Long Chen[4,5]

[1] Gansu Key Laboratories of Crop Genetic and Germplasm Enhancement and Aridland Crop Science, College of Agronomy, Gansu Agricultural University, Lanzhou, China
[2] Department of Chemistry, School of Chemistry and Chemical Engineering, Nanjing University, Nanjing, China
[3] State Key Laboratory of Grassland Agro-ecosystems, College of Pastoral Agriculture Science and Technology, Lanzhou University, Lanzhou, China
[4] Plant Nutrition and Soil Science and UWA Institute of Agriculture, School of Earth and Environment, The University of Western Australia, Perth, WA, Australia
[5] State Key Laboratory of Soil Erosion and Dryland Farming on the Loess Plateau, Institute of Soil and Water Conservation, Chinese Academy of Sciences and Ministry of Education, Northwest A&F University, Yangling, China

Edited by:
Mary Jane Beilby, The University of New South Wales, Australia

Reviewed by:
Huazhong Shi, Texas Tech University, USA
Vadim Volkov, London Metropolitan University, UK

***Correspondence:**
Di Wang, Gansu Key Laboratories of Crop Genetic and Germplasm Enhancement and Aridland Crop Science, College of Agronomy, Gansu Agricultural University, 1 Yingmen Village, Anning District, Lanzhou 730070, Gansu, China
e-mail: wangd@gsau.edu.cn;
Jin-Lin Zhang, State Key Laboratory of Grassland Agro-ecosystems, College of Pastoral Agriculture Science and Technology, Lanzhou University, 768 West Jiayuguan Road, Chengguan District, Lanzhou 730020, Gansu, China
e-mail: jlzhang@lzu.edu.cn

† These authors have contributed equally to this work.

Salinity is one of the major abiotic stresses that impacts plant growth and reduces the productivity of field crops. Compared to field plants, test tube plantlets offer a direct and fast approach to investigate the mechanism of salt tolerance. Here we examined the ultrastructural and physiological responses of potato (*Solanum tuberosum* L. c.v. "Longshu No. 3") plantlets to gradient saline stress (0, 25, 50, 100, and 200 mM NaCl) with two consequent observations (2 and 6 weeks, respectively). The results showed that, with the increase of external NaCl concentration and the duration of treatments, (1) the number of chloroplasts and cell intercellular spaces markedly decreased, (2) cell walls were thickened and even ruptured, (3) mesophyll cells and chloroplasts were gradually damaged to a complete disorganization containing more starch, (4) leaf Na and Cl contents increased while leaf K content decreased, (5) leaf proline content and the activities of catalase (CAT) and superoxide dismutase (SOD) increased significantly, and (6) leaf malondialdehyde (MDA) content increased significantly and stomatal area and chlorophyll content decline were also detected. Severe salt stress (200 mM NaCl) inhibited plantlet growth. These results indicated that potato plantlets adapt to salt stress to some extent through accumulating osmoprotectants, such as proline, increasing the activities of antioxidant enzymes, such as CAT and SOD. The outcomes of this study provide ultrastructural and physiological insights into characterizing potential damages induced by salt stress for selecting salt-tolerant potato cultivars.

Keywords: potato plantlets, saline stress, ultrastructure, antioxidant defense system, ion distribution

INTRODUCTION

As a major abiotic stresses, salinity affects plant growth and significantly reduces crop yield (Zhang et al., 2010; Zhang and Shi, 2013; Deinlein et al., 2014; Shabala et al., 2014). High soil salinity can lead to osmotic imbalance, ion-specific toxicity, alteration of composition and structure of membranes, and disruption of photosynthesis (Hasegawa et al., 2000; Zhang and Shi, 2013; Maathuis et al., 2014; Cabot et al., 2014; Zhang et al., 2014). Plants generally develop salt resistance mechanism and unique structures to survive under high saline-stress conditions (Deinlein et al., 2014; Gupta and Huang, 2014; Roy et al., 2014; Shabala et al., 2014). Therefore, a better understanding of the structural variations, ion distribution and physiological changes in crop plants induced by salinity should facilitate the identification of saline tolerance mechanisms (Roy et al., 2014).

Potato (*Solanum tuberosum* L.), as the fourth most important food crop in the world, has been identified as moderately salt-sensitive or salt-tolerant (Katerji et al., 2000). Under 50 mM NaCl treatment, potato growth decreased and tuber yield reduced to about 50%, while the growth of plants is completely inhibited at 150 mM NaCl (Hmida-Sayari et al., 2005). Bruns and Hecht-Buchholz (1990) found that the salt-induced changes were mainly observed in the chloroplasts, especially in the thylakoids. Different potato cultivars reacted differently to salt stress. Mitsuya et al. (2000) found the degradation of thylakoid membranes of chloroplast of sweet potato *in vitro* resulting from salt-induced oxidative stress (0 and 80 mM). In addition, ultrastructural changes at the cellular level in a salt-adapted potato callus lines grown in 150 mM NaCl (Queirós et al., 2011) demonstrated that salt-adapted potato cell line contained more large starch, reduced membrane system and no vesicles. Although the ultrastructural alterations induced by saline have been reported in many plant cells (Yamane et al., 2004; Miyake et al., 2006; Ferreira and Lima-Costa, 2008; Bennici and Tani, 2009, 2012), information regarding the effects of salinity on potato cells cultured *in vitro* is not specified and is incomplete.

Plants could sense changes of external environment and adapt to new conditions (Vij and Tyagi, 2007; Cabot et al., 2014; Deinlein et al., 2014). Plants have developed complex physiological and biochemical mechanisms to maintain a stable intracellular environment through accumulating various antioxidant enzymes and solute under salt stress (Wang et al., 2007; Zhang and Shi, 2013; Gupta and Huang, 2014; Roy et al., 2014). The osmotic adjustment in plant can maintain water uptake and cell turgor, allowing regular physiological metabolism (Serraj and Sinclair, 2002; Han et al., 2014). Proline, as an important osmosis protective agent, contributes to osmotic adjustment, protecting cells from damage (Silva-Ortega et al., 2008; Abrahám et al., 2010; Hou et al., 2013; Bojorquez-quintal et al., 2014; Gupta and Huang, 2014). Salt stress also caused overproduction of reactive oxygen species (ROS), leading to secondary oxidative stress (Nounjan et al., 2012; Mishra et al., 2011). ROS mainly generated from chloroplasts and mitochondria (Munns and Tester, 2008), attributed to membrane damage (Abdullahil-Baque et al., 2010), decrease of protein synthesis and inactivation of enzymes, seriously disrupting cell normal metabolism and inducing lipid peroxidation (Csiszár et al., 2012). Malondialdehyde (MDA) as a product of membrane lipid peroxidation could reflects oxidative damage to cell membrane (Koca et al., 2006; Yazici et al., 2007; Han et al., 2014). To avoid ROS-induced oxidative damage, plants could form antioxidant defense system to remove free radical and effectively avoid oxidative damage.Therefore, the increase of catalase (CAT) and superoxide dismutase (SOD) activity is correlated to the tolerance of plant to abiotic stresses (Hernåndez et al., 1993; Hossain et al., 2004; Daneshmand et al., 2010). Salt-tolerant potato could evolve a better protective mechanisms to detoxifying ROS by increasing the activity of antioxidant enzymes and content of proline (Arbona et al., 2008; Cho et al., 2012).

Higher accumulation of salt ions in leaves is very harmful for plant growth (Neocleous and Vasilakakis, 2007; Sabra et al., 2012; Khayyat et al., 2014; Liu et al., 2014a). Naeini et al. (2006) reported that more Na^+ accumulated in roots and more Cl^- in leaves of pomegranates (*Punica granatum*) exposed to salt stress. Soil salinity usually reduces K^+ uptake by roots of higher plants (Zhang et al., 2010; Maathuis et al., 2014). Recent research suggests that maintaining a high level of K^+/Na^+ ratio is important to salt tolerance in glycophytes (Maathuis and Amtmann, 1999; Carden et al., 2003; Peng et al., 2004; Lv et al., 2011; Maathuis et al., 2014). A number of studies have demonstrated that salinity also reduced Ca^{2+} absorption and transportation in plant (Tattini and Traversi, 2009; Evelin et al., 2012; Zhang and Shi, 2013; Liu et al., 2014a). Ca^{2+} has vital signal transduction function triggered by various environmental stresses. Especially, Ca^{2+} could alleviate Na^+ toxicity on plants and has a regulation effect on ion selectivity absorption and transport (Zhu, 2002; Ben-Amor et al., 2010). Ca^{2+} is an essential component of the middle lamella and cell walls which participates in maintaining the stability of cell membrane, cell wall, and membrane-bound proteins, preventing membrane damage and leakage, and stabilizing wall structure (Maathuis and Amtmann, 1999; Liu et al., 2014a). Scanning electron microscope (SEM) equipped with energy dispersive X-ray Spectroscopy (EDX) has been extensively utilized for analysis of the elements distributed in plant tissues. Moreover, ion concentrations analyzed by EDX is comparable to that derived from atomic absorption or flame photometry of whole samples (Ebrahimi and Bhatla, 2011, 2012).

The present study was to investigate the anatomical response, ion distribution and physiological changes of potato plants to gradient salt (NaCl). Test tube plantlets were used in this study to allow a direct and fast approach to examine the physiological and biochemical mechanisms of salt tolerance. The present study will provide the insight of the anatomical response, in addition to physiological response, of *in vitro* propagated potato plantlets exposed to saline stress, and develop a useful method for screening salt-tolerant cultivars.

MATERIALS AND METHODS

PLANT MATERIAL AND TREATMENTS

A local potato cultivar "Longshu No. 3," released in 2002 by Gansu Academy of Agricultural Sciences, China, was used in this study. This cultivar has been largely grown in Northwestern China because of its moderate resistance to low temperature, drought and salinity. Potato plantlets were propagated in solidified Murashige and Skoog (MS) medium. A total of 6 plantlets were cultured in each triangular flask under 16/8 h photoperiods at 200 μmol/m^2/s and 23 $\pm$ 2°C. For salt stress treatment, plantlet stems with at least two leaves were transferred to the MS medium containing NaCl at concentrations of 0 (control), 25, 50, 100, and 200 mM, respectively. Root, stem and leaf samples were collected 2 or 6 weeks after treatments for analysis. There were six plantlets in six triangular flasks for each treatment.

TRANSMISSION ELECTRON MICROSCOPY

At each sampling time, the fully expanded uppermost leaves of potato plantlets were collected and fixed for 3 h at room temperature with 2% glutaraldehyde in 100 mM sodium cacodylate buffer with a pH value of 7.4 (Sabatini et al., 1963). Samples were post-treated in 1% (w/v) OsO_4, similarly buffered for 6 h at room temperature, dehydrated in a graded ethanol series and propylene oxide, and infiltrated and embedded in Spurr's epoxy resin (Spurr, 1969). Ultrasections were obtained using a LKBV ultramicrotome and stained with uranyl acetate and lead phosphate. Images were observed and generated using a transmission electron microscope (JEM-1230 JEOL, Japan). The size of the intercellular space and cell wall was measured manually on the printed micrographs.

X-RAY MICROANALYSIS OF IONS

Root, shoot and leaf samples of each treatment were washed with distilled water, respectively. The middle sections of plant tissues were dipped in 5% agar, inserted to a depth of 1.0 cm in a copper holder, and sliced freehand with a razor blade to obtain transverse sections, and immediately frozen in liquid nitrogen. The samples were freeze-dried in vacuum and stored in a desiccator, followed by carbon coated with a high vacuum sputter coater and sputter-coated with gold in an argon atmosphere. Samples were analyzed in an scanning electron microscope (JSM-5600LV, JEOL, Japan) equipped with energy dispersive X-ray spectroscopy (INCA X-Max 80, Oxford Instruments) detector. The accelerating voltage was 10 kV. The counting time for each analysis was 60 s and the data were expressed as counts per second (cps) of an element peak after subtraction of the background. Then, these spectra were transformed to normalized data. All the detectable elements were transformed into the relative element weight. Counts per second of K, Na, and Cl were discerned by weight percentage in tissues. Five location spots of the same tissue of each section were analyzed.

PHYSIOLOGICAL ASSAYS

Free proline and malondialdehyde content from plantlet were extracted and quantified following the ninhydrin-based colorimetric assays (Delauney et al., 1992) and thiobarbituric acid (Hodges et al., 2014), respectively. Activities of SOD and CAT were determined according to the ultraviolet absorption method assays of Giannopotitis and Ries (1977) and Stewart and Bewley (1980). To measure the stomatal aperture, leaf samples (2 × 2 mm) were collected from plantlets treated with or without NaCl stress. The lower epidermis of leaves was collected by scotch tape and examined under a compound Digital Microscope (Motic) after stained with 0.1% I-KI. The morphological parameters of stomata [guard cell length—L (μM) and guard cell width—W (μM)] magnified 200 ×, were measured with Motic Images Advanced 3.2. Stomatal area (S) was calculated as the product of L and W. Leaf chlorophyll content was determined spectrophotometrically in 80% acetone as described by Arnon (1949).

DATA ANALYSIS

Parameter data were presented as means with standard deviations ($n = 6$). Data were subjected to One-Way ANOVA and Duncan's multiple range tests for each parameter at $P < 0.05$ using SPSS 13.0.

RESULTS

EFFECTS OF SALINE STRESS ON THE ULTRASTRUCTURE OF LEAF MESOPHYLL CELLS

For 2 weeks of control plantlets (without salt stress), the ultrastructural distortion of mesophyll cells and chloroplasts was not observed. The structure of mesophyll cell was intact and the cell membrane was in close contact with the cell wall. Moreover, there was large intercellular space in mesophyll cells (**Figure 1A**). After 6 weeks growth, integrated chloroplasts of control plantlets were still closely arranged along plasma membrane (**Figure 1B, Table 1**).

For plantlets with 2 weeks of 25 mM NaCl treatment, mesophyll cell walls were twisted and plasma membrane crimpled remarkably. A small proportion of the chloroplasts with distended thylakoids were apart from the cell wall and membranous invagination was observed (**Figure 1C**). After 6-week treatment more starch grains were attached to the chloroplasts (**Figure 1D**) and intercellular space decreased (**Table 1**). For plantlets grown in 50 mM NaCl for 2 weeks, mesophyll cells showed some alterations (**Figure 1E**). The number of chloroplast decreased dramatically. Plasmolysis in some cells was accompanied by a reduction in mesophyll intercellular spaces. Six weeks later, chloroplasts showed irregular shape and complex vesiculation in the vacuoles was observed. Moreover, a number of cells appeared to be linked together without space (**Figure 1F, Table 1**). When plantlets

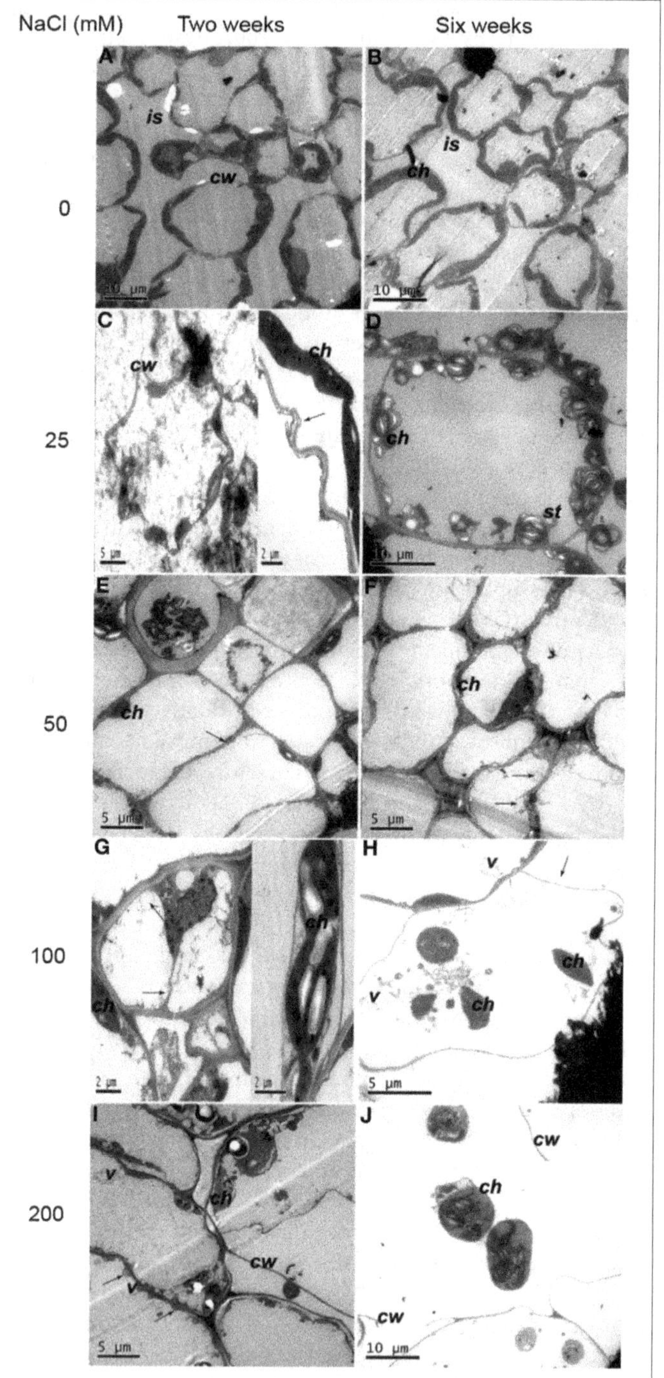

FIGURE 1 | Ultrastructural changes of mesophyll cells. (A) Two weeks of non-salinity: intact mesophyll cells 2 weeks of non-salinity treatment. **(B)** Six weeks of non-salinity: more chloroplasts were present in mesophyll cells and cellular intercellular spaces increased for 6 weeks of 25 mM NaCl treatment. **(C)** Two weeks of 25 mM NaCl: cell walls were twisted, and the plasma membrane was apparently crimpled. Note chloroplasts were apart from the cell walls with membranous invaginations (black arrows). **(D)** Six weeks of 25 mM NaCl: mesophyll cell-contained chloroplasts have more starch grains. **(E)** Two weeks of 50 mM NaCl: mesophyll cells displayed plasmolysis (white arrow) and reduced intercellular spaces (black arrow). **(F)** Six weeks of 50 mM NaCl: complex vesiculation (black arrows), and dramatically reduced numbers of chloroplasts. **(G)** Two weeks of 100 mM

(Continued)

FIGURE 1 | Continued
NaCl: plasmolysis (white arrow), numerous vesicles (black arrows) and embedded chloroplasts. **(H)** Six weeks of 100 mM NaCl: cells showed severe plasmolysis (black arrows) and more vesicles and chloroplasts moved toward the cell center. **(I)** Two weeks of 200 mM NaCl: cells displayed severely damaged membrane systems, with severe membranous invagination (black arrow). **(J)** Six weeks of 200 mM NaCl: cell walls ruptured, and whole cells disintegrated. Note: ch, chloroplast; g, grana; pl, plastoglobuli; st, starch grains; w, cell wall; is, intercellular space; v, vesicle.

Table 1 | Size of the Intercellular space and cell wall of the Mesophyll cell.

NaCl (mM)	0	25	50	100	200
Intercellular space (μm)	6.41 ± 0.57a	2.34 ± 0.07b	0 ± 0c	0 ± 0c	NA
Cell wall (μm)	0.18 ± 0.02a	0.19 ± 0.01a	0.18 ± 0.00a	0.26 ± 0.02b	NA

Values are means ± standard deviation (n = 6). Means in each line followed by different letters were statistically different (P < 0.05) by Duncan's multiple range tests. NA, not available. At 200 mM, parameters could not be obtained due to cell wall rupture and cell disintegration.

were exposed to 100 mM NaCl for 2 weeks, serious plasmolysis was observed. Membranous invaginations resulted in numerous vesicles. Some chloroplasts embedded together (**Figure 1G**). Six weeks later, plasmolysis occurred severely accompanied by the presence of more vesicles in the vacuole. Chloroplasts moved toward the center of the cell (**Figure 1H**). The most dramatic alterations were observed in plantlets treated with 200 mM NaCl for 2 weeks. Membrane structure was severely damaged, characterized by severe membranous invagination (**Figure 1I**). After 6 weeks of 200 mM NaCl treatment, cell walls ruptured and the whole cell disorganized (**Figure 1J**).

EFFECTS OF SALINE STRESS ON THE ULTRASTRUCTURE OF CHLOROPLASTS

For 2 weeks of control plantlets, integrated chloroplasts with few and small starch, containing compactly arranged thylakoids and well compartmentalized grana stacks with distinct grana lamellaes parallel to the chloroplasts' long axes, were observed (**Figure 2A**). Six weeks later, the membrane system was complete. The grana and stromal lamellae of chloroplast closely arranged and compacted thylakoids (**Figure 2B**).

When exposed to 25 mM NaCl for 2 weeks, the cell walls were thickened (**Figure 2C**, **Table 1**). The outer membrane of the chloroplast was vague. After 6 weeks of 25 mM NaCl treatment, the swelling of the thylakoids became obvious. The arrangement of lamella remained consistent, but showed a slight bend (**Figure 2D**). After 2 weeks of 50 mM NaCl treatment, chloroplast envelope was partially fragmented and evaginated to form complex vesicles (**Figure 2E**). Six weeks later, chloroplast envelopes disrupted with outer membranes disorganized. Grana lamella loosened with severely swollen thylakoids and space between lamella increased (**Figure 2F**). For plantlets treated with 100 mM NaCl for 2 weeks, the cell walls were much thicker (**Table 1**).

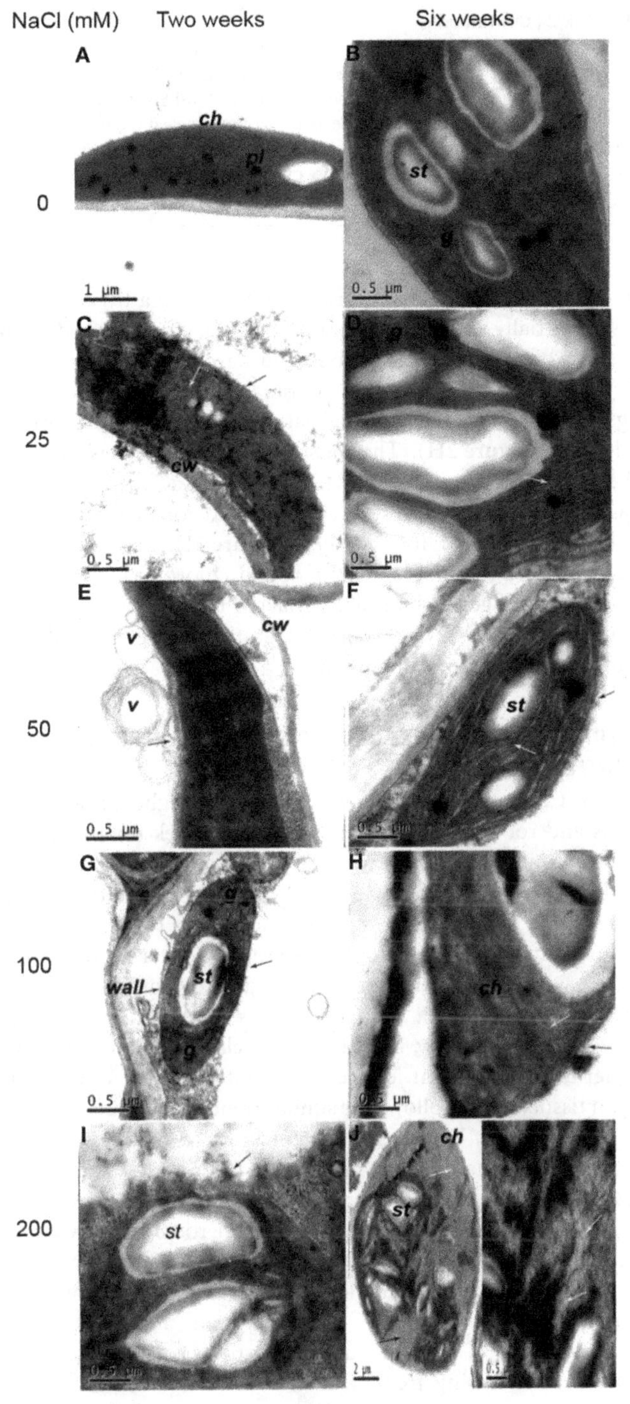

FIGURE 2 | Ultrastructural changes of chloroplast in mesophyll cell. (A) Two weeks of non-salinity: ellipse- or spindle-shaped chloroplast with few and small starch. **(B)** Six weeks of non-salinity: chloroplast structure was complete. **(C)** Two weeks of 25 mM NaCl: chloroplast with vague outer membranes (black arrows) showed distended thylakoids (white arrows). **(D)** Six weeks of 25 mM NaCl: obvious swelling of the thylakoid (white arrow). **(E)** Two weeks of 50 mM NaCl: chloroplast envelope evagination, forming vesicles (black arrow). **(F)** Two weeks of 50 mM NaCl: chloroplast envelope disruption (black arrow) and distorted lamella (white arrow). **(G)** Two weeks of 100 mM NaCl: chloroplast envelope disintegration (black arrow) and thicker cell walls and partially dissolved grana thylakoid. **(H)** Six *(Continued)*

FIGURE 2 | Continued
weeks of 100 mM NaCl: envelope (black arrow) and lamellar structure (white arrow) partly dissolved. **(I)** Two weeks of 200 mM NaCl: chloroplast disintegrated with inclusions effused (black arrows). **(J)** Six weeks of 200 mM NaCl: the grana and stromal lamella of chloroplast digest basically (black arrow), while thylakoids disintegrate and cavitate gradually (white arrows). Note: ch, chloroplast; g, grana; pl, plastoglobuli; st, starch grains; w, cell wall; is, intercellular space; v, vesicle.

Chloroplast envelope disintegrated and the grana thylakoid dissolved partially with reduced grana stacking, characterized by the presence of enlarged plastoglobuli and starch grains (**Figure 2G**). Six week later, the orientation of grana changed. Lamellar stacking decreased and dissolved dramatically. Membrane system was indistinct (**Figure 2H**). The most serious impact was observed when plantlets were treated with 200 mM NaCl. Some chloroplasts disintegrated with inclusions effused for plantlets treated with 200 mM NaCl for 2 weeks (**Figure 2I**). Six weeks later, the grana and stromal lamella of round chloroplasts with some starch grains digested basically, thylakoid membranes adhered to each other, while thylakoids disintegrated, cavitated, and even gradually disappeared (**Figure 2J**).

EFFECTS OF SALINE STRESS ON ION DISTRIBUTION IN POTATO PLANTLET TISSUES

Na and Cl contents in leaves were relatively higher than that in stems and roots for all treatments. After 2 week treatments, Na relative content in leaves was 5.1, 4.2, 3.4, 3.0, and 1.9 times of that in roots at 0, 25, 50, 100, 200 mM NaCl treatments, respectively; Cl relative content in leaves was 1.2, 4.4, 2.5, 6.4, and 5.0 times of that in roots, respectively. After 6 week treatments, with the increase of NaCl in growth environment, the relative contents of Na and Cl in tissues were higher than those at 2 weeks, respectively. In addition, Cl relative content remained higher than Na content for the same treatment and for the same organ tissue, which follows the similar trend as at 2 weeks. After 6 week treatments, Na relative content in leaves was 1.7, 1.6, 2.0, 1.7, and 1.5 times of that in roots at 0, 25, 50, 100, 200 mM NaCl treatments, respectively; Cl relative content in leaves were 2.3, 1.7, 1.8, 2.0, and 1.2 times of that in roots at corresponding NaCl treatments, respectively. These results indicated that Na and Cl were mainly distributed in leaves of potato plantlets. (**Figures 3A–F**).

In contrast, K relative content in roots, stems and leaves showed a decreasing trend with the increase of external NaCl concentration. Accumulation of K in stems was reduced, particularly in leaves. After 2 weeks of salt treatment, K relative content in roots was 1.1, 1.3, 1.3, 3.0, and 2.1 times of that in leaves at 0, 25, 50, 100, 200 mM NaCl treatments, respectively. Six weeks later, K relative content in roots, stems and leaves decreased compared to that at 2 weeks. K relative content in roots was 1.3, 1.5, 1.6, 2.7, and 1.8 times of that in leaves at 0, 25, 50, 100, 200 mM NaCl treatments, respectively (**Figures 3G–I**). The comparison of K distribution in the different parts of potato plantlets showed that salinity seriously reduced K allocation toward leaves.

The Na/K ratio dramatically increased, especially in leaves after treated with various concentrations of NaCl. After 2 weeks of treatments, Na/K ratio significantly increased by 2.0, 4.3, 6.0, and 19.0 times in roots, 1, 2, 3.1, and 5.1 times in stems, and 1.6, 2.6, 8.9, and 12.1 times in leaves, at 25, 50, 100, 200 mM NaCl treatments, respectively, compared to that in control tissues, After 6-week treatment, compared to the corresponding organs of control plantlets, Na/K ratio significantly increased by 1.7, 2.1, 5.5, and 7.9 times in roots, 1.3, 1, 7, and 9.1 times in stems, and 1.8, 3.3, 11.7, and 9.7 times in leaves at corresponding NaCl treatments, respectively. Potato plantlets treated with salt for 6 weeks had higher Na/K ratio in the relevant organs than those treated for 2 weeks except for leaf Na/K ratio at 200 mM NaCl concentration (**Figures 3J–L**).

EFFECTS OF SALINE STRESS ON LEAF FREE PROLINE CONTENT, CAT AND SOD ACTIVITIES AND MDA CONTENT

Salt stress significantly increased free proline levels in leaves (**Figure 4**). After 2 weeks of treatment, proline content significantly increased by 1.6, 1.9, 3.4, and 4.5 times at 25, 50, 100, and 200 mM NaCl treatments, respectively, compared to control ($P < 0.05$). After 6 weeks of treatments, proline significantly content increased by 0.8, 3.1, 4.7, and 3.7 times, respectively ($P < 0.05$). Proline content decreased significantly at 200 mM NaCl compared to that at 100 Mm NaCl ($P < 0.05$). Leaf proline content in plantlets treated for 6 weeks by 50, 100, and 200 mM NaCl was significant higher than that in plantlets treated for 2 weeks ($P < 0.05$).

Salt stress increased the activity of the antioxidant enzymes. After 2 week treatment, compared to control, CAT activity significantly increased by 28.9, 57.9, 96.8, and 63.4% at 25, 50, 100, and 200 mM NaCl, respectively; while SOD activity significantly increased by 18.6, 41.2, 38.4, and 52.9%, respectively ($P < 0.05$). After 6 weeks, CAT and SOD activities significantly increased by 50.0, 80.5, 102.6, and 13.6%, and 13.1, 29.5, 29.6, and 23.9% at 25, 50, 100, and 200 mM NaCl, respectively, compared to corresponding control ($P < 0.05$). Leaf CAT activity in plantlets treated with 200 mM NaCl for 2 and 6 weeks and SOD activity for 6 weeks decreased significantly compared to that in plantlets treated with 100 mM NaCl ($P < 0.05$). Also, activities of leaf CAT and SOD in plantlets treated for 6 weeks were significantly higher than those in plantlets treated for 2 weeks except for 200 mM NaCl treatment ($P < 0.05$) (**Figure 5**).

Leaf MDA content was used as an indicator of oxidative damage by salt stresses. After 2 week treatment, MDA content significantly increased by 0.8, 1.0, 1.8, and 2.0 times with the increase of external NaCl concentration compared to control plantlets; after 6 week treatment, MDA content sharply increased by 0.7, 1.1, 1.7, and 2.4 times with the increase of salinity ($P < 0.05$). Leaf MDA content in plantlets treated for 6 weeks were significantly higher than that in plantlets treated for 2 weeks ($P < 0.05$) (**Figure 6**).

EFFECTS OF SALINITY STRESS ON LEAF STOMATAL AREA AND CHLOROPHYLL CONTENT

Two weeks of salt treatment reduced stomatal area significantly by 18.0, 35.4, 61.5, and 86.7% at 25, 50, 100, and 200 mM NaCl

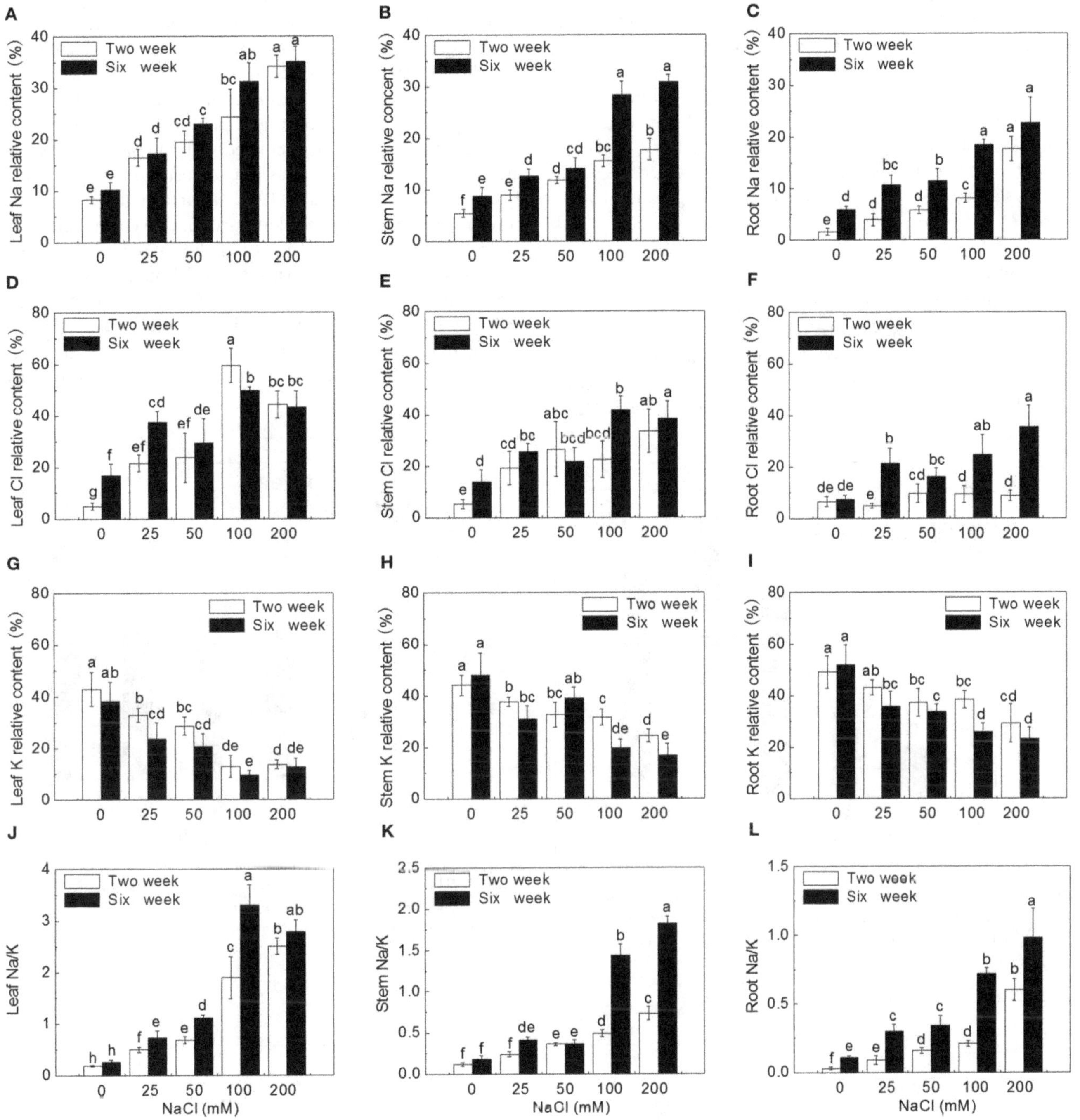

FIGURE 3 | Ion relative content and Na/K ratio under different concentrations of NaCl using SEM-EDS. **(A)** Leaf Na relative content, **(B)** stem Na relative content, **(C)** root Na relative content, **(D)** leaf Cl relative content, **(E)** stem Cl relative content, **(F)** root Cl relative content, **(G)** leaf K relative content, **(H)** stem K relative content, **(I)** root K relative content, **(J)** ratio of Na to K in leaf, **(K)** ratio of Na to K in stem, **(L)** ratio of Na to K in root. Values are means and bars indicate SDs ($n = 6$). Columns with different letters indicate significant difference by Duncan's multiple range tests at $P < 0.05$ (Duncan test).

concentrations, respectively, compared to control ($P < 0.05$). Six weeks of salt treatment dramatically reduced stomatal area by 70.3, 88.2, 91.6, and 99.4% with the increase of NaCl concentration ($P < 0.05$). Stoma was almost closed after 6 weeks of 200 mM NaCl treatment (**Figure 7A**).

The trend of changes for chlorophyll content was similar to that for stomatal area. After 2 weeks of salt treatment, leaf chlorophyll content decreased gradually by 24.8, 44.2, 65.5, and 70.8% with the increase of NaCl concentration, compared to control ($P < 0.05$). After 6 weeks of salt treatment, chlorophyll content sharply decreased by 33.9, 68.3, 88.1, and 93.6% with the increase of NaCl concentration ($P < 0.05$), and was much lower than that at 2 weeks under corresponding salt stresses (**Figure 7B**).

At the whole plantlet level, NaCl treatments inhibited potato plantlet growth. The height of seedlings gradually decreased with increase of external NaCl concentration. After 6 weeks of treatment, severe salt stress (200 mM NaCl) induced a greater decline in shoot growth and root development of potato plantlets (**Figure S1**).

DISCUSSION

SALINITY INDUCED ULTRASTRUCTURAL CHANGES OF LEAF MESOPHYLL CELLS AND CHLOROPLASTS

In present study, high levels of Na and Cl, and low level of K were distributed in leaves. The changes in chemical contents could result in ultrastructural alteration in leaf cells. Three salt-stress related alterations were observed. Firstly, the number of chloroplasts displaying swelled and distorted thylakoids decreased, accompanied by chloroplasts moving to the cell center. This chloroplast change is a typical effect of salinity as previously observed in salt-stressed *Cucumis sativus* L. (Shu et al., 2013). Secondly, cell walls thickened and plasmolysis occurred and the intercellular spaces of cell decreased with the increase of external salt concentration, which was also reported in potato cultivars (Bruns and Hecht-Buchholz, 1990; Navarro et al., 2007). Thirdly, lamella became disordered, loosened, and even indistinct, with reduced grana stacking because of inhibition of protein synthesis. Krzesłowska (2010) has reported that thickened cell wall could be as a barrier, protecting cell from toxicity of trace metals. So cell wall may function and limit passive Na and Cl enter into protoplast, maintaining structural integrity of the cell in the early low salt stress. It has been known salt stress can lead to osmotic

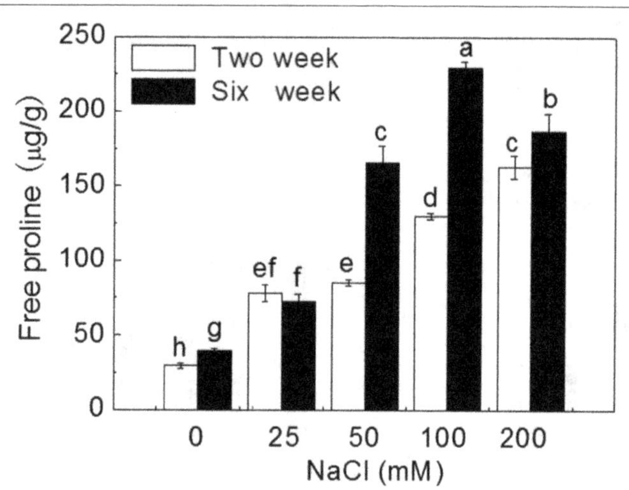

FIGURE 4 | Effects of NaCl treatment on free proline content. Values are means and bars indicate SDs ($n = 6$). Columns with different letters indicate significant difference by Duncan's multiple range tests at $P < 0.05$.

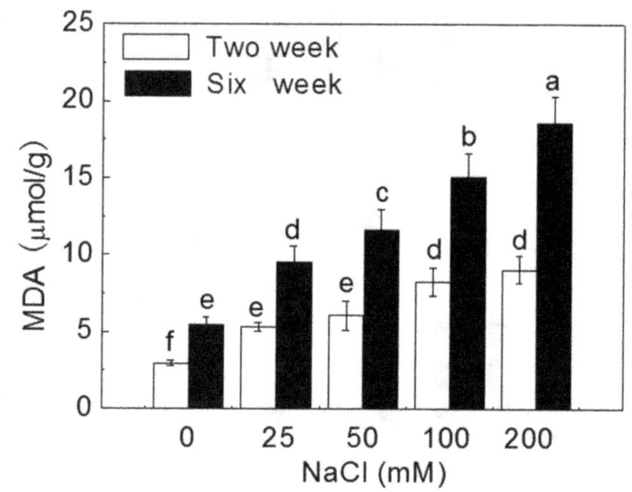

FIGURE 6 | Effects of NaCl treatment on malondialdehyde (MDA) content. Values are means and bars indicate SDs ($n = 6$). Columns with different letters indicate significant difference by Duncan's multiple range tests at $P < 0.05$.

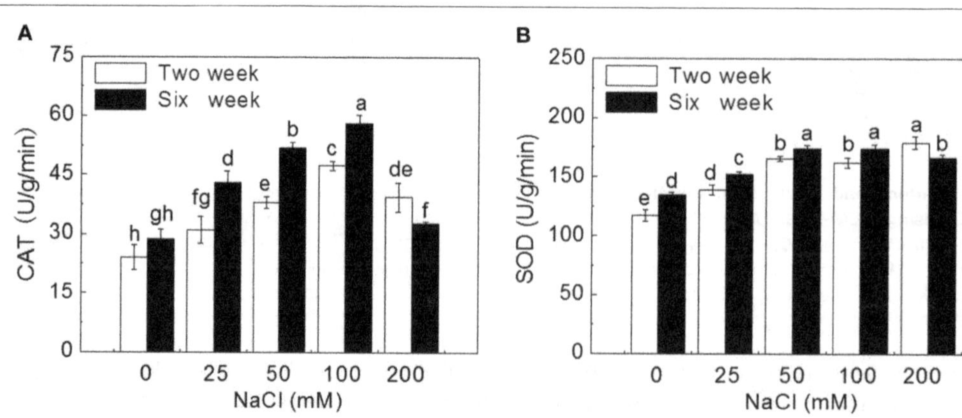

FIGURE 5 | Effects of NaCl treatment on activities of catalase (CAT) and superoxide dismutase (SOD). (A) CAT activity, **(B)** SOD activity. Values are means and bars indicate SDs ($n = 6$). Columns with different letters indicate significant difference by Duncan's multiple range tests at $P < 0.05$.

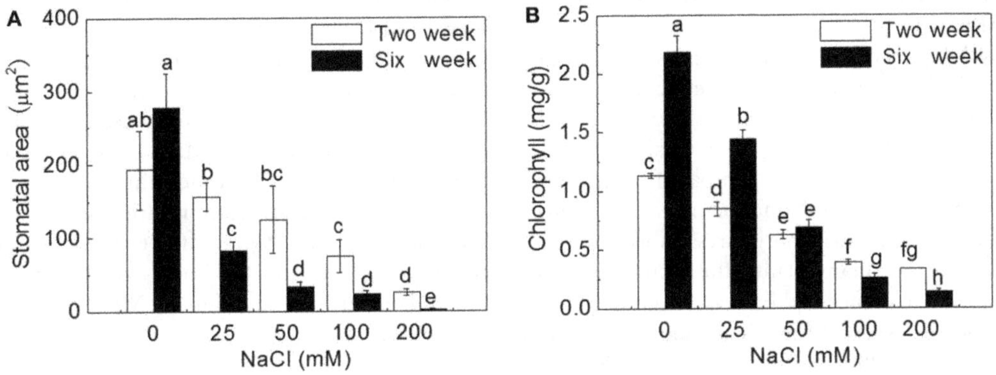

FIGURE 7 | Effects of NaCl treatment on stomatal area (A) and chlorophyll content (B). Values are means and bars indicate SDs ($n = 6$). Columns with different letters indicate significant difference by Duncan's multiple range tests at $P < 0.05$.

damage. Na^+ could be used directly for osmotic adjustment to maintain cell turgor and photosynthetic activity under low external salt concentration (Yousfi et al., 2010; Ebrahimi and Bhatla, 2012; Ma et al., 2012). However, with the increase of salt levels (NaCl concentration >50 mM), high concentrations of Na and Cl accumulated in leaf apoplast, leading to water loss of cell, plasmolysis and decrease of intercellular spaces in the leaves of potato plantlets. The present study observed invaginated membrane system forming numerous vesicles under salt treatments supporting observations by Kim and Park (2010), whilst contrary to Queirós et al. (2011) in which no vesicle was found in salt-adapted potato cell line. Vacuolation may be a response to membrane system damage induced by ROS caused by toxicity of Na and Cl (Kim and Park, 2010). ROS lead to the increase of plasma membrane permeability and extravasations of soluble substances, causing osmotic water imbalance, aggravating plasmolysis. Since membrane vesicles have Na^+/H^+ antiporter (Blumwald et al., 2000) and cell can sequester ion into vacuole (Kim and Park, 2010), vesicles may compartmentalize Na and Cl and migrate to walls. When plants were exposed to high NaCl concentration (100 mM), membrane disappeared. Salt inhibits absorption of Ca^{2+}, further leading to instability of cell membrane and cell wall. Integral of membrane is essential in ions absorption and distribution. The destruction of the membrane structure inevitably disrupted ion homeostasis, affecting osmotic potential and inducing ion toxicity.

Disorganization of whole cells was accompanied by disintegrated chloroplasts having more starch and dissolved stroma lamella under 200 mM NaCl. It was speculated that starch synthesis plays a role in lessening the hyperosmotic stress as osmoticum. A total disorganization of the protoplast in callus cells was reported in other plants, possibly caused by dehydration (Bennici and Tani, 2012). Disintegration of chloroplasts and mesophyll cells end the photosynthesis, thus, maintaining structural integrity is necessary in plant growth (Bennici and Tani, 2012).

SALINITY CHANGED ION HOMEOSTASIS IN POTATO PLANTLETS

It has been known that the total Na^+ and Cl^- content increased under salt in potato cell line, and K^+/Na^+ ratio was a little higher in the adapted line (Queirós et al., 2011). Ruan et al. (2005) showed that Na^+ accumulation decreased from the roots to leaves in *Kosteletzkya virginica*. Higher Na^+ distributed in roots than in leaves in maize under salt stress (Azevedo-Neto and Prisco, 2004). In *Capsicum chinense*, more Na^+ was restricted in roots (Bojorquez-quintal et al., 2014). Higher levels of Na^+ in roots can maintain the normal osmotic potential and prevent it from being transported to the leaves, therefore avoiding the accumulation of Na^+ in the leaves (Tester and Davenport, 2003; Munns and Tester, 2008; Xue et al., 2013). Queirós et al. (2009) reported that higher Na^+ distributed in roots, inhibiting Na^+ transport to leaves in potato cell. In present study, the distribution of Na and Cl increased from roots to stems and leaves in potato plantlets, indicating that potato is not a salt exclusion plant and has lower capacity to retain saline ions in their roots. High ions in leaves leaded to osmotic damage and oxidative stress, affecting physiological and biochemical metabolism. In addition, as a whole more Cl accumulated in potato tissue than Na, indicating the absorption of Cl^- was higher than Na, which is similar to the findings in sunflower (Ebrahimi and Bhatla, 2011) and in Clions (Greenway and Munns, 1980). Higher Cl^- accumulation lead to more serious and instant damage under salt stress (Yao and Fang, 2008). In our study, the absorption of Na and Cl in roots, stems and leaves of potato plantlet was enhanced with the increases of NaCl concentration, and the relative contents of Na and Cl were the highest in leaves, and lowest in roots.

K^+ participates in many cellular functions, such as protein synthesis, enzyme activation and osmotic regulation (Peng et al., 2004; Takahashi et al., 2007; Amtmann et al., 2008). Therefore, the regulation of K^+ homeostasis plays a critical role in plant tolerance to abiotic stresses (Ashley et al., 2006; Wang and Wu, 2010; Demidchik, 2014; Anschütz et al., 2014; Shabala and Pottosin, 2014). Salinity induced plant nutritional disorders, such as the suppression of K^+ uptake (Kader and Lindberg, 2005; Kronzucker et al., 2006; Shabala and Cuin, 2008). Bojorquez-quintal et al. (2014) suggested that more K^+ accumulated in roots is correlated with the salt tolerance of *Capsicum chinense*. In present study, salt stress dramatically reduced K^+ uptake and accumulation, especially in leaves, resulting in increased Na/K ratio in all tissues with the increase of external salt concentration and the duration of treatments.

PHYSIOLOGICAL MECHANISM OF POTATO PLANTLETS ADAPTING TO GRADIENT SALINE STRESS

Salinity leads to physiological changes in plant, especially osmotic and oxidative stress (Zhang and Shi, 2013). The accumulation of osmoprotectants is important for plant to adapt to osmotic stress (Apse and Blumwald, 2002; Waditee et al., 2007; Chan et al., 2011; Rivero et al., 2014). Proline, an important compatible osmolyte in plants, could maintain cell turgor and function in osmotic adjustment to improve plant tolerance to osmotic stress (Abrahám et al., 2010; Huang et al., 2013). In many plants, the accumulation of proline could lead to salt tolerance and has even been used as an important trait in selecting tolerant species or genotypes (Ashraf and Harris, 2004; Khelil et al., 2007; Ruffino et al., 2010). Recently, Bojorquez-quintal et al. (2014) found that more proline was accumulated in leaves of salt-tolerant habanero pepper (*Capsicum chinense* Jacq.) cultivar (Rex) than in salt-sensitive one (Chichen-Itza) under 150 mM NaCl treatment. In our study, the levels of free proline increased significantly with the increase of external salt concentration and with the duration of treatments except for a little decline at 200 mM NaCl after 6-week treatment (**Figure 3**). The reason may be that 200 mM induced excessive damage to plant cells and inhibited proline synthesis.

Antioxidant enzymes in plant can remove ROS and alleviate oxidative damage (Krantev et al., 2008; Mishra et al., 2011). It has been known that the higher activities of CAT and SOD could improve plant tolerance to salinity and K^+-deficiency conditions (Wang et al., 2010; Zhou et al., 2014). It was found that SOD activity was significantly higher in the leaves of salt-tolerant wild tomato (*Lycopersicon pennellii*) than that of salt-sensitive cultivated tomato (*Lycopersicon esculentum*) after 12 and 84 d of salt treatment (140 mM NaCl) (Koca et al., 2006). Similarly, salt-tolerant *Plantago maritima* showed a better protection mechanism against oxidative damage caused by salt stress by its higher induced activities of CAT, SOD, glutathione reductase (GR) and peroxidase (POX) than the salt-sensitive *P. media* (Sekmen et al., 2007). Co-expression of the *Suaeda salsa* CAT and glutathione S-transferase (GST) genes enhanced the active oxygen-scavenging system that led to improved salt tolernace in transgenic rice, resulting from not only increased CAT and GST activities but also the combined increase in SOD activity (Zhao and Zhang, 2006). Jing et al. (2014) reported that overexpression of mangrove (*Kandelia candel*) copper/zinc superoxide dismutase gene (*KcCSD*) enhanced salinity tolerance in tobacco: *KcCSD*-transgenic lines were more Na^+ tolerant than wild-type (WT) tobacco in terms of lipid peroxidation, root growth, and survival rate; Na^+ injury to chloroplast was less pronounced in transgenic tobacco plants due to enhanced SOD activity by an increment in SOD isoenzymes under 100 mM NaCl stress from 24 h to 7 d; catalase activity rose in *KcCSD* overexpressing tobacco plants and transgenic plants better scavenged NaCl-elicited ROS compared to WT ones. In present study, the activities of CAT and SOD in leaves of potato plantlets significantly increased with the increase of NaCl concentration (0~100 mM) in medium. When exposed to 200 mM NaCl, especially after 6 weeks, leaf cells were severely damaged, even disorganized (**Figure 1**), leading to the damage of cellular structure or alterations of metabolism, and reducing the synthesis of CAT and SOD.

Soil salinity is known to increase the level of ROS in plant leaves and MDA is a major product of membrane lipid peroxidation (Mittova et al., 2004; Koca et al., 2006; Yazici et al., 2007). Therefore, leaf MDA content could represent the degree of cell membrane damage and is usually used to evaluate plant salt tolerance (Luna et al., 2000; Miao et al., 2010; Han et al., 2014). In our study, leaf MDA content increased significantly with the increase of external salt concentration after 2-week treatment and even increased more rapidly after 6-week treatment. However, the activities of SOD and CAT may not enough to eliminate ROS, resulted in large production of MDA under higher salt stress (200 mM).

SALINITY REDUCED LEAF STOMATAL AREA AND CHLOROPHYLL CONTENT

Chlorophyll is essential for photosynthesis, and the increase of chlorophyll content can reflect the increase of photosynthetic activity (Yamori et al., 2006). Ben et al. (2010) and Su et al. (2011) suggested that the accumulation of chlorophyll content could enhance plant salt tolerance. In the present study, leaf chlorophyll content gradually decreased with the increase of NaCl treatment and duration, which could result from the inhibition of chlorophyll synthesis caused by chloroplast damage.

Gas exchange through stoma play important role in carbon assimilation (Wilkinson and Davies, 2002). Salt stress decreases leaf stomatal area by reducing leaf water content and leaf turgor induced by ABA signal (Wilkinson and Davies, 2002). Therefore, stomatal conductance was correlated to salinity stress (Liu et al., 2014b). In our study, salt stress seriously induced stomatal closure. Reduced CO_2 diffusion caused by stomatal closure lead to suppression of photosynthesis, affecting plant growth (**Figure S1**).

In conclusion, the adaptation of plants to salt stress is a complex process at cellular, biochemical and physiological levels. In the present study, several parameters were analyzed to demonstrate ultrastructural and physiological responding mechanisms of potato (*Solanum tuberosum* L.) plantlets to gradient saline stress (**Figure 8**). We found that with the increase of external NaCl concentration and the duration of treatments, the number of chloroplasts and cell intercellular space markedly decreased, cell wall thickened and even ruptured, and mesophyll cells and chloroplasts were gradually damaged to a complete disorganization. Above ultrastructural changes may be induced by the increased concentrations of Na^+ that was transported into cytosol probably through non-selective cation channels (NSCCs), high-affinity K^+ transporters (HKTs, probably HKT1;2; HKT1;4; HKT1;5 and HKT2;1) and permeated directly across plasma membrane, and Cl^- that was probably transported by cation-Cl^- cotransporter (CCC) (Apse and Blumwald, 2007; Plett and Moller, 2010; Zhang et al., 2010; Zhang and Shi, 2013; Almeida et al., 2014a,b; Maathuis, 2014; Maathuis et al., 2014). More and more K^+ was probably transported out of the cell by K^+ outward-rectifying channels (KORs) activated by membrane depolarization (DPZ) (Chen et al., 2007; Sun et al., 2009; Lu et al., 2013; Demidchik, 2014; Demidchik et al., 2014; Lai et al., 2014). Leaf MDA content increased significantly

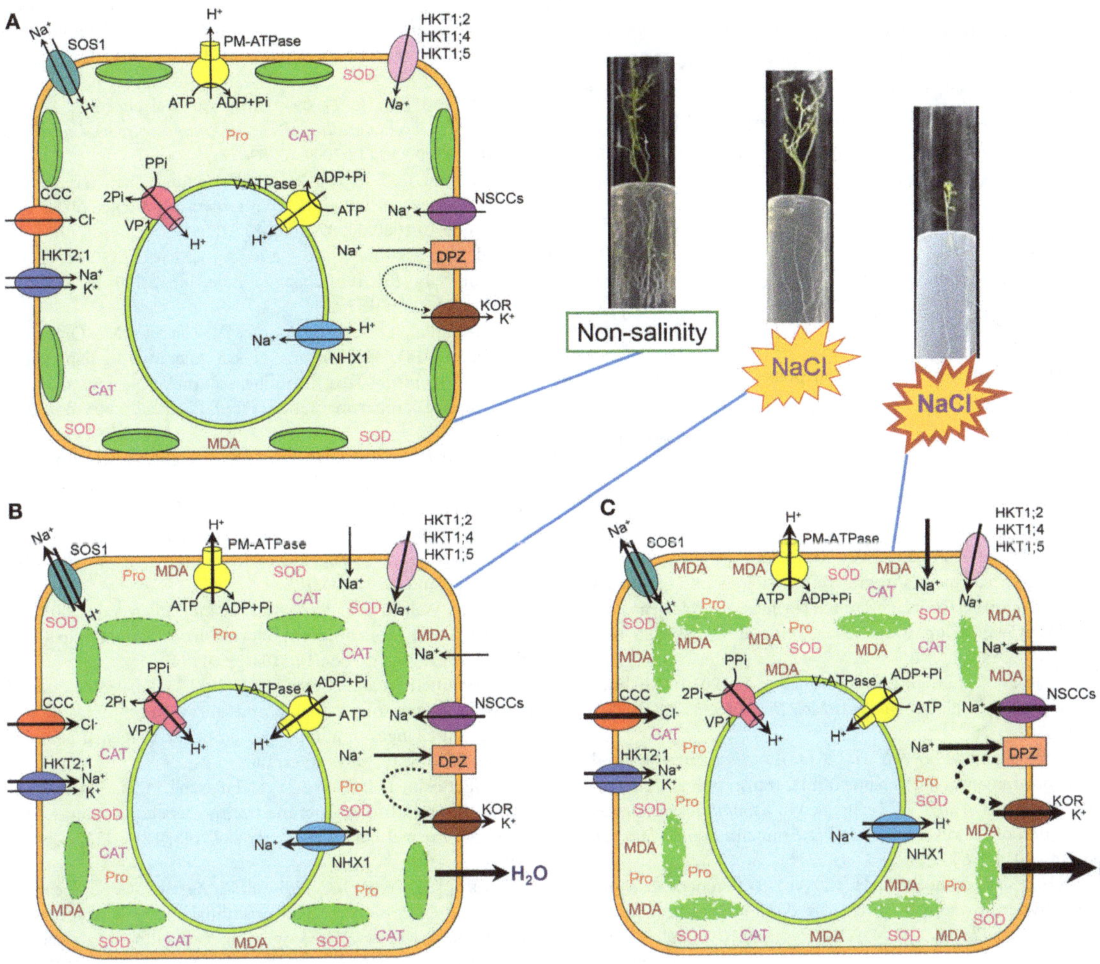

FIGURE 8 | Schematic model of ultrastructural and physiological responding mechanisms of the mesophyll cells from potato (*Solanum tuberosum* L.) plantlets to gradient saline stress. (A) Under non-salinity condition, water and ions was maintained at a balance status, only little proline (Pro), CAT, SOD, and MDA were accumulated within cytosol, and integrated chloroplasts were closely arranged along plasma membrane. **(B)** Under moderate salinity condition, abundant Na^+ was transported into cytosol probably through non-selective cation channels (NSCCs), high-affinity K^+ transporters (HKTs, probably HKT1;2, HKT1;4, HKT1;5, and HKT2;1) and a little permeated directly across plasma membrane, and Cl^- was probably transported by cation-Cl^- cotransporter (CCC). Some K^+ was transported out of the cell by K^+ outward-rectifying channels (KORs) activated by membrane depolarization (DPZ). The membrane system was damaged resulting in the increase of MDA and damaged chloroplasts were not closely arranged along plasma membrane. Stoma closed because of water loss and chlorophyll content decreased because of chloroplast damage. For adaptation to moderate salinity, Na^+ efflux or extrusion by plasma membrane Na^+/H^+ antiporter (salt overly sensitive, SOS1) motivated by plasma membrane ATPase (PM-ATPase) and vacuolar Na^+ compartmentation by tonoplast Na^+/H^+ antiporter (NHX1) motivated by vacuolar ATPase (V-ATPase) and H^+-pyrophosphatase (VP1) functioned to reduce Na^+ toxicity in cytosol, at the same time osmoprotectants such as proline were accumulated and the activities of antioxidant enzymes (CAT and SOD) increased. **(C)** Under high salinity condition, more and more Na^+ was transported into cytosol probably through NSCCs and permeated directly across plasma membrane although the amount of Na^+ transported by HKTs did not increase, and more Cl^- was probably transported by CCC. More and more K^+ was transported out of the cell by KOR. The membrane system was seriously damaged resulting in the rapid increase of MDA and disintegrated chloroplasts appeared. Stoma closed completely because of increasing water loss and chlorophyll content decreased dramatically because of severe chloroplast damage. However, the ability of Na^+ efflux or extrusion by SOS1 and vacuolar Na^+ compartmentation by NHX1 were not enhanced because of serious damage to membrane system, at the same time osmoprotectant content and the activities of antioxidant enzymes (CAT and SOD) did not increased any more, but even decreased. Therefore, the growth of potato plantlets was inhibited.

due to all membrane lipid peroxidation induced by increasing and continuous salt stress, which also induced stomata closure and chlorophyll content decline. Potato plantlets showed adaptation ability to moderate salt stress through Na^+ efflux or extrusion by plasma membrane Na^+/H^+ antiporter (salt overly sensitive, SOS1) motivated by plasma membrane ATPase (PM-ATPase), vacuolar Na^+ compartmentation by tonoplast Na^+/H^+ antiporter (NHX1) driven by vacuolar ATPase (V-ATPase) and H^+-pyrophosphatase (VP1), accumulating osmoprotectants such as proline, and improving the activities of antioxidant enzymes (CAT and SOD). This work provided both anatomical and physiological data for characterization of damages induced by salinity and the method could be used for selecting salt-tolerant potato cultivars.

ACKNOWLEDGMENTS

This research was supported by Program for Changjiang Scholars and Innovative Research Team in University (IRT13019), International Science & Technology Cooperation Program of China (2014DFG31570), Gansu S&T Foundation (1308RJZA131 and 1308RJIA005), Lanzhou S&T Research Project (2013-4-156 and GSCS-2012-04) and NSFC (31222053).

SUPPLEMENTARY MATERIAL

The Supplementary Material for this article can be found online at: http://www.frontiersin.org/journal/10.3389/fpls.2014.00787/abstract

Figure S1 | Growth of potato plantlets in MS agar plates. Plantlets grown on MS were transferred to new solid agar MS supplemented with various concentrations of NaCl (0, 25, 50, 100, and 200 mM) for 2 and 6 weeks, respectively.

REFERENCES

Abdullahil-Baque, M., Lee, E., j., Paek, K, Y., Ashley, M. K., and Grabov, A. (2010). Medium salt strength induced changes in growth, physiology and secondary metabolite content in adventitious roots of *Morinda citrifolia*: the role of antioxidant enzymes and phenylalanine amMonia lyase. *Plant Cell Rep.* 29, 685–694. doi: 10.1007/s00299-010-0854-4

Abrahám, E., Hourton-Cabassa, C., Erdei, L., and Szabados, L. (2010). Methods for determination of proline in plants. *Methods Mol. Biol.* 639, 317–331. doi: 10.1007/978-1-60761-702-0_20

Almeida, P. M., de Boer, G. J., and de Boer, A. H. (2014a). Assessment of natural variation in the first pore domain of the tomato HKT1;2 transporter and characterization of mutated versions of *SlHKT1;2* expressed in *Xenopus laevis* oocytes and via complementation of the salt sensitive *athkt1;1* mutant. *Front. Plant Sci.* 5:600. doi: 10.3389/fpls.2014.00600

Almeida, P., de Boer, G.-J., and de Boer, A. H. (2014b). Differences in shoot Na^+ accumulation between two tomato species are due to differences in ion affinity of HKT1;2. *J. Plant Physiol.* 171, 438–447. doi: 10.1016/j.jplph.2013.12.001

Amtmann, A., Troufflard, S., and Armengaud, P. (2008). The effect of potassium nutrition on pest and disease resistance in plants. *Physiol. Plant.* 133, 682–691. doi: 10.1111/j.1399-3054.2008.01075.x

Anschütz, U., Becker, D., and Shabala, S. (2014). Going beyond nutrition: regulation of potassium homoeostasis as a common denominator of plant adaptive responses to environment. *J. Plant Physiol.* 171, 670–687. doi: 10.1016/j.jplph.2014.01.009

Apse, M. P., and Blumwald, E. (2002). Engineering salt tolerance in plants. *Curr. Opin. Biotechnol.* 13, 146–150. doi: 10.1016/S0958-1669(02)00298-7

Apse, M. P., and Blumwald, E. (2007). Na^+ transport in plants. *FEBS Lett.* 581, 2247–2254. doi: 10.1016/j.febslet.2007.04.014

Arbona, V., Hossain, Z., López-Climent, M. F., Pérez-Clemente, R. M., and Gómez Cadenas, A. (2008). Antioxidant enzymatic activity is linked to waterlogging stress tolerance in citrus. *Physiol. Plant.* 132, 452–466. doi: 10.1111/j.1399-3054.2007.01029.x

Arnon, D. I. (1949). Copper enzymes in isolated chloroplasts. *Plant Physiol.* 24, 1–15. doi: 10.1104/pp.24.1.1

Ashley, M. K., Grant, M., and Grabov, A. (2006). Plant responses to potassium deficiencies: a role for potassium transport proteins. *J. Exp. Bot.* 57, 425–436. doi: 10.1093/jxb/erj034

Ashraf, M., and Harris, P. J. C. (2004). Potential biochemical indicators of salinity tolerance in plants. *Plant Sci.* 166, 3–16. doi: 10.1016/j.plantsci.2003.10.024

Azevedo-Neto, A. D., and Prisco, J. T. (2004). Effects of salt stress on plant growth, stomatal response and solute accumulation of different maize genotypes. *Braz. J. Plant Physiol.* 16, 31–38. doi: 10.1590/s1677-04202004000100005

Ben, A. C., Ben, R. B., Sensoy, S., Boukhriss, M., and Ben, A. F. (2010). Exogenous proline effects on photosynthetic performance and antioxidant defense system of young olive tree. *J. Agric. Food Chem.* 58, 4216–4222. doi: 10.1021/jf9041479

Ben-Amor, N., Megdiche, W., Jiménez, A., Sevilla, F., and Abdelly, C. (2010). The effect of calcium on the antioxidant systems in the halophyte *Cakile maritima* under salt stress. *Acta Physiol. Plant.* 32, 453–461. doi: 10.1007/s11738-009-0420-2

Bennici, A., and Tani, C. (2009). Ultrastructural effects of salinity in *Nicotiana bigelovii* var. *bigelovii callus cells and Allium cepa roots. Caryologia* 62, 124–133. doi: 10.1080/00087114.2004.10589677

Bennici, A., and Tani, C. (2012). Ultrastructural characteristics of callus cells of *Nicotiana tabacum* L. var. BELW3 grown in presence of NaCl. *Caryologia* 65, 72–81. doi: 10.1080/00087114.2012.678091

Blumwald, E., Aharon, G. S., and Apse, M. P. (2000). Sodium transport in plant cells. *Biochim. Biophys. Acta* 1465, 140–151. doi: 10.1016/S0005-2736citation(00)00135-8

Bojorquez-quintal, J., Velarde, A., Ku, A., Carrillo, M., Ortega, D., Echevarria, I., et al. (2014). Mechanisms of salt tolerance in habanero pepper plants (*Capsicum chinense* Jacq.): proline accumulation, ions dynamics, root-shoot partition and compartmentation. *Front. Plant Sci.* 5:605. doi: 10.3389/fpls.2014.00605

Bruns, S., and Hecht-Buchholz, C. (1990). Light and electron microscope studies on the leaves of several potato cultivars after application of salt at various development stages. *Potato Res.* 33, 33–41. doi: 10.1007/BF02358128

Cabot, C., Sibole, J. V., Barceló, J., and Poschenrieder, C. (2014). Lessons from crop plants struggling with salinity. *Plant Sci.* 226, 2–13. doi: 10.1016/j.plantsci.2014.04.013

Carden, D. E., Walker, D. J., Flowers, T. J., and Miller, A. J. (2003). Single-cell measurements of the contributions of cytosolic Na^+ and K^+ to salt tolerance. *Plant Physiol.* 131, 676–683. doi: 10.1104/pp.011445

Chan, Z., Grumet, R., and Loescher, W. (2011). Global gene expression analysis of transgenic, mannitol-producing, and salt-tolerant *Arabidopsis thaliana* indicates widespread changes in abiotic and biotic stress-related genes. *J. Exp. Bot.* 62, 4787–4803. doi: 10.1093/jxb/err130

Chen, Z., Pottosin, I. I., Cuin, T. A., Fuglsang, A. T., Tester, M., Jha, D., et al. (2007). Root plasma membrane transporters controlling K^+/Na^+ homeostasis in salt-stressed barley. *Plant Physiol.* 145, 1714–1725. doi: 10.1104/pp.107.110262

Cho, K., Kim, Y. C., Woo, J. C., Rakwal, R., Agrawal, G. K., Yoeun, S., et al. (2012). Transgenic expression of dual positional maize lipoxygenase-1 leads to the regulation of defense-related signaling molecules and activation of the antioxidative enzyme system in rice. *Plant Sci.* 238–245. doi: 10.1016/j.plantsci.2011.10.016

Csiszár, J., Gallé, A., Horváth, E., Dancsó, P., Gombos, M., Váry, Z., et al. (2012). Different peroxidase activities and expression of abiotic stress-related peroxidases in apical root segments of wheat genotypes with different drought stress tolerance under osmotic stress. *Plant Physiol. Biochem.* 52, 119–129. doi: 10.1016/j.plaphy.2011.12.006

Daneshmand, F., Arvin, M., and Kalantari, K. (2010). Physiological responses to NaCl stress in three wild species of potato *in vitro*. *Acta Physiol. Plant.* 32, 91–101. doi: 10.1007/s11738-009-0384-2

Deinlein, U., Stephan, A. B., Horie, T., Luo, W., Xu, G., and Schroeder, J. I. (2014). Plant salt-tolerance mechanisms. *Trends Plant Sci.* 19, 371–379. doi: 10.1016/j.tplants.2014.02.001

Delauney, A. J., Hu, C. A., and Verma, D. P. (1992). A bifunctional enzyme (delta 1-pyrroline-5-carboxylate synthetase) catalyzes the first two steps in proline biosynthesis in plants. *Proc. Natl. Acad. Sci. U.S.A.* 89, 9354–9358. doi: 10.1073/pnas.89.19.9354

Demidchik, V., Straltsova, D., Medvedev, S. S., Pozhvanov, G. A., Sokolik, A., and Yurin, V. (2014). Stress-induced electrolyte leakage: the role of K^+-permeable channels and involvement in programmed cell death and metabolic adjustment. *J. Exp. Bot.* 65, 1259–1270. doi: 10.1093/jxb/eru004

Demidchik, V. (2014). Mechanism and physiological roles of K^+ efflux from root cells. *J. Plant Physiol.* 171, 696–707. doi: 10.1016/j.jplph.2014.01.015

Ebrahimi, R., and Bhatla, S. (2012). Ion distribution measured by electron probe X-ray microanalysis in apoplastic and symplastic pathways in root cells in sunflower plants grown in saline medium. *J. Bioscience.* 37, 713–721. doi: 10.1007/s12038-012-9246-y

Ebrahimi, R., and Bhatla, S. C. (2011). Effect of sodium chloride levels on growth, water status, uptake, transport, and accumulation pattern of sodium and chloride Ions in young sunflower plants. *Commun. Soil Sci. Plant.* 42, 815–831. doi: 10.1080/00103624.2011.552657

Evelin, H., Giri, B., and Kapoor, R. (2012). Contribution of Glomus intraradices inoculation to nutrient acquisition and mitigation of ionic imbalance in NaCl-stressed *Trigonella foenum-graecum*. *Mycorrhiza* 22, 203–217. doi: 10.1007/s00572-011-0392-0

Ferreira, A., and Lima-Costa, M. (2008). Growth and ultrastructural characteristics of *Citrus* cells grown in medium containing NaCl. *Biol. Plant.* 52, 129–132. doi: 10.1007/s10535-008-0026-3

Giannopotitis, C. N., and Ries, S. K. (1977). Superoxide dismutase in higher plants. *Plant Physiol.* 59, 309–314. doi: 10.1104/pp.59.2.309

Greenway, H., and Munns, R. (1980). Mechanisms of Salt Tolerance in Nonhalophytes. *Annu. Rev. Plant Physiol.* 31, 149–190. doi: 10.1146/annurev.pp.31.060180.001053

Gupta, B., and Huang, B. (2014). Mechanism of salinity tolerance in plants: physiological, biochemical, and molecular characterization. *Int. J. Genom.* 2014:701596. doi: 10.1155/2014/701596

Han, Q. Q., Lü, X.P., Bai, J. P., Qiao, Y., Paré, P. W., Wang, S. M., et al. (2014). Beneficial soil bacterium *Bacillus subtilis* (GB03) augments salt tolerance of white clover. *Front. Plant Sci.* 5:525. doi: 10.3389/fpls.2014.00525

Hasegawa, P. M., Bressan, R. A., Zhu, J. K., and Bohnert, H. J. (2000). Plant cellular and molecular responses to high salinity. *Annu. Rev. Plant Biol.* 51, 463–499. doi: 10.1146/annurev.arplant.51.1.463

Hernández, J. A., Corpas, F. J., Gómez, M., Del-Río, L. A., and Sevilla, F. (1993). Salt-induced oxidative stress mediated by activated oxygen species in pea leaf mitochondria. *Physiol. Plant.* 89, 103–110. doi: 10.1111/j.1399-3054.1993.tb01792.x

Hmida-Sayari, A., Gargouri-Bouzid, R., Bidani, A., Jaoua, L., Savouré, A., and Jaoua, S. (2005). Overexpression of $^{\Delta}$1-pyrroline-5-carboxylate synthetase increases proline production and confers salt tolerance in transgenic potato plants. *Plant Sci.* 169, 746–752. doi: 10.1016/j.plantsci.2005.05.025

Hodges, D. M., John, M. D., Charles, F. F., and Robert, K. P. (2014). Improving the thiobarbituric acid-reactive-substances assay for estimating lipid peroxidation in plant tissues containing anthocyanin and other interfering compounds. *Planta* 207, 604–611. doi: 10.1007/s004250050524

Hossain, Z., Mandal, A. K. A., Shukla, R., and Datta, S. K. (2004). NaCl stress—its chromotoxic effects and antioxidant behavior in roots of *Chrysanthemum morifolium* Ramat. *Plant Sci.* 166, 215–220. doi: 10.1016/j.plantsci.2003.09.009

Hou, X., Liang, Y., He, X., Shen, Y., and Huang, Z. (2013). A novel ABA-responsive *TaSRHP* gene from wheat contributes to enhanced resistance to salt stress in *Arabidopsis thaliana*. *Plant Mol. Biol. Rep.* 31, 791–801. doi: 10.1007/s11105-012-0549-9

Huang, Z., Zhao, L., Chen, D., Liang, M., Liu, Z., Shao, H., et al. (2013). Salt stress encourages proline accumulation by regulating proline biosynthesis and degradation in *Jerusalem artichoke* plantlets. *PloS ONE* 8:e62085. doi: 10.1371/journal.pone.0062085

Jing, X., Hou, P., Lu, Y., Deng, S., Li, N., Zhao, R., et al. (2014). Overexpression of copper/zinc superoxide dismutase from mangrove *Kandelia candel* in tobacco enhances salinity tolerance by the reduction of reactive oxygen species in chloroplast. *Front. Physiol.* 5:515. doi: 10.3389/fphys.2014.00515

Kader, M. A., and Lindberg, S. (2005). Uptake of sodium in protoplasts of salt-sensitive and salt-tolerant cultivars of rice, *Oryza sativa* L. determined by the fluorescent dye SBFI. *J. Exp. Bot.* 56, 3149–3158. doi: 10.1093/jxb/eri312

Katerji, N., Van-Hoorn, J., Hamdy, A., and Mastrorilli, M. (2000). Salt tolerance classification of crops according to soil salinity and to water stress day index. *Agric. Water Manage.* 43, 99–109. doi: 10.1016/S0378-3774citation(99)00048-7

Khayyat, M., Tehranifar, A., Davarynejad, G. H., and Sayyari-Zahan, M. H. (2014). Vegetative growth, compatible solute accumulation, ion partitioning and chlorophyll fluorescence of 'Malas-e-Saveh' and 'Shishe-Kab' pomegranates in response to salinity stress. *Photosynthetica* 52, 301–312. doi: 10.1007/s11099-014-0034-9

Khelil, A., Menu, T., and Ricard, B. (2007). Adaptive response to salt involving carbohydrate metabolism in leaves of a salt-sensitive tomato cultivar. *Plant Physiol. Biochem.* 45, 551–559. doi: 10.1016/j.plaphy.2007.05.003

Kim, I., and Park, S. (2010). Ultrastructural characteristics of three chenopod halophytes lacking salt excretion structures. *J. Plant Biol.* 53, 314–320. doi: 10.1007/s12374-010-9119-6

Koca, H., Ozdemir, F., and Turkan, I. (2006). Effect of salt stress on lipid peroxidation and superoxide dismutase and peroxidase activities of *Lycopersicon esculentum* and *L. pennellii*. *Biol. Plant.* 50, 745–748. doi: 10.1007/s10535-006-0121-2

Krantev, A., Yordanova, R., Janda, T., Szalai, G., and Popova, L. (2008). Treatment with salicylic acid decreases the effect of cadmium on photosynthesis in maize plants. *J. Plant Physiol.* 165, 920–931. doi: 10.1016/j.jplph.2006.11.014

Kronzucker, H. J., Szczerba, M. W., Moazami-Goudarzi, M., and Britto, D. T. (2006). The cytosolic Na^+: K^+ ratio does not explain salinity-induced growth impairment in barley: a dual-tracer study using 42 $K^+$and 24 Na^+. *Plant Cell Environ.* 29, 2228–2237. doi: 10.1111/j.1365-3040.2006.01597.x

Krzesłowska, M. (2010). The cell wall in plant cell response to trace metals: polysaccharide remodeling and its role in defense strategy. *Acta Physiol. Plant.* 33, 35–51. doi: 10.1007/s11738-010-0581-z

Lai, D., Mao, Y., Zhou, H., Li, F., Wu, M., and Zhang, J., et al. (2014). Endogenous hydrogen sulfide enhances salt tolerance by coupling the reestablishment of redox homeostasis and preventing salt-induced K^+ loss in seedlings of *Medicago sativa*. *Plant Sci.* 225, 117–129. doi: 10.1016/j.plantsci.2014.06.006

Liu, W., Yuan, X., Zhang, Y., Xuan, Y., and Yan, Y. (2014a). Effects of salt stress and exogenous Ca^{2+} on Na^+ compartmentalization, ion pump activities of tonoplast and plasma membrane in *Nitraria tangutorum* Bobr. leaves. *Acta Physiol. Plant.* 36, 2183–2193. doi: 10.1007/s11738-014-1595-8

Liu, X., Mak, M., Babla, M., Wang, F., Chen, G., Veljanoski, F., et al. (2014b). Linking stomatal traits and expression of slow anion channel genes *HvSLAH1* and *HvSLAC1* with grain yield for increasing salinity tolerance in barley. *Front. Plant Sci.* 5:634. doi: 10.3389/fpls.2014.00634

Lu, Y., Li, N., Sun, J., Hou, P., Jing, X., and Zhu, H., et al. (2013). Exogenous hydrogen peroxide, nitric oxide and calcium mediate root ion fluxes in two non-secretor mangrove species subjected to NaCl stress. *Tree Physiol.* 33, 81–95. doi: 10.1093/treephys/tps119

Luna, C., Seffino, L. G., Arias, C., and Taleisnik, E. (2000). Oxidative stress indicators as selection tools for salt tolerance in *Chloris gayana*. *Plant Breed.* 119, 341–345. doi: 10.1046/j.1439-0523.2000.00504.x

Lv, S., Nie, L., Fan, P., Wang, X., Jiang, D., Chen, X., et al. (2011). Sodium plays a more important role than potassium and chloride in growth of *Salicornia europaea*. *Acta Physiol. Plant.* 34, 503–513. doi: 10.1007/s11738-011-0847-0

Ma, Q., Yue, L. J., Zhang, J. L., Wu, G. Q., Bao, A. K., and Wang, S. M. (2012). Sodium chloride improves photosynthesis and water status in the succulent xerophyte *Zygophyllum xanthoxylum*. *Tree Physiol.* 32, 4–13. doi: 10.1093/treephys/tpr098

Maathuis, F. J. M. (2014). Sodium in plants: perception, signalling, and regulation of sodium fluxes. *J. Exp. Bot.* 65, 849–858. doi: 10.1093/jxb/ert326

Maathuis, F. J. M., Ahmad, I., and Patishtan, J. (2014). Regulation of Na^+ fluxes in plants. *Front. Plant Sci.* 5:467. doi: 10.3389/fpls.2014.00467

Maathuis, F. J. M., and Amtmann, A. (1999). K^+ nutrition and Na^+ toxicity: the basis of cellular K^+/Na^+ ratios. *Ann. Bot.* 84, 123–133. doi: 10.1006/anbo.1999.0912

Miao, B. H., Han, X. G., and Zhang, W. H. (2010). The ameliorative effect of silicon on soybean seedlings grown in potassium-deficient medium. *Ann. Bot.* 105, 967–973. doi: 10.1093/aob/mcq063

Mishra, P., Bhoomika, K., and Dubey, R. (2011). Differential responses of antioxidative defense system to prolonged salinity stress in salt-tolerant and salt-sensitive Indica rice (*Oryza sativa* L.) seedlings. *Protoplasma* 250, 3–19. doi: 10.1007/s00709-011-0365-3

Mitsuya, S., Takeoka, Y., and Miyake, H. (2000). Effects of sodium chloride on foliar ultrastructure of sweet potato (*Ipomoea batatas* Lam.) plantlets grown under light and dark conditions *in vitro*. *J. Plant Physiol.* 157, 661–667. doi: 10.1016/S0176-1617citation(00)80009-7

Mittova, V., Guy, M., Tal, M., and Volokita, M. (2004). Salinity up-regulates the antioxidative system in root mitochondria and peroxisomes of the wild salt-tolerant tomato species *Lycopersicon pennellii*. *J. Exp. Bot.* 55, 1105–1113. doi: 10.1093/jxb/erh113

Miyake, H., Mitsuya, S., and Rahman, M. S. (2006). "Ultrastructural effects of salinity stress in higher plants," in *Abiotic Stress Tolerance in Plants*, eds R. Ashwanik and T. Teruhiro (Dordrecht: Springer), 215–226.

Munns, R., and Tester, M. (2008). Mechanisms of salinity tolerance. *Annu. Rev. Plant Biol.* 59, 651–681. doi: 10.1146/annurev.arplant.59.032607.092911

Naeini, M. R., Khoshgoftarmanesh, A. H., and Fallahi, E. (2006). Partitioning of chlorine, sodium, and potassium and shoot growth of three pomegranate cultivars under different levels of salinity. *J. Plant Nutr.* 29, 1835–1843. doi: 10.1080/01904160600899352

Navarro, A., Bañon, S., Olmos, E., and Sánchez-Blanco, M. D. J. (2007). Effects of sodium chloride on water potential components, hydraulic conductivity, gas exchange and leaf ultrastructure of *Arbutus unedo* plants. *Plant Sci.* 172, 473–480. doi: 10.1016/j.plantsci.2006.10.006

Neocleous, D., and Vasilakakis, M. (2007). Effects of NaCl stress on red raspberry (*Rubus idaeus* L. *'Autumn Bliss'*). *Sci. Horticult.* 112, 282–289. doi: 10.1016/j.scienta.2006.12.025

Nounjan, N., Nghia, P. T., and Theerakulpisut, P. (2012). Exogenous proline and trehalose promote recovery of rice seedlings from salt-stress and differentially modulate antioxidant enzymes and expression of related genes. *J. Plant Physiol.* 169, 596–604. doi: 10.1016/j.jplph.2012.01.004

Peng, Y. H., Zhu, Y. F., Mao, Y. Q., Wang, S. M., Su, W. A., and Tang, Z. C. (2004). Alkali grass resists salt stress through high $[K^+]$ and an endodermis barrier to Na^+. *J. Exp. Bot.* 55, 939–949. doi: 10.1093/jxb/erh071

Plett, D. C., and Moller, I. S. (2010). Na^+ transport in glycophytic plants: what we know and would like to know. *Plant Cell Environ.* 33, 612–626. doi: 10.1111/j.1365-3040.2009.02086.x

Queirós, F., Fontes, N., Silva, P., Almeida, D., Maeshima, M., Gerós, H., et al. (2009). Activity of tonoplast proton pumps and Na^+/H^+ exchange in potato cell cultures is modulated by salt. *J. Exp. Bot.* 60, 1363–1374. doi: 10.1093/jxb/erp011

Queirós, F., Rodrigues, J. A., Almeida, J. M., Almeida, D. P., and Fidalgo, F. (2011). Differential responses of the antioxidant defence system and ultrastructure in a salt-adapted potato cell line. *Plant Physiol. Biochem.* 49, 1410–1419. doi: 10.1016/j.plaphy.2011.09.020

Rivero, R. M., Mestre, T. C., Mittler, R., Rubio, F., Garcia-Sanchez, F., and Martinez, V. (2014). The combined effect of salinity and heat reveals a specific physiological, biochemical and molecular response in tomato plants. *Plant Cell Environ.* 37, 1059–1073. doi: 10.1111/pce.12199

Roy, S. J., Negrão, S., and Tester, M. (2014). Salt resistant crop plants. *Curr. Opin. Biotechnol.* 26, 115–124. doi: 10.1016/j.copbio.2013.12.004

Ruan, C. J., Qin, P., He, Z. X., and Xie, M. (2005). Concentrations of major and minor mineral elements in different organs of *Kosteletzkya virginica* and saline soils. *J. Plant Nutr.* 28, 1191–1200. doi: 10.1081/PLN-200063228

Ruffino, A., Rosa, M., Hilal, M., González, J., and Prado, F. (2010). The role of cotyledon metabolism in the establishment of quinoa (*Chenopodium quinoa*) seedlings growing under salinity. *Plant Soil* 326, 213–224. doi: 10.1081/pln-200063228

Sabatini, D. D., Bensch, K., and Barrnett, R. J. (1963). Cytochemistry and electron microscopy. The preservation of cellular ultrastructure and enzymatic activity by aldehyde fixation. *J. Cell Biol.* 17, 19–58. doi: 10.1083/jcb.17.1.19

Sabra, A., Daayf, F., and Renault, S. (2012). Differential physiological and biochemical responses of three *Echinacea* species to salinity stress. *Sci. Horticult.* 135, 23–31. doi: 10.1016/j.scienta.2011.11.024

Sekmen, A. H., Turkan, I., and Takio, S. (2007). Differential responses of antioxidative enzymes and lipid peroxidation to salt stress in salt-tolerant *Plantago maritima* and salt-sensitive *Plantago media*. *Physiol. Plant.* 131, 399–411. doi: 10.1111/j.1399-3054.2007.00970.x

Serraj, R., and Sinclair, T. (2002). Osmolyte accumulation: can it really help increase crop yield under drought conditions? *Plant Cell Environ.* 25, 333–341. doi: 10.1046/j.1365-3040.2002.00754.x

Shabala, S., and Cuin, T. A. (2008). Potassium transport and plant salt tolerance. *Physiol. Plant.* 133, 651–669. doi: 10.1111/j.1399-3054.2007.01008.x

Shabala, S., and Pottosin, I. (2014). Regulation of potassium transport in plants under hostile conditions: implications for abiotic and biotic stress tolerance. *Physiol. Plant.* 151, 257–279. doi: 10.1111/ppl.12165

Shabala, S., Bose, J., and Hedrich, R. (2014). Salt bladders: do they matter? *Trends Plant Sci.* 19, 687–691. doi: 10.1016/j.tplants.2014.09.001

Shu, S., Yuan, L. Y., Guo, S. R., Sun, J., and Yuan, Y. H. (2013). Effects of exogenous spermine on chlorophyll fluorescence, antioxidant system and ultrastructure of chloroplasts in *Cucumis sativus* L. under salt stress. *Plant Physiol. Biochem.* 63, 209–216. doi: 10.1016/j.plaphy.2012.11.028

Silva-Ortega, C. O., Ochoa-Alfaro, A. E., Reyes-Agüero, J. A., Aguado-Santacruz, G. A., and Jiménez-Bremont, J. F. (2008). Salt stress increases the expression of p5cs gene and induces proline accumulation in cactus pear. *Plant Physiol. Biochem.* 46, 82–92. doi: 10.1016/j.plaphy.2007.10.011

Spurr, A. R. (1969). A low-viscosity epoxy resin embedding medium for electron microscopy. *J. Ultrastruct. Res.* 26, 31–43. doi: 10.1016/s0022-5320citation(69)90033-1

Stewart, R. R., and Bewley, J. D. (1980). Lipid peroxidation associated with accelerated aging of soybean axes. *Plant Physiol.* 65, 245–248. doi: 10.1104/pp.65.2.245

Su, X., Chu, Y., Li, H., Hou, Y., Zhang, B., Huang, Q., et al. (2011). Expression of multiple resistance genes enhances tolerance to environmental stressors in transgenic poplar (*Populus × euramericana* 'Guariento'). *PLoS ONE* 6:e24614. doi: 10.1371/journal.pone.0024614

Sun, J., Dai, S., Wang, R., Chen, S., Li, N., and Zhou, X., et al. (2009). Calcium mediates root K^+/Na^+ homeostasis in poplar species differing in salt tolerance. *Tree Physiol.* 29, 1175–1186. doi: 10.1093/treephys/tpp048

Takahashi, R., Nishio, T., Ichizen, N., and Takano, T. (2007). Salt-tolerant reed plants contain lower Na^+ and higher K^+ than salt-sensitive reed plants. *Acta Physiol. Plant.* 29, 431–438. doi: 10.1007/s11738-007-0052-3

Tattini, M., and Traversi, M. L. (2009). On the mechanism of salt tolerance in olive (*Olea europaea* L.) under low or high Ca^{2+} supply. *Environ. Exp. Bot.* 65, 72–81. doi: 10.1016/j.envexpbot.2008.01.005

Tester, M., and Davenport, R. (2003). Na^+ tolerance and Na^+ transport in higher plants. *Ann. Bot.* 91, 503–527. doi: 10.1093/aob/mcg058

Vij, S., and Tyagi, A. K. (2007). Emerging trends in the functional genomics of the abiotic stress response in crop plants. *Plant Biotechnol. J.* 5, 361–380. doi: 10.1111/j.1467-7652.2007.00239.x

Waditee, R., Bhuiyan, N. H., Hirata, E., Hibino, T., Tanaka, Y., Shikata, M., et al. (2007). Metabolic engineering for betaine accumulation in microbes and plants. *J. Biol. Chem.* 282, 34185–34193. doi: 10.1074/jbc.M704939200

Wang, Y. C., Qu, G. Z., Li, H. Y., Wu, Y. J., Wang, C., Liu, G. F., et al. (2010). Enhanced salt tolerance of transgenic poplar plants expressing a manganese superoxide dismutase from *Tamarix androssowii*. *Mol. Biol. Rep.* 37, 1119–1124. doi: 10.1007/s11033-009-9884-9

Wang, Y., and Wu, W. H. (2010). Plant sensing and signaling in response to K^+ deficiency. *Mol. Plant* 3, 280–287. doi: 10.1093/mp/ssq006

Wang, Z. Q., Yuan, Y. Z., Ou, J. Q., Lin, Q. H., and Zhang, C. F. (2007). Glutamine synthetase and glutamate dehydrogenase contribute differentially to proline accumulation in leaves of wheat (*Triticum aestivum*) seedlings exposed to different salinity. *J. Plant Physiol.* 164, 695–701. doi: 10.1016/j.jplph.2006.05.001

Wilkinson, S., and Davies, W. J. (2002). ABA-based chemical signalling: the coordination of responses to stress in plants. *Plant Cell Environ.* 25, 195–210. doi: 10.1046/j.0016-8025.2001.00824.x

Xue, Z., Zhao, S., Gao, H., and Sun, S. (2013). The salt resistance of wild soybean (*Glycine soja* Sieb. et Zucc. ZYD 03262) under NaCl stress is mainly determined by Na^+ distribution in the plant. *Acta Physiol. Plant.* 36, 61–70. doi: 10.1007/s11738-013-1386-7

Yamane, K., Rahman, M. S., Kawasaki, M., Taniguchi, M., and Miyake, H. (2004). Pretreatment with antioxidants decreases the effects of salt stress on chloroplast ultrastructure in rice leaf segments (*Oryza sativa* L.). *Plant Prod. Sci.* 7, 292–300. doi: 10.1626/pps.7.292

Yamori, W., Suzuki, K., Noguchi, K., Nakai, M., and Terashima, I. (2006). Effects of Rubisco kinetics and Rubisco activation state on the temperature dependence of the photosynthetic rate in spinach leaves from contrasting growth temperatures. *Plant Cell Environ.* 29, 1659–1670. doi: 10.1111/j.1365-3040.2006.01550.x

Yao, R., and Fang, S. (2008). Effect of NaCl stress on ion distribution in roots and growth of *Cyclocarya paliurus* seedlings. *Front. Forest. China* 4, 208–215. doi: 10.1007/s11461-009-0007-5

Yazici, I., Tuerkan, I., Sekmen, A. H., and Demiral, T. (2007). Salinity tolerance of purslane (*Portulaca oleracea* L.) is achieved by enhanced antioxidative system, lower level of lipid peroxidation and proline accumulation. *Environ. Exp. Bot.* 61, 49–57. doi: 10.1016/j.envexpbot.2007.02.010

Yousfi, S., Rabhi, M., Hessini, K., Abdelly, C., and Gharsalli, M. (2010). Differences in efficient metabolite management and nutrient metabolic regulation between wild and cultivated barley grown at high salinity. *Plant Biol.* 12, 650–658. doi: 10.1111/j.1438-8677.2009.00265.x

Zhang, J. L., and Shi, H. (2013). Physiological and molecular mechanisms of plant salt tolerance. *Photosynth. Res.* 115, 1–22. doi: 10.1007/s11120-013-9813-6

Zhang, J. L., Flowers, T. J., and Wang, S. M. (2010). Mechanisms of sodium uptake by roots of higher plants. *Plant Soil* 326, 45–60. doi: 10.1007/s11104-009-0076-0

Zhang, X., Lu, G., Long, W., Zou, X., Li, F., and Nishio, T. (2014). Recent progress in drought and salt tolerance studies in Brassica crops. *Breed. Sci.* 64, 60–73. doi: 10.1270/jsbbs.64.60

Zhao, F. Y., and Zhang, H. (2006). Salt and paraquat stress tolerance results from co-expression of the *Suaeda salsa* glutathione S-transferase and catalase in transgenic rice. *Plant Cell Tiss. Org.* 86, 349–358. doi: 10.1007/s11240-006-9133-z

Zhou, J., Wang, J. J., and Bi, Y. F. (2014). Overexpression of *PtSOS2* enhances salt tolerance in transgenic poplars. *Plant Mol. Biol. Rep.* 32, 185–197. doi: 10.1007/s11105-013-0640-x

Zhu, J. K. (2002). Salt and drought stress signal transduction in plants. *Annu. Rev. Plant Biol.* 53, 247–273. doi: 10.1146/annurev.arplant.53.091401.143329

14

Making plants break a sweat: The structure, function, and evolution of plant salt glands

Maheshi Dassanayake* and John C. Larkin*

Department of Biological Sciences, Louisiana State University, Baton Rouge, LA, USA

Edited by:
Vadim Volkov,
London Metropolitan University, UK
Reviewed by:
Sergey Shabala,
University of Tasmania, Australia
Edward Glenn,
University of Arizona, USA
Wee Kee Tan,

Salt stress is a complex trait that poses a grand challenge in developing new crops better adapted to saline environments. Some plants, called recretohalophytes, that have naturally evolved to secrete excess salts through salt glands, offer an underexplored genetic resource for examining how plant development, anatomy, and physiology integrate to prevent excess salt from building up to toxic levels in plant tissue. In this review we examine the structure and evolution of salt glands, salt gland-specific gene expression, and the possibility that all salt glands have originated via evolutionary modifications of trichomes. Salt secretion via salt glands is found in more than 50 species in 14 angiosperm families distributed in caryophyllales, asterids, rosids, and grasses.The salt glands of these distantly related clades can be grouped into four structural classes. Although salt glands appear to have originated independently at least 12 times, they share convergently evolved features that facilitate salt compartmentalization and excretion. We review the structural diversity and evolution of salt glands, major transporters and proteins associated with salt transport and secretion in halophytes, salt gland relevant gene expression regulation, and the prospect for using new genomic and transcriptomic tools in combination with information from model organisms to better understand how salt glands contribute to salt tolerance. Finally, we consider the prospects for using this knowledge to engineer salt glands to increase salt tolerance in model species, and ultimately in crops.

Keywords: salt glands, halophytes, trichomes, salt secretion, convergent evolution

INTRODUCTION

Plants face many challenges from the abiotic world, and among the most significant of these is salt stress. Salt water intrusion due to rising sea levels in coastal regions, extensive irrigation in arid regions, and widespread erosion contribute to increasing soil salinity, limiting agricultural productivity and preventing the use of much needed marginal lands (IPCC, 2014). Indeed, it is no exaggeration to say that breeding crops with increased salt tolerance is among the most significant challenges facing 21st century agriculture. Virtually all major crops, with a few exceptions (e.g., *Chenopodium quinoa* and *Gossypium hirsutum*), are naturally sensitive to salt stress. Only about 0.25% of all flowering plants are reportedly able to complete their lifecycle in saline soils (Flowers et al., 2010) and are hence considered to be halophytes. Although halophytes have evolved independently in a variety of taxonomically diverse lineages, they exhibit many examples of convergently evolved adaptations to salt stress (Flowers et al., 2010; Bromham, 2015).

The capacity to generate high-throughput genomic and transcriptomic data from non-model plant species has catalyzed the growth of comparative, functional and evolutionary genomics, and this new knowledge base provides opportunities for understanding the mechanisms underpinning the halophytic lifestyle and also provides opportunities for adapting these lessons to improving the salt tolerance of agricultural crops.

A significant proportion of halophytes have evolved specialized epidermal structures called salt glands to store and exclude salt (Flowers and Colmer, 2015; Santos et al., 2016). The epidermis is the surface through which a plant interacts with its environment, and thus the epidermis has a wide variety of functional specializations at the cellular level. Some of these, including stomates for gas exchange and cuticle-covered pavement cells that prevent dehydration and pathogen attack, are shared by most land plants and all angiosperms. In addition, plants have developed a myriad of epidermal structural adaptations to defend themselves from or to exploit their environments, such as trichomes, nectaries, prickles, and hydathodes, which range in complexity from specialized single cells to multicellular structures consisting of several cell types (Esau, 1965). Although all salt glands function to increase salt tolerance, they differ in structural complexity and mechanism of salt exclusion, suggesting that salt glands have multiple evolutionary origins (Flowers et al., 2010).

Salt uptake, signaling, transport, detoxification, and storage mechanisms are among the integral biological processes we need to understand in solving the puzzle of salt adaptation (see reviews Hasegawa et al., 2000; Deinlein et al., 2014). The use of halophytes to study these processes is rare (see reviews Flowers et al., 2010; Shabala et al., 2015; Volkov, 2015), and the targeted use of specialized structures such as salt glands to study salt exclusion in a molecular genetic framework is even less common. The scarcity of genetic, cellular, or biochemical research on salt glands could be due to their occurrence on diverse taxa in plant families that are ecologically important, but not economically valued as crops. Limited research focusing on salt glands also may have arisen from the difficulty in studying salt glands as an isolated system consisting of just a few cells in the leaf epidermis. The magnitude of such barriers is, however, declining as new molecular genetic tools become available that make non-model organisms and rare cell types more tractable to study (Schwab and Ossowski, 2006; Deal and Henikoff, 2011; Olofsson et al., 2012; Etalo et al., 2015). Salt glands are found mostly on leaves of plants that grow on dry saline soils, on salt marsh grasses, and in a variety of mangroves, which are woody plants that inhabit tropical and subtropical intertidal zones (Flowers et al., 1986; Tomlinson, 1986). Therefore, most of the salt gland baring plants are also considered as halophytes, but a few exceptions are found throughout land plants (Mooney et al., 1980; Chen and Chen, 2005; Maricle et al., 2009; Peng et al., 2016). Although plant models such as *Arabidopsis* and rice are devoid of salt glands, they still have the analogous cell structures and the orthologous gene families that are likely key effectors in sensing, transporting, and compartmentalizing salt in halophytes that carry salt glands. We are now at a point where a comparison between the extensive information available from models such as *Arabidopsis* and new genomic resources from halophytes naturally selected for salt stress adaptation can illuminate key aspects of this important adaptation (Oh et al., 2012). Therefore, in this review, we attempt to evaluate the structure and development of salt glands, as well as the existing genetic resources that have been largely underexplored in plants equipped with salt glands, and we also assess the practicality of using model systems to effectively study them. Finally, we consider the feasibility of improving salt tolerance by engineering existing trichomes on *Arabidopsis* to function as salt glands and challenges associated with the gap in our knowledge to develop engineered salt glands in candidate crops.

SALT GLANDS ARE STRUCTURALLY DIVERSE

The term "salt gland" is quite broad, and has been applied to a wide variety of structures with different anatomical features and functional mechanisms. Halophytes with salt glands are collectively termed salt secretors (Liphschitz et al., 1974) or recretohalophytes (Breckle, 1990). From a structural perspective, all salt glands appear to be largely epidermal in origin and thus are in essence specialized trichomes (Esau, 1965). From a functional perspective, there are two types of salt glands, those that directly secrete salts to the surface of the leaf (exo-recretohalophytes), and those that collect salt in the vacuole of a specialized bladder cell

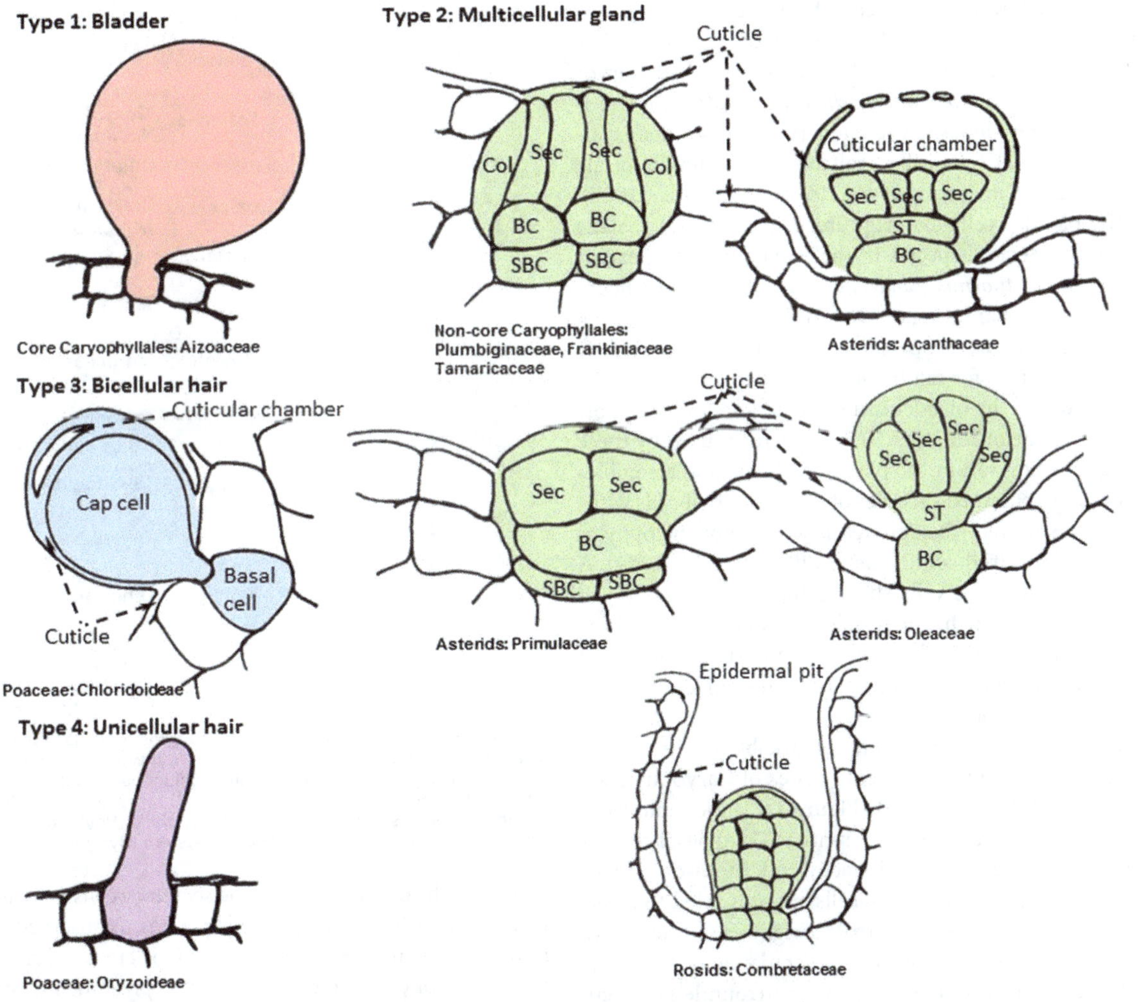

FIGURE 1 | Representative cellular organization of distinct salt gland structures found in angiosperms. Drawings are based on consensus representations of species specific salt gland structures. References used to create consensus figures for each type are given in **Table 1**. The cells that constitute the salt gland are colored while the adjacent cells are kept blank. The continuous cuticle around the salt gland is also colored and changed to blank when the cuticle overlays the surrounding epidermis. The dynamic intracellular structures such as vacuoles, vesicles, and laminated membranes are not depicted in the representative figures. Collecting cell (Col), secretory cell (Sec), basal cell (BC), sub-basal cell (SBC), stalk cell (ST).

(endo-recretohalophytes) (Breckle, 1990; Ding et al., 2010b). Although few species of plants have salt glands, they are distributed among four major divisions of flowering plants: Caryophyllales, asterids, rosids, and Poaceae (Santos et al., 2016). This broad phylogenetic distribution suggests that salt glands have originated independently multiple times as previously proposed for halophyte origins (Flowers et al., 2010). Yet the salt glands of widely divergent species have many phenotypic similarities, providing some striking examples of convergent evolution that give insight into the mechanisms through which salt glands protect plants. The similarities among salt glands enable categorization into four broad structural groups: (1) salt bladders consisting of a large vacuolated cell with or without 1 to 2 stalk cells, found only in Aizoaceae and Amaranthaceae (**Figure 1**, Type 1), (2) multicellular salt glands varying from 4 to 40 cells, with cells typically differentiated into collecting and secretory cells in a cuticle lined structure, widely distributed phylogenetically (**Figure 1**, Type 2), (3) bicellular secretory hair-like structures with a basal cell and a cap cell, found in chloridoid grasses (**Figure 1**, Type 3), and (4) unicellular highly vacuolated secretory hairs (found in *Porteresia*) (**Figure 1**, Type 4). The first two structural types are found in eudicots while the third and fourth types are found in monocots (**Figure 2**).

Among eudicots the structurally simplest form of salt glands, called salt bladders, are found in two families in the order Caryophyllales (**Figure 1**). In *Mesembryanthemum crystallinum* (Aizoaceae) salt is simply deposited in the large vacuole of specialized swollen epidermal cells called salt bladders (Steudle et al., 1975; Lüttge et al., 1978; Adams et al., 1998; Agarie et al., 2007). Eventually the bladder cells may rupture, depositing salt on the epidermal surface. Several species in the Amaranthaceae, exemplified by *Atriplex lentiformis, Bienertia sinuspersici,* and *Chenopodium quinoa* (Karimi and Ungar, 1989; Akhani et al., 2005; Park et al., 2009; Adolf et al., 2013; Shabala et al., 2014), have a slightly more elaborate structure for salt bladders compared to that of *M. crystallinum*, in which the bladder cell is located on top of a short stalk consisting of one or few cells. The mechanism used by these plants for sequestering salt in the bladder cell vacuole resembles the storage of salt in enlarged vacuoles of the mesophyll cells within succulent leaves in many halophytes as well as non-halophytes upon salt stress (Longstreth and Nobel, 1979; Park et al., 2009). A mutant line lacking bladder cells showed high sensitivity to salt and severely limited growth under salt stress compared to the wild type *M. crystallinum*, establishing the important role of salt compartmentalization and ion homeostasis achieved through salt bladders (Agarie et al., 2007).

The level of convergence is quite remarkable in the second type of salt glands spanning the diverse clades of Caryophyllales, asterids, and rosids (Shi et al., 2005) (**Figure 1**). These multicellular glands typically have cell types differentiated into basal collecting cells and distal secretory cells (Faraday and Thomson, 1986b; Thomson et al., 1988). The collecting cells are presumed to create a salt efflux gradient to collect salt from neighboring mesophyll cells and transport it to secretory cells (Faraday and Thomson, 1986a,b). The secretory cells are completely surrounded by a cuticle, with the exception of where they contact the subtending basal collecting cells, a feature which appears to channel the flow of salt through the secretory cells and prevent leakage back into the neighboring tissue via the apoplast (Thomson and Liu, 1967; Campbell et al., 1974; Tan et al., 2013). It is not uncommon to see the cuticle layer wrapped around the basal collecting cell if the collecting cell is partially above the epidermal layer (Thomson, 1975; Thomson et al., 1988). The secretory cells are cytoplasmically dense, possessing many mitochondria and endomembranes, and have internal projections of the cell wall (Vassilyev and Stepanova, 1990), resembling those in phloem transfer cells, which are presumed to increase surface area (Gunning and Pate, 1969).

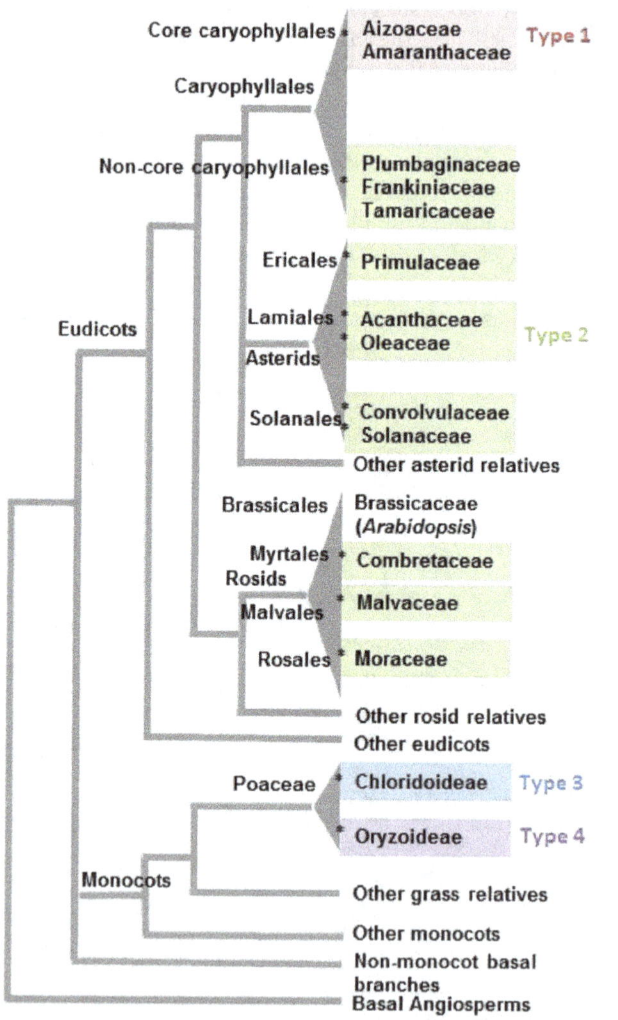

FIGURE 2 | Phylogenetic distribution of angiosperm clades reported to have salt glands. Families associated with the four types of salt glands discussed in this review are grouped in the same colors used to distinguish salt gland types in Figure 1. The asterisk symbols (*) represent 12 likely independent introductions of salt glands into a family/clade. Phylogenetic relationships are based on the APG IV classification system (Byng et al., 2016).

Although the outer surface of the secretory cells is covered with cuticle, this cuticle is either pierced by pores, as observed in *Limonium bicolor* salt glands (Feng et al., 2015), or creates a cuticular chamber on top of the secretory cells that is presumed to store secreted salts, as observed in salt glands of *Avicennia marina* (Campbell and Thomson, 1976; Naidoo, 2016) and *Aeluropus littoralis* (Barhoumi et al., 2008) (**Figure 1** type 2 and 3). Contrasting the secretory cells, the collecting cells have numerous plasmodesmata connections amongst surrounding mesophyll cells. Thus it appears that salt is actively transported through the symplast from the collecting cells into the secretory cells, and then the salt solution is deposited outside the cell via the pores in the cuticle (Campbell and Stong, 1964; Campbell and Thomson, 1976). These salt glands are organized into a bulbous or discoid structure where salt is extruded from the top of the dome or cup-shaped center. The entire structure is often sunken into the epidermis, such that the cuticle overlaying the secretory cells is at or slightly below the level of the ground epidermal cells. This type of salt gland is represented by plants in the Tamaricaceae (Campbell and Stong, 1964; Xue and Wang, 2008) (e.g., *Tamarix* and *Reaumuria*), Frankeniaceae (Campbell and Thomson, 1976) (e.g., *Frankenia* spp.), and Plumbaginaceae (Faraday and Thomson, 1986b) (e.g., *Limonium, Aegialitis,* and *Limoniastrum*), all of which are closely related families in Caryophyllales (Byng et al., 2016). The rest of the eudicot salt glands share the same core structure with slight modifications.

The Type 2 multicellular salt glands of asterids (**Figure 1**), which are distributed among five families (**Figure 2; Table 1**), tend to have one or two stalk cells connecting the secretory cells to the basal collecting cells contrasting the structure of the Tamarix-type salt glands (Shimony et al., 1973; Drennan et al., 1987; Das, 2002). While maintaining the overall similarity of the structure with a cuticular envelope covering the salt gland, the number of secretory cells compared to the number of collecting cells varies between species in the asterids. For example, *Aegiceras corniculatum* and *Glaux maritima* (Primulaceae) have salt glands consisting of 8–40 radially arranged secretory cells atop a single basal cell (Cardale and Field, 1971; Rozema et al., 1977) while the mangroves, *Avicennia* and *Acanthus* spp. (Acanthaceae), have salt glands consisting of two to four collecting cells connected by one or two stalk cells to eight to twelve radially arranged secretory cells (Shimony et al., 1973; Drennan et al., 1987; Das, 2002). Similar to *Tamarix*, the cuticle of the secretory cells contains pores through which the saline solution is secreted; the secretory cells are cytoplasmically dense and rich in mitochondria and endomembranes, and the basal cell is highly vacuolated. Plasmodesmata connect the basal cell to the secretory cells and to the underlying sub-basal cells. The less studied *Cressa cretica* (Convolvulaceae) and *Phillyrea latifolia* (Oleaceae) also produce multicellular salt glands consisting of multiple secreting cells connected by a stalk cell to vacuolated basal collecting cells, similar to the other asterid salt glands (Weiglin and Winter, 1988).

Only a few species reportedly have salt glands in the large rosid clade. The mangrove *Laguncularia racemosa* in Combretaceae has multicellular salt glands located in deep adaxial epidermal pits of the leaf (Francisco et al., 2009). The pit is likely lined by a thick cuticle and the secretory cells at the base of the pit are dense in cytoplasm. Salt is extruded as a chain of crystals from the narrow mouth of the pit (Stace, 1965; Tomlinson, 1986; Francisco et al., 2009). Although the anatomy of these glands has not been described in detail, they have been shown to secrete salt (Sobrado, 2004). Mangrove species in two other genera in the family Combretaceae, *Lumnitzera* and *Conocarpus*, have similar structures, but there is no direct evidence to confirm that these glands function as salt glands (Tomlinson, 1986; Parida and Jha, 2010). Despite their diverse phylogenetic origins, all mangrove salt glands appear to have a similar structural organization spanning asterids and rosids. Additionally, two non-halophytic species in the rosids, *Gossypium hirsutum* (Malvaceae) and *Ficus formosana* (Moraceae), develop salt secreting glandular trichomes. Going by the broad definition of salt glands, these species show the capacity to extrude salt through salt glands on leaves and the structures described are similar to multicellular glandular trichomes described for salt glands in halophytes (Chen and Chen, 2005; Peng et al., 2016). In *Gossypium hirsutum* the ability to exclude NaCl under salt stress via leaf salt glands is thought to be an adaptation shared with ancestral genotypes from coastal regions (Peng et al., 2016). Excretion of salt through salt glands in non-halophytes may represent a facultative trait in response to salt stress derived from halophytic ancestral traits.

TABLE 1 I Halophytes reported with salt glands, their salt gland structural organization, and availability of sequence resources.

Clade/Family	Species	Structure	References for publicly available cDNA/RNAseq data
Asterids			
Acanthaceae	*Acanthus ebracteatus**, *A. ilicifolius**	Organized into secretory, stalk, and basal cells (Das, 2002; Ong and Gong, 2013)	ESTs (Nguyen et al., 2006, 2007); RNAseq (Yang et al., 2015)
	*Avicennia germinans**, *A. officinalis**, *A. marina**	Organized into secretory, stalk, and basal cells (Shimony et al., 1973; Drennan et al., 1987; Balsamo and Thomson, 1993; Tan et al., 2013; Naidoo, 2016)	RNAseq (Huang et al., 2014); ESTs (Mehta et al., 2004; Jyothi-Prakash et al., 2014)
Convolvulaceae	*Cressa cretica*	Multiple secretory cells on top of a single stalk cell subtended by a basal cell (Weiglin and Winter, 1988)	N/F
Oleaceae	*Phillyrea latifolia*	Several secretory cells, a stalk cell, and a basal cell formed in an epidermal pit (Gucci et al., 1997)	N/F
Primulaceae	*Aegiceras corniculatum**	24–40 secretory cells connect to a single basal cell on top of sub-basal cells (Cardale and Field, 1971)	ESTs (Fu et al., 2005)
	Glaux maritima	A large vacuolated basal cell, a stalk cell, and 4–8 cytoplasm dense secretory cells in an epidermal pit (Rozema et al., 1977)	N/F

	Samolus repens	6–12 unequally sized secretory cells arranged on a single stalk and basal cell in an epidermal pit (Adam and Wiecek, 1983)	N/F
Solanaceae	*Nolana mollis*[†]	Structure undefined, but presence of glands confirmed (Mooney et al., 1980)	N/F
Caryophyllales			
Aizoaceae	*Mesembryanthemum crystallinum*[†], *M. nodiflorum*[†]	Large highly vacuolar bladder cell (Steudle et al., 1975; Agarie et al., 2007; Grigore et al., 2014)	cDNA (Roeurn et al., 2016), ESTs (Cushman et al., 2008); RNAseq (Oh 2015; Tsukagoshi et al., 2015); miRNAseq (Chiang et al., 2016)
	Aizoon canariense	Large bladder cells (Grigore et al., 2014)	
Amaranthaceae	*Atriplex amnicola, A. canescens, A. lentiformis, A. semilunaris*	Stalked bladder cell forms a bicellular gland (Malcolm et al., 2003; Shabala et al., 2014; Pan et al., 2016)	ESTs (Li et al., 2014); cDNA (Adair et al., 1992)
	Bienertia sinuspersici[‡]	Stalked bladder cell forms a bicellular gland (Akhani et al., 2005; Park et al., 2009)	454 cDNA (Offermann et al., 2015)
	Chenopodium quinoa, C. album	A highly vacuolated bladder cell is connected to a cytoplasm dense stalk cell (Reimann and Breckle, 1988; Adolf et al., 2013; Shabala et al., 2014)	ESTs (Coles et al., 2005; Stevens et al., 2006; Gu et al., 2011); RNAseq (Zhang et al., 2012); genome (Yasui et al., 2016)
Frankeniaceae	*Frankenia grandifolia*	Organized into two highly vacuolar collecting cells and six largely cytoplasmic secretory cells (Balsamo and Thomson, 1993)	N/F
Plumbaginaceae	*Aegialitis annulata**, *A. rotundifolia**	Organized into three concentric rings. Inner two rings contain palisade cells with large vacuoles and outer ring has smaller cells and cytoplasm dense basal cells (Atkinson et al., 1967; Das, 2002)	N/F
	Armeria canescens	Organized into 12 gland cells and 4 subsidiary cells with a structure similar to other salt glands in the family (Scassellati et al., 2016)	N/F
	Limoniastrum guyonianum, L. monopetalum	Organized as an embedded cup of multiple cells (Ioannidou-Akoumianaki et al., 2015; Zouhaier et al., 2015)	N/F
	Limonium bicolor, L. delicatulum, L. furfuraceum, L. gmelinii, L. linifolium, L. perezii, L. platyphyllum	4 types of cells in a total of 16 cells organized into secretory, accessory, inner cup, outer cup, and basal cells (Faraday and Thomson, 1986a; Vassilyev and Stepanova, 1990; Daraban et al., 2013; Grigore et al., 2014; Yuan et al., 2015b; Aymen et al., 2016)	ESTs (Wang et al., 2008), RNAseq (Yuan et al., 2015b, 2016b)
Tamaricaceae	*Reaumuria soongorica, R. trigyna*	Inner and outer secretory cells arranged in a cuticle lined cup arranged on top of a basal cell (Weiglin and Winter, 1991; Wang et al., 2016)	RNAseq (Dang et al., 2013; Shi et al., 2013; Liu et al., 2014, 2015)
	Tamarix androssowii, T. ahylla, T. hispida, T. minoa, T. pentandra, T. usneoides	Highly vacuolar two basal cells and mostly cytoplasmic dense six secretory cells (Campbell and Stong, 1964; Thomson and Liu, 1967; Villar et al., 2015; Wilson et al., 2016)	cDNA (Gao et al., 2014; Wang L. et al., 2014; Yang et al., 2014); ESTs (Wang et al., 2006; Gao et al., 2008); RNAseq (Wang C. et al., 2014)
Rosids			
Combretaceae	*Laguncularia racemosa**	Multicellular gland in a pit (Francisco et al., 2009; Pelozo et al., 2016)	N/F
Malvaceae	*Gossypium hirsutum*	The salt gland structure is not described in detail but resembles a multicellular glandular trichome (Peng et al., 2016)	Genome (Li et al., 2015); microarray (Rodriguez-Uribe et al., 2011; Yin et al., 2012), RNAseq (Peng et al., 2014; Lin et al., 2015); micro-RNAseq (Xie et al., 2014) [Only a few selected references are given for *G. hirsutum* genetic resources]
Moraceae	*Ficus formosana*	Multicellular glandular trichome (Chen and Chen, 2005)	N/F
Poaceae			
Chloridoideae		Organized as a bicellular gland with a basal collecting cell and a secretory cap cell	
Cynodonteae	*Aeluropus littoralis*[‡]	(Zouari et al., 2007; Barhoumi et al., 2008)	ESTs (Zouari et al., 2007)
	Buchloe dactyloides[‡]	(Liphschitz and Waisel, 1974; Marcum, 2006, 2008)	RNAseq (Wachholtz et al., 2013), cDNA (Budak et al., 2005, 2006)

	Bouteloua spp.‡	Céccoli et al., 2015	Wachholtz et al., 2013; Amaradasa and Amundsen, 2016
	Chloris gayana‡	Amarasinghe and Watson, 1988; Takao et al., 2012	N/F
	Cynodon dactylon‡	Oross and Thomson, 1982; Amarasinghe and Watson, 1988; Marcum, 2006, 2008	RNAseq (Hu et al., 2015), cDNA (Peña-Castro et al., 2006; Kim et al., 2008)
	Dactyloctenium aegyptium‡	Liphschitz and Waisel, 1974	N/F
	Diplachne fusca‡	Céccoli et al., 2015	N/F
	Distichlis spicata‡	Liphschitz and Waisel, 1974; Oross and Thomson, 1982; Marcum, 2006; Semenova et al., 2010; Céccoli et al., 2015	cDNA (Zhao et al., 1989)
	Eleusine indica‡	Liphschitz and Waisel, 1974	N/F
	Leptochloa digitata‡, *L. fusca*‡	Wieneke et al., 1987; Amarasinghe and Watson, 1988	N/F
	Odyssea paucinervis‡	Somaru et al., 2002	N/F
	Pappophorum philippianum‡	Taleisnik and Anton, 1988; Céccoli et al., 2015	N/F
	Munroa argentina‡	Céccoli et al., 2015	N/F
Zoysieae	*Spartina* spp.‡	Levering and Thomson, 1971; Liphschitz and Waisel, 1974	RNAseq (Baisakh et al., 2008; Ferreira de Carvalho et al., 2013; Bedre et al., 2016; Nah et al., 2016), miRNA (Qin et al., 2015; Zandkarimi et al., 2015)
	Sporobolus virginicus‡	Amarasinghe and Watson, 1988; Marcum, 2006, 2008	RNAseq (Yamamoto et al., 2015)
	Zoysia spp.‡	Amarasinghe and Watson, 1988; Marcum et al., 1998; Marcum and Murdoch, 1990	Genomes (Tanaka et al., 2016), RNAseq (Ahn et al., 2015; Wei et al., 2015; Xie et al., 2015), cDNA (Chen et al., 2015), ESTs (Cheng et al., 2009; Ko et al., 2010)
Oryzoideae	*Porteresia coarctata*	Unicellular finger shaped or peg shaped hairs (Flowers et al., 1990; Sengupta and Majumder, 2010)	RNAseq (Garg et al., 2013); miRNAseq (Mondal et al., 2015)

* *mangroves;* † *CAM species;* ‡ *C4 species; N/F none found.*

The last two types of salt glands are found in Chloridoideae and Oryzoideae subfamilies in Poaceae (Amarasinghe and Watson, 1988; Flowers et al., 1990). A recent review by Céccoli et al. (2015) provides a detailed report of chloridoid type salt gland structures and their physiological features (Type 3 in **Figure 1**). Although somewhat similar to the salt-secreting glands of eudicots, the salt glands of grasses differ in three important ways. First, they are simpler in structure, consisting of only one or two cells. Second, they lack the cuticular boundary surrounding the secretory and basal cells that appears to channel the flow of salt in the eudicot salt glands. Finally, the basal cell is not vacuolated, contrasting the vacuolated basal collecting cells of eudicots. The Chloridoideae salt glands are two-celled trichomes differentiated into a basal and a cap cell. Both the basal cell and the cap cell are cytoplasmically dense and rich in mitochondria, plastids, and vesicles. Wall protrusions and the associated plasma membrane extend from the cap cell deep into the basal cell, increasing surface area. These are often found in epidermal depressions, within the folds of the leaf laminar structure, sunken in the epidermis, or placed above the epidermis (Liphschitz and Waisel, 1974; Céccoli et al., 2015). The continuous cuticle on the epidermis in some species thickens on top of the cap cell and forms a cuticular chamber that stores secreted salts as seen for some eudicot salt glands (Amarasinghe and Watson, 1988). The thick cuticle extends from the top of the cap cell to the side walls of the basal cell where adjacent epidermal cells connect and where the side walls of the basal cell are often lignified (Liphschitz and Waisel, 1974).

The fourth type of salt glands is found in the wild rice species *Porteresia coarctata*, closely related to the cultivated rice in Oryzoideae. These salt glands are unicellular hairs (Type 4 in **Figure 1**). The finger-shaped adaxial salt hairs in *P. coarctata* continue to secrete salt even at high soil NaCl levels, but the peg-shaped shorter salt hairs on the abaxial surface rupture as intracellular NaCl accumulates, and regrow when soil salt levels decline (Sengupta and Majumder, 2010). It appears that *P. coarctata* can modulate the type and number of salt hairs, adjusting to external salt levels. These unicellular hairs seem to lack organelles and appear to be completely filled with vacuoles in contrast to the bladder cells in eudicot glands (Flowers et al., 1990; Oh et al., 2015).

SALT GLANDS HAVE EVOLVED INDEPENDENTLY MANYTIMES

It is more than likely that glandular adaptations to salt have developed multiple times in the angiosperms, using distinct mechanisms involving either sequestration of salt in vacuoles or secretion. A conservative estimation of multiple independent origins proposed by Flowers et al. (2010) and Flowers and Colmer (2015) suggests a minimum of three origins for salt glands among the angiosperms, one in monocots, one in rosids, and one in the joint clade of asterids and Caryophyllales. However, given that only a fraction of a percent of flowering plants are halophytes, and only a small percentage of halophytes have salt-secreting glands, it seems exceedingly unlikely that the common ancestor of the Caryophyllales and the asterids had salt glands that were subsequently lost in the vast majority of the species in the relevant clade. Although the asterids are one of the largest flowering plant groups, encompassing nearly one-third of all angiosperm species classified in 144 families (Soltis et al., 2005), salt glands are only reported in five families distributed among the three orders Ericales, Lamiales, and Solanales. It is likely that salt glands were independently acquired within each of the individual asterid families containing salt gland-bearing species: Acanthaceae, Convolvulaceae, Oleaceae, Primulaceae, and Solanaceae (indicated by an asterisk in **Figure 2** for each independent introduction). Similarly, rosids include more than a quarter of angiosperm species classified into about 140 families (Soltis et al., 2005). Yet, salt glands are recorded for only three families (Combretaceae, Malvaceae, and Moraceae) in the three diverse orders of Myrtales, Malvales, and Rosales. Thus, rosid salt glands likely represent three additional events of salt gland evolution in angiosperms.

Closer inspection of salt gland structure and function supports the hypothesis of many independent origins for salt glands. Within the Caryophyllales there are two structurally and functionally distinct types of salt glands. It is likely that the sister groups of Aizoaceae and Amaranthaceae had a shared ancestor with salt bladders (**Figure 1**, Type 1). These families are in a monophyletic clade known as the core Caryophyllales, and their salt glands are all of the salt bladder type (**Figure 1**). In contrast the Tamaricaceae, Frankeniaceae, and Plumbaginaceae families, which are in a clade termed the non-core Caryophyllales, sister to the core Caryophyllales, all have type 2 multicellular salt-secreting glands that are structurally similar to each other, with a number of cytoplasmically dense secretory cells overlying several vacuolated collecting basal cells (**Figure 1**, Type 2). These cells are very different from the bladder cells of the Aizoaceae and Amaranthaceae, and in fact are structurally more similar to the multicellular salt glands found among the asterids. Similarly, the two salt gland types in grasses likely present two additional events of acquiring salt glands independently. Thus it would be reasonable to assume that salt glands have originated independently 12 times or more in angiosperms. Even if it is assumed that the most closely related pairs of asterid families (Acanthaceae-Oleaceae and Convolvulaceae-Solanaceae) each share a single origin, salt glands can have arisen no less than 10 times.

These different evolutionary origins present compelling examples of convergent evolution in the structure of salt glands. Species located in a wide range of clades have cytoplasmically dense secretory cells overlying vacuolate collecting cells, a pattern seen in all asterid salt glands and in the salt glands of non-core Caryophyllales (Plumbaginaceae, Tamaricaceae, and Frankeniaceae), although the numbers of secretory and collecting cells vary (**Table 1**). In a number of cases, cuticular material extends down the sides of the secretory and/or the basal collecting cells. While these parallels are striking, glands of similar structure that secrete volatile secondary metabolites, nectar, mucilage, and digestive enzymes are widespread throughout the asterids and Caryophyllales. Indeed, the salt glands of various families tend to greatly resemble the structure of secretory glands of related plants that lack salt glands. For example, both *Acanthus* and *Avicennia* have a short stalk composed of 1–2 cells bearing a globular head consisting of secretory cells (Shimony et al., 1973), while similar short stalked gland functions are ubiquitous in Acanthaceae (Immelman, 1990; Tripp and Fekadu, 2014; Bhogaonkar and Lande, 2015). The Acanthaceae (Lamiales) salt glands also bear a strong resemblance to the glandular trichomes that secrete essential oils in the closely related Lamiaceae. These trichomes have a basal cell embedded in the epidermis, a one or two celled stalk, and a globular head of secretory cells with a sub-cuticular space where oils containing volatile secondary metabolites accumulate. This structural feature is redolent to the cuticular chambers with salt on top of salt glands (Werker et al., 1993; Ascensão et al., 1995; Serrato-Valenti et al., 1997; Giuliani and Bini, 2008). Glandular trichomes are common among the other clades in asterids as well. For example, Solanaceae short stalked globular trichomes (Type VI) that secrete defensive proteins (Shepherd et al., 2005) and other secondary metabolites are also structurally similar to the asterid salt glands with respect to the cellular organization of a basal cell, 1–2 stalk cells, and a few secretory cells on top (Reis et al., 2002; Glas et al., 2012; Munien et al., 2015).

The non-core Caryophyllales families, Plumbaginaceae, Tamaricaceae and Frankeniaceae, are sister to a clade of mostly carnivorous plants consisting of the families Droseraceae, Drosophyllaceae, Nepanthaceae, Dioncophyllaceae, and Ancistrocladaceae, which have glands that secrete digestive enzymes and mucilage. Although the carnivorous plants have a variety of elaborate glandular morphologies that show secretory as well as absorption functions, these are thought to be derived from an ancestral character state for glands that are very similar to the salt glands of *Tamarix* and *Frankinia* (Cameron et al., 2002; Heubl et al., 2006; Renner and Specht, 2013). The digestive glands of *Dionaea muscipula* (Venus fly-trap), which consist of two layers of secretory cells above a pair of stalk cells and several basal cells that are embedded in the epidermis, may be taken as an example close to the ancestral state (Scala et al., 1968; Robins and Juniper, 1980). Like the salt gland secretory cells of *Tamarix*, these secretory cells have projections of cell wall material that increase the surface area of the secretory cell plasma membrane. The pattern of convergent evolution of the secretory-type salt glands (**Figure 1**, Type 2) described here, combined with the resemblance of these salt glands to other types of glands on closely related plants, and in conjunction with the overall low frequency of plants bearing salt glands, suggests that these Type 2 salt glands have evolved independently multiple times from a common type of multicellular secretory gland found widely throughout eudicots.

A similar trend is observed for salt glands in monocots. Liphschitz and Waisel (1974) previously have suggested a common halophytic ancestor for the Chloridoideae species with salt glands. The Chloridoideae-type bicellular glands that secrete salts are found in a number of species in Cynodonteae and Zoysieae, but not all grasses in these subclades are halophytes. For example, the bicellular glands in *Eleusine indica* and *Sporobolus elongatus* in Cynodonteae and Zoysieae, respectively, do not secrete salts and are not known as halophytes even if they carry glands with the same ultrastructure shared with Cynodonteae and Zoysieae halophytes (Amarasinghe and Watson, 1989). Interestingly, the glandular organization consisting of a basal and cap cell is not limited to the Chloridoideae species, but it is also observed in more than 5000 species in the sister clade of panicoid grasses (includes sorghum and corn). However, these lack the plasma membrane invaginations in the basal cell characteristic of the halophytes in

Chloridoideae (Amarasinghe and Watson, 1988). Some of these non-halophytes that do not develop "salt glands" retain the capacity to secrete NaCl to some extent and also induce the rate of microhair formation under salt stress (Ramadan and Flowers, 2004). Although salt glands are generally associated with halophytes, several *Spartina* spp. from freshwater habitats also carry salt glands at a level similar to their relatives from saltmarshes (Maricle et al., 2009). This could be a derived trait from an ancestral halophytic lifestyle of *Spartina* from saltmarshes and also coincides with the view presented by Bennett et al. (2013) wherein it is inferred that the salt tolerance trait evolved more than 70 times independently in diverse grass lineages with multiple events of loss of trait in some genera. Collectively, we see that the ubiquitous bicellular glands in grasses can differentiate to salt secreting glands, microhairs without secretions, or glands that secrete other substances. The salt secretory unicellular hairs reported for *Porteresia coarctata* show close resemblance to microhairs found in cultivated rice (both in Oryzoideae), but rice microhairs do not show salt secretory functions detectable at significant levels (Flowers et al., 1990). Some convergent trends occur multiple times in subsets of eudicot and monocot recretohalophytes separated by large evolutionary distances, indicative of the selective pressures driving salt gland evolution. For example, cell wall projections resulting in an increase in plasma membrane surface area are seen in both the Poaceae (Levering and Thomson, 1971; Amarasinghe and Watson, 1989) and in the Tamaricaceae-Frankeniaceae-Plumbaginaceae clade (Campbell and Thomson, 1976; Faraday and Thomson, 1986b), although in Poaceae these projections protrude into the basal cell, while in Caryophyllales the protrusions occur into the secreting cell. Such wall protrusions are characteristic of a wide variety of transfer cells that are involved in the intercellular transport of solutes (Gunning and Pate, 1969). In another common trend, secretory-type salt glands are often located in pits or depressions on the leaf surface (Tamaricaceae, Frankeniaceae, Plumbaginaceae, Primulaceae, Acanthaceae, Combretaceae, and Poaceae). Perhaps these depressions collect dew into which salts can be efficiently secreted. This trait may have been further developed in *Nolana mollis* (Solanaceae) salt glands that primarily secreted NaCl, where excreted salts were used to condense water from unsaturated atmospheres as an adaptation to retrieve water for survival in the Atacama Desert (Mooney et al., 1980). This may suggest a trait highlighting adaptations to extreme drought tolerance from a preadapted halophytic trait.

The density of salt glands is highly species specific. For example, salt gland density generally ranges from 20 to 50 salt glands mm^{-2} in leaves of *Limonium* and *Zoysia* species (Ding et al., 2010a; Yamamoto et al., 2016). The structural integrity of the salt glands may also depend on soil salinity and leaf age. For instance, the abaxial peg-like salt hairs on *Porteresia coarctata* tend to burst with increasing soil salinity where the adaxial more elongated salt hairs increase in density (Sengupta and Majumder, 2009). In *Ficus formosana* the salt glands near hydathodes get dropped as the leaf ages removing compartmentalized excess salts more efficiently (Chen and Chen, 2005).

The functional significance provided by salt glands also changes with leaf development. NaCl sequestration capacity may be the most critical function of salt bladders in young leaves of Aizoaceae and Amaranthaceae halophytes (Agarie et al., 2007; Bonales-Alatorre et al., 2013; Barkla et al., 2016), but as the leaf matures and the salt bladders reach their maximum volume, salt sequestration rate needs to be paused (Adams et al., 1998; Jou et al., 2007; Barkla and Vera-Estrella, 2015; Oh et al., 2015). Other functions including providing a secondary epidermal layer to protect against water loss, UV stress, and also serving as reserves for ROS scavenging metabolites and organic osmoprotectants may contribute more to plant survival under abiotic stress as the leaf matures (Adolf et al., 2013; Barkla and Vera-Estrella, 2015; Ismail et al., 2015; Oh et al., 2015). The corresponding increased rate of salt secretion as a response to increasing concentrations of soil NaCl is also observed for salt glands in other plant clades (Marcum et al., 1998; Mishra and Das, 2003). The maximum rate of salt secretion, however, is dependent on the species. For example, *Spartina anglica* has been reported to secrete up to 60% of absorbed salts while *Limonium vulgare* and *Glaux maritima* showed 33 to 20%, respectively, in a comparative study (Rozema et al., 1981).

NEW GENETIC RESOURCES ANDTOOLS PROVIDE INSIGHTS INTOTHE MOLECULAR COMPONENTS INVOLVED IN SALT GLAND FUNCTION

Model Species Studies

Because salt glands represent only a small proportion of the cells on the leaves of salt gland-bearing plants, studies regarding the cellular physiology and molecular genetics of salt glands have been limited in the past. However, new methods are increasing our ability to study the detailed function of salt glands at the cellular level. The most accessible salt glands for study until recently have been bladder cells. The salt tolerant extremophiles *Mesembryanthemum crystallinum* (ice plant) has been the focus of the greatest number of biochemical, physiological, and genetic studies among halophytes with salt glands. Steudle et al. (1975) first measured bladder cell membrane potential (between −10 and −40 mV), hydraulic conductivity (LP of the bladder cell membrane was on average 2 $\times 10^{-6}$ cm $s^{-1}bar^{-1}$) and demonstrated high bladder membrane salt permeability, consistent with their role in compartmentalizing excess NaCl in the vacuoles (Steudle et al., 1975; Lüttge et al., 1978). The critical role played by salt bladders in *M. crystallinum* for development and survival under high NaCl was further confirmed by the creation of growth impaired mutant plants without bladder cells (Agarie et al., 2007). The remarkable salt and drought tolerance capacity exhibited by *M. crystallinum* has also led to its use as a model halophyte in multiple gene expression studies using ESTs and RNAseq from bulk tissues to discover gene regulatory mechanisms related to salt tolerance (Bohnert and Cushman, 2000; Cushman et al., 2008; Tsukagoshi et al., 2015; Chiang et al., 2016). Additionally, the recent cell specific targeted transcriptome, proteome, and metabolome analyses have reported the type of genes, proteins, and metabolites expressed specifically in salt glands in *M. crystallinum* (Barkla et al., 2012; Barkla and Vera-Estrella, 2015; Oh et al., 2015). These studies have helped to establish the importance of salt glands and their distinct functions from other leaf cells in a model halophyte. With the recent cell type specific experiments, we know that epidermal bladder cells of *M. crystallinum* are not just passive storage organs for salts as perceived before, but they also carry out active metabolism related to energy generation, UV protection, organic osmolyte accumulation, and stress signaling. A significant number of lineage-specific genes of unknown function in response to salt stress were detected in these bladder cells. Some of the lineage specific transcripts are easily detected in the epidermal bladder cell transcriptomes at high expression levels, but appear to be expressed at low levels or are undetected in whole shoot transcriptomes, indicating the importance of studies of individual salt gland cell types (Oh et al., 2015). Genes specific to bladder cell function and development that were identified using a suppression subtractive hybridization library construction between wild type *M. crystallinum* and mutant plants without bladder cells also revealed a significant number of lineage specific genes with unknown functions (Roeurn et al., 2016). One such gene of unknown function detected via the comparison between wildtype and mutant plants was subsequently overexpressed in *Arabidopsis*, resulting in a phenotype with an increased number of trichomes on leaves, and this gene was inferred to regulate trichome initiation via regulating *GL2* in the trichome development pathway (Roeurn et al., 2016). The availability of a reference genome for *M. crystallinum* will facilitate new comprehensive investigations of the critical role of salt glands in the survival of the whole plant under salt stress.

Chenopodium quinoa (Amaranthaceae), is an emerging model halophyte and a seed crop with several salt tolerant cultivars adapted to salt levels that are as high as that of sea water (Adolf et al., 2012; Ruiz et al., 2016). Its genomic complexity and polyploid nature have made molecular genetic analyses of the genetic mechanisms underlying its salt tolerance traits challenging. However, the draft genome of *C. quinoa* that was recently made available will be an excellent resource opening new paths to explore its stress adapted biology (Yasui et al., 2016). Also, the genome of the closely related non-halophyte *Beta vulgaris* (Amaranthaceae) and additional transcriptomes of the halophytic but non-salt gland subspecies *B. vulgaris* ssp. *maritima* (Dohm et al., 2014; Skorupa et al., 2016) should further facilitate comparative genomic analyses of the role of salt glands in salt tolerance in the Amaranthaceae.

A number of electrophysiological studies performed on quinoa leaf cells and salt bladders suggest that a polar organization of Na^+ transporters and anion channels mediates NaCl net influx into the bladder cell vacuoles, while the small stalk cell serves as an intracellular ion transport controller between the epidermal and bladder cells (reviewed in Adolf et al., 2013; Shabala et al., 2014). The entry of Na^+ and Cl^- into the bladder cell vacuole are likely dependent on the NHX1 transporter, CLC-type chloride channels, and the electrochemical proton gradient provided by the vacuolar H^+-ATPases and vacuolar H^+-pyrophosphatases, while plasma membrane Na^+/K^+ transporters like HKT1 may play a major role in getting Na^+ into the bladder cell cytoplasm. The importance of the vacuolar proton pumps in sequestering Na^+ in the vacuolar lumen is supported by transcriptomic, proteomic, and biochemical studies done on ice plant and quinoa bladder cell systems (Barkla et al., 2012; Adolf et al., 2013; Oh et al., 2015).

Recently, *Limonium bicolor* has been developed as a model for the study of secretory multicellular salt glands. Transcriptomic analysis of developing *Limonium bicolor* leaves while monitoring salt gland developmental stages suggests that salt gland development might be regulated by transcription factors homologous to those regulating trichome development in *Arabidopsis thaliana*, however, this suggestion is based solely on correlated expression patterns and weakly documented evidence for orthology (Yuan et al., 2015b, 2016b). Yuan et al. (2014) have further developed a transformation system for *L. bicolor* to enable validation of predicted gene functions within the native genome. Additionally, the same group has optimized gamma radiation mutagenesis to create large mutant populations of *L. bicolor* (Yuan et al., 2015a) and has developed an autofluorescence-based screen to identify mutants in salt gland function (Yuan et al., 2013). The efforts to create a molecular toolbox for forward and reverse genetics in order to study the multicellular salt gland functions in *Limonium bicolor* are exemplary, given its status as a non-model organism in plant genetics.

Transport of Na^+ through a multicellular gland that ultimately excretes salt outside the leaf is a far more complex process than understanding valuolar compartmentalization in salt bladders. In a salt gland, when certain cells take up the role of absorbing salt from neighboring cells and intercellular spaces (main function proposed for collecting cells, basal cells, an sub-basal cells found in Type 2 and 3 salt glands in **Figure 1**), other cells in the gland would need to export salts (secretory and cap cells in Type 2 and 3 glands from **Figure 1**). Given that there are several channels and transporters that can transport Na^+ exclusively or together with other organic and inorganic ions in plant cells (reviewed in Maathuis, 2014; Maathuis et al., 2014), this process needs to be coordinated between multiple membrane systems to avoid futile cycling of Na^+ and other ions including K^+ or toxic accumulation of NaCl. Salt tolerance is also tightly linked to K^+ homeostasis in plant tissues. Halophytes are known to accumulate high K^+ levels or prevent loss of K^+ when treated with high Na^+ (Flowers et al., 2015). For example, *Limonium* salt glands increase K^+ retention upon high Na^+ treatments (Feng et al., 2015). Additionally, there are a number of aquaporins that transport water and other molecules that need to be integrated into the Na^+ transport systems when we attempt to understand salt transport management in plant tissues (reviewied in Maurel et al., 2015). A plasma membrane aquaporin was among the highest membrane transporters/channels in the cell specific salt bladder transcriptome of *M. crystallinum* (Oh et al., 2015), further supporting the idea that suites of transporters, including water channels and K^+ transporters, need to be considered in addition to Na^+ transporters and membrane proton pumps to accurately model salt secretion.

Salt from collecting basal cells can also be bulk transported via vesicles that fuse to the plasma membranes of collecting and secretory cells (or cap cells in grasses), releasing salt to the extracellular space. A few studies have looked into the significance of vesicle transport in delivery of NaCl to secretory cells or extracellular spaces (cuticle lined chamber in most multicellular salt glands and bicellular glands in grasses). These studies have reported the formation of extra vesicles and fusion with the plasma membrane between basal cells and mesophyll cells and also basal and secretory cells upon salt treatment (Thomson and Liu, 1967; Shimony et al., 1973; Barhoumi et al., 2008). Faraday and Thomson (1986a) reported ion efflux rates in *Limonium perezii* salt glands that were significantly higher than rates possible exclusively via transmembrane transport. Congruently, Yuan et al. (2016b) have reported genes associated with vesicle function enriched in *Limonium bicolor* leaves upon NaCl treatment. Vesicle-mediated NaCl transport may provide the energy efficiency required for transporting salts through the salt glands that may not be feasible via transmembrane ion channels alone. Physiological and molecular studies have attempted to model the unidirectional flux of Na^+ and Cl^- in multicellular salt glands of *Limonium* and *Avicennia* (Tan et al., 2013; Yuan et al., 2016b), but the details of the cell-specific roles in any multicellular salt gland remain largely unknown.

The fiber crop *Gossypium hirsutum* (Malvaceae), although is not considered a halophyte, is among the crop species most adapted to salt stress, and some cultivars also develop functional salt glands (Gossett et al., 1994; Du et al., 2016; Peng et al., 2016). The availability of a reference genome, multiple large scale transcriptome datasets, genetic transformation techniques, and genetic diversity estimates for a large group of cultivars make *G. hirsutum* an attractive candidate for studying salt gland functions between salt adapted and sensitive cultivars (Shen et al., 2006; Khan et al., 2010; TianZi et al., 2010; Rodriguez-Uribe et al., 2011; Rahman et al., 2012; Xie et al., 2014; Li et al., 2015; Lin et al., 2015). However, the role of salt glands in adapting to salt stress in cotton has not been explored much until recent work published by Peng et al. (2016). High levels of activity inferred for the plasma membrane H^+-ATPase and the Na^+/H^+ antiporter to compartmentalize more Na^+ into the apoplast or the vacuole were suggested as key transporters in extruding excess salt from the young cotton leaves.

Among the monocot recretohalophytes, studies on *Spartina* spp. offer multiple snapshots into the leaf transcriptomics that investigate how salt glands contribute to salt tolerance (Baisakh et al., 2008; Ferreira de Carvalho et al., 2013; Bedre et al., 2016). *Spartina* is among the few recretohalophytes where both RNASeq and microRNASeq profiles are available (Qin et al., 2015; Zandkarimi et al., 2015). In addition, the genus *Spartina* offers an interesting evolutionary context where one can study the relaxed selection on genes important in salt gland functions when salt glands do not provide a fitness advantage to species that occupy freshwater habitats. Freshwater species including *S. cynosuroide*, *S. gracilis*, and *S. pectinata* show no difference in their salt gland distribution compared to the closely related salt marsh species *S. alterniflora, S. anglica,* and *S. densiflora* (Maricle et al., 2009). The development of salt glands in the freshwater species may be a result of a recent speciation event from ancestral salt marsh species. This provides an excellent set of plants with natural replicates for comparative genomics in search of salt gland associated genes and their recruitment driven by salt stress (or loss of recruitment in the absence of the selection pressure).

Genome Wide Data and Tools for Salt Gland Specific Expression

Salt gland specific transcriptomic, proteomic, or metabolic datasets as genetic resources are challenging to obtain, often due to the tight integration of salt glands in leaf or other photosynthetic tissue. **Table 1** lists all genome wide molecular studies reporting datasets from plants with salt glands available at present (October 2016). Several of these studies provide RNAseq-based experiments that target tissues enriched in salt glands. A few studies have focused on enrichment of salt gland cell types or isolation of exclusive salt gland populations. Due to the structural diversity of these species, a method optimized for one species is difficult to implement in others. Barkla et al. (2012) accomplished this task for ice plant epidermal bladder cells by vacuum aspiration of the cell sap using a fine needle attached to a collecting tube. This technique is able to provide clean cell specific sap, but is impractical for multicellular salt glands. Techniques developed using pressure probes and picoliter osmometers to measure water potential and osmotic potential in single plant cells (reviewed in Fricke, 2013) often used in crop plants (Malone et al., 1989; Fricke, 1997; Fricke and Peters, 2002; Volkov et al., 2006) offer additional tools to test salt gland cell specific traits. The use of epidermal peels enriched in salt glands is an alternative solution, although this technique introduces molecular signatures of regular epidermal cells to the sample, as contaminants are difficult to avoid (Tan et al., 2015). Use of enzymatic digestion and subsequent grinding of epidermal peels has also been shown to be effective in isolating mangrove salt glands devoid of neighboring epidermal cells (Tan et al., 2010). However, enzymatic digestion adds a significant amount of time that may lower the feasibility of using salt glands isolated through such techniques to detect transcript profiles dependent on plant treatments and conditions. Treating epidermal peels with clearing solutions and detecting salt glands based on their autofluorescence has been successfully demonstrated for *Limonium* and *Avicennia* in identifying the salt gland structure and organization, but this method too would not allow time-sensitive assessments of salt gland-specific transcripts or proteins (Tan et al., 2010; Yuan et al., 2013).

Effective methods shown successful in capturing multicellular gland-specific transcripts do not exist for halophytes at present. However, this can be attempted using current molecular techniques. For example, fluorescent tags labeling entire cells, nuclei, or polysomes allow capture of cell-type specific transcripts in model plants (Mustroph et al., 2009; Deal and Henikoff, 2011; Rogers et al., 2012). Creating targeted transgenic lines for non-model halophytes could be a greater challenge than optimizing methods for cell-type specific tagging. One may need to explore *Agrobacterium*-independent transformation techniques if certain recretohalophytes prove to be recalcitrant to widely used transformation protocols (Altpeter et al., 2016). Furthermore, such methods require the identification of salt gland-specific promoter sequences. Candidate promoters might be deduced from promoters functioning in glandular trichome gene expression of related plants (Choi et al., 2012; Spyropoulou et al., 2014), given the evidence presented above that multicellular salt glands in eudicots are likely derived from multicellular secretory trichomes. Alternatively, physically isolating multicellular glandular structures before extracting the cell sap for RNA, protein, or metabolite profiling has been established using laser capture microdissection methods (Olofsson et al., 2012; Soetaert et al., 2013).

COULD WE ENGINEER WORKING SALT GLANDS IN A MODEL SYSTEM?

Is *Arabidopsis* trichome development a suitable model for engineering bladder cell-like salt glands? Salt glands provide an end destination for excess salts, and understanding the function of these specialized structures may ultimately play a role in producing salt-tolerant crops. Although the engineered expression of individual genes involved in salt tolerance has had some success in increasing salt tolerance in artificial situations, this has not translated to increased salt tolerance under field conditions (Flowers and Colmer, 2015; Mickelbart et al., 2015; Polle and Chen, 2015). Salt tolerance under real-world conditions is likely to require careful attention to cell and tissue-type specific expression of multiple proteins involved in salt tolerance. As noted above, virtually all salt glands are similar in structure, and likely homologous, to the trichomes of closely related plants. The trichomes of *Arabidopsis thaliana* are one of the most well-studied models for plant development at the cellular level, and it was recently suggested that knowledge from *Arabidopsis* trichome development could be used to guide the engineering of bladder cell-type salt glands in crop plants (Shabala et al., 2014). This is a striking proposal that deserves serious consideration. A first step would be attempting to engineer *Arabidopsis* trichomes to function as bladder cells.

The trichomes of *Arabidopsis thaliana* are unicellular and branched, and like bladder cells, they have a large cell volume in comparison with other epidermal cells, most of it being occupied by a large vacuole (Hülskamp et al., 1994; Mathur et al., 2003). Trichome development is initiated by a transcription factor complex containing the R2–R3 MYB protein GLABRA1 (GL1), the bHLH protein GLABRA3 (GL3), and the WD-repeat protein TRANSPARENT TESTA GLABRA (TTG), and is restrained by several inhibitory single-repeat R3 MYBs, typified by TRIPTYCHON (TRY) (Larkin et al., 2003). Many direct downstream targets of this transcription factor complex have been identified, and mutations and gene-expression manipulations are established that alter the density of trichomes on leaves, trichome cell shape, and cell wall properties. A number of direct downstream target genes of the trichome development transcription complex are known, and several relatively trichome-specific promoters are noted, e.g., for *GLABRA2* (*GL2*), *GL3*, *TRY* and *NOEK* (*NOK*) (Schnittger et al., 1998; Szymanski et al., 1998; Jakoby et al., 2008). The putative transcription factor identified in wild type *M. crystallinum* compared to the mutant without bladder cells expressed in *Arabidopsis* was proposed to act upstream of the GL1-GL3 complex via positively regulating *GL2* (Roeurn et al., 2016). A functional homolog of *GL1* in cotton, *GaMYB2*, was shown to have trichome specific expression in *Arabidopsis*, but in cotton both fiber cells and trichomes showed *GaMYB2* promoter driven GUS expression. Interestingly, the *GaMYB2* promoter directed GUS expression exclusively in glandular cells of glandular secreting trichomes in tobacco where different types of trichomes exist (Shangguan et al., 2008). This suggests that complex tissue specific signals may exist for trichome specific expression in different halophytes even when the genetic components are well described in the model species.

This detailed knowledge of *Arabidopsis* trichome development, in combination with new large-scale gene assembly tools that aid in transferring whole pathways to plant genomes such as BioBrick, Golden Gate, and Gibson assembly methods (reviewed in Patron, 2014), suggest that attempting to modify *Arabidopsis* trichomes to function as salt glands may be feasible. As a start, one might engineer expression of the plasma membrane/vacuolar H^+-ATPase and/or the vacuolar H^+-pyrophosphatase, the tonoplast Na^+/H^+ antiporter *NHX1* in trichomes, along with the *P5CS* and *P5CR*, proline biosynthesis genes to increase the proline concentration to act as an organic osmolyte, myo-inositol-1-phosphate synthase (INPS), and myoinositol O-methyltransferase 1 (IMT1) that are key enzymes in polyol synthesis pathways important in ROS scavenging. It should be noted that some of the key target proteins involved in the salt response may include multiple subunits from different polypeptides and therefore, multiple genes need to be coordinately expressed to get the desired level of expression of the holoenzyme. The vacuolar H^+-ATPase is encoded by multiple genes coding for distinct essential subunits while the vacuolar H^+-pyrophosphatase can be generated by a single gene. Additionally, both of these may have variable gene copy numbers for each subunit or protein in different species (Silva and Gerós, 2009; Fuglsang et al., 2011; Volkov, 2015). For example, salt gland bladder cells in *Mesembryanthemum crystallinum* in response to salt stress showed significantly higher expression for 10 transcripts coding for different subunits of the vacuolar H^+-ATPase, while two transcripts likely encoding two copies for the vacuolar H^+-pyrophosphatase showed downregulation (Oh et al., 2015). The coordinated regulation of the vacuolar H^+-ATPase and the vacuolar H^+-pyrophosphatase can be complex and recent research suggests that the combined activity of these proton pumps is required for vacuolar acidification (Kriegel et al., 2015). If salt excretion to the leaf surface as opposed to salt sequestration in a vacuole of bladder cells is envisioned, plasma membrane transporters and proton pumps that govern Na^+ influx into and efflux out of the salt gland should be carefully orchestrated. For example, Na^+ transporters, including SOS1, would need to be regulated together with plasma membrane proton ATPases to excrete salt to the surface against an electrochemical gradient while Na^+/K^+ membrane transporters like HKT1 would be useful for the influx of Na^+ into the secretory bladder cell from neighboring cells. Additional membrane transporters associated with Na^+ and Cl^- transport that may play an important role in developing functional salt glands are reviewed in Shabala et al. (2015) and Yuan et al. (2016a). Further refinements could be made by taking advantage of the knowledge that increased *GL3* expression increases trichome density on leaves (Payne et al., 2000; Morohashi et al., 2007). Thus, introducing a copy of *GL3* under the control of an ABA-inducible, salt-responsive promoter would be expected to increase the number of bladder cell-modified trichomes on the leaf in response to salt stress.

Although the prospect of engineering trichomes of a non-halophyte into functional bladder cells is exciting, there are naturally some serious caveats. First, salt glands of any sort are only one line of defense against salt, and this is achieved via the sequestration of salt that has reached photosynthetic shoot tissues to ameliorate the effects. Truly salt-tolerant plants are likely to require engineering of gene expression in multiple tissues. Much evidence indicates that for plants, the initial line of defense is to prevent the accumulation of salt in the roots in the first place (reviewed in Flowers and Colmer, 2015). Thus, for example, it would likely be necessary to engineer increased *SOS1* expression in root hairs to pump Na^+ out from the root epidermis, limiting salt intake, as well to increase expression of *SOS1* in the endodermis to feed Na^+ that does enter the plant into the transpiration stream for transport to the shoot. Fortunately, well-characterized promoters are now available for engineering cell type-specific expression in *Arabidopsis* roots. A second caveat is that this approach has ignored the roles of signaling by Ca^{2+} and reactive oxygen species in salt tolerance. The incorporation of tissue-specificity through the use of tissue-specific promoters is still ultimately too simplistic and likely will fail to capture the dynamic nature of true halophyte responses to saline conditions.

The final caveat to this approach is that the engineering of bladder cell-type salt glands based on *Arabidopsis* trichomes as a model is likely to be limited phylogenetically to plants sharing the same trichome initiation regulatory network. While the transcription complex that regulates *Arabidopsis* trichome development is clearly homologous to the transcription factors that regulate anthocyanin biosynthesis in plants as distantly related as the grasses, it appears that asterids regulate trichome development via the *MIXTA*-like MYB proteins, which lack the ability to bind GL3-like bHLH proteins (Payne et al., 1999; Serna and Martin, 2006). Furthermore, expression of *Antirrhinum MIXTA* does not affect *Arabidopsis* trichome development, and expression of *Arabidopsis* trichome regulators in *Nicotiana* also does not affect trichome formation. Thus, trichome development appears to be regulated independently in the rosids and the asterids. In this light, it is interesting to note that in *Mesembryanthemum crystallinum*, a putative ortholog of the trichome development gene *GL2*, exhibits increased expression in bladder cells in response to salt (Oh et al., 2015), and that in *Limonium bicolor*, the expression of putative orthologs of several trichome development transcription factors is correlated with the development of salt glands. Both of these plants are in the Caryophyllales. Thus, among dicotyledonous crops, approaches to salt gland engineering based on *Arabidopsis* trichomes may be limited to crops in the rosids, such as *Brassica* spp. and legumes, and perhaps to crops in the Caryophyllales.

More significant to the engineering of crop plants, the limited evidence to date on trichomes in the grasses gives no support for the involvement of any MYB, basic-helix-loop-helix, or WD-repeat proteins in trichome development. In maize, the mutant *macrohairless1* lacks the large single-celled trichomes known as macrohairs, but the gene product is not known (Moose et al., 2004). In rice, mutants of *glabrous leaf and hull1* (*gl1*) lack both macrohairs and microhairs, two classes of unicellular trichomes, but do not affect the development of the glandular trichomes. The mutations defining this locus are in the 5[1] untranslated region of a gene of unknown function, Os05g0118900 (Li et al., 2010). Thus what we learn from manipulating *Arabidopsis* trichomes to function as salt glands may not be readily applied to some of our most important crops, although crops in the rosids include not only the *Brassica* spp. (e.g., canola), which are very closely related to *Arabidopsis*, but the legumes, which include soybeans.

If engineering multicellular salt glands into a crop prior to establishing a proof of concept protocol in *Arabidopsis* is envisioned, Solanaceae crops provide alternative candidates. For example, engineering potato or tomato could take advantage of substantial molecular resources that are already available. These crops have reference genomes available for both the main commercial cultivars and also for more stress tolerant wild relatives (Xu et al., 2011; Bolger et al., 2014; Aversano et al., 2015). Solanaceae crops also have cultivars more tolerant to moderate salt levels (Shahbaz et al., 2012; Watanabe, 2015), have naturally developed secretory trichomes with structural features shared with recretohalophytes, have well-developed protocols to study gene expression exclusive to glandular trichomes, and have established transformation protocols (Butler et al., 2015; Čermák et al., 2015; Kortbeek et al., 2016). The idea of converting a glandular trichome to a salt secreting trichome bypasses the need to engineer cellular structural features needed for liquid excretion. Still, this endeavor requires the knowledge of coupling stress signaling and coordination of salt transport from roots to shoots and finally to the modified glandular trichomes at a metabolic energy cost (or yield penalty) applicable or tolerable for a crop species.

If a cereal crop model is chosen for engineering salt glands, rice would naturally be a top candidate, given the genetic resources available for rice as the prominent monocot model. This essential crop that feeds more than 3 billion people is being increasingly threatened by salinity stress caused by climate induced salt water intrusion, thus endangering the nutrition of the billions that consume rice. However, more targeted functional genomic studies have to be conducted to identify its trichome development pathway as discussed above. Comparative transcriptome-based studies on *Porteresia coarctata* salt hairs can further facilitate identification of the candidate orthologous genes one would need to introduce to selected rice cultivars. Alternatively, given the availability of genetic resources, including a reference genome, for sorghum, its relatively high capacity for abiotic stress tolerance as a C4 grass, and its phylogenetic proximity to almost all the grass species that are known to secrete salt through salt glands makes sorghum another attractive model for salt gland engineering in cereals (Paterson et al., 2009). It should be noted that all reported salt-secreting grasses also happen to be C4 grasses, with the exception of *Porteresia* (**Table 1**). The bicellular microhairs in *Zea mays* that are not considered to be salt glands show an increase in microhair density on leaves in response to increasing soil salinity (Ramadan and Flowers, 2004). This suggests the possibility of shared regulatory pathways in microhair initiation between salt secreting grasses and non-secretors. Notably, *Zea mays* has a significant amount of genomic resources, optimized genetic engineering tools, diverse germplasm from wild relatives, and cell type specific metabolic data (Liang et al., 2014; Nannas and Dawe, 2015; Dwivedi et al., 2016; Wen et al., 2016). Such factors, in conjunction with the importance as a major food and as a biofuel crop, make it another candidate for engineering salt hairs with significant secretion capacity upon problematic soil salt levels. Inarguably, a significant amount of functional, evolutionary, and comparative genomics studies need to be initiated to understand the organization and coordination of molecular networks that could transform a non-salt secreting species to a salt secreting plant. If we succeed with a non-crop model, success in the exercise would be a substantial test of our skills in combining – omics data, cell biology, and classical whole plant physiology to understand and manipulate a plant's response to environmental stress, a seemingly worthy objective in itself.

CONCLUDING REMARKS

Salt-stress is a substantial challenge for agriculture in the 21st century. One mechanism used by a wide variety of plants to deal with saline conditions is the use of epidermal salt glands that sequester or excrete salt. Salt glands have independently evolved likely twelve or more times and exist in at least four distinct morphological types. Despite these diverse origins, significant shared features due to convergent evolution give insight into the selective forces that have shaped their evolution and function. Although salt glands are challenging to study at the cellular and molecular level, new resources and tools have begun to elucidate the mechanisms by which salt glands alleviate salt stress. The time is now ripe to begin applying lessons from salt gland physiology to improving the salt tolerance of agricultural crops.

AUTHOR CONTRIBUTIONS

MD and JL developed, wrote, and edited the manuscript.

FUNDING

MD was supported by a National Science Foundation award (MCB 1616827) and the National Research Foundation of Korea (BioGreen21 program 2012R1A2A1A01003133). JL was supported by a National Science Foundation award (MCB 1615782).

ACKNOWLEDGMENTS

We like to thank the LSU Interlibrary Loan services for efficiently retrieving many publications unavailable electronically that were used for this review. We also like to thank Kieu-Nga Tran for help with processing **Figure 1** and John Johnson and Lauren Taylor for assistance with proofreading. We thank the four reviewers whose critical comments helped improve this manuscript.

REFERENCES

Adair, L. S., Andrews, D. L., Cairney, J., Funkhouser, E. A., Newton, R. J., and Aldon, E. F. (1992). Characterizing gene responses to drought stress in fourwing saltbush (*Atriplex canescens* (Pursh) Nutt). *J. Range Manag.* 45, 454–461. doi: 10.2307/4002902

Adam, P., and Wiecek, B. M. (1983). The salt glands of *Samolus repens*. *Wetl. Aust. J.* 3, 2–11.

Adams, P., Nelson, D. E., Yamada, S., Chmara, W., Jensen, R. G., Bohnert, H. J., et al. (1998). Growth and development of *Mesembryanthemum crystallinum* (Aizoaceae). *New Phytol.* 138, 171–190. doi: 10.1046/j.1469-8137.1998.00111.x

Adolf, V. I., Jacobsen, S. E., and Shabala, S. (2013). Salt tolerance mechanisms in quinoa (*Chenopodium quinoa* Willd.). *Environ. Exp. Bot.* 92, 43–54. doi: 10.1016/j.envexpbot.2012.07.004

Adolf, V. I., Shabala, S., Andersen, M. N., Razzaghi, F., and Jacobsen, S. E. (2012). Varietal differences of quinoa's tolerance to saline conditions. *Plant Soil* 357, 117–129. doi: 10.1007/s11104-012-1133-7

Agarie, S., Shimoda, T., Shimizu, Y., Baumann, K., Sunagawa, H., Kondo, A., et al. (2007). Salt tolerance, salt accumulation, and ionic homeostasis in an epidermal bladder-cell-less mutant of the common ice plant *Mesembryanthemum crystallinum*. *J. Exp. Bot.* 58, 1957–1967. doi: 10.1093/jxb/erm057

Ahn, J. H., Kim, J. S., Kim, S., Soh, H. Y., Shin, H., Jang, H., et al. (2015). De novo transcriptome analysis to identify anthocyanin biosynthesis genes responsible for tissue-specific pigmentation in Zoysiagrass (*Zoysia japonica* Steud.). *PLoS ONE* 10:e0124497. doi: 10.1371/journal.pone.0124497

Akhani, H., Barroca, J., Koteeva, N., Voznesenskaya, E., Franceschi, V., Edwards, G., et al. (2005). *Bienertia sinuspersici* (Chenopodiaceae): a new species from southwest asia and discovery of a third terrestrial C4 plant without Kranz anatomy. *Syst. Bot.* 30, 290–301. doi: 10.1600/0363644054223684

Altpeter, F., Springer, N. M., Bartley, L. E., Blechl, A. E., Brutnell, T. P., Citovsky, V., et al. (2016). Advancing crop transformation in the era of genome editing. *Plant Cell* 28, 1510–1520. doi: 10.1105/tpc.16.00196

Amaradasa, B. S., and Amundsen, K. (2016). Transcriptome profiling of buffalograss challenged with the leaf spot pathogen *Curvularia inaequalis*. *Front. Plant Sci.* 7:715. doi: 10.3389/fpls.2016.00715

Amarasinghe, V., and Watson, L. (1988). Comparative ultrastructure of microhairs in grasses. *Bot. J. Linn. Soc.* 98, 303–319. doi: 10.1111/j.1095-8339.1988.tb01705.x

Amarasinghe, V., and Watson, L. (1989). Variation in salt secretory activity of microhairs in grasses. *Austr. J. Plant Physiol.* 16, 219–229. doi: 10.1071/ PP9890219

Ascensão, L., Marques, N., and Pais, M. S. (1995). Glandular trichomes on vegetative and reproductive organs of *Leonotis leonurus* (Lamiaceæ). *Ann. Bot.* 75, 619–626. doi: 10.1006/anbo.1995.1067

Atkinson, M. R., Findlay, G. P., Hope, A. B., Pitman, M. G., Saddler, H. D. W., and West, K. R. (1967). Salt regulation in the mangroves *Rhizophora mucronata* Lam. and *Aegialitis annulata* Rbr. *Austr. J. Biol. Sci.* 20, 589–600. doi: 10.1071/BI9670589

Aversano, R., Contaldi, F., Ercolano, M. R., Grosso, V., Iorizzo, M., Tatino, F., et al. (2015). The *Solanum commersonii* genome sequence provides insights into adaptation to stress conditions and genome evolution of wild potato relatives. *Plant Cell* 27, 954–968. doi: 10.1105/tpc.114.135954

Aymen, S., Morena, G., Vincenzo, L., Laura, P., Lorenza, B., Abderrazak, S., et al. (2016). Salt tolerance of the halophyte *Limonium delicatulum* is more associated with antioxidant enzyme activities than phenolic compounds. *Funct. Plant Biol.* 43, 607–619. doi: 10.1071/FP15284

Baisakh, N., Subudhi, P. K., and Varadwaj, P. (2008). Primary responses to salt stress in a halophyte, smooth cordgrass (*Spartina alterniflora* Loisel.). *Funct. Integr. Genom.* 8, 287–300. doi: 10.1007/s10142-008-0075-x

Balsamo, R. A., and Thomson, W. W. (1993). Ultrastructural features associated with secretion in the salt glands of *Frankenia grandifolia* (Frankeniaceae) and *Avicennia germinans* (Avicenniaceae). *Am. J. Bot.* 80, 1276–1283. doi: 10.2307/2445711

Barhoumi, Z., Djebali, W., Abdelly, C., Chaïbi, W., and Smaoui, A. (2008). Ultrastructure of *Aeluropus littoralis* leaf salt glands under NaCl stress. *Protoplasma* 233, 195–202. doi: 10.1007/s00709-008-0003-x

Barkla, B. J., and Vera-Estrella, R. (2015). Single cell-type comparative metabolomics of epidermal bladder cells from the halophyte *Mesembryanthemum crystallinum*. *Front. Plant Science* 6:435. doi: 10.3389/fpls. 2015.00435

Barkla, B. J., Vera-Estrella, R., and Pantoja, O. (2012). Protein profiling of epidermal bladder cells from the halophyte *Mesembryanthemum crystallinum*. *Proteomics* 12, 2862–2865. doi: 10.1002/pmic.201200152

Barkla, B. J., Vera-Estrella, R., and Raymond, C. (2016). Single-cell-type quantitative proteomic and ionomic analysis of epidermal bladder cells from the halophyte model plant *Mesembryanthemum crystallinum* to identify salt-responsive proteins. *BMC Plant Biol.* 16:110. doi: 10.1186/s12870-016-0797-1

Bedre, R., Mangu, V. R., Srivastava, S., Sanchez, L. E., and Baisakh, N. (2016). Transcriptome analysis of smooth cordgrass (*Spartina alterniflora* Loisel), a monocot halophyte, reveals candidate genes involved in its adaptation to salinity. *BMC Genomics* 17:657. doi: 10.1186/s12864-016-3017-3

Bennett, T. H., Flowers, T. J., and Bromham, L. (2013). Repeated evolution of salt-tolerance in grasses. *Biol. Lett.* 9:20130029. doi: 10.1098/rsbl.2013.0029

Bhogaonkar, P. Y., and Lande, S. K. (2015). Anatomical characterization of *Lepidagathis cristata* Willd. a ethnomedicinal herb. *J. Glob. Biosci.* 4, 2282–2288.

Bohnert, H. J., and Cushman, J. C. (2000). The ice plant cometh: lessons in abiotic stress tolerance. *J. Plant Growth Regul.* 19, 334–346. doi: 10.1007/s003440000033

Bolger, A., Scossa, F., Bolger, M. E., Lanz, C., Maumus, F., Tohge, T., et al. (2014). The genome of the stress-tolerant wild tomato species *Solanum pennellii*. *Nat. Genet.* 46, 1034–1038. doi: 10.1038/ng.3046

Bonales-Alatorre, E., Shabala, S., Chen, Z., and Pottosin, I. I. (2013). Reduced tonoplast FV and SV channels activity is essential for conferring salinity tolerance in a facultative halophyte, *Chenopodium quinoa*. *Plant Physiol.* 162, 940–952. doi: 10.1104/pp.113.216572

Breckle, S. W. (1990). "Salinity tolerance of different halophyte types," in *Genetic Aspects of Plant Mineral Nutrition*, eds N. El Bassam, M. Dambroth, and B. C. Loughman (Berlin: Springer), 167–175. doi: 10.1007/978-94-009-2053-8_26

Bromham, L. (2015). Macroevolutionary patterns of salt tolerance in angiosperms. *Ann. Bot.* 115, 333–341. doi: 10.1093/aob/mcu229

Budak, H., Kasap, Z., Shearman, R. C., Dweikat, I., Sezerman, U., and Mahmood, A. (2006). Molecular characterization of cDNA encoding resistance gene-like sequences in *Buchloe dactyloides*. *Mol. Biotechnol.* 34, 293–301. doi: 10.1385/MB:34:3:293

Budak, H., Shearman, R. C., and Dweikat, I. (2005). Evolution of *Buchloe dactyloides* based on cloning and sequencing of matK, rbcL, and cob genes from plastid and mitochondrial genomes. *Genome* 48, 411–416. doi: 10.1139/g05-002

Butler, N. M., Atkins, P. A., Voytas, D. F., and Douches, D. S. (2015). Generation and inheritance of targeted mutations in potato (*Solanum tuberosum* L.) using the CRISPR/Cas system. *PLoS ONE* 10:e0144591. doi: 10.1371/journal.pone.0144591

Byng, J. W., Chase, M. W., Christenhusz, M. J. M., Fay, M. F., Judd, W. S., Mabberley,

D. J., et al. (2016). An update of the angiosperm phylogeny group classification for the orders and families of flowering plants: APG IV. *Bot. J. Linn. Soc.* 181, 1–20. doi: 10.1111/boj.12385

Cameron, K. M., Wurdack, K. J., and Jobson, R. W. (2002). Molecular evidence for the common origin of snap-traps among carnivorous plants. *Am. J. Bot.* 89, 1503–1509. doi: 10.3732/ajb.89.9.1503

Campbell, C. J., and Stong, J. E. (1964). Salt gland anatomy in *Tamarix penandra* (Tamaricaceae). *Southwest. Nat.* 9, 232–238. doi: 10.2307/3669691

Campbell, N., and Thomson, W. W. (1976). The ultrastructure of *Frankenia* salt glands. *Ann. Bot.* 40, 681–686. doi: 10.1093/oxfordjournals.aob.a085181

Campbell, N., Thomson, W. W., and Platt, K. (1974). The apoplastic pathway of transport to salt glands. *J. Exp. Bot.* 25, 61–69. doi: 10.1093/jxb/25.1.61

Cardale, S., and Field, C. D. (1971). The structure of the salt gland of *Aegiceras corniculatum*. *Planta* 99, 183–191. doi: 10.1007/BF00386836

Céccoli, G., Ramos, J., Pilatti, V., Dellaferrera, I., Tivano, J. C., Taleisnik, E., et al. (2015). Salt glands in the Poaceae family and their relationship to salinity tolerance. *Bot. Rev.* 81, 162–178. doi: 10.1007/s12229-015-9153-7

Čermák, T., Baltes, N. J., Čegan, R., Zhang, Y., and Voytas, D. F. (2015). High-frequency, precise modification of the tomato genome. *Genome Biol.* 16, 232. doi: 10.1186/s13059-015-0796-9

Chen, C.-C., and Chen, Y.-R. (2005). Study on laminar hydathodes of *Ficus formosana* (Moraceae) I. Morphology and ultrastructure. *Bot. Bull.* 46, 205–215.

Chen, Y., Zong, J., Tan, Z., Li, L., Hu, B., Chen, C., et al. (2015). Systematic mining of salt-tolerant genes in halophyte *Zoysia matrella* through cDNA expression library screening. *Plant Physiol. Biochem.* 89, 44–52. doi: 10.1016/j.plaphy.2015.02.007

Cheng, X., Dai, X., Zeng, H., Li, Y., Tang, W., and Han, L. (2009). Gene expression involved in dark-induced leaf senescence in zoysiagrass (*Zoysia japonica*). *Plant Biotechnol. Rep.* 3, 285–292. doi: 10.1007/s11816-009-0104-9

Chiang, C. P., Yim, W. C., Sun, Y. H., Ohnishi, M., Mimura, T., Cushman, J. C., et al. (2016). Identification of ice plant (*Mesembryanthemum crystallinum* L.) microRNAs using RNA-seq and their putative roles in high salinity responses in seedlings. *Front. Plant Sci.* 7:1143. doi: 10.3389/fpls.2016.01143

Choi, Y. E., Lim, S., Kim, H. J., Han, J. Y., Lee, M. H., Yang, Y., et al. (2012). Tobacco NtLTP1, a glandular-specific lipid transfer protein, is required for lipid secretion from glandular trichomes. *Plant J.* 70, 480–491. doi: 10.1111/j.1365-313X.2011. 04886.x

Coles, N. D., Coleman, C. E., Christensen, S. A., Jellen, E. N., Stevens, M. R., Bonifacio, A., et al. (2005). Development and use of an expressed sequenced tag library in quinoa (*Chenopodium quinoa* Willd.) for the discovery of single nucleotide polymorphisms. *Plant Sci.* 168, 439–447. doi: 10.1016/j.plantsci.2004.09.007

Cushman, J. C., Tillett, R. L., Wood, J., Branco, J., and Schlauch, K. (2008). Large-scale mRNA expression profiling in the common ice plant, *Mesembryanthemum crystallinum*, performing C3 photosynthesis and Crassulacean acid metabolism (CAM). *J. Exp. Bot.* 59, 1875–1894. doi: 10.1093/jxb/ern008

Dang, Z., Zheng, L., Wang, J., Gao, Z., Wu, S., Qi, Z., et al. (2013). Transcriptomic profiling of the salt-stress response in the wild recretohalophyte *Reaumuria trigyna*. *BMC Genomics* 14:29. doi: 10.1186/1471-2164-14-29

Daraban, I.-N., Mihali, C. V., Turcus, V., Ardelean, A., and Arsene, G. G. (2013). ESEM and EDAX observations on leaf and stem epidermal structures (stomata and salt glands) in *Limonium gmelinii* (Willd.) Kuntze. *Ann. Rom. Soc. Cell Biol.* 18, 123–130.

Das, S. (2002). On the ontogeny of stomata and glandular hairs in some Indian mangroves. *Acta Bot. Croat.* 61, 199–205.

Deal, R. B., and Henikoff, S. (2011). The INTACT method for cell type-specific gene expression and chromatin profiling in *Arabidopsis thaliana*. *Nat. Protoc.* 6, 56–68. doi: 10.1038/nprot.2010.175

Deinlein, U., Stephan, A. B., Horie, T., Luo, W., Xu, G., and Schroeder, J. I. (2014). Plant salt-tolerance mechanisms. *Trends Plant Sci.* 19, 371–379. doi: 10.1016/j.tplants.2014.02.001

Ding, F., Chen, M., Sui, N., and Wang, B. S. (2010a). $Ca2^+$ significantly enhanced development and salt-secretion rate of salt glands of *Limonium bicolor* under NaCl treatment. *S. Afr. J. Bot.* 76, 95–101. doi: 10.1016/j.sajb.2009.09.001

Ding, F., Yang, J.-C., Yuan, F., and Wang, B.-S. (2010b). Progress in mechanism of salt excretion in recretohalopytes. *Front. Biol.* 5:164–170. doi: 10.1007/s11515-010-0032-7

Dohm, J. C., Minoche, A. E., Holtgrawe, D., Capella-Gutierrez, S., Zakrzewski, F., Tafer, H., et al. (2014). The genome of the recently domesticated crop plant sugar beet (*Beta vulgaris*). *Nature* 505, 546–549. doi: 10.1038/nature 12817

Drennan, P. M., Berjak, P., Lawton, J. R., and Pammenter, N. (1987). Ultrastructure of the salt glands of the mangrove, *Avicennia marina* (Forssk.) Vierh., as indicated by the use of selective membrane staining. *Planta* 172, 176–183. doi: 10.1007/BF00394585

Du, L., Cai, C., Wu, S., Zhang, F., Hou, S., and Guo, W. (2016). Evaluation and exploration of favorable QTL alleles for salt stress related traits in cotton cultivars (*G. hirsutum* L.). *PLoS ONE* 11:e0151076. doi: 10.1371/journal.pone. 0151076

Dwivedi, S. L., Ceccarelli, S., Blair, M. W., Upadhyaya, H. D., Are, A. K., and Ortiz, R. (2016). Landrace germplasm for improving yield and abiotic stress adaptation. *Trends Plant Sci.* 21, 31–42. doi: 10.1016/j.tplants.2015. 10.012 Esau, K. (1965). *Plant Anatomy*. New York, NY: John Wiley & Sons, Inc.

Etalo, D. W., Vos, R. C., Joosten, M. H., and Hall, R. D. (2015). Spatially resolved plant metabolomics: some potentials and limitations of laser-ablation electrospray ionization mass spectrometry metabolite imaging. *Plant Physiol.* 169, 1424–1435. doi: 10.1104/pp.15.01176

Faraday, C. D., and Thomson, W. W. (1986a). Morphometric analysis of *Limonium* salt glands in relation to ion efflux. *J. Exp. Bot.* 37:471. doi: 10.1093/jxb/37.4.471

Faraday, C.D., and Thomson, W.W. (1986b). Structural aspects of the salt glands of the Plumbaginaceae. *J. Exp. Bot.* 37, 461–470. doi: 10.1093/jxb/37.4.461

Feng, Z., Deng, Y., Zhang, S., Liang, X., Yuan, F., Hao, J., et al. (2015). K^+ accumulation in the cytoplasm and nucleus of the salt gland cells of *Limonium bicolor* accompanies increased rates of salt secretion under NaCl treatment using NanoSIMS. *Plant Sci.* 238, 286–296. doi: 10.1016/j.plantsci.2015. 06.021

Ferreira de Carvalho, J., Poulain, J., Da Silva, C., Wincker, P., Michon-Coudouel, S., Dheilly, A., et al. (2013). Transcriptome de novo assembly from next-generation sequencing and comparative analyses in the hexaploid salt marsh species *Spartina maritima* and *Spartina alterniflora* (Poaceae). *Heredity* 110, 181–193. doi: 10.1038/hdy.2012.76

Flowers, T. J., and Colmer, T. D. (2015). Plant salt tolerance: adaptations in halophytes. *Ann. Bot.* 115, 327–331. doi: 10.1093/aob/mcu267

Flowers, T. J., Flowers, S. A., Hajibagheri, M. A., and Yeo, A. R. (1990). Salt tolerance in the halophytic wild rice, *Porteresia coarctata* Tateoka. *New Phytol.* 114, 675–684. doi: 10.1111/j.1469-8137.1990.tb00439.x

Flowers, T. J., Galal, H. K., and Bromham, L. (2010). Evolution of halophytes: multiple origins of salt tolerance in land plants. *Funct. Plant Biol.* 37, 604–612. doi: 10.1071/FP09269

Flowers, T. J., Hajibagheri, M. A., and Clipson, N. J. W. (1986). Halophytes. *Q. Rev. Biol.* 61, 313–337. doi: 10.1086/415032

Flowers, T. J., Munns, R., and Colmer, T. D. (2015). Sodium chloride toxicity and the cellular basis of salt tolerance in halophytes. *Ann. Bot.* 115, 419–431. doi: 10.1093/aob/mcu217

Francisco, A. M., Diaz, M., Romano, M., and Sanchez, F. (2009). Morphoanatomical description of leaves glands types in the white mangrove *Laguncularia racemosa* (L.) Gaertn (f.). *Acta Microsc.* 18, 237–252.

Fricke, W. (1997). Cell turgor, osmotic pressure and water potential in the upper epidermis of barley leaves in relation to cell location and in response to NaCl and air humidity. *J. Exp. Bot.* 48, 45–58. doi: 10.1093/jxb/48.1.45

Fricke, W. (2013). "Plant single cell sampling," in *Plant Mineral Nutrients: Methods and Protocols*, ed. F. J. M. Maathuis (Totowa, NJ: Humana Press), 209–231. doi: 10.1007/978-1-62703-152-3_14

Fricke, W., and Peters, W. S. (2002). The biophysics of leaf growth in salt-stressed barley. A study at the cell level. *Plant Physiol.* 129, 374–388. doi: 10.1104/pp. 001164

Fu, X., Huang, Y., Deng, S., Zhou, R., Yang, G., and Ni, X. (2005). Construction of a SSH library of *Aegiceras corniculatum* under salt stress and expression analysis of four transcripts. *Plant Sci.* 169, 147–154. doi: 10.1016/j.plantsci.2005.03.009

Fuglsang, A. T., Paez-Valencia, J., and Gaxiola, R. A. (2011). "Plant proton pumps: regulatory circuits involving H+-ATPase and H+-PPase," in *Transporters and Pumps in Plant Signaling*, eds M. Geisler and K. Venema (Berlin: Springer), 39–64. doi: 10.1007/978-3-642-14369-4_2

Gao, C., Liu, Y., Wang, C., Zhang, K., and Wang, Y. (2014). Expression profiles of 12 late embryogenesis abundant protein genes from *Tamarix hispida* in response to abiotic stress. *Sci. World J.* 2014:868391. doi: 10.1155/2014/868391

Gao, C., Wang, Y., Liu, G., Yang, C., Jiang, J., and Li, H. (2008). Expression profiling of salinity-alkali stress responses by large-scale expressed sequence tag analysis in *Tamarix hispida*. *Plant Mol. Biol.* 66, 245–258. doi: 10.1007/s11103-007-9266-4

Garg, R., Verma, M., Agrawal, S., Shankar, R., Majee, M., and Jain, M. (2013). Deep transcriptome sequencing of wild halophyte rice, *Porteresia coarctata*, provides novel insights into the salinity and submergence tolerance factors. *DNA Res.* 21, 69–84. doi: 10.1093/dnares/dst042

Giuliani, C., and Bini, L. M. (2008). Insight into the structure and chemistry of glandular trichomes of Labiatae, with emphasis on subfamily Lamioideae. *Plant Syst. Evol.* 276, 199–208. doi: 10.1007/s00606-008-0085-0

Glas, J. J., Schimmel, B. C. J., Alba, J. M., Escobar-Bravo, R., Schuurink, R. C., and Kant, M. R. (2012). Plant glandular trichomes as targets for breeding or engineering of resistance to herbivores. *Int. J. Mol. Sci.* 13:17077. doi: 10.3390/ijms131217077

Gossett, D. R., Millhollon, E. P., and Lucas, M. C. (1994). Antioxidant response to NaCl stress in salt-tolerant and salt-sensitive cultivars of cotton. *Crop Sci.* 34, 706–714. doi: 10.2135/cropsci1994.0011183X003400030020x

Grigore, M. N., Ivanescu, L., and Toma, C. (2014). *Halophytes: An Integrative Anatomical Study.* Cham: Springer, doi: 10.1007/978-3-319-05729-3_10

Gu, L., Xu, D., You, T., Li, X., Yao, S., Chen, S., et al. (2011). Analysis of gene expression by ESTs from suppression subtractive hybridization library in *Chenopodium album* L. under salt stress. *Mol. Biol. Rep.* 38, 5285–5295. doi: 10.1007/s11033-011-0678-5

Gucci, R., Aronne, G., Lombardini, L., and Tattini, M. (1997). Salinity tolerance in *Phillyrea* species. *New Phytol.* 135, 227–234. doi: 10.1046/j.1469-8137.1997.00644.x

Gunning, B. E. S., and Pate, J. S. (1969). "Transfer cells" plant cells with wall ingrowths, specialized in relation to short distance transport of solutes -their occurrence, structure, and development. *Protoplasma* 68, 107–133. doi: 10. 1007/BF01247900

Hasegawa, P. M., Bressan, R. A., Zhu, J.-K., and Bohnert, H. J. (2000). Plant cellular and molecular responses to high salinity. *Annu. Rev. Plant Physiol. Plant Mol. Biol.* 51, 463–499. doi: 10.1146/annurev.arplant.51.1.463

Heubl, G., Bringmann, G., and Meimberg, H. (2006). Molecular phylogeny and character evolution of carnivorous plant families in Caryophyllales -revisited. *Plant Biol.* 8, 821–830. doi: 10.1055/s-2006-924460

Hu, L., Li, H., Chen, L., Lou, Y., Amombo, E., and Fu, J. (2015). RNA-seq for gene identification and transcript profiling in relation to root growth of bermudagrass (*Cynodon dactylon*) under salinity stress. *BMC Genomics* 16:575. doi: 10.1186/s12864-015-1799-3

Huang, J., Lu, X., Zhang, W., Huang, R., Chen, S., and Zheng, Y. (2014). Transcriptome sequencing and analysis of leaf tissue of *Avicennia marina* using the Illumina platform. *PLoS ONE* 9:e108785. doi: 10.1371/journal.pone. 0108785

Hülskamp, M., Miséra, S., and Jürgens, G. (1994). Genetic dissection of trichome cell development in Arabidopsis. *Cell* 76, 555–566. doi: 10.1016/0092-8674(94)90118-X

Immelman, K. (1990). Studies in the southern African species of *Justicia* and *Siphonoglossa* (Acanthaceae): indumentum. *Bothalia* 20, 61–66. doi: 10.4102/ abc. v20i1.894

Ioannidou-Akoumianaki, A., Spentza, R. P., and Fasseas, C. (2015). *Limoniastrum monopetalum* (L.) Boiss, a candidate plant for use in urban and suburban areas with adverse conditions. An anatomical and histochemical study. *Bull. Univ. Agric. Sci. Vet. Med. Cluj-Napoca. Hortic.* 72, 438–440. doi: 10.15835/ buasvmcn-hort:11347

IPCC (2014). "Climate change 2014: synthesis report," in *Contribution of Working Groups I, II and III to the Fifth Assessment Report of the Intergovernmental Panel on Climate Change*, eds R. K. Pachauri and L. A. Meyer Geneva (Geneva: IPCC).

Ismail, H., Dragišic Maksimovic, J., Maksimovic, V., Shabala, L., Živanovic, B. D., Tian, Y., et al. (2015). Rutin, a flavonoid with antioxidant activity, improves plant salinity tolerance by regulating K^+ retention and Na^+ exclusion from leaf mesophyll in quinoa and broad beans. *Funct. Plant Biol.* 43, 75–86.

Jakoby, M. J., Falkenhan, D., Mader, M. T., Brininstool, G., Wischnitzki, E., Platz, N., et al. (2008). Transcriptional profiling of mature Arabidopsis trichomes reveals that NOECK encodes the MIXTA-like transcriptional regulator MYB106. *Plant Physiol.* 148, 1583–1602. doi: 10.1104/pp.108.126979

Jou, Y., Wang, Y., and Yen, H. (2007). Vacuolar acidity, protein profile, and crystal composition of epidermal bladder cells of the halophyte *Mesembryanthemum crystallinum*. *Funct. Plant Biol.* 34, 353–359. doi: 10.1071/FP06269

Jyothi-Prakash, P. A., Mohanty, B., Wijaya, E., Lim, T. M., Lin, Q., Loh, C. S., et al. (2014). Identification of salt gland-associated genes and characterization of a dehydrin from the salt secretor mangrove *Avicennia officinalis*. *BMC Plant Biol.* 14:291. doi: 10.1186/s12870-014-0291-6

Karimi, S. H., and Ungar, I. A. (1989). Development of epidermal salt hairs in Atriplex triangularis Willd. in response to salinity, light intensity, and aeration. *Bot. Gazette* 150, 68–71. doi: 10.1086/337749

Khan, T., Reddy, V. S., and Leelavathi, S. (2010). High-frequency regeneration via somatic embryogenesis of an elite recalcitrant cotton genotype (*Gossypium hirsutum* L.) and efficient *Agrobacterium*-mediated transformation. *Plant Cell Tissue Organ. Cult.* 101, 323–330. doi: 10.1007/s11240-010-9691-y

Kim, C., Jang, C. S., Kamps, T. L., Robertson, J. S., Feltus, F. A., and Paterson, A. H. (2008). Transcriptome analysis of leaf tissue from Bermudagrass (*Cynodon dactylon*) using a normalised cDNA library. *Funct. Plant Biol.* 35, 585–594. doi: 10.1071/FP08133

Ko, S., Kang, H., Kim, J., Seo, J., Kim, M., Sin, H., et al. (2010). Transcriptomic analysis of zoysiagrass using expressed sequence tags. *Hortic. Environ. Biotechnol.* 51, 562–565.

Kortbeek, R. W. J., Xu, J., Ramirez, A., Spyropoulou, E., Diergaarde, P., Otten-Bruggeman, I., et al. (2016). "Chapter twelve -Engineering of tomato glandular trichomes for the production of specialized metabolites," in *Synthetic Biology and Metabolic Engineering in Plants and Microbes Part B: Metabolism in Plants Methods in Enzymology*, ed. S. E. O'Connor (Cambridge, MA: Academic Press), 305–331. doi: 10.1016/bs.mie.2016.02.014

Kriegel, A., Andrés, Z., Medzihradszky, A., Krüger, F., Scholl, S., Delang, S., et al. (2015). Job sharing in the endomembrane system: vacuolar acidification requires the combined activity of V-ATPase and V-PPase. *Plant Cell* 27, 3383–3396. doi: 10.1105/tpc.15.00733

Larkin, J. C., Brown, M. L., and Schiefelbein, J. (2003). How do cells know what they want to be when they grow up? Lessons from epidermal patterning in Arabidopsis. *Annu. Rev. Plant Biol.* 54, 403–430. doi: 10.1146/annurev.arplant.54.031902.134823

Levering, C. A., and Thomson, W. W. (1971). The ultrastructure of the salt gland of *Spartina foliosa*. *Planta* 97, 183–196. doi: 10.1007/BF00389200

Li, F., Fan, G., Lu, C., Xiao, G., Zou, C., Kohel, R. J., et al. (2015). Genome sequence of cultivated Upland cotton (*Gossypium hirsutum* TM-1) provides insights into genome evolution. *Nat. Biotechnol.* 33, 524–530. doi: 10.1038/nbt.3208

Li, J., Sun, X., Yu, G., Jia, C., Liu, J., and Pan, H. (2014). Generation and analysis of expressed sequence tags (ESTs) from halophyte *Atriplex canescens* to explore salt-responsive related genes. *Int. J. Mol. Sci.* 15, 11172–11189. doi: 10.3390/ijms150611172

Li, W., Wu, J., Weng, S., Zhang, D., Zhang, Y., and Shi, C. (2010). Characterization and fine mapping of the glabrous leaf and hull mutants (gl1) in rice (*Oryza sativa* L.). *Plant Cell Rep.* 29, 617–627. doi: 10.1007/s00299-010-0848-2

Liang, Z., Zhang, K., Chen, K., and Gao, C. (2014). Targeted mutagenesis in *Zea mays* using TALENs and the CRISPR/Cas system. *J. Genet. Genom.* 41, 63–68. doi: 10.1016/j.jgg.2013.12.001

Lin, M., Pang, C., Fan, S., Song, M., Wei, H., and Yu, S. (2015). Global analysis of the *Gossypium hirsutum* L. Transcriptome during leaf senescence by RNA-Seq. *BMC Plant Biol.* 15:43. doi: 10.1186/s12870-015-0433-5

Liphschitz, N., Adiva-Shomer-Ilan, Eshel, A., and Waisel, Y. (1974). Salt glands on leaves of rhodes grass (*Chloris gayana* Kth.). *Ann. Bot.* 38, 459–462. doi: 10.1093/oxfordjournals.aob.a084829

Liphschitz, N., and Waisel, Y. (1974). Existence of salt glands in various genera of the Gramineae. *New Phytol.* 73, 507–513. doi: 10.1111/j.1469-8137.1974. tb02129.x

Liu, M., Li, X., Liu, Y., Shi, Y., and Ma, X. (2015). Analysis of differentially expressed genes under UV-B radiation in the desert plant *Reaumuria soongorica*. *Gene* 574, 265–272. doi: 10.1016/j.gene.2015.08.026

Liu, Y., Liu, M., Li, X., Cao, B., and Ma, X. (2014). Identification of differentially expressed genes in leaf of *Reaumuria soongorica* under PEG-induced drought stress by digital gene expression profiling. *PLoS ONE* 9:e94277. doi: 10.1371/journal.pone.0094277

Longstreth, D. J., and Nobel, P. S. (1979). Salinity effects on leaf anatomy: consequences for photosynthesis. *Plant Physiol.* 63, 700–703. doi: 10.1104/pp. 63.4.700

Lüttge, U., Fischer, E., and Steudle, E. (1978). Membrane potentials and salt distribution in epidermal bladders and photosynthetic tissue of *Mesembryanthemum crystallinum* L. *Plant Cell Environ.* 1, 121–129. doi: 10.1111/j.1365-3040.1978.tb00753.x

Maathuis, F. J. M. (2014). Sodium in plants: perception, signalling, and regulation of sodium fluxes. *J. Exp. Bot.* 65, 849–858. doi: 10.1093/jxb/ert326

Maathuis, F. J. M., Ahmad, I., and Patishtan, J. (2014). Regulation of Na+ fluxes in plants. *Front. Plant Sci.* 5:467. doi: 10.3389/fpls.2014.00467

Malcolm, C. V., Lindley, V. A., O'Leary, J. W., Runciman, H. V., and Barrett-Lennard, E. G. (2003). Halophyte and glycophyte salt tolerance at germination and the establishment of halophyte shrubs in saline environments. *Plant Soil* 253, 171–185. doi: 10.1023/A:1024578002235

Malone, M., Leigh, R. A., and Tomos, A. D. (1989). Extraction and analysis of sap from individual wheat leaf cells: the effect of sampling speed on the osmotic pressure of extracted sap. *Plant Cell Environ.* 12, 919–926. doi: 10.1111/j.1365-3040.1989.tb01971.x

Marcum, K. B. (2006). Use of saline and non-potable water in the turfgrass industry: constraints and developments. *Agric. Water Manag.* 80, 132–146. doi: 10.1016/ j.agwat.2005.07.009

Marcum, K. B. (2008). "Saline tolerance physiology in grasses," in *Ecophysiology of High Salinity Tolerant Plants*, eds M. A. Khan and D. J. Weber (Berlin: Springer), 157–172.

Marcum, K. B., Anderson, S. J., and Engelke, M. C. (1998). Salt gland ion secretion: a salinity tolerance mechanism among five Zoysiagrass species. *Crop Science* 38, 806–810. doi: 10.2135/cropsci1998.0011183X003800030031x

Marcum, K. B., and Murdoch, C. L. (1990). Salt glands in the Zoysieae. *Ann. Bot.* 66, 1–7. doi: 10.1093/oxfordjournals.aob.a087991

Maricle, B. R., Koteyeva, N. K., Voznesenskaya, E. V., Thomasson, J. R., and Edwards, G. E. (2009). Diversity in leaf anatomy, and stomatal distribution and conductance, between salt marsh and freshwater species in the C4 genus *Spartina* (Poaceae). *New Phytol.* 184, 216–233. doi: 10.1111/j.1469-8137.2009. 02903.x

Mathur, J., Mathur, N., Kernebeck, B., and Hülskamp, M. (2003). Mutations in actin-related proteins 2 and 3 affect cell shape development in Arabidopsis. *Plant Cell* 15, 1632–1645. doi: 10.1105/tpc.011676

Maurel, C., Boursiac, Y., Luu, D.-T., Santoni, V., Shahzad, Z., and Verdoucq, L. (2015). Aquaporins in plants. *Physiol. Rev.* 95, 1321–1358. doi: 10.1152/physrev. 00008.2015

Mehta, P. A., Sivaprakash, K., Parani, M., Venkataraman, G., and Parida, A. K. (2004). Generation and analysis of expressed sequence tags from the salt-tolerant mangrove species *Avicennia marina* (Forsk) Vierh. *Theor. Appl. Genet.* 110, 416–424. doi: 10.1007/s00122-004-1801-y

Mickelbart, M. V., Hasegawa, P. M., and Bailey-Serres, J. (2015). Genetic mechanisms of abiotic stress tolerance that translate to crop yield stability. *Nat. Rev. Genet.* 16, 237–251. doi: 10.1038/nrg3901

Mishra, S., and Das, A. B. (2003). Effect of NaCl on leaf salt secretion and antioxidative enzyme level in roots of a mangrove, *Aegiceras corniculatum*. *Indian J. Exp. Biol.* 41, 160–166.

Mondal, T. K., Ganie, S. A., and Debnath, A. B. (2015). Identification of novel and conserved miRNAs from extreme halophyte, *Oryza coarctata*, a wild relative of rice. *PLoS ONE* 10:e0140675. doi: 10.1371/journal.pone.0140675

Mooney, H. A., Gulmon, S. L., Ehleringer, J., and Rundel, P. W. (1980). Atmospheric water uptake by an Atacama Desert shrub. *Science* 209, 693–694. doi: 10.1126/ science.209.4457.693

Moose, S. P., Lauter, N., and Carlson, S. R. (2004). The maize macrohairless1 locus specifically promotes leaf blade macrohair initiation and responds to factors regulating leaf identity. *Genetics* 166, 1451–1461. doi: 10.1534/genetics.166. 3.1451

Morohashi, K., Zhao, M., Yang, M., Read, B., Lloyd, A., Lamb, R., et al. (2007). Participation of the Arabidopsis bHLH Factor GL3 in trichome initiation regulatory events. *Plant Physiol.* 145, 736–746. doi: 10.1104/pp.107.104521

Munien, P., Naidoo, Y., and Naidoo, G. (2015). Micromorphology, histochemistry and ultrastructure of the foliar trichomes of *Withania somnifera* (L.) Dunal (Solanaceae). *Planta* 242, 1107–1122. doi: 10.1007/s00425-015-2341-1

Mustroph, A., Zanetti, M. E., Jang, C. J. H., Holtan, H. E., Repetti, P. P., Galbraith, D. W., et al. (2009). Profiling translatomes of discrete cell populations resolves altered cellular priorities during hypoxia in *Arabidopsis*. *Proc. Natl. Acad. Sci. U.S.A.* 106, 18843–18848. doi: 10.1073/pnas.0906131106

Nah, G., Lee, M., Kim, D. S., Rayburn, A. L., Voigt, T., and Lee, D. K. (2016). Transcriptome analysis of *Spartina pectinata* in response to freezing stress. *PLoS ONE* 11:e0152294. doi: 10.1371/journal.pone.0152294

Naidoo, G. (2016). The mangroves of South Africa: an ecophysiological review. *S. Afr. J. Bot.* 107, 101–113. doi: 10.1016/j.sajb.2016.04.014

Nannas, N. J., and Dawe, R. K. (2015). Genetic and genomic toolbox of *Zea mays*. *Genetics* 199, 655–669. doi: 10.1534/genetics.114.165183

Nguyen, P., Ho, C. L., Harikrishna, J. A., Wong, M. L., and Rahim, R. A. (2007). Functional screening for salinity tolerant genes from *Acanthus ebracteatus* Vahl using *Escherichia coli* as a host. *Trees* 21, 515. doi: 10.1007/s00468-007-0144-0

Nguyen, P. D., Ho, C. L., Harikrishna, J. A., Wong, M. C., and Rahim, R. A. (2006). Generation and analysis of expressed sequence tags from the mangrove plant, *Acanthus ebracteatus* Vahl. *Tree Genet. Genomes* 2, 196–201. doi: 10.1007/ s11295-006-0044-2

Offermann, S., Friso, G., Doroshenk, K. A., Sun, Q., Sharpe, R. M., Okita, T. W., et al. (2015). Developmental and subcellular organization of single-cell C4 photosynthesis in *Bienertia sinuspersici* determined by large-scale proteomics and cDNA assembly from 454 DNA sequencing. *J. Proteome Res.* 14, 2090–2108. doi: 10.1021/pr5011907

Oh, D., Barkla, B. J., Vera-estrella, R., Pantoja, O., Lee, S., Bohnert, H. J., et al. (2015). Cell type-specific responses to salinity – the epidermal bladder cell transcriptome of *Mesembryanthemum crystallinum*. *New Phytol.* 207, 627–644. doi: 10.1111/nph.13414

Oh, D., Dassanayake, M., Bohnert, H. J., and Cheeseman, J. M. (2012). Life at the extreme: lessons from the genome. *Genome Biol.* 13:241. doi: 10.1186/gb-2012-13-3-241

Olofsson, L., Lundgren, A., and Brodelius, P. E. (2012). Trichome isolation with and without fixation using laser microdissection and pressure catapulting followed by RNA amplification: expression of genes of terpene metabolism in apical and sub-apical trichome cells of *Artemisia annua* L. *Plant Sci.* 183, 9–13. doi: 10.1016/j.plantsci.2011.10.019

Ong, J. E., and Gong, W. K. (2013). "Structure, function and management of mangrove ecosystems," in *ISME Mangrove Educational Book Series No. 2*, ed. H. T. Chan (Yokohama: International Tropical Timber Organization).

Oross, J. W., and Thomson, W. W. (1982). The ultrastructure of the salt glands of *Cynodon* and *Distichlis* (Poaceae). *Am. J. Bot.* 69, 939–949. doi: 10.2307/2442890

Pan, Y., Guo, H., Wang, S., Zhao, B., Zhang, J., Ma, Q., et al. (2016). The photosynthesis, Na+/K+ homeostasis and osmotic adjustment of *Atriplex* canescens in response to salinity. *Front. Plant Sci.* 7:848. doi: 10.3389/fpls.2016. 00848

Parida, A. K., and Jha, B. (2010). Salt tolerance mechanisms in mangroves: a review. *Trees* 24, 199–217. doi: 10.1007/s00468-010-0417-x

Park, J., Okita, T. W., and Edwards, G. E. (2009). Salt tolerant mechanisms in single-cell C4 species *Bienertia sinuspersici* and *Suaeda aralocaspica* (Chenopodiaceae). *Plant Sci.* 176, 616–626. doi: 10.1016/j.plantsci.2009. 01.014

Paterson, A. H., Bowers, J. E., Bruggmann, R., Dubchak, I., Grimwood, J., Gundlach, H., et al. (2009). The *Sorghum* bicolor genome and the diversification of grasses. *Nature* 457, 551–556. doi: 10.1038/nature07723

Patron, N. J. (2014). DNA assembly for plant biology: techniques and tools. *Curr. Opin. Plant Biol.* 19, 14–19. doi: 10.1016/j.pbi.2014.02.004

Payne, C. T., Zhang, F., and Lloyd, A. M. (2000). GL3 encodes a bHLH protein that regulates trichome development in Arabidopsis through interaction with GL1 and TTG1. *Genetics* 156, 1349–1362.

Payne, T., Clement, J., Arnold, D., and Lloyd, A. (1999). Heterologous myb genes distinct from GL1 enhance trichome production when overexpressed in *Nicotiana tabacum*. *Development* 126, 671–682.

Pelozo, A., Boeger, M. R. T., Sereneski-de-Lima, C., and Soffiatti, P. (2016). Leaf morphological strategies of seedlings and saplings of *Rhizophora mangle* (Rhizophoraceae), *Laguncularia racemosa* (Combretaceae) and *Avicennia schaueriana* (Acanthaceae) from Southern Brazil. *Rev. Biol. Trop.* 64, 321–333. doi: 10.15517/ rbt.v64i1.17923

Peña-Castro, J. M., Barrera-Figueroa, B. E., Fernández-Linares, L., Ruiz-Medrano, R., and Xoconostle-Cázares, B. (2006). Isolation and identification of up-regulated genes in bermudagrass roots (*Cynodon dactylon* L.) grown under petroleum hydrocarbon stress. *Plant Sci.* 170, 724–731. doi: 10.1016/j.plantsci. 2005.11.004

Peng, Z., He, S., Gong, W., Sun, J., Pan, Z., Xu, F., et al. (2014). Comprehensive analysis of differentially expressed genes and transcriptional regulation induced by salt stress in two contrasting cotton genotypes. *BMC Genomics* 15:760. doi: 10.1186/1471-2164-15-760

Peng, Z., He, S., Sun, J., Pan, Z., Gong, W., Lu, Y., et al. (2016). Na^+ compartmentalization related to salinity stress tolerance in upland cotton (*Gossypium hirsutum*) seedlings. *Sci. Rep.* 6:34548. doi: 10.1038/ srep34548

Polle, A., and Chen, S. (2015). On the salty side of life: molecular, physiological and anatomical adaptation and acclimation of trees to extreme habitats. *Plant Cell Environ.* 38, 1794–1816. doi: 10.1111/pce.12440

Qin, Z., Chen, J., Jin, L., Duns, G. J., and Ouyang, P. (2015). Differential expression of miRNAs under salt stress in *Spartina alterniflora* leaf tissues. *J. Nanosci. Nanotechnol.* 15, 1554–1561. doi: 10.1166/jnn.2015.9004

Rahman, M., Shaheen, T., Tabbasam, N., Iqbal, M. A., Ashraf, M., Zafar, Y., et al. (2012). Cotton genetic resources. A review. *Agron. Sustain. Dev.* 32, 419–432. doi: 10.1007/s13593-011-0051-z

Ramadan, T., and Flowers, T. J. (2004). Effects of salinity and benzyl adenine on development and function of microhairs of *Zea mays* L. *Planta* 219, 639–648. doi: 10.1007/s00425-004-1269-7

Reimann, C., and Breckle, S. (1988). Anatomy and development of bladder hairs in *Chenopodium* species. *Flora* 180, 275–288. doi: 10.1016/S0367-2530(17) 30323-7

Reis, C., Sajo, M. G., and Stehmann, J. R. (2002). Leaf structure and taxonomy of *Petunia* and *Calibrachoa* (Solanaceae). *Braz. Arch. Biol. Technol.* 45, 59–66. doi: 10.1590/S1516-89132002000100010

Renner, T., and Specht, C. D. (2013). Inside the trap: gland morphologies, digestive enzymes, and the evolution of plant carnivory in the Caryophyllales. *Curr. Opin. Plant Biol.* 16, 436–442. doi: 10.1016/j.pbi.2013.06.009

Robins, R. J., and Juniper, B. E. (1980). The secretory cycle of *Dionaea muscipula* Ellis. *New Phytol.* 86, 279–296. doi: 10.1111/j.1469-8137.1980.tb00789.x

Rodriguez-Uribe, L., Higbie, S. M., Stewart, J. M., Wilkins, T., Lindemann, W., Sengupta-Gopalan, C., et al. (2011). Identification of salt responsive genes using comparative microarray analysis in upland cotton (*Gossypium hirsutum* L.). *Plant Sci.* 180, 461–469. doi: 10.1016/j.plantsci.2010.10.009

Roeurn, S., Hoshino, N., Soejima, K., Inoue, Y., Cushman, J. C., and Agarie, S. (2016). Suppression subtractive hybridization library construction and identification of epidermal bladder cell related genes in the common ice plant, *Mesembryanthemum crystallinum* L. *Plant Prod. Sci.* 19, 552–561. doi: 10.1080/1343943X.2016.1221320

Rogers, E. D., Jackson, T., Moussaieff, A., Aharoni, A., and Benfey, P. N. (2012). Cell type-specific transcriptional profiling: implications for metabolite profiling. *Plant J.* 70, 5–17. doi: 10.1111/j.1365-313X.2012.04888.x

Rozema, J., Gude, H., and Pollak, G. (1981). An ecophysiological study of the salt secretion of four halophytes. *New Phytol.* 89, 201–217. doi: 10.1111/j.1469-8137.1981.tb07483.x

Rozema, J., Riphagen, I., and Sminia, T. (1977). A light and electron -microscopical study on the structure and function of the salt gland of *Glaux maritima* L. *New Phytol.* 79, 665–671. doi: 10.1111/j.1469-8137.1977.tb02251.x

Ruiz, K. B., Biondi, S., Martínez, E. A., Orsini, F., Antognoni, F., and Jacobsen, S.-E. (2016). Quinoa – a model crop for understanding salt-tolerance mechanisms in halophytes. *Plant Biosyst.* 150, 357–371. doi: 10.1080/11263504.2015.1027317

Santos, J., Al-Azzawi, M., Aronson, J., and Flowers, T. J. (2016). eHALOPH a database of salt-tolerant plants: helping put halophytes to work. *Plant Cell Physiol.* 57:e10. doi: 10.1093/pcp/pcv155

Scala, J., Schwab, D., and Simmons, E. (1968). The fine structure of the digestive gland of venus's flytrap. *Am. J. Bot.* 55, 649–657. doi: 10.2307/2440522

Scassellati, E., Pasqua, G., Valletta, A., and Abbate, G. (2016). Salt glands of *Armeria canescens* (Host) Boiss: morphological and functional aspects. *Plant Biosyst.* 150, 1134–1139. doi: 10.1080/11263504.2016.1186126

Schnittger, A., Jurgens, G., and Hulskamp, M. (1998). Tissue layer and organ specificity of trichome formation are regulated by GLABRA1 and TRIPTYCHON in Arabidopsis. *Development* 125, 2283–2289.

Schwab, R., and Ossowski, S. (2006). Highly specific gene silencing by artificial microRNAs in *Arabidopsis*. *Plant Cell* 18, 1121–1133. doi: 10.1105/tpc.105.039834.1

Semenova, G. A., Fomina, I. R., and Biel, K. Y. (2010). Structural features of the salt glands of the leaf of *Distichlis spicata* "Yensen 4a" (Poaceae). *Protoplasma* 240, 75–82. doi: 10.1007/s00709-009-0092-1

Sengupta, S., and Majumder, A. L. (2009). Insight into the salt tolerance factors of a wild halophytic rice, *Porteresia coarctata*: a physiological and proteomic approach. *Planta* 229, 911–929. doi: 10.1007/s00425-008-0878-y

Sengupta, S., and Majumder, A. L. (2010). Porteresia coarctata (Roxb.) Tateoka, a wild rice: a potential model for studying salt-stress biology in rice. *Plant Cell Environ.* 33, 526–542. doi: 10.1111/j.1365-3040.2009.02054.x

Serna, L., and Martin, C. (2006). Trichomes: different regulatory networks lead to convergent structures. *Trends Plant Sci.* 11, 274–280. doi: 10.1016/j.tplants.2006.04.008

Serrato-Valenti, G., Bisio, A., Cornara, L., and Ciarallo, G. (1997). Structural and histochemical investigation of the glandular trichomes of *Salvia aurea* L. leaves, and chemical cnalysis of the essential Oil. *Ann. Bot.* 79, 329–336. doi: 10.1006/anbo.1996.0348

Shabala, S., Bose, J., Fuglsang, A. T., and Pottosin, I. (2015). On a quest for stress tolerance genes: membrane transporters in sensing and adapting to hostile soils. *J. Exp. Bot.* 67, 1015–1031. doi: 10.1093/jxb/erv465

Shabala, S., Bose, J., and Hedrich, R. (2014). Salt bladders: do they matter? *Trends Plant Sci.* 19, 687–691. doi: 10.1016/j.tplants.2014.09.001

Shahbaz, M., Ashraf, M., Al-Qurainy, F., and Harris, P. J. C. (2012). Salt tolerance in selected vegetable crops. *Crit. Rev. Plant Sci.* 31, 303–320. doi: 10.1080/07352689.2012.656496

Shangguan, X.-X., Xu, B., Yu, Z.-X., Wang, L.-J., and Chen, X.-Y. (2008). Promoter of a cotton fibre MYB gene functional in trichomes of *Arabidopsis* and glandular trichomes of tobacco. *J. Exp. Bot.* 59, 3533–3542. doi: 10.1093/jxb/ern204

Shen, X., Zhang, T., Guo, W., Zhu, X., and Zhang, X. (2006). Mapping fiber and yield QTLs with main, epistatic, and QTL × environment interaction effects in recombinant inbred lines of upland cotton. *Crop Sci.* 46, 61–66. doi: 10.2135/cropsci2005.0056

Shepherd, R. W., Bass, W. T., Houtz, R. L., and Wagner, G. J. (2005). Phylloplanins of tobacco are defensive proteins deployed on aerial surfaces by short glandular trichomes. *Plant Cell* 17, 1851–1861. doi: 10.1105/tpc.105.031559

Shi, S., Huang, Y., Zeng, K., Tan, F., He, H., and Huang, J. (2005). Molecular phylogenetic analysis of mangroves: independent evolutionary origins of vivipary and salt secretion. *Mol. Phylogenet. Evol.* 34, 159–166. doi: 10.1016/j.ympev.2004.09.002

Shi, Y., Yan, X., Zhao, P., Yin, H., Zhao, X., Xiao, H., et al. (2013). Transcriptomic analysis of a Tertiary relict plant, extreme xerophyte *Reaumuria soongorica* to identify genes related to drought adaptation. *PLoS ONE* 8:e63993. doi: 10.1371/journal.pone.0063993

Shimony, C., Fahm, A., and Reinhold, L. (1973). Ultrastructure and ion gradients in the salt glands of *Avicennia marina* (Forssk.) Vierh. *New Phytol.* 72, 27–36. doi: 10.1111/j.1469-8137.1973.tb02006.x

Silva, P., and Gerós, H. (2009). Regulation by salt of vacuolar H+-ATPase and H+-pyrophosphatase activities and Na+/H+ exchange. *Plant Signal. Behav.* 4, 718–726. doi: 10.4161/psb.4.8.9236

Skorupa, M., Gołębiewski, M., Domagalski, K., Kurnik, K., Nahia, K. A., Złoch, M., et al. (2016). Transcriptomic profiling of the salt stress response in excised leaves of the halophyte *Beta vulgaris* ssp. *maritima*. *Plant Sci.* 243, 56–70. doi: 10.1016/j.plantsci.2015.11.007

Sobrado, M. A. (2004). Influence of external salinity on the osmolality of xylem sap, leaf tissue and leaf gland secretion of the mangrove *Laguncularia racemosa* (L.) Gaertn. *Trees* 18, 422–427. doi: 10.1007/s00468-004-0320-4

Soetaert, S. S. A., Van Neste, C. M. F., Vandewoestyne, M. L., Head, S. R., Goossens, A., Van Nieuwerburgh, F. C. W., et al. (2013). Differential transcriptome analysis of glandular and filamentous trichomes in *Artemisia annua*. *BMC Plant Biol.* 13:220. doi: 10.1186/1471-2229-13-220

Soltis, D., Soltis, P., and Edwards, C. (2005). *Core Eudicots. The Tree of Life Web Project.* Available at: http://tolweb.org/Core_Eudicots/20714/2005.01.01

Somaru, R., Naidoo, Y., and Naidoo, G. (2002). Morphology and ultrastructure of the leaf salt glands of *Odyssea paucinervis* (Stapf) (Poaceae). *Flora* 197, 67–75. doi: 10.1078/0367-2530-00016

Spyropoulou, E. A., Haring, M. A., and Schuurink, R. C. (2014). Expression of Terpenoids 1, a glandular trichome-specific transcription factor from tomato that activates the terpene synthase 5 promoter. *Plant Mol. Biol.* 84, 345–357. doi: 10.1007/s11103-013-0142-0

Stace, C. A. (1965). The significance of the leaf epidermis in the taxonomy of the Combretaceae. *J. Linn. Soc. Lond. Bot.* 59, 229–252. doi: 10.1111/j.1095-8339.1965.tb00060.x

Steudle, E., Lüttge, U., and Zimmermann, U. (1975). Water relations of the epidermal bladder cells of the halophytic species *Mesembryanthemum crystallinum*: direct measurements of hydrostatic pressure and hydraulic conductivity. *Planta* 126, 229–246. doi: 10.1007/BF00388965

Stevens, M. R., Coleman, C. E., Parkinson, S. E., Maughan, P. J., Zhang, H.-B., Balzotti, M. R., et al. (2006). Construction of a quinoa (*Chenopodium quinoa* Willd.) BAC library and its use in identifying genes encoding seed storage proteins. *Theor. Appl. Genet.* 112, 1593–1600. doi: 10.1007/s00122-006-0266-6

Szymanski, D. B., Jilk, R. A., Pollock, S. M., and Marks, M. D. (1998). Control of GL2 expression in Arabidopsis leaves and trichomes. *Development* 125, 1161–1171.

Takao, O., Mitsutaka, T., and Hiroshi, M. (2012). Morphology and ultrastructure of the salt glands on the leaf surface of Rhodes grass (*Chloris gayana*Kunth). *Int. J. Plant Sci.* 173, 454–463. doi: 10.1086/665588

Taleisnik, E. L., and Anton, A. M. (1988). Salt glands in *Pappophorum* (Poaceae). *Ann. Bot.* 62, 383–388. doi: 10.1093/oxfordjournals.aob.a087671

Tan, W. K., Lim, T. K., Loh, C. S., Kumar, P., and Lin, Q. (2015). Proteomic characterisation of the salt gland-enriched tissues of the mangrove tree species *Avicennia officinalis*. *PLoS ONE* 10:e0133386. doi: 10.1371/journal.pone. 0133386

Tan, W. K., Lim, T. M., and Loh, C. S. (2010). A simple, rapid method to isolate salt glands for three-dimensional visualization, fluorescence imaging and cytological studies. *Plant Methods* 6:24. doi: 10.1186/1746-4811-6-24

Tan, W. K., Lin, Q., Lim, T. M., Kumar, P., and Loh, C. S. (2013). Dynamic secretion changes in the salt glands of the mangrove tree species *Avicennia officinalis* in response to a changing saline environment. *Plant Cell Environ.* 36, 1410–1422. doi: 10.1111/pce.12068

Tanaka, H., Hirakawa, H., Kosugi, S., Nakayama, S., Ono, A., Watanabe, A., et al. (2016). Sequencing and comparative analyses of the genomes of zoysiagrasses. *DNA Res.* 23, 171–180. doi: 10.1093/dnares/dsw006

Thomson, W., Faraday, C. D., and Oross, J. W. (1988). "Salt glands," in *In Solute Transport in Plant Cells and Tissues*, eds D. A. Baker and J. L. Hall (Harlow: Longman Scientific & Technical), 498–537.

Thomson, W. W. (1975). "The structure and function of salt glands," in *Plants in Saline Environments, Ecological Studies*, Vol. 15, eds A. Poljakoff-Mayber and J. Gale (Berlin: Springer), 118–146.

Thomson, W. W., and Liu, L. L. (1967). Ultrastructural features of the salt gland of *Tamarix aphylla* L. *Planta* 73, 201–220. doi: 10.1007/BF00387033

TianZi, C., ShenJie, W., Jun, Z., WangZhen, G., and TianZhen, Z. (2010). Pistil drip following pollination: a simple in planta *Agrobacterium*-mediated transformation in cotton. *Biotechnol. Lett.* 32, 547–555. doi: 10.1007/s10529-009-0179-y

Tomlinson, P. (1986). *The Botany of Mangroves.* New York, NY: Cambridge University Press.

Tripp, E. A., and Fekadu, M. (2014). Comparative leaf and stem anatomy in selected species of Ruellieae (Acanthaceae) representative of all major lineages. *Kew Bull.* 69, 1–8. doi: 10.1007/s12225-014-9543-8

Tsukagoshi, H., Suzuki, T., Nishikawa, K., Agarie, S., Ishiguro, S., and Higashiyama, T. (2015). RNA-seq analysis of the response of the halophyte, *Mesembryanthemum crystallinum* (ice plant) to high salinity. *PLoS ONE* 10:e0118339. doi: 10.1371/journal.pone.0118339

Vassilyev, A. E., and Stepanova, A. A. (1990). The ultrastructure of ion-secreting and non-secreting salt glands of *Limonium platyphyllum*. *J. Exp. Bot.* 41, 41–46. doi: 10.1093/jxb/41.1.41

Villar, J. L., Turland, N. J., Juan, A., Gaskin, J. F., Alonso, M. Á, and Crespo, M. B. (2015). *Tamarix minoa* (Tamaricaceae), a new species from the island of Crete (Greece) based on morphological and plastid molecular sequence data. *Willdenowia* 45, 161–172. doi: 10.3372/wi.45.45201

Volkov, V. (2015). Salinity tolerance in plants. Quantitative approach to ion transport starting from halophytes and stepping to genetic and protein engineering for manipulating ion fluxes. *Front. Plant Sci.* 6:873. doi: 10.3389/ fpls.2015.00873

Volkov, V., Hachez, C., Moshelion, M., Draye, X., Chaumont, F., and Fricke, W. (2006). Water permeability differs between growing and non-growing barley leaf tissues. *J. Exp. Bot.* 58, 377–390. doi: 10.1093/jxb/erl203

Wachholtz, M., Heng-Moss, T., Twigg, P., Baird, L., Lu, G., and Amundsen, K. (2013). Transcriptome analysis of two buffalograss cultivars. *BMC Genomics* 14:613. doi: 10.1186/1471-2164-14-613

Wang, C., Gao, C., Wang, L., Zheng, L., Yang, C., and Wang, Y. (2014). Comprehensive transcriptional profiling of NaHCO3 stressed *Tamarix hispida* roots reveals networks of responsive genes. *Plant Mol. Biol.* 84, 145–157. doi: 10.1007/s11103-013-0124-2

Wang, L., Wang, C., Wang, D., and Wang, Y. (2014). Molecular characterization and transcript profiling of NAC genes in response to abiotic stress in *Tamarix hispida*. *Tree Genet. Genomes* 10, 157–171. doi: 10.1007/s11295-013-0672-2

Wang, X., Xiao, H., Cheng, Y., and Ren, J. (2016). Leaf epidermal water-absorbing scales and their absorption of unsaturated atmospheric water in *Reaumuria soongorica*, a desert plant from the northwest arid region of China. *J. Arid Environ.* 128, 17–29. doi: 10.1016/j.jaridenv.2016.01.005

Wang, Y., Ma, H., Liu, G., Zhang, D., Ban, Q., Zhang, G., et al. (2008). Generation and analysis of expressed sequence tags from a NaHCO3-treated *Limonium bicolor* cDNA library. *Plant Physiol. Biochem.* 46, 977–986. doi: 10.1016/j. plaphy.2008.06.001

Wang, Y. C., Yang, C. P., Liu, G. F., Jiang, J., and Wu, J. H. (2006). Generation and analysis of expressed sequence tags from a cDNA library of *Tamarix androssowii*. *Plant Sci.* 170, 28–36. doi: 10.1016/j.plantsci.2005.07.027

Watanabe, K. (2015). Potato genetics, genomics, and applications. *Breed. Sci.* 65, 53–68. doi: 10.1270/jsbbs.65.53

Wei, S., Du, Z., Gao, F., Ke, X., Li, J., Liu, J., et al. (2015). Global transcriptome profiles of "Meyer" zoysiagrass in response to cold stress. *PLoS ONE* 10:e0131153. doi: 10.1371/journal.pone.0131153

Weiglin, C., and Winter, E. (1988). Studies on the ultrastructure and development of the glandular trichomes of *Cressa cretica* L. *Flora* 181, 19–27. doi: 10.1016/ S0367-2530(17)30346-8

Weiglin, C., and Winter, E. (1991). Leaf structures of xerohalophytes from an East Jordanian salt pan. *Flora* 185, 405–424. doi: 10.1016/S0367-2530(17) 30508-X

Wen, W., Brotman, Y., Willmitzer, L., Yan, J., and Fernie, A. R. (2016). Broadening our portfolio in the genetic improvement of maize chemical composition. *Trends Genet.* 32, 459–469. doi: 10.1016/j.tig.2016.05.003

Werker, E., Putievsky, E., Ravid, U., Dudai, N., and Katzir, I. (1993). Glandular hairs and essential oil in developing leaves of *Ocimum basilicum* L. (Lamiaceae). *Ann. Bot.* 71, 43–50. doi: 10.1006/anbo.1993.1005

Wieneke, J., Sarwar, G., and Roeb, M. (1987). Existence of salt glands on leaves of kallar grass (*Leptochloa fusca* L. Kunth). *J. Plant Nutr.* 10, 805–819. doi: 10.1080/01904168709363611

Wilson, H., Mycock, D., and Weiersbye, I. M. (2016). The salt glands of *Tamarix usneoides* E. Mey. ex Bunge (South African Salt Cedar). *Int. J. Phytoremediation* 19, 587–595. doi: 10.1080/15226514.2016.1244163

Xie, F., Wang, Q., Sun, R., and Zhang, B. (2014). Deep sequencing reveals important roles of microRNAs in response to drought and salinity stress in cotton. *J. Exp. Bot.* 66, 789–804. doi: 10.1093/jxb/eru437

Xie, Q., Niu, J., Xu, X., Xu, L., Zhang, Y., Fan, B., et al. (2015). De novo assembly of the Japanese lawngrass (*Zoysia japonica* Steud.) root transcriptome and identification of candidate unigenes related to early responses under salt stress. *Front. Plant Sci.* 6:610. doi: 10.3389/fpls.2015.00610

Xu, X., Pan, S., Cheng, S., Zhang, B., Mu, D., Ni, P., et al. (2011). Genome sequence and analysis of the tuber crop potato. *Nature* 475, 189–195. doi: 10.1038/nature10158

Xue, Y., and Wang, Y. C. (2008). Study on characters of ions secretion from *Reaumuria trigyna*. *J. Desert Res.* 28, 437–442.

Yamamoto, A., Hashiguchi, M., Akune, R., Masumoto, T., Muguerza, M., Saeki, Y., et al. (2016). The relationship between salt gland density and sodium accumulation/secretion in a wide selection from three *Zoysia* species. *Austr. J. Bot.* 64, 277–284. doi: 10.1071/BT15261

Yamamoto, N., Takano, T., Tanaka, K., Ishige, T., Terashima, S., Endo, C., et al. (2015). Comprehensive analysis of transcriptome response to salinity stress in the halophytic turf grass *Sporobolus virginicus*. *Front. Plant Sci.* 6:241. doi: 10.3389/fpls.2015.00241

Yang, G., Wang, Y., Zhang, K., and Gao, C. (2014). Expression analysis of nine small heat shock protein genes from *Tamarix hispida* in response to different abiotic stresses and abscisic acid treatment. *Mol. Biol. Rep.* 41, 1279–1289. doi: 10.1007/ s11033-013-2973-9

Yang, Y., Yang, S., Li, J., Deng, Y., Zhang, Z., Xu, S., et al. (2015). Transcriptome analysis of the holly mangrove *Acanthus ilicifolius* and its terrestrial relative, *Acanthus leucostachyus*, provides insights into adaptation to intertidal zones. *BMC Genomics* 16:605. doi: 10.1186/s12864-015-1813-9

Yasui, Y., Hirakawa, H., Oikawa, T., Toyoshima, M., Matsuzaki, C., Ueno, M., et al. (2016). Draft genome sequence of an inbred line of *Chenopodium quinoa*, an allotetraploid crop with great environmental adaptability and outstanding nutritional properties. *DNA Res.* 23, 535–546. doi: 10.1093/dnares/ dsw037

Yin, Z., Li, Y., Yu, J., Liu, Y., Li, C., Han, X., et al. (2012). Difference in miRNA expression profiles between two cotton cultivars with distinct salt sensitivity. *Mol. Biol. Rep.* 39, 4961–4970. doi: 10.1007/s11033-011-1292-2

Yuan, F., Chen, M., Leng, B. Y., and Wang, B. S. (2013). An efficient autofluorescence method for screening *Limonium bicolor* mutants for abnormal salt gland density and salt secretion. *S. Afr. J. Bot.* 88, 110–117. doi: 10.1016/j. sajb.2013.06.007

Yuan, F., Chen, M., Yang, J., Leng, B., and Wang, B. (2014). A system for the transformation and regeneration of the recretohalophyte *Limonium bicolor*. *In Vitro Cell. Dev. Biol. Plant* 50, 610–617. doi: 10.1007/s11627-014-9611-7

Yuan, F., Chen, M., Yang, J., and Wang, B. (2015a). The optimal dosage of 60 co gamma irradiation for obtaining salt gland mutants of exo-recretohalophyte *Limonium bicolor* (Bunge) O. Kuntze. *Pakistan J. Bot.* 47, 71–76.

Yuan, F., Leng, B., and Wang, B. (2016a). Progress in studying salt secretion from the salt glands in recretohalophytes: how do plants secrete salt? *Front. Plant Sci.* 7:977. doi: 10.3389/fpls.2016.00977

Yuan, F., Lyu, M. A., Leng, B., Zheng, G. Y., Feng, Z. T., Li, P., et al. (2015b). Comparative transcriptome analysis of developmental stages of the *Limonium bicolor* leaf generates insights into salt gland differentiation. *Plant Cell Environ.* 38, 1637–1657. doi: 10.1111/pce.12514

Yuan, F., Lyu, M. A., Leng, B., Zhu, X., and Wang, B. (2016b). The transcriptome of NaCl-treated *Limonium bicolor* leaves reveals the genes controlling salt secretion of salt gland. *Plant Mol. Biol.* 91, 241–256. doi: 10.1007/s11103-016-0460-0

Zandkarimi, H., Bedre, R., Solis, J., Mangu, V., and Baisakh, N. (2015). Sequencing and expression analysis of salt-responsive miRNAs and target genes in the halophyte smooth cordgrass (*Spartina alternifolia* Loisel). *Mol. Biol. Rep.* 42, 1341–1350. doi: 10.1007/s11033-015-3880-z

Zhang, Y., Pei, X., Zhang, C., Lu, Z., Wang, Z., Jia, S., et al. (2012). De novo foliar transcriptome of *Chenopodium amaranticolor* and analysis of its gene expression during virus-induced hypersensitive response. *PLoS ONE* 7:e45953. doi: 10.1371/journal.pone.0045953

Zhao, Z., Heyser, J. W., and Bohnert, H. J. (1989). Gene expression in suspension culture cells of the halophyte *Distichlis spicata* during adaptation to high salt. *Plant Cell Physiol.* 30, 861–867. doi: 10.1093/oxfordjournals.pcp.a077817

Zouari, N., Ben Saad, R., Legavre, T., Azaza, J., Sabau, X., Jaoua, M., et al. (2007). Identification and sequencing of ESTs from the halophyte grass *Aeluropus littoralis*. *Gene* 404, 61–69. doi: 10.1016/j.gene.2007.08.021

Zouhaier, B., Abdallah, A., Najla, T., Wahbi, D., Wided, C., Aouatef, B. A., et al. (2015). Scanning and transmission electron microscopy and X-ray analysis of leaf salt glands of *Limoniastrum guyonianum* Boiss. under NaCl salinity. *Micron* 78, 1–9. doi: 10.1016/j.micron.2015.06.001

15

Light as stress factor to plant roots: Case of root halotropism

Ken Yokawa[1,2], Rossella Fasano[3], Tomoko Kagenishi[1] and František Baluška[1]*

[1] Department of Plant Cell Biology, Institute of Cellular and Molecular Botany, University of Bonn, Bonn, Germany
[2] Department of Biological Sciences, Tokyo Metropolitan University, Tokyo, Japan
[3] Department of Pharmacy, University of Salerno, Fisciano, Italy

Edited by:
Vadim Volkov, London Metropolitan University, UK

Reviewed by:
Francois Bouteau, Université Paris Diderot, France
Zheng-Hui He, San Francisco State University, USA

***Correspondence:**
František Baluška, Department of Plant Cell Biology, Institute of Cellular and Molecular Botany, University of Bonn, Kirschallee 1, 53115 Bonn, Germany
e-mail: baluska@uni-bonn.de

Despite growing underground, largely in darkness, roots emerge to be very sensitive to light. Recently, several important papers have been published which reveal that plant roots not only express all known light receptors but also that their growth, physiology and adaptive stress responses are light-sensitive. In *Arabidopsis*, illumination of roots speeds-up root growth via reactive oxygen species-mediated and F-actin dependent process. On the other hand, keeping *Arabidopsis* roots in darkness alters F-actin distribution, polar localization of PIN proteins as well as polar transport of auxin. Several signaling components activated by phytohormones are overlapping with light-related signaling cascade. We demonstrated that the sensitivity of roots to salinity is altered in the light-grown *Arabidopsis* roots. Particularly, light-exposed roots are less effective in their salt-avoidance behavior known as root halotropism. Here we discuss these new aspects of light-mediated root behavior from cellular, physiological and evolutionary perspectives.

Keywords: root, light response, plant hormones, reactive oxygen species, root tropism

LIGHT AS IMPORTANT ENVIRONMENTAL FACTOR FOR ROOTS

In nature, sessile plants have to respond to diurnal change in the light environment. One of main roles of light in plant's life is to provide energy for photosynthesis and for the regulation of plant development at different stages such as seed germination, vegetative growth, tropisms and flowering. It is known that plant photoreceptors and related light-sensitive signaling molecules participate in the regulation of physiological conditions and morphological plasticity in response to the light environment. Darwin (1879) has discovered negative phototropism of plant roots. One year later, Francis Darwin and his father, Charles Darwin, published the book, "The Power of Movements in Plants." They described both root and shoot tropisms. In addition, they also proposed that some form of long-distance signaling connect the sensory organ apices with the actively growing basal parts (Darwin, 1879, 1880). Since then, dedicated research work in plant physiology has discovered the long-distance signaling molecule, auxin, resulting in insights into plant photoreception. This directional growth response to incoming light is called phototropism. Positive phototropism, observed in shoots, is growth toward a light source; whereas negative phototropism, seen in roots, is bending away from the light source. We demonstrated that short (10 s), but strong (the photon flux was 82 μmol m^{-2} s^{-1}), blue light illumination of *Arabidopsis* roots induces the immediate generation of reactive oxygen species (ROS) in root apex region, resulting in rapid increase of the root growth rate (Yokawa et al., 2011, 2013). This active response of light-stimulated root growth is termed escape tropism (Xu et al., 2013; Yokawa et al., 2013; for maize roots see Burbach et al., 2012). This tropism would allow *Arabidopsis* roots to escape from unfavorable light conditions if growing outside of our laboratories in the nature.

PHOTORECEPTORS IN ROOTS

It has been shown that *Arabidopsis* plant expresses 14 photoreceptors, most of which are also present in roots (Briggs and Lin, 2012; Jeong and Choi, 2013; Briggs, 2014). Roots grow in the dark soil to anchor the plant and to absorb mineral nutrients and water. It has been reported that light can penetrate less than several millimeters due to the rather high absorbance of soil (Woolley and Stoller, 1978). Nevertheless, small cracks or mechanical impacts can often happen which allows light to penetrate deeper. For instance, roots may be exposed to light due to sudden temperature changes, earthquake, heavy rain, wind, and so on. In addition, it is very important for emerging radicle to increase the root growth rate shortly after seed germination on the ground. It was necessary to evolve the ability of roots to respond to environmental light when the first flowering plants with modern root system emerged in land plant evolution. In the next section, intriguing interplays between phytohormones and light-related signaling pathways will be discussed.

FROM ACTIN CYTOSKELETON, VIA PIN2 RECYCLING, TO SALT AVOIDANCE TROPISMS OF ROOTS

At the cellular level, it was reported that PIN2 proteins (PIN-FORMED 2; auxin efflux carrier) in root apices respond to the light environment (Laxmi et al., 2008). Wan et al. (2012) demonstrated that the basipetal (shootward) PIN2-based polar auxin transport is subject to blue light control, which regulates the negative phototropism of *Arabidopsis* roots (Wan et al., 2012). Moreover, Dyachok et al. (2011) reported that light-activated COP1, E3 ubiquitin ligase, promotes actin polymerization and F-actin bundling, through regulation of the downstream ARP2/3-SCAR pathway in root cells. It results in increased root growth under the illuminated conditions (Dyachok et al., 2011). It was also reported that light controls bundling of F-actin in maize coleoptiles (Waller and Nick, 1997), changing sensitivity of cells to auxin, which is feeding back to control F-actin as well as cell growth (Nick et al., 2009). The interplays between F-actin and polar auxin transport mediated by endocytic vesicle recycling, especially in the transition zone of root apex, control root tropisms (Baluška et al., 1996, 2004, 2005, 2010; Baluška and Mancuso, 2013). Interestingly, precursor of endogenous auxin, indole-3-acetaldehyde (IAAld), is produced non-enzymatically *in vitro* by illumination of tryptophan in the presence of flavin which is abundant in living plant cells (Koshiba et al., 1993). Recently, we have proposed close links between the redox status and auxin (IAA) biosynthesis in plants (Yokawa et al., 2014). Taken together, it is obvious that roots are extraordinarily sensitive to light exposure due to their inherent evolutionary optimization for the underground life. Therefore, it is not surprising that illuminated roots of young *Arabidopsis* seedlings enhance their growth with the concomitant phototropism.

A few yeas ago, salt-stressed roots of *Arabidopsis* have been shown to alter root growth direction in order to avoid high salt areas via so-called salt avoidance tropism (Li and Zhang, 2008; Sun et al., 2008). This active root tropism requires ion gradient sensing pathway which would then control the PIN2 abundance, recycling and degradation (Li and Zhang, 2008; Sun et al., 2008). This unique *Arabidopsis* root behavior was linked to phospholipase D Zeta2 (PLDζ2) activity which stimulates clathrin-mediated endocytosis of PIN2, and this tropism was also termed root halotropism (Galvan-Ampudia et al., 2013; Rosquete and Kleine-Vehn, 2013; Pierik and Testerink, 2014). Interestingly, similarly, as in halotropism, PLDζ2 is involved in root hydrotropism through the PIN2-mediated suppression of root gravitropism (Taniguchi et al., 2010). Moreover, PLDζ2 is crucial for brefeldin A-sensitive endocytic recycling driving PIN2 recycling (Li and Xue, 2007) and polar auxin transport in the transition zone (Mancuso et al., 2007). Because PIN2 is crucial in this respect, as well as in adaptive responses of roots to light (Laxmi et al., 2008; Sassi et al., 2012; Wan et al., 2012), it is very important to test whether light condition affect the response of *Arabidopsis* roots to the salt stress.

PLANT HORMONES ARE INTEGRATED WITH LIGHT SIGNALING PATHWAYS

Karrikin molecule was isolated from the smoke of combusted plant materials and found to potently stimulate the germination of plant seeds (Flematti et al., 2004). This compound was identified as 3-methyl-2H-furo[2,3-c]pyran-2-one, karrikin-1 or karrikinolide-1 (KAR_1) shown in **Figure 1A**-a, and another analogs of karrikins (KAR_2-KAR_6) commonly possess butenolide structure. The *Arabidopsis* mutant *kai2* lacking *KAI2* genes insensitive to KAR_1 triggered promotion of seed germination. KAI2 is thought to be a putative receptor of karrikins (Nelson et al., 2010; Waters et al., 2012). KAI2 is a member of α/β-hydrolase family and is also known as HYPOSENSITIVE TO LIGHT (HTL). Role of KAI2/HTL in light responses (**Figure 1B**) of roots will be described in the next section.

Arabidopsis thaliana AtD14 (ortholog of DWARF14) is a paralog of KAI2/HTL, which belongs to α/β-hydrolase family and plays a role in strigolactone perception (Waters et al., 2012).

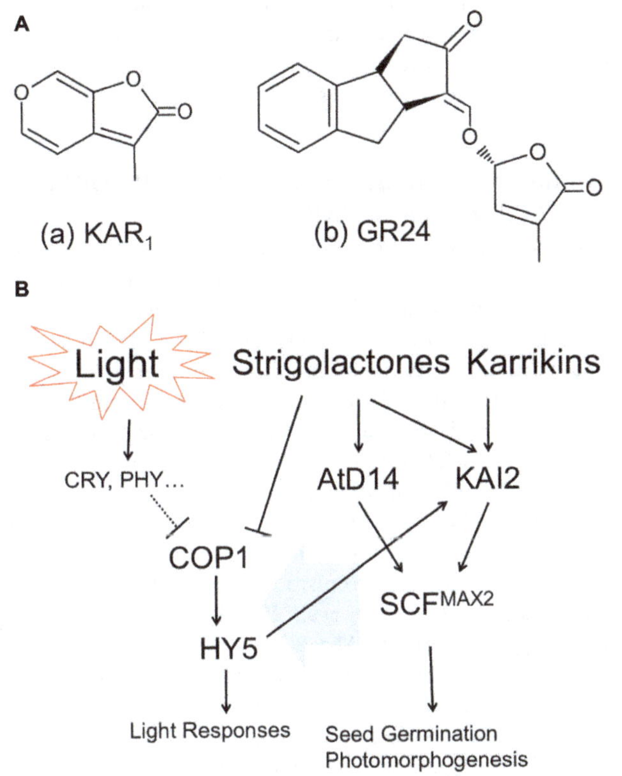

FIGURE 1 | Cross talk between butenolide plant hormones and light signaling pathways. (A) Chemical structures of (a) Karrikin1, KAR_1 and (b) Strigolactone analog, GR24. **(B)** Interaction of hormonal and light signaling pathways.

Strigolactones are synthesized from carotenoids and released into root exudates to promote germination of parasitic weeds (Yoneyama et al., 2007), as well as to stimulate hyphal branching in arbuscular mycorrhizal fungi (Akiyama et al., 2005). Strigolactones act also as a new class of plant hormone that inhibit the growth of axillary buds (Umehara et al., 2008) and alter root architecture (Koltai, 2011). Strigolactones are categorized as a sesquiterpenes lactone and have two moieties of butenolide (shown in **Figure 1A**-b as synthesized analog, GR24). As described above, karrikins have butenolide moieties as well. Therefore, because of the structural similarity of two molecules, karrikin receptor KAI2/HTL can respond to strigolactones. However, in contrast, the proposed strigolactone receptor AtD14 is insensitive to karrikins; indicating that the activity of KAI2/HTL as α/β-hydrolase is more flexible to detect butenolide structure than AtD14, probably due to evolutionary issues (Waters et al., 2012).

It was reported that F-box protein, MORE AXILLARY BRANCHES2 (MAX2), is located in the downstream of both KAI2/HTL and AtD14 and thus necessary for plants to respond to karrikin and strigolactone (Nelson et al., 2010; Waters et al., 2012). MAX2 forms SCF E3 ubiquitin ligase complex (Skp1, Cullin1, F-box), which modulates further downstream transcriptions (Stirnberg et al., 2007; Shen et al., 2012). However, molecular functions of KAI2/HTL or AtD14-mediated SCF^{MAX2} regulation of plant morphogenesis are not known yet. Interestingly in this respect, *max2* mutants are hyper-sensitive to drought and osmotic stress, including high NaCl, mannitol, and glucose (Bu et al., 2014). Furthermore, strigolactones exert positive roles in plant adaptation to drought and salt stress (Ha et al., 2014). We will discuss these newly emerging aspects in the final section.

LIGHT SIGNAL TRANSDUCTION VIA COP1 AND HY5 IS INTEGRATED WITH HORMONAL SIGNALING

In principle, plants recognize light as specific wavelengths by specific photoreceptors. Activated (excited) photoreceptors convert the light information into physiological signaling in diverse manners. CONSTITUTIVE PHOTOMORPHOGENIC1(COP1), E3 ubiquitin ligase, participates in the light-perceiving signaling cascade via affecting ubiquitination of target proteins. Once the activity of COP1 is inhibited, the downstream bZIP transcription factor, ELONGATED HYPOCOTYL5 (HY5) is freed from ubiquitination by COP1 and starts specific gene transcriptions related to light responses (**Figure 1B** left). COP1 and HY5 control root growth, lateral root formation, root hair tip growth, as well as root touch-responses and gravitropism (Oyama et al., 1997; Ang et al., 1998; Cluis et al., 2004; Datta et al., 2006, 2007; Sibout et al., 2006). Importantly, SALT TOLERANCE HOMOLOG2 (STH2) is interacting partner of COP1 and HY5 which controls roots and their anthocyanin levels (Datta et al., 2007).

Interestingly, exposure of plants to strigolactone inhibits the COP1 activity, suggesting that strigolactones can mimick light perception in plants (Tsuchiya et al., 2010). They also demonstrated that *max2* mutant of rice produces excess of strigolactones, resulting in the inhibition of COP1 and expression of light-responsive genes (Tsuchiya et al., 2010). In addition, it was reported that *max2* mutant is hypersensitive to red, far-red and blue light (Shen et al., 2007). Auxin interaction with the signaling pathway of strigolactones via MAX2 was already reported (Hayward et al., 2009). Finally, the key transcription factor of light signaling HY5 requires strigolactones in order to stimulate *Arabidopsis* seed germination during thermoinhibiton (Toh et al., 2012). Light also induces auxin biosynthesis via photoexcitation of flavins (Koshiba et al., 1993; Yokawa et al., 2014).

Karrikin receptor KAI2 is also known as HTL. KAI2/HTL expression was strongly increased by blue, red and far-red light and HY5 binds to the promoter region of HTL, indicating the expression of KAI2/HTL is regulated in response to light environment (Sun and Ni, 2011). Furthermore, KAI2/HTL is located downstream of HY5 (**Figure 1B**) and induced by light treatment (Sun and Ni, 2011). Meanwhile, it was reported that the treatment of seeds with KAR_1 (chemical structure is shown in **Figure 1A**-a) improved light responses in germination and early development of seedlings (Nelson et al., 2010). It implies that karrikins can affect the array of gene expression regarding light responses.

Plant roots exposed to salinity stress increase abscisic acid (ABA) and ABA-related gene expressions (Raghavendra et al., 2010). ABA and strigolactones are synthesized from carotenoids and the levels of these two plant hormones are affecting each other (López-Ráez et al., 2010). Salt stress increases strigolactone levels in roots which then promote arbuscular mycorrhizal symbiosis, resulting in changing ABA contents and alleviating stress response (Aroca et al., 2013). Strigolactones, ABA and cytokinin signaling pathways are integrated to allow plants to cope with high salinity environment very effectively (Ha et al., 2014). Many players of the phytohormone signaling pathways are overlapping and integrating with light-response pathways.

Taken together, these findings suggest that hormonal and light signaling pathway utilizes same junction for plant morphogenesis as well as for adaptation to abiotic stresses. The signaling pathways proposed (**Figure 1B**) are highly integrated, interacting with each other and, obviously, they represent just a 'tip of the iceberg.' In the case of roots, we should consider light solely as information (not as source of energy) into our careful consideration for our understanding of plant photomorphogenesis. In the next session, we are discussing how the root sensitivity to salinity is altered via exposure of roots to light.

UV LIGHT AND UVR8 IN DROUGHT AND SALINITY RESPONSES OF ROOTS

It has been reported that the reduction of plant growth under water deficit is driven by the UV-B photoreceptor UV RESISTANCE LOCUS 8 (UVR8; Kliebenstein et al., 2002; Brown et al., 2005; Favory et al., 2009; Fasano et al., 2014). Fasano et al. (2014) previously shown that the *UVR8* gene complements the osmosensitive yeast mutant *mpk1 ppz1*. The expression of *UVR8* was found upregulated in *Arabidopsis* plants grown under salt or osmotic stress conditions. Furthermore, the ectopic expression of *UVR8* causes pleiotropic effects on plant growth, such as a general reduction of plant organ size, leaves with smaller cells, reduced root growth, and the accumulation of flavonoids. This suggests that the UV-B morphogenic responses are enhanced in the *UVR8*-overexpressing plants grown under low levels of UV-B light. The growth defects of the *UVR8*-overexpressing plants are even more severe under osmotic and salt stress. In contrast, the inactivation of *UVR8* expression does not affect shoot or root growth under standard or mild drought stress conditions. Thus, the hypersensitive response to osmotic stress of the *35SUVR8* plants is strictly UVR8-dependent (Fasano et al., 2014).

There are extensive evidences that osmotic stress as well as UV-B light impacts are more prominent on shoots than on roots. Under mild osmotic stress, roots continue to grow (Bartels and Sunkar, 2005; Pardo, 2010; Werner et al., 2010). Moreover, enhanced root growth was observed in transgenic plants with a higher drought or salt-stress tolerance (Bartels and Sunkar, 2005). Whereas the impact of osmotic stress on root development is well known (recently reviewed in Liu et al., 2014), effects of UV-B on root growth are poorly understood. In general UV-B reduces primary root growth of *Arabidopsis* seedling (Kim et al., 1998; Tong et al., 2008). In the adult plant, increased allocation of biomass to roots has been reported to occur under UV-B stress (Bussell et al., 2012).

Development of a larger root system is considered as a drought-avoidance strategy that plants adopt to improve the uptake of water and nutrients when their availability in the soil is limited (Fukai and Cooper, 1995; Liao et al., 2001; Sharp et al., 2004; Pierik and Testerink, 2014). The UV-B photoreceptor UVR8 is expressed in roots of wild type plants (Rizzini et al., 2011; reviewed in Yokawa and Baluška, 2014). As shown in **Figure 2**, UVR8::GFP is

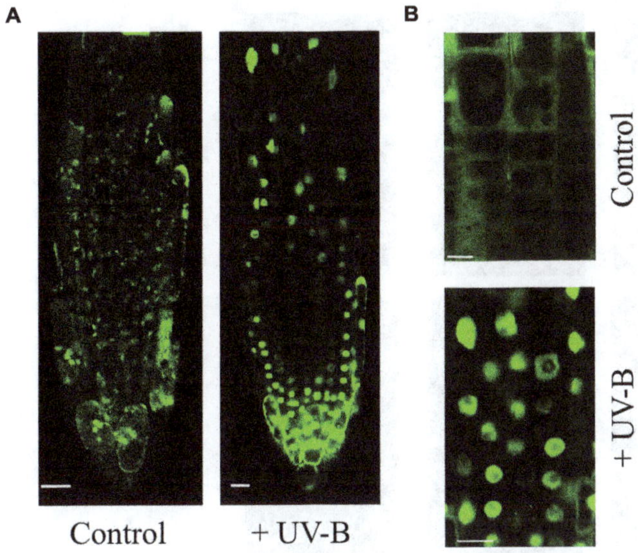

FIGURE 2 | Relocation of UVR8::GFP to nuclei in *Arabidopsis* root cells. **(A)** Root apex region. **(B)** Magnified view of cells in the transition zone. Bar indicates 10 μm. *Arabidopsis* roots were exposed to 1 h UV-B treatment (3.9 μmol m^{-2} s^{-1}).

expressing in *Arabidopsis* roots and is transported into nuclei upon irradiation of roots with UV-B. Over-expression of UVR8 greatly reduces root growth under light exposure and aggravates their weaker performance under the osmotic stress (Fasano et al., 2014). In comparison to the control plants, the primary root and lateral root densities were 13 and 60%, respectively. This indicates that auxin-dependent lateral root growth was most hampered. Indeed the root-phenotypes of the *35S-UVR8* plants are reminiscent that one of auxin mutant (Zhao, 2010). The analysis of lateral roots showed that the number of lateral root primordia and emerged lateral roots of the *35S-UVR8* plants was 12 and 68%, respectively, reduced in comparison to the control plants (Fasano et al., 2014). Flavonoids accumulation was found to be 2.2-fold increased in the root of the *UVR8*-over-expressing plants, whereas the content of IAA-conjugates showed a tendency to decrease (Fasano et al., 2014). Thus, the defects in cell expansion of the *35S-UVR8* roots could be associated to the increased levels of flavonoids which, in turn, alter polar auxin transport and/or auxin homeostasis (e.g., Santelia et al., 2008).

Flavonoids are synthesized in roots, regulating root branching, gravitropism, and stress adaptation (Brown et al., 2001; Buer and Muday, 2004; Agati et al., 2011; Emiliani et al., 2013). Flavonoid concentrations are significantly higher in combined stress-treated plants than in those treated with UV-B alone (Hughes et al., 2010; Comont et al., 2012). Together, all these reports suggest that *UVR8*, via the control of flavonoid accumulation, might be a common intermediate in light and hormone signaling pathways to regulate root growth and development under abiotic stress challenges. Besides UV-B, drought and salinity have also been associated with anthocyanin accumulations in various tissues including roots (Agati et al., 2011; Emiliani et al., 2013; Meng, 2014). Importantly in this respect, the light-induced stimulation of *Arabidopsis* root growth is mediated via the COP1-mediated accumulation of anthocyanins (Meng, 2014).

IMPACT OF LIGHT ON SALINITY AVOIDANCE VIA ROOT HALOTROPISM

Although root apices growing in soil are known to be at a front line that can be exposed to salt stress, an impact of light on modulation of the sensitivity of roots to salinity has not been considered to date. Importantly, most experiments using the laboratory grown *Arabidopsis* seedlings are using roots exposed to light, although root apices outside in the nature are mostly in darkness.

As shown in the **Figure 3**, we have demonstrated that light exposure affects the root response to the salinity stress. Columbia WT seedlings were grown on agar plates containing NaCl only at their bottom parts. With this system, root tropism responses to salinity, halotropism, can be observed. Our methods followed the protocol described in (Galvan-Ampudia et al., 2013; Rosquete and Kleine-Vehn, 2013). After replacing bottom part of agar with the NaCl-agar, plastic dishes were covered with aluminum foil to allow dark treatment. Light/dark period was 16 h/8 h and light intensity was about the 120 μmol m^{-2} s^{-1}, comparable to conditions in Galvan-Ampudia et al. (2013). Both the dark- and light-grown roots grew straight to the boundary of two agar media in control experiments with 0 mM NaCl (**Figure 3**). On the other hand, roots growing in darkness have higher sensitivity to NaCl than that roots exposed to light (**Table 1**). Dark-grown roots showed halotropism by avoiding salt-enriched agar areas at all concentrations of NaCl (**Figure 3**; **Table 1**). At the 100 mM concentration of NaCl, light-grown roots kept growing into the salty part of agar across the borderline between two agar parts even after several days. This suggests that light either prevents roots to accomplish the halotropism or that light lowers the sensitivity of *Arabidopsis* roots to perceive a gradient of salinity. The fact that light grown roots performed halotropism at higher concentrations of NaCl (**Table 1**) suggests that light interferes with the elusive root's ability to sense the Na^+ gradients (Maathuis, 2014). As root apices of roots growing outside in the nature are typically located deep in the soil in darkness, it can be expected that their halotropism is more efficient than for roots of laboratory grown seedlings.

LIGHT-INDUCED ROOT GROWTH IS EXPENSIVE AND ALTERS PHYSIOLOGY AND MORPHOLOGY OF WHOLE SEEDLINGS

It is important to be aware that light-induced stimulation of root growth is changing the physiology of whole seedlings/plants. **Figure 4** summarizes light-induced root growth mediated via COP1 interactions with the actin cytoskeleton (Dyachok et al., 2011) and anthocyanin biosynthesis (Meng, 2014). Both SCAR and PAP mediated pathways require sucrose as energy source (Kurata and Yamamoto, 1997; Kircher and Schopfer, 2012; Maier et al., 2013; Meng, 2014). Roots lacking ANGUSTIFOLIA3 (AN3) transcription coactivator have reduced anthocyanin levels and longer roots in light than WT roots (Meng, 2014). In contrast, the *an3* mutant roots are shorter that WT roots when grown in darkness. Interestingly, AN3 binds to COP1 promoter to inhibit the light-induced root elongation (Meng, 2014). Besides anthocyanins, light-exposed roots show also rather dramatic increase of

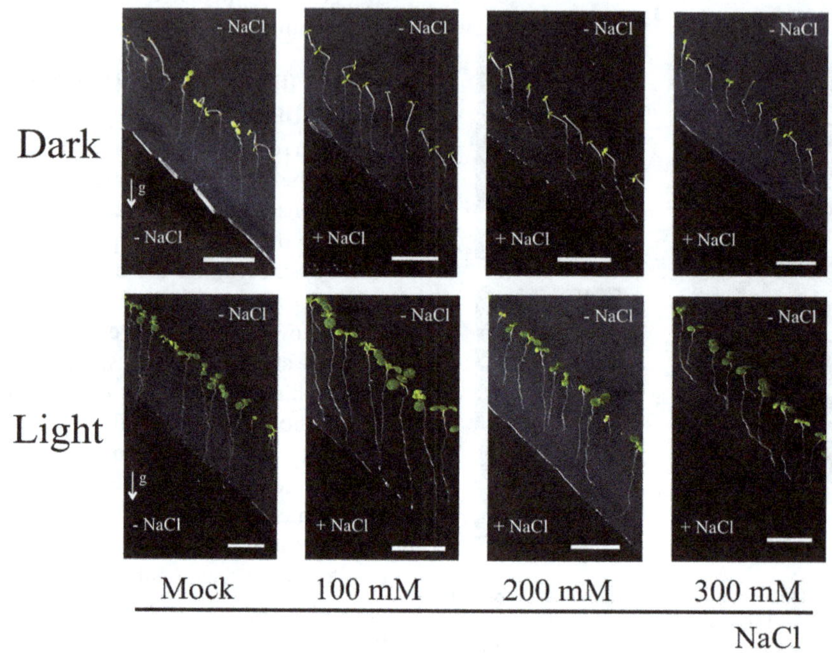

FIGURE 3 | Light environment affects sensitivity of *Arabidopsis* roots to salinity. Seedlings of Col-0 were firstly grown on 1/2 MS media for 5 days. Bottom part of agar was removed diagonally and new 1/2 MS media containing NaCl was added. Petri dishes were placed either in light or dark condition for 3 days. Roots growing on agar plates in dark/light conditions with 0, 100, 200, and 300 mM NaCl media in the bottom half of Petri dishes. The method used for imposing the root halotropism is following exactly the protocol as published by Galvan-Ampudia et al. (2013). White arrows indicate the gravity vector. All seedlings were grown in vertical position. Bar indicates 1 cm.

Table 1 | Halotropic bendings of *Arabidopsis* roots grown in different light conditions.

	Mock	100 mM	200 mM	300 mM
Dark	0% ($n = 25$)	26% ($n = 23$)	90% ($n = 30$)	92.3% ($n = 26$)
Light	0% ($n = 29$)	18.5% ($n = 27$)	16.7% ($n = 30$)	46.4% ($n = 28$)

Roots have been considered halotropic when scored root apex angles are greater than 45° with respect of the gravity vector (= same angle with the boundary of agar).

phenylpropanoid metabolism, inducing not only flavonoids but also monolignol glucosides (Hemm et al., 2004). Possible impacts of light on the Casparian bands formation in endodermis can be expected and easily tested.

Light-induced actin polymerization, auxin polar transport, root growth and anthocyanin biosynthesis are energetically demanding processes. The increased demand of sucrose for the root photomorphogenesis (Costigan et al., 2011; Warnasooriya and Montgomery, 2011; Yokawa et al., 2013) is going at the expense of other potential sinks of the plant body (Kurata and Yamamoto, 1997; Yazdanbakhsh et al., 2011; Kircher and Schopfer, 2012). Moreover, salt stress perceived locally at the root apex is rapidly spread throughout the plant body via systemic signaling (Choi et al., 2014; Gilroy et al., 2014). Therefore, it is important to be aware that illumination of roots has rather dramatic consequences for the whole seedling's physiology and morphology.

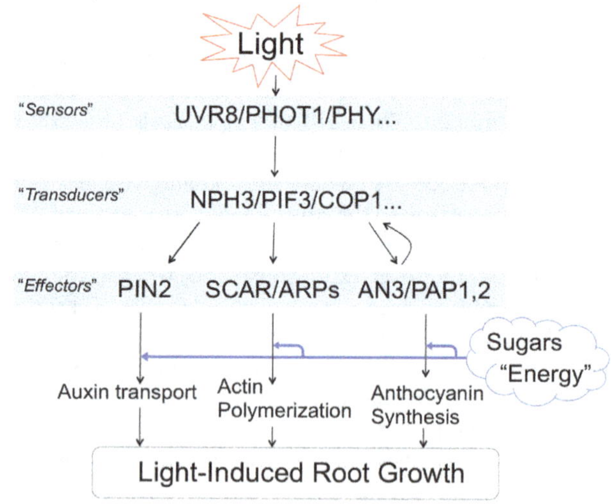

FIGURE 4 | Light-induced root growth is expensive. Light-induced root elongation is mediated via COP1 interactions. Both SCAR and PAP mediated pathways require enough sucrose as an energy source. This figure is based on following references: Dyachok et al. (2011), Wan et al. (2012), Bai et al. (2014), Fasano et al. (2014), and Meng (2014).

CONCLUSIONS AND PERSPECTIVES

Our results using salt stress reveal that light exposed roots show different behavior and responses under salt stress. One possible scenario, based on reports that illuminated roots enhance

Table 2 | *Arabidopsis* root lengths in different salt and light conditions.

	Mock	100 mM	200 mM	300 mM
Dark	10.2 ± 2.9 mm	11.6 ± 2.6 mm	9.3 ± 1.8 mm	10.4 ± 2.4 mm
Light	21.1 ± 4.4 mm	20.5 ± 2.7 mm	15.9 ± 2.9 mm	14.5 ± 3.1 mm

Root lengths were measured after 3-days-incubation on NaCl treatment plates. 20–30 roots in each plate were measured and standard deviations are presented next to the average length (mm).

strigolactone levels (Koltai et al., 2011) and strigolactone increases drought and salt tolerance of roots (Ha et al., 2014), is that the light-exposed roots are less sensitive to salt stress. Whereas the dark-grown roots show halotropism (salt avoidance tropism) by growing around the salt-enriched agar parts, the light-exposed roots grow into the salt-enriched agar if NaCl levels are not too high (**Tables 1** and **2**). As light-induced strigolactone levels modify also root responses to low phosphate (Mayzlish-Gati et al., 2012), one can propose that exposure of *Arabidopsis* roots to light is modifying their whole physiology. In fact, our preliminary data reveal that *Arabidopsis* roots exposed to light show changes the sensitivity also to the aluminum toxicity (Kagenishi et al., in preparation). In addition, IAA levels in light exposed (1 h) maize roots were more than doubled compared to control levels (Yokawa et al., in preparation). It is intriguing that enhanced strigolactone levels resemble light-induced effects on roots. It was shown that light impacts on the actin cytoskeleton, also increased abundance and recycling of PIN2 auxin transporter (Laxmi et al., 2008; Dyachok et al., 2011; Wan et al., 2012; Pandya-Kumar et al., 2014). Moreover, salt stress, drought stress, cold stress, alkaline stress, aluminum toxicity; all these challenges target the PIN2 auxin transporter which is expressed in the root apex transition zone (discussed in Baluška et al., 2010; Baluška and Mancuso, 2013). In order to avoid light exposure of roots, we recommend using of the partially darkened Petri dishes (Yokawa et al., 2011, 2013; Xu et al., 2013). This is important as stressing roots with light is modifying not only root responses to diverse stresses but also the overall physiology of such seedlings. For example, it has been reported that circadian rhythms in *Arabidopsis* roots are governed by the shoot part (James et al., 2008). Importantly in this respect, ROS signaling and homeostasis was shown to be a major modulator of circadian rhythms both in prokaryotes and eukaryotes (Edgar et al., 2012; Lai et al., 2012; Stangherlin and Reddy, 2013). It is possible that seedlings will change even their circadian clock when roots are illuminated.

Although plant roots are heterotrophic plant organs, they require active suppression of photosynthetic gene expression. Tyrosylprotein sulfotransferase (TPST) protein HPS7 emerged recently as crucial player controlling this active suppression (Kang et al., 2014). Interestingly enough, *hsp7* mutant roots show enhanced photosynthesis-related effects under phosphate deficiency stress. Previous studies identified Golden-Like transcription factors GLK1 and GLK2 involved in activation of photosynthetic genes in roots (Waters et al., 2009; Kobayashi et al., 2013). Both *hsp7* and GLK-ox mutant roots do not show stunted root growth and any other phenotypes when grown in darkness (Kang et al., 2014).

Finally, the plant physiology perspective of the light-induced root growth in *Arabidopsis* is considering the high root growth rate as a sign of optimal root growth conditions. In stark contrast, the plant neurobiology interpretation of this increased root growth rate is that the light-exposed roots are experiencing stress and that such stressed roots are trying to escape from this unfavorable situation (Yokawa et al., 2011, 2013; Xu et al., 2013). Alpi et al. (2007), critics of the plant neurobiology initiative claimed that '. . . plant neurobiology does not add to our understanding of plant physiology, plant cell biology or signaling.' Now, some 7 years later, it is getting obvious that plants and their roots are behaviorally much more complex than envisioned in the framework of classical plant physiology (Trewavas, 2005, 2009, 2014; Trewavas and Baluška, 2011; Baluška and Mancuso, 2013; Pierik and Testerink, 2014). Illumination of the roots is common laboratory praxis ever since *Arabidopsis thaliana* was introduced as model organism to plant sciences. However, it is important to be aware that light exposure of roots emerges as stress factor for the laboratory grown *Arabidopsis* seedlings. This is clear example for the usefulness of viewing plants as actively living and sensitive organisms solving their own plant-specific problems in intelligent manner.

ACKNOWLEDGMENTS

Ken Yokawa was supported by the JSPS (Japanese Society for the Promotion of Science) Postdoctoral Fellowship. We thank Prof. Gareth I. Jenkins (University of Glasgow) for providing us with seeds of the UVR8::GFP line.

REFERENCES

Agati, G., Biricolti, S., Guidi, L., Ferrini, F., Fini, A., and Tattini, M. (2011). The biosynthesis of flavonoids is enhanced similarly by UV radiation and root zone salinity in *L. vulgare* leaves. *J. Plant Physiol.* 168, 204–212. doi: 10.1016/j.jplph.2010.07.016

Akiyama, K., Matsuzaki, K., and Hayashi, H. (2005). Plant sesquiterpenes induce hyphal branching in arbuscular mycorrhizal fungi. *Nature* 435, 824–827. doi: 10.1038/nature03608

Alpi, A., Amrhein, N., Bertl, A., Blatt, M. R., Blumwald, E., Cervone, F., et al. (2007). Plant neurobiology: no brain, no gain? *Trends Plant Sci.* 12, 135–136. doi: 10.1016/j.tplants.2007.03.002

Ang, L. H., Chattopadhyay, S., Wei, N., Oyama, T., Okada, K., Batschauer, A., et al. (1998). Molecular interaction between COP1 and HY5 defines a regulatory switch for light control of *Arabidopsis* development. *Mol. Cell* 1, 213–222. doi: 10.1016/S1097-2765(00)80022-2

Aroca, R., Ruiz-Lozano, J. M., Zamarreño, A. M., Paz, J. A., García-Mina, J. M., Pozo, M. J., et al. (2013). Arbuscular mycorrhizal symbiosis influences strigolactone production under salinity and alleviates salt stress in lettuce plants. *J. Plant Physiol.* 170, 47–55. doi: 10.1016/j.jplph.2012.08.020

Bai, S., Yao, T., Li, M., Guo, X., Zhang, Y., Zhu, S., et al. (2014). PIF3 is involved in the primary root growth inhibition of *Arabidopsis* induced by nitric oxide in the light. *Mol. Plant* 7, 616–625. doi: 10.1093/mp/sst142

Baluška, F., and Mancuso, S. (2013). Root apex transition zone as oscillatory zone. *Front. Plant Sci.* 4:354. doi: 10.3389/fpls.2013.00354

Baluška, F., Mancuso, S., Volkmann, D., and Barlow, P. W. (2004). Root apices as plant command centres: the unique 'brain-like' status of the root apex transition zone. *Biologia* 59, 9–17.

Baluška, F., Mancuso, S., Volkmann, D., and Barlow, P. W. (2010). Root apex transition zone: a signalling – response nexus in the root. *Trends Plant Sci.* 15, 402–408. doi: 10.1016/j.tplants.2010.04.007

Baluška, F., Volkmann, D., and Barlow, P. W. (1996). Specialized zones of development in roots: view from the cellular level. *Plant Physiol.* 112, 3–4.

Baluška, F., Volkmann, D., and Menzel, D. (2005). Plant synapses: actin-based domains for cell-to-cell communication. *Trends Plant Sci.* 10, 106–111. doi: 10.1016/j.tplants.2005.01.002

Bartels, D., and Sunkar, R. (2005). Drought and salt tolerance in plants. *Crit. Rev. Plant Sci.* 24, 23–58. doi: 10.1080/07352680590910410

Briggs, W. R. (2014). Phototropism: some history, some puzzles, and a look ahead. *Plant Physiol.* 164, 13–23. doi: 10.1104/pp.113.230573

Briggs, W., and Lin, C. T. (2012). Photomorphogenesis – from one photoreceptor to 14: 40 years of progress. *Mol. Plant* 3, 531–532. doi: 10.1093/mp/sss059

Brown, B. A., Cloix, C., Jiang, G. H., Kaiserli, E., Herzyk, P., Kliebenstein, D. J., et al. (2005). A UV-B-specific signaling component orchestrates plant UV protection. *Proc. Natl. Acad. Sci. U.S.A.* 102, 18225–18230. doi: 10.1073/pnas.0507187102

Brown, D. E., Rashotte, A. M., Murphy, A. S., Normanly, J., Tague, B. W., Peer, W. A., et al. (2001). Flavonoids act as negative regulators of auxin transport in vivo in *Arabidopsis. Plant Physiol.* 126, 524–535. doi: 10.1104/pp.126.2.524

Bu, Q., Lv, T., Shen, H., Luong, P., Wang, J., Wang, Z., et al. (2014). Regulation of drought tolerance by the F-box protein MAX2 in *Arabidopsis. Plant Physiol.* 164, 424–439. doi: 10.1104/pp.113.226837

Buer, C. S., and Muday, G. K. (2004). The transparent testa4 mutation prevents flavonoid synthesis and alters auxin transport and the response of *Arabidopsis* roots to gravity and light. *Plant Cell* 16, 1191–1205. doi: 10.1105/tpc.020313

Burbach, C., Markus, K., Zhang, Y., Schlicht, M., and Baluška, F. (2012). Photophobic behavior of maize roots. *Plant Signal. Behav.* 7, 874–878. doi: 10.4161/psb.21012

Bussell, J. S., Gwynn-Jones, D., Griffith, G. W., and Scullion, J. (2012). Above- and below-ground responses of *Calamagrostis purpurea* to UV-B radiation and elevated CO_2 under phosphorus limitation. *Physiol. Plant.* 145, 619–628. doi: 10.1111/j.1399-3054.2012.01595.x

Choi, W. G., Toyota, M., Kim, S. H., Hilleary, R., and Gilroy, S. (2014). Salt stress-induced Ca^{2+} waves are associated with rapid, long-distance root-to-shoot signaling in plants. *Proc. Natl. Acad. Sci. U.S.A.* 111, 6497–6502. doi: 10.1073/pnas.1319955111

Cluis, C. P., Mouchel, C. F., and Hardtke, C. S. (2004). The *Arabidopsis* transcription factor HY5 integrates light and hormone signaling pathways. *Plant J.* 38, 332–347. doi: 10.1111/j.1365-313X.2004.02052.x

Comont, D., Winters, A., and Gwynn-Jones, D. (2012). Acclimation and interaction between drought and elevated UV-B in *A. thaliana*: differences in response over treatment, recovery and reproduction. *Ecol. Evol.* 2, 2695–2709. doi: 10.1002/ece3.387

Costigan, S. E., Warnasooriya, S. N., Humphries, B. A., and Montgomery, B. L. (2011). Root-localized phytochrome chromophore synthesis is required for photoregulation of root elongation and impacts root sensitivity to jasmonic acid in *Arabidopsis. Plant Physiol.* 157, 1138–1150. doi: 10.1104/pp.111.184689

Darwin, C. R. (1880). *The Power of Movements in Plants.* London: John Murray.

Darwin, F. (1879). Über das negativ wachstum heliotropischer wurzeln im licht und im finstern. *Arb. D. Bot. Inst. Würzburg* 2, 521–528.

Datta, S., Hettiarachchi, C., Johansson, H., and Holm, M. (2007). SALT TOLERANCE HOMOLOG2, a B-box protein in *Arabidopsis* that activates transcription and positively regulates light-mediated development. *Plant Cell* 19, 3242–3255. doi: 10.1105/tpc.107.054791

Datta, S., Hettiarachchi, G. H., Deng, X. W., and Holm, M. (2006). *Arabidopsis* CONSTANS-LIKE3 is a positive regulator of red light signaling and root growth. *Plant Cell* 18, 70–84. doi: 10.1105/tpc.105.038182

Dyachok, J., Zhu, L., Liao, F., He, J., Huq, E., and Blancaflor, E. B. (2011). SCAR mediates light-induced root elongation in *Arabidopsis* through photoreceptors and proteasomes. *Plant Cell* 23, 3610–3626. doi: 10.1105/tpc.111.088823

Edgar, R. S., Green, E. W., Zhao, Y., van Ooijen, G., Olmedo, M., Qin, X., et al. (2012). Peroxiredoxins are conserved markers of circadian rhythms. *Nature* 485, 459–464. doi: 10.1038/nature11088

Emiliani, J., Grotewold, E., Falcone Ferreyra, M. L., and Casati, P. (2013). Flavonols protect *Arabidopsis* plants against UV-B deleterious effects. *Mol. Plant* 6, 1376–1379. doi: 10.1093/mp/sst021

Fasano, R., Gonzalez, N., Tosco, A., Piaz, F. D., Docimo, T., Serrano, R., et al. (2014). Role of *Arabidopsis*–UV RESISTANCE LOCUS 8 in plant growth reduction under osmotic stress and low levels of UV-B. *Mol. Plant* 5, 773–791. doi: 10.1093/mp/ssu002

Favory, J. J., Stec, A., Gruber, H., Rizzini, L., Oravecz, A., Funk, M., et al. (2009). Interaction of COP1 and UVR8 regulates UV-B-induced photomorphogenesis and stress acclimation in *Arabidopsis. EMBO J.* 28, 591–601. doi: 10.1038/emboj.2009.4

Flematti, G. R., Ghisalberti, E. L., Dixon, K. W., and Trengove, R. D. (2004). A compound from smoke that promotes seed germination. *Science* 305:977. doi: 10.1126/science.1099944

Fukai, S., and Cooper, M. (1995). Development of drought-resistant cultivars using physio-morphological traits in rice. *Field Crops Res.* 40, 67–86. doi: 10.1016/0378-4290(94)00096-U

Galvan-Ampudia, C. S., Julkowska, M. M., Darwish, E., Gandullo, J., Korver, R. A., Brunoud, G., et al. (2013). Halotropism is a response of plant roots to avoid a saline environment. *Curr. Biol.* 23, 2044–2050. doi: 10.1016/j.cub.2013.08.042

Gilroy, S., Suzuki, N., Miller, G., Choi, W. G., Toyota, M., Devireddy, A. R., et al. (2014). A tidal wave of signals: calcium and ROS at the forefront of rapid systemic signaling. *Trends Plant Sci.* 19, 623–630. doi: 10.1016/j.tplants.2014.06.013

Guo, Y., Zheng, Z., La Clair, J. J., Chory, J., and Noel, J. P. (2013). Smoke-derived karrikin perception by the α/β-hydrolase KAI2 from *Arabidopsis. Proc. Natl. Acad. Sci. U.S.A.* 110, 8284–8289. doi: 10.1073/pnas.1306265110

Ha, C. V., Leyva-González, M. A., Osakabe, Y., Tran, U. T., Nishiyama, R., Watanabe, Y., et al. (2014). Positive regulatory role of strigolactone in plant responses to drought and salt stress. *Proc. Natl. Acad. Sci. U.S.A.* 111, 851–856. doi: 10.1073/pnas.1322135111

Hayward, A., Stirnberg, P., Beveridge, C., and Leyser, O. (2009). Interactions between auxin and strigolactone in shoot branching control. *Plant Physiol.* 151, 400–412. doi: 10.1104/pp.109.137646

Hemm, M. R., Rider, S. D., Ogas, J., Murry, D. J., and Chapple, C. (2004). Light induces phenylpropanoid metabolism in *Arabidopsis* roots. *Plant J.* 38, 765–778. doi: 10.1111/j.1365-313X.2004.02089.x

Hughes, N. M., Reinhardt, K., Feild, T. S., Gerardi, A. R., and Smith, W. K. (2010). Association between winter anthocyanin production and drought stress in angiosperm evergreen species. *J. Exp. Bot.* 61, 1699–1709. doi: 10.1093/jxb/erq042

James, A. B., Monreal, J. A., Nimmo, G. A., Kelly, C. L., Herzyk, P., Jenkins, G. I., et al. (2008). The circadian clock in *Arabidopsis* roots is a simplified slave version of the clock in shoots. *Science* 322, 1832–1835. doi: 10.1126/science.1161403

Jeong, J., and Choi, G. (2013). Phytochrome-interacting factors have both shared and distinct biological roles. *Mol. Cells* 35, 371–380. doi: 10.1007/s10059-013-0135-5

Kang, J., Yu, H., Tian, C., Zhou, W., Li, C., Jiao, Y., et al. (2014). Suppression of photosynthetic gene expression in roots is required for sustained root growth under phosphate deficiency. *Plant Physiol.* 165, 1156–1170. doi: 10.1104/pp.114.238725

Kim, B. C., Tennessen, D. J., and Last, R. L. (1998). UV-B-induced photomorphogenesis in *Arabidopsis thaliana. Plant J.* 15, 667–674. doi: 10.1046/j.1365-313x.1998.00246.x

Kircher, S., and Schopfer, P. (2012). Photosynthetic sucrose acts as cotyledon-derived long-distance signal to control root growth during early seedling development in *Arabidopsis. Proc. Natl. Acad. Sci. U.S.A.* 109, 11217–11221. doi: 10.1073/pnas.1203746109

Kliebenstein, D. J., Lim, J. E., Landry, L. G., and Last, R. L. (2002). *Arabidopsis* UVR8 regulates ultraviolet-B signal transduction and tolerance and contains sequence similarity to human regulator of chromatin condensation. *Plant Physiol.* 130, 234–243. doi: 10.1104/pp.005041

Kobayashi, K., Sasaki, D., Noguchi, K., Fujinuma, D., Komatsu, H., Kobayashi, M., et al. (2013). Photosynthesis of root chloroplasts developed in *Arabidopsis* lines overexpressing GOLDEN2-LIKE transcription factors. *Plant Cell Physiol.* 54, 1365–1377. doi: 10.1093/pcp/pct086

Koltai, H. (2011). Strigolactones are regulators of root development. *New Phytol.* 190, 545–549. doi: 10.1111/j.1469-8137.2011.03678.x

Koltai, H., Cohen, M., Chesin, O., Mayzlish-Gati, E., Bécard, G., Puech, V., et al. (2011). Light is a positive regulator of strigolactone levels in tomato roots. *J. Plant Physiol.* 168, 1993–1996. doi: 10.1016/j.jplph.2011.05.022

Koshiba, T., Yamauchi, K., Matsuyama, H., Miyakado, M., Sori, I., and Satô, M. (1993). Flavin-photosensitized production of indole-3-acetaldehyde from tryptophan. *Tetrahedron Lett.* 34, 7603–7606. doi: 10.1016/S0040-4039(00)60411-2

Kurata, T., and Yamamoto, K. T. (1997). Light stimulated root elongation in *Arabidopsis thaliana. J. Plant Physiol.* 151, 346–351. doi: 10.1016/S0176-1617(97)80263-5

Lai, A. G., Doherty, C. J., Mueller-Roeber, B., Kay, S. A., Schippers, J. H., and Dijkwel, P. P. (2012). CIRCADIAN CLOCK-ASSOCIATED 1 regulates ROS homeostasis and oxidative stress responses. *Proc. Natl. Acad. Sci. U.S.A.* 109, 17129–17134. doi: 10.1073/pnas.1209148109

Laxmi, A., Pan, J., Morsy, M., and Chen, R. (2008). Light plays an essential role in intracellular distribution of auxin efflux carrier PIN2 in *Arabidopsis thaliana. PLoS ONE* 3:e1510. doi: 10.1371/journal.pone.0001510

Li, G., and Xue, H. W. (2007). *Arabidopsis* PLDzeta2 regulates vesicle trafficking and is required for auxin response. *Plant Cell* 19, 281–295. doi: 10.1105/tpc.106.041426

Li, X., and Zhang, W. (2008). Salt-avoidance tropism in *Arabidopsis thaliana. Plant Signal. Behav.* 3, 351–353. doi: 10.4161/psb.3.5.5371

Liao, H., Rubio, G., Yan, X., Cao, A., Brown, K. M., and Lynch, J. P. (2001). Effect of phosphorus availability on basal root shallowness in common bean. *Plant Soil* 232, 69–79. doi: 10.1023/A:1010381919003

Liu, J., Rowe, J., and Lindsey, K. (2014). Hormonal crosstalk for root development: a combined experimental and modeling perspective. *Front. Plant Sci.* 5:116. doi: 10.3389/fpls.2014.00116

López-Ráez, J. A., Kohlen, W., Charnikhova, T., Mulder, P., Undas, A. K., Sergeant, M. J., et al. (2010). Does abscisic acid affect strigolactone biosynthesis? *New Phytol.* 187, 343–354. doi: 10.1111/j.1469-8137.2010.03291.x

Maathuis, F. J. (2014). Sodium in plants: perception, signalling, and regulation of sodium fluxes. *J. Exp. Bot.* 65, 849–858. doi: 10.1093/jxb/ert326

Maier, A., Schrader, A., Kokkelink, L., Falke, C., Welter, B., Iniesto, E., et al. (2013). Light and the E3 ubiquitin ligase COP1/SPA control the protein stability of the MYB transcription factors PAP1 and PAP2 involved in anthocyanin accumulation in *Arabidopsis. Plant J.* 74, 638–651. doi: 10.1111/tpj.12153

Mancuso, S., Marras, A. M., Mugnai, S., Schlicht, M., Zársky, V., Li, G., et al. (2007). Phospholipase dzeta2 drives vesicular secretion of auxin for its polar cell-cell transport in the transition zone of the root apex. *Plant Signal. Behav.* 2, 240–244. doi: 10.4161/psb.2.4.4566

Mayzlish-Gati, E., De-Cuyper, C., Goormachtig, S., Beeckman, T., Vuylsteke, M., Brewer, P. B., et al. (2012). Strigolactones are involved in root response to low phosphate conditions in *Arabidopsis. Plant Physiol.* 160, 1329–1341. doi: 10.1104/pp.112.202358

Meng, L. S. (2014). Transcription coactivator *Arabidopsis* ANGUSTIFOLIA3 modulates anthocyanin accumulation and light-induced root elongation through transrepression of constitutive photomorphogenic 1. *Plant Cell Environ.* doi: 10.1111/pce.12456 [Epub ahead of print].

Nelson, D. C., Flematti, G. R., Riseborough, J. A., Ghisalberti, E. L., Dixon, K. W., and Smith, S. M. (2010). Karrikins enhance light responses during germination and seedling development in *Arabidopsis thaliana. Proc. Natl. Acad. Sci. U.S.A.* 107, 7095–7100. doi: 10.1073/pnas.0911635107

Nick, P., Han, M. J., and An, G. (2009). Auxin stimulates its own transport by shaping actin filaments. *Plant Physiol.* 151, 155–167. doi: 10.1104/pp.109.140111

Oyama, T., Shimura, Y., and Okada, K. (1997). The *Arabidopsis* HY5 gene encodes a bZIP protein that regulates stimulus-induced development, of root and hypocotyl. *Genes Dev.* 11, 2983–2995. doi: 10.1101/gad.11.22.2983

Pandya-Kumar, N., Shema, R., Kumar, M., Mayzlish-Gati, E., Levy, D., Zemach, H., et al. (2014). Strigolactone analog GR24 triggers changes in PIN2 polarity, vesicle trafficking and actin filament architecture. *New Phytol.* 202, 1184–1196. doi: 10.1111/nph.12744

Pardo, J. M. (2010). Biotechnology of water and salinity stress tolerance. *Curr. Opin. Biotechnol.* 21, 185–196. doi: 10.1016/j.copbio.2010.02.005

Pierik, R., and Testerink, C. (2014). The art of being fexible: how to escape from shade, salt, and drought. *Plant Physiol.* 166, 5–22. doi: 10.1104/pp.114.239160

Raghavendra, A. S., Gonugunta, V. K., Christmann, A., and Grill, E. (2010). ABA perception and signalling. *Trends Plant Sci.* 15, 395–401. doi: 10.1016/j.tplants.2010.04.006

Rizzini, L., Favory, J. J., Cloix, C., Faggionato, D., O'Hara, A., Kaiserli, E., et al. (2011). Perception of UV-B by the *Arabidopsis* UVR8 protein. *Science* 332, 103–106. doi: 10.1126/science.1200660

Rosquete, M. R., and Kleine-Vehn, V. (2013). Halotropism: turning down the salty date. *Curr. Biol.* 23, R927–R929. doi: 10.1016/j.cub.2013.08.020

Santelia, D., Henrichs, S., Vincenzetti, V., Sauer, M., Bigler, L., Klein, M., et al. (2008). Flavonoids redirect PIN-mediated polar auxin fluxes during root gravitropic responses. *J. Biol. Chem.* 283, 31218–31226. doi: 10.1074/jbc.M710122200

Sassi, M., Lu, Y., Zhang, Y., Wang, J., Dhonukshe, P., Blilou, I., et al. (2012). COP1 mediates the coordination of root and shoot growth by light through modulation of PIN1 and PIN2-dependent auxin transport in *Arabidopsis. Development* 139, 3402–3412. doi: 10.1242/dev.078212

Sharp, R. E., Poroyko, V., Hejlek, L. G., Spollen, W. G., Springer, G. K., Bohnert, H. J., et al. (2004). Root growth maintenance during water deficits: physiology to functional genomics. *J. Exp. Bot.* 55, 2343–2351. doi: 10.1093/jxb/erh276

Shen, H., Luong, P., and Huq, E. (2007). The F-box protein MAX2 functions as a positive regulator of photomorphogenesis in *Arabidopsis. Plant Physiol.* 145, 1471–1483. doi: 10.1104/pp.107.107227

Shen, H., Zhu, L., Bu, Q. Y., and Huq, E. (2012). MAX2 affects multiple hormones to promote photomorphogenesis. *Mol. Plant* 5, 750–762. doi: 10.1093/mp/sss029

Sibout, R., Sukumar, P., Hettiarachchi, C., Holm, M., Muday, G. K., and Hardtke, C. S. (2006). Opposite root growth phenotypes of *hy5* versus *hy5 hyh* mutants correlate with increased constitutive auxin signaling. *PLoS Genet.* 2:e202. doi: 10.1371/journal.pgen.0020202

Stangherlin, A., and Reddy, A. B. (2013). Regulation of circadian clocks by redox homeostasis. *J. Biol. Chem.* 288, 26505–26511. doi: 10.1074/jbc.R113.457564

Stirnberg, P., Furner, I. J., and Leyser, H. M. O. (2007). MAX2 participates in an SCF complex which acts locally at the node to suppress shoot branching. *Plant J.* 50, 80–94. doi: 10.1111/j.1365-313X.2007.03032.x

Sun, F., Zhang, W., Hu, H., Li, B., Wang, Y., Zhao, Y., et al. (2008). Salt modulates gravity signaling pathway to regulate growth direction of primary roots in *Arabidopsis. Plant Physiol.* 146, 178–188. doi: 10.1104/pp.107.109413

Sun, X. D., and Ni, M. (2011). HYPOSENSITIVE TO LIGHT, an alpha/beta fold protein, acts downstream of ELONGATED HYPOCOTYL 5 to regulate seedling de-etiolation. *Mol. Plant* 4, 116–126. doi: 10.1093/mp/ssq055

Taniguchi, Y. Y., Taniguchi, M., Tsuge, T., Oka, A., and Aoyama, T. (2010). Involvement of *Arabidopsis thaliana* phospholipase Dzeta2 in root hydrotropism through the suppression of root gravitropism. *Planta* 231, 491–497. doi: 10.1007/s00425-009-1052-x

Toh, S., McCourt, P., and Tsuchiya, Y. (2012). *HY5* is involved in strigolactone-dependent seed germination in Arabidopsis. *Plant Signal. Behav.* 7, 556–558. doi: 10.4161/psb.19839

Tong, H., Leasure, C. D., Hou, X., Yuen, G., Briggs, W., and He, Z.-H. (2008). Role of root UV-B sensing in *Arabidopsis* early seedling development. *Proc. Natl. Acad. Sci. U.S.A.* 105, 21039–21044. doi: 10.1073/pnas.0809942106

Trewavas, A. (2005). Plant intelligence. *Naturwissenschaften* 92, 401–413. doi: 10.1007/s00114-005-0014-9

Trewavas, A. (2009). What is plant behaviour? *Plant Cell Environ.* 32, 606–616. doi: 10.1111/j.1365-3040.2009.01929.x

Trewavas, A. (2014). *Plant Behaviour and Intelligence.* Oxford: Oxford University Press. doi: 10.1093/acprof:oso/9780199539543.001.0001

Trewavas, A., and Baluška, F. (2011). The ubiquity of consciousness. The ubiquity of consciousness, cognition and intelligence in life. *EMBO Rep.* 12, 1221–1225. doi: 10.1038/embor.2011.218

Tsuchiya, Y., Vidaurre, D., Toh, S., Hanada, A., Nambara, E., Kamiya, Y., et al. (2010). A small-molecule screen identifies new functions for the plant hormone strigolactone. *Nat. Chem. Biol.* 6, 741–749. doi: 10.1038/nchembio.435

Umehara, M., Hanada, A., Yoshida, S., Akiyama, K., Arite, T., Takeda-Kamiya, N., et al. (2008). Inhibition of shoot branching by new terpenoid plant hormones. *Nature* 455, 195–200. doi: 10.1038/nature07272

Waller, F., and Nick, P. (1997). Response of actin microfilaments during phytochrome-controlled growth of maize seedlings. *Protoplasma* 200, 154–162. doi: 10.1007/BF01283291

Wan, Y., Jasik, J., Wang, L., Hao, H., Volkmann, D., Menzel, D., et al. (2012). The signal transducer NPH3 integrates the phototropin1 photosensor with PIN2-based polar auxin transport in *Arabidopsis* root phototropism. *Plant Cell* 24, 551–565. doi: 10.1105/tpc.111.094284

Warnasooriya, S. N., and Montgomery, B. L. (2011). Spatial-specific regulation of root development by phytochromes in *Arabidopsis thaliana. Plant Signal. Behav.* 6, 2047–2050. doi: 10.4161/psb.6.12.18267

Waters, M. T., Nelson, D. C., Scaffidi, A., Flematti, G. R., Sun, Y. K., Dixon, K. W., et al. (2012). Specialisation within the DWARF14 protein family confers distinct responses to karrikins and strigolactones in *Arabidopsis. Development* 139, 1285–1295. doi: 10.1242/dev.074567

Waters, M. T., Wang, P., Korkaric, M., Capper, R. G., Saunders, N. J., and Langdale, J. A. (2009). GLK transcription factors coordinate expression of the photosynthetic apparatus in *Arabidopsis. Plant Cell* 21, 1109–1128. doi: 10.1105/tpc.108.065250

Werner, T., Nehnevajova, E., Köllmer, I., Novák, O., Strnad, M., Krämer, U., et al. (2010). Root-specific reduction of cytokinin causes enhanced root growth, drought tolerance, and leaf mineral enrichment in *Arabidopsis* and tobacco. *Plant Cell* 22, 3905–3920. doi: 10.1105/tpc.109.072694

Woolley, J. T., and Stoller, E. W. (1978). Light penetration and light-induced seed germination in soil. *Plant Physiol.* 61, 597–600. doi: 10.1104/pp.61.4.597

Xu, W., Ding, G., Yokawa, K., Baluška, F., Li, Q. F., Liu, Y., et al. (2013). An improved agar-plate method for studying root growth and response of *Arabidopsis thaliana. Sci. Rep.* 3:1273. doi: 10.1038/srep01273

Yazdanbakhsh, N., Sulpice, R., Graf, A., Stitt, M., and Fisahn, J. (2011). Circadian control of root elongation and C partitioning in *Arabidopsis thaliana. Plant Cell Environ.* 34, 877–894. doi: 10.1111/j.1365-3040.2011.02286.x

Yokawa, K., and Baluška, F. (2014). Pectins, ROS homeostasis and UV-B responses in plant roots. *Phytochemistry* doi: 10.1016/j.phytochem.2014.08.016 [Epub ahead of print].

Yokawa, K., Kagenishi, T., and Baluška, F. (2013). Root photomorphogenesis in laboratory-maintained *Arabidopsis* seedlings. *Trends Plant Sci.* 18, 117–119. doi: 10.1016/j.tplants.2013.01.002

Yokawa, K., Kagenishi, T., Kawano, T., Mancuso, S., and Baluška, F. (2011). Illumination of *Arabidopsis* roots induces immediate burst of ROS production. *Plant Signal. Behav.* 6, 1460–1464. doi: 10.4161/psb.6.10.18165

Yokawa, K., Koshiba, T., and Baluška, F. (2014). Light-dependent control of redox balance and auxin biosynthesis in plants. *Plant Signal. Behav.* 9:e29522. doi: 10.4161/psb.29522

Yoneyama, K., Yoneyama, K., Takeuchi, Y., and Sekimoto, H. (2007). Phosphorus deficiency in red clover promotes exudation of orobanchol, the signal for mycorrhizal symbionts and germination stimulant for root parasites. *Planta* 225, 1031–1038. doi: 10.1007/s00425-006-0410-1

Zhao, Y. (2010). Auxin biosynthesis and its role in plant development. *Annu. Rev. Plant Biol.* 61, 49–64. doi: 10.1146/annurev-arplant-042809-112308

16

Salinity tolerance in plants: Quantitative approach to ion transport starting from halophytes and stepping to genetic and protein engineering for manipulating ion fluxes

Vadim Volkov*

Faculty of Life Sciences and Computing, London Metropolitan University, London, UK

Edited by:
Bjoern Usadel,
RWTH Aachen University, Germany
Reviewed by:
Camila Caldana,
Brazilian Bioethanol Science and Technology Laboratory – Centro Nacional de Pesquisa em Energia e Materiais/ABTLuS, Brazil
Federico Valverde, Instituto de Bioquímica Vegetal y Fotosíntesis, Spain
*Correspondence:
Vadim Volkov
vadim.s.volkov@gmail.com

Ion transport is the fundamental factor determining salinity tolerance in plants. The Review starts from differences in ion transport between salt tolerant halophytes and salt-sensitive plants with an emphasis on transport of potassium and sodium via plasma membranes. The comparison provides introductory information for increasing salinity tolerance. Effects of salt stress on ion transport properties of membranes show huge opportunities for manipulating ion fluxes. Further steps require knowledge about mechanisms of ion transport and individual genes of ion transport proteins. Initially, the Review describes methods to measure ion fluxes, the independent set of techniques ensures robust and reliable basement for quantitative approach. The Review briefly summarizes current data concerning Na+ and K+ concentrations in cells, refers to primary thermodynamics of ion transport and gives special attention to individual ion channels and transporters. Simplified scheme of a plant cell with known transport systems at the plasma membrane and tonoplast helps to imagine the complexity of ion transport and allows choosing specific transporters for modulating ion transport. The complexity is enhanced by the influence of cell size and cell wall on ion transport. Special attention is given to ion transporters and to potassium and sodium transport by HKT, HAK, NHX, and SOS1 proteins. Comparison between non-selective cation channels and ion transporters reveals potential importance of ion transporters and the balance between the two pathways of ion transport. Further on the Review describes in detail several successful attempts to overexpress or knockout ion transporters for changing salinity tolerance. Future perspectives are questioned with more attention given to promising candidate ion channels and transporters for altered expression. Potential direction of increasing salinity tolerance by modifying ion channels and transporters using single point mutations is discussed and questioned. An alternative approach from synthetic biology is to create new regulation networks using novel transport proteins with desired properties for transforming agricultural crops. The approach had not been widely used earlier; it leads also to theoretical and pure scientific aspects of protein chemistry, structure-function relations of membrane proteins, systems biology and physiology of stress and ion homeostasis. Summarizing, several potential ways are aimed at required increase in salinity tolerance of plants of interest.

Keywords: salinity tolerance, halophyte, ion channel, ion transporter, ion fluxes, systems biology, synthetic biology, protein engineering

Salinity is among the most serious problems for modern agriculture with the estimated annual losses nowadays being over USD 12 billion (Pitman and Läuchli, 2002; Shabala, 2013). More than 20% of irrigated lands and up to 50% (Flowers, 1998) are affected by salinity; it essentially reduces the yield of agricultural crops since most of them are salt-sensitive glycophytes (Munns and Tester, 2008). The collapse of Sumer civilization about 4000 years ago was caused by improper agricultural techniques, which led to soil salinization and drop in the agricultural productivity in the area (Serrano, 1996; Pitman and Läuchli, 2002). However, progress in modern agriculture and science rather allowed to set problems and pose questions than to get clear answers. We still do not know how to deal with salinity and to grow plants under salinization.

Effects of salinity on living plants include osmotic stress, toxic effects of high salt, mostly sodium ions, and corresponding oxidative stress (Flowers, 2004; Munns and Tester, 2008; Shabala, 2013). Osmotic stress is proportional to concentration of external salt solution, usually with osmotic pressures over 1 MPa. Toxic effect of sodium ions results from their rise in cytoplasm of plant cells. Developing in plant cells oxidative stress adds to the negative effects of salinity on the whole plants.

Ion transport via cell membranes is the basic factor determining salinity tolerance. Ion fluxes control ion concentrations; finally the values and regulation of the fluxes are essential for salinity tolerance. The review starts from brief description of ion transport in halophytes. Then it provides basic details and features of ion transport aiming to understand what could be altered and the potential results of the changes. Further discussion is directed to specific transporters and to genetically modified and artificially designed transport proteins for modulating ion transport under conditions of salinization.

IONTRANSPORT IN HALOPHYTES AND EFFECTS OF SALT STRESS ON MEMBRANETRANSPORT IN GLYCOPHYTES

Brilliant examples of salinity tolerance are represented by halophytes, which are able to grow at high concentrations of salt (**Figure 1**), under irrigation by seawater and even under several times higher salt concentrations than in seawater (over 2 M of NaCl) (Flowers et al., 1977; Flowers and Colmer, 2008). Halophytes are usually considered like sodium tolerant plants. Indeed, NaCl is the main source of salinity in most areas, though chloride, sulfate, calcium, magnesium, and the other ions are involved constituting sometimes the main salt for soil and water salinization (Flowers et al., 1977). "Domestication" of halophytes is one more way to use salt-affected and salinized territories.

Halophytes is an undisputable example and proof that salinity tolerance in plants is achievable. The main questions are:

- what are the salinity tolerance traits in halophytes;
- how to bring the relevant trait or multiple traits to agricultural plants;
- to which extent the transfer could be realized without essentially influencing the growth rate, agricultural productivity and quality of grain, fruits, edible parts, flowers or the other economically important elements/features/ traits of agricultural plants.

The simplest idea under salinity is to decrease sodium net influx and increase potassium net uptake via plasma membrane of epidermal root cells to alter mineral nutrition. Surprisingly, in most cases the pure straightforward and strict approach is not realized neither in halophytes, nor in glycophytes (Flowers, 2004; Maathuis et al., 2014). Halophytes imply several strategies to cope with high concentrations of salts including (1) sodium exclusion from roots, (2) accumulation of high sodium in shoots, (3) shedding specialized leaves, (4) localizing salts in vacuoles, (5) excreting salts via salt glands etc. The role and contribution of each strategy depend on the habitat and type of a halophyte plant (Flowers et al., 1977; Breckle, 2002; Lüttge, 2002; Flowers and Colmer, 2008; Munns and Tester, 2008; Shabala, 2013). So far the known transport systems in halophytes are basically the same like in glycophytes due to common ancestry and evolution (Flowers et al., 2010): the trait of salt tolerance emerged independently over 70 times in different groups of grasses (Bennett et al., 2013). The knowledge facilitates potential transfer of salinity tolerance traits to agriculturally important plants.

It is important to consider living cell as a complex system. There are networks of signaling events and metabolic reactions. Under salinity the increase in cytoplasmic Na^+ and reduction of K^+ result in changes of membrane potential, osmotic pressure, turgor pressure, calcium signaling, reactive oxygen species signaling, transcriptional regulation, alteration of gene expression, modification of protein expression pattern and spectra of siRNAs, signaling molecules, phytohormones and metabolites. The set of parameters including ion fluxes characterizes cell in a definite physiological state. Transition from one physiological state to the other one could be described by a trajectory in a multidimensional space, while the stable physiological states are considered as dynamic attractors. The volume and shape of the attractor in multidimensional space could be registered using means of "omics" that is RNA expression microarrays, proteomics, ionomics, metabolomics etc. (**Figure 2**). Properties of the attractors are slowly studied and understood for biological systems (Spiller et al., 2010; Breeze et al., 2011), though the idea is initially well developed in physics, especially in plasma physics for non-equilibrium thermodynamic systems (Akhromeeva et al., 1989, 1992). Obviously and intuitively the biological systems with tens of thousands of participating expressed genes are more complex than physical systems. Salt treatment up- and down-regulates tens and hundreds of genes including transcription factors, genes of transporters and regulatory proteins (e.g., Taji et al., 2004; Volkov et al., 2004; Liu et al., 2013; Takahashi et al., 2015) when the cells are moving to a new distinct physiological state of gene expression, metabolic control and ion transport.

FIGURE 1 | Halophytes *Salicornia sp.* (larger and bright green plants at the picture) and *Suaeda maritima* (smaller and grayish-green plants at the picture) are growing at the salt-affected soil near river Medway in the UK, where the area is flooded with mixed sea and river water under high tides (Beginning of June 2014). The size of the largest specimen is about 15 cm.

Comparison of ion fluxes via membranes between halophytes and glycophytes often demonstrates lower sodium uptake for halophytes (reviewed in: Flowers and Colmer, 2008). However, an evident problem in comparison is the high variability in ion transport between plant species because of growth rate, tissue-specific variability and the other physiological factors. It is reasonable to consider similar plants and achieve comparable values. Recent results on ion fluxes in glycophyte *Arabidopsis thaliana* and similar from the point of genome and morphology halophyte *Thellungiella halophila* demonstrated lower Na^+ fluxes and higher K^+/Na^+ selectivity of ion currents in the roots and root protoplasts of the halophyte under salt treatment (**Figures 3** and **4**, and in: Volkov et al., 2004; Volkov and Amtmann, 2006; Wang, 2006; Wang et al., 2006; Amtmann, 2009; resembling results for roots of the two plants: Alemán et al., 2009). Lower Na^+ accumulation and higher K^+/Na^+ ratio under salt treatment were also found in roots of leguminous halophyte *Melilotus indicus* compared to similar herbaceous glycophyte *Medicago intretexta* (Zahran et al., 2007).

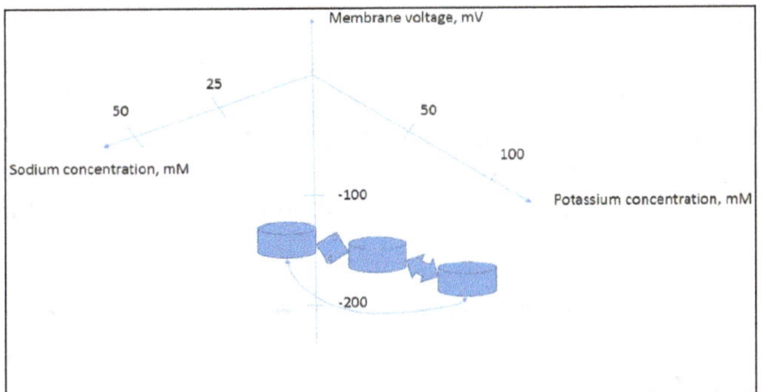

FIGURE 2 | Proposed simplified model of several attractors for a plant cell with specific metabolic and regulatory networks determined by cytoplasmic ion concentrations and membrane potential. Changes in extracellular ion concentrations result in transition of cells from one state of concentrations-membrane potential-protein and DNA–RNA expression levels and activity pattern to another one. Stability of the physiological states and trajectories of transitions could be studied in more detail using complex approaches of ionomics–metabolomics–proteomics–nucleic acids expression arrays and biophysical methods.

Other strategies may be involved depending on the level of salinity tolerance, plant morphology, habitat and the other environmental factors and evolutionary history. For example, salt tolerant *Plantago maritima* had similar sodium uptake rates by roots compared to salt-sensitive *P. media* (Erdei and Kuiper, 1979; de Boer, 1985); the salt-tolerance in the pair is rather associated with xylem transport and sodium accumulation in vacuoles of leaf cells.

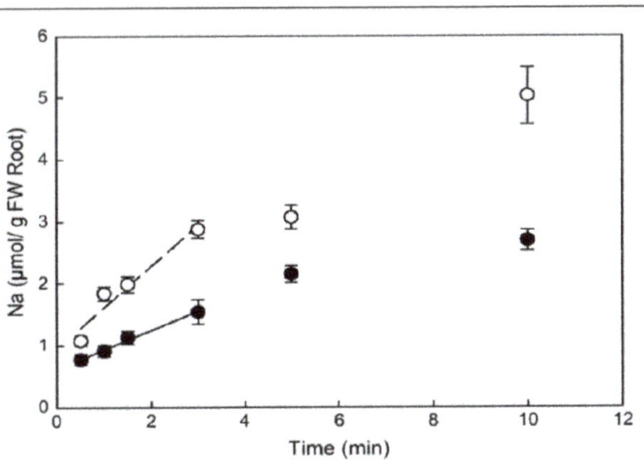

FIGURE 3 | Kinetics of initial unidirectional Na^+ influx into roots of *Arabidopsis thaliana* (open circles) and *Thellungiella halophila* (closed circles) as determined from $^{22}Na^+$ accumulation of individual plants from $^{22}Na^+$ labeled nutrient solution with 100 mM NaCl and 0.1 mM $CaCl_2$. Error bars are SE ($n = 4$). Reproduced from Wang et al. (2006) with the permission from the publisher Oxford University Press.

Vacuolar membranes of several halophytes were also a subject of special investigation. Electrophysiological patch clamp study of vacuoles from leaves of *Suaeda maritima* did not find any unusual features to support the high salt tolerance of the halophyte (Maathuis et al., 1992). Patch clamp experiments to compare vacuoles from roots of *P. maritima* and *P. media* also did not reveal striking differences apart from an extra smaller ion channel conductance in the tonoplast of the halophyte; salt stress essentially reduced the open probability of larger non-selective between K^+ and Na^+ ion channel conductance in both species but did not change the properties of the conductance (Maathuis and Prins, 1990). However, comparison of tonoplast from suspension culture cells of halophytic sugar beet with glycophytic tomato revealed rectification properties of ion channels in vacuolar membrane of the halophyte (Pantoja et al., 1989). At positive voltages in outside-out configuration (corresponding to ion currents out of vacuole) the ion-channel- like conductance dropped 6.5 times presumably preventing the transport of ions from vacuole to cytoplasm; the conductance was not selective for K^+ over Na^+ in both species (Pantoja et al., 1989).

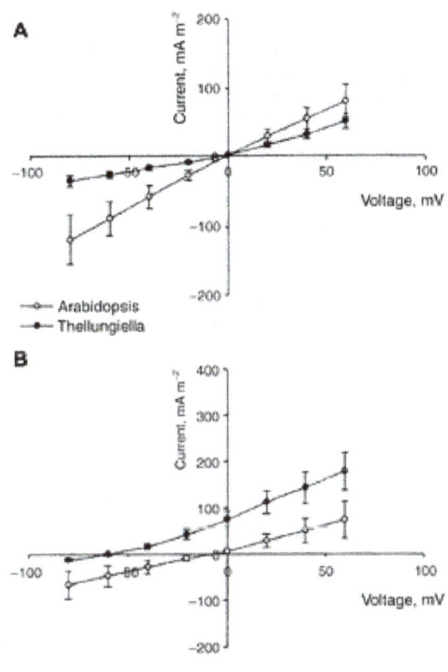

FIGURE 4 | Comparison of current–voltage curves for whole-cell instantaneous currents in root protoplasts of *A. thaliana* (open symbols) and *T. halophila* (closed symbols). Currents are normalized to protoplast surface. The pipette solution was always 100 mM KCl. The bath solution was 100 KCl **(A)** or 100 NaCl **(B)**. Data are given as means ± SE ($n = 6$ for *A. thaliana*, $n = 13$ for *T. halophila*). Reproduced from Volkov and Amtmann (2006) with the permission from the publisher John Wiley and Sons. Note that IV curve for instantaneous current in *T. halophila* resembles (though not completely obeys) the expected one from Goldmann–Hodkin–Katz equation and the shift in reversal potential of ion current in 100 mM KCl/100 mM NaCl is reflected in slope of IV curve for voltages above and below the reversal potential. The electric current characterizes the ion transport properties of plasma membranes and their selectivity for K^+ over Na^+.

Complexity of ion transport within the whole plant under salinity is confirmed also in experiments with glycophytes. Ion transport could be essentially influenced by salt stress in a cell-specific manner. The information is required to consider the whole plant responses to salinity and to alter them in a desired direction. An example of changes in membrane electric conductance after salt stress in several types of cells from barley leaves is given in **Figure 5**.

BASIC ASSUMPTIONS FOR REGULATING IONTRANSPORT IN PLANT CELLS

Before pondering ways to modulate ion transport it is helpful to describe techniques to measure ion fluxes and then introduce basic important parameters, which could be influenced and manipulated to alter ion fluxes and intracellular ion concentrations. The knowledge is also required to explain, understand and predict the results with overexpression/knockout of individual transporters and ion pumps (see below).

Comparison between Different Methods for Measuring Ion Fluxes

Total activity of proton pumps, ion channels, transporters and potential "leak" pathways of ion transport (see below) is reflected in changing and regulated ion fluxes. The coinciding values of ion fluxes measured by several independent techniques are needed to provide reliable and robust basis for the fur-

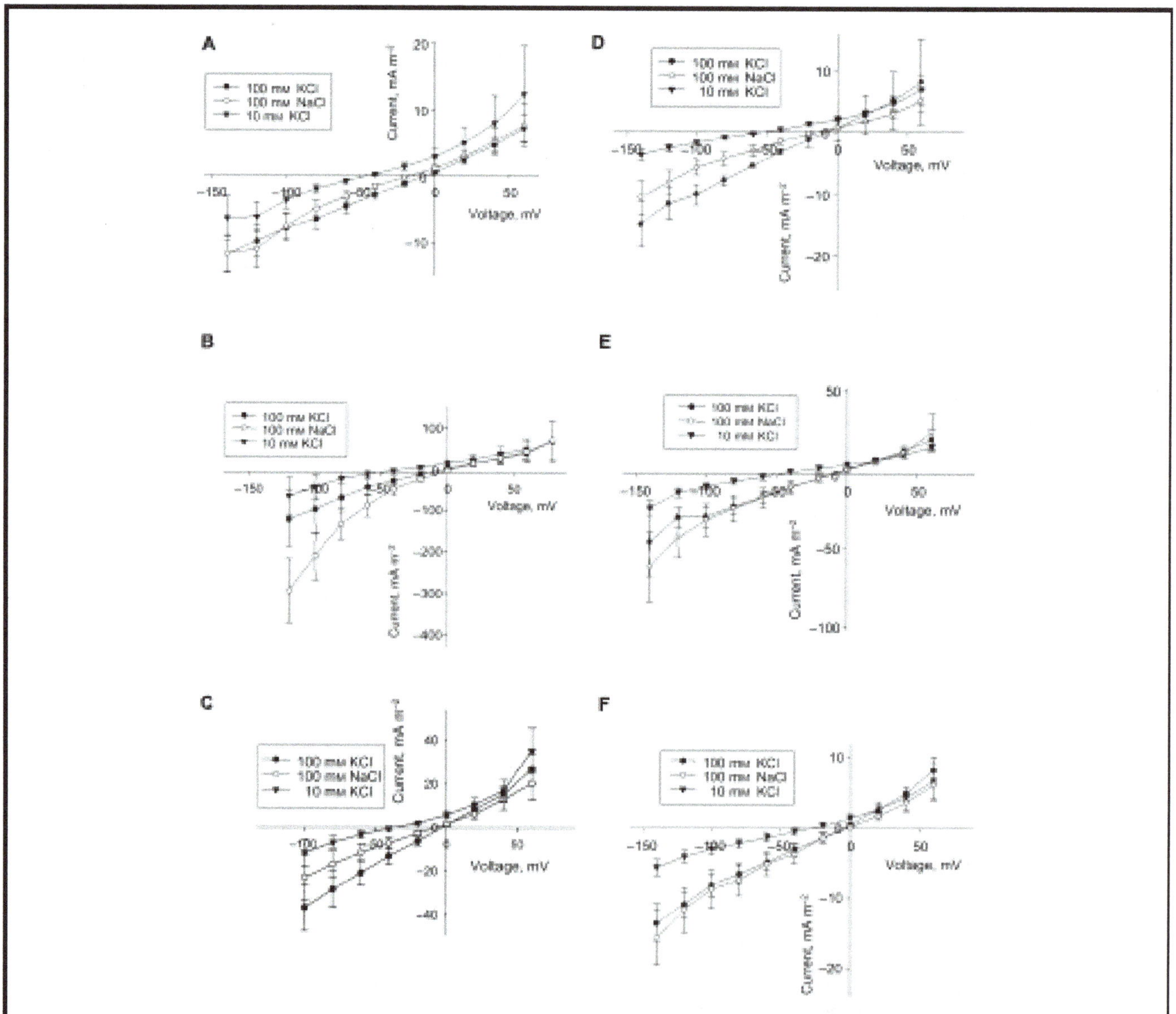

FIGURE 5 | Effect of salt stress on instantaneous current in protoplasts from the elongation zone and emerged blade portion of the developing leaf 3 of barley (*Hordeum vulgare* L.) (A–F). Averaged current-voltage (I–V) relationship in protoplasts from mesophyll **(A,D)** and epidermis **(B,E)** of the emerged blade, and from protoplasts of the elongation zone **(C,F)**. Control **(A–C)** and salt treatment **(D–F)**, $n = 5$–8 for control protoplasts and $n = 3$–6 for protoplasts for salt treatment; error bars are standard errors. Concentrations of KCl and NaCl in the bath are indicated. The pipette solution was always 100 mM KCl. Plants had been exposed to 100 mM NaCl for 3 days prior to protoplast isolation. Reproduced from Volkov et al. (2009), composed from Figures 1, 2, 4 and 5 with the permission from the publisher John Wiley and Sons.

ther conclusions. Direct and indirect methods include (1) estimates based on kinetic measurements of ion concentrations (both in plants and in nutrient solution), (2) electrophysiological methods, (3) technique of measuring ion fluxes using vibrating ion-selective electrodes [microelectrode ion flux estimate/measurements (MIFE)], (4) measurements of unidirectional ion fluxes (e.g., radioactive isotope $^{22}Na^+$ for sodium, while Rb^+ is often used to imitate K^+ fluxes) and (5) several other methods including ion-selective fluorescent and non-fluorescent indicators, potentially NMR spectroscopy, ion-conductance scanning microscopy etc. (**Table 1**).

Estimates from ion concentrations are the sum of influx and efflux of ions from plant cells. The measurements finally provide the important physiological information about averaged net changes and fluxes. However, the existing concentrations are often already high, therefore it takes at least hours and sometimes days to determine kinetic changes of ion concentrations and obtain kinetic curves, usually without detailed peculiarities and high temporal resolution of minutes. For example, after 25 h of 100 mM NaCl treatment sodium contents in roots of glycophyte *A. thaliana* rose from 0.2 to 2% of dry weight (DW) of the roots (0.087–0.87 mmole Na^+ per g of DW), while in halophyte *T. halophila* sodium in roots increased from 0.15 to 1.2% (0.065–0.52 mmole Na^+ per g of DW) (Volkov et al., 2004). Fast initial rates of net Na^+ uptake by roots during the first 6 h were 0.064 μmole/(g root DW∗min) in *Thellungiella* and 0.048 μmole/(g root DW∗min) in *Arabidopsis*. After 24 h of salt treatment, the rates dropped to 0.0004 nmole/(g root DW∗min) in *Thellungiella* and 0.003 nmole/(g root DW∗min) in *Arabidopsis* (Wang, 2006).

To elucidate detailed mechanisms of ion transport it is essential to add the other methods. One more requirement is to compare ion fluxes measured by different techniques: changes of ion concentrations are expressed in moles/g of fresh or dry weight (FW or DW), while electrophysiological and MIFE measurements are normalized per a unit of surface area, A/m^2 and moles/(m^2∗s), correspondingly. Rough estimate for conversion is that DW is about 10–15% of FW; for recalculation per surface area the size of roots or protoplasts is required. Estimates relating potassium flux in epidermal cells to ion electric currents were done for rye roots (White and Lemtiri-Chlieh, 1995). The net K^+ flux was estimated 1.0–1.9 μmole/(g root WF∗h) and unidirectional K^+ flux was about 7 μmole/(g root WF∗h) from mineral solution with 0.6 mM K^+ (White et al., 1991). Epidermal cells with diameter 26 μm were considered 8.3% of root volume; then the net fluxes in μmole/(g root WF∗h) corresponded to 0.11–0.21 pmole/(cell∗hour) or 3.1–5.9 pA/cell. Unidirectional flux of 7 μmole/(g root WF∗h) corresponded to 0.77 pmole/(cell∗hour) or 21.7 pA/cell (White and Lemtiri-Chlieh, 1995). Converting per surface area of the whole protoplasts, the fluxes of 0.11–0.21 pmole/(cell∗ hour) correspond to 14–27 nmole/(m^2 ∗s) and 0.77 pmole/(cell∗ hour) ≈ 100 nmole/(m^2 ∗s), the values reasonably confirmed by the other methods (see below). Conversion of the flux in pmoles to number of ions per a cell gives 0.77 pmole/(cell∗ hour) ≈ $1.3 * 10^8$ ions/(cell∗ second) (see below).

TABLE 1 | Comparative summary for methods to measure ion fluxes in plants and their tissues and cells.

Method or group of methods	Principle of the method	Spatial and temporal resolution, advantages	Disadvantages
Kinetic measurements of ion concentrations	Ion concentrations in plant tissues or in nutrient solution are measured in time, the concentration differences along the time points are plotted against time.	Resolution is at the level of whole plant or plant tissues, usually tens of minutes and hours are needed to register changes. Simple methods without a need of special equipment, convenient for most ions.	Measures net fluxes, not separate influx and flux outside of plant tissues. Low level of resolution.
MIFE: microelectrode ion flux estimate/ measurements	Tiny ion-selective microelectrode with tip around a μmeter vibrates within seconds in the vicinity of a cell or plant tissue and measures ion concentrations. Changing in time difference in concentrations is recalculated to ion fluxes.	Resolution at the level of individual cells within seconds, measurements may last for hours.	Requires special equipment. Ion-selective electrodes often interfere with organic compounds. Reliable ion-selective resins available for a few major ions only.
Measurements of unidirectional ion fluxes	Plant tissues or organs are loaded with radioactive ions or rare ions to imitate ions of interest. Unidirectional usually outward ion fluxes of the isotope or rare ion are measured then as changes in concentrations against time.	Spatial resolution at the level of whole plant or plant tissues, recordings from minutes lasting to hours are needed to register changes.	Requires radioactive isotopes and often complicated calculations with several proposed pools of ions.
Electrophysiological methods	Isolated cell membrane, piece of membrane or single cell within a plant preparation are subjected to set of physiological voltages and ion current is registered in the form of electric current.	Spatial resolution of single molecules or single cells. Temporal resolution from μseconds to minutes. High accuracy and possibility to find out specific molecules for transport of definite ions.	Indirect measurements, measure total electric current carried out by several ion species.
Fluorescent indicators	Fluorescent ion selective indicators	Resolution at the level of individual cells within tens of seconds to minutes, recordings over tens of minutes to hours.	Require specific protocols for loading, intrinsic autofluorescence and non-specific adsorption of fluorescent indicators could be drawbacks.

The distinction between the methods is partially arbitrary.

Electrophysiological techniques of two-electrode voltage clamp or patch clamp measure ion currents across membrane of plant cells (protoplasts for patch clamp) under determined applied voltage via the membrane and provide current–voltage curves with resolution up to pA and μseconds (**Figures 4** and **5**). For the example of patch clamp study with root protoplasts of *Arabidopsis* and *Thellungiella*, the inward K^+ fluxes in external 100 mM KCl at −80 mV were 120 mA/m^2 = 120 fA/μm^2 ≈ 1.2 10−18 mole/(μm^2∗s) = 1.2 μmole/(m^2∗s) for *Arabidopsis* and 30 mA/m^2 ≈ 300 nmole/(m^2 ∗s) for *Thellungiella*, correspondingly. Inward Na^+ fluxes under the same conditions, but in 100 mM NaCl in the external medium are 70 mA/m^2 ≈ 700 nmole/(m^2∗s) for *Arabidopsis* and 15 mA/m^2 ≈ 150 nmole/(m^2∗s) for *Thellungiella* (Volkov and Amtmann, 2006).

Comparisons of electric currents and ion fluxes were performed in a series of simultaneous measurements using patch clamp and MIFE for wheat root protoplasts, when MIFE measured H^+ fluxes (Tyerman et al., 2001). Proton fluxes in the experiments basically correlated to electric currents, though with large variation between protoplasts (Tyerman et al., 2001). Further on MIFE proved the lack of "dark" electroneutral fluxes for K^+ transport in the wheat root protoplasts (the situation was different for Ca^{2+}, large Ca^{2+} fluxes were not electrogenic): K^+ ion currents via outward and inward potassium channels nearly exactly corresponded to the K^+ ion fluxes measured by ion selective electrodes of MIFE (Tyerman et al., 2001; Gilliham et al., 2006). Some deviations from 1:1 ratio and variation between protoplasts were explained by uneven distribution of ion channels (Gilliham et al., 2006). Potentially atomic force microscopy or ion-conductance scanning microscope with further ion channel recordings (e.g., Hansma et al., 1989; Korchev et al., 1997; Lab et al., 2013 for ion-conductance scanning microscope) could be useful to explore the distribution of lipid rafts and clusters of ion channels at nanometre resolution.

Values of ion fluxes measured by patch clamp in protoplasts are usually higher than fluxes measured by MIFE for intact roots (e.g., Shabala and Lew, 2002; Chen et al., 2013), but coincide within orders of magnitude and essentially depend on composition of ambient medium, concentration of ions, membrane potential of cells and on multiple physiological factors (e.g., Ivashikina et al., 2001; Shabala et al., 2009). Experiments with MIFE are non-invasive and simpler; therefore provide huge opportunities with temporal resolution of seconds and spatial resolution within less than tens of microns for exploring physiological factors and conditions, which influence ion fluxes to and out of cells (Newman et al., 1987; Newman, 2001; Shabala, 2006; Sun et al., 2009; Shabala and Bose, 2012). Ion fluxes of K^+, Na^+, and other ions in the vicinity of roots were measured by MIFE and compared for different salt-tolerant and salt- sensitive cultivars and agricultural species (Chen et al., 2005; Cuin et al., 2008, 2012), along root zones of several plants (Garnett et al., 2001; Chen et al., 2005; Pang et al., 2006), for mutants in specific ion channels or transporters (e.g., Shabala et al., 2005; Demidchik et al., 2010), under treatment by physiologically active compounds (Cuin and Shabala, 2007a; Shabala et al., 2009; Pandolfi et al., 2010; Demidchik et al., 2011; Ordoñez et al., 2014), after generation of reactive oxygen species and salt stress (Cuin and Shabala, 2007b; Demidchik et al., 2010) and for numerous other conditions. MIFE is also a good method for studying ion transport in cell biology (Lew et al., 2006; Valencia-Cruz et al., 2009; Demidchik et al., 2010). The results provided huge volume of information about characteristics, kinetics and physiological features of ion fluxes under salt stress, helped to develop fast tests for salinity tolerance (Chen et al., 2005; Cuin et al., 2008). The results are described in hundreds of publications and several reviews (e.g., Newman, 2001; Shabala, 2006; Sun et al., 2009; Shabala and Bose, 2012), so are not covered here in more detail. Among the limitations of MIFE is the selectivity of ion-selective electrodes, which is influenced by interfering ions (e.g., Knowles and Shabala, 2004) and sometimes affected by physiologically active compounds and proteins in the medium for measurements after interaction with the material of ion-selective electrodes (e.g., Chen et al., 2005), hence demanding more control checks.

Unidirectional fluxes of $^{22}Na^+$ for sodium and $^{42}K^+$ or Rb^+ (a tracer for K^+) provide fast kinetics of ion transport within tens of seconds as determined by the speed of sampling and changes in concentrations at the background of initial level without $^{22}Na^+$ and $^{42}K^+/Rb^+$ (e.g., **Figure 5** for $^{22}Na^+$ fluxes). The method is used widely and had already provided essential advances in plant ion transport (e.g., MacRobbie and Dainty, 1958; Rains and Epstein, 1965; Epstein, 1966). The outward fluxes could be measured after loading plants with $^{22}Na^+$ or $^{42}K^+/Rb^+$ and transferring then to different chemical solutions without the ions (e.g., Wang, 2006; Wang et al., 2006 for $^{22}Na^+$ for roots of *Arabidopsis* and *Thellungiella*). Details and the methodical procedures are well described with different modifications to get more information about compartmentation of the absorbed ions and to exclude potential sources of errors (Cheeseman, 1986 for analysis of compartmentation based on efflux kinetics; Britto et al., 2006; Wang, 2006 for $^{42}K^+$ in application for the tracer efflux by barley roots; Britto and Kronzucker, 2013 for comprehensive practical description of experimental procedures to measure potassium fluxes). The values of measured fluxes differ significantly, sometimes over 100 times depending on plant species, physiological conditions and ion concentrations (e.g., summarized for Na^+ fluxes in: Kronzucker and Britto, 2011).

Analysis of unidirectional fluxes is often complemented by the other methods to obtain better understanding of the processes. An interesting comparison for influx of K^+ ($^{86}Rb^+$) from 0.1 mM K_2SO_4 solution and net K^+ influx measured by external K^+ microelectrodes (prototype of MIFE) gave nearly the same values of fluxes: 2.6 μmole/(g FW∗h) and 2.5 μmole/(g FW∗h), correspondingly (Newman et al., 1987). Measurements of unidirectional $^{24}Na^+$, $^{42}K^+$ fluxes in barley together with membrane potential measurements and pharmacological profiling of the fluxes allowed to study high affinity transport of sodium from low μM – 50 mM solutions and provided predictions about possible mechanisms for the transport (Schulze et al., 2012).

Among methods to determine ion fluxes is the use of ion- selective fluorescent indicators for estimating cytoplasmic and vacuolar concentrations of Na^+ and K^+ and kinetics of their changes (e.g., mostly for protoplasts: Lindberg, 1995; Halperin and Lynch, 2003; D'Onofrio et al., 2005; Kader and Lindberg, 2005), measurements by intracellular ion-selective electrodes (e.g., Carden et al., 2003 and references there), ^{23}Na-NMR spectroscopy (e.g., Bental et al., 1988) and several others; the methods are not discussed in detail here.

Ion Concentrations in Cells

Ion concentrations are the reflection of net ion fluxes via plasma membrane. Certain range of ion concentrations, especially of K^+, Na^+, and Ca^{2+}, is vital for cell physiology and function of proteins. Typical potassium concentrations in cytoplasm of plant cells were measured independently by several methods (including ion-selective electrodes, fluorescent dyes and X-ray microanalysis) and range around 60–140 mM (Pitman et al., 1981; Hajibagheri et al., 1988; Hajibagheri and Flowers, 1989;

Walker et al., 1995, 1998; Korolev et al., 2000; Cuin et al., 2003; Halperin and Lynch, 2003; Shabala et al., 2006; Hammou et al., 2014), though concentrations above 200 mM were estimated by efflux analysis (reviewed in Britto and Kronzucker, 2008) and as low as 12 mM and even lower K^+ concentrations were measured by ion-selective electrodes in root cells of potassium-deprived *Arabidopsis* plants (Armengaud et al., 2009). Higher potassium concentrations of 200–350 mM measured in cell sap by X-ray microanalysis or capillary electrophoresis are rather attributed to vacuolar compartment under sufficient potassium supply (Malone et al., 1991; Fricke et al., 1994; Bazzanella et al., 1998; Volkov et al., 2004). It is worth mentioning that in animal cells potassium seems to be among regulators of apoptotic enzymes activating them at K^+ concentrations below 50 mM (Hughes and Cidlowski, 1999).

Sodium cytoplasmic concentrations of plant cells are usually low reaching about 20-50 mM after several days of NaCl treatment (Carden et al., 2001, 2003; Halperin and Lynch, 2003). Higher sodium concentrations had also been measured depending on duration of salt stress, external sodium, concentration of the other ions and on plant species: cytoplasmic sodium concentrations over 200 mM were reported in salt- tolerant halophytes (Hajibagheri and Flowers, 1989; Halperin and Lynch, 2003; Flowers and Colmer, 2008; Kronzucker and Britto, 2011).

An interesting example is halotolerant alga *Dunaliella salina*, which is a good unicellular eukaryotic model for studying salinity tolerance within the range of 0.05–5.5 M NaCl (e.g., Katz and Avron, 1985). Cytoplasmic sodium concentrations about 90 mM (88 ± 28 mM) were reported in the alga using ^{23}Na- NMR spectroscopy (Bental et al., 1988). Na^+ concentrations were nearly the same (within the error of a few measurements) in the algal cells adapted to a wide range of external Na^+, from 0.1 to 4 M (Bental et al., 1988). Similar or even lower sodium concentrations below 100 mM were measured by the other methods for the alga under 0.5–4 M or 1–4 M sodium treatment (Katz and Avron, 1985; Pick et al., 1986). The small alga *D. salina* has length about 10–11 μm, width of 6 μm and volume around 200 fL (Masi and Melis, 1997) or even smaller dimensions with volume around 90–100 fL then (Katz and Avron, 1985). It probably possesses specific transport system to exclude Na^+, similar to Na^+-ATPase expected to function in the plasma membrane of the marine unicellular alga *Platymonas viridis* (Balnokin and Popova, 1994), alga *D. maritima* (Popova et al., 2005) and several other marine algae (reviewed in: Balnokin, 1993; Gimmler, 2000).

High sodium concentrations over 100 mM often have inhibiting effect on protein synthesis at least in salt-sensitive glycophytes (Hall and Flowers, 1973; Wyn Jones and Pollard, 1983; Flowers and Dalmond, 1992). Sodium is also (1) competing with potassium for allosteric sites of enzymes and (2) interacting with ion channels (for example, sodium ions change the gating of potassium outward rectifying currents in root protoplasts of halophyte plant *Thellungiella*, which are most likely carried by Shaker type potassium channels: Volkov and Amtmann, 2006). Moreover at the cellular level salt stress induces apoptosis (Katsuhara and Kawasaki, 1996; Huh et al., 2002; shortly reviewed in Shabala, 2009; Demidchik et al., 2010).

Much higher sodium concentrations could be tolerated in vacuoles, one of the functions of the organelle is to sequester and isolate sodium. Concentrations of sodium in vacuoles may exceed 0.5–1 M being up to ten times over the cytoplasmic sodium concentrations (eg Flowers et al., 1977; Zhao et al., 2005; Flowers and Colmer, 2008) at the expenses of activity of specific ion- transport systems (reviewed in: Martinoia et al., 2012). Under salt treatment of 2 M NaCl for 85 days shoot tissue concentrations of sodium in halophytes *Tecticornia* were about 2 M, so presumably vacuolar Na^+ concentrations could be over 2 M in the halophytes (English and Colmer, 2013).

The reasons for sodium competing with potassium are that the ions have (1) the same electric charge, $1.6*10^{-19}$ coulombs, (2) similar cation radii in non-hydrated, about 0.1 nm = 1.0 Å for sodium cation and 0.14 nm = 1.4 Å for potassium (diameter being nearly 2–3% of cell membrane thickness) and (3) hydrated forms, about 3.6 Å for sodium and 3.2 Å for potassium ions (Nightingale, 1959; Collins, 1997; Mähler and Persson, 2012) and, hence, (4) similar surface electric charge densities, which differ about two times (twice higher for non- hydrated sodium according to charge and diameter of the ion).

Therefore interactions of Na^+ and K^+ with amino acids of protein surfaces, active centers of enzymes, pockets of allosteric regulation or binding of proteins, selectivity filters of ion channels are similar and often differ only several times in selectivity of the interactions. The selectivity depends on the nature and number of interacting amino acids and their spatial location. Molecular dynamics simulations together with conductivity measurements for several proteins and oligopeptides demonstrated up to five times higher affinity of sodium over potassium to the protein surfaces (especially with numerous carboxyl groups; Vrbka et al., 2006). The phenomenon could explain higher destabilizing effect of sodium over potassium on proteins ("salting them out"), which was initially discovered with white proteinaceous part of hen's eggs by Hofmeister in 1888 (Hofmeister, 1888; Kunz et al., 2004). The effect could be the reason why potassium and not sodium is chosen and naturally selected for being the major intracellular monovalent cation, pumped into cells while pumping out sodium cations (Collins, 1997) though sea and ocean water contains more than 40 times higher concentration of sodium.

Under salt stress, for plants it is important to keep higher K^+/Na^+ ratio (Maathuis and Amtmann, 1999). It is essential, however, to mention that some proteins (due to specific amino acid composition or structural peculiarities) and processes from halophytes are able to withstand higher sodium concentrations without loss of activity (eg. Flowers and Dalmond, 1992; Premkumar et al., 2005); it seems to be the secondary evolutionary adaptation. It is also interesting that cell wall proteins of studied halophytes and also glycophytes did not change their activity within wide range of sodium concentrations, often from 0 to over 0.5–1 M (Thiyagarajah et al., 1996).

Driving Forces and Pathways for IonTransport to Cells

Transport of ions is driven by physico-chemical forces including differences of ion concentrations (to be more precise, activities of ions) and differences in electric potential at the sides of membranes.

Membrane potential of plant cells is routinely measured by microelectrodes with tiny sharp tips around 0.1 μm in diameter after impalement of a plant cell of interest (e.g., described in: Blatt, 1991). Recently developed voltage-sensitive fluorescent proteins and dyes (reviewed in: Mutoh et al., 2012) could also be used for at least indications of membrane potential in cell tissues and populations of cells (Matzke and Matzke, 2013). Membrane potentials below −70 mV and above −200 to −220 mV are recorded by microelectrodes though values around −300 mV were also reported; more negative values are often measured in root cells compared to leaf ones (apart from leaf guard cells) (Higinbotham, 1973; L'Roy and Hendrix, 1980; Blatt, 1987; Walker et al., 1995, 1998; Carden et al., 2001, 2003; Shabala and Lew, 2002; Fricke et al., 2006; Murthy and Tester, 2006; Shabala et al., 2006; Volkov and Amtmann, 2006; Armengaud et al., 2009; Hammou et al., 2014). Vacuolar membrane potential is the same or 10–40 mV above the values for cytoplasm with pH about 2 or over units lower, about 5.0–6.1 or less in vacuoles compared to pH = 7.0–7.7 in cytoplasm (e.g., Walker et al., 1995; Carden et al., 2003; Cuin et al., 2003; Martinoia et al., 2012) (**Figure 6**).

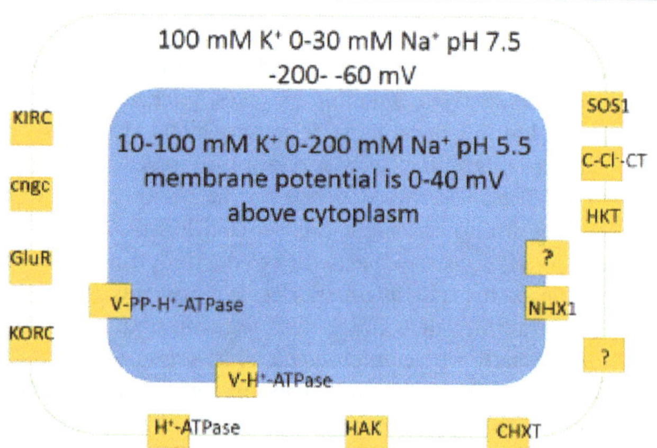

FIGURE 6 | Basic scheme of membrane potentials, potassium K^+ and sodium Na^+ concentrations and pH values in a generalized plant cell together with main ion transport systems ensuring potassium and sodium transport according to electrochemical forces. Different cell types usually have less transporters, though specialized for more determined ion transport functions. The concentrations and membrane potentials are rather indicative and change depending on conditions of mineral nutrition and are not the same for different cell types (see text and references for more details). KIRC are inward rectifying potassium channels (e.g., Hirsch et al., 1998); KORC are outward rectifying potassium channels; GluR are glutamate receptors; cngc are cyclic nucleotide gated ion channels; HAK is high affinity potassium transporter; CHXT are cation H^+ exchange transporters (e.g., Evans et al., 2012); HKT are high K^+ affinity transporters; C-Cl^--CT are cation chloride contrasnporters; SOS1 is well studied sodium-proton antiporter; H^+-ATPase is proton pump of plasma membrane; V-H^+-ATPase is vacuolar proton pump; V-PP-H^+-ATPase is vacuolar pyrophosphatase, another vacuolar proton pump; NHX1 is vacuolar sodium (cation)/proton antiporter. For more details and description see text.

Thermodynamics of ion transport is described by several equations. Nernst equation applied to selectively permeable membrane links ion concentrations at the sides of membrane to the electric potential via the membrane under equilibrium conditions (when net flux of ions via the membrane is absent):

$$E_S = E_1 - E_2 = R^*T/(Z_S{}^*F)^*\ln([S_2]/[S_1])$$

(Hille, 2001). Here E is electric potential, R is universal gas constant equal to 8.31 J/(K∗ mole), T is temperature in K, Z_S is the charge of ion S, F is Faraday constant equal to 96,500 s∗A/mole, $[S_1]$ and $[S_2]$ are concentrations of ion S at the sides of the membrane. Basically, the diffusion of ion S due to different concentrations is equilibrated by distinction in electric potential, which is about ± 60 mV (slightly depending on temperature) per 10-fold difference in concentrations with sign determined by the ion charge. For potassium with typical 100 mM in cytoplasm of epidermal root cell and low 100 μM in soil solution the membrane potential to ensure uptake of K^+ should be below −180 mV to satisfy the electrochemically downhill transport of K^+ ions. Lower concentrations of potassium outside the cells may require co-transport of K^+ with the other ions (e.g., with H^+).

For several ion species with specific permeabilities via the membrane a more complicated Goldman–Hodgkin–Katz voltage equation is applicable; it takes into account permeabilities of ions. For sodium, potassium and chloride (obviously more ions to be considered and more components should be added) the equation will be:

$$E_{membrane} = R^*T/F^*\ln((P_{Na+}[Na^+]_{out} + P_{K+}[K^+]_{out} + P_{Cl-}[Cl^-]_{in})/(P_{Na+}[Na^+]_{in} + P_{K+}[K^+]_{in} + P_{Cl-}[Cl^-]_{out}))$$

where $E_{membrane}$ is membrane potential via the membrane (or $E_{reversal}$ with zero net ion current via the membrane), P are permeabilities of the corresponding ions and [] stands for concentrations of the ions (Hille, 2001). Usually potassium permeability is dominating and membrane potential is close to $E_{reversal}$ of K^+, though may change and depend on cell type. Active plasma membrane proton pump H^+-ATPase (reviewed e.g., in: Palmgren, 2001) shifts membrane potential to more negative values compared to the calculated Ereversal for K^+.

Transport of most ions including Na^+ and K^+ in plants occurs passively (following the electrochemical forces) via ion-selective proteinaceous pores of ion channels. Most ion channels can change their conformation from open to close states and vice versa (so called "gating") under applied voltages or after binding ligands and regulators. Another pathway is via proteinaceous transporters with slower transport rates. Ion transport via ion channels is electrogenic since ions carry electric charge, while transporters realize electrogenic or non-electrogenic transport, transporting one ion or co-transporting/antiporting several charged ions in one or opposite directions, correspondingly.

Co-transport of several ions or even small molecules may add extra energy for transport (**Figure 7**). For example, HAK transporters presumably co-transport K^+ together with H^+ (Banuelos et al., 1995; Rodriguez-Navarro, 2000; Grabov, 2007), which gives several orders of concentration differences extra due to transport of H^+ according to electric charge. Membrane potential of −180 mV potentially allows potassium uptake, when co-transported with H^+ with, e.g., stoichiometry 1:1 under similar external pH to pH of cytoplasm, against 10^6 differences in K^+ concentrations (e.g., Rodriguez-Navarro, 2000). Higher concentrative capacity could be achieved using also pH differences or higher number of protons per K^+; cytoplasmic pH is about 7.5 and external low pH of 4 will add over three orders of concentration more. The surprising example is described for yeast *Schwanniomyces occidentalis*, which was reported to deplete external potassium to 0.03 μM, presumably taking up K^+ against 3,000,000 differences in concentrations (assuming over 100 mM of cytoplasmic K^+) due to HAK transporters (Banuelos et al., 1995). Much higher concentrations, around 80 μM, arising from K^+ contamination from agar and the other chemicals (Armengaud et al., 2009; Kellermeier et al., 2014) (though contamination from agar was estimated at 1–3 μM: Kellermeier et al., 2014) resulted in symptoms of severe potassium deficiency in *Arabidopsis* and essentially changed transcription profile in roots and shoots of the plants (Armengaud et al., 2004), so more detailed examination with special attention to transport systems of different species (e.g., Coskun and Kronzucker, 2013) is required. Potentially new transporters and ion channels from different organisms could be a source of diversity and comparison for the existing pool of membrane transport proteins and for creating novel artificial transporters.

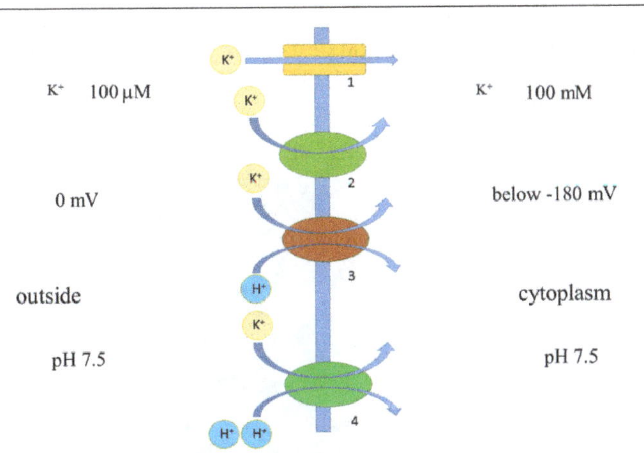

FIGURE 7 | Simplified scheme demonstrating principles of ion transport via membrane. Voltage difference below −180 mV allows potassium transport against 1000-fold concentration difference via potassium-selective pore of ion channel 1. Similar thermodynamically favorable potassium transport with lower rates and specific mechanism is facilitated by potassium transporter 2. Transporters 2 and 3 are H^+/K^+-cotransporters, they co-transport one or two H^+ per K^+; H^+ is transported according to voltage difference, hence adding energy for K^+ transport. Transporters 2 and 3 can potentially lead to inward transport of K^+ against over 1,000,000 concentration difference; their functioning depends also on pH difference across membrane. More details are in the text.

A set of transporters and ion channels is specific for cell types and organisms, includes tens and more distinct characterized so far proteins, which often form heteromers with variable properties and regulation (**Figure 6**; e.g., reviewed in: Isayenkov, 2012). Detailed analysis of genome sequences of salt-sensitive model plant *Arabidopsis* revealed that about 5% of about 25,000 genes of the plant potentially encode membrane transport proteins; the genes of about 880 proteins are classified in 46 unique groups, while genes of cation channels/transporters predict for coding over 150 proteins (Mäser et al., 2001). Special databases include information about transport proteins, e.g., plant membrane transport database http://aramemnon.uni-koeln.de/; http://www.yeastgenome.org/ is a useful source of information for yeast proteins including yeast membrane transport proteins.

Cell Size-Volume-Surface/Volume Ratio and Effects of Cell Wall

Interestingly, cell surface/volume ratio has an effect on ion concentration under the same ion fluxes (**Figure 8**). The sizes of cells within a plant differ orders of magnitude. Cells of xylem parenchyma in roots of dicot *Arabidopsis* are less than 5 μm in diameter with length often below 20–30 μm (eg: Dolan et al., 1993; Ivanov, 1997; Kurup et al., 2005; Verbelen et al., 2006; Ivanov and Dubrovsky, 2013); the cells could be isolated and result in protoplasts of about 10 μm in diameter compared to larger 20 μm epidermal protoplasts from root elongation zone or 15–25 μm protoplasts from root cortex parenchyma cells and the other root tissues (Demidchik and Tester, 2002; Demidchik et al., 2002a; Volkov and Amtmann, 2006). The volumes would correspond to $4/3*\pi*R^3$ that is about 500 fL for 10 μm protoplasts and about 8 pL for 25 μm protoplasts.

Cells in leaf epidermis of several monocots are quite large. Barley epidermal leaf cells could be up to 2 mm long and about 25–30 μm wide with nearly square cross-section, thus reaching volume over 1000 pL (Volkov et al., 2007). Several studies involved isolated protoplasts from barley leaf epidermal cells and reported 60 μm (100 pL) protoplasts (Dietz et al., 1992), 40 μm (from 20 to 80 μm) (30 pL, from to 4 to 250 pL) protoplasts (Karley et al., 2000), 25 pL protoplasts with large variations (over 10 times) in volume (Volkov et al., 2007). It is worth mentioning that about 99% of large epidermal leaf protoplasts could be occupied by vacuole (Winter et al., 1994).

Assuming volumes of usual plant cells within 500 fL to 1000 pL, the calculated surface to volume ratios will differ about 10 times for the cells: from about 0.9 μm−1 to 0.16 μm−1 for oblong cells within plant tissues. The larger cells need higher ion fluxes or larger time for the same concentration changes.

Effects of cell walls on ion transport are well known though not often remembered and studied in detail (**Figure 9**). They include changes of cell membrane potential depending on ion composition of ambient medium due to fixed electric charges in cell walls, assumed ion-rectifying properties of cell walls and expected effect of shrinking and swelling of cell walls on mechanosensitive ion channels. Effects of cell walls on ion buffering and ion concentrations have more experimental evidence. For example, protoplasts isolated from bean leaf mesophyll did not show NaCl-induced calcium efflux compared to mesophyll tissue (Shabala and Newman, 2000). Salt-induced H^+ efflux also differed between protoplasts compared to tissue, so, presumably, all the $Ca2^+$ efflux over an hour of measurement was from calcium ion exchange stores of cell walls (Shabala and Newman, 2000).

Ion Channels vs. IonTransporters: More about Pathways of Ion Transport to Cells

One of disputable questions of ion transport is about the relative role of ion channels and transporters in transport. It is commonly accepted that ion channels in an open state/conformation allow passage of over 106–108 ions per second via a selective pore formed within a protein molecule. The diameter of the pore is determined by the molecular structure of ion channel, from 12 Å for potassium channel KcsA with narrow part of 4 Å in diameter (e.g., Doyle et al., 1998; Jiang et al., 2002; MacKinnon, 2004) to 15 Å and even 28 Å diameters of pores for the general bacterial porins with low selectivity and permitted passage for small hydrophilic molecules (about 6 Å pores for the highly selective porins) (e.g., Galdiero et al., 2012). The diameter of the pore and amino acids lining it essentially determine the ion selectivity of ion channel and potential number of passing ions per unit of time. The selectivity could be, for example, over 1,000 for K^+ over Na^+ in potassium selective ion channels or over 10 for Na^+ over K^+ in sodium selective channels due to special selectivity filters with conserved amino acids for specific channel types. Often amino acid sequence glycine-tyrosine- glycine (GYG) indicates selectivity for K^+, introducing mutations into the pore to change the amino acids converted potassium selective ion channels to non-selective ones (Heginbotham et al., 1992). The interactions of ions with the protein molecule of ion channel are not well understood yet and probably involve non-electrostatic ion–ion interactions, van der Waals forces, interaction with water molecules and numerous other interactions. Several methods of modeling and simulations of molecular dynamics are applied within at least the last 30 years; sharp increase in computing power allowed to include the lipid environment of membranes, pH and the known biochemical factors and regulators to the models (e.g., reviewed in: Maffeo et al., 2012).

Direct measurements are the basement for investigating ion fluxes via ion channels; they provide information about permeating ions, number of ions per second, selectivity and complex transitions of protein molecules of ion channels (gating) during the transport processes. Indeed, a small current of 1 pA corresponds to 10^{-12} A/(1.6*10^{-19} coulombs) ≈ 6*10^6 ions/second (1.6*10^{-19} coulombs is elementary charge, a charge of monovalent cation), while most ion channels demonstrate much larger electric currents with complex voltage-dependent patterns of open–closed states (**Figure 10**).

Transporters could be considered as enzymes where conformational changes of a protein molecule are required for a complete transport cycle of ions (e.g., Gadsby, 2009). Turnover rate of the transporter is the number of complete transport cycles performed per second (e.g., Longpré and Lapointe, 2011). The lower estimate for turnover rate of transporters is the activity of ion pumps. Plant plasma membrane H^+-ATPase pumps about 100 ions per a second (Sze et al., 1999). The value is comparable to 20–100 H^+ per second by yeast plasma membrane ATPase Pma1 (Serrano, 1988) and turnover rate of 160 s−1 of animal Na^+/K^+-ATPase (Skou, 1998). Similar or even lower turnover rates, from 3 to 60 s−1, were shown for sodium/glucose cotransporter (Longpré and Lapointe, 2011 and references therein), while turnover rate about 500 s−1 was estimated for sucrose/H^+ co-transporter from maize ZmSUT1 (Carpaneto et al., 2010). The highest possible turnover rate for activity of ion transporters could be assessed from protein structure studies and frequency of conformational changes with estimated

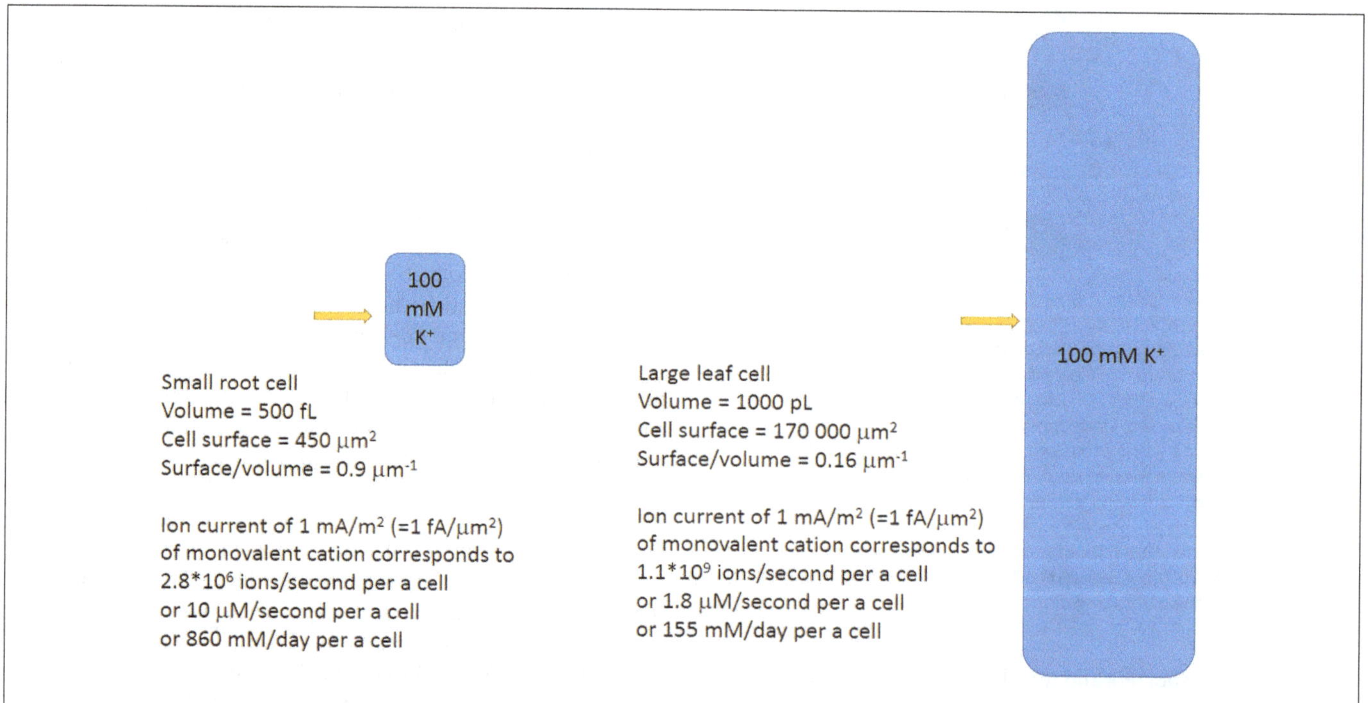

FIGURE 8 | Simplified scheme of different plant cells and the ion fluxes for the cells. The same ion fluxes result in different changes of ion concentrations for cells of different sizes and surface/volume ratios. Ion fluxes are directly recalculated from number of ions/second to concentrations.

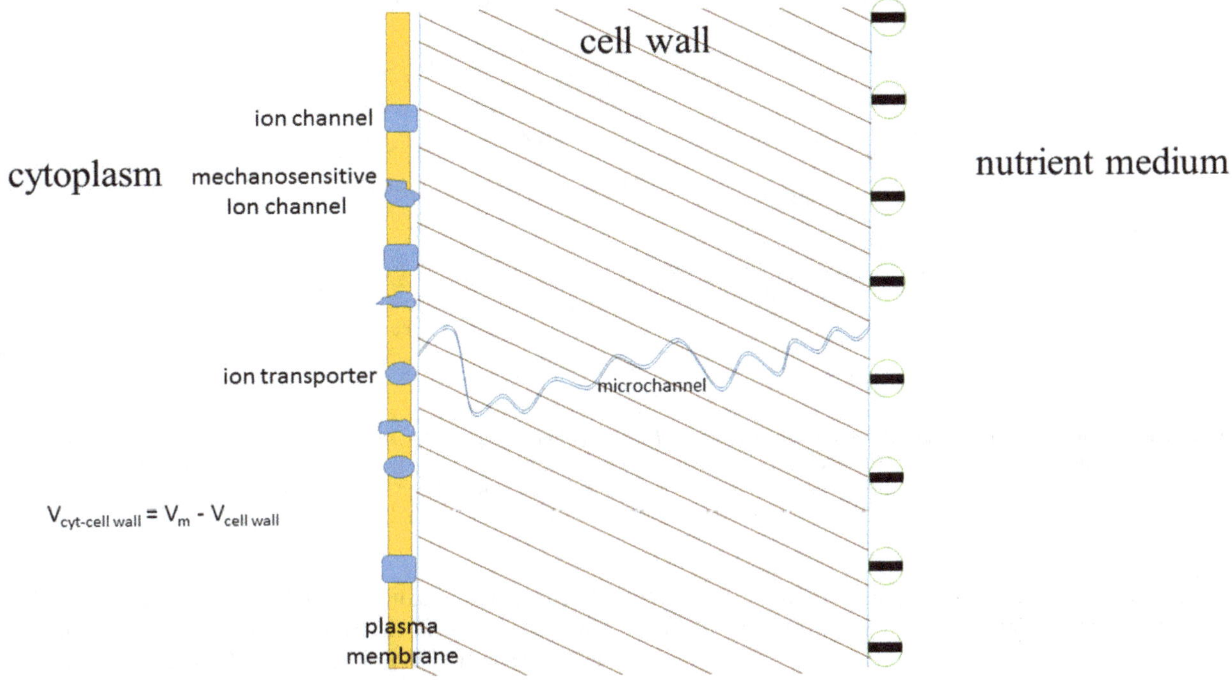

FIGURE 9 | The scheme indicating expected and studied effects of cell walls on ion fluxes and cell plasma membrane potential. Plasma membrane of approximately 10 nm thickness is surrounded by the cell wall of about 250 nm thickness, the interior of the cell is under hydrostatic turgor pressure of several bars (1 bar = 0.1 MPa) due to concentration differences and, hence, differences in osmotic pressures inside the cell and outside of it. Cell wall has negative surface charge at pH around 6–7, hence influencing the voltage difference between the sides of plasma membrane. One of micro- or nanochannels in the porous cell wall is depicted with charged inner walls and potential ion-rectifying properties. More details are given in the text.

upper limit of 106 s–1 (Chakrapani and Auerbach, 2005), which seems to be an overestimated value. A more realistic value for the fast transporters is around 10,000 ions/second, when they are mutated and have accelerated turnover rate (Gadsby, 2009). A higher value of 100,000 ions per second was reported for Cl^-/H^+ antiporter ClC-5, which is rather a unique type of transporter similar to ClC channels (Zdebik et al., 2008).

Mechanisms of ion transport by transporters are less understood than transport via pores of ion channels, though the genes of transporters are sequenced and well-studied. Briefly, several mechanisms are expected for different transporters and described below.

HKT transporters with at least eight transmembrane domains could be similar to ion channels, they form a specific ion-selective pore with properties distinct from the pore of ion channels according to basic crystal structure analysis (Cao et al., 2011; reviewed in: Yamaguchi et al., 2013). HKT transporters can electrically resemble ion channels with similar IV curve, reversal potential of ion current mediated by HKT can shift following ion concentrations inside and outside the cell. The phenomenon is observed in *Xenopus* oocytes heterologously expressing different HKT transporters (Jabnoune et al., 2009; Almeida et al., 2014; de Almeida, 2014), where rectification and Nernstian shift (according to Nernst equation for ion concentrations and voltages) in reversal potential were measured. Pore of HKT transporters has selectivity filter in the first transmembrane domain with conserved glycine for K^+-selective and serine for Na^+-selective HKT transporters (reviewed in Yamaguchi et al., 2013; de Almeida, 2014). An extra amino acid constriction with arginine residue in the last transmembrane domain makes an additional energy barrier for ion transport (Kato et al., 2007; Cao et al., 2011; reviewed in: Yamaguchi et al., 2013; Benito et al., 2014; de Almeida, 2014). Selectivity of HKT transporters for Na^+ or K^+ could be altered by amino acid substitutions, while K^+/Na^+ symport or Na^+ uniport are exhibited by different HKT transporters and even varied from K^+/Na^+ symport to Na^+ uniport depending on the cation concentrations (Mian et al., 2011; Almeida et al., 2014; reviewed in Almeida et al., 2013; Waters et al., 2013; de Almeida, 2014). Some HKT transporters symport K^+ with Na^+ at low K^+ concentrations (reviewed in Waters et al., 2013); presumably Na^+ ion adds energy for co-transport "pushing" K^+ via the amino acid constrictions (Benito et al., 2014), though the exact mechanisms of transport are still to be elucidated.

HAK transporters are not found in animals and Protista (Grabov, 2007). Presumably they co-transport K^+ together with H^+ (Banuelos et al., 1995; Rodriguez-Navarro, 2000; Grabov, 2007), but had not been crystallized so far while attempts to express them in Xenopus oocytes failed; therefore mechanisms of ion transport by HAK transporters are not well studied yet. Gene sequences and comparison of HAK transporters predict for 10–14 transmembrane domains (Greiner et al., 2011). Amino acid substitutions within the region between the second and third putative transmembrane domains of *Arabidopsis* HAK5 transporter essentially changed ion transport selectivity indicating for selectivity filter within the region (Alemán et al., 2014). Lack of putative specific

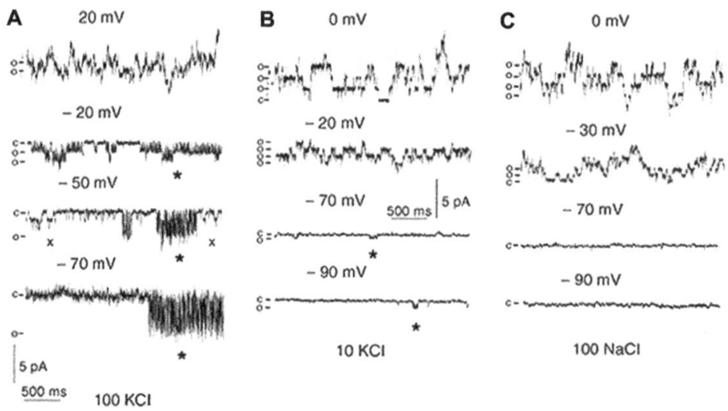

FIGURE 10 | Single-channel recordings in outside-out patches of *T. halophila* root protoplasts. The pipette solution was 100 mM KCl. The bath solution contained 100 mM KCl **(A)**, 10 mM KCl **(B)** or 100 mM NaCl **(C)**. c, current level with no open channels; o, current levels of single-channel openings. Spiky openings of outward-rectifying channels allowing inward current are indicated with asterisks. Openings of a second type of channel are indicated with crosses. Reproduced from Volkov and Amtmann (2006) with the permission from the publisher John Wiley and Sons.

pore similar to HKT transporters suggests that HAK transporters realize specific mechanism for K^+/H^+ symport (Alemán et al., 2014).

Nhx1 and SOS1 are cation/H^+ antiporters. SOS1 has 12 predicted transmembrane domains at the N-terminal part and long C-terminal tail composed of 700 amino acids (Shi et al., 2000), the protein forms homodimers (Núñez-Ramírez et al., 2012). Molecular mechanisms of Na^+/H^+ antiport by SOS1 are under investigation, though the known crystal structure of bacterial Na^+/H^+ antiporter NhaA (Hunte et al., 2005) could be a potential basement for understanding transport by both Nhx1 and SOS1. NhaA has 12 transmembrane domains, exists as a dimer, amino acid helices of the protein form negatively charged funnel-like structure, which leads to cytoplasm from the center of membrane and selectively attracts cations. Cation- bound Nha1 follows transformation and releases cation to the outer side being protonated at the same time at aspartate moieties. Deprotonation, release of H^+ to cytoplasmic side and return to the initial conformation completes the transport cycle (Hunte et al., 2005). The mechanism with conformation changes limited within a part of the protein makes NhaA one of the fastest transporters; NhaA has activity of catalytic center (turnover rate) about 89,000 s–1 (reviewed in Hunte et al., 2005).

Tonoplast Nhx1 could share the same transport mechanism. Discovery of Nhx1 initially provided a molecular basis for Na^+/H^+ antiport activity of vacuolar membrane. Na^+/H^+ antiport was measured in tonoplast of several plants: it included amiloride-sensitive transport in vesicles from *Beta vulgaris* (Blumwald and Poole, 1985) and from mesophyll cells of halophytic plant *Mesembryanthemum crystallinum* (Barkla et al., 1995), vesicles from roots of NaCl-grown salt-tolerant *P. maritima*, but not salt-sensitive *P. media* (Staal et al., 1991), in preparations from salt-grown barley roots (Garbarino and DuPont, 1988). Gene *AtNHX1* of the transporter was cloned from *Arabidopsis* and rescued some of the salt-sensitive yeast phenotypes under heterologous expression (Gaxiola et al., 1999). Moreover, overexpression of *AtNHX1* conferred salinity tolerance to *Arabidopsis* and significantly increased Na^+/H^+ antiport activity of vacuolar membrane (Apse et al., 1999). Expression of *AtNHX1* in yeast resulted in increased amiloride-sensitive electroneutral Na^+/H^+ exchange in yeast vacuolar vesicles (Darley et al., 2000). However, AtNHX1 with 9–11 putative transmembrane domains (reviewed in Rodríguez-Rosales et al., 2009) demonstrated also high K^+/H^+ exchange capacity depending on regulation by luminal C-terminal domain; the ratio of maximal rates of K^+ to Na^+ transport rose upon binding calmodulin in calcium and pH- dependent manner (Yamaguchi et al., 2003, 2005). Further evidence for role of *AtNHX1* in K^+ transport came from transgenic plants. Overexpression of *AtNHX1* in tomato plants conferred higher vacuolar K^+ under different growth conditions and increased salinity tolerance via retaining intracellular K^+ without influencing vacuolar Na^+ accumulation (Leidi et al., 2010). The simple model of AtNHX1 transporting and localizing excess Na^+ in vacuole was modified to more complex schemes. It was suggested that AtNHX1 is more important for K^+ transport to vacuole thus stimulating K^+ uptake by roots, then K^+ ions recycle between cytoplasm and vacuole, while Na^+ is transported to vacuole and "locked" there (Jiang et al., 2010). Different effects of overexpressed *NHX* genes on vacuolar K^+ and Na^+ concentrations under salt stress and increase in salt tolerance led to conclusions that NHX plays role in both Na^+ and K^+ vacuolar transport and K^+ homeostasis (reviewed in: Rodríguez-Rosales et al., 2009; Bassil et al., 2012; Yamaguchi et al., 2013; Bassil and Blumwald, 2014).

Unfortunately, ion fluxes via a single transporter (order of several fA or much lower) are below the resolution for electrophysiological measurements. However, potentially the ion currents via at least hundreds and rather thousands or millions of electrogenic ion transporters could be measured under specific conditions. A report about unitary conductance of Cl^-/H^+ antiporter ClC-5 is an exception, the conductance of 0.45 pS (ion current about 63 fA at 140 mV) for the transporter was determined from noise analysis of recordings with hundreds/thousands of transporters (Zdebik et al., 2008).

To study their properties, transporters are routinely heterologously expressed in large *Xenopus* oocytes. Detectable ion currents or fluxes of radioactive tracers are reported following the expression at the background of usually small endogenous electric currents of the oocytes. The successful expression in *Xenopus* oocytes was reported for several HKT transporters and cation-chloride cotransporters, attempts to record activity of HAK, SOS1, or Nhx transporters were less fruitful so far (Rodriguez-Navarro, 2000; Liu et al., 2001; Colmenero-Flores et al., 2007; Jabnoune et al., 2009; Rodríguez-Rosales et al., 2009). Mature *Xenopus* oocytes used for heterologous expression are quite even and have diameter of 1 mm, so measurements could be easily normalized per surface area. A typical surface area of *Xenopus* oocyte is 4*π*500*500 μm2 ≈ 4*3.14*250000 ≈ 3,100,000 μm2. Assuming that the very high recorded values of about −10 μA at −150 mV for heterologous expression of rice transporter HKT in *Xenopus* oocytes (Jabnoune et al., 2009) are reasonable and not due to incorrect folding/partial proteolysis/interaction with endogenous transport systems of *Xenopus* oocytes, we can recalculate the current per a unit of oocyte surface and compare with recordings from plant protoplasts: −10 μA/3,100,000 μm2 ≈ 3 $pA/\mu m^2$ = 3,000 mA/m^2 or about 1 nA per a small root protoplast with diameter around 10 μm. The values are very high and comparable or even much higher (see below) than the ones recorded from activity of ion channels using patch clamp.

Another theoretical estimate is useful for assessing activity of ion transporters expressed in *Xenopus* oocytes. Assuming the high turnover rate for the transporter around 10,000 and the very high expression of about 10,000 transporters/μm2 (means a transporter per 10*10 nm2, nearly maximal density due to the physical and steric limitations) we get the possible presumed current per oocyte of *Xenopus*: 10,000 ions per transporter per second * 3,100,000 μm2 * 10,000 transporters/μm2 = 3,1*1014 ions/second, then taking elementary charge of monovalent ion being 1.6*10−19 coulombs, the estimate gives 1.6*10−19 coulombs/ion * 3,1*1014 ions/second ≈ 50 μA per an oocyte. An excessive order of magnitude could be reasonably explained by a lower level of expression and lower transport rate of a transporter, so gives a reasonable agreement with experimental data and leads to several conclusions.

(1) Transporters can provide sufficient ion currents for registered ion transport under conditions of salinity.
(2) Ion channels are not the only essential pathway for ion transport under salinity.
(3) Balance between relative share of ion transport by ion channels or by ion transporters depends on abundances of the corresponding proteins (ion channels or transporters), their regulation and the other factors (composition of ion solutions, membrane potential etc.). Transporters can provide coupled transport of several ions and potentially may ensure fine-tuning of ion transport, while ion channels provide large ion fluxes when required.

The estimates indicate that total ion current via thousands/millions of electrogenic transporters could be measured and characterized in plant protoplasts using patch clamp in whole-cell configuration (recording sum of all ion currents via the whole membrane). Well-studied non-selective cation channels with low conductance carry small instantaneous currents; potential total current via numerous transporters could be of the same range. Ion current via ion-selective ion channels is described by Goldmann–Hodgkin–Katz equations based on assumption of independent passage of ions via channel pore (or constant electric field along the diffusion zone) (Hille, 2001), additional charges of lipid bilayer and the surface of ion channel can modify the ideal curves (Alcaraz et al., 2004). However, similar to Goldmann–Hodgkin–Katz curves were recorded for HKT transporters expressed in *Xenopus* oocytes (Jabnoune et al., 2009; de Almeida, 2014).

Recently sodium currents via AtHKT1;1 transporters were presumably measured in *Arabidopsis* root stelar protoplasts overexpressing AtHKT1;1 (Møller et al., 2009; Xue et al., 2011). Patch clamp experiments recorded about 30 mA/m2 lower (more negative) currents in 10 mM external Na^+ and about 50 mA/m2 lower currents in 25 mM external Na^+ at −120 mV, when compared to control protoplasts. The currents in AtHKT1;1 overexpressing protoplasts shifted the reversal potential according to external Na^+ concentrations, so confirmed Na^+ selectivity. Further study (Xue et al., 2011) compared Na^+ and K^+ currents in root stelar cell protoplasts from wild type (control) and *athtk1;1-4* mutant plants lacking AtHKT1;1. Potassium currents were similar (about −50 pA at −120 mV in 5 mM internal/50 mM external K^+), while sodium currents were about −50 pA at −120 mV in control protoplasts at 50 mM internal/50 mM external Na^+ compared to about −10 pA in *athtk1;1-4* mutant ones and demonstrated Nernstian shift according to reversal potential of sodium (Xue et al., 2011). The results pose questions for future study. Earlier it had been demonstrated that non-selective cation currents in similar root protoplasts of *Arabidopsis* (e.g., Demidchik and Tester, 2002 and see below) are slightly (1.5 times) more selective for potassium over sodium, so predictably *athtk1;1-4* mutant protoplasts should have 2–4 times higher sodium currents than measured in (Xue et al., 2011) due to expected non-selective cation currents.

The non-selective cation currents are studied well for root protoplasts, especially in *Arabidopsis* and carried by cyclic nucleotide gated channels (about 20 genes for *Arabidopsis*) and glutamate receptors (about 20 genes for *Arabidopsis*) (see below). One of the possible explanations for the paradox is to assume that nonselective cation currents in root stelar protoplasts of *Arabidopsis* are highly selective for potassium over sodium; the selectivity was shown for root protoplasts of *Thellungiella* (Volkov and Amtmann, 2006), salt-tolerant relative of *Arabidopsis* (Bressan et al., 2001; Inan et al., 2004; Amtmann, 2009). It is important to understand specific tissue and cell type expression of genes and proteins for non-selective cation channels and HKT transporters for characterizing their role in total ion currents. So far, more electrophysiological studies were performed with non-selective cation currents, while HKT transporters were mostly studied using molecular biology with the known genes in heterologous expression systems.

Special modeling for different pores of ion channels will help to understand better the peculiarities of ion currents via ion channels and HKT-like transporters. Pharmacological analysis and profiling of ion currents is also essential together with the further use of mutants (knock out or overexpression) and heterologous expression of the genes of interest in cell culture or in *Xenopus* oocytes. Non-selective ion currents are well characterized electrophysiologically and pharmacologically, especially for root protoplasts (White, 1993; White and Lemtiri- Chlieh, 1995; Roberts and Tester, 1997; Tyerman et al., 1997; Tyerman and Skerrett,

1998; Demidchik and Tester, 2002; reviewed in: Demidchik et al., 2002b; Volkov and Amtmann, 2006; reviewed in: Demidchik and Maathuis, 2007), less is known for transporters. Recently quinine (500 μM) was shown to have slight inhibiting effect on ion currents induced by HKT1;4 transporters from durum wheat after heterologous expression in *Xenopus* oocytes; $Zn2^+$, $La3^+$, $Gd3^+$, or amiloride had no effect (Ben Amar et al., 2014). For cation-chloride co-transporter from *A. thalina*, which was expressed in *Xenopus* oocytes and presumably transported Na^+:K^+:2Cl–100 μM bumetanide had an inhibiting effect on uptake of radioactive ions similar to analogous animal co-transporters (Colmenero-Flores et al., 2007). Amiloride is shown to inhibit Nhx1 vacuolar Na^+(cation)/H^+ antiporter (Barkla et al., 1990; Darley et al., 2000). Experiments with heterologous expression of rice HKT transporter OsHKT2;4 in *Xenopus* oocytes demonstrated channel-like behavior with single channel traces and inhibition by $Ba2^+$, $La3^+$, and $Gd3^+$ (Lan et al., 2010), however, the properties were not typical for a transporter and further on the results were not confirmed and attributed to endogenous currents of the expression system (Sassi et al., 2012).

A further complication for understanding the pathways of membrane ion transport comes from electroporation experiments (e.g., Pakhomov et al., 2009; Wegner et al., 2011, 2013; Wegner, 2013, 2014). Nanopores with diameter about 1.0 –1.8 nm are formed in lipid membrane bilayer of several animal cell lines and plant protoplasts after (1) voltage pulses of 1 V and lower to –350 to $^+$250 mV and (2) also during patch clamp experiments after applied holding voltages above $^+$200– 250 mV or below –300 to –250 mV (Pakhomov et al., 2009; Wegner et al., 2011, 2013; Wegner, 2013). The voltage-induced nanopores existed for minutes and demonstrated selectivity for cations (including even TEA^+) over anions (Cl–) and slight selectivity for different cations ($Ca2^+$ and Li^+ were the most permeable); certain similarity to behavior of non-selective cation channels was found (Wegner et al., 2011, 2013; Wegner, 2013). It is a question whether the nanopores appear under physiological conditions in plants (voltages below –300 mV were recorded for plant cells). Heterologous expression of specific ion channels together with use of mutants, elucidating detailed properties, pharmacological profiles and peculiarities of ion fluxes are required to exclude misinterpretation.

GENETIC ENGINEERING WITH NON-SPECIFIC OR TISSUE-SPECIFIC OVEREXPRESSION/KNOCKOUT/ DISRUPTION OF SPECIFIC TRANSPORTERS MODIFIES SALINITY TOLERANCE

Several obvious ways to achieve salinity tolerance include: (1) decreasing sodium conductance and increasing potassium/sodium selectivity of plasma membrane of root epidermal cells; (2) increasing sodium efflux by root epidermal cells; (3) increasing sodium accumulation in vacuoles; (4) altering sodium and potassium loading and unloading to xylem and phloem depending on plant strategy to cope with salinity. The strategies had been realized by salt-tolerant plants or revealed in plants overexpressing genes of specific transporters.

Modification of gene activity started after the essential rise and success of molecular methods together with the identification and characterisation of individual ion channels and transporters. Lower sodium conductance and higher K^+/Na^+ selectivity of oot epidermis discovered in halophytes (see above) could be potentially reached in agriculturally important plants by RNA silencing of non-selective cation channels or modifying their expression pattern and regulation. However, still not much is known about the exact genes for the ion channels and importance of the individual genes in sodium uptake. Successful attempts to overexpress or knockout genes of vacuolar proton pump H^+-PPase, NHX, HKT, or SOS1-like transporters and to modulate the salinity tolerance of plants had already been reported.

Overexpression of the vacuolar H^+-pump would enhance the proton pumping activity at vacuolar membrane and thus permit to accumulate more Na^+ in vacuoles due to activity of Na^+(cation)/ H^+ antiporters NHX. The choice of H^+- pyrophosphatase is explained by a single gene required for the protein, while the other vacuolar H^+-ATPase is composed of several subunits and needs correct overexpression of several genes (reviewed in e.g., Silva and Gerós, 2009). Overexpression of vacuolar H^+-PPase under control of strong non-specific viral *35S* promoter sharply increased salinity tolerance in *Arabidopsis*, to 250 mM of NaCl (Gaxiola et al., 2001). Overexpressing plants accumulated more sodium and potassium in their leaves and also demonstrated higher drought resistance. Further attempts to overexpress vacuolar H^+-PPases from different microbial (D'yakova et al., 2006) and plant species increased salinity tolerance in tobacco (D'yakova et al., 2006; Gao et al., 2006; Li et al., 2014), transgenic rice overexpressing also vacuolar transporter NHX1 (Zhao et al., 2006), in alfalfa (Bao et al., 2009), cotton (Pasapula et al., 2011), tomato (Bhaskaran and Savithramma, 2011), and sugarcane (Kumar et al., 2014). The experiments used *35S* promoter, NaCl concentration of 150–400 mM and reported higher Na^+ concentrations in leaves of overexpressing plants under salt treatment, while K^+ changes were not consistent between the species (Gao et al., 2006; Zhao et al., 2006; Bao et al., 2009; Bhaskaran and Savithramma, 2011; Li et al., 2014). Gene of vacuolar H^+-pyrophosphatase was among salinity tolerance determinants in barley (Shavrukov et al., 2013). Overexpression of vacuolar H^+-PPase had also an effect on the whole physiology of plants, for example, increasing root growth via probably auxin transport-associated genes, antioxidant enzymes activities and photosynthetic rate in tobacco (Li et al., 2014). Without salt stress the transgenic plants overexpressing H^+-pyrophosphatase under control of *35S* promoter demonstrated phenotypes either similar to non-transformed plants (Zhao et al., 2006; Bao et al., 2009; Bhaskaran and Savithramma, 2011), or exhibited improved morphological features (Kumar et al., 2014; Li et al., 2014) or lower osmotic potential in leaves (Gao et al., 2006). Salt treatment for non- transgenic plants resulted in both up- and down-regulation of vacuolar H^+-PPase in different species, therefore suggesting an important role of vacuolar H^+-ATPase in responses to the stress factor (reviewed in: Silva and Gerós, 2009).

Another candidates for overexpression are vacuolar *NHX* genes. Overexpression of *AtNHX1* increased salinity tolerance in *Arabidopsis* to 200 mM NaCl, the overexpressing plants accumulated more Na^+ compared to wild type and demonstrated higher Na^+/H^+ exchange activity in isolated leaf vacuoles (Apse et al., 1999). The approach of overexpressing *AtNHX1* to improve salinity tolerance proved to be successful for tomato; the transgenic plants accumulated more sodium in leaves but not in fruits at

200 mM NaCl (Zhang and Blumwald, 2001). Cotton plants with *AtNHX1* from *Arabidopsis* (He et al., 2005), rice overexpressing *SsNHX1* from halophyte *Suaeda salsa* (Zhao et al., 2006), tomato with heterologous NHX from *Pennisetum glaucum* (Bhaskaran and Savithramma, 2011) also showed increased salinity tolerance. Overexpression of *NHX* did not influence the phenotype of plants under control conditions (Apse et al., 1999; Zhang and Blumwald, 2001; He et al., 2005; Zhao et al., 2006; Bhaskaran and Savithramma, 2011).The results with heterologous expression or overexpression of NHX transporters lead to conclusions that the gene is among determinants and potential candidates for engineering salinity tolerance (e.g., Rodríguez-Rosales et al., 2009; Peleg et al., 2011 with more references for successful overexpression of *NHX* to increase salinity tolerance in sugar beet, wheat, maize and the other plants). However, the overexpression of *NHX* was not tissue- specific and under the control of strong promoters, one report did not confirm increase in salinity tolerance in *Arabidopsis* overexpressing *AtNHX1* (Yang et al., 2009). Expression in a tissue-specific manner could be the next step for using NHX to increase salinity tolerance.

The amazing simplicity of the idea to play with the expression of known and functionally well characterized transporters and get salt tolerant or salt sensitive plants is applied to plasma membrane SOS1 Na^+/H^+ antiporters and Na^+ or Na^+/K^+ HKT transporters. SOS1 is expressed in (1) epidermal root cells where it participates in sodium efflux and in (2) xylem parenchyma cells where SOS1 may load Na^+ to xylem under moderate salinity and unloads Na^+ under high salinity or has more complex mode of xylem loading/unloading (Shi et al., 2000, 2002; Pardo et al., 2006; Oh et al., 2007, 2009a; Olías et al., 2009). *Arabidopsis* mutants with defects in gene of SOS1 exhibited strong growth inhibition under salt treatment (Wu et al., 1996), which was rescued in *sos1* mutant by overexpression of *SOS1* gene under *35S* promoter (Shi et al., 2000). Overexpression of *SOS1* gene in wild type plants under *35S* promoter enhanced salinity tolerance of *Arabidopsis* at 100–200 mM NaCl (Shi et al., 2003; Yang et al., 2009), reduced sodium accumulation in shoots and sodium concentration in xylem sap (Shi et al., 2003). Further on overexpression of *SOS1* from *A. thaliana* increased salinity tolerance in transgenic tobacco (Yue et al., 2012) and in transgenic tall fescue (Ma et al., 2014). *SOS1* gene from durum wheat conferred salinity tolerance to *sos1* mutant of *Arabidopsis* (Feki et al., 2014). Interestingly, the effects of overexpression were observed under salt treatment, while in the absence of stress no differences were observed in growth or morphology between wild-type plants and the transgenic lines. Disruption of SOS1 activity by RNA interference in *Thellungiella* on the opposite resulted in the loss of tolerance of the halophyte indicating importance of Na^+ efflux and essential role of SOS1 in salinity tolerance (Oh et al., 2009a). RNA interference of *SOS1* significantly changed the whole transcriptome of *Thellungiella* (Oh et al., 2007) and vacuolar pH under salt treatment (Oh et al., 2009b) proving the complex nature of metabolic and regulatory networks in plants (**Figures 2** and **6**) and yet the probabilistic chances of success in strict overexpression of specific transporters for salinity tolerance improvement. A more complicated situation emerges due to tissue-specific expression. SOS1 is important for long-distance ion transport and xylem loading/unloading in *Arabidopsis* (Shi et al., 2002; discussed in: de Boer and Volkov, 2003), sodium partioning between plant organs in tomato (Olías et al., 2009) and ion fluxes in root meristem zone (Guo et al., 2009), therefore attempts to express it in specific tissues could increase salinity tolerance to a higher extent.

Genetic modification of salinity tolerance using HKT transporters was also successful. Analysis of *Arabidopsis* plants with mutated *HKT* gene revealed higher salt sensitivity of the mutants under long term stress, higher sodium accumulation in their shoots under mild salinity treatment (Mäser et al., 2002) and suggested that HKT is involved in recirculation of sodium within plants (Berthomieu et al., 2003). Further study confirmed increased sodium in the shoots of *Arabidopsis hkt1;1* mutants and clarified that HKT is important for root accumulation of Na^+ and Na^+ uptake from xylem in *Arabidopsis* (Davenport et al., 2007). The next step was to create plants overexpressing *HKT* (Møller et al., 2009). *Arabidopsis* plants overexpressing *AtHKT* under the control of *35S* promoter were compared with plants specifically overexpressing *HKT* in cells of root stele. *Pro35S:HKT1;1* plants were salt sensitive probably due to higher Na^+ uptake by roots while tissue specific overexpression of *HKT* in stele increased salinity tolerance and reduced sodium accumulation in shoots (Møller et al., 2009). The approach was applied to rice where gene from *Arabidopsis AtHKT1;1* was heterologously expressed in root cortex. It resulted in lower shoot Na^+ concentrations, improved salinity tolerance and involved up- and down-regulation of several membrane transport genes including vacuolar H^+-pyrophosphatases (Plett et al., 2010). Overexpression of HKT had none (Møller et al., 2009; Plett et al., 2010; Mian et al., 2011) or slight inhibiting pleiotropic effect on growth without NaCl depending on type of promotor for expression and on plant line studied (Møller et al., 2009; Plett et al., 2010). HKT transporters proved to be important for Na^+ exclusion in wheat and were transferred from durum wheat to bread wheat by interspecific crossing; the genes gave beneficial effects including higher K^+/Na^+ ratio in leaves under saline conditions (James et al., 2011). Some plants including barley accumulate Na^+ in shoots; overexpression of barley *HvHKT2;1* under *35S* promoter in barley increased salinity tolerance at 100 mM NaCl, but opposite to *Arabidopsis* increased Na^+ concentration in xylem and Na^+ accumulation in barley leaves (Mian et al., 2011). Taken together the results set HKT transporters to potential candidates for engineering salinity tolerance and among the determinants of the trait (reviewed in: Horie et al., 2009; Almeida et al., 2013; Maathuis et al., 2014) together with the above mentioned NHX1, SOS1 and presumably new studied transporters, e.g., similar to CHX21 from *Arabidopsis* (Hall et al., 2006).

Genes which are important for salinity tolerance in the other groups of organisms and not present in higher plants could also be potential candidates for engineering the trait. Sodium pumping ATPase from moss *Physcomitrella patens* was cloned and expressed under *35S* promoter in rice; plasma membrane expression resulted in higher biomass of transgenic plants compared to control ones after 2 weeks of 50 mM salt treatment. Surprisingly, expression of Na^+-ATPase did not influence Na^+ and K^+ concentrations in transgenic compared to control plants under any tested conditions, so needs more investigation (Jacobs et al., 2011). Recently a new bacterial rhodopsin from *Krokinobacter eikastus* was discovered; it is the first light-driven Na^+ pump. This rhodopsin was crystallized and resolved using X-ray; the structural basis for Na^+ transport was revealed (Gushchin et al., 2015; Kato et al., 2015). The photo switchable sodium pump seems a good simple molecular tool. The initial results are already promising and involve the diverse pool of Na^+-ATPases and transporters from yeast, algae and microbes for improving salinity tolerance in plants.

PERSPECTIVES OF PROTEIN ENGINEERING. STRUCTURE-FUNCTION STUDIES AND POTENTIAL FUTURE FOR EXPRESSION OF NOVEL ION CHANNELS, PUMPS, AND TRANSPORTERS

Novel opportunities for increasing salinity tolerance in plants are arising with the development of new methods of molecular biology, understanding regulation networks from synthetic biology and growing knowledge about single point mutations changing specific amino acids within molecules of an ion channel or a transporter.

Single amino acid substitutions, e.g., within K^+ selectivity filter GYG of potassium channels, may change selectivity of the ion channels rendering them from K^+ selective to non-selective ones (e.g., Heginbotham et al., 1992). Amino acid substitutions within a presumed pore region of HKT transporters are able to alter them from Na^+ selective to Na^+ and K^+ permeable (e.g., Mäser et al., 2002; de Almeida, 2014). Moreover, the single point mutations could be determining for salinity tolerance, e.g., amino acid substitution V395L in rice transporter OsHKT1;5 presumably explained the salt sensitivity of the rice cultivar (Cotsaftis et al., 2012). Specific single point substitution in *Arabidopsis* HAK5 transporter (F130S) over 100 times increased affinity for K^+ under heterologous expression and reduced inhibition constants for Na^+ and Cs^+ (Alemán et al., 2014). Effects of single point mutations on the whole pattern of physiology and on phenotype are well known and better studied in human biomedical science when inherited diseases cystic fibrosis and sickle cell anemia are caused by amino acid substitutions in transport protein CFTR and in hemoglobin, correspondingly.

The knowledge about structure-function correlations of proteins allows to modify the selectivity and create the required properties of ion transport proteins. The way in the direction is to employ the existing and growing information about structure-function of different ion channels and transporters. Next step is to change their ion selectivity and gating properties according to the requirements using single amino acids mutations and to transform the plants of interest in a tissue-specific or cell-specific manner. Examples of tissue-specific transformation already exist (e.g., Møller et al., 2009; Plett et al., 2010) while the new methods and opportunities are progressing enormously (e.g., Brand et al., 2006; Oszvald et al., 2008 etc.). Recent opportunities to directly edit genome using CRISPR–Cas system could overcome some difficulties and directly modify the expressed genes of ion transport proteins (reviewed in e.g., Shan et al., 2013; Kumar and Jain, 2015).

An alternative approach from synthetic biology is not to modify the existing membrane transport proteins, but to create new ones with desired properties for the further cell-specific transformation (**Figure 11**). The idea is different from what could be assumed at a first glance. Existing biological organisms emerged over the process of long evolution, when previous "building blocks" and elements were used for the future development and often could not be essentially modified due to intrinsic links within organisms and biological systems. It leaves out the question of ideal design, which is mostly not present in biological organisms. Indeed, they are largely predetermined by the previous evolutional history with intrinsic evolutionary trajectories and evolved under multifactor environment (composition of atmosphere, illumination, temperature, salinity and mineral nutrients, water availability etc., while interactions with the other organisms and biotic interactions are often the most important). A simple example is related to temperature. Ion channels in homoeothermic animals like mammals or birds evolved over hundreds of millions of years under stable conditions and hence differ in many properties from ion channels in plants. Sodium and calcium selective ion channels had not been found in plants while in animals they ensure action potentials in neurons and cardiomyocytes. Specialized highly temperature-sensitive ion channel in animals provide temperature sensation (for example transient receptor potential channels, e.g., Ramsey et al., 2006; Myers et al., 2009). Plants are different, they rely on calcium signaling via non-selective cation channels, may have distinct groups of transporters with specific properties, have no known sodium-selective ion channels and use ion channels with low temperature sensitivity (e.g., plant potassium channels and their regulation are reviewed in: Dreyer and Uozumi, 2011; more general review: Hedrich, 2012). Obviously, cell signaling and regulation in plants have numerous specific peculiarities compared to homoeothermic animals. The differences could be reflected in protein structures, therefore simple comparisons may be misleading. Moreover, plant ion channels and transporters are often functioning at membrane voltages below −200 mV while animal membranes are usually not experiencing voltages below −100 mV. Structure and functioning of ion transport proteins from yeast, algae, Protista, and microbes could provide more insights for synthetic proteins for plants.

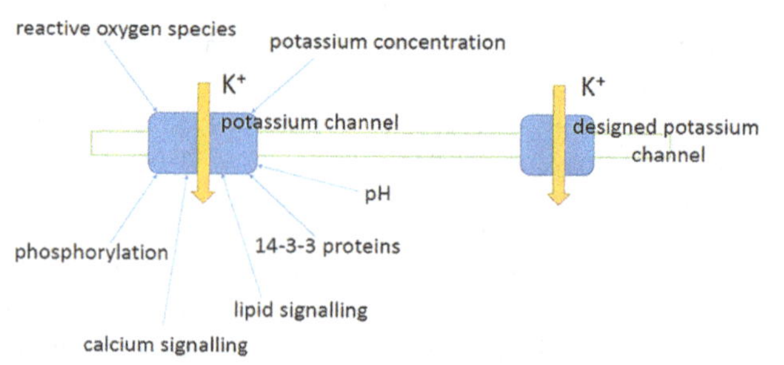

FIGURE 11 | Novel artificially designed ion channels and ion transporters potentially provide an opportunity to avoid the present evolved protein regulation networks and may be useful for altering ion concentrations, membrane potential and signaling without essentially interacting with the fine-tuning of the existing regulatory networks. A schematic generalized plant potassium channel in a plasma membrane with several potentially known regulation factors and a model of artificially designed ion channel lacking the evolved known regulation feedbacks.

From the point of the above mentioned it seems that attempts to design novel artificial ion channels and transporters with known characterized ion selectivity filters and voltage sensors adding or excluding specific interacting regulatory elements for the proteins might be productive. The ion channels and transporters when expressed in cell-specific manner under controlled conditions and in defined numbers may avoid fine tuning of regulation and potentially could provide shortcuts in natural signaling networks. The appearing opportunities offer new chances to design salt tolerant plants with previously unknown features and wider ranges of regulation circuits and networks. The potential strategy to increase salinity tolerance includes (1) choice of plant, (2) understanding ion transport and features of salinity tolerance for the plant, (3) determining ion fluxes and ion conductances important for Na^+ and K^+ accumulation and compartmentation, (4) modeling ion fluxes and adding/removing in the models ion fluxes and conductances to ensure better nutrient supply and rise of salinity tolerance under salt stress, (5) correlating required changes in ion transport with potential membrane ion transport proteins to realize the ion fluxes, (6) designing specific membrane proteins, (7) expressing the membrane proteins in a tissue-specific manner, (8) checking the salinity tolerance of transgenic plants in laboratory and field experiments. The proposed sequence of events is hypothetical so far and had not been realized yet. However, progress in molecular biology and new ways of thinking from synthetic biology may bring it to fruition and provide new discoveries on the path.

Design of transmembrane proteins is at the very beginning nowadays. Certain structural blocks of the proteins are known better and used for several applications. Voltage-sensors are intensively studied for voltage-gated ion channels and for transporters (Bezanilla, 2008). Moreover, optogenetics needs and already efficiently applies voltage-sensing structural blocks of proteins (Mutoh et al., 2012; Cao et al., 2013; Zou et al., 2014). Selectivity filters of ion channels and transporters are also known and under investigation. Addition of regulatory domains for binding depends on our current information and new endeavors to understand protein–protein interactions and regulation. The further tissue-specific expression and cell-specific localisation of the proteins to be realized via signal peptides and tissue-specific promoters. Problems, strategies and perspectives for engineering novel membrane proteins and re-engineering the existing ones are widely discussed and reviewed (e.g., Grosse et al., 2011; Subramanyam and Colecraft, 2015). *De novo* design, synthesis confirmed by X-ray structure of Zn2 $^+$-transporting four-helix transmembrane protein bundle is a recent experimental advance (Joh et al., 2014). The $Zn2^+$-transport activity of the novel protein was confirmed in functional assays (Joh et al., 2014) proving feasibility of the approach and setting engineering of membrane proteins to a new higher level.

CONCLUSION AND PERSPECTIVES

Development of agriculture often coincides with salinization of the used land, which is later not available for agricultural plants and inhabited by natural halophyte plants. The problem exists since collapse of Sumer civilization about 4000 years ago due to improper agricultural techniques. Recent decades added modern plant physiology, biophysics, molecular and systems biology to pure agriculture and breeding for creating salt-tolerant crops and agriculturally important plants. New information about individual ion transport systems gave solid physical basis for improving salinity tolerance of plants. Emerging opportunities to overexpress individual genes allowed to sharply increase salinity tolerance in laboratory trials. Growing knowledge in protein engineering and synthetic biology sets novel aims and horizons for producing artificial proteins with predefined transport properties and for designing new regulation networks. Potentially the progress in the direction may lead to partially artificial plants with desired salinity tolerance. The next step would be to fill the huge gap between rapid success in laboratory experiments and field practice. Expectedly the future advances will help to release the problem of salinization of agricultural lands.

ACKNOWLEDGMENTS

The Author would like to thank colleagues who we worked together over several projects summarized in the Review, Dr Svetlana Bagirova for critical reading and comments on the Manuscript and to bring apologies for not citing all the relevant literature sources due to the limitations by the volume of the Review. The earlier preprint version of the Review entitled “Salinity tolerance in plants: attempts to manipulate ion transport” was uploaded to repository of electronic preprints arXiv at http://arxiv.org/abs/1411.1553 on the 6th of November 2014.

REFERENCES

Akhromeeva, T. S., Kurdyumov, S. P., Malinetskii, G. G., and Samarskii, A. A. (1989). Nonstationary dissipative structures and diffusion-induced chaos in nonlinear media. *Physics. Rep.* 176, 5–6. doi: 10.1016/0370-1573(89)9 0001-X

Akhromeeva, T. S., Kurdyumov, S. P., Malinetskij, G. G., and Samarskij, A. (1992). *Nonstationary Structures and Diffusion Chaos (Nestatsionarnye Struktury i Diffuznyj Khaos).* Moskva: Nauka Publisher, 544.

Alcaraz, A., Nestorovich, E. M., Aguilella-Arzo, M., Aguilella, V. M., and Bezrukov, S. M. (2004). Salting out the ionic selectivity of a wide channel: the asymmetry of OmpF. *Biophys. J.* 87, 943–957. doi: 10.1529/biophysj.104/043414

Alemán, F., Caballero, F., Ródenas, R., Rivero, R. M., Martínez, V., and Rubio, F. (2014). The F130S point mutation in the *Arabidopsis* high-affinity K^+ transporter AtHAK5 increases K^+ over Na^+ and Cs^+ selectivity and confers Na^+ and Cs^+ tolerance to yeast under heterologous expression. *Front. Plant Sci.* 5:430. doi: 10.3389/fpls.2014.00430

Alemán, F., Nieves-Cordones, M., Martínez, V., and Rubio, F. (2009). Potassium/sodium steady-state homeostasis in *Thellungiella halophila* and *Arabidopsis thaliana* under long-term salinity conditions. *Plant Sci.* 176, 768– 774. doi: 10.1016/j.plantsci.2009.02.020

Almeida, P., Katschnig, D., and de Boer, A. H. (2013). HKT transporters-state of the art. *Int. J. Mol. Sci.* 14, 20359–20385. doi: 10.3390/ijms141020359

Almeida, P. M. F., de Boer, G. J., and de Boer, A. H. (2014). Assessment of natural variation in the first pore domain of the tomato HKT1;2 transporter and characterization of mutated versions of SlHKT1;2 expressed in *Xenopus laevis* oocytes and via complementation of the salt sensitive athkt1;1 mutant. *Front. Plant Sci.* 5:600. doi: 10.3389/fpls.2014.00600

Amtmann, A. (2009). Learning from evolution: thellungiella generates new knowledge on essential and critical components of abiotic stress tolerance in plants. *Mol. Plant* 2, 3–12. doi: 10.1093/mp/ssn094

Apse, M. P., Aharon, G. S., Snedden, W. A., and Blumwald, E. (1999). Salt tolerance conferred by overexpression of a vacuolar Na^+/H^+ antiport in *Arabidopsis. Science* 285, 1256–1258. doi: 10.1126/science.285.5431.1256

Armengaud, P., Breitling, R., and Amtmann, A. (2004). The potassium-dependent transcriptome of *Arabidopsis* reveals a prominent role of jasmonic acid in nutrient signaling. *Plant Physiol.* 136, 2556–2576. doi: 10.1104/pp.104.046482

Armengaud, P., Sulpice, R., Miller, A. J., Stitt, M., Amtmann, A., and Gibon, Y. (2009). Multilevel analysis of primary metabolism provides new insights into the role of potassium nutrition for glycolysis and nitrogen assimilation in *Arabidopsis* roots. *Plant Physiol.* 150, 772–785. doi: 10.1104/pp.108.1 33629

Balnokin, Y. V. (1993). Ionic homeostasis and osmoregulation in halotolerant microalgae. *Russian J. Plant Physiol.* 40, 567–576.

Balnokin, Y. V., and Popova, L. G. (1994). The ATP-driven Na^+ -pump in the plasma membrane of the marine unicellular alga, *Platymonas viridis. FEBS Lett.* 343, 61–64. doi: 10.1016/0014-5793(94)80607-1

Banuelos, M. A., Klein, R. D., Alexander-Bowman, S. J., and Rodriguez-Navarro, A. (1995). A potassium transporter of the yeast Schwanniomyces occidentalis homologous to the Kup system of *Escherichia coli* has a high concentrative capacity. *EMBO J.* 14, 3021–3027.

Bao, A. K., Wang, S. M., Wu, G. Q., Xi, J. J., Zhang, J. L., and Wang, C. M. (2009). Overexpression of the Arabidopsis H^+-PPase enhanced resistance to salt and drought stress in transgenic alfalfa (*Medicago sativa* L.). *Plant Sci.* 176, 232–240. doi: 10.1016/j.plantsci.2008.10.009

Barkla, B. J., Charuk, J. H., Cragoe, E. J., and Blumwald, E. (1990). Photolabeling of tonoplast from sugar beet cell suspensions by [3H]5-(N-Methyl-N-Isobutyl)- Amiloride, an inhibitor of the vacuolar Na^+/H^+ antiport. *Plant Physiol.* 93, 924–930. doi: 10.1104/pp.93.3.924

Barkla, B. J., Zingarelli, L., Blumwald, E., and Smith, J. (1995). Tonoplast Na^+/H^+ antiport activity and its energization by the vacuolar h^+-atpase in the halophytic plant *Mesembryanthemum crystallinum* L. *Plant Physiol.* 109, 549–556.

Bassil, E., and Blumwald, E. (2014). The ins and outs of intracellular ion homeostasis: NHX-type cation/H^+ transporters. *Curr. Opin. Plant Biol.* 22, 1–6. doi: 10.1016/j.pbi.2014.08.002

Bassil, E., Coku, A., and Blumwald, E. (2012). Cellular ion homeostasis: emerging roles of intracellular NHX Na^+/H^+ antiporters in plant growth and development. *J. Exp. Bot.* 63, 5727–5740. doi: 10.1093/jxb/ers250

Bazzanella, A., Lochmann, H., Tomos, A. D., and Bächmann, K. (1998). Determination of inorganic cations and anions in single plant cells by capillary zone electrophoresis. *J. Chromatogr.* 809, 231–239. doi: 10.1016/S0021- 9673(98)00144-7

Ben Amar, S., Brini, F., Sentenac, H., Masmoudi, K., and Véry, A. A. (2014). Functional characterization in *Xenopus oocytes* of Na^+ transport systems from durum wheat reveals diversity among two HKT1;4 transporters. *J. Exp. Bot.* 65, 213–222. doi: 10.1093/jxb/ert361

Benito, B., Haro, R., Amtmann, A., Cuin, T. A., and Dreyer, I. (2014). The twins K^+ and Na^+ in plants. *J. Plant Physiol.* 171, 723–731. doi: 10.1016/j.jplph.2013.10.014

Bennett, T. H., Flowers, T. J., and Bromham, L. (2013). Repeated evolution of salt-tolerance in grasses. *Biol. Lett.* 9, 20130029. doi: 10.1098/rsbl.2013.0029

Bental, M., Degani, H., and Avron, M. (1988). 23Na-NMR studies of the intracellular sodium ion concentration in the halotolerant alga *Dunaliella salina. Plant Physiol.* 87, 813–817. doi: 10.1104/pp.87.4.813

Berthomieu, P., Conéjéro, G., Nublat, A., Brackenbury, W. J., Lambert, C., Savio, C., et al. (2003). Functional analysis of AtHKT1 in *Arabidopsis* shows that Na^+ recirculation by the phloem is crucial for salt tolerance. *EMBO J.* 22, 2004–2014. doi: 10.1093/emboj/cdg207

Bezanilla, F. (2008). How membrane proteins sense voltage. *Nat. Rev. Mol. Cell Biol.* 9, 323–332. doi: 10.1038/nrm2376

Bhaskaran, S., and Savithramma, D. L. (2011). Co-expression of *Pennisetum glaucum* vacuolar Na^+/H^+ antiporter and *Arabidopsis* H^+-pyrophosphatase enhances salt tolerance in transgenic tomato. *J. Exp. Bot.* 62, 5561–5570. doi: 10.1093/jxb/err237

Blatt, M. R. (1987). Electrical characteristics of stomatal guard cells: the ionic basis of the membrane potential and the consequence of potassium chlorides leakage from microelectrodes. *Planta* 170, 272–287. doi: 10.1007/BF00397898

Blatt, M. R. (1991). "A primer in plant electrophysiological methods," in *Methods in Plant Biochemistry*, ed. K. Hostettmann (London: Academic Press), 281–321.

Blumwald, E., and Poole, R. J. (1985). Na^+/H^+ antiport in isolated tonoplast vesicles from storage tissue of *Beta vulgaris. Plant Physiol.* 78, 163–167. doi: 10.1104/pp.78.1.163

Brand, L., Hörler, M., Nüesch, E., Vassalli, S., Barrell, P., Yang, W., et al. (2006). A versatile and reliable two-component system for tissue-specific gene induction in *Arabidopsis. Plant Physiol.* 141, 1194–1204. doi: 10.1104/pp.106.081299

Breckle, S.-W. (2002). "Salinity, halophytes and salt affected natural ecosystems," in *Salinity: Environment-Plants-Molecules*, eds R. Läuchli and U. Lüttge (The Netherlands: Kluwer Academic Publishers), 53–77.

Breeze, E., Harrison, E., McHattie, S., Hughes, L., Hickman, R., Hill, C., et al. (2011). High-resolution temporal profiling of transcripts during *Arabidopsis* leaf senescence reveals a distinct chronology of processes and regulation. *Plant Cell* 23, 873–894. doi: 10.1105/tpc.111.083345

Bressan, R. A., Zhang, C., Zhang, H., Hasegawa, P. M., Bohnert, H. J., and Zhu, J. K. (2001). Learning from the *Arabidopsis* experience. The next gene search paradigm. *Plant Physiol.* 127, 1354–1360. doi: 10.1104/pp.010752

Britto, D. T., and Kronzucker, H. J. (2008). Cellular mechanisms of potassium transport in plants. *Physiol. Plant.* 133, 637–650. doi: 10.1111/j.1399-3054.2008.01067.x

Britto, D. T., and Kronzucker, H. J. (2013). Flux measurements of cations using radioactive tracers. *Methods Mol. Biol.* 953, 161–170. doi: 10.1007/978-1-62703-152-3_10

Britto, D. T., Szczerba, M. W., and Kronzucker, H. J. (2006). A new, non- perturbing, sampling procedure in tracer exchange measurements. *J. Exp. Bot.* 57, 1309–1314. doi: 10.1093/jxb/erj105

Cao, G., Platisa, J., Pieribone, V. A., Raccuglia, D., Kunst, M., and Nitabach, M. N. (2013). Genetically targeted optical electrophysiology in intact neural circuits. *Cell* 154, 904–913. doi: 10.1016/j.cell.2013.07.027

Cao, Y., Jin, X., Huang, H., Derebe, M. G., Levin, E. J., Kabaleeswaran, V., et al. (2011). Crystal structure of a potassium ion transporter, TrkH. *Nature* 471, 336–340. doi: 10.1038/nature09731

Carden, D. E., Diamond, D., and Miller, A. J. (2001). An improved Na^+ -selective microelectrode for intracellular measurements in plant cells. *J. Exp. Bot.* 52, 1353–1359. doi: 10.1093/jexbot/52.359.1353

Carden, D. E, Walker, D. J., Flowers, T. J., and Miller, A. J. (2003). Single-cell measurements of the contributions of cytosolic Na^+ and K^+ to salt tolerance. *Plant Physiol.* 131, 676–683. doi: 10.1104/pp.011445

Carpaneto, A., Koepsell, H., Bamberg, E., Hedrich, R., and Geiger, D. (2010). Sucrose- and H^+-dependent charge movements associated with the gating of sucrose transporter ZmSUT1. *PLoS ONE* 5:e12605. doi: 10.1371/journal.pone.0012605

Chakrapani, S., and Auerbach, A. (2005). A speed limit for conformational change of an allosteric membrane protein. *Proc. Natl. Acad. Sci. U.S.A.* 102, 87–92. doi: 10.1073/pnas.0406777102

Cheeseman, J. M. (1986). Compartmental efflux analysis: an evaluation of the technique and its limitations. *Plant Physiol.* 80, 1006–1011. doi: 10.1104/pp.80.4.1006

Chen, J., Xiong, D. Y., Wang, W. H., Hu, W. J., Simon, M., Xiao, Q., et al. (2013). Nitric oxide mediates root K^+/Na^+ balance in a mangrove plant. *PLoS ONE* 8:e71543. doi: 10.1371/journal.pone.0071543

Chen, Z., Newman, I., Zhou, M., Mendham, N., Zhang, G., and Shabala, S. (2005). Screening plants for salt tolerance by measuring K^+ flux: a case study for barley. *Plant Cell Environ.* 28, 1230–1246. doi: 10.1111/j.1365-3040.2005.01364.x

Collins, K. D. (1997). Charge density-dependent strength of hydration and biological structure. *Biophys. J.* 72, 65–76. doi: 10.1016/S0006-3495(97)78647-8

Colmenero-Flores, J. M., Martínez, G., Gamba, G., Vázquez, N., Iglesias, D. J., Brumós, J., et al. (2007). Identification and functional characterization of cation-chloride cotransporters in plants. *Plant J.* 50, 278–292. doi: 10.1111/j.1365-313X.2007.03048.x

Coskun, D., and Kronzucker, H. J. (2013). Complexity of potassium acquisition: how much flows through channels? *Plant Signal Behav.* 8:e24799. doi: 10.4161/psb.24799

Cotsaftis, O., Plett, D., Shirley, N., Tester, M., and Hrmova, M. (2012). A two-staged model of Na^+ exclusion in rice explained by 3D modeling of HKT transporters and alternative splicing. *PLoS ONE* 7:e39865. doi: 10.1371/journal.pone.0039865

Cuin, T. A., Betts, S. A., Chalmandrier, R., and Shabala, S. (2008). A root's ability to retain K^+ correlates with salt tolerance in wheat. *J. Exp. Bot.* 59, 2697–2706. doi: 10.1093/jxb/ern128

Cuin, T. A., Miller, A. J., Laurie, S. A., and Leigh, R. A. (2003). Potassium activities in cell compartments of salt-grown barley leaves. *J. Exp. Bot.* 54, 657–661. doi: 10.1093/jxb/erg072

Cuin, T. A., and Shabala, S. (2007a). Amino acids regulate salinity-induced potassium efflux in barley root epidermis. *Planta* 225, 753–761. doi: 10.1007/s00425-006-0386-x

Cuin, T. A., and Shabala, S. (2007b). Compatible solutes reduce ROS-induced potassium efflux in *Arabidopsis* roots. *Plant Cell Environ.* 30, 875–885. doi: 10.1111/j.1365-3040.2007.01674.x

Cuin, T. A., Zhou, M., Parsons, D., and Shabala, S. (2012). Genetic behaviour of physiological traits conferring cytosolic K^+/Na^+ homeostasis in wheat. *Plant Biol.* 14, 438–446. doi: 10.1111/j.1438-8677.2011.00526.x

Darley, C. P., van Wuytswinkel, O. C., van der Woude, K., Mager, W. H., and de Boer, A. H. (2000). *Arabidopsis thaliana* and *Saccharomyces cerevisiae* NHX1 genes encode amiloride sensitive electroneutral Na^+/H^+ exchangers. *Biochem. J.* 351, 241–249. doi: 10.1042/0264-6021:3510241

Davenport, R. J., Muñoz-Mayor, A., Jha, D., Essah, P. A., Rus, A., and Tester, M. (2007). The Na^+ transporter AtHKT1;1 controls retrieval of Na^+ from the xylem in *Arabidopsis. Plant Cell Environ.* 30, 497–507. doi: 10.1111/j.1365-3040.2007.01637.x

de Almeida, P. M. F. (2014). *The Role of HKT Transporters in Salinity Tolerance of*

Tomato. Ph.D. thesis, Vrije Universiteit, Amsterdam, 219.

de Boer, A. H. (1985). *Xylem/Symplast Ion Exchange: Mechanism and Function in Salt Tolerance and Growth*. Ph.D. thesis, Rijksuniversiteit Groningen, Groningen.

de Boer, A. H., and Volkov, V. (2003). Logistics of water and salt transport through the plant: structure and functioning of the xylem. *Plant Cell Environ.* 26, 87–101. doi: 10.1046/j.1365-3040.2003.00930.x

Demidchik, V., Bowen, H. C., Maathuis, F. J., Shabala, S. N., Tester, M. A., White, P. J., et al. (2002a). *Arabidopsis thaliana* root non-selective cation channels mediate calcium uptake and are involved in growth. *Plant J.* 32, 799–808. doi: 10.1046/j.1365-313X.2002.01467.x

Demidchik, V., Davenport, R. J., and Tester, M. (2002b). Nonselective cation channels in plants. *Annu. Rev. Plant Biol.* 53, 67–107. doi: 10.1146/annurev.arplant.53.091901.161540

Demidchik, V., Cuin, T. A., Svistunenko, D., Smith, S. J., Miller, A. J., Shabala, S., et al. (2010). *Arabidopsis* root K^+-efflux conductance activated by hydroxyl radicals: single-channel properties, genetic basis and involvement in stress-induced cell death. *J. Cell Sci.* 123, 1468–1479. doi: 10.1242/jcs.064352

Demidchik, V., and Maathuis, F. J. (2007). Physiological roles of nonselective cation channels in plants: from salt stress to signalling and development. *New Phytol.* 175, 387–404. doi: 10.1111/j.1469-8137.2007.02128.x

Demidchik, V., Shang, Z., Shin, R., Colaço, R., Laohavisit, A., Shabala, S., et al. (2011). Receptor-like activity evoked by extracellular ADP in *Arabidopsis* root epidermal plasma membrane. *Plant Physiol.* 156, 1375–1385. doi: 10.1104/pp.111.174722

Demidchik, V., and Tester, M. (2002). Sodium fluxes through nonselective cation channels in the plasma membrane of protoplasts from *Arabidopsis* roots. *Plant Physiol.* 128, 379–387. doi: 10.1104/pp.010524

Dietz, K. J., Schramm, M., Betz, M., Busch, H., Dürr, C., and Martinoia, E. (1992). Characterization of the epidermis from barley primary leaves: I. Isolation of epidermal protoplasts. *Planta* 187, 425–430. doi: 10.1007/BF00199959

Dolan, L., Janmaat, K., Willemsen, V., Linstead, P., Poethig, S., Roberts, K., et al. (1993). Cellular organisation of the *Arabidopsis thaliana* root. *Development* 119, 71–84.

D'Onofrio, C., Kader, A., and Lindberg, S. (2005). Uptake of sodium in quince, sugar beet, and wheat protoplasts determined by the fluorescent sodium- binding dye benzofuran isophthalate. *J. Plant Physiol.* 162, 421–428. doi: 10.1016/j.jplph.2004.07.017

Doyle, D. A., Cabral, M. J., Pfuetzner, R. A., Kuo, A., Gulbis, J. M., Cohen, S. L., et al. (1998). The structure of the potassium channel: molecular basis of K^+ conduction and selectivity. *Science* 280, 69–77. doi: 10.1126/science.280.5360.69

Dreyer, I., and Uozumi, N. (2011). Potassium channels in plant cells. *FEBS J.* 278, 4293–4303. doi: 10.1111/j.1742-4658.2011.08371.x

D'yakova, E. V., Rakitin, A. L., Kamionskaya, A. M., Baikov, A. A., Lahti, R., Ravin, N. V., et al. (2006). A study of the effect of expression of the gene encoding the membrane H^+-pyrophosphatase of *Rhodospirillum rubrum* on salt resistance of transgenic tobacco plants. *Doklady Biol. Sci.* 409, 346–348. doi: 10.1134/S0012496606040247

English, J. P., and Colmer, T. D. (2013). Tolerance of extreme salinity in two stem-succulent halophytes (*Tecticornia* species). *Funct. Plant Biol.* 40, 897–912.

Epstein, E. (1966). Dual pattern of ion absorption by plant cells and by plants. *Nature* 212, 1324–1327. doi: 10.1038/2121324a0

Erdei, L., and Kuiper, P. J. C. (1979). The effect of salinity on growth. Cation Content, Na^+ -uptake and translocation in salt-sensitive and salt- tolerant *Plantago* Species. *Physiol. Plant.* 47, 95–99. doi: 10.1111/j.1399-3054.1979.tb03197.x

Evans, A., Hall, D., Pritchard, J., and Newbury, H. J. (2012). The roles of the cation transporters CHX21 and CHX23 in the development of *Arabidopsis thaliana*. *J. Exp. Bot.* 63, 59–67. doi: 10.1093/jxb/err271

Feki, K., Quintero, F. J., Khoudi, H., Leidi, E. O., Masmoudi, K., Pardo, J. M., et al. (2014). A constitutively active form of a durum wheat Na^+ /H^+ antiporter SOS1 confers high salt tolerance to transgenic *Arabidopsis*. *Plant Cell Rep.* 33, 277–288. doi: 10.1007/s00299-013-1528-9

Flowers, T. J. (1998). Salinisation and horticultural production. *Sci. Hortic.* 78, 1–4.

Flowers, T. J. (2004). Improving crop salt tolerance. *J. Exp. Bot.* 55, 307–319. doi: 10.1093/jxb/erh003

Flowers, T. J., and Colmer, T. D. (2008). Salinity tolerance in halophytes. *New Phytol.* 179, 945–963. doi: 10.1111/j.1469-8137.2008.02531.x

Flowers, T. J., and Dalmond, D. (1992). Protein synthesis in halophytes: the influence of potassium, sodium and magnesium in vitro. *Plant Soil* 146, 153– 161. doi: 10.1007/BF00012008

Flowers, T. J., Galal, H. K., and Bromham, L. (2010). Evolution of halophytes: multiple origins of salt tolerance in land plants. *Funct. Plant Biol.* 37, 604–612. doi: 10.1071/FP09269

Flowers, T. J., Troke, P. F., and Yeo, A. R. (1977). The mechanism of salt tolerance in halophytes. *Annu. Rev. Plant Physiol.* 28, 89–121. doi: 10.1146/annurev.pp.28.060177.000513

Fricke, W., Akhiyarova, G., Wei, W., Alexandersson, E., Miller, A., Kjellbom, P. O., et al. (2006). The short-term growth response to salt of the developing barley leaf. *J. Exp. Bot.* 57, 1079–1095. doi: 10.1093/jxb/erj095

Fricke, W., Leigh, R. A., and Tomos, A. D. (1994). Epidermal solute concentrations and osmolality in barley leaves studied at the single-cell level. *Planta* 192, 317–323. doi: 10.1007/BF00198566

Gadsby, D. C. (2009). Ion channels versus ion pumps: the principal difference, in principle. *Nat. Rev. Mol. Cell Biol.* 10, 344–352. doi: 10.1038/nrm2668

Galdiero, S., Falanga, A., Cantisani, M., Tarallo, R., Della Pepa, M. E., D'Oriano, V., et al. (2012). Microbe-host interactions: structure and role of Gram-negative bacterial porins. *Curr. Protein Pept. Sci.* 13, 843–854. doi: 10.2174/1389203711213080012

Gao, F., Gao, Q., Duan, X., Yue, G., Yang, A., and Zhang, J. (2006). Cloning of an H^+-PPase gene from *Thellungiella halophila* and its heterologous expression to improve tobacco salt tolerance. *J. Exp. Bot.* 57, 3259–3270. doi: 10.1093/jxb/erl090

Garbarino, J., and DuPont, F. M. (1988). NaCl Induces a Na^+ /H^+ Antiport in tonoplast vesicles from barley roots. *Plant Physiol.* 86, 231–236. doi: 10.1104/pp.86.1.231

Garnett, T. P., Shabala, S. N., Smethurst, P. J., and Newman, I. A. (2001). Simultaneous measurements of ammonium, nitrate and proton fluxes along the length of eucalypt roots. *Plant Soil* 236, 55–62. doi: 10.1023/A:1011951413917

Gaxiola, R. A., Li, J., Undurraga, S., Dang, L. M., Allen, G. J., Alper, S. L., et al. (2001). Drought- and salt-tolerant plants result from overexpression of the AVP1 H^+-pump. *Proc. Natl. Acad. Sci. U.S.A.* 98, 11444–11449. doi: 10.1073/pnas.191389398

Gaxiola, R. A., Rao, R., Sherman, A., Grisafi, P., Alper, S. L., and Fink, G. R. (1999). The *Arabidopsis thaliana* proton transporters. AtNhx1 and Avp1, can function in cation detoxification in yeast. *Proc. Natl. Acad. Sci. U.S.A.* 96, 1480–1485. doi: 10.1073/pnas.96.4.1480

Gilliham, M., Sullivan, W., Tester, M., and Tyerman, S. D. (2006). Simultaneous flux and current measurement from single plant protoplasts reveals a strong link between K^+ fluxes and current, but no link between Ca2 $^+$ fluxes and current. *Plant J.* 46, 134–144. doi: 10.1111/j.1365-313X.2006.02676.x

Gimmler, H. (2000). Primary sodium plasma membrane ATPases in salt- tolerant algae: facts and fictions. *J. Exp. Bot.* 51, 1171–1178. doi: 10.1093/jexbot/51.348.1171

Grabov, A. (2007). Plant KT/KUP/HAK potassium transporters: single family - multiple functions. *Ann. Bot.* 99, 1035–1041. doi: 10.1093/aob/mcm066

Greiner, T., Ramos, J., Alvarez, M. C., Gurnon, J. R., Kang, M., Van Etten, J. L., et al. (2011). Functional HAK/KUP/KT-like potassium transporter encoded by chlorella viruses. *Plant J.* 68, 977–986. doi: 10.1111/j.1365-313X.2011.04748.x

Grosse, W., Essen, L. O., and Koert, U. (2011). Strategies and perspectives in ion-channel engineering. *Chembiochem* 12, 830–839. doi: 10.1002/cbic.201000793

Guo, K. M., Babourina, O., and Rengel, Z. (2009). Na^+ /H^+ antiporter activity of the SOS1 gene: lifetime imaging analysis and electrophysiological studies on *Arabidopsis* seedlings. *Physiol. Plant.* 137, 155–165. doi: 10.1111/j.1399-3054.2009.01274.x

Gushchin, I., Shevchenko, V., Polovinkin, V., Kovalev, K., Alekseev, A., Round, E., et al. (2015). Crystal structure of a light-driven sodium pump. *Nat. Struct. Mol. Biol.* 22, 390–395. doi: 10.1038/nsmb.3002

Hajibagheri, M. A., and Flowers, T. J. (1989). X-ray microanalysis of ion distribution within root cortical cells of the halophyte *Suaeda maritima* (L.) Dum. *Planta* 177, 131–134. doi: 10.1007/BF00392163

Hajibagheri, M. A., Flowers, T. J., Collins, J. C., and Yeo, A. R. (1988). A comparison of the methods of X-ray microanalysis. *J. Exp. Bot.* 39, 279–290. doi: 10.1093/jxb/39.3.279

Hall, D., Evans, A. R., Newbury, H. J., and Pritchard, J. (2006). Functional analysis of CHX21: a putative sodium transporter in *Arabidopsis*. *J. Exp. Bot.* 57, 1201–1210. doi: 10.1093/jxb/erj092

Hall, J. L., and Flowers, T. J. (1973). The effect of salt on protein synthesis in the halophyte *Suaeda maritima*. *Planta* 110, 361–368. doi: 10.1007/BF00387064

Halperin, S. J., and Lynch, J. P. (2003). Effects of salinity on cytosolic Na^+ and K^+ in root hairs of *Arabidopsis thaliana*: in vivo measurements using the fluorescent dyes SBFI and PBFI. *J. Exp. Bot.* 54, 2035–2043. doi: 10.1093/jxb/erg219

Hammou, K. A., Rubio, L., Fernández, J. A., and García-Sánchez, M. J. (2014). Potassium uptake in the halophyte *Halimione portulacoides* L. Aellen *Environ. Exp. Bot.* 107, 15–24. doi: 10.1016/j.envexpbot.2014.05.001

Hansma, P. K., Drake, B., Marti, O., Gould, S. A., and Prater, C. B. (1989). The scanning ion-conductance microscope. *Science* 243, 641–643. doi: 10.1126/science.2464851

He, C., Yan, J., Shen, G., Fu, L., Holaday, A. S., Auld, D., et al. (2005). Expression of an Arabidopsis vacuolar sodium/proton antiporter gene in cotton improves photosynthetic performance under salt conditions and increases fiber yield in the field. *Plant Cell Physiol.* 46, 1848–1854. doi: 10.1093/pcp/pci201

Hedrich, R. (2012). Ion Channels in Plants. *Physiol. Rev.* 92, 1777–1811. doi: 10.1152/physrev.00038.2011

Heginbotham, L., Abramson, T., and MacKinnon, R. (1992). A functional connection between the pores of distantly related ion channels as revealed by mutant K^+ channels. *Science* 258, 1152–1155. doi: 10.1126/science.1279807

Higinbotham, N. (1973). Electropotentials of Plant Cells. *Annu. Rev. Plant Physiol.* 24, 25–46. doi: 10.1146/annurev.pp.24.060173.000325

Hille, B. (2001). *Ion Channels and Excitable Membranes*, 3rd Edn. (Sunderland, MA: Sinauer Associates), 814.

Hirsch, R. E., Lewis, B. D., Spalding, E. P., and Sussman, M. R. (1998). A role for the AKT1 potassium channel in plant nutrition. *Science* 280, 918–921. doi: 10.1126/science.280.5365.918

Hofmeister, F. (1888). Zur Lehre von der Wirkung der Salze. *Archiv Exp. Pathol. Pharmakol.* 24, 247–260. doi: 10.1007/BF01918191

Horie, T., Hauser, F., and Schroeder, J. I. (2009). HKT transporter-mediated salinity resistance mechanisms in *Arabidopsis* and monocot crop plants. *Trends Plant Sci.* 14, 660–668. doi: 10.1016/j.tplants.2009.08.009

Hughes, F. M. Jr., and Cidlowski, J. A. (1999). Potassium is a critical regulator of apoptotic enzymes in vitro and in vivo. *Adv. Enzyme Regul.* 39, 157–171. doi: 10.1016/S0065-2571(98)00010-7

Huh, G. H., Damsz, B., Matsumoto, T. K., Reddy, M. P., Rus, A. M., Ibeas, J. I., et al. (2002). Salt causes ion disequilibrium-induced programmed cell death in yeast and plants. *Plant J.* 29, 649–659. doi: 10.1046/j.0960-7412.2001.01247.x

Hunte, C., Screpanti, E., Venturi, M., Rimon, A., Padan, E., and Michel, H. (2005). Structure of a Na^+/H^+ antiporter and insights into mechanism of action and regulation by pH. *Nature* 435, 1197–1202. doi: 10.1038/nature03692

Inan, G., Zhang, Q., Li, P., Wang, Z., Cao, Z., Zhang, H., et al. (2004). Salt cress. A halophyte and cryophyte *Arabidopsis* relative model system and its applicability to molecular genetic analyses of growth and development of extremophiles. *Plant Physiol.* 135, 1718–1737. doi: 10.1104/pp.104.041723

Isayenkov, S. V. (2012). Physiological and molecular aspects of salt stress in plants. *Cytol. Genet.* 46, 302–318. doi: 10.3103/S0095452712050040

Ivanov, V. B. (1997). Relationship between cell proliferation and transition to elongation in plant roots. *Int. J. Dev. Biol.* 41, 907–915.

Ivanov, V. B., and Dubrovsky, J. G. (2013). Longitudinal zonation pattern in plant roots: conflicts and solutions. *Trends Plant Sci.* 18, 237–243. doi: 10.1016/j.tplants.2012.10.002

Ivashikina, N., Becker, D., Ache, P., Meyerhoff, O., Felle, H. H., and Hedrich, R. (2001). K^+ channel profile and electrical properties of *Arabidopsis* root hairs. *FEBS Lett.* 508, 463–469. doi: 10.1016/S0014-5793(01)03114-3

Jabnoune, M., Espeout, S., Mieulet, D., Fizames, C., Verdeil, J. L., Conéjéro, G., et al. (2009). Diversity in expression patterns and functional properties in the rice HKT transporter family. *Plant Physiol.* 150, 1955–1971. doi: 10.1104/pp.109.138008

Jacobs, A., Ford, K., Kretschmer, J., and Tester, M. (2011). Rice plants expressing the moss sodium pumping ATPase PpENA1 maintain greater biomass production under salt stress. *Plant Biotechnol. J.* 9, 838–847. doi: 10.1111/j.1467-7652.2011.00594.x

James, R. A., Blake, C., Byrt, C. S., and Munns, R. (2011). Major genes for Na^+ exclusion. Nax1 and Nax2 (wheat HKT1;4 and HKT1;5), decrease Na^+ accumulation in bread wheat leaves under saline and waterlogged conditions. *J. Exp. Bot.* 62, 2939–2947. doi: 10.1093/jxb/err003

Jiang, X., Leidi, E. O., and Pardo, J. M. (2010). How do vacuolar NHX exchangers function in plant salt tolerance? *Plant Signal Behav.* 5, 792–795. doi: 10.4161/psb.5.7.11767

Jiang, Y., Lee, A., Chen, J., Cadene, M., Chait, B. T., and MacKinnon, R. (2002). The open pore conformation of potassium channels. *Nature* 417, 523–526. doi: 10.1038/417523a

Joh, N. H., Wang, T., Bhate, M. P., Acharya, R., Wu, Y., Grabe, M., et al. (2014). De novo design of a transmembrane Zn2 ⁺-transporting four-helix bundle. *Science* 6216, 1520–1524. doi: 10.1126/science.1261172

Kader, M. A., and Lindberg, S. (2005). Uptake of sodium in protoplasts of salt- sensitive and salt-tolerant cultivars of rice. *J. Exp. Bot.* 56, 3149–3158. doi: 10.1093/jxb/eri312

Karley, A. J., Leigh, R. A., and Sanders, D. (2000). Differential ion accumulation and ion fluxes in the mesophyll and epidermis of barley. *Plant Physiol.* 122, 835–844. doi: 10.1104/pp.122.3.835

Kato, H. E., Inoue, K., Abe-Yoshizumi, R., Kato, Y., Ono, H., Konno, M., et al. (2015). Structural basis for Na^+ transport mechanism by a light-driven Na^+ pump. *Nature* 521, 48–53. doi: 10.1038/nature14322

Kato, N., Akai, M., Zulkifli, L., Matsuda, N., Kato, Y., Goshima, S., et al. (2007). Role of positively charged amino acids in the M2D transmembrane helix of Ktr/Trk/HKT type cation transporters. *Channels (Austin)* 1, 161–171. doi: 10.4161/chan.4374

Katsuhara, M., and Kawasaki, T. (1996). Salt stress induced nuclear and DNA degradation in meristematic cells of barley roots. *Plant Cell Physiol.* 37, 169–173. doi: 10.1093/oxfordjournals.pcp.a028928

Katz, A., and Avron, M. (1985). Determination of intracellular osmotic volume and sodium concentration in dunaliella. *Plant Physiol.* 78, 817–820. doi: 10.1104/pp.78.4.817

Kellermeier, F., Armengaud, P., Seditas, T. J., Danku, J., Salt, D. E., and Amtmann, A. (2014). Analysis of the root system architecture of arabidopsis provides a quantitative readout of crosstalk between nutritional signals. *Plant Cell* 26, 1480–1496. doi: 10.1105/tpc.113.122101

Knowles, A., and Shabala, S. (2004). Overcoming the problem of non-ideal liquid ion exchanger selectivity in microelectrode ion flux measurements. *J. Membr. Biol.* 202, 51–59. doi: 10.1007/s00232-004-0719-2

Korchev, Y. E., Bashford, C. L., Milovanovic, M., Vodyanoy, I., and Lab, M. J. (1997). Scanning ion conductance microscopy of living cells. *Biophys. J.* 73, 653–658. doi: 10.1016/S0006-3495(97)78100-1

Korolev, A. V., Tomos, A. D., Bowtell, R., and Farrar, J. F. (2000). Spatial and temporal distribution of solutes in the developing carrot taproot measured at single-cell resolution. *J. Exp. Bot.* 51, 567–577. doi: 10.1093/jexbot/51.344.56

Kronzucker, H. J., and Britto, D. T. (2011). Sodium transport in plants: a critical review. *New Phytol.* 189, 54–81. doi: 10.1111/j.1469-8137.2010.03540.x

Kumar, T., Uzma, A. S., Khan, M. R., Abbas, Z., and Ali, G. M. (2014). Genetic Improvement of sugarcane for drought and salinity stress tolerance using arabidopsis vacuolar pyrophosphatase (AVP1) gene. *Mol. Biotechnol.* 56, 199– 209. doi: 10.1007/s12033-013-9695-z

Kumar, V., and Jain, M. (2015). The CRISPR-Cas system for plant genome editing: advances and opportunities. *J. Exp. Bot.* 66, 47–57. doi: 10.1093/jxb/eru429

Kunz, W., Henle, J., and Ninham, B. W. (2004). Zur Lehre von der Wirkung der Salze' (about the science of the effect of salts): franz Hofmeister's historical papers. *Curr. Opin. Colloid Interface Sci.* 9, 19–37. doi: 10.1016/j.cocis.2004.05.005

Kurup, S., Runions, J., Köhler, U., Laplaze, L., Hodge, S., and Haseloff, J. (2005). Marking cell lineages in living tissues. *Plant J.* 42, 444–453. doi: 10.1111/j.1365-313X.2005.02386.x

Lab, M. J., Bhargava, A., Wright, P. T., and Gorelik, J. (2013). The scanning ion conductance microscope for cellular physiology. *Am. J. Physiol. Heart Circ. Physiol.* 304, H1–H11. doi: 10.1152/ajpheart.00499.2012

Lan, W. Z., Wang, W., Wang, S. M., Li, L. G., Buchanan, B. B., Lin, H. X., et al. (2010). A rice high-affinity potassium transporter (HKT) conceals a calcium- permeable cation channel. *Proc. Natl. Acad. Sci. U.S.A.* 107, 7089–7094. doi: 10.1073/pnas.1000698107

Leidi, E. O., Barragán, V., Rubio, L., El-Hamdaoui, A., Ruiz, M. T., Cubero, B., et al. (2010). The AtNHX1 exchanger mediates potassium compartmentation in vacuoles of transgenic tomato. *Plant J.* 61, 495–506. doi: 10.1111/j.1365-313X.2009.04073.x

Lew, R. R., Levina, N. N., Shabala, L., Anderca, M. I., and Shabala, S. N. (2006). Role of a mitogen-activated protein kinase cascade in ion flux-mediated turgor regulation in fungi. *Eukaryot. Cell* 5, 480–487. doi: 10.1128/EC.5.3.480- 487.2006

Li, X., Guo, C., Gu, J., Duan, W., Zhao, M., Ma, C., et al. (2014). Overexpression of VP, a vacuolar H^+-pyrophosphatase gene in wheat (*Triticum aestivum L.),* improves tobacco plant growth under Pi and N deprivation, high salinity, and drought. *J. Exp. Bot.* 65, 683–696. doi: 10.1093/jxb/ert442

Lindberg, S. (1995). In-situ determination of intracellular concentrations of K^+ in barley (*Hordeum vulgare* L. cv. Kara) using the K^+-binding fluorescent dye benzofuran isophthalate *Planta* 195, 525–529. doi: 10.1007/BF00195710

Liu, W., Fairbairn, D. J., Reid, R. J., and Schachtman, D. P. (2001). Characterization of two HKT1 homologues from *Eucalyptus camaldulensis* that display intrinsic osmosensing capability. *Plant Physiol.* 127, 283–294.

Liu, Y., Ji, X., Zheng, L., Nie, X., and Wang, Y. (2013). Microarray analysis of transcriptional responses to abscisic acid and salt stress in *Arabidopsis thaliana. Int. J. Mol. Sci.* 14, 9979–9998. doi: 10.3390/ijms14059979

Longpré, J. P., and Lapointe, J. Y. (2011). Determination of the Na^+/glucose cotransporter (SGLT1) turnover rate using the ion-trap technique. *Biophys. J.* 100, 52–59. doi: 10.1016/j.bpj.2010.11.012

L'Roy, A., and Hendrix, D. L. (1980). Effect of salinity upon cell membrane potential in the marine halophyte, *Salicornia bigelovii* Torr. *Plant Physiol.* 65, 544–549. doi: 10.1104/pp.65.3.544

Lüttge, U. (2002). "Mangroves," in *Salinity: Environment-Plants-Molecules*, eds R. Läuchli and U. Lüttge (The Netherlands: Kluwer Academic Publishers), 113–135.

Ma, D. M., Xu, W. R., Li, H. W., Jin, F. X., Guo, L. N., Wang, J., et al. (2014). Co-expression of the *Arabidopsis* SOS genes enhances salt tolerance in transgenic tall fescue (*Festuca arundinacea* Schreb.). *Protoplasma* 251, 219–231. doi: 10.1007/s00709-013-0540-9

Maathuis, F. J., and Prins, H. B. (1990). Patch clamp studies on root cell vacuoles of a salt-tolerant and a salt-sensitive *Plantago* species. *Plant Physiol.* 92, 23–28. doi: 10.1104/pp.92.1.23

Maathuis, F. J. M., Ahmad, I., and Patishtan, J. (2014). Regulation of Na^+ fluxes in plants. *Front. Plant Sci.* 5:467. doi: 10.3389/fpls.2014.00467

Maathuis, F. J. M., and Amtmann, A. (1999). K^+ nutrition and Na^+ toxicity: the basis of cellular K^+/Na^+ ratios. *Ann. Bot.* 84, 123–133. doi: 10.1006/anbo.1999.0912

Maathuis, F. J. M., Flowers, T. J., and Yeo, A. R. (1992). Sodium-chloride compartmentation in leaf vacuoles of the halophyte *Suaeda maritima* (L) Dum. and its relation to tonoplast permeability. *J. Exp. Bot.* 43, 1219–1223. doi: 10.1093/jxb/43.9.1219

MacKinnon, R. (2004). Potassium channels and the atomic basis of selective ion conduction (Nobel Lecture). *Angew. Chem. Int. Ed. Engl.* 43, 4265–4277. doi: 10.1002/anie.200400662

MacRobbie, E. A., and Dainty, J. (1958). Ion transport in Nitellopsis obtusa. *J. Gen. Physiol.* 42, 335–353. doi: 10.1085/jgp.42.2.335

Maffeo, C., Bhattacharya, S., Yoo, J., Wells, D., and Aksimentiev, A. (2012). Modeling and simulation of ion channels. *Chem. Rev.* 112, 6250–6284. doi: 10.1021/cr3002609

Mähler, J., and Persson, I. (2012). A study of the hydration of the alkali metal ions in aqueous solution. *Inorg. Chem.* 51, 425–438. doi: 10.1021/ic2018693

Malone, M., Leigh, R. A., and Tomos, A. D. (1991). Concentrations of vacuolar inorganic ions in individual cells of intact wheat leaf epidermis. *J. Exp. Bot.* 42, 305–309. doi: 10.1093/jxb/42.3.305

Martinoia, E., Meyer, S., De Angeli, A., and Nagy, R. (2012). Vacuolar transporters in their physiological context. *Annu. Rev. Plant Biol.* 63, 183–213. doi: 10.1146/annurev-arplant-042811-105608

Mäser, P., Eckelman, B., Vaidyanathan, R., Horie, T., Fairbairn, D. J., Kubo, M., et al. (2002). Altered shoot/root Na^+ distribution and bifurcating salt sensitivity in *Arabidopsis* by genetic disruption of the Na^+ transporter AtHKT1. *FEBS Lett.* 531, 157–161. doi: 10.1016/S0014-5793(02)03488-9

Mäser, P., Thomine, S., Schroeder, J. I., Ward, J. M., Hirschi, K., Sze, H., et al. (2001). Phylogenetic relationships within cation transporter families of *Arabidopsis*. *Plant Physiol.* 126, 1646–1667. doi: 10.1104/pp.126.4.1646

Masi, A., and Melis, A. (1997). Morphological and molecular changes in the unicellular green alga *Dunaliella salina* grown under supplemental UV-B radiation: cell characteristics and Photosystem II damage and repair properties. *BBA Bioenerget.* 1321, 183–193. doi: 10.1016/S0005-2728(97)00054-6

Matzke, A. J., and Matzke, M. (2013). Membrane "potential-omics": toward voltage imaging at the cell population level in roots of living plants. *Front. Plant Sci.* 4:311. doi: 10.3389/fpls.2013.00311

Mian, A., Oomen, R. J., Isayenkov, S., Sentenac, H., Maathuis, F. J., and Véry, A. A. (2011). Over-expression of an Na^+ -and K^+-permeable HKT transporter in barley improves salt tolerance. *Plant J.* 68, 468–479. doi: 10.1111/j.1365-313X.2011.04701.x

Møller, I. S., Gilliham, M., Jha, D., Mayo, G. M., Roy, S. J., Coates, J. C., et al. (2009). Shoot Na^+ exclusion and increased salinity tolerance engineered by cell type-specific alteration of Na^+ transport in *Arabidopsis*. *Plant Cell* 21, 2163–2178. doi: 10.1105/tpc.108.064568

Munns, R., and Tester, M. (2008). Mechanisms of salinity tolerance. *Annu. Rev. Plant Biol.* 59, 651–681. doi: 10.1146/annurev.arplant.59.032607.092911

Murthy, M., and Tester, M. (2006). Cation currents in protoplasts from the roots of a Na^+ hyperaccumulating mutant of *Capsicum annuum*. *J. Exp. Bot.* 57, 1171–1180. doi: 10.1093/jxb/erj115

Mutoh, H., Akemann, W., and Knöpfel, T. (2012). Genetically engineered fluorescent voltage reporters ACS. *Chem. Neurosci.* 3, 585–592. doi: 10.1021/cn300041b

Myers, B. R., Sigal, Y. M., and Julius, D. (2009). Evolution of thermal response properties in a cold-activated TRP channel. *PLoS ONE* 4:e5741. doi: 10.1371/journal.pone.0005741

Newman, I. A. (2001). Ion transport in roots: measurement of fluxes using ion-selective microelectrodes to characterize transporter function. *Plant Cell Environ.* 24, 1–14. doi: 10.1046/j.1365-3040.2001.00661.x

Newman, I. A., Kochian, L. V., Grusak, M. A., and Lucas, W. J. (1987). Fluxes of H^+ and K^+ in corn roots: characterization and stoichiometries using ion-selective microelectrodes. *Plant Physiol.* 84, 1177–1184. doi: 10.1104/pp.84.4.1177

Nightingale, E. R. Jr. (1959). Phenomenological theory of ion solvation. Effective radii of hydrated ions. *J. Phys. Chem.* 63, 1381–1387. doi: 10.1021/j150579a011

Núñez-Ramírez, R., Sánchez-Barrena, M. J., Villalta, I., Vega, J. F., Pardo, J. M., Quintero, F. J., et al. (2012). Structural insights on the plant salt- overly-sensitive 1 (SOS1) Na^+/H^+ antiporter. *J. Mol. Biol.* 424, 283–294. doi: 10.1016/j.jmb.2012.09.015

Oh, D. H., Gong, Q., Ulanov, A., Zhang, Q., Li, Y., Ma, W., et al. (2007). Sodium stress in the halophyte *Thellungiella halophila*, and transcriptional changes in a thsos1-RNA iterference line. *J. Integr. Plant Biol.* 49, 1484–1496. doi: 10.1111/j.1672-9072.2007.00548.x

Oh, D. H., Leidi, E., Zhang, Q., Hwang, S. M., Li, Y., Quintero, F. J., et al. (2009a). Loss of halophytism by interference with SOS1 expression. *Plant Physiol.* 151, 210–222. doi: 10.1104/pp.109.137802

Oh, D. H., Zahir, A., Yun, D. J., Bressan, R. A., and Bohnert, H. J. (2009b). SOS1 and halophytism. *Plant Signal. Behav.* 4, 1081–1083. doi: 10.4161/psb.4.11.9786

Olías, R., Eljakaoui, Z., Li, J., De Morales, P. A., Marín-Manzano, M. C., Pardo, J. M., et al. (2009). The plasma membrane Na^+/H^+ antiporter SOS1 is essential for salt tolerance in tomato and affects the partitioning of Na^+ between plant organs. *Plant Cell Environ.* 32, 904–916. doi: 10.1111/j.1365-3040.2009.01971.x

Ordoñez, N. M., Marondedze, C., Thomas, L., Pasqualini, S., Shabala, L., Shabala, S., et al. (2014). Cyclic mononucleotides modulate potassium and calcium flux responses to H2O2 in *Arabidopsis* roots. *FEBS Lett.* 588, 1008–1015. doi: 10.1016/j.febslet.2014.01.062

Oszvald, M., Gardonyi, M., Tamas, C., Takacs, I., Jenes, B., Tamas, L., et al. (2008). Development and characterization of a chimaeric tissue-specific promoter in wheat and rice endosperm. *In Vitro Cell. Dev. Biol. Plant* 44, 1–7. doi: 10.1007/s11627-007-9082-1

Pakhomov, A. G., Bowman, A. M., Ibey, B. L., Andre, F. M., Pakhomova, O. N., and Schoenbach, K. H. (2009). Lipid nanopores can form a stable, ion channel-like conduction pathway in cell membrane. *Biochem. Biophys. Res. Commun.* 385, 181–186. doi: 10.1016/j.bbrc.2009.05.035

Palmgren, M. G. (2001). PLANT PLASMA MEMBRANE H^+-ATPases: powerhouses for nutrient uptake. *Annu. Rev. Plant. Physiol. Plant Mol. Biol.* 52, 817–845. doi: 10.1146/annurev.arplant.52.1.817

Pandolfi, C., Pottosin, I., Cuin, T., Mancuso, S., and Shabala, S. (2010). Specificity of polyamine effects on Nacl-induced ion flux kinetics and salt stress amelioration in plants. *Plant Cell Physiol.* 51, 422–434. doi: 10.1093/pcp/pcq007

Pang, J. Y., Newman, I., Mendham, N., Zhou, M., and Shabala, S. (2006). Microelectrode ion and O2 fluxes measurements reveal differential sensitivity of barley root tissues to hypoxia. *Plant Cell Environ.* 29, 1107–1121. doi: 10.1111/j.1365-3040.2005.01486.x

Pantoja, O., Dainty, J., and Blumwald, E. (1989). Ion channels in vacuoles from halophytes and glycophytes. *FEBS Lett.* 255, 92–96. doi: 10.1016/0014-5793(89)81067-1

Pardo, J. M., Cubero, B., Leidi, E. O., and Quintero, F. J. (2006). Alkali cation exchangers: roles in cellular homeostasis and stress tolerance. *J. Exp. Bot.* 57, 1181–1199. doi: 10.1093/jxb/erj114

Pasapula, V., Shen, G., Kuppu, S., Paez-Valencia, J., Mendoza, M., Hou, P., et al. (2011). Expression of an *Arabidopsis* vacuolar H^+-pyrophosphatase gene (AVP1) in cotton improves drought- and salt tolerance and increases fibre yield in the field conditions. *Plant Biotechnol. J.* 9, 88–99. doi: 10.1111/j.1467-7652.2010.00535.x

Peleg, Z., Apse, M. P., and Blumwald, E. (2011). Engineering salinity and water-stress tolerance in crop plants: getting closer to the field. *Adv. Bot. Res.* 57, 405–443. doi: 10.1016/B978-0-12-387692-8.00012-6

Pick, U., Karni, L., and Avron, M. (1986). Determination of ion content and ion fluxes in the halotolerant alga *Dunaliella salina*. *Plant Physiol.* 81, 92–96. doi: 10.1104/pp.81.1.92

Pitman, M. G., and Läuchli, A. (2002). "Global impact of salinity and agricultural ecosystems," in *Salinity: Environment-Plants-Molecules*, eds R. Läuchli and U. Lüttge (The Netherlands: Kluwer Academic Publishers), 3–20.

Pitman, M. G., Läuchli, A., and Stelzer, R. (1981). Ion distribution in roots of barley seedlings measured by electron probe x-ray microanalysis. *Plant Physiol.* 68, 673–679. doi: 10.1104/pp.68.3.673

Plett, D., Safwat, G., Gilliham, M., Møller, I. S., Roy, S., Shirley, N., et al. (2010). Improved salinity tolerance of rice through cell type-specific expression of AtHKT1;1. *PLoS ONE* 5:e12571. doi: 10.1371/journal.pone.0012571

Popova, L. G., Shumkova, G. A., Andreev, I. M., and Balnokin, Y. V. (2005). Functional identification of electrogenic Na^+ -translocating ATPase in the plasma membrane of the halotolerant microalga *Dunaliella maritima*. *FEBS Lett.* 579, 5002–5006. doi: 10.1016/j.febslet.2005.07.087

Premkumar, L., Greenblatt, H. M., Bagashwar, U. K., Savchenko, T., Gokhman, I., Sussman, J. L., et al. (2005). 3D structure of a halotolerant algal carbonic anhydrase predicts halotolerance of a mammalian homolog. *Proc. Natl. Acad. Sci. U.S.A.* 102, 7493–7498. doi: 10.1073/pnas.0502829102

Rains, D. W., and Epstein, E. (1965). Transport of sodium in plant tissue. *Science*

148, 1611. doi: 10.1126/science.148.3677.1611

Ramsey, I. S., Delling, M., and Clapham, D. E. (2006). An introduction to TRP channels. *Annu. Rev. Physiol.* 68, 619–647. doi: 10.1146/annurev. physiol.68.040204.100431

Roberts, S. K., and Tester, M. (1997). A patch clamp study of Na^+ transport in maize roots. *J. Exp. Bot.* 48, 431–440. doi: 10.1093/jxb/48

Rodriguez-Navarro, A. (2000). Potassium transport in fungi and plants. *BBA Rev. Biomembr.* 1469, 1–30. doi: 10.1016/S0304-4157(99)00013-1

Rodríguez-Rosales, M. P., Gálvez, F. J., Huertas, R., Aranda, M. N., Baghour, M., Cagnac, O., et al. (2009). Plant NHX cation/proton antiporters. *Plant Signal. Behav.* 4, 265–276. doi: 10.4161/psb.4.4.7919

Sassi, A., Mieulet, D., Khan, I., Moreau, B., Gaillard, I., Sentenac, H., et al. (2012). The rice monovalent cation transporter OsHKT2;4: revisited ionic selectivity. *Plant Physiol.* 160, 498–510. doi: 10.1104/pp.112.194936

Schulze, L. M., Britto, D. T., Li, M., and Kronzucker, H. J. (2012). A pharmacological analysis of high-affinity sodium transport in barley (*Hordeum vulgare* L.): a $24Na^+/42K^+$ study. *J. Exp. Bot.* 63, 2479–2489. doi: 10.1093/jxb/err419

Serrano, R. (1988). Structure and function of proton translocating ATPase in plasma membranes of plants and fungi. *Biochim. Biophys. Acta* 947, 1–28. doi: 10.1016/0304-4157(88)90017-2

Serrano, R. (1996). Salt tolerance in plants and microorganisms: toxicity targets and defense mechanisms. *Int. Rev. Cytol.* 165, 1–52. doi: 10.1016/S0074-7696(08)62219-6

Shabala, L., Cuin, T. A., Newman, I. A., and Shabala, S. (2005). Salinity-induced ion flux patterns from the excised roots of *Arabidopsis* sos mutants. *Planta* 222, 1041–1050. doi: 10.1007/s00425-005-0074-2

Shabala, S. (2006). "Non-invasive microelectrode ion flux measurements in plant stress physiology," in *Plant Electrophysiology—Theory and Methods*, ed. A. G. Volkov (Berlin: Springer-Verlag), 35–71.

Shabala, S. (2009). Salinity and programmed cell death: unravelling mechanisms for ion specific signalling. *J. Exp. Bot.* 60, 709–712. doi: 10.1093/jxb/erp013

Shabala, S. (2013). Learning from halophytes: physiological basis and strategies to improve abiotic stress tolerance in crops. *Ann. Bot.* 112, 1209–1221. doi: 10.1093/aob/mct205

Shabala, S., and Bose, J. (2012). "Application of non-invasive microelectrode flux measurements in plant stress physiology," in *Plant Electrophysiology—Methods and Cell Electrophysiology*, ed. A. G. Volkov (Berlin: Springer-Verlag), 91–126.

Shabala, S., Demidchik, V., Shabala, L., Cuin, T. A., Smith, S. J., Miller, A. J., et al. (2006). Extracellular $Ca2^+$ ameliorates NaCl-induced K^+ loss from *Arabidopsis* root and leaf cells by controlling plasma membrane K^+ -permeable channels. *Plant Physiol.* 141, 1653–1665. doi: 10.1104/pp.106.082388

Shabala, S., and Newman, I. (2000). Salinity effects on the activity of plasma membrane H^+ and $Ca2^+$ transporters in bean leaf mesophyll: masking role of the cell wall. *Ann. Bot.* 85, 681–686. doi: 10.1006/anbo.2000.1131

Shabala, S., Pang, J., Zhou, M., Shabala, L., Cuin, T. A., Nick, P., et al. (2009). Electrical signalling and cytokinins mediate effects of light and root cutting on ion uptake in intact plants. *Plant Cell Environ.* 32, 194–207. doi: 10.1111/j.1365-3040.2008.01914.x

Shabala, S. N., and Lew, R. R. (2002). Turgor regulation in osmotically stressed *Arabidopsis* epidermal root cells. Direct support for the role of inorganic ion uptake as revealed by concurrent flux and cell turgor measurements. *Plant Physiol.* 129, 290–299. doi: 10.1104/pp.020005

Shan, Q., Wang, Y., Li, J., Zhang, Y., Chen, K., Liang, Z., et al. (2013). Targeted genome modification of crop plants using a CRISPR-Cas system. *Nat. Biotechnol.* 31, 686–688. doi: 10.1038/nbt.2650

Shavrukov, Y., Bovill, J., Afzal, I., Hayes, J. E., Roy, S. J., Tester, M., et al. (2013). HVP10 encoding V-PPase is a prime candidate for the barley HvNax3 sodium exclusion gene: evidence from fine mapping and expression analysis. *Planta* 237, 1111–1122. doi: 10.1007/s00425-012-1827-3

Shi, H., Ishitani, M., Kim, C., and Zhu, J. K. (2000). The *Arabidopsis thaliana* salt tolerance gene SOS1 encodes a putative Na^+ /H^+ antiporter. *Proc. Natl. Acad. Sci. U.S.A.* 97, 6896–6901. doi: 10.1073/pnas.120170197

Shi, H., Lee, B. H., Wu, S. J., and Zhu, J. K. (2003). Overexpression of a plasma membrane Na^+/H^+ antiporter gene improves salt tolerance in *Arabidopsis thaliana*. *Nat. Biotechnol.* 21, 81–85. doi: 10.1038/nbt766

Shi, H., Quintero, F. J., Pardo, J. M., and Zhu, J. K. (2002). The putative plasma membrane Na^+/H^+ antiporter SOS1 controls long-distance Na^+ transport in plants. *Plant Cell* 14, 465–477. doi: 10.1105/tpc.010371

Silva, P., and Gerós, H. (2009). Regulation by salt of vacuolar H^+-ATPase and H^+-pyrophosphatase activities and Na^+ /H^+ exchange. *Plant Signal. Behav.* 4, 718–726. doi: 10.4161/psb.4.8.9236

Skou, J. C. (1998). Nobel Lecture. The identification of the sodium pump. *Biosci. Rep.* 18, 155–169.

Spiller, D. G., Wood, C. D., Rand, D. A., and White, M. R. (2010). Measurement of single-cell dynamics. *Nature* 465, 736–745. doi: 10.1038/nature09232

Staal, M., Maathuis, F. J. M., Elzenga, J. T. M., Overbeek, J. H. M., and Prins, H. B. A. (1991). Na^+/H^+ antiport activity in tonoplast vesicles from roots of the salt-tolerant *Plantago maritima* and the salt-sensitive *Plantago media*. *Physiol. Plant.* 82, 179–184. doi: 10.1111/j.1399-3054.1991.tb00078.x

Subramanyam, P., and Colecraft, H. M. (2015). Ion channel engineering: perspectives and strategies. *J. Mol. Biol.* 427, 190–204. doi: 10.1016/j. jmb.2014.09.001

Sun, J., Chen, S. L., Dai, S. X., Wang, R. G., Li, N. Y., Shen, X., et al. (2009). Ion flux profiles and plant ion homeostasis control under salt stress. *Plant Signal. Behav.* 4, 261–264. doi: 10.4161/psb.4.4.7918

Sze, H., Li, X., and Palmgren, M. G. (1999). Energization of plant cell membranes by H^+-pumping ATPases. Regulation and biosynthesis. *Plant Cell* 11, 677–690.

Taji, T., Seki, M., Satou, M., Sakurai, T., Kobayashi, M., Ishiyama, K., et al. (2004). Comparative genomics in salt tolerance between *Arabidopsis* and *Arabidopsis*-related halophyte salt cress using *Arabidopsis* microarray. *Plant Physiol.* 135, 1697–1709. doi: 10.1104/pp.104.039909

Takahashi, F., Tilbrook, J., Trittermann, C., Berger, B., Roy, S. J., Seki, M., et al. (2015). Comparison of leaf sheath transcriptome profiles with physiological traits of bread wheat cultivars under salinity stress. *PLoS ONE* 10:e0133322. doi: 10.1371/journal.pone.0133322

Thiyagarajah, M., Fry, S. C., and Yeo, A. R. (1996). In vitro salt tolerance of cell wall enzymes from halophytes and glycophytes. *J. Exp. Bot.* 47, 1717–1724. doi: 10.1093/jxb/47.11.1717

Tyerman, S. D., Beilby, M., Whittington, J., Juswono, U., Newman, I., and Shabala, S. (2001). Oscillations in proton transport revealed from simultaneous measurements of net current and net proton fluxes from isolated root protoplasts: MIFE meets patch-clamp. *Austr. J. Plant Physiol.* 28, 591–606. doi: 10.1071/PP01030

Tyerman, S. D., and Skerrett, I. M. (1998). Root ion channels and salinity. *Sci. Hortic.* 78, 175–235. doi: 10.1016/S0304-4238(98)00194-0

Tyerman, S. D., Skerrett, M., Garrill, A., Findlay, G. P., and Leigh, R. A. (1997). Pathways for the permeation of Na^+ and Cl– into protoplasts derived from the cortex of wheat roots. *J. Exp. Bot.* 48, 459–480. doi: 10.1093/jxb/48.Special_Issue.459

Valencia-Cruz, G., Shabala, L., Delgado-Enciso, I., Shabala, S., Bonales-Alatorre, E., Pottosin, I. I., et al. (2009). Kbg and Kv1.3 channels mediate potassium efflux in the early phase of apoptosis in Jurkat T lymphocytes. *Am. J. Physiol. Cell Physiol.* 297, C1544–C1553. doi: 10.1152/ajpcell.00064.2009

Verbelen, J. P., De Cnodder, T., Le, J., Vissenberg, K., and Baluska, F. (2006). The root apex of arabidopsis thaliana consists of four distinct zones of growth activities: meristematic zone. *Plant Signal. Behav.* 1, 296–304. doi: 10.4161/psb.1.6.3511

Volkov, V., and Amtmann, A. (2006). *Thellungiella halophila*, a salt-tolerant relative of *Arabidopsis thaliana*, has specific root ion-channel features supporting K^+/Na^+ homeostasis under salinity stress. *Plant J.* 48, 342–353. doi: 10.1111/j.1365-313X.2006.02876.x

Volkov, V., Boscari, A., Clément, M., Miller, A. J., Amtmann, A., and Fricke, W. (2009). Electrophysiological characterization of pathways for K^+ uptake into growing and non-growing leaf cells of barley. *Plant Cell Environ.* 32, 1778–1790. doi: 10.1111/j.1365-3040.2009.02034.x

Volkov, V., Hachez, C., Moshelion, M., Draye, X., Chaumont, F., and Fricke, W. (2007). Water permeability differs between growing and non-growing barley leaf tissues. *J. Exp. Bot.* 58, 377–390. doi: 10.1093/jxb/erl203

Volkov, V., Wang, B., Dominy, P. J., Fricke, W., and Amtmann, A. (2004). *Thellungiella halophila*, a salt-tolerant relative of *Arabidopsis thaliana*, possesses effective mechanisms to discriminate between potassium and sodium. *Plant Cell Environ.* 27, 1–14. doi: 10.1046/j.0016-8025.2003.01116.x

Vrbka, L., Vondrásek, J., Jagoda-Cwiklik, B., Vácha, R., and Jungwirth, P. (2006). Quantification and rationalization of the higher affinity of sodium over potassium to protein surfaces. *Proc. Natl. Acad. Sci. U.S.A.* 103, 15440–15444. doi: 10.1073/pnas.0606959103

Walker, D. J., Black, C. R., and Miller, A. J. (1998). The role of cytosolic potassium and pH in the growth of barley roots. *Plant Physiol.* 118, 957–964. doi: 10.1104/pp.118.3.957

Walker, D. J., Smith, S. J., and Miller, A. J. (1995). Simultaneous measurement of intracellular pH and K^+ or NO3 - in barley root cells using triple-barreled, ion-selective microelectrodes. *Plant Physiol.* 108, 743–751.

Wang, B. (2006) *Physiological and Molecular Strategies for Salt Tolerance in Thellungiella halophila, a Close Relative of Arabidopsis thaliana*. Ph.D. thesis, Division of Biochemistry and Molecular Biology, Institute of Biomedical and Life Sciences, University of Glasgow, Glasgow, 235.

Wang, B., Davenport, R. J., Volkov, V., and Amtmann, A. (2006). Low unidirectional sodium influx into root cells restricts net sodium accumulation in *Thellungiella halophila*, a salt-tolerant relative of *Arabidopsis thaliana*. *J. Exp. Bot.* 57, 1161–

1170. doi: 10.1093/jxb/erj116

Waters, S., Gilliham, M., and Hrmova, M. (2013). Plant high-affinity potassium (HKT) transporters involved in salinity tolerance: structural insights to probe differences in ion selectivity. *Int. J. Mol. Sci.* 14, 7660–7680. doi: 10.3390/ijms14047660

Wegner, L. H. (2013). Cation selectivity of the plasma membrane of tobacco protoplasts in the electroporated state. *BBA Biomembr.* 1828, 1973–1981. doi: 10.1016/j.bbamem.2013.04.010

Wegner, L. H. (2014). The conductance of cellular membranes at supra-physiological voltages. *Bioelectricity* 103, 34–38. doi: 10.1016/j. bioelechem.2014.08.005

Wegner, L. H., Flickinger, B., Eing, C., Berghöfer, T., Hohenberger, P., Frey, W., et al. (2011). A patch clamp study on the electro-permeabilization of higher plant cells: supra-physiological voltages induce a high-conductance. *BBA Biomembr.* 1808, 1728–1736. doi: 10.1016/j.bbamem.2011.01.016

Wegner, L. H., Frey, W., and Schönwälder, S. (2013). A critical evaluation of whole cell patch clamp studies on electroporation using the voltage sensitive dye ANNINE-6. *Bioelectrochemistry* 92, 42–46. doi: 10.1016/j.bioelechem.2013.03.002

White, P. J. (1993). Characterization of a high-conductance, voltage-dependent cation channel from the plasma membrane of rye roots in planar lipid bilayers. *Planta* 191, 541–551. doi: 10.1007/BF00195756

White, P. J., Cooper, H. D., Clarkson, D. T., Earnshaw, M. J., and Loughman, B. C. (1991). Effects of low temperature on growth and nutrient accumulation in rye (*Secale cereale*) and wheat (*Triticum aestivum*). *Ann. Bot.* 68, 23–31.

White, P. J., and Lemtiri-Chlieh, F. (1995). Potassium currents across the plasma membrane of protoplasts derived from rye roots: a patch clamp study. *J. Exp. Bot.* 46, 497–511. doi: 10.1093/jxb/46.5.497

Winter, H., Robinson, D. G., and Heldt, H. W. (1994). Subcellular volumes and metabolite concentrations in spinach leaves. *Planta* 193, 530–535. doi: 10.1007/BF02411558

Wu, S. J., Ding, L., and Zhu, J. K. (1996). SOS1, a genetic locus essential for salt tolerance and potassium acquisition. *Plant Cell* 8, 617–627. doi: 10.1105/tpc.8.4.617

Wyn Jones, R. G. A., and Pollard, A. (1983). "Proteins, enzymes and inorganic ions," in *Encyclopedia of Plant Physiology, New Series*, Vol. 15B, eds A. Lauchli and A. Person (New York: Springer), 528–562.

Xue, S., Yao, X., Luo, W., Jha, D., Tester, M., Horie, T., et al. (2011). AtHKT1;1 mediates Nernstian sodium channel transport properties in *Arabidopsis* root stelar cells. *PLoS ONE* 6:e24725. doi: 10.1371/journal.pone.0024725

Yamaguchi, T., Aharon, G. S., Sottosanto, J. B., and Blumwald, E. (2005). Vacuolar Na^+ /H^+ antiporter cation selectivity is regulated by calmodulin from within the vacuole in a Ca2⁺- and pH-dependent manner. *Proc. Natl Acad. Sci. U.S.A.* 102, 16107–16112. doi: 10.1073/pnas.0504437102

Yamaguchi, T., Apse, M. P., Shi, H., and Blumwald, E. (2003). Topological analysis of a plant vacuolar Na^+ /H^+ antiporter reveals a luminal C terminus that regulates antiporter cation selectivity. *Proc. Natl. Acad. Sci. U.S.A.* 100, 12510–12515. doi: 10.1073/pnas.2034966100

Yamaguchi, T., Hamamoto, S., and Uozumi, N. (2013). Sodium transport system in plant cells. *Front. Plant Sci.* 4:410. doi: 10.3389/fpls.2013.00410

Yang, Q., Chen, Z. Z., Zhou, X. F., Yin, H. B., Li, X., Xin, X. F., et al. (2009). Overexpression of SOS (Salt Overly Sensitive) genes increases salt tolerance in transgenic *Arabidopsis*. *Mol. Plant* 2, 22–31. doi: 10.1093/mp/ssn058. doi: 10.1093/mp/ssn058

Yue, Y., Zhang, M., Zhang, J., Duan, L., and Li, Z. (2012). SOS1 gene overexpression increased salt tolerance in transgenic tobacco by maintaining a higher K^+/Na^+ ratio. *J. Plant Physiol.* 169, 255–261. doi: 10.1016/j.jplph.2011.10.007

Zahran, H. H., Marín-Manzano, M. C., Sánchez-Raya, A. J., Bedmar, E. J., Venema, K., and Rodríguez-Rosales, M. P. (2007). Effect of salt stress on the expression of NHX-type ion transporters in *Medicago intretexta* and *Melilotus indicus* plants. *Physiol. Plant.* 131, 122–130. doi: 10.1111/j.1399- 3054.2007.00940.x

Zdebik, A. A., Zifarelli, G., Bergsdorf, E. Y., Soliani, P., Scheel, O., Jentsch, T. J., et al. (2008). Determinants of anion-proton coupling in mammalian endosomal CLC proteins. *J. Biol. Chem.* 283, 4219–4227. doi: 10.1074/jbc.M708368200

Zhang, H. X., and Blumwald, E. (2001). Transgenic salt-tolerant tomato plants accumulate salt in foliage but not in fruit. *Nat. Biotechnol.* 19, 765–768. doi: 10.1038/90824

Zhao, F. Y., Zhang, X. J., Li, P. H., Zhao, Y. X., and Zhang, H. (2006). Co-expression of the *Suaeda salsa* SsNHX1 and *Arabidopsis* AVP1 confer greater salt tolerance to transgenic rice than the single SsNHX1. *Mol. Breed.* 17, 341–353. doi: 10.1007/s11032-006-9005-6

Zhao, K. F., Fan, H., Song, J., Sun, M. X., Wang, B. Z., Zhang, S. Q., et al. (2005). Two Na^+ and Cl− Hyperaccumulators of the Chenopodiaceae. *J. Integr. Plant Biol.* 47, 311–318. doi: 10.1111/j.1744-7909.2005.00057.x

Zou, P., Zhao, Y., Douglass, A. D., Hochbaum, D. R., Brinks, D., Werley, C. A., et al. (2014). Bright and fast multicoloured voltage reporters via electrochromic FRET. *Nat. Commun.* 5, 4625. doi: 10.1038/ncomms5625

17

Beneficial soil bacterium *Bacillus subtilis* (GB03) augments salt tolerance of white clover

Qing-Qing Han[1†], Xin-Pei Lü[1†], Jiang-Ping Bai[2], Yan Qiao[2], Paul W. Paré[3], Suo-Min Wang[1], Jin-Lin Zhang[1*], Yong-Na Wu[1], Xiao-Pan Pang[1], Wen-Bo Xu[1] and Zhi-Liang Wang[1]

[1] State Key Laboratory of Grassland Agro-ecosystems, College of Pastoral Agriculture Science and Technology, Lanzhou University, Lanzhou, China
[2] Gansu Key Laboratory of Crop Genetic and Germplasm Enhancement, Gansu Provincial Key Laboratory of Arid Land Crop Science, College of Agronomy, Gansu Agricultural University, Lanzhou, China
[3] Department of Chemistry and Biochemistry, Texas Tech University, Lubbock, TX, USA

Edited by:
Vadim Volkov, London Metropolitan University, UK

Reviewed by:
Vadim Volkov, London Metropolitan University, UK
Salman Gulzar, University of Karachi, Pakistan
Yinglong Chen, The University of Western Australia, Australia

***Correspondence:**
Jin-Lin Zhang, State Key Laboratory of Grassland Agro-ecosystems, College of Pastoral Agriculture Science and Technology, Lanzhou University, 768 West Jiayuguan Road, Chengguan District, Lanzhou 730020, Gansu, China
e-mail: jlzhang@lzu.edu.cn

† These authors have contributed equally to this work.

Soil salinity is an increasingly serious problem worldwide that reduces agricultural output potential. Selected beneficial soil bacteria can promote plant growth and augment tolerance to biotic and abiotic stresses. *Bacillus subtilis* strain GB03 has been shown to confer growth promotion and abiotic stress tolerance in the model plant *Arabidopsis thaliana*. Here we examined the effect of this beneficial soil bacterium on salt tolerance in the legume forage crop, white clover. Plants of white clover (*Trifolium repens* L. cultivar Huia) were grown from seeds with or without soil inoculation of the beneficial soil bacterium *Bacillus subtilis* GB03 supplemented with 0, 50, 100, or 150 mM NaCl water into soil. Growth parameters, chlorophyll content, malondialdehyde (MDA) content and osmotic potential were monitored during the growth cycle. Endogenous Na^+ and K^+ contents were determined at the time of harvest. White clover plants grown in GB03-inoculated soil were significantly larger than non-inoculated controls with respect to shoot height, root length, plant biomass, leaf area and chlorophyll content; leaf MDA content under saline condition and leaf osmotic potential under severe salinity condition (150 mM NaCl) were significantly decreased. Furthermore, GB03 significantly decreased shoot and root Na^+ accumulation and thereby improved K^+/Na^+ ratio when GB03-inoculated plants were grown under elevated salt conditions. The results indicate that soil inoculation with GB03 promotes white clover growth under both non-saline and saline conditions by directly or indirectly regulating plant chlorophyll content, leaf osmotic potential, cell membrane integrity and ion accumulation.

Keywords: *Bacillus subtilis*, white clover, salt tolerance, MDA, osmotic potential, Na^+ accumulation

INTRODUCTION

Soil salinity has a significant negative impact on global agricultural productivity, and is a particularly acute issue in both irrigated and non-irrigated areas of the world (Flowers, 2004; Zhang et al., 2010a; Kronzucker and Britto, 2011; Zhang and Shi, 2013). Most crop and forage plants that feed the global population are sensitive to high salt concentration in soils (Rengasamy, 2010; Kronzucker et al., 2013). Soil salinity promotes osmotic stress, water deficit, stomatal closure and reduced leaf expansion (Rahnama et al., 2010; James et al., 2011); moreover, soil salinity causes deficiency of essential nutrients such as K^+. Elevated Na^+ inside plants can decrease plant photosynthetic rates and biomass accumulation (Mahajan and Tuteja, 2005; Munns and Tester, 2008; Zhang et al., 2010a; Zhang and Shi, 2013). Therefore, there is a need to find new ways to cope with the threat of global soil salinization to agriculture.

Beneficial interactions between bacteria and plants accelerate seed germination, promote growth, increase crop yields and secondary metabolites, and augment reproductive success. More recently plant-microbe interactions have been attributed to increased biotic and abiotic stress tolerance, including plant disease resistance and salt and drought tolerance (van Hulten et al., 2006; Zhang et al., 2008a; Hayat et al., 2010; Marques et al., 2010; Rudrappa et al., 2010; Medeiros et al., 2011; Cappellari et al., 2013; Diagne et al., 2013; de Zelicourta et al., 2013). *Bacillus subtilis* that is not toxic to humans widely exists in soils and can produce a wealth of antibacterial substances including lipopeptides, polypeptides, and phospholipids (Stein et al., 2002). Recently, Medeiros et al. (2011) found that transcriptional profiling in cotton was linked with *Bacillus subtilis* (UFLA285) which promoted biotic-stress tolerance. Zhang et al. (2011) found that a bioorganic fertilizer could effectively control banana wilt by strong colonization of *Bacillus subtilis* N11. *B. subtilis* strain GB03 that emits a complex blend of volatile organic compounds (VOCs) enhances growth and abiotic stress tolerance in *Arabidopsis* (Ryu et al., 2004; Farag et al., 2006). A bouquet of over 25 bacterial volatile odors have been identified that trigger differential expression of approximately 600 *Arabidopsis* transcripts related to cell wall modifications, primary and secondary metabolism, stress responses, hormone regulation and other expressed proteins (Zhang et al., 2007). In *Arabidopsis*, GB03 regulates auxin homeostasis and cell expansion (Zhang et al., 2007), augments photosynthesis by decreasing glucose sensing and ABA levels (Zhang et al., 2008b); promotes salt tolerance as well as reduces total Na^+ by regulating tissue specific expression of *AtHKT1* (Zhang et al., 2008a). GB03 has also been shown to stimulate iron acquisition and increased photosynthetic capacity (Zhang et al., 2009). More recently GB03 has been found to improve osmotic-stress tolerance by elevating levels of endogenous osmoprotectants (Zhang et al., 2010b). With such beneficial plant responses activated by GB03, a legume forage crop is now being examined as to how it responds to this beneficial microbe.

White clover (*Trifolium repens* L.) is an important forage crop worldwide. Like many forage crops, white clover is sensitive to soil salinity (Rogers et al., 1997). Salt stress can decrease both above-ground growth and the number of root nodules that in turn compromise nitrogen fixation and soil fertility (Acharya et al., 2011). Since GB03 promotes growth and salt tolerance in *Arabidopsis*, the question of how GB03 improves growth and salt tolerance in the salt-sensitive forage crop white clover was addressed. The aim of current work was to evaluate the efficiency of *Bacillus subtilis* GB03 for growth promotion and salt tolerance in white clover.

MATERIALS AND METHODS

BACTERIAL SUSPENSION CULTURE

Bacillus subtilis strain GB03 was streaked onto LB agar plates and incubated at 28°C without light for 24 h (The bacterium strain was presented by Professor Paul W. Paré at Texas Tech University, USA). Bacterial cells were then harvested from LB agar plates, transferred into liquid LB media and cultured at 28°C with 250 rpm rotation to yield 10^9 colony forming units (CFU) mL^{-1}, as determined by optical density and serial dilutions (Zhang et al., 2008a).

PLANT GROWTH AND TREATMENTS

White clover (*Trifolium repens* L. cultivar Huia) seeds were surface sterilized with 2% NaClO for 1 min followed by 70% ethanol for 10 min, and rinsed with sterile water five times (The white clover seeds was presented by Dr. Wanhai Zhou at Gansu Agricultural University, China). Seeds were then sown in presterilized plastic pots (diameter 20 cm, depth 22 cm, 10 seeds/pot with eight replicates) containing 600 g of heat-sterilized (95°C, 48 h) vermiculite and soil mix (1:1) watered with half strength Hoagland's nutrient solution (5 mM KNO_3, 1 mM $NH_4H_2PO_4$, 0.5 mM $Ca(NO_3)_2$, 0.5 mM $MgSO_4$, 60 μM Fe-citrate, 92 μM H_3BO_3, 18 μM $MnCl_2{\cdot}4H_2O$, 1.6 μM $ZnSO_4{\cdot}7H_2O$, 0.6 μM $CuSO_4{\cdot}5H_2O$, and 0.7 μM $(NH_4)_6Mo_7O_{24}{\cdot}4H_2O$). Each pot was watered with nutrient solution (400 mL) once per week. After germination, six uniform seedlings per pot were selected for continued growth and inoculated directly to the soil with 1 mL bacterial suspension culture as inoculation treatment or 1 mL liquid LB medium as a control. For salt stress treatments, seedlings were watered with 0, 50, 100, and 150 mM NaCl present in the nutrient solution. Each treatment contained eight pots as replications. Plants were grown in a glasshouse at the temperature regulated to around 28°C during the day and 22°C at night. Relative humidity averaged 65 and 75% for day and night periods, respectively, with natural photoperiod and light intensity.

PLANT BIOMASS AND PHYSIOLOGICAL MEASUREMENTS

Sixty-day old plants were used for plant growth measurements and physiological index determination. Whole plant leaves were harvested to count leaf number per plant, measure leaf area using a leaf area meter (Epson Perfection 4870 Photo scanner, Epson America Inc., Long Beach, CA, USA) (Ma et al., 2012). Average area per leaf was calculated from leaf area per plant divided by leaf number per plant.

Plants were removed from the pots and roots were water rinsed to remove attached soil. Shoot height and root length (root depth) were measured by a ruler. Then, root and shoot were separated and blotted gently. Fresh weights were determined immediately and samples were oven dried at 80°C for 2 day for dry weight measurements.

Leaf chlorophyll content was estimated according to Porra et al. (1989). Fresh leaf samples were ground thoroughly with 80% acetone in the dark and centrifuged at 9000 g for 10 min at 4°C. Absorbance reading (UV-2102C Spectrophotometer, Unico Instrument Co., Ltd, Shanghai, China) at 645 and 663 nm for collected supernatant was used to estimate total chlorophyll content based on chlorophyll equations of Porra et al. (1989).

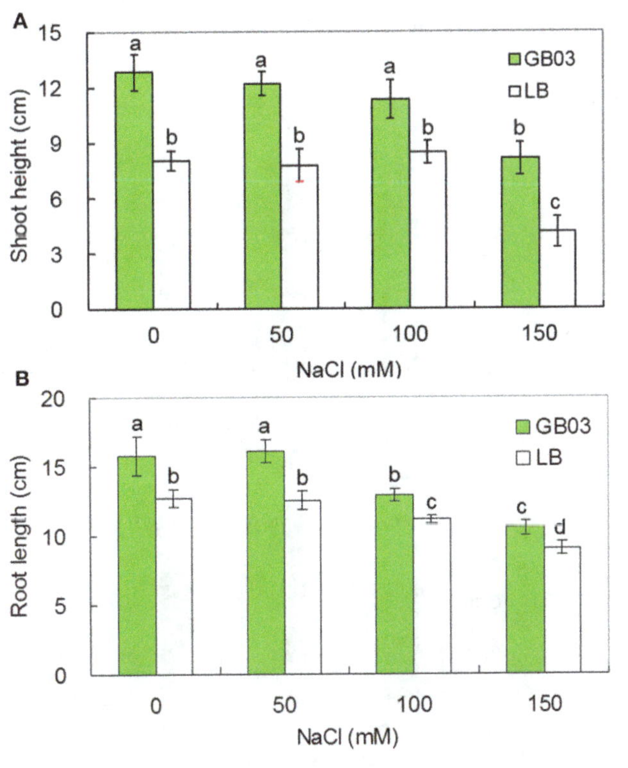

FIGURE 1 | Effects of GB03 bacterization on shoot height (A) and root length (B) of white clover under various concentrations of NaCl. Values are means and bars indicate SDs ($n = 8$). Columns with different letters indicate significant difference at $P < 0.05$ (Duncan test).

To probe oxidative stress the biomarker malondialdehyde (MDA) was extracted and measured spectrophotometrically using a thiobarbituric acid (TBA) protocol (Bao et al., 2009). Absorbance was determined at 450, 532, and 600 nm using a UV spectrophotometer (UV-2102C, Unico Instrument Co., Ltd, Shanghai, China).

Leaf osmotic potential (Ψs) was measured according to Ma et al. (2012). Fresh leaf samples were frozen in liquid nitrogen. Cell sap was collected by thawing slowly and then Ψs was determined using a cryoscopic osmometer (Osmomat-030, Gonotec GmbH, Berlin, Germany) at 25°C. The readings ($mol \cdot L^{-1}$) were used to calculate the solute potential (Ψs) in MPa with the formula $\Psi s = -$ The readings $\times R \times T$, here $R = 0.008314\ MPa \cdot L \cdot mol^{-1} \cdot K^{-1}$ and $T = 298.8$ K (Ma et al., 2012).

ION ANALYSIS

Na^+ and K^+ contents were measured according to the method described by Wang et al. (2007). Roots were washed twice for eight min in ice-cold 20 mM $CaCl_2$ to exchange cell wall-bound K^+ and Na^+ and the shoots were rinsed in deionized water to remove surface salts. Root and shoot were separated and samples oven dried at 70°C for 2 day. Na^+ and K^+ were extracted from dried plant tissues in 100 mM acetic acid at 90°C for 2 h. Ion analysis was performed using an atomic absorption spectrophotometer (2655-00, Cole-Parmer Instrument Co., Vernon Hills, IL, USA).

DATA ANALYSIS

Results of the growth, physiological index, ion contents and ion ratio were presented as means with standard deviations ($n = 8$). Statistical analyses, One-Way ANOVA and Duncan's multiple range tests, were performed.

RESULTS

BACILLUS SUBTILIS GB03 PROMOTED WHITE CLOVER GROWTH

Bacillus subtilis GB03 enhanced both shoot height and root length of white clover under both non-saline conditions and salinity stress (**Figure 1** and **Supplementary Figure 1**). Shoot height was increased by 60.1% ($P < 0.05$) under non-saline conditions; compared to corresponding media control, GB03 significantly improved shoot height by 57.4% ($P < 0.01$), 33.7% ($P < 0.05$), and 95.6% ($P < 0.01$) at 50, 100, and 150 mM NaCl treatments, respectively (**Figure 1A**). Root length was increased by 23.9% ($P < 0.05$) under non-saline conditions; compared to corresponding media control, GB03 significantly improved root length by 28.0% ($P < 0.01$), 15.4% ($P < 0.01$), and 16.9% ($P < 0.05$) with 50, 100, and 150 mM NaCl treatments, respectively (**Figure 1B**).

GB03 increased plant biomass of white clover under both non-saline conditions and salinity stress (**Figure 2**). Shoot fresh weight was increased 4.1-fold ($P < 0.05$) under non-saline conditions; compared to corresponding media control, GB03 significantly improved shoot fresh weight by 5.5-fold ($P < 0.05$), 6.9-fold ($P < 0.01$), and 3.0-fold ($P < 0.05$) with 50, 100, and 150 mM NaCl treatment, respectively (**Figure 2A**). Shoot dry weight was increased by 4.0-fold ($P < 0.05$) under non-saline conditions; compared to corresponding media control, GB03 significantly improved shoot dry weight by 4.9-fold ($P < 0.05$), 6.4-fold ($P < 0.01$), and 3.2-fold ($P < 0.05$) with 50, 100, and

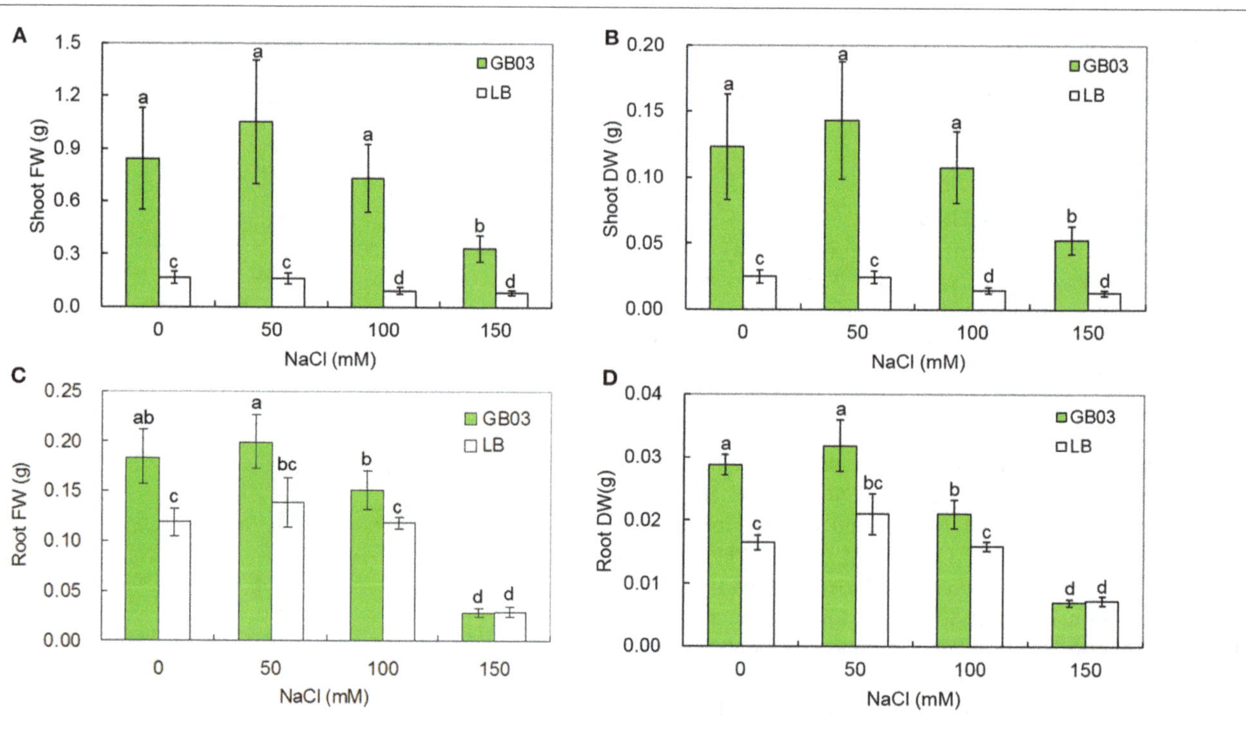

FIGURE 2 | Effects of GB03 bacterization on plant growth of white clover under various concentrations of NaCl. (A) shoot fresh weight, **(B)** shoot dry weight, **(C)** root fresh weight, and **(D)** root dry weight. Values are means and bars indicate SDs ($n = 8$). Columns with different letters indicate significant difference at $P < 0.05$ (Duncan test).

150 mM NaCl treatment, respectively (**Figure 2B**). Root fresh weight was increased by 55.0% ($P < 0.05$) under non-saline conditions; compared to corresponding media control, GB03 significantly improved shoot dry weight by 44.1% ($P < 0.05$) and 27.7% ($P < 0.05$) at 50 and 100 mM NaCl treatment, respectively (**Figure 2C**). Root dry weight was increased by 74.7% ($P < 0.01$) under non-saline conditions; compared to corresponding media control, GB03 significantly improved shoot dry weight by 51.7% ($P < 0.01$) and 32.5% ($P < 0.05$) at 50 and 100 mM NaCl treatment, respectively (**Figure 2D**). GB03 had no significant effects on root fresh and dry weights under 150 mM NaCl stress.

GB03 INCREASED LEAF AREA AND CHLOROPHYLL CONTENT

GB03 promoted leaf growth under both non-saline and salinity stress conditions (**Figure 3**). Leaf area per plant was increased by 3.1-fold ($P < 0.05$) under non-saline conditions; compared to corresponding media control, GB03 significantly increased leaf area per plant by 5.7-fold ($P < 0.05$), 7.8-fold ($P < 0.01$), and 1.5-fold ($P < 0.05$) with 50, 100, and 150 mM NaCl treatment, respectively (**Figure 3A**). Leaf number per plant was increased by 84.2 % ($P < 0.05$) under non-saline conditions; compared to corresponding media control, GB03 significantly elevated leaf number per plant by 142.1% ($P < 0.05$) and 107.7% ($P < 0.05$) with 50 and 100 mM NaCl treatments, respectively (**Figure 3B**). Average area per leaf was increased by 125.6% ($P < 0.05$) under non-saline conditions; compared to corresponding media control, GB03 significantly improved average area per leaf by 159.4% ($P < 0.01$), 334.5% ($P < 0.01$), and 81.5% ($P < 0.01$) with 50, 100, and 150 mM NaCl treatments, respectively (**Figure 3C**).

In addition to promoting leaf growth, *Bacillus subtilis* GB03 increased leaf chlorophyll content under both non-saline and salinity stress (**Figure 4**). Leaf chlorophyll content rose by 36.0% ($P < 0.01$) under non-saline condition; compared to corresponding media control, GB03 significantly increased leaf chlorophyll content by 34.3% ($P < 0.01$), 37.5% ($P < 0.01$), and 57.4% ($P < 0.01$) with 50, 100, and 150 mM NaCl treatments, respectively (**Figure 4**).

GB03 REDUCED LEAF OSMOTIC POTENTIAL UNDER HIGHER SALINITY CONDITION

At lower salt concentration (0, 50, and 100 mM NaCl), compared with the controls, GB03 had no significant effect on leaf osmotic potential; however, under higher salt concentration (150 mM NaCl), GB03 significantly decreased leaf osmotic potential of white clover by 46.7% ($P < 0.05$) (**Figure 5**).

GB03 REDUCED LEAF MALONDIALDEHYDE (MDA) CONTENT UNDER SALINITY CONDITIONS

The results indicated that GB03 reduced leaf MDA content under both non-saline conditions and salinity stress (**Figure 6**). Leaf MDA content was not changed under non-saline conditions; however, compared to corresponding media control, GB03 significantly reduced leaf MDA content by 52.7% ($P < 0.01$), 56.8% ($P < 0.01$), and 53.8% ($P < 0.05$) with 50, 100, and 150 mM NaCl treatments, respectively (**Figure 6**), thus, oxidative stress tolerance and integrity of the cell membrane as measured by the biomarker MDA were both favorably regulated by GB03.

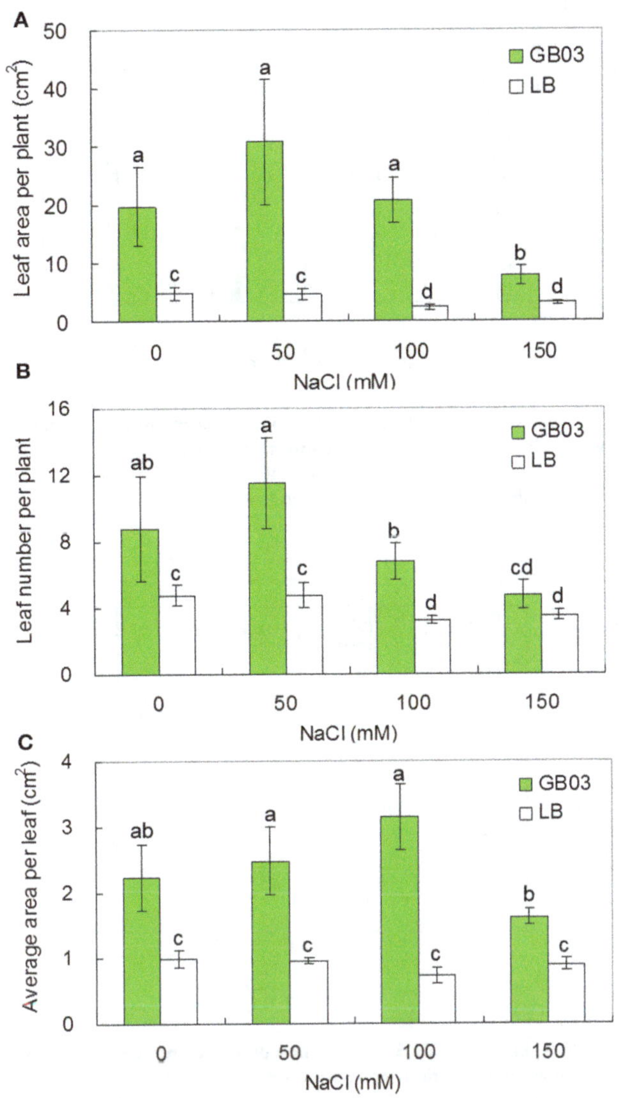

FIGURE 3 | Effects of GB03 bacterization on leaf development and growth of white clover under various concentrations of NaCl. (A) leaf area per plant, **(B)** leaf number per plant, and **(C)** average area per leaf. Values are means and bars indicate SDs ($n = 8$). Columns with different letters indicate significant difference at $P < 0.05$ (Duncan test).

GB03 REDUCED SODIUM ACCUMULATION UNDER SALINITY CONDITIONS

To test whether GB03 inoculation resulted in altered ion accumulation in white clover under various salinity conditions, endogenous Na^+ and K^+ content was measured in the 0, 50, 100, and 150 mM NaCl treatments. K^+ accumulation was uneffected by GB03 inoculation (**Figures 7A,B**). However, endogenous Na^+ accumulation was reduced significantly. Compared to the media control, GB03 significantly decreased shoot Na^+ content by 40.7% ($P < 0.01$), 22.5% ($P < 0.01$), and 26.3% ($P < 0.01$), as well as root Na^+ content by 27.1% ($P < 0.05$), 39.7% ($P < 0.01$), and 40.7% ($P < 0.01$) with 50, 100, and 150 mM NaCl treatments, respectively (**Figures 7C,D**).

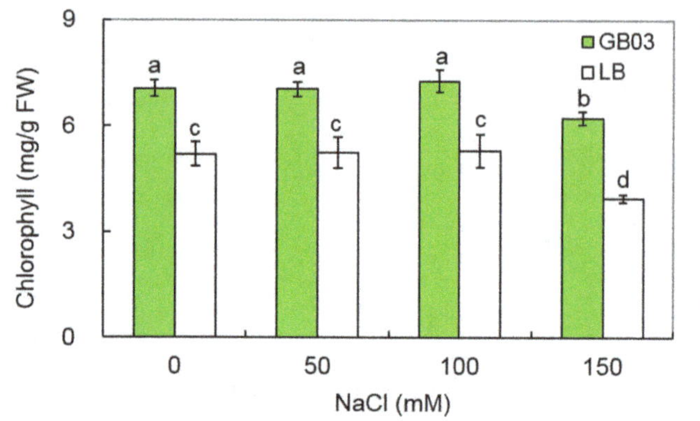

FIGURE 4 | Effects of GB03 bacterization on total chlorophyll content of white clover under various concentrations of NaCl. Values are means and bars indicate SDs ($n = 8$). Columns with different letters indicate significant difference at $P < 0.05$ (Duncan test).

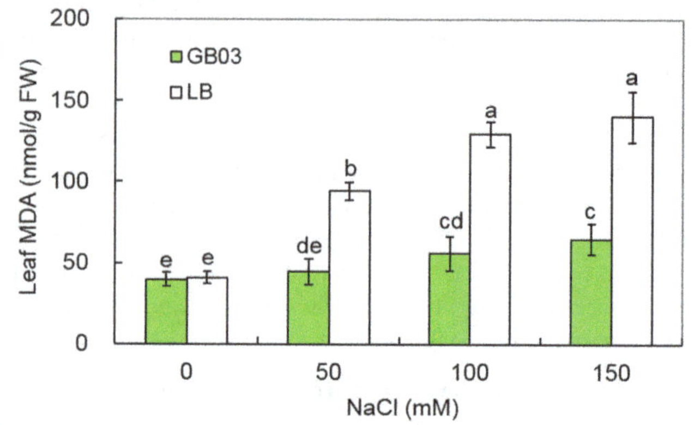

FIGURE 6 | Effects of GB03 bacterization on leaf malondialdehyde (MDA) content of white clover under various concentrations of NaCl. Values are means and bars indicate SDs ($n = 8$). Columns with different letters indicate significant difference at $P < 0.05$ (Duncan test).

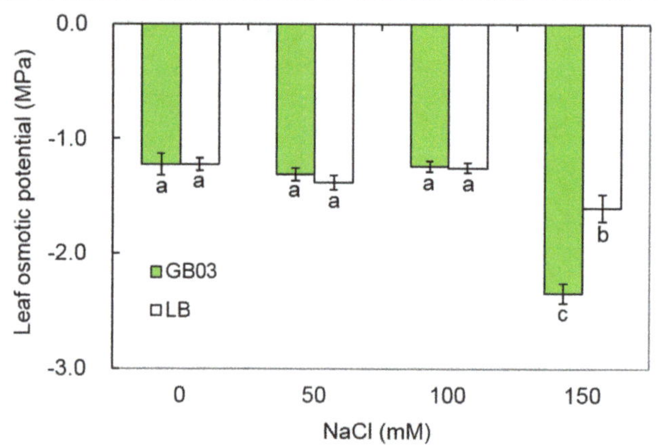

FIGURE 5 | Effects of GB03 bacterization on leaf osmotic potential of white clover under various concentrations of NaCl. Values are means and bars indicate SDs ($n = 8$). Columns with different letters indicate significant difference at $P < 0.05$ (Duncan test).

K^+/Na^+ ratios were also significantly improved by GB03 inoculation with 50, 100, and 150 mM NaCl treatments. Compared to corresponding media control, Shoot K^+/Na^+ ratios increased by 78.6% ($P < 0.01$), 32.0% ($P < 0.05$), and 39.7% ($P < 0.05$) as well as root K^+/Na^+ ratios by 34.6% ($P < 0.05$), 82.2% ($P < 0.01$), and 66.2% ($P < 0.01$) under soil salinity conditions of 50, 100, and 150 mM NaCl, respectively (**Table 1**).

DISCUSSION

INFLUENCE OF *BACILLUS SUBTILIS* GB03 ON GROWTH OF WHITE CLOVER UNDER SALINE CONDITION

Beneficial soil bacteria promoted plant growth of many plant species (Bashan et al., 2000; Gray and Smith, 2005; Xie et al., 2009; Paré et al., 2011). Under salinity stress, inducible plant growth promotion mediated by beneficial soil bacteria has also been observed in several cultivated and wild plant species including dwarf saltwort (*Salicornia bigelovii*) (Bashan et al., 2000), tomato (*Lycopersicon esculentum*) (Mayak et al., 2004), chickpea (*Cicer arietinum*) (Mhadhbi et al., 2004), alfalfa (*Medicago sativa*) (Ibragimova et al., 2006), common glasswort (*Salicornia europea*) (Ozawa et al., 2007), maize (*Zea mays*) (Bano and Fatima, 2009), and wheat (*Triticum aestivum*) (Tiwari et al., 2011). The growth promotion of white clover by GB03 inoculation in soil evaluated in the present study is consistent with previous reports in *Arabidopsis* (Zhang et al., 2007, 2008a,b, 2010b; Xie et al., 2009; Paré et al., 2011). Interestingly, our results indicated that GB03 is a more efficient promoter of shoot growth than root growth in white clover, especially under salt exposure (50, 100, and 150 mM NaCl). Whether GB03 promotes growth of white clover by enhancing nitrogen fixation and the coordination between nitrogen-fixing bacteria and *Bacillus subtilis* (GB03) in the rhizosphere remain to be investigated.

Leaf development plays an important role in plant production since it affects the area available for photosynthesis, which is strongly related to plant growth and biomass accumulation (Gutierrez-Boem and Thomas, 1998; Battie-Laclau et al., 2013). In our study, GB03 inoculation significantly increased leaf area per plant, leaf number per plant, and thereby average area per leaf in white clover. Leaf chlorophyll content is also an important physiological trait directly linked to photosynthesis rate in plants (Ma et al., 2012). Previous studies showed that plants grown under salinity conditions produced less chlorophyll and dry matter than those without salinity stress due to chlorophyll peroxidation (Hernandez et al., 1995; Zayed and Zeid, 1998; Lunde et al., 2007; Tuna et al., 2008; Barry, 2009). Zhang et al. (2008b) found that GB03 augments photosynthetic capacity by increasing photosynthetic efficiency and chlorophyll content in *Arabidopsis*. With white clover, GB03 significantly improved leaf chlorophyll content under both non-saline and salinity conditions (50, 100, and 150 mM NaCl).

GB03 AMELIORATED LEAF OSMOTIC ADJUSTMENT ABILITY AND ALLEVIATED CELL MEMBRANE DAMAGE UNDER SALINE CONDITIONS

When plants are exposed to saline or drought conditions, osmotic stress rapidly follows (Munns and Tester, 2008). Salt stress reduces

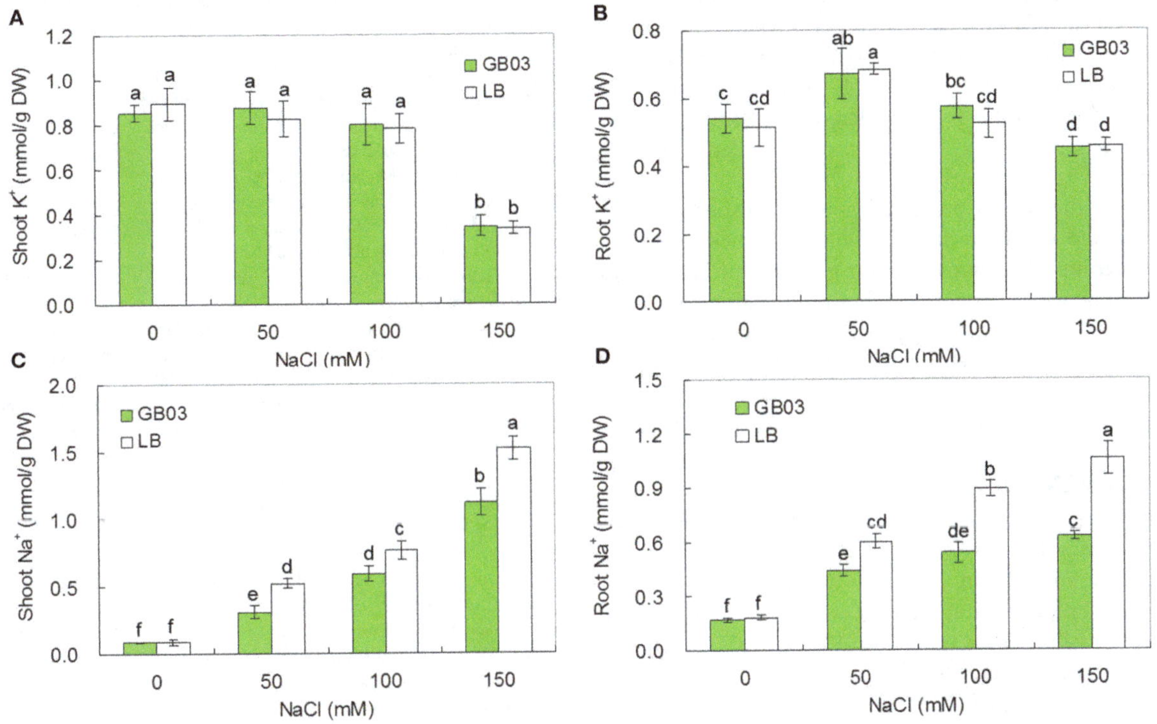

FIGURE 7 | Effects of GB03 bacterization on ion contents of white clover under various concentrations of NaCl. (A) shoot K^+ content, **(B)** root K^+ content, **(C)** shoot Na^+ content, **(D)** root Na^+ content. Values are means and bars indicate SDs ($n = 8$). Columns with different letters indicate significant difference at $P < 0.05$ (Duncan test).

Table 1 | Effects of GB03 bacterization on K^+/Na^+ ratio in shoot and root of white clover under various concentrations of NaCl.

NaCl (mM)	Shoots		Roots	
	GB03	LB	GB03	LB
0	9.87 ± 0.42a	9.81 ± 0.80a	3.13 ± 0.25a	2.81 ± 0.30a
50	2.81 ± 0.24b	1.58 ± 0.15c	1.53 ± 0.17b	1.14 ± 0.03c
100	1.35 ± 0.15cd	1.03 ± 0.09d	1.07 ± 0.07c	0.59 ± 0.05e
150	0.31 ± 0.04e	0.22 ± 0.02e	0.72 ± 0.05d	0.43 ± 0.02f

Values are means ± SDs ($n = 8$). Different letters indicate significant difference at $P < 0.05$ (Duncan test) within shoots and roots, individually.

the rates of photosynthesis and transpiration, stomatal conductance and leaf area etc. (Wang et al., 2005; Ramani et al., 2006). Leaves could accumulate abundant osmoprotectant and adjust their osmotic potential (≈leaf water potential) below that of the apoplast and soil in order to ensure that the plants can continue to absorb moisture from the soil and maintain turgor pressure, thus improving their stress tolerance (Bao et al., 2009; Jha et al., 2011; Janz and Polle, 2012; Ma et al., 2012).

Zhang et al. (2010b) reported that GB03 enhanced *Arabidopsis* choline and glycine betaine synthesis associated with enhanced osmolyte content, resulting in increased plant tolerance to osmotic stress. In the case of white clover, GB03 significantly reduced leaf osmotic potential under severe salt stress conditions (150 mM NaCl).

Soil salinity is known to increase the level of reactive oxygen species in plant leaves, which are well recognized for membrane lipid peroxidation and cause an increase in leaf malondialdehyde (MDA), a product of membrane lipid peroxidation (Koca et al., 2006; Yazici et al., 2007). Therefore, leaf MDA content, representing the degree of cell membrane damage, is usually used to evaluate plant tolerance to salinity and drought (Luna et al., 2000; Miao et al., 2010). In this work, GB03 significantly decreased leaf MDA content of white clover under saline conditions (50, 100, and 150 mM NaCl).

GB03 REDUCED Na^+ CONTENT IN PLANTS

Saline soils contain a varied and complex array of cation-anion pairs (e.g., Na_2SO_4, $MgSO_4$, $CaSO_4$, $MgCl_2$, KCl, and Na_2CO_3), with Na^+ often the dominant species (Zhang et al., 2010a). Growth inhibition, a common plant response to soil salinity, is correlated with high internal Na^+ concentration and low K^+/Na^+ ratio in the plant (Zhang et al., 2010a). Plants grown in saline conditions can minimize Na^+ toxicity by restricting Na^+ uptake and Na^+ xylem loading, extruding Na^+ from root cells and redirecting Na^+ from shoots to roots (Tester and Davenport, 2003; Munns and Tester, 2008; Zhang et al., 2010a; Kronzucker and Britto, 2011; Zhang and Shi, 2013). Zhang et al. (2008a) found that in *Arabidopsis* GB03 decreased whole plant Na^+ content to 54% of that in control plants by down-regulating *HKT1* expression in roots to decrease root Na^+ uptake and up-regulating *HKT1* expression in shoots to enhance shoot-to-root Na^+ recirculation, respectively. In this study, GB03 soil inoculation significantly decreased Na^+ accumulation and increased K^+/Na^+ in both shoots and roots of white clover under salt stress with no measurable effect on K^+ content in the whole plant. Whether GB03 improves salt tolerance of white clover by tissue specific regulation of *HKT* gene expression remains to be investigated.

In summary, the results presented here established that the inoculation of the soil bacterium *Bacillus subtilis* GB03 to the rhizosphere significantly increases plant growth and biomass of the forage legume white clover under both non-saline and saline conditions. GB03-regulated plant processes include chlorophyll abundance, leaf osmotic potential, cell membrane integrity, and ion accumulation. This study provides insight for the application of selected bacteria to cultivated legumes to combat saline toxicity.

ACKNOWLEDGMENTS

This research was supported by the National Basic Research Program of China (grant No. 2014CB138701), the National Natural Science Foundation of China (grant No. 31172256, 31222053), the Programs for New Century Excellent Talents, Ministry of Education, China (grant No. NCET-11-0217) and the Fundamental Research Funds for the Central Universities (grant No. LZUJBKY-2013-K08, BT05, and 2014-M01).

SUPPLEMENTARY MATERIAL

The Supplementary Material for this article can be found online at: http://www.frontiersin.org/journal/10.3389/fpls.2014.00525/abstract

Supplementary Figure 1 | Effects of bacterization on growth of white clover plants in pots under various concentrations of NaCl. Here GB03 represents *Bacillus subtilis* GB03 suspension culture in LB medium, LB Luria Broth medium without bacteria and 0, 50, 100, and 150 the concentrations of NaCl (mM).

REFERENCES

Acharya, A. R., Wofford, D. S., Kenworthy, K., and Quesenberry, K. H. (2011). Combining ability analysis of resistance in white clover to southern root-knot nematode. *Crop Sci.* 51, 1928–1934. doi: 10.2135/cropsci2010.12.0695

Bano, A., and Fatima, M. (2009). Salt tolerance in *Zea mays* (L.) following inoculation with Rhizobium and Pseudomonas. *Biol. Fert. Soils* 45, 405–413. doi: 10.1007/s00374-008-0344-9

Bao, A. K., Wang, S. M., Wu, G. Q., Xi, J. J., Zhang, J. L., and Wang, C. M. (2009). Overexpression of the *Arabidopsis* H^+-PPase enhanced the salt and drought tolerance in transgenic alfalfa (*Medicago sativa* L.). *Plant Sci.* 176, 232–240. doi: 10.1016/j.plantsci.2008.10.009

Barry, C. S. (2009). The stay-green revolution: recent progress in deciphering the mechanisms of chlorophyll degradation in higher plants. *Plant Sci.* 176, 325–333. doi: 10.1016/j.plantsci.2008.12.013

Bashan, Y., Moreno, M., and Troyo, E. (2000). Growth promotion of the seawater-irrigated oilseed halophyte *Salicornia bigelovii* inoculated with mangrove rhizosphere bacteria and halotolerant *Azospirillum* spp. *Biol. Fert. Soils* 32, 265–272. doi: 10.1007/s003740000246

Battie-Laclau, P., Laclau, J. P., Piccolo, M., Arenque, B., Beri, C., Mietton, L., et al. (2013). Influence of potassium and sodium nutrition on leaf area components in *Eucalyptus grandis* trees. *Plant Soil* 371, 19–35. doi: 10.1007/s11104-013-1663-7

Cappellari, L. R., Santoro, M. V., Nievas, F., Giordano, W., and Banchio, E. (2013). Increase of secondary metabolite content in marigold by inoculation with plant growth-promoting rhizobacteria. *Appl. Soil Ecol.* 70, 16–22. doi: 10.1016/j.apsoil.2013.04.001

de Zelicourta, A., Al-Yousif, M., and Hirta, H. (2013). Rhizosphere microbes as essential partners for plant stress tolerance. *Mol. Plant* 6, 242–245. doi: 10.1093/mp/sst028

Diagne, N., Thioulouse, J., Sanguin, H., Prin, Y., Krasova-Wade, T., Sylla, S., et al. (2013). Ectomycorrhizal diversity enhances growth and nitrogen fixation of *Acacia mangium* seedlings. *Soil Biol. Biochem.* 57, 468–476. doi: 10.1016/j.soilbio.2012.08.030

Farag, M. A., Ryu, C. M., Sumner, L. W., and Paré, P. W. (2006). GC-MS SPME profiling of rhizobacterial volatiles reveals prospective inducers of growth promotion and induced systemic resistance in plants. *Phytochemistry* 67, 2262–2268. doi: 10.1016/j.phytochem.2006.07.021

Flowers, T. J. (2004). Improving crop salt tolerance. *J. Exp. Bot.* 55, 307–319. doi: 10.1093/jxb/erh003

Gray, E. J., and Smith, D. L. (2005). Intracellular and extracellular PGPR: commonalities and distinctions in the plant-bacterium signaling processes. *Soil Biol. Biochem.* 37, 395–412. doi: 10.1016/j.soilbio.2004.08.030

Gutierrez-Boem, F. H., and Thomas, G. W. (1998). Phosphorus nutrition affects wheat response to water deficit. *Agron. J.* 90, 166–171. doi: 10.2134/agronj1998.00021962009000020008x

Hayat, R., Ali, S., Amara, U., Khalid, R., and Ahmed, I. (2010). Soil beneficial bacteria and their role in plant growth promotion: a review. *Ann. Microbiol.* 60, 579–598. doi: 10.1007/s13213-010-0117-1

Hernandez, J. A., Olmos, E., Corpas, F. J., Sevill, F., and Delrio, L. A. (1995). Salt-induced oxidative stress in chloroplasts of pea plants. *Plant Sci.* 105, 151–167. doi: 10.1016/0168-9452(94)04047-8

Ibragimova, M. V., Rumyantseva, M. L., Onishchuk, O. P., Belova, V. S., Kurchak, O. N., Andronov, E. E., et al. (2006). Symbiosis between the root-nodule bacterium *Sinorhizobium meliloti* and alfalfa (*Medicago sativa*) under salinization conditions. *Microbiology* 75, 77–81. doi: 10.1134/S0026261706010140

James, R. A., Blake, C., Byrt, C. S., and Munns, R. (2011). Major genes for Na^+ exclusion, *Nax1* and *Nax2* (wheat *HKT1;4* and *HKT1;5*), decrease Na^+ accumulation in bread wheat leaves under saline and waterlogged conditions. *J. Exp. Bot.* 62, 2939–2947. doi: 10.1093/jxb/err003

Janz, D., and Polle, A. (2012). Harnessing salt for woody biomass production. *Tree Physiol.* 32, 1–3. doi: 10.1093/treephys/tpr127

Jha, Y., Subramanian, R. B., and Patel, S. (2011). Combination of endophytic and rhizospheric plant growth promoting rhizobacteria in *Oryza sativa* shows higher accumulation of osmoprotectant against saline stress. *Acta Physiol. Plant.* 33, 797–802. doi: 10.1007/s11738-010-0604-9

Koca, H., Ozdemir, F., and Turkan, I. (2006). Effect of salt stress on lipid peroxidation and superoxide dismutase and peroxidase activities of *Lycopersicon esculentum* and *L. pennellii*. *Biol. Plant.* 50, 745–748. doi: 10.1007/s10535-006-0121-2

Kronzucker, H. J., and Britto, D. T. (2011). Sodium transport in plants: a critical review. *New Phytol.* 189, 54–81. doi: 10.1111/j.1469-8137.2010.03540.x

Kronzucker, H. J., Coskun, D., Schulze, L. M., Wong, J. R., and Britto, D. T. (2013). Sodium as nutrient and toxicant. *Plant Soil* 369, 1–23. doi: 10.1007/s11104-013-1801-2

Luna, C., Seffino, L. G., Arias, C., and Taleisnik, E. (2000). Oxidative stress indicators as selection tools for salt tolerance in *Chloris gayana*. *Plant Breeding* 119, 341–345. doi: 10.1046/j.1439-0523.2000.00504.x

Lunde, C., Drew, D. P., Jacobs, A. K., and Tester, M. (2007). Exclusion of Na^+ via sodium ATPase (PpENA1) ensures normal growth of *Physcomitrella patens* under moderate salt stress. *Plant Physiol.* 144, 1786–1796. doi: 10.1104/pp.106.094946

Ma, Q., Yue, L. J., Zhang, J. L., Wu, G. Q., Bao, A. K., and Wang, S. M. (2012). Sodium chloride improves photosynthesis and water status in the succulent xerophyte *Zygophyllum xanthoxylum*. *Tree Physiol.* 32, 4–13. doi: 10.1093/treephys/tpr098

Mahajan, S., and Tuteja, N. (2005). Cold, salinity and drought stresses: an overview. *Arch. Biochem. Biophys.* 444, 139–158. doi: 10.1016/j.abb.2005.10.018

Marques, A. P., Pires, C., Moreira, H., Rangel, A. O., and Castro, P. M. (2010). Assessment of the plant growth promotion abilities of six bacterial isolates using *Zea mays* as indicator plant. *Soil Biol. Biochem.* 42, 1229–1235. doi: 10.1016/j.soilbio.2010.04.014

Mayak, S., Tirosh, T., and Glick, B. R. (2004). Plant growth-promoting bacteria confer resistance in tomato plants to salt stress. *Plant Physiol. Biochem.* 42, 565–572. doi: 10.1016/j.plaphy.2004.05.009

Medeiros, F. H., Souza, R. M., Medeiros, F. C., Zhang, H., Wheeler, T., Payton, P., et al. (2011). Transcriptional profiling in cotton associated with *Bacillus subtilis* (UFLA285) induced biotic-stress tolerance. *Plant Soil* 347, 327–337. doi: 10.1007/s11104-011-0852-5

Mhadhbi, H., Jebara, M., Limam, F., and Aouani, M. E. (2004). Rhizobial strain involvement in plant growth, nodule protein composition and antioxidant enzyme activities of chickpea-rhizobia symbioses: modulation by salt stress. *Plant Physiol. Biochem.* 42, 717–722. doi: 10.1016/j.plaphy.2004.07.005

Miao, B. H., Han, X. G., and Zhang, W. H. (2010). The ameliorative effect of silicon on soybean seedlings grown in potassium-deficient medium. *Ann. Bot.* 105, 967–973. doi: 10.1093/aob/mcq063

Munns, R., and Tester, M. (2008). Mechanisms of salinity tolerance. *Annu. Rev. Plant Biol.* 59, 651–681. doi: 10.1146/annurev.arplant.59.032607.092911

Ozawa, T., Wu, J., and Fujii, S. (2007). Effect of inoculation with a strain of *Pseudomonas pseudoalcaligenes* isolated from the endorhizosphere of *Salicornia europea* on salt tolerance of the glasswort. *Soil Sci. Plant Nutr.* 53, 12–16. doi: 10.1111/j.1747-0765.2007.00098.x

Paré, P. W., Zhang, H. M., Aziz, M., Xie, X. T., Kim, M. S., Shen, X., et al. (2011). Beneficial rhizobacteria induce plant growth: mapping signaling networks in *Arabidopsis. Biocommun. Soil Microorg. Soil Biol.* 23, 403–412. doi: 10.1007/978-3-642-14512-4_15

Porra, R. J., Thompson, W. A., and Kriendemann, P. E. (1989). Determination of accurate extinction coefficients and simultaneous equations for assaying chlorophyll a and b extracted with four different solvents: verification of the concentration of chlorophyll standards by atomic absorption spectroscopy. *Biochim. Biophys. Acta* 975, 384–394. doi: 10.1016/S0005-2728(89)80347-0

Rahnama, A., James, R. A., Poustini, K., and Munns, R. (2010). Stomatal conductance as a screen for osmotic stress tolerance in durum wheat growing in saline soil. *Funct. Plant Biol.* 37, 255–263. doi: 10.1071/FP09148

Ramani, B., Reeck, T., Debez, A., Stelzer, R., Huchzermeyer, B., Schmidt, A., et al. (2006). *Aster tripolium* L. and *Sesuvium portulacastrum* L.: two halophytes, two strategies to survive in saline habitats. *Plant Physiol. Biochem.* 44, 395–408. doi: 10.1016/j.plaphy.2006.06.007

Rengasamy, P. (2010). Soil processes affecting crop production in salt-affected soils. *Funct. Plant Biol.* 37, 613–620. doi: 10.1071/FP09249

Rogers, M. E., Noble, C. L., Halloran, G. M., and Nicolas, M. E. (1997). Selecting for salt tolerance in white clover (*Trifolium repens*): chloride ion exclusion and its heritability. *New Phytol.* 135, 645–654. doi: 10.1046/j.1469-8137.1997.00685.x

Rudrappa, T., Biedrzychi, M. L., Kunjeti, S. G., Donofrio, N. M., Czymmek, K. J., Paré, P. W., et al. (2010). The rhizobacterial elicitor acetoin induces systemic resistance in *Arabidopsis thaliana*. *Commun. Integr. Biol.* 3, 130–138. doi: 10.4161/cib.3.2.10584

Ryu, C. M., Farag, M. A., Hu, C. H., Reddy, M. S., Kloepper, J. W., and Paré, P. W. (2004). Bacterial volatiles induce systemic resistance in *Arabidopsis*. *Plant Physiol.* 134, 1017–1026. doi: 10.1104/pp.103.026583

Stein, T., Borchert, S., Conrad, B., Feesche, J., Hofemeister, B., Hofemeister, J., et al. (2002). Two different lantibiotic like peptides originate from the ericin gene cluster of *Bacillus subtilis* A1/3. *J. Bacteriol.* 184, 1703–1711. doi: 10.1128/JB.184.6.1703-1711.2002

Tester, M., and Davenport, R. (2003). Na^+ tolerance and Na^+ transport in higher plants. *Ann. Bot.* 91, 503–527. doi: 10.1093/aob/mcg058

Tiwari, S., Singh, P., Tiwari, R., Meena, K. K., Yandigeri, M., Singh, D. P., et al. (2011). Salt-tolerant rhizobacteria-mediated induced tolerance in wheat (*Triticum aestivum*) and chemical diversity in rhizosphere enhance plant growth. *Biol. Fert. Soils* 47, 907–916. doi: 10.1007/s00374-011-0598-5

Tuna, A. L., Kaya, C., Higgs, D., Murillo-Amador, B., Aydemir, S., and Girgin, A. R. (2008). Silicon improves salinity tolerance in wheat plants. *Environ. Exp. Bot.* 62, 10–16. doi: 10.1016/j.envexpbot.2007.06.006

van Hulten, M., Pelser, M., van Loon, L. C., Pieterse, C. M., and Ton, J. (2006). Costs and benefits of priming for defense in *Arabidopsis*. *Proc. Natl. Acad. Sci. U.S.A.* 103, 5602–5607. doi: 10.1073/pnas.0510213103

Wang, L. W., Showalter, A. M., and Ungar, I. A. (2005). Effects of intraspecific competition on growth and photosynthesis of *Atriplex prostrata*. *Aquat. Bot.* 83, 187–192. doi: 10.1016/j.aquabot.2005.06.005

Wang, S. M., Zhang, J. L., and Flowers, T. J. (2007). Low-affinity Na^+ uptake in the halophyte *Suaeda maritima*. *Plant Physiol.* 145, 559–571. doi: 10.1104/pp.107.104315

Xie, X., Zhang, H., and Paré, P. W. (2009). Sustained growth promotion in *Arabidopsis* with long-term exposure to the beneficial soil bacterium *Bacillus subtilis* (GB03). *Plant Signal Behav.* 4, 948–953. doi: 10.4161/psb.4.10.9709

Yazici, I., Tuerkan, I., Sekmen, A. H., and Demiral, T. (2007). Salinity tolerance of purslane (*Portulaca oleracea* L.) is achieved by enhanced antioxidative system, lower level of lipid peroxidation and proline accumulation. *Environ. Exp. Bot.* 61, 49–57. doi: 10.1016/j.envexpbot.2007.02.010

Zayed, M. A., and Zeid, I. M. (1998). Effect of water and salt stresses on growth, chlorophyll, mineral ions and organic solutes contents, and enzymes activity in mung bean seedlings. *Biol. Plant.* 40, 351–356. doi: 10.1023/A:1001057728794

Zhang, H., Kim, M. S., Krishnamachari, V., Payton, P., Sun, Y., Grimson, M., et al. (2007). Rhizobacterial volatile emissions regulate auxin homeostasis and cell expansion in *Arabidopsis*. *Planta* 226, 839–851. doi: 10.1007/s00425-007-0530-2

Zhang, H., Kim, M. S., Sun, Y., Dowd, S. E., Shi, H., and Paré, P. W. (2008a). Soil bacteria confer plant salt tolerance by tissue-specific regulation of the sodium transporter *HKT1*. *Mol. Plant Microbe Interact.* 21, 737–744. doi: 10.1094/MPMI-21-6-0737

Zhang, H., Murzello, C., Sun, Y., Kim, M. S., Xie, X., Jeter, R. M., et al. (2010b). Choline and osmotic-stress tolerance induced in *Arabidopsis* by the soil microbe *Bacillus subtilis* (GB03). *Mol. Plant Microbe Interact.* 23, 1097–1104. doi: 10.1094/MPMI-23-8-1097

Zhang, H., Sun, Y., Xie, X., Kim, M. S., Dowd, S. E., and Paré, P. W. (2009). A soil bacterium regulates plant acquisition of iron via deficiency-inducible mechanisms. *Plant J.* 58, 568–577. doi: 10.1111/j.1365-313X.2009.03803.x

Zhang, H., Xie, X., Kim, M. S., Kormyeyev, D. A., Holaday, S., and Paré, P. W. (2008b). Soil bacteria augment *Arabidopsis* photosynthesis by decreasing glucose sensing and abscisic acid levels in planta. *Plant J.* 56, 264–273. doi: 10.1111/j.1365-313X.2008.03593.x

Zhang, J. L., Flowers, T. J., and Wang, S. M. (2010a). Mechanisms of sodium uptake by roots of higher plant. *Plant Soil* 326, 45–60. doi: 10.1007/s11104-009-0076-0

Zhang, J. L., and Shi, H. Z. (2013). Physiological and molecular mechanisms of plant salt tolerance. *Photosynth. Res.* 115, 1–22. doi: 10.1007/s11120-013-9813-6

Zhang, N., Wu, K., He, X., Li, S. Q., Zhang, Z. H., Shen, B., et al. (2011). A new bioorganic fertilizer can effectively control banana wilt by strong colonizaton with *Bacillus subtilis* N11. *Plant Soil* 344, 87–97. doi: 10.1007/s11104-011-0729-7

Permissions

All chapters in this book were first published in STPMRIT, by Frontiers Media SA; hereby published with permission under the Creative Commons Attribution License or equivalent. Every chapter published in this book has been scrutinized by our experts. Their significance has been extensively debated. The topics covered herein carry significant findings which will fuel the growth of the discipline. They may even be implemented as practical applications or may be referred to as a beginning point for another development.

The contributors of this book come from diverse backgrounds, making this book a truly international effort. This book will bring forth new frontiers with its revolutionizing research information and detailed analysis of the nascent developments around the world.

We would like to thank all the contributing authors for lending their expertise to make the book truly unique. They have played a crucial role in the development of this book. Without their invaluable contributions this book wouldn't have been possible. They have made vital efforts to compile up to date information on the varied aspects of this subject to make this book a valuable addition to the collection of many professionals and students.

This book was conceptualized with the vision of imparting up-to-date information and advanced data in this field. To ensure the same, a matchless editorial board was set up. Every individual on the board went through rigorous rounds of assessment to prove their worth. After which they invested a large part of their time researching and compiling the most relevant data for our readers.

The editorial board has been involved in producing this book since its inception. They have spent rigorous hours researching and exploring the diverse topics which have resulted in the successful publishing of this book. They have passed on their knowledge of decades through this book. To expedite this challenging task, the publisher supported the team at every step. A small team of assistant editors was also appointed to further simplify the editing procedure and attain best results for the readers.

Apart from the editorial board, the designing team has also invested a significant amount of their time in understanding the subject and creating the most relevant covers. They scrutinized every image to scout for the most suitable representation of the subject and create an appropriate cover for the book.

The publishing team has been an ardent support to the editorial, designing and production team. Their endless efforts to recruit the best for this project, has resulted in the accomplishment of this book. They are a veteran in the field of academics and their pool of knowledge is as vast as their experience in printing. Their expertise and guidance has proved useful at every step. Their uncompromising quality standards have made this book an exceptional effort. Their encouragement from time to time has been an inspiration for everyone.

The publisher and the editorial board hope that this book will prove to be a valuable piece of knowledge for researchers, students, practitioners and scholars across the globe.

List of Contributors

Fang Yuan, Bingying Leng and Baoshan Wang
Key Lab of Plant Stress Research, College of Life Science, Shandong Normal University, Ji'nan, China

Frans J. M. Maathuis, Izhar Ahmad and Juan Patishtan
Department of Biology, University of York, York, UK

Vadim Volkov
Faculty of Life Sciences, School of Human Sciences, London Metropolitan University, London, UK

Bernardo Duarte and Isabel Caçador
Centre of Oceanography, Faculty of Sciences, University of Lisbon, Lisbon, Portugal
Marine and Environmental Sciences Centre, Faculty of Sciences, University of Lisbon, Lisbon, Portugal

Noomene Sleimi
UR: MaNE, Faculté des Sciences de Bizerte, Université de Carthage, Bizerte, Tunisie

Pedro M. F. Almeida and Albertus H. de Boer
Department of Structural Biology, Faculty Earth and Life Sciences, Vrije Universiteit Amsterdam, Amsterdam, Netherlands

Gert-Jan de Boer
R&D Department, Enza Zaden, Enkhuizen, Netherlands

Ya-Qing Pan, Huan Guo, Suo-Min Wang, Jin-Lin Zhang, Qing Ma, Hong-Ju Yin and Ai-Ke Bao
State Key Laboratory of Grassland Agro-ecosystems, College of Pastoral Agriculture Science and Technology, Lanzhou University, Lanzhou, China,

Bingyu Zhao
Department of Horticulture, Virginia Polytechnic Institute and State University, Blacksburg, VA, USA

Fernando Alemán, Fernando Caballero, Reyes Ródenas, Rosa M. Rivero, Vicente Martínez and Francisco Rubio
Centro de Edafología y Biología Aplicada del Segura-CSIC, Murcia, Spain

Michelle Mak, Mohammad Babla, Filip Veljanoski and Zhong-Hua Chen
School of Science and Health, University of Western Sydney, Penrith, NSW, Australia

Xiaohui Liu
School of Science and Health, University of Western Sydney, Penrith, NSW, Australia
School of Chemical Engineering and Technology, Tianjin University, Tianjin, China

Feifei Wang, Sergey Shabala and Meixue Zhou
School of Land and Food, University of Tasmania, Hobart, TAS, Australia

Guang Chen
College of Agriculture and Biotechnology, Zhejiang University, Hangzhou, China

Gang Wang
School of Environmental Science and Engineering, Tianjin University, Tianjin, China

Xiaoshu Jing, Yanjun Lu, Shurong Deng, Rui Zhao, Yang Wang, Yansha Han, Tao Lang, Xin Shen and Shaoliang Chen
Department of Plant Science, College of Biological Sciences and Technology, Beijing Forestry University, Beijing, China

Peichen Hou
Department of Bio-Instruments, National Engineering Research Center for Information Technology in Agriculture, Beijing, China

Niya Li
Department of Biology, College of Life Science, Hainan Normal University, Haikou, China

Jian Sun
Department of Plant Science, College of Life Science, Jiangsu Normal University, Xuzhou, China

Mingquan Ding
Department of Crop Science, College of Agricultural and Food Science, Zhejiang Agricultural and Forestry University, Hangzhou, China

Emanuel Bojórquez-Quintal, Ángela Ku-González, Mildred Carillo-Pech, Ileana Echevarría-Machado and Manuel Martínez-Estévez
Unidad de Bioquímica y Biología Molecular de Plantas, Centro de Investigación Científica de Yucatán, Yucatán, México

Ana Velarde-Buendía and Igor Pottosin
Centro Universitario de Investigaciones Biomédicas, Universidad de Colima, Colima, México

Daniela Ortega-Camacho
Unidad de Ciencias del Agua, Centro de Investigación Científica de Yucatán, Yucatán, México

Mary J. Beilby
Plant Membrane Biophysics, Physics/Biophysics, School of Physics, University of New South Wales, Sydney, NSW, Australia

Honghong Wu, Lana Shabala, Meixue Zhou, Jayakumar Bose and Sergey Shabala
Faculty of Science, Engineering and Technology, School of Land and Food, University of Tasmania, Hobart, TAS, Australia

Elisa Azzarello, Camilla Pandolfi and Stefano Mancuso
Department of Agrifood Production and Environmental Sciences, University of Florence, Florence, Italy

Hui-Juan Gao, Hong-Yu Yang, Jiang-Ping Bai, Yan Lou, Jun-Lian Zhang and Di Wang
Gansu Key Laboratories of Crop Genetic and Germplasm Enhancement and Aridland Crop Science, College of Agronomy, Gansu Agricultural University, Lanzhou, China

Xin-Yue Liang
Department of Chemistry, School of Chemistry and Chemical Engineering, Nanjing University, Nanjing, China

Shu-Qi Niu
State Key Laboratory of Grassland Agro-ecosystems, College of Pastoral Agriculture Science and Technology, Lanzhou University, Lanzhou, China

Maheshi Dassanayake and John C. Larkin
Department of Biological Sciences, Louisiana State University, Baton Rouge, LA, USA

Ying-Long Chen
Plant Nutrition and Soil Science and UWA Institute of Agriculture, School of Earth and Environment, The University of Western Australia, Perth, WA, Australia
State Key Laboratory of Soil Erosion and Dryland Farming on the Loess Plateau, Institute of Soil and Water Conservation, Chinese Academy of Sciences and Ministry of Education, Northwest A&F University, Yangling, China

Tomoko Kagenishi and František Baluška
Department of Plant Cell Biology, Institute of Cellular and Molecular Botany, University of Bonn, Bonn, Germany

Ken Yokawa
Department of Plant Cell Biology, Institute of Cellular and Molecular Botany, University of Bonn, Bonn, Germany
Department of Biological Sciences, Tokyo Metropolitan University, Tokyo, Japan

Rossella Fasano
Department of Pharmacy, University of Salerno, Fisciano, Italy

Qing-Qing Han, Xin-Pei Lü, Yong-Na Wu, Xiao-Pan Pang, Wen-Bo Xu and Zhi-Liang Wang
State Key Laboratory of Grassland Agro-ecosystems, College of Pastoral Agriculture Science and Technology, Lanzhou University, Lanzhou, China

Yan Qiao
Gansu Key Laboratory of Crop Genetic and Germplasm Enhancement, Gansu Provincial Key Laboratory of Arid Land Crop Science, College of Agronomy, Gansu Agricultural University, Lanzhou, China

Paul W. Paré
Department of Chemistry and Biochemistry, Texas Tech University, Lubbock, TX, USA

Index

N

O

P

R

S

V

Y

www.ingramcontent.com/pod-product-compliance
Lightning Source LLC
LaVergne TN
LVHW060558210926
848485LV00005B/1333